# Environmental Engineering:
## Fundamentals, Sustainability, Design

## Authors and Editors

### James R. Mihelcic
*University of South Florida*

### Julie Beth Zimmerman
*Yale University*

## Contributing Authors

Martin T. Auer
*Michigan Technological University*

David W. Hand
*Michigan Technological University*

Richard E. Honrath, Jr.
*Michigan Technological University*

Alex S. Mayer
*Michigan Technological University*

Mark W. Milke
*University of Canterbury*

Kurtis G. Paterson
*Michigan Technological University*

Judith A. Perlinger
*Michigan Technological University*

Michael E. Penn
*University of Wisconsin-Platteville*

Noel R. Urban
*Michigan Technological University*

Brian E. Whitman
*Wilkes University*

Qiong Zhang
*University of South Florida*

**John Wiley & Sons, Inc.** WILEY

## About the Cover

Richard Buckminster Fuller (1895-1983) was an engineer, architect, poet, and designer. During his life, he pondered the question, "Does humanity have a chance to survive lastingly and successfully on planet Earth, and if so, how?" To begin to answer this question, Fuller ascribed to the "Spaceship Earth" worldview that expresses concern over the use of limited global resources and the behavior of everyone on it to act as a harmonious crew working toward the greater good.

In 1969 Fuller wrote and published a book entitled "Operating Manual for Spaceship Earth." The following quotation from this book reflects his worldview: "Fossil fuels can make all of humanity successful through science's world-engulfing industrial evolution provided that we are not so foolish as to continue to exhaust in a split second of astronomical history the orderly energy savings of billions of years' energy conservation aboard our Spaceship Earth. These energy savings have been put into our Spaceship's life-regeneration-guaranteeing bank account for use only in self-starter functions."

To further communicate his ideas, Fuller developed the Dymaxion Map, shown on the cover. This map is a projection of a World map onto the surface of a polyhedron. The projection can be unfolded in many different ways and flattened out to form a two-dimensional map that retains the look and integrity of a globe map. Importantly, the Dymaxion map has no "right way up." Fuller believed that in the universe there was no "up" and "down" or "north" and "south": only "in" and "out." He linked the north-up-superior/south-down-inferior presentation of most other world maps to cultural bias.

| | |
|---|---|
| PUBLISHER | Don Fowley |
| ACQUISITIONS EDITOR | Jennifer Welter |
| SENIOR PRODUCTION EDITOR | Valerie A. Vargas |
| MARKETING MANAGER | Christopher Ruel |
| CREATIVE DIRECTOR | Harry Nolan |
| DESIGN DIRECTOR | Jeof Vita |
| SENIOR DESIGNER | Kevin Murphy |
| INTERIOR DESIGNER | Amy Rosen |
| COVER DESIGN | David Levy |
| PRODUCTION MANAGEMENT SERVICES | Elm Street Publishing Services, Inc. |
| PHOTO EDITOR | Sheena Goldstein |
| EDITORIAL ASSISTANT | Mark Owens |
| MEDIA EDITOR | Lauren Sapira |
| COVER PHOTO | © 2002 Buckminster Fuller Institute and Jim Knighton. Coordinate transformation software written by Robert W. Gray and modified by Jim Knighton. |

This book was set in Palatino by Thomson Digital and printed and bound by Hamilton Printing. The cover was printed by Lehigh Phoenix.

ISBN-13   978-0-470-16505-8

Printed in the United States of America

10 9 8 7 6 5 4 3

# Preface

By the time you finish reading this sentence, ten people will be added to the population of the planet, one thousand tonnes of carbon dioxide will be added to the atmosphere, and ten acres of land will be deforested. Unless something dramatically changes about the way humans interact with the world and the products, processes, and systems we design, we will end up exactly where we are headed—an unsustainable future. Sustainability, as a concept, means making those transformative changes needed to ensure that the current generation and future generations have the ability meet their needs. How we go about accomplishing this tremendous challenge is the central purpose of this book.

Stark and egregious environmental problems gave rise to the field of environmental engineering more than five decades ago. Unhealthy air in Los Angeles and Pittsburgh, rivers on fire in Ohio, and entire communities evacuated because of toxics in New York and Missouri were some of the problems that environmental engineers had to address as a nascent field. Creativity and diligence of the field resulted in tremendous benefits to the environment and improved quality of life. Knowledge of how to deal with those egregious problems is foundational and critical to the field of environmental engineering as it continues to evolve. Sadly, today new challenges are arising that are more subtle, more complex, more global, and potentially more devastating to people and the planet.

For these and other reasons, environmental engineering has now emerged as a distinct discipline. Environmental engineering is a recognized specialty on professional engineering licensing exams, and the number of environmental engineers employed in the United States alone is estimated to range from 54,000 to 100,000. As a profession, this places environmental engineering in the United States as a significantly larger profession than biomedical, materials, and chemical engineering. The discipline is also one of only two engineering disciplines the U.S. Bureau of Labor Statistics predicts will have "much faster than average growth" over the next ten years.

The discipline of environmental engineering, though, is in constant motion. The problems that have been created as a result of the manner in which the human society and economy have interacted with the natural environment since the industrial revolution have changed over the past decades. Some general characterizations of the types of problems environmental engineers are dealing with are shown in the table below.

| 20th Century Environmental Issues | 21st Century Environmental Issues |
|---|---|
| Local | Global |
| Acute | Chronic |
| Obvious | Subtle |
| Immediate | Multigenerational |
| Discrete | Complex |

The evolution of the problems themselves and the level of understanding we have about these problems will require engineers to take on new skills, capabilities, and perspectives about how we approach our work. It is not that the skills previously learned are antiquated and need to be replaced. It is that the traditional skills need to be augmented, complemented, and enhanced with new knowledge, new perspectives, and new awareness. The melding of old and new fundamentals and design skills is the purpose of this text. It is our hope that this text provides engineers with the knowledge and confidence to deal with 21st century challenges as well as they dealt with the daunting challenges of the 20th century.

## Hallmark Features

### A FOCUS ON SUSTAINABLE DESIGN

Perhaps one of the most important aspects of the textbook is that it will focus the student on the elements of *design*. Design of products, processes, and systems will be essential not only in responding to the environmental issues in ways that our profession has done historically but also in informing the design of new products, processes, and systems to reduce or eliminate problems from occurring in the first place.

To use the tools of green engineering design truly to design for sustainability, students need a command of the framework for this design. The framework perhaps can be summarized in the *four I's*: (1) Inherency, (2) Integration, (3) Interdisciplinary, and (4) International.

**Inherency** As a reader proceeds through the text, it will become obvious that we are not merely looking at how to change the conditions or circumstances that make a product, process, or system a problem. Readers will understand the *inherent* nature of the material and energy inputs and outputs so that they are able to understand the fundamental basis of the hazard and the root causes of the adverse consequence they seek to address. Only through this inherency approach can we begin to design for sustainability rather than generating elegant technological bandages for flawed conceptions.

**Integration** Our historical approaches toward many environmental issues have been fragmented—often by media, life cycle, culture, or geographic region. Understanding that energy is inextricably linked to water, water to climate change, climate change to food production, food production to health care, health care to societal development, and so on will be essential in the new paradigm of sustainable design. It is equally necessary to understand that we cannot think about approaching any environmental problem without looking at the problem across all elements of its life cycle. There have been countless attempts to improve environmental circumstances that have resulted in unintended problems that have often been worse than the problem they intended to fix. Attempts to increase drinking water supply in Bangladesh resulted in widespread arsenic poisoning. Attempts to increase crop yields through the production of pesticides in Bhopal, India, resulted in one of the greatest chemical tragedies of our time. Understanding the complex interconnections and ensuring the *integration* of multiple factors in the development of solutions is something that 21st century environmental engineering requires.

**Interdisciplinary** To achieve the goals of sustainable design, environmental engineers will be working increasingly with a wide array of other disciplines. Technical disciplines of chemistry and biology and other engineering disciplines will be essential but so will the disciplines of economics, systems analysis, health, sociology, and anthropology. This text seeks to introduce the *interdisciplinary* dimensions that will be important to the successful environmental engineer in this century.

**International** Many well-intentioned engineering solutions of the 20th century would fail by not considering the very different context found in the diversity of nations around the world. Although water purification or municipal waste may seem like they can be dealt with through identical processes anywhere in the world, it has been shown repeatedly that the local factors—geographic, climatic, cultural, socioeconomic, political, ethnic, and historical—can all play a role in the success or failure of an environmental engineering solution. The *international* perspective is an important one this textbook emphasizes and incorporates into the fundamentals of the training of environmental engineers.

## MATERIAL AND ENERGY BALANCES AND LIFE CYCLE THINKING

The book provides a rigorous development of mass and energy and mass balance concepts with numerous easy-to-follow example problems. It then applies mass and energy balance concepts to a wide range of natural and engineered systems and different environmental media. The book has appropriate coverage of life cycle assessment with an in-depth example problem and provides a life cycle–thinking approach in discussion throughout other chapters.

## PEDAGOGY AND ASSESSMENT

Beyond including the elements mentioned previously to prepare engineers for the 21st century, this book also incorporates changes in pedagogy and assessment that provide structure for delivering this new information in a meaningful education experience.

**Fink's Taxonomy of Significant Learning** One such element is the use of Fink's taxonomy of significant learning in guiding the development of learning objectives for each chapter as well as in example and homework problems. Fink's taxonomy recognizes six domains including the traditionally considered foundational knowledge, including: foundational knowledge; application of knowledge; integration of knowledge; human dimensions of learning and caring; and learning how to learn. Without much background on the taxonomy, it is clear from these knowledge domain headings alone that these areas recognized by Fink are critical to an engineer tasked with designing solutions to many of today's sustainability challenges.

**Web Modules** Icons in the margin indicate when Web modules are available on the book Web site to enhance and expand on the concepts presented in the book. Modules include animations, video clips, spreadsheets, document and PDF files, and executables. This platform provides students with a visual and interactive learning environment in which they can explore fundamentals, design, and sustainability by visually seeing how changes in numerical inputs affect the output of design and operation.

www.wiley.com/college/mihelcic

**Important Equations** Boxes around important equations indicate for students which are most critical.

**Learning Exercises** Learning exercises through end-of-chapter problems not only ask students to solve traditional numerical problems of assessment and design but also challenge students to research problems and innovate solutions at different levels: campus, apartment, home, city, region, state, or world.

**Discussion Topics** To further emphasize the importance of the domains of knowledge discussed in the previous paragraph, the book encourages classroom discussions and interaction between students as well as between the students and the instructor. These discussion topics are noted by a symbol in the margin.

**Resources for Further Learning** Online resources for further learning and exploration are listed in the margin where appropriate. These resources give students the opportunity to explore topics in much greater detail and learn of geographical commonalities and uniqueness to specific environmental engineering issues. More important, use of these online resources prepares students better for professional practice by expanding their knowledge of information available at government and nongovernment Web sites.

## BOOK WEB SITE

The following resources for students and instructors are available on the book Web site, located at **www.wiley.com/college/mihelcic.**

**Web Modules** Martin Auer had the vision and then provided the leadership to oversee development of 60 interactive Web modules that are integrated throughout the text. The goal of the Web modules is to enhance and expand on concepts presented in the book. The modules also provide students with the opportunity to reach out to practice with government, communities, and the world as they explore concepts in greater depth. Because the Web modules are an ever-evolving product, additional Web modules will be added over time. These Web modules are flagged in the text by an icon that alerts a student to the availability of Web support for a particular topic. Instructors are encouraged to use the modules during class as an additional tool for student motivation, education, and engagement.

**Classroom Materials for Instructors** Through an NSF Course, Curriculum, and Laboratory Improvement grant awarded to three of this book's authors (Qiong Zhang, Julie Zimmerman, and James Mihelcic) and to Linda Vanasupa (California Polytechnic State University), we have developed in-depth educational materials on the following six topics:

1. Systems Thinking
2. (Introduction to) Sustainability
3. Systems Thinking: Population
4. Systems Thinking: Energy
5. Systems Thinking: Material
6. Systems Thinking: Water

Each set provides an array of classroom materials whose design aligns with educational research on how to foster more significant learning. Each set includes:

- **Learning objectives** within each of five critical areas of learning (foundational knowledge, application of knowledge, integration of knowledge, human dimensions of learning and caring, and learning how to learn)
- A set of editable and notated **slides for faculty to present lecture material**
- **Active learning exercises** that range from two-minute to three-hour investments; notated guides for faculty using the exercises
- A set of **assessment activities** that includes learning objectives, criteria for assessment, and standards for judging the criteria

## ADDITIONAL RESOURCES FOR INSTRUCTORS

Additional resources for instructors to support this text include:

- **Solutions Manual** containing solutions for all the end-of-chapter problems in the text.
- **Image Gallery** with illustrations from the text appropriate for use in lecture slides.

These resources are available only to instructors who adopt the text. Please visit the instructor section of the Web site at **www.wiley.com/college/mihelcic** to register for a password.

## Genesis of the Book

In 1999, we published a book titled *Fundamentals of Environmental Engineering* (John Wiley & Sons). That book was an outgrowth of a team-teaching effort required at that time for environmental and civil engineering undergraduates at Michigan Technological University. One strength of *Fundamentals of Environmental Engineering* is that it provides in-depth coverage of the basic environmental engineering fundamentals required for design, operation, analysis, and modeling of both natural and engineered systems. This new book you are reading now, *Environmental Engineering: Fundamentals, Sustainability, Design,* not only includes updated chapters on those same fundamentals with continued strong emphasis on material and energy balances and inclusion of issues of energy and climate, it also includes application of those fundamental skills to design and operation of treatment and prevention systems. Importantly, this new book provides chapters on treatment and management of multimedia environmental problems (e.g., drinking water treatment and distribution, wastewater collection and treatment, solid waste management, and air resources engineering) while also providing chapters on sustainability, environmental risk, green engineering, and the built environment. The important issues of energy and climate are incorporated into every chapter and are not considered as stand-alone issues. Critical concepts of ethics and justice are included and readers are provided a global perspective. The engineering applications are also updated; for example, the water quality chapter not only includes management of lakes and rivers but also the design of low-impact development infrastructure for stormwater management. The wastewater chapter includes issues of reuse and reclamation, and the air resources chapter includes material on indoor air quality. Last, the built environment chapter includes topics such as context-sensitive design, LEED certification of buildings, and issues related to engineering livable communities and more sustainable transportation systems.

## Acknowledgements

As we marvel and appreciate all those who have dedicated themselves to leaving the world a better place than they found it—environmental engineers and others—we are grateful for all the talented people who have helped make this book possible and are poised to change the very nature of the field of environmental engineering.

Besides all the individuals who contributed content to the book, the following faculty provided high-quality review and insight through several stages of the book development:

Zuhdi Aljobeh, *Valparaiso University*

Robert W. Fuessle, *Bradley University*

Keri Hornbuckle, *University of Iowa*

Benjamin S. Magbanua Jr., *Mississippi State University*

Taha F. Marhaba, *New Jersey Institute of Technology*

William F. McTernan, *Oklahoma State University*

Gbekeloluwa B. Oguntimein, *Morgan State University*

Joseph Reichenberger, *Loyola Marymount University*

Sukalyan Sengupta, *University of Massachusetts*

Thomas Soerens, *University of Arkansas*

Linda Vanasupa (California Polytechnic State University) reviewed the chapters and assisted in developing learning objectives in the context of Fink's taxonomy of significant learning. Linda Phillips (University of South Florida) provided her international perspective, especially regarding integrating service learning with practitioner involvement. The editorial team of Jennifer Welter, Mark Owens, and Heather Johnson that assisted us on this project has also been a key to success. Their early vision of the book's purpose and attention and contributions to detail, style, and pedagogy have made this a fulfilling and equal partnership.

The following students at the University of South Florida reviewed every chapter of the book and provided valuable comments during the editing process: Jonathan Blanchard, Justin Meeks, Colleen Naughton, Kevin Orner, Duncan Peabody, and Steven Worrell. Ezekiel Fugate and Jennifer Ace (Yale University) and Helen E. Muga (University of South Florida) helped us obtain permissions and search for materials. We are especially grateful to Ziad Katirji (Michigan Technological University) and Heather E. Wright Wendel (University of South Florida), who assisted efforts to create, assemble, and proof the *Solutions Manual*.

Finally, thanks to Karen, Paul, Raven, and Kennedy for embracing the vision of this project over the past several years.

James R. Mihelcic

Julie Beth Zimmerman

# About the Authors

**James R. Mihelcic** is a professor of civil and environmental engineering and a State of Florida 21st Century World Class Scholar at the University of South Florida. He is founder of the Peace Corps Master's International Program in Civil and Environmental Engineering. His teaching and research interests are in sustainability, water and wastewater engineering, design and operation of appropriate technology, and green engineering. Dr. Mihelcic is a past president of the Association of Environmental Engineering and Science Professors and a board-certified member of the American Academy of Environmental Engineers. He has traveled extensively in the developing world to serve and conduct research on development issues related to water supply and treatment, sanitation, shelter, and global health.

**Julie Beth Zimmerman** is an assistant professor jointly appointed in the School of Engineering and Applied Science in the Environmental Engineering program and the School of Forestry and Environmental Studies at Yale University. She also serves as Associate Director for Research for the Center for Green Chemistry and Green Engineering at Yale. Her research interests include green chemistry and engineering, systems dynamics modeling of natural and engineered water systems, environmentally benign design and manufacturing, the fate and impacts of anthropogenic compounds in the environment, and appropriate water treatment technologies for the developing world. She also conducts research on corporate environmental behavior and governance interventions to enhance the integration of sustainability in industry and academia.

**Martin T. Auer** is a professor of civil and environmental engineering at Michigan Technological University. He teaches introductory courses in environmental engineering and advanced coursework in surface water–quality engineering and mathematical modeling of lakes, reservoirs, and rivers. Dr. Auer's research interests involve field and laboratory studies and mathematical modeling of water quality in lakes and rivers.

**David W. Hand** is a professor of civil and environmental engineering at Michigan Technological University. He teaches senior-level and graduate courses in drinking water treatment, wastewater treatment, and physical chemical processes in environmental engineering. Dr. Hand's research interests include physical-chemical treatment processes, mass transfer, adsorption, air stripping, homogeneous and heterogeneous advanced oxidation processes, process modeling of water treatment and wastewater treatment processes, and development of engineering software design tools for pollution prevention practice.

**Richard E. Honrath** was a professor of geological and mining engineering and Sciences and of civil and environmental engineering at Michigan Technological University, where he also directed the Atmospheric Sciences graduate program. He taught courses in introductory environmental engineering, advanced air quality engineering and science, and atmospheric chemistry. His research activities involved studies of the large-scale impacts of air pollutant emissions from anthropogenic sources and from wildfires, with a focus on the interaction between transport processes and chemical processing. He also studied photochemistry in ice and snow, including field studies of the interactions among snow, air, and sunlight.

**Alex S. Mayer** is a professor of environmental and geological engineering and Director of the Center for Water and Society at Michigan Technological University. His teaching and research interests focus on human-biophysical interactions in water systems; watershed management and modeling; and groundwater flow, transport, and remediation. Dr. Mayer has served in editorial positions for several professional journals: *Water Resources Research*, *Journal of Contaminant Hydrology*, and *Advances in Water Resources*. He has consulted for engineering companies, legal firms, and nonprofit groups and is a registered professional engineer.

**Mark W. Milke** is an associate professor/reader at the Department of Civil and Natural Resources Engineering, University of Canterbury, New Zealand, where he has worked since 1991. His research and teaching interests are in solid-waste management, groundwater, and uncertainty analysis. He is a chartered professional engineer in New Zealand and serves as an associate editor for the journal *Waste Management*.

**Kurtis G. Paterson** is an assistant professor of civil and environmental engineering at Michigan Technological University. His research and teaching interests are in appropriate technology design, international project-based service learning, and engineering education reform. Dr. Paterson currently serves as Director of Michigan Tech's D80 Center, a consortium of research, education, and service programs dedicated to creating sustainable development solutions for the poorest 80 percent of humanity. He is currently involved in several international initiatives at the American Society for Engineering Education and numerous educational activities at Engineers Without Borders-USA.

**Michael R. Penn** is a professor of civil and environmental engineering at the University of Wisconsin-Platteville. He teaches undergraduate courses in introductory environmental engineering, fluid mechanics, hydrology, groundwater, wastewater and drinking water treatment, and solid and hazardous waste management. Dr. Penn's research interests focus on involving undergraduates in studies of agricultural runoff, nutrient cycling in lakes, and urban infrastructure management. Dr. Penn is currently writing a new Wiley textbook, *Introduction to Infrastructure: Civil Engineering, Environmental Engineering, and the Built Environment*, intended for first- and second-year undergraduates.

**Judith Perlinger** is an associate professor of civil and environmental engineering at Michigan Technological University. Her interests are in air and water quality, and her expertise is in environmental transport and transformation of organic chemicals and the effects of these on human health, ecosystem health, and climate. Dr. Perlinger teaches courses in the environmental engineering and atmospheric sciences programs, including Fundamentals of Environmental Engineering, Applications of Sustainability Principles and Environmental Regulations to Engineering Practice, Transport and Transformation of Organic Pollutants, and Boundary Layer Meteorology. Her research is oriented toward development and verification of predictive models to assess risk.

**Noel R. Urban** is a professor of civil and environmental engineering at Michigan Technological University. His teaching interests focus on environmental chemistry and surface water–quality modeling. His research interests include environmental cycles of major and trace elements, sediment diagenesis and stratigraphy, chemistry of natural organic matter, wetland biogeochemistry, environmental impact and fate of pollutants, influence of organisms on the chemical environment, and the role of the chemical environment in controlling populations.

**Brian E. Whitman** is an associate professor of environmental engineering at Wilkes University. He teaches courses in water distribution and wastewater collection system design, hydrology, water resources engineering, and water and wastewater treatment process design. Dr. Whitman's research interests include hydraulic modeling of water distribution and wastewater collection systems, environmental microbiology, and the development of lubrication systems for combustion engines. He is the recipient of two Wilkes University Outstanding Faculty Awards and has co-authored three books in the areas of water distribution and wastewater collection system modeling and design.

**Qiong Zhang** is an assistant professor of civil and environmental engineering at the University of South Florida. She was previously the Operations Manager of the Sustainable Futures Institute at Michigan Technological University. Her teaching interests are in green engineering, water treatment, and environmental assessment for sustainability. Dr. Zhang's research interests include developing embodied energy models for water systems, water treatment cost modeling and the associated software tools to optimize the design of water-air treatment systems, incorporating green engineering principles in engineering design and engineering curriculum, and integrating environmental assessment and chemical fate with life cycle assessment.

# Brief Table of Contents

# Detailed Table of Contents

# chapter/One Engineering and Sustainable Development

James R. Mihelcic
and Julie Beth Zimmerman

*This chapter defines sustainable development and reviews the global history of sustainability over the past few decades. Several emerging issues related to population, urbanization, water, energy, health, and the built environment are reviewed—challenges and opportunities for future engineers.*

## Learning Objectives

1. Identify ten emerging environmental issues that have local and global significance.
2. Define sustainable development and sustainable engineering in your own words and according to others.
3. Redefine engineering problems in a balanced social, economic, and environmental context.
4. Relate *The Limits to Growth*, "The Tragedy of the Commons," and the definition of carrying capacity to sustainable development.
5. Identify concerns faced by citizens of developing countries.
6. Summarize the eight Millennium Development Goals (MDGs) and targets set forth by the global community.
7. Identify one of the MDGs that interests you, and discuss how you as an engineer could contribute to achieving that goal.
8. Relate each of the eight MDGs to engineering practice.
9. Identify the types and magnitude of environment risk that exist for people living in the developing world.
10. Discuss problems caused by drivers of global change (such as population, urbanization, land use, and climate) and ways that engineering can offer solutions.
11. Define sustainable development, and explain how it relates to issues of population growth, urbanization, health, sanitation, water scarcity, water conflict, energy consumption, climate, toxic chemicals, material use, and the built environment.
12. Identify Internet resources related to global issues of sustainability, the environment, and health.
13. Describe the historical context of the sustainability revolution and the related changes in commonly used vocabulary.

## 1.1 Background

Engineers play a crucial role in improving living standards throughout the world. As a result, engineers can have a significant impact on progress towards sustainable development.

World Federation of Engineering Organizations (2002)

The environmental movement in the United States began in earnest in the late 1960s and early 1970s with the creation of the Environmental Protection Agency (EPA). This consolidated in one agency a variety of federal research, monitoring, standard-setting, and enforcement activities. During the same decades, Congress passed key environmental regulations such as the National Environmental Protection Act (NEPA), the Clean Air Act, the Water Pollution Control Act, and the Endangered Species Act. These acts were designed to address glaring environmental challenges such as the Cuyahoga River catching on fire in 1969 and the toxic waste and subsequent health problems in neighborhoods such as Love Canal in Niagara Falls, New York.

While we have made tremendous strides in addressing the most egregious environmental insults and maintained a growing economy, the environmental challenges of today are more subtle and more complex. They involve clear connections between emissions to air, land, and water and come from highly distributed sources. We also have a much higher level of understanding of the linkages among society, the economy, and the environment. In this case, scientific, technological, and policy innovations are recognized as powerful tools for advancing these areas for mutual benefit.

---

**Box / 1.1** **Rachel Carson and the Modern Environmental Movement**

**Rachel Carson** at Hawk Mountain, Pennsylvania (photograph taken ca. 1945 by Shirley Briggs, courtesy of the Lear/Carson Collection, Connecticut College).

**Rachel Carson** is considered one the leaders of the modern environmental movement. She was born 15 miles northeast of Pittsburgh in 1907. Educated at the undergraduate and graduate levels in science and zoology, she first worked for the government agency that became the U.S. Fish and Wildlife Service. As a scientist, she excelled at communicating complex scientific concepts to the public through clear and accurate writing. She wrote several books, including *The Sea around Us* (published in 1951) and *Silent Spring* (published in 1962).

*Silent Spring* was a commercial success soon after its publication. It dramatically captured the fact that songbirds were facing reproductive failure and early death because of pollution associated with manufacturing and prolific use of chemicals such as DDT, which had bioaccumulated in their small bodies. Some historians believe that *Silent Spring* was the initial catalyst that led to the creation of the modern environmental movement in the United States, including establishment of the U.S. Environmental Protection Agency.

It is through new scientific, technological, and policy innovation that we can maintain economic prosperity while also improving the quality of life for our citizens. This goal of creating and maintaining a prosperous society needs to be met without the negative impacts that have historically harmed our natural resources, the environment, and communities. This requires a new perspective and new understanding of the environmental damages that have been traditionally associated with development.

As Albert Einstein stated, "We can't solve problems by using the same kind of thinking we used when we created them." Through awareness of sustainability, defined in the next section, we can simultaneously advance society, the environment, and the economy for the long-term prosperity of future generations. Engineers, in particular, have a unique role to play, because they have a direct effect on the design and development of products, processes, and systems, as well as on natural systems through material selection, project siting, and the end-of-life handling of products.

The world's population exceeds 6 billion, and 80 million people are added each year. Resource consumption per capita also is on the rise. For example, over 25 percent of the possible terrestrial and aquatic solar energy captured in photosynthesis by primary producers (plants and cyanobacteria) is now appropriated by humans. Just two more doublings of the human impact on the world's natural resources—through a combination of population increase and consumption-fueled economic growth—would result in 100 percent of the net primary production being utilized by humans. This ecological impossibility would leave ecosystems with nothing. The results could also have catastrophic implications for humans, because of our well-established reliance on ecosystems for economic prosperity and health (Daly, 1986).

**Laws + Regulations**
http://www.epa.gov/lawsRegs

© Steve Geer/iStockphoto.

**The Story of Stuff**
http://www.storyofstuff.com

---

### Box / 1.2    Tragedy of the Commons

**"The Tragedy of the Commons"** describes the relationship where individuals or organizations consume shared resources (for example, freshwater, fish from the ocean) and then return their wastes back into the shared resource (air, land). In this way, the individual or organization receives all of the benefit of the shared resource but distributes the cost across anyone who also uses that resource. The tragedy arises when each individual or organization fails to recognize that everyone else is acting in the same way.

It is this logic that has led to the current situation in ocean fisheries, the Amazon rain forest, and global climate change. In each case, the consumptive behavior of a few has led to a significant impact on the many—and the destruction of the integrity of the shared resource.

---

As the world's population and per capita consumption increases, so does the urgency for engineers to protect and enhance the environments and communities where people reside. This, however, will present numerous challenges to engineers. The United Nations Environment Programme, described in the next section, lists ten existing or emerging environmental issues (Table 1.1). Engineers are—or soon will be—engaged in developing sustainable solutions for all these issues.

## Table / 1.1

### Existing and Emerging Environmental Issues

1. Globalization, trade, and development
2. Coping with climate change and variability
3. Growth of megacities
4. Human vulnerability to climate change
5. Freshwater depletion and degradation
6. Marine and coastal degradation
7. Population growth
8. Rising consumption in developing countries
9. Biodiversity depletion
10. Biosecurity

SOURCE: United Nations Environment Programme, 2002.

## 1.2 Defining Sustainability

If you Google the words *sustainability, sustainable development*, and *sustainable engineering*, you will get more than 300 definitions. Try it! The abundance of varying definitions makes grasping the concept of sustainability difficult for some.

> **Sustainable engineering** *is defined as the design of human and industrial systems to ensure that humankind's use of natural resources and cycles do not lead to diminished quality of life due either to losses in future economic opportunities or to adverse impacts on social conditions, human health, and the environment. (Mihelcic et al., 2003)*

Under this definition, sustainability requires integrating the three elements of the triple bottom line (environment, economy, society). Most definitions incorporate the triple bottom line, along with the aim of meeting the needs of current and future generations. This definition, unlike many in the sustainability literature, explicitly describes a role for engineers by highlighting the design of human-made systems.

Worldwide, numerous discussions have taken place over the past several decades, yielding significant contributions to the concept of sustainability. Engineers should understand the historical context of these discussions. For example, the United Nations (UN) Conference on the Human Environment, held in Stockholm in 1972, is significant because for the first time, it added the environment to the list of global problems. As evidence, Principle 1 of the conference's Stockholm Declaration states:

> *Man has the fundamental right to freedom, equality, and adequate conditions of life, environment of quality that permits a life of dignity*

**Global Environmental Outlook**
http://www.unep.org/GEO

and well-being, and he bears a solemn responsibility to protect and improve the environment for present and future generations.

Principle 2 states:

*The natural resources of the earth including air, water, land, flora, and fauna and especially representative samples of natural ecosystems must be safeguarded for the benefit of present and future generations through careful planning and management, as appropriate.*

The Stockholm conference also resulted in the creation of the **United Nations Environment Programme (UNEP)**. See www.unep.org to learn more about this organization and global environmental problems. The UNEP's stated mission is

*to provide leadership and encourage partnership in caring for the environment by inspiring, informing, and enabling nations and peoples to improve their quality of life without compromising that of future generations. (www.unep.org)*

Also in 1972, an influential book was published by the Club of Rome, a group of 30 individuals from ten countries who organized in 1968 to discuss the present and future predicament of the human race. The book, *The Limits to Growth* (Meadows et al., 1972), warned of the limitations of the world's resources and pointed out there might not be enough resources remaining for the developing world to industrialize. In *The Limits to Growth*, the authors used mathematical models to demonstrate that "the basic behavior mode of the world system is exponential growth of population and capital, followed by collapse."

Carrying capacity is a way to think of resource limitations. **Carrying capacity** refers to the upper limit to population or community size (for example, biomass) imposed through environmental resistance. In nature, this resistance is related to the availability of renewable resources, such as food, and nonrenewable resources, such as space, as they affect biomass through reproduction, growth, and survival.

One solution to the world's environmental problems is to use technological advances to solve the issue of dwindling or harder-to-extract resources. However, as demonstrated in *The Limits to Growth*, in the past, society has "evolved around the principle of fighting against limits rather than learning to live with them." This is demonstrated in Figure 1.1 for the whaling industry. Historically, humans could live within a system of finite resources. Not only did they have access to a relatively large amount of resources and available land, but they also had a limited population that produced a limited amount of pollutants. However, with population increasing and industrial production and consumption on the rise, this historical trend of a world that can moderate the environmental impact of humans might not be feasible in the long term.

In 1987, the UN World Commission on Environment and Development released *Our Common Future*. This book is also referred to as the **Brundtland Commission** report because Gro Brundtland, a former prime minister of Norway, chaired the commission. This influential report not only adopted the concept of sustainable development

**Class Discussion**
How is the discipline of environmental engineering grounded in these two principles?

**Class Discussion**
Is it better to live within a determined limit by accepting some restrictions on consumption-fueled growth?

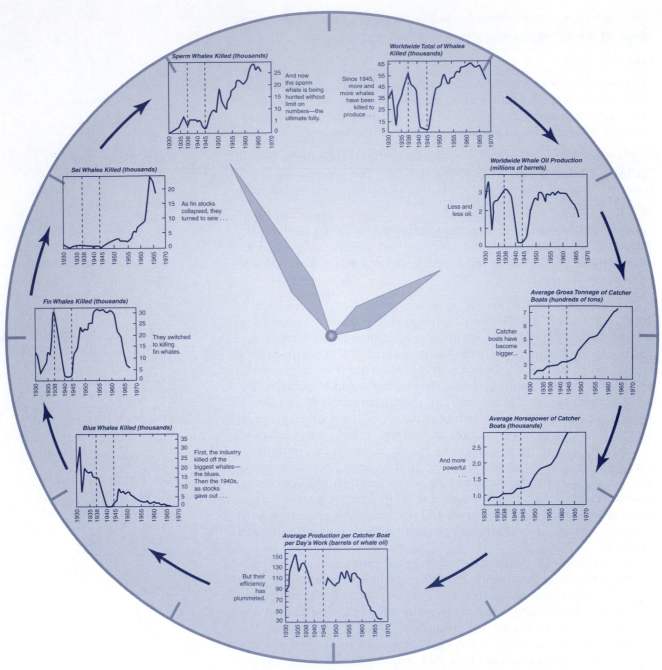

The following labels and text appear within the figure:

**Sperm Whales Killed (thousands)**
And now the sperm whale is being hunted without limit on numbers—the ultimate folly.

**Worldwide Total of Whales Killed (thousands)**
Since 1945, more and more whales have been killed to produce . . .

**Sei Whales Killed (thousands)**
As fin stocks collapsed, they turned to seis . . .

**Worldwide Whale Oil Production (millions of barrels)**
Less and less oil.

**Fin Whales Killed (thousands)**
They switched to killing fin whales.

**Average Gross Tonnage of Catcher Boats (hundreds of tons)**
Catcher boats have become bigger...

**Blue Whales Killed (thousands)**
First, the industry killed off the biggest whales—the blues. Then the 1940s, as stocks gave out . . .

**Average Horsepower of Catcher Boats (thousands)**
And more powerful . . .

**Average Production per Catcher Boat per Day's Work (barrels of whale oil)**
But their efficiency has plummeted.

**Figure 1.1 Limits to Growth and Technology of the Whaling Industry** Maintaining growth in a limited system by advances in technology will eventually result in extinction for both whales and the whaling industry. As wild herds of whales are destroyed, finding the survivors has become more difficult and has required more effort. As larger whales are killed off, smaller species are exploited to keep the industry alive. Without species limits, large whales are always taken wherever and whenever encountered. Thus, small whales subsidize the extermination of large ones.

Adapted from Payne, 1968.

but also provided the stimulus for the 1992 UN Conference on Environment and Development, known as the Earth Summit. The Brundtland Commission report defined **sustainable development** as "development which meets the needs of the present without compromising the ability of the future to meet its needs."

The 1992 Earth Summit, held in Rio de Janeiro, Brazil, was the first global conference to specifically address the environment. It also integrated for the first time environmental and economic issues. One outcome of this Rio summit was the nonbinding agenda for the 21st century, titled, *Agenda 21*, which set forth goals and recommendations related to environmental, economic, and social issues. In addition, the UN Commission on Sustainable Development was created to oversee implementation of *Agenda 21*. The complete document is available at the UNEP Web site (www.unep.org).

At the 2002 World Summit on Sustainable Development in Johannesburg, South Africa, world leaders reaffirmed the principles of sustainable development adopted at the Earth Summit ten years earlier. They also adopted the **Millennium Development Goals** (MDGs), listed in Table 1.2. The MDGs are an ambitious agenda for reducing poverty and improving lives based on what world leaders agreed on at the Millennium Summit in September 2000. For each goal, one or more targets have been set, most for achievement by 2015, using 1990 as a benchmark. These goals are lofty and far-reaching, and the follow-up commitments, including financial and human capital, are uncertain at this time.

Regardless, the eight MDGs as policy goals present a vision of a better world that can guide engineering innovation and practice for the next several decades. They represent commitments to reduce poverty and hunger and to tackle ill health, gender inequality, lack of access to

**Class Discussion**

What relationship can you see between this legal argument and the definition of sustainable development proposed by the Brundtland Commission?

Investigate the public trust doctrine as it applies to the public's access to navigable waters and shores in your state. What type of access is legally guaranteed to fishers, boaters, and beach users?

**Millennium Development Goals**

**Millennium Development Goals**

You can go to www.un.org to learn more about progress towards meeting the MDGs.

**Millennium Development Goals (MDGs)** MDGs are an ambitious agenda embraced by the world community for reducing poverty and improving lives of the global community. Learn more at www.un.org/millenniumgoals/.

| Millennium Development Goal | Background | Example Target(s) (of 21 total targets) |
|---|---|---|
| 1. Eradicate extreme poverty and hunger. | More than a billion people still live on less than $1 a day. | 1a) Halve the proportion of people living on less than $1 a day and those who suffer from hunger. |
| 2. Achieve universal primary education. | As many as 113 million children do not attend school. | 2a) Ensure that all boys and girls complete primary school. |
| 3. Promote gender equality and empower women. | Two-thirds of illiterates are women, and the rate of employment among women is two-thirds that of men. | 3a) Eliminate gender disparities in primary and secondary education preferably by 2005, and at all levels by 2015. |
| 4. Reduce child mortality. | Every year, nearly 11 million young children die before their 5th birthday, mainly from preventable illnesses. | 4a) Reduce by two-thirds the mortality rate among children under 5. |
| 5. Improve maternal health. | In the developing world, the risk of dying in childbirth is one in 48. | 5a) Reduce by three-quarters the ratio of women dying in childbirth. |
| 6. Combat HIV/AIDS, malaria, and other diseases. | 40 million people are living with HIV, including 5 million newly infected in 2001. | 6a and 6c) Halt and begin to reverse the spread of HIV/AIDS and the incidence of malaria and other major diseases. |
| 7. Ensure environmental sustainability. | More than 1 billion people lack access to safe drinking water and more than 2 billion people lack sanitation. | 7a) Integrate the principles of sustainable development into country policies and programs and reverse the loss of environmental resources. 7b) Reduce by half the proportion of people without access to safe drinking water. 7c) Achieve significant improvement in the lives of at least 100 million slum dwellers. |
| 8. Develop a global partnership for development. | | 8a) Develop further an open, rule-based, predictable, nondiscriminatory trading and financial system. 8b) Address the special needs of the least-developed countries. 8c) Address the special needs of landlocked countries and small island developing states. 8d) Deal comprehensively with the debt problems of developing countries through national and international measures to make debt sustainable in the long term. 8e) In cooperation with pharmaceutical companies, provide access to affordable, essential drugs in developing countries. 8f) In cooperation with the private sector, make available the benefits of new technologies, especially information and communications. |

SOURCE: www.un.org/millenniumgoals/

clean water, and environmental degradation. This is a good example of the link between policy and engineering; policy can drive engineering innovation, and new engineering advancements can encourage the development of policies with advanced standards that redefine "best available technologies."

## 1.3 Issues That Will Affect Engineering Practice in the Future

As noted in the previous section, during the past 40 years, there has been increased attention to global sustainability, with a growing consensus that the world faces serious challenges in terms of long-term economic growth, societal prosperity, and environmental protection. These challenges arise from current scientific, technical, and policy approaches and from the behavior of individuals, communities, corporations, and government.

In the ongoing debate over the major challenges to sustainability, key problems and most engaged in solutions involve engineering systems related to water quality, climate, air quality, sanitation, waste management, health, energy, food production, chemicals and materials, and the built environment. These issues pose local and global challenges that uniquely affect communities located in every part of the world and are closely related to population and demographics. Solutions will require an integrated approach that combines technology, governance, and economics. With an understanding of these broader issues, current engineering design can be engaged more effectively to advance the goal of local, regional, and global sustainability. This section provides an overview of some challenges that engineers will face in this century.

### 1.3.1 POPULATION AND URBANIZATION

The current global population of 6 billion is expected to reach 9 billion to 10 billion people during this century (Figure 1.2). The impact of **population growth** has long been understood as one of the grand challenges to mutually advancing environmental, economic, and societal goals and creating a sustainable future. It also has a great impact on how we manage natural resources and design and invest in engineering infrastructure. Most population growth is occurring in the developing world, especially in urban areas, while population is stagnant—and in some cases declining—in much of the industrialized world.

This pattern of population growth suggests that—within the complexities of growing populations, including birth and mortality rates, sociopolitical pressures, access to health care and education, gender equality, and cultural norms—an empirical correlation exists between the rate of population growth and the level of economic development, which often is equated with quality of life. This relation would mean that meeting the challenges of stabilizing population growth and advancing the goal of sustainability is possible through improved quality of life and expanded development that is equitable and thus sustainable. Historically, however, increases in development and quality

Population

**Figure 1.2**  **Global Population, 1750–2000, and Projected Increases to 2150**
Increases in the population are attributed to developed or developing countries. For the first time in history, urban population exceeds rural population, and most of the population growth expected over the next century will be added to urban environments.

United Nations, 2006.

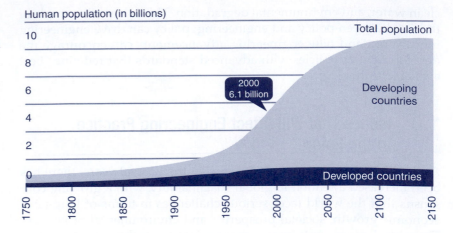

Human population (in billions)

Total population

2000
6.1 billion

Developing countries

Developed countries

---

| Box / 1.4 | Defining Developed and Developing Countries |

While there is no single definition of a **developed country**, the generally accepted concept refers to countries that have reached relatively high levels of economic achievement through advanced production, increased per capita income and **consumption**, and continued utilization of natural and human resources. Japan, Canada, the United States, Australia, New Zealand, and most countries in northern and western Europe are considered developed countries.

A **developing country** is one that has not reached the stage of economic development characterized by the growth of industrialization. In developing countries, the national income is less than the amount of money needed for basic infrastructure and human services, leading to a relatively low standard of living, an undeveloped industry base, and substandard per capita income.

---

of life have been inextricably linked with consumption and associated resource depletion and environmental degradation. A significant amount of evidence suggests that an increasing human population places additional strain on natural resources as society begins to develop its infrastructure. The opportunity for the engineering community is to continue to develop and enhance quality of life through the protection and restoration of ecosystems and to design, develop, implement, and maintain infrastructure that does not have the historical consequences of environmental degradation, resource consumption, and adverse and unjust impacts on society.

One of the environmental issues listed in Table 1.1 is the growth of megacities, a process called **urbanization**. For the first time in human history, urban population exceeds rural population. In fact, by 2030, 61 percent of the global population is expected to live in urban areas. Urbanization is widely recognized to be a source of health problems. For example, 30 to 60 percent of the urban population in the developing world lacks adequate sanitary facilities, drainage systems, and piping for clean water.

### 1.3.2  HEALTH

The **World Health Organization (WHO)** (see www.who.org) estimates that poor environmental quality contributes to 25 percent of all

## Table / 1.3

**Distribution of Global Population Not Served with Improved Water Supply and Sanitation**

| Region* | % of Population Lacking Improved Water Supply | % of Population Lacking Improved Sanitation | 2000 Population (in millions) |
|---|---|---|---|
| Asia | 19 | 52 | 3,683 |
| Africa | 38 | 40 | 784 |
| Latin America and Caribbean | 15 | 22 | 519 |
| Oceania | 12 | 7 | 30 |
| Europe | 4 | 8 | 729 |

\* Coverage in the United States and Canada approaches 100%.

SOURCE: Data from WHO and UNICEF, 2000.

preventable illnesses in the world. In addition, WHO reports that 900 million people lack access to an **improved water supply**—a household connection, public standpipe, borehole, protected dug well, protected spring, or rainwater collection (see Table 1.3). (Bottled water is not considered an improved water supply.) Access to adequate sanitation is even worse, with 2.5 billion people lacking access to any type of sanitation equipment. One consequence is devastating ecological impact on surface waters that receive domestic water processed by households and businesses, because more than 90 percent of the wastewater in developing countries and 33 percent in developed countries is not treated (WHO, 1999). This has led to dire consequences for downstream communities' water supplies and for fishing communities dependent on aquatic ecosystems for their economic livelihood.

Because many disease-causing vectors are transmitted through contact with water, air, and solid waste, health issues are critical to the environmental engineering profession. As WHO points out, health is inextricably linked to sustainable development:

*Health is both a resource for, as well as an outcome of, sustainable development. The goals of sustainable development cannot be achieved when there is a high prevalence of debilitating illness and poverty, and the health of a population cannot be maintained without a responsive health system and a healthy environment. Environmental degradation, mismanagement of natural resources, and unhealthy consumption patterns and lifestyles impact health. Ill-health, in turn, hampers poverty alleviation and economic development (WHO, 2005).*

HIV/AIDS, tuberculosis, and malaria are among the world's largest killers. All have their greatest impact on developing nations, interact in ways that make their combined impact worse, and create an enormous economic burden on families and communities—especially those where economic livelihood depends on good health (UNESA, 2004). Figure 1.3 shows the types of **environmental risk** that lead to the greatest loss of disability-free days of a person's life. Much of the burden of

**Figure 1.3** Environmental Risk and Loss of Disability-Free Days

Data from Ezzati et al., 2004.

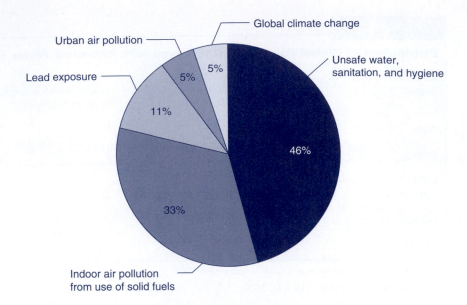

this risk is assumed by people living in the developing world. Note that almost half of the risk is associated with poor access to drinking water and sanitation, and much of the other half is due to exposure to indoor and outdoor air pollution.

WHO (2004) explains that health problems become economic problems:

*For people living in poverty, illness and disability translate directly into loss of income. This can be devastating for individuals and their families who are dependent on their health for household income.*

The effects of ill health have significant ramifications at the macroeconomic scale as well. For instance, a significant portion of Africa's economic shortfall is attributed to climate and disease burden.

Environmental degradation can have an even more direct effect on household income. The income derived from ecosystems (that is, **environmental income**) provides a "fundamental stepping stone in the economic empowerment of the rural poor" (WRI, 2005). This "natural capital" provided by the environment is the stock that yields the flow of natural resources. Those resources may be either renewable (for example, fish, trees) or nonrenewable (for example, petroleum). Nonrenewable natural capital can be depleted, while renewable natural capital can either be left alone to regenerate or be cultivated with the use of human-made capital, such as fish ponds, cattle herds, or forest plantations.

### 1.3.3 WATER SCARCITY, CONFLICT, AND RESOLUTION

Water scarcity is a situation where there is insufficient water to satisfy normal human requirements. Normal human requirements are perhaps visualized best by the World Health Organization definition for *reasonable access* to a water source: availability of at least 20 L/capita-day from a source within 1 km of the user's dwelling. Figure 1.4 depicts a daily occurrence in the developing world, young children collecting water from a source located far from home.

**Figure 1.4** Daily Activity of Collecting Water that is Seen in much of the World

(Photo courtesy of James R. Mihelcic).

A country is defined as experiencing **water stress** when annual water supplies drop below 1,700 m$^3$ per person. When annual water supplies drop below 1,000 m$^3$ per person, the country is defined as **water scarce**. By one measure, nearly 2 billion people now suffer from severe water scarcity. Furthermore, of the additional one billion people expected to face water scarcity by the year 2025, 20% will be associated with direct effects of climate change (Vörösmarty et al. 2000).

Figure 1.5 shows the countries currently experiencing water stress or water scarcity. By the year 2025, additional countries will encounter

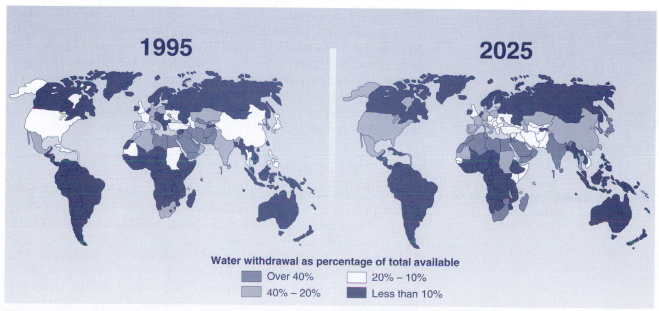

**Figure 1.5** Countries Facing Water Stress or Scarce Conditions, 1995 and 2025 (Projected)

Data from World Meteorological Organization and figure adapted from United Nations Environment Programme, 2007.

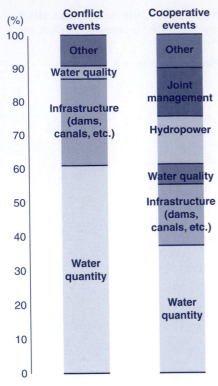

(%)

**Conflict events**

**Cooperative events**

**Figure 1.6 Water Cooperation or Water Conflict?** Shown are the percent of events that have either caused water conflicts or lead to water cooperation (Redrawn with permission from: UNEP/GRID-Arendal, Water–cooperation or conflict?, *UNEP/GRID-Arendal Maps and Graphics Library*, http://maps.grida.no/go/graphic/water-cooperation-or-conflict).

**Class Discussion**

Research and discuss a specific **water conflict** in your region or globally.

**How Clean Is the Energy You Use?**

http://www.epa.gov/cleanenergy

water scarcity (see Figure 1.5). The water stress indicator in these maps measures the proportion of water withdrawal with respect to total renewable resources. It is a criticality ratio, which implies that water stress depends on the variability of resources. Water stress causes deterioration of fresh water resources in terms of quantity (over-exploitation of groundwater, dry rivers, etc.) and quality (organic matter pollution, eutrophication, saltwater intrusion, etc.).

Water is expected to be a source of both tension and cooperation in the future. This is because more than 215 major rivers and 300 groundwater aquifers are shared by two or more countries. The Organisation for Economic Co-operation and Development (OECD) consists of 30 member countries. The OECD Development Assistant Committee writes that *"Water-related tensions can emerge on various geographical scales. The international community can help address factors that determine whether these tensions will lead to violent conflict. Water can also be the focus of measures to improve trust and cooperation."*

The following website chronicles water conflict going back to 3,000 B.C. (http://www.worldwater.org/conflictchronology.pdf). History shows that most water conflict is resolved peacefully. In fact, there have been 507 recorded water conflicts and 1,228 recorded water cooperative events. However, there have been less than 40 recorded reports of violence over water. This shows that water conflict is perhaps not as sensational as popularized in movies like *Chinatown* and books like *Cadillac Desert*. Figure 1.6 shows the specific events that historic water conflict or cooperation was related to. As seen in this figure, most of the documented events associated with water conflict and cooperation are related to changes in the quantity of water flow and design and construction of infrastructure like dams and canals.

Finding sustainable economic solutions to water infrastructure problems is another challenge. The funding gap between current levels and needed levels of investment in the U.S.'s water and wastewater infrastructure is estimated to be equally split and cost hundreds of billions of dollars. UN-Habitat estimates the cost of meeting the Millennium Development Goal, Target 11, committed to improving lives of at least 100 million slum dwellers, is $67 billion. However, another 400 million people will be in slums by 2020. These people will require an additional $300 billion if they are to have access to basic services and decent housing. One of these basic services is providing improved water and sanitation.

## 1.3.4 ENERGY AND CLIMATE

U.S. **energy consumption** in all sectors has increased in the past 30 years and is projected to increase in the future (Figure 1.7a). Much of the energy consumption is in sectors designed, constructed, and managed by engineers (for example, transportation and residential and commercial buildings). Figure 1.7b shows the breakdown of fuels that provide electricity in the United States, including the small percentage of U.S. energy needs currently—and projected to be—provided by renewable energy sources.

Figure 1.8 illustrates the energy consumption of North America (including the United States, Canada, and Mexico) on a per capita basis

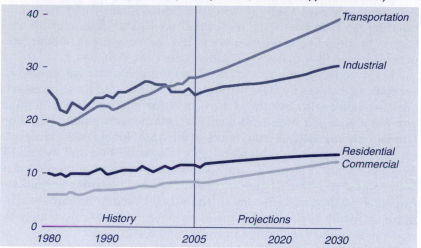

**(a) Delivered energy consumption by sector, 1980–2030 (quadrillion Btu)**

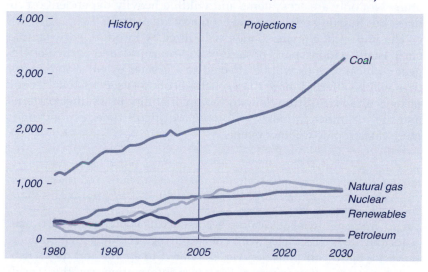

**(b) Electricity generation by fuel, 1980–2030 (billion kilowatt-hours)**

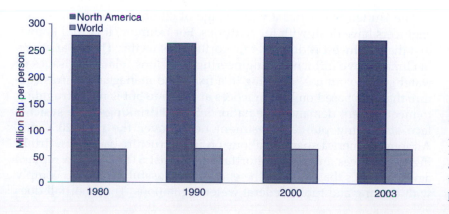

**Figure 1.8 Annual Energy Consumption per Capita in North America and the Rest of the World**
North America's energy consumption is approximately four times higher than the rest of the world over this 23-year period

compared with the rest of the world since 1980. In 2003, North America's per capita energy consumption was approximately four times higher than that of the rest of the world. This demonstrates that engineers should be concerned about the source and use of energy in every decision they make, especially in the United States.

Energy consumption is one reason why **greenhouse gas** emissions are causing changes in global **climate**. The majority of these emissions are associated with burning fossil fuels for energy, with a smaller amount associated with land use. The **Intergovernmental Panel on Climate Change (IPCC)** (cowinners of the 2007 Nobel Peace Prize) was established by the World Meteorological Organiza-tion and the UNEP to assess scientific, technical, and socioeconomic information related to better understanding of climate change. (For more information, see http://www.ipcc.ch/.) The more than 2,000 notable scientists who make up the IPCC predict that the likely range of temperature increase in the next century will range from 2.4°C to 6.4°C.

**Global Climate Change**

The global consequences of warming will be significant. Figure 1.9 shows the expected impacts on water, ecosystems, food, coastal areas, and health as they relate to the specific increase in global mean temperature. Not only are ecosystems and wildlife heavily dependent on climate, but human health and the economy are as well.

The impact of climate change will differ by location. For example, small island nations, parts of the developing world, and particular U.S. geographical regions will be affected to a greater extent. Some industries will be affected more than others. Economic sectors that depend on agriculture will struggle with more variability in weather patterns, and the insurance industry will have a difficult time responding to more catastrophic weather events.

---

**Box / 1.5**   **Climate and Health**

Climate change also has a relationship to human health. Increases in temperature have a documented relationship to illness and death. Common vector-born diseases like malaria and dengue are sensitive to temperature. The impact on health and infrastructure from extreme weather events such as flooding and hurricane associated with climate change could be great. Indoor air pollution issues related to spores and mold also may increase. For more information, see the World Health Organization Web site at www.who.int/en.

---

The United States has a wide range of climate conditions, which engineers have dealt with for centuries. For example, it is well known that the Southwest is dry and the Northeast is wetter. These variations in climate have influenced engineering decisions related to issues of water supply and use, resulting in a fixed and manageable infrastructure that was based on best practices at the time but is now struggling to meet current demand. The nation's 54,000 drinking-water systems face staggering public investment needs over the next 20 years. Although America spends billions on infrastructure each year, drinking water faces an annual shortfall of at least $11 billion to replace aging facilities that are near the end of their useful life and to comply with existing and future federal water regulations. That shortfall does

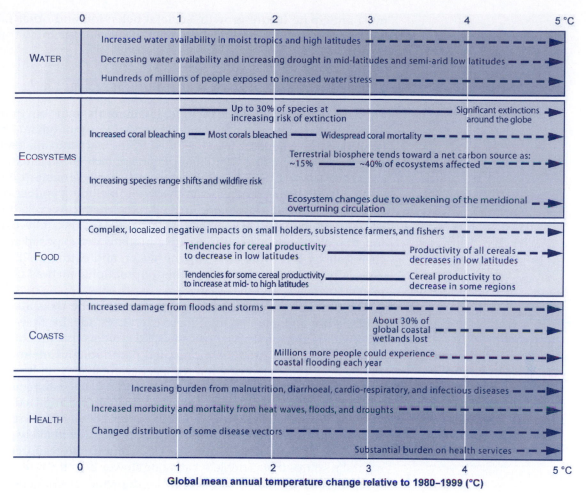

**Figure 1.9** **Expected Impacts from Climate Change** The type and extent of the impact will be influenced by the magnitude of the temperature increases.

With permission of Intergovernmental Panel on Climate Change, Climate Change 2007: Impacts, Adaption and Vulnerability, Summary for Policymakers, Table SPM.2, 2007.

not account for any growth in the demand for drinking water over the next 20 years (ASCE, 2005).

As climate, population, and demographics change in the future, engineers must not only incorporate technological advances to reduce energy and water usage, but also make use of renewable sources of energy and materials. Engineers designing infrastructure

**Read the World Health Report**

http://www.who.int/whr/en/index.html

---

**Box / 1.6** **The 2 Percent Solution**

Some scientists say the most dangerous impacts of global warming can be curbed if overall emissions are reduced by 80 percent by 2050. This might seem like a lot, but it would only require that everyone cut his or her carbon dioxide emissions by 2 percent a year from now until 2050. If you search on the Internet for keywords like *two percent* and *climate change*, you will find links to information about how you can implement voluntary emission reductions and be part of the 2 percent solution.

must anticipate future growth, societal behavior, and other factors that affect demand during the intended lifetime of the infrastructure projects.

### 1.3.5  TOXIC CHEMICALS AND FINITE RESOURCES

The use, generation, and release of **toxic chemicals** to the environment remains a global issue. In the United States alone, more than 4 billion pounds of toxic chemicals were released by industry into air, land, and water in 2004, including 72 million pounds of recognized carcinogens, according to the EPA's Toxics Release Inventory. **Persistent organic pollutants (POPs)** and other toxic chemicals, including **endocrine disruptors**, are serious global concerns. As these chemicals cycle through natural and human systems, they pose significant risks to ecosystem function and human health, because humans are exposed to these chemicals by breathing air, drinking water, and eating food. This is especially important for susceptible populations such as children, pregnant women, and the elderly.

Engineers play a significant role in reducing the risks associated with the use and generation of these chemicals. Ideally, they achieve this by designing products, processes, and systems that do not specify these chemicals in production, repair, operation, and maintenance. Another important contribution engineers can make is understanding the fate and transport of these chemicals so that damage to natural systems and exposure to humans can be eliminated or minimized.

When it comes to materials, another concern beyond toxicity is our current reliance on **nonrenewable resources**, which will likely grow in magnitude as the population increases. A **renewable resource** is any natural resource that is depleted at a rate slower than the rate at which it regenerates or that is unlikely to be depleted in the conceivable future. For the current population of our planet to live at the same quality of life as we do in America, it would require the resources of four Earths (Rees, 2006).

Engineers can contribute to meeting this challenge in several ways. The first is to incorporate renewable resources into designs and specifications. The second is to design products, processes, and systems for high material efficiency, reducing the amount of material acquired, manufactured, and later wasted. There is also significant opportunity to improve our current material efficiency. Recent analysis has found that, of all raw materials used in manufacturing processes, 94 percent ends up as waste. In addition, 99 percent of the original material used in the production of or contained in U.S. goods becomes waste within six weeks of sale (Lovins, 1997). The majority of those discarded materials are from nonrenewable resources, particularly petroleum, which adds to environmental and human health impacts.

### 1.3.6  MATERIAL FLOWS AND THE BUILT ENVIRONMENT

The **built environment** is where we live, work, shop, study, and play. It includes everything that is built: buildings, roads, bridges, and harbors. While only 2 to 3 percent of North America's land area is built on,

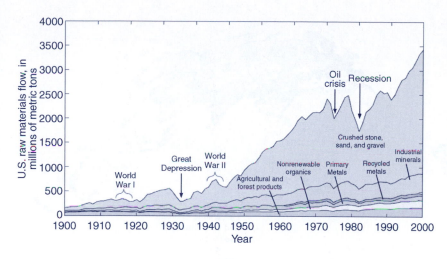

**Figure 1.10** **Flow of Raw Materials in the U.S. by Weight, 1900–1998** The use of raw materials has increased extensively in the 20th century and much of it is used in the engineering profession as aggregates, cement (which is included in industrial minerals), and steel reinforcement (which is included in primary metals).

From U.S. Geological Survey, Wagner, 2002.

approximately 60 percent of this land area is now affected by the built environment (UNEP, 2002).

The built environment also requires a tremendous amount of water, energy, and natural resources for its construction and operation. An analysis of **materials flow** in the United States (Figure 1.10) shows that approximately 85 percent of the U.S. material flow by weight is associated with items such as aggregate, cement, steel reinforcement, and wood—materials incorporated into engineering infrastructure. (Note that water is not included in this analysis, but if included, it would be the single largest material flow.)

Aggregates are used in foundations and in production of concrete, which is made up of cement, sand, stone, and water. Industrial materials include Portland cement and wallboard used in residential and commercial construction. The **embodied energy**—the amount of energy required in the life cycle stages of acquisition of raw materials, manufacturing, use, and end of life—of concrete has a significant impact on our current energy flows (Table 1.4). The transportation of aggregates and cement to job sites accounts for more than 10 percent of the total embodied energy. In addition, production of 1 kg of Portland cement results in production of approximately 1 kg of $CO_2$. When

| Table / 1.4 | | |
|---|---|---|
| **Embodied Energy Associated with Components of Concrete** | | |
| **Material** | **% Material by Weight** | **% of Embodied Energy** |
| Cement | 12 | 94 |
| Sand | 34 | 1.7 |
| Crushed stone | 48 | 5.9 |
| Water | 6 | 0 |

SOURCE: Horvath, 2004, reprinted with permission from *Annual Review of Environment and Resources*, Volume 29, © 2004 by Annual Reviews, www.annualreviews.org.

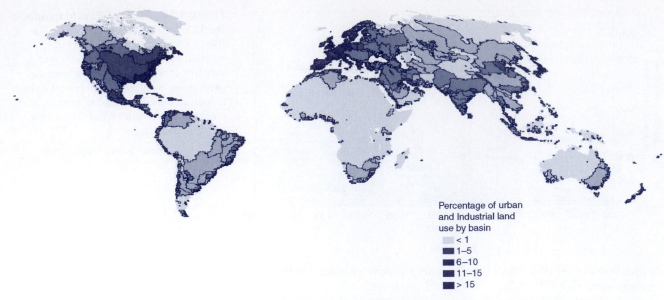

Percentage of urban
and Industrial land
use by basin
■ < 1
■ 1–5
■ 6–10
■ 11–15
■ > 15

**Figure 1.11   World Urban and Industrial Land Use by River Basins**   Highly urbanized watersheds are concentrated along the east coast of the United States, Western Europe, and Japan. Lesser urban and industrial land use occurs in coastal China, India, Central America, most of the United States, Western Europe, and the Persian Gulf. (Redrawn with permission of World Resources Institute, 2000. Pilot Analysis of Global Ecosystems: Freshwater systems. World Resources Institute, Washington, D.C.).

considering the end-of-life stage for the life cycle of engineering materials, 13 to 19 percent of solid waste is construction and demolition debris, half of this is concrete (by volume), and only 20 to 30 percent of this material is recycled (Horvath, 2004).

The built environment also affects the local heating of urban areas—termed the **urban heat island**—as well as the quantity and quality of water that cycles through it. Figure 1.11 shows the intensity of **land use** around the world. Many highly urbanized watersheds are located along the East Coast of the United States, and similar land use intensity exists in Western Europe and Japan. There are less-dense urban concentrations in the remainder of the United States and coastal areas of China, India, Central America, and the Persian Gulf. These urban areas were built up by changing and removing rivers, lakes, forests, and wetlands. This has had great impact on the infiltration properties, transpiration rates, and associated runoff from these watersheds. Impervious surfaces (discussed in Chapters 8 and 14) that cover our roads increase not only the volume, but also the rate at which water runs off. This has diminished the recharge of groundwater resources as well as the quality of receiving water bodies. Loss of wetlands in these areas has exacerbated the effects.

Moving out from urban centers and creating suburban developments, also known as "sprawl," creates problems as well. This type of growth requires significant material and energy to develop and maintain; causes people to commute greater distances, often in personal cars rather than on public transit; and can contribute to the fragmentation of communities, wilderness, and open space. One strategy to mitigate the growing concerns with sprawl is through initiatives such as **smart growth** or **new urbanism**. Both of these approaches to urban development are focused on designing communities that preserve natural lands, protect water and air quality, and reuse developed land. By designing mixed-use neighborhoods (residential, shops, offices, and schools), communities are creating environments where residents can walk or bike, use public transportation, and drive their car—with a focus on living and working within the same community. The high quality of life in these communities makes them economically competitive, creates business opportunities, and improves the local tax base (EPA, 2007).

## 1.4 The Sustainability Revolution

Society and the engineering community are at the onset of the **sustainability revolution**, one of several revolutions occurring over the past 10,000 years that have changed how humans interact with natural systems. The first two revolutions were the agricultural and industrial revolutions, which took place over many decades and centuries. Although the pressures on the environment may be greater because of population and technology, we are near the beginning of this newest revolution.

When the agricultural revolution began 10,000 years ago, world population was approximately 10 million. By 1750, it had grown to an estimated 800 million. An agriculture-based society was not necessarily more productive or efficient. In fact, some believe it was simply the basis to accommodate an increasing population. For example, the agricultural revolution resulted in food of a lower nutritional value on a per acre basis; however, it limited challenges to wildlife and wilderness. The agricultural revolution also brought forward concepts of land ownership, feudalism, wealth, status, trade, money, power, guilds, temples, armies, and cities (Meadows et al., 2004).

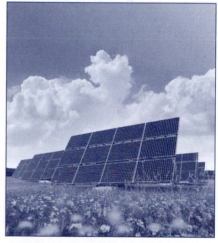

iStockphoto.

Over the relatively recent time frame of the industrial revolution, the global population increased to more than 6 billion. The industrial revolution brought machines, capitalism, roads, railroads, combustion, smokestacks, factories, and large urban areas. Along with the agricultural revolution, it brought the world the current and emerging environmental issues listed in Table 1.1. The industrial revolution also contributed to our vocabulary, with commonly used words that describe many modern environmental challenges and opportunities. Table 1.5 lists some of the words common to our vocabulary as a result of the industrial revolution. It also lists words that are becoming common during the sustainability revolution.

Engineers can significantly contribute to the success of the sustainability revolution through their power and potential to design and manage the future through innovative, sustainable design. Engineers can play a significant role in achieving a sustainable future by thinking beyond incremental improvements to leapfrog technologies, providing services without physical entities, and designing and engineering with intent. As mentioned earlier, Einstein said that a new level of awareness is needed to create solutions—in this case, to design a better tomorrow.

## Table / 1.5

**Vocabulary of the Industrial Revolution and Sustainability Revolution**

| Industrial Revolution | Sustainability Revolution |
|---|---|
| Nonrenewable energy | Renewable energy |
| Waste | Efficiency |
| Climate change | Ecological restoration |
| Consumption | Resource equity |
| Accumulation | Social and environmental justice |
| Toxicity, smog, persistent organic pollutants, endocrine disruptors | Green chemistry |
| Transportation | Accessibility |
| Concrete hydraulic channels | Low-impact storm water development, rain gardens |
| Urban heat island effect | Green roofs |
| Bioaccumulation | Biodiversity |
| Industrial design | Green design |
| Gross national product (GNP) | Index of sustainable economic welfare, environmental sustainability index, genuine progress indicator |

## Key Terms

- Brundtland Commission
- built environment
- carrying capacity
- Carson, Rachel
- climate
- consumption
- developed country
- developing country
- embodied energy
- endocrine disruptors
- energy consumption
- environmental income
- environmental risk
- greenhouse gas
- improved water supply

- Intergovernmental Panel on Climate Change (IPCC)
- land use
- *The Limits to Growth*
- materials flow
- Millennium Development Goals (MDGs)
- new urbanism
- nonrenewable resources
- persistent organic pollutants (POPs)
- population growth
- public trust doctrine
- renewable resources
- smart growth
- sustainability

- sustainability revolution
- sustainable development
- sustainable engineering
- toxic chemicals
- "The Tragedy of the Commons"
- triple bottom line
- United Nations Environment Programme (UNEP)
- urban heat island
- urbanization
- water conflict
- water scarcity
- water stress
- World Health Organization (WHO)

**1.1** Identify three definitions of sustainability from three sources (for example, local, state, or federal government; industry; environmental organization; international organization; financial or investment organization). Compare and contrast those definitions with the Brundtland Commission definition. How do the definitions reflect their sources?

**1.2** Write your own definition of sustainable development as it applies to your engineering profession. Explain its appropriateness and applicability in 2–3 sentences.

**1.3** Presume you were born in a developing country. Your community's health and prosperity are threatened by climate changes caused by anthropogenic $CO_2$ emissions generated primarily by developed countries. This situation is so extreme that people in your community do not name their children until they live past 5 years old, since many die before that age. If you had a chance to talk to an engineer in a developed country, what would you say?

**1.4** Research the progress that two countries of your choice have made in meeting each of the eight Millennium Development Goals (MDGs). Summarize the results in a table. Among other sources, you might consult the UN's MDG Web site, www.un.org/millenniumgoals/.

**1.5** Research two examples of persistent organic pollutants (POPs), and write a short paragraph that describes each chemical, its most common uses and applications, and its known or suspected impacts on human health and the environment. What are some economic, societal, and environmental issues associated with your chemicals?

**1.6** Construct a plot that shows world population on the $y$-axis (since 1960) and a time line identifying major global conferences related to the environment and sustainability on the $x$-axis. Indicate any major global events (e.g., disaster, war, famine) that have contributed to the loss of environmental or social attributes that you personally value.

**1.7** Relate "The Tragedy of the Commons" to a local environmental issue. Be specific on what you mean in terms of the "commons" for this particular example, and carefully explain how these "commons" are being damaged for current and future generations.

**1.8** Research a current global environmental problem at the UNEP Web site, www.unep.org. What is the current state of the environmental resource associated with the problem? Are any future projections provided? If so, what are they?

**1.9** At the UNEP Web site, many graphs depict environmental issues on a global scale. Select one graph related to one of the issues discussed in Section 1.3. Reproduce this graph, and discuss how different regions of the world (North America, Central America and the Caribbean, South America, Europe, Asia, Oceania) are affected by this particular issue. How might future issues of population growth, urbanization, and climate change affect this issue (positively or negatively)? Present a sustainable solution to the problem that includes a balance of societal, environmental, and economic issues.

**1.10** Research the World Health Organization's Web site (www.who.org), and write a one-page essay on how health and sustainability are linked. Provide some specific examples.

**1.11** Familiarize yourself with the Web site of the Intergovernmental Panel on Climate Change (www.ipcc.ch/). Write a one-page essay on how changes in climate may affect two aspects of your engineering profession in the future.

**1.12** Research a water conflict issue in and outside the United States. What similarities and differences do you see between each conflict?

**1.13** Identify a nearby watershed that crosses state or national boundaries. What are some possible conflicts over management of this watershed? Identify at least three stakeholders.

**1.14** Go to the U.S. Department of Energy's Web site (www.doe.gov), and research energy consumption in the household, commercial, industrial, and transportation sectors. Develop a table on how this specific energy consumption relates to the percent of U.S. and global $CO_2$ emissions. Identify a sustainable solution for each sector that would reduce energy use and $CO_2$ emissions.

**1.15** Go to the EPA's Web site (www.epa.gov), and research "smart growth" or "urban heat island." Write a one-page essay that defines your topic and, importantly, relates it to engineering practice.

**1.16** Research the definition of civil society on the United Nations Web site (www.un.org). Discuss why members of civil society should be actively engaged in a sustainable solution to a common problem that engineers work on in your community (for example, water quality, air quality, transportation, accessibility, or land use).

**1.17** Go to the Web site of the American Society of Civil Engineers (www.asce.org), and research and discuss the state of infrastructure in the United States.

**1.18** At the World Health Organization Web site (www.who.org), research a health issue related to water, indoor air, or climate. In one page, discuss the issue and provide your own sustainable solution to solving this problem, including its societal, environmental, and economic components.

**1.19** In your own words, write in one paragraph how health and sustainability are related. Is your discussion influenced by whether you live in the developed or developing world? Are current transportation systems designed with health and sustainability in mind?

**1.20** Go to the Web site of the World Resources Institute (www.wri.org) and access the report *World Resources 2005: The Wealth of the Poor; Managing Ecosystems to Fight Poverty*. Select one case study, and demonstrate how people living in poverty depend on the environment for a substantial portion of their economic livelihood. Be specific on the actual amount of economic wealth the environment provides in this study.

# References

American Society of Civil Engineers (ASCE). 2005. "Report Card for America's Infrastructure, 2005," www.asce.org/reportcard/2005/index.cfm, accessed April 1, 2008.

Daly, H. E. 1986. *Beyond Growth*. Boston: Beacon Press.

Environmental Protection Agency (EPA). 2007. "About Smart Growth." Office of Policy Economics and Innovation, www.epa.gov/livability/ about_sg.htm, accessed February 8, 2007.

Ezzati, M., A. Rodgers, A. D. Lopez, et al. 2004. "Mortality and Burden of Disease Attributable to Individual Risk Factors." In *Comparative Quantification of Health Risks: Global and Regional Burden of Disease Attributable to Selected Major Risk Factors*, vol. 2, ed. M. Ezzati, A. D. Lopez, et al. Geneva, World Health Organization.

Horvath, A. 2004. "Construction Materials and the Environment." *Annual Review of Environment and Resources* 29:181–204.

Intergovernmental Panel on Climate Change (IPCC). 2007. "Summary for Policy Makers." In *Climate Change 2007: Impacts, Adaptation and Vulnerability*, ed. M. L. Parry, O. F. Canziani, J. P. Palutikof, P. J. van der Linden, and C. E. Hanson, 7–22. Cambridge: Cambridge University Press.

Lovins, A. B., L. H. Lovins, and E. U. Weizsacker. 1997. *Factor Four: Doubling Wealth, Halving Resource Use; The New Report to the Club of Rome*. London: James & James/Earthscan.

Meadows, D. H., D. L. Meadows, J. Randers, and W. W. Behrens III. 1972. *The Limits to Growth*. London: Earth Island Limited.

Meadows, D., J. Randers, and D. Meadows. 2004. *The Limits to Growth: The 30-Year Update*. White River Junction, Vt.: Chelsea Green Publishing Co.

Mihelcic, J. R., J. C. Crittenden, M. J. Small, D. R. Shonnard, D. R. Hokanson, Q. Zhang, H. Chen, S. A. Sorby, V. U. James, J. W. Sutherland, and J. L. Schnoor. 2003. "Sustainability Science and Engineering: Emergence of a New Metadiscipline." *Environmental Science & Technology* 37 (23): 5314–5324.

Payne, R. 1968. "Among Wild Whales." *New York Zoological Society Newsletter* (November).

Rees, W. E. 2006. "Ecological Footprints and Bio-Capacity: Essential Elements in Sustainability Assessment." In *Renewables-Based Technology: Sustainability Assessment*, ed. Jo Dewulf and Herman Van Langenhove, 143–158. Chichester, UK: John Wiley and Sons.

United Nations. 2006. *World Population Prospects*, 2006 revision, esa.un.org/unpp/.

United Nations Department of Economic and Social Affairs (UNESA). 2004. *World Urbanization Prospects: The 2003 Revision*, UN Sale No. E.04.XIII.6, New York: UNESA Population Division.

United Nations Environment Programme (UNEP). 2002. *Global Environmental Outlook 3*. London: Earthscan.

UNEP. 2007. *Global Environmental Outlook 4*. Valletta, Malta: Progress Press.

U.S. Department of Energy. 2007. Annual Energy Review. Energy Information Administration Web site, www.eia.doe.gov/emeu/aer/contents.html, accessed August 15, 2007.

Vörösmarty, C. J., P. Green, J. Salisbury, and R. B. Lammers. 2000. "Global Water Resources: Vulnerability from Climate Change and Population Growth." *Science* 289:284–288.

Wagner, L. A. 2002. "Materials in the Economy: Material Flows, Scarcity and the Economy," U.S. Geological Survey Circular 1221, U.S. Department of the Interior, U.S. Geological Survey.

WFEO – ComTech (World Federation of Engineering Organizations and ComTech) Engineers and Sustainable Development, 2002.

World Commission on Environment and Development. 1987. *Our Common Future*. Oxford: Oxford University Press.

World Health Organization (WHO). 1999. "The World Health Report 1995: Bridging the Gaps." Geneva: WHO.

World Health Organization (WHO). 2004. Global and Regional Burden of Diseases Attributable to Selected Risk Factors, vol. 1 and 2. Geneva, WHO.

WHO. 2005. "World Summit on Sustainable Development." WHO Web site, www.who.int/wssd/en/, accessed June 10, 2005.

WHO and United Nations Children's Fund (UNICEF). 2000. *Global Water Supply and Sanitation Assessment 2000 Report*. Geneva: WHO and UNICEF.

World Resources Institute (WRI). 2000. "Pilot Analysis of Global Ecosystems (PAGE): Freshwater Systems." Washington, D.C.: WRI.

World Resources Institute (WRI) in collaboration with United Nations Development Programme, United Nations Environment Programme, and World Bank 2005. "World Resources 2005: The Wealth of the Poor-Managing Ecosystems to Fight Poverty." Washington, D.C.

# chapter/Two Environmental Measurements

James R. Mihelcic, Richard E.
Honrath Jr., Noel R. Urban,
Julie Beth Zimmerman

*In this chapter, readers become familiar with the different units used to measure pollutant levels in aqueous (water), soil/sediment, atmospheric, and global systems.*

## Major Sections

## Learning Objectives

1. Calculate chemical concentration in mass/mass, mass/volume, volume/volume, mole/mole, mole/volume, and equivalent/volume units.

2. Convert chemical concentration to a parts per million or parts per billion basis.

3. Calculate chemical concentration in units of partial pressure.

4. Calculate chemical concentration in common constituent units such as hardness, nitrogen, phosphorus, global warming potential, carbon equivalents, and carbon dioxide equivalents.

5. Use the ideal gas law to convert between units of $ppm_v$ and $\mu g/m^3$.

6. Calculate particle concentrations in air and water.

7. Represent specific chemical concentration in mixtures to a direct effect such as oxygen depletion to express units of biochemical oxygen demand and chemical oxygen demand.

## 2.1 Mass Concentration Units

Chemical concentration is one of the most important determinants in almost all aspects of chemical fate, transport, and treatment in both natural and engineered systems. This is because concentration is the driving force that controls the movement of chemicals within and between environmental media, as well as the rate of many chemical reactions. In addition, concentration often determines the severity of adverse effects, such as toxicity, bioconcentration, and climate change.

Concentrations of chemicals are routinely expressed in a variety of units. The choice of units to use in a given situation depends on the chemical, where it is located (air, water, or soil/sediments), and how the measurement will be used. It is therefore necessary to become familiar with the units used and methods for converting between different sets of units. Representation of concentration usually falls into one of the categories listed in Table 2.1. Important prefixes to know include pico ($10^{-12}$, abbreviated as p), nano ($10^{-9}$, abbreviated as n), micro ($10^{-6}$, abbreviated as $\mu$), milli ($10^{-3}$, abbreviated as m), and kilo ($10^{+3}$, abbreviated as k).

Concentration units based on chemical mass include mass chemical per total mass and mass chemical per total volume. In these descriptions, $m_i$ is used to represent the mass of the chemical referred to as chemical $i$.

### 2.1.1 MASS/MASS UNITS

**Mass/mass concentrations** are commonly expressed as parts per million, parts per billion, parts per trillion, and so on. For example, 1 mg of a solute placed in 1 kg of solvent equals 1 $ppm_m$. **Parts per million by mass** (referred to as **ppm** or **$ppm_m$**) is defined as the number of units of mass of chemical per million units of total mass. Thus, we can express the previous example mathematically:

$$ppm_m = \text{g of } i \text{ in } 10^6 \text{ g total} \qquad \textbf{(2.1)}$$

**Clear Water Act Analytical Methods**
http://www.epa.gov/waterscience/methods

**Air Pollution Monitoring Techniques**
http://www.epa.gov/ebtpages/airair pollutiontesting.html

## Table / 2.1

**Common Units of Concentration Used in Environmental Measurements**

| Representation | Example | Typical Units |
|---|---|---|
| mass chemical/total mass | mg/kg in soil | mg/kg, $ppm_m$ |
| mass chemical/total volume | mg/L in water or air | mg/L, $\mu g/m^3$ |
| volume chemical/total volume | volume fraction in air | $ppm_v$ |
| moles chemical/total volume | moles/L in water | M |

SOURCE: Mihelcic (1999), reprinted with permission of John Wiley & Sons, Inc.

This definition is equivalent to the following general formula, which is used to calculate $ppm_m$ concentration from measurements of chemical mass in a sample of total mass $m_{total}$:

$$ppm_m = \frac{m_i}{m_{total}} \times 10^6 \qquad (2.2)$$

Note that the factor $10^6$ in Equation 2.2 is really a conversion factor. It has the implicit units of $ppm_m$/mass fraction (mass fraction = $m_i/m_{total}$), as shown in Equation 2.3:

$$ppm_m = \frac{m_i}{m_{total}} \times 10^6 \frac{ppm_m}{mass\ fraction} \qquad (2.3)$$

In Equation 2.3, $m_i/m_{total}$ is defined as the mass fraction, and the conversion factor of $10^6$ is similar to the conversion factor of $10^2$ used to convert fractions to percentages. For example, the expression $0.25 = 25\%$ can be thought of as:

$$0.25 = 0.25 \times 100\% = 25\% \qquad (2.4)$$

© Anthony Rosenberg/iStockphoto.

Similar definitions are used for the units $ppb_m$, $ppt_m$, and percent by mass. That is, 1 **ppb_m** equals 1 **part per billion** or 1 g of a chemical per billion ($10^9$) g total, so that the number of $ppb_m$ in a sample is equal to $m_i/m_{total} \times 10^9$. And 1 **ppt_m** usually means 1 **part per trillion** ($10^{12}$). However, be cautious about interpreting ppt values, because they may refer to either parts per thousand or parts per trillion.

Mass/mass concentrations can also be reported with the units explicitly shown (for example, mg/kg or μg/kg). In soils and sediments,

---

example/2.1 Concentration in Soil

A 1 kg sample of soil is analyzed for the chemical solvent trichloroethylene (TCE). The analysis indicates that the sample contains 5.0 mg of TCE. What is the TCE concentration in $ppm_m$ and $ppb_m$?

solution

$$[TCE] = \frac{5.0\ mg\ TCE}{1.0\ kg\ soil} = \frac{0.005\ g\ TCE}{10^3\ g\ soil}$$

$$= \frac{5 \times 10^{-6}\ g\ TCE}{g\ soil} \times 10^6 = 5\ ppm_m = 5{,}000\ ppb_m$$

Note that in soil and sediments, mg/kg equals $ppm_m$, and μg/kg equals $ppb_m$.

---

1 ppm$_m$ equals 1 mg of pollutant per kg of solid (mg/kg), and 1 ppb$_m$ equals 1 µg/kg. **Percent by mass** is analogously equal to the number of grams of pollutant per 100 g total.

## 2.1.2 MASS/VOLUME UNITS: mg/L AND µg/m³

In the atmosphere, it is common to use concentration units of mass per volume of air, such as mg/m³ and µg/m³. In water, mass/volume concentration units of mg/L and µg/L are common. In most aqueous systems, ppm$_m$ is equivalent to mg/L. This is because the density of pure water is approximately 1,000 g/L (demonstrated in Example 2.2). The density of pure water is actually 1,000 g/L at 5°C. At 20°C the density has decreased slightly to 998.2 g/L. This equality is strictly true only for *dilute* solutions, in which any dissolved material does not contribute significantly to the mass of the water, and the total density remains approximately 1,000 g/L. Most wastewaters and natural waters can be considered dilute, except perhaps seawaters, brines, and some recycled streams.

example/2.2 **Concentration in Water**

One liter of water is analyzed and found to contain 5.0 mg of TCE. What is the TCE concentration in mg/L and ppm$_m$?

solution

$$[TCE] = \frac{5.0 \text{ mg TCE}}{1.0 \text{ L H}_2\text{O}} = \frac{5.0 \text{ mg}}{L}$$

To convert to ppm$_m$, a mass/mass unit, it is necessary to convert the volume of water to mass of water. To do this, divide by the density of water, which is approximately 1,000 g/L:

$$[TCE] = \frac{5.0 \text{ mg TCE}}{1.0 \text{ L H}_2\text{O}} \times \frac{1.0 \text{ L H}_2\text{O}}{1,000 \text{ g H}_2\text{O}}$$

$$= \frac{5.0 \text{ mg TCE}}{1,000 \text{ g total}} = \frac{5.0 \times 10^{-6} \text{ g TCE}}{\text{g total}} \times \frac{10^6 \text{ ppm}_m}{\text{mass fraction}}$$

$$= 5.0 \text{ ppm}_m$$

In most dilute aqueous systems, mg/L is equivalent to ppm$_m$.

In this example, the TCE concentration is well above the allowable U.S. drinking water standard for TCE, 5 µg/L (or 5 ppb), which was set to protect human health. Five ppb is a small value. Think of it this way: Earth's population exceeds 6 billion people, meaning that 30 individuals in one of your classes constitute a human concentration of approximately 5 ppb!

example/2.3 Concentration in Air

What is the carbon monoxide (CO) concentration expressed in $\mu g/m^3$ of a 10 L gas mixture that contains $10^{-6}$ mole of CO?

solution

In this case, the measured quantities are presented in units of moles of the chemical per total volume. To convert to mass of the chemical per total volume, convert the moles of chemical to mass of chemical by multiplying moles by CO's molecular weight. The molecular weight of CO (28 g/mole) is equal to 12 (atomic weight of C) plus 16 (atomic weight of O).

$$[CO] = \frac{1.0 \times 10^{-6} \text{ mole CO}}{10 \text{ L total}} \times \frac{28 \text{ g CO}}{\text{mole CO}}$$

$$= \frac{28 \times 10^{-6} \text{ g CO}}{10 \text{ L total}} \times \frac{10^6 \ \mu g}{g} \times \frac{10^3 \text{ L}}{m^3} = \frac{2,800 \ \mu g}{m^3}$$

## 2.2 Volume/Volume and Mole/Mole Units

Units of volume fraction or mole fraction are frequently used for gas concentrations. The most common volume fraction units are **parts per million by volume** (referred to as **ppm** or **ppm$_v$**), defined as:

$$ppm_v = \frac{V_i}{V_{\text{total}}} \times 10^6 \qquad (2.5)$$

where $V_i/V_{\text{total}}$ is the volume fraction and $10^6$ is a conversion factor, with units of $10^6$ ppm$_v$ per volume fraction.

Other common units for gaseous pollutants are **parts per billion ($10^9$) by volume (ppb$_v$)**.

The advantage of volume/volume units is that gaseous concentrations reported in these units do not change as a gas is compressed or expanded. Atmospheric concentrations expressed as mass per volume (for example, $\mu g/m^3$) decrease as the gas expands, since the pollutant mass remains constant but the volume increases. Both mass/volume units, such as $\mu g/m^3$, and ppm$_v$ units are frequently used to express gaseous concentrations. (See Equation 2.9 for conversion between $\mu g/m^3$ and ppm$_v$.)

### 2.2.1 USING THE IDEAL GAS LAW TO CONVERT ppm$_v$ TO $\mu g/m^3$

The ideal gas law can be used to convert gaseous concentrations between mass/volume and volume/volume units. The **ideal gas law** states that *pressure (P)* times *volume occupied (V)* equals *the number of moles (n)* times the *gas constant (R)* times the *absolute temperature (T)*

in degrees Kelvin or Rankine. This is written in the familiar form
of

$$PV = nRT \qquad (2.6)$$

In Equation 2.6, the **universal gas constant**, $R$, may be expressed in
many different sets of units. Some of the most common values for $R$ are
listed here:

0.08205 L-atm/mole-K

$8.205 \times 10^{-5}$ m$^3$-atm/mole-K

82.05 cm$^3$-atm/mole-K

$1.99 \times 10^{-3}$ kcal/mole-K

8.314 J/mole-K

1.987 cal/mole-K

62,358 cm$^3$-torr/mole-K

62,358 cm$^3$-mm Hg/mole-K

Because the gas constant may be expressed in different units, always
be careful of its units and cancel them out to ensure you are using the
correct value of $R$.

The ideal gas law also states that the volume occupied by a given
number of molecules of any gas is the same, no matter what the molecular weight or composition of the gas, as long as the pressure and temperature are constant. The ideal gas law can be rearranged to show the
volume occupied by $n$ moles of gas:

$$V = n\frac{RT}{P} \qquad (2.7)$$

At standard conditions ($P = 1$ atm and T = 273.15 K), 1 mole of any
pure gas will occupy a volume of 22.4 L. This result can be derived by
using the corresponding value of $R$ (0.08205 L-atm/mole-K) and the
form of the ideal gas law provided in Equation 2.7. At other temperatures and pressures, this volume varies as determined by Equation 2.7.

In Example 2.4, the terms $RT/P$ cancel out. This demonstrates an
important point that is useful in calculating volume fraction or mole
fraction concentrations: *For gases, volume ratios and mole ratios are equivalent*. This is clear from the ideal gas law, because at constant temperature and pressure, the volume occupied by a gas is proportional to the
number of moles. Therefore, Equation 2.5 is equivalent to Equation 2.8:

$$\text{ppm}_v = \frac{\text{moles } i}{\text{moles total}} \times 10^6 \qquad (2.8)$$

The solution to Example 2.4 could have been found simply by using
Equation 2.8 and determining the mole ratio. Therefore, in any given
problem, you can use either units of volume or units of moles to

example/2.4 Gas Concentration in Volume Fraction

A gas mixture contains 0.001 mole of sulfur dioxide ($SO_2$) and 0.999 mole of air. What is the $SO_2$ concentration, expressed in units of ppm$_v$?

solution

The concentration in ppm$_v$ is determined using Equation 2.5.

$$[SO_2] = \frac{V_{SO_2}}{V_{total}} \times 10^6$$

To solve, convert the number of moles of $SO_2$ to volume using the ideal gas law (Equation 2.6) and the total number of moles to volume. Then divide the two expressions:

$$V_{SO_2} = 0.001 \text{ mole } SO_2 \times \frac{RT}{P}$$

$$V_{total} = (0.999 + 0.001) \text{ mole total} \times \frac{RT}{P}$$

$$= (1.000) \text{ mole total} \times \frac{RT}{P}$$

Substitute these volume terms for ppm$_v$:

$$\text{ppm}_v = \frac{0.001 \text{ mole } SO_2 \times \dfrac{RT}{P}}{1.000 \text{ mole total} \times \dfrac{RT}{P}} \times 10^6$$

$$\text{ppm}_v = \frac{0.001 \text{ L } SO_2}{1.000 \text{ L total}} \times 10^6 = 1,000 \text{ ppm}_v$$

calculate ppm$_v$. Being aware of this will save unnecessary conversions between moles and volume.

The **mole ratio** (moles $i$/moles total) is sometimes referred to as the **mole fraction**, $X$.

Example 2.5 and Equation 2.9 show how to use the ideal gas law to convert concentrations between $\mu g/m^3$ and ppm$_v$.

Example 2.5 demonstrates a useful way to write the conversion for air concentrations between units of $\mu g/m^3$ and ppm$_v$:

$$\frac{\mu g}{m^3} = \text{ppm}_v \times MW \times \frac{1,000 \, P}{RT} \qquad (2.9)$$

where MW is the molecular weight of the chemical species, $R$ equals 0.08205 L-atm/mole-K, $T$ is the temperature in degrees K, and 1,000 is example/2.5 **Conversion of Gas Concentration between ppb$_v$ and $\mu$g/m$^3$**

The concentration of $SO_2$ is measured in air to be 100 ppb$_v$. What is this concentration in units of $\mu$g/m$^3$? Assume the temperature is 28°C and pressure is 1 atm. Remember that $T$ expressed in °K is equal to $T$ expressed in °C plus 273.15.

solution

To accomplish this conversion, use the ideal gas law to convert the volume of $SO_2$ to moles of $SO_2$, resulting in units of moles/L. This can be converted to $\mu$g/m$^3$ using the molecular weight of $SO_2$, (which equals 64). This method will be used to develop a general formula for converting between ppm$_v$ and $\mu$g/m$^3$.

First, use the definition of ppb$_v$ to obtain a volume ratio for $SO_2$:

$$100 \text{ ppb}_v = \frac{100 \text{ m}^3 \text{ SO}_2}{10^9 \text{ m}^3 \text{ air solution}}$$

Now convert the volume of $SO_2$ in the numerator to units of mass. This is done in two steps. First, convert the volume to a number of moles, using a rearranged format of the ideal gas law (Equation 2.6), $n/V = P/RT$, and the given temperature and pressure:

$$\frac{100 \text{ m}^3 \text{ SO}_2}{10^9 \text{ m}^3 \text{ air solution}} \times \frac{P}{RT} = \frac{100 \text{ m}^3 \text{ SO}_2}{10^9 \text{ m}^3 \text{ air solution}}$$

$$\times \frac{1 \text{ atm}}{8.205 \times 10^{-5} \frac{\text{m}^3\text{-atm}}{\text{mole-K}}(301 \text{ K})} = \frac{4.05 \times 10^{-6} \text{ mole SO}_2}{\text{m}^3 \text{ air}}$$

In the second step, convert the moles of $SO_2$ to mass of $SO_2$ by using the molecular weight of $SO_2$:

$$\frac{4.05 \times 10^{-6} \text{ mole SO}_2}{\text{m}^3 \text{ air}} \times \frac{64 \text{ g SO}_2}{\text{mole SO}_2} \times \frac{10^6 \text{ }\mu\text{g}}{\text{g}} = \frac{260 \text{ }\mu\text{g}}{\text{m}^3}$$

a conversion factor (1,000 L = m$^3$). Note that for 0°C, $RT$ has a value of 22.4 L-atm/mole, while at 20°C, $RT$ has a value of 24.2 L-atm/mole.

## 2.3 Partial-Pressure Units

In the atmosphere, concentrations of chemicals in the gas and particulate phases may be determined separately. A substance will exist in the gas phase if the atmospheric temperature is above the substance's boiling (or sublimation) point or if its concentration is below the saturated vapor pressure of the chemical at a specified temperature (vapor

## Table / 2.2

### Composition of the Atmosphere

| Compound | Concentration (% volume or moles) | Concentration ($ppm_v$) |
|---|---|---|
| Nitrogen ($N_2$) | 78.1 | 781,000 |
| Oxygen ($O_2$) | 20.9 | 209,000 |
| Argon (Ar) | 0.93 | 9,300 |
| Carbon dioxide ($CO_2$) | 0.038 | 379 |
| Neon (Ne) | 0.0018 | 18 |
| Helium (He) | 0.0005 | 5 |
| Methane ($CH_4$) | 0.00018 | 1.774 |
| Krypton (Kr) | 0.00011 | 1.1 |
| Hydrogen ($H_2$) | 0.00005 | 0.50 |
| Nitrous Oxide ($N_2O$) | 0.000032 | 0.319 |
| Ozone ($O_3$) | 0.000002 | 0.020 |

SOURCE: Adapted from Mihelcic (1999), with permission of John Wiley & Sons, Inc.

pressure is defined in Chapter 3). The major and minor gaseous constituents of the atmosphere all have boiling points well below atmospheric temperatures. Concentrations of these species typically are expressed either as volume fractions (for example, percent, $ppm_v$, or $ppb_v$) or as partial pressures (units of atmospheres).

Table 2.2 summarizes the concentrations of the most abundant atmospheric gaseous constituents, including carbon dioxide and methane. Carbon dioxide is the largest human contributor to greenhouse gases. The **Intergovernmental Panel on Climate Change** (IPCC; see www.ipcc.ch) reports that the global atmospheric concentration of carbon dioxide has increased to 379 $ppm_v$ in 2005 from pre-industrial revolution levels of 280 $ppm_v$. Global atmospheric concentrations of methane recorded in 2005 have reached 1,774 ppb. This recorded methane concentration greatly exceeds the natural range of 320 to 790 $ppb_v$, measured in ice cores, that dates over the past 650,000 years. According to the IPCC, it is very likely that this increase in methane concentration is due to agricultural land use, population growth, and energy use associated with burning fossil fuels.

The total pressure exerted by a gas mixture may be considered as the sum of the partial pressures exerted by each component of the mixture. The **partial pressure** of each component is equal to the pressure that would be exerted if all of the other components of the mixture were suddenly removed. Partial pressure is commonly written as $P_i$, where $i$ refers to a particular gas. For example, the partial pressure of oxygen in the atmosphere $P_{O_2}$ is 0.21 atm.

Remember, the ideal gas law states that, at a given temperature and volume, pressure is directly proportional to the number of moles of gas present; therefore, pressure fractions are identical to mole fractions (and volume fractions). For this reason, partial pressure can be calculated as the product of the mole or volume fraction and the total pressure. For example:

$$P_i = [\text{volume fraction}_i \text{ or mole fraction}_i \times P_{\text{total}}]$$
$$= [(\text{ppm}_v)_i \times 10^{-6} \times P_{\text{total}}] \qquad \text{(2.10)}$$

In addition, rearranging Equation 2.10 shows that $\text{ppm}_v$ values can be calculated from partial pressures as follows:

$$\text{ppm}_v = \frac{P_i}{P_{\text{total}}} \times 10^6 \qquad \text{(2.11)}$$

Partial pressure can thus be added to the list of unit types that can be used to calculate $\text{ppm}_v$. That is, either volume (Equation 2.5), moles (Equation 2.8), or partial pressures (Equation 2.11) can be used in $\text{ppm}_v$ calculations.

Example 2.6 applies these principles to the partial pressure of a formerly popular family of chemical compounds known as polychlorinated biphenyls (PCBs), illustrated in Figure 2.1.

---

### example/2.6 Concentration as Partial Pressure

The concentration of gas-phase polychlorinated biphenyls (PCBs) in the air above Lake Superior was measured to be 450 picograms per cubic meter ($\text{pg/m}^3$). What is the partial pressure (in atm) of PCBs? Assume the temperature is 0°C, the atmospheric pressure is 1 atm, and the average molecular weight of PCBs is 325.

### solution

The partial pressure is defined as the mole or volume fraction times the total gas pressure. First, find the number of moles of PCBs in a liter of air. Then use the ideal gas law (Equation 2.7) to calculate that 1 mole of gas at 0°C and 1 atm occupies 22.4 L. Substitute this value into the first expression to determine the mole fraction of PCBs:

$$450 \frac{\text{pg}}{\text{m}^3 \text{ air}} \times \frac{\text{mole}}{325 \text{ g}} \times 10^{-12} \frac{\text{g}}{\text{pg}} \times 10^{-3} \frac{\text{m}^3}{\text{L}} = 1.38 \times 10^{-15} \frac{\text{mole PCB}}{\text{L air}}$$

$$1.38 \times 10^{-15} \frac{\text{mole PCB}}{\text{L air}} \times \frac{22.4 \text{ L}}{\text{mole air}} = 3.1 \times 10^{-14} \frac{\text{mole PCB}}{\text{mole air}}$$

Multiplying the mole fraction by the total pressure (1 atm) (see Equation 2.10) yields the PCB partial pressure of $3.1 \times 10^{-14}$ atm.

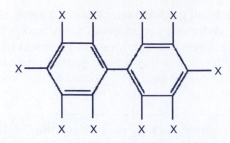

X = chlorine or hydrogen

**Figure 2.1  Chemical Structure of Polychlorinated Biphenyls (PCBs)**  PCBs are a family of compounds produced commercially by chlorinating biphenyl. Chlorine atoms can be placed at any or all of ten available sites, with 209 possible PCB congeners. The great stability of PCBs caused them to have a wide range of uses, including serving as coolants in transformers and as hydraulic fluids and solvents. However, the chemical properties that resulted in this stability also resulted in a chemical that did not degrade easily, bioaccumulated in the food chain, and was hazardous to humans and wildlife. The 1976 Toxic Substances Control Act (TSCA) banned the manufacture of PCBs and PCB-containing products. TSCA also established strict regulations regarding the future use and sale of PCBs. PCBs typically were sold as mixtures commonly referred to as Arochlors. For example, the Arochlor 1260 mixture consists of 60 percent chlorine by weight, meaning the individual PCBs in the mixture primarily are substituted with 6 to 9 chlorines per biphenyl molecule. In contrast, Arochlor 1242 consists of 42 percent chlorine by weight; thus, it primarily consists of PCBs with 1 to 6 substituted chlorines per biphenyl molecule.

From Mihelcic (1999). Reprinted with permission of John Wiley & Sons, Inc.

example/2.7 Concentration as Partial Pressure Corrected for Moisture

What would be the partial pressure (in atm) of carbon dioxide ($CO_2$) when the barometer reads 29.0 inches of Hg, the relative humidity is 80 percent, and the temperature is 70°F? Use Table 2.2 to obtain the concentration of $CO_2$ in dry air.

solution

The partial-pressure concentration units in Table 2.2 are for dry air, so the partial pressure must first be corrected for the moisture present in the air. In dry air, the $CO_2$ concentration is 379 $ppm_v$. The partial pressure will be this volume fraction times the total pressure of dry air. The total pressure of dry air is the total atmospheric pressure (29.0 in. Hg) minus the contribution of water vapor. The vapor pressure of water at 70°F is 0.36 lb./in.$^2$ Thus, the total pressure of dry air is

$$P_{total} - P_{water} = 29.0 \text{ in. Hg} - \left[ 0.36 \frac{\text{lb.}}{\text{in.}^2} \times \frac{29.9 \text{ in. Hg}}{14.7 \text{ lb./in.}^2} \times 0.8 \right]$$

$$= 28.4 \text{ in. Hg}$$

## example/2.7 Continued

The partial pressure of $CO_2$ would be:

$$\text{vol. fraction} \times P_{\text{total}} = 379\ \text{ppm}_v \times \frac{10^{-6}\ \text{vol. fraction}}{\text{ppm}_v} \times \left[ 28.4\ \text{in. Hg} \times \frac{1\ \text{atm}}{29.9\ \text{in. Hg}} \right] = 3.6 \times 10^{-4}\ \text{atm}$$

## 2.4 Mole/Volume Units

Units of **moles per liter** (molarity, M) are used to report concentrations of compounds dissolved in water. **Molarity** is defined as the number of moles of compound per liter of solution. Concentrations expressed in these units are read as **molar**.

Molarity, M, should not be confused with molality, m. Molarity is usually used in equilibrium calculations and throughout the remainder of this book. **Molality** is the number of moles of a solute added to exactly 1 L of solvent. Thus, the actual volume of a molal solution is slightly larger than 1 L. Molality is more likely to be used when properties of the solvent, such as boiling and freezing points, are a concern. Therefore, it is rarely used in environmental situations.

**The Mole**

## example/2.8 Concentration as Molarity

The concentration of trichloroethylene (TCE) is 5 ppm. Convert this to units of molarity. The molecular weight of TCE is 131.5 g/mole.

### solution

Remember, in water, $\text{ppm}_m$ is equivalent to mg/L, so the concentration of TCE is 5.0 mg/L. Conversion to molarity units requires only the molecular weight:

$$\frac{5.0\ \text{mg TCE}}{L} \times \frac{1\ \text{g}}{10^3\ \text{mg}} \times \frac{1\ \text{mole}}{131.5\ \text{g}} = \frac{3.8 \times 10^{-5}\ \text{moles}}{L}$$

$$= 3.8 \times 10^{-5}\ \text{M}$$

Often, concentrations below 1 M are expressed in units of millimoles per liter, or millimolar (1 mM = $10^{-3}$ moles/L), or in micromoles per liter, or micromolar (1 μM = $10^{-6}$ moles/L). Thus, the concentration of TCE could be expressed as 0.038 mM or 38 μM.

## example/2.9 Concentration as Molarity

The concentration of alachlor, a common herbicide, in the Mississippi River was found to range from 0.04 to 0.1 μg/L. What is the concentration range in nanomoles/L? The molecular formula for alachlor is $C_{14}H_{20}O_2NCl$, and its molecular weight is 270.

## 2.5    Other Types of Units

Concentrations can also be expressed as normality, as a common constituent, or represented by effect.

### 2.5.1    NORMALITY

**Normality** (equivalents/L) typically is used in defining the chemistry of water, especially in instances where acid–base and oxidation-reduction reactions are taking place. Normality is also used frequently in the laboratory during the analytical measurement of water constituents. For example, "Standard Methods for the Examination of Water and Wastewater" (Eaton et al., 2005) has many examples where concentrations of chemical reagents are prepared and reported in units of normality and not molarity.

Reporting concentration on an **equivalent basis** is useful because if two chemical species react and the two species reacting have the same strength on an equivalent basis, a 1 mL volume of reactant number 1 will react with a 1 mL volume of reactant number 2. In acid–base chemistry, the number of equivalents per mole of acid equals the number of moles of $H^+$ the acid can potentially donate. For example, HCl has 1 equivalent/mole, $H_2SO_4$ has 2 equivalents/mole, and $H_3PO_4$ has 3 equivalents/mole. Likewise, the number of equivalents per mole of a base equals the number of moles of $H^+$ that will react with 1 mole of the base. Thus, NaOH has 1 equivalent/mole, $CaCO_3$ has 2 equivalents/ mole, and $PO_4^{3-}$ has 3 equivalents/mole.

In oxidation-reduction reactions, the number of equivalents is related to how many electrons a species donates or accepts. For example, the number of equivalents of $Na^+$ is 1 (where $e^-$ equals an electron) because $Na \rightarrow Na^+ + e^-$. Likewise, the number of equivalents for $Ca^{2+}$ is 2 because $Ca \rightarrow Ca^{2+} + 2e^-$. The **equivalent weight** (in grams (g) per equivalent (eqv)) of a species is defined as the molecular weight of the species divided by the number of equivalents in the species (g/mole divided by eqv/mole equals g/eqv).

All aqueous solutions must maintain charge neutrality. Another way to state this is that the sum of all cations on an equivalent basis must

© Nadezda Pyastolova/iStockphoto.

## example/2.10 Calculations of Equivalent Weight

What are the equivalent weights of HCl, $H_2SO_4$, NaOH, $CaCO_3$, and aqueous $CO_2$?

## solution

To find the equivalent weight of each compound, divide the molecular weight by the number of equivalents:

$$\text{eqv wt of HCl} = \frac{(1 + 35.5)\text{g/mole}}{1 \text{ eqv/mole}} = \frac{36.5 \text{ g}}{\text{eqv}}$$

$$\text{eqv wt of } H_2SO_4 = \frac{(2 \times 1) + 32 + (4 \times 16)\text{g/mole}}{2 \text{ eqv/mole}} = \frac{49 \text{ g}}{\text{eqv}}$$

$$\text{eqv wt of NaOH} = \frac{(23 + 16 + 1)\text{g/mole}}{1 \text{ eqv/mole}} = \frac{40 \text{ g}}{\text{eqv}}$$

$$\text{eqv wt of } CaCO_3 = \frac{40 + 12 + (3 \times 16)\text{g/mole}}{2 \text{ eqv/mole}} = \frac{50 \text{ g}}{\text{eqv}}$$

Determining the equivalent weight of aqueous $CO_2$ requires additional information. Aqueous carbon dioxide is not an acid until it hydrates in water and forms carbonic acid ($CO_2 + H_2O \rightarrow H_2CO_3$). So aqueous $CO_2$ really has 2 eqv/mole. Thus, one can see that the equivalent weight of aqueous carbon dioxide is

$$\frac{12 + (2 \times 16)\text{g/mole}}{2 \text{ eqv/mole}} = \frac{22 \text{ g}}{\text{eqv}}$$

## example/2.11 Calculation of Normality

What is the normality (N) of 1 M solutions of HCl and $H_2SO_4$?

## solution

$$1 \text{ M HCl} = \frac{1 \text{ mole HCl}}{L} \times \frac{1 \text{ eqv}}{\text{mole}} = \frac{1 \text{ eqv}}{L} = 1 \text{ N}$$

$$1 \text{ M } H_2SO_4 = \frac{1 \text{ mole } H_2SO_4}{L} \times \frac{2 \text{ eqv}}{\text{mole}} = \frac{2 \text{ eqv}}{L} = 2 \text{ N}$$

Note that on an equivalent basis, a 1 M solution of sulfuric acid is twice as strong as a 1 M solution of HC1.

equal the sum of all anions on an equivalent basis. Thus, water samples can be checked to determine whether something is incorrect in the analyses or a constituent is missing. Example 2.12 shows how this is done.

example/2.12 Use of Equivalents in Determining the Accuracy of a Water Analysis

The label on a bottle of New Zealand mineral water purchased in the city of Dunedin states that a chemical analysis of the mineral water resulted in the following cations and anions being identified with corresponding concentrations (in mg/L):

$$[Ca^{2+}] = 2.9 \quad [Mg^{2+}] = 2.0 \quad [Na^+] = 11.5 \quad [K^+] = 3.3$$

$$[SO_4^{2-}] = 4.7 \quad [Fl^-] = 0.09 \quad [Cl^-] = 7.7$$

Is the analysis correct?

solution

First, convert all concentrations of major ions to an equivalent basis. To do this, multiply the concentration in mg/L by a unit conversion (g/1,000 mg), and then divide by the equivalent weight of each substance (g/eqv). Then sum the concentrations of all cations and anions on an equivalent basis. A solution with less than 5 percent error generally is considered acceptable.

**Cations**

$$[Ca^{2+}] = \frac{1.45 \times 10^{-4} \text{ eqv}}{L}$$

$$[Mg^{2+}] = \frac{1.67 \times 10^{-4} \text{ eqv}}{L}$$

$$[Na^+] = \frac{5 \times 10^{-4} \text{ eqv}}{L}$$

$$[K^+] = \frac{8.5 \times 10^{-5} \text{ eqv}}{L}$$

**Anions**

$$[SO_4^{2-}] = \frac{9.75 \times 10^{-5} \text{ eqv}}{L}$$

$$[Fl^-] = \frac{4.73 \times 10^{-6} \text{ eqv}}{L}$$

$$[Cl^-] = \frac{2.17 \times 10^{-4} \text{ eqv}}{L}$$

The total amount of cations equals $9.87 \times 10^{-4}$ eqv/L, and the total amount of anions equals $3.2 \times 10^{-4}$ eqv/L.

The analysis is not within 5 percent. The analysis resulted in more than three times more cations than anions on an equivalent basis. Therefore, either of two conclusions is possible: (1) One or more of the reported concentrations are incorrect, assuming all major cations and anions are accounted for. (2) One or more important anions were not accounted for by the chemical analysis. (Bicarbonate, $HCO_3^-$, would be a good guess for the missing anion, as it is a common anion in most natural waters.)

## 2.5.2 CONCENTRATION AS A COMMON CONSTITUENT

Concentrations can be reported as a **common constituent** and can therefore include contributions from a number of different chemical compounds. Greenhouse gases, nitrogen, and phosphorus are chemicals that have their concentration typically reported as a common constituent.

For example, the phosphorus in a lake or wastewater may be present in inorganic forms called orthophosphates ($H_3PO_4$, $H_2PO_4^-$, $HPO_4^{2-}$, $PO_4^{3-}$ $HPO_4^{2-}$ complexes), polyphosphates (for example, $H_4P_2O_7$ and $H_3P_3O_{10}^{2-}$), metaphosphates (for example, $HP_3O_9^{2-}$), and/or organic phosphates. Because phosphorus can be chemically converted between these forms and can thus be found in several of these forms, it makes sense at some times to report the total P concentration, without specifying which form(s) are present. Thus, each concentration for every individual form of phosphorus is converted to mg P/L using the molecular weight of the individual species, the molecular weight of P (which is 32), and simple stoichiometry. These converted concentrations of each individual species can then be added to determine the total phosphorus concentration. The concentration is then reported in units of mg/L as phosphorus (written as mg P/L, mg/L as P, or mg/L P).

---

example/2.13 **Nitrogen Concentrations as a Common Constituent**

A water contains two nitrogen species. The concentration of $NH_3$ is 30 mg/L $NH_3$, and the concentration of $NO_3^-$ is 5 mg/L $NO_3^-$. What is the total nitrogen concentration in units of mg N/L?

solution

Use the appropriate molecular weight and stoichiometry to convert each individual species to the requested units of mg N/L, and then add the contribution of each species:

$$\frac{30 \text{ mg NH}_3}{L} \times \frac{\text{mole NH}_3}{17 \text{ g}} \times \frac{\text{mole N}}{\text{mole NH}_3} \times \frac{14 \text{ g}}{\text{mole N}}$$

$$= \frac{24.7 \text{ mg NH}_3 - N}{L}$$

$$\frac{5 \text{ mg NO}_3^-}{L} \times \frac{\text{mole NO}_3^-}{62 \text{ g}} \times \frac{\text{mole N}}{\text{mole NO}_3^-} \times \frac{14 \text{ g}}{\text{mole N}}$$

$$= \frac{1.1 \text{ mg NO}_3^- - N}{L}$$

$$\text{total nitrogen concentration} = 24.7 + 1.1 = \frac{25.8 \text{ mg N}}{L}$$

The alkalinity and hardness of a water typically are reported by determining all of the individual species that contribute to either alkalinity or hardness, then converting each of these species to **units of mg $CaCO_3$/L**, and finally summing up the contribution of each species. Hardness is thus typically expressed as mg/L as $CaCO_3$.

The **hardness** of a water is caused by the presence of divalent cations in water. $Ca^{2+}$ and $Mg^{2+}$ are by far the most abundant divalent cations in natural waters, though $Fe^{2+}$, $Mn^{2+}$, and $Sr^{2+}$ may contribute as well. In Michigan, Wisconsin, and Minnesota, untreated waters usually have a hardness of 121 to 180 mg/L as $CaCO_3$. In Illinois and Iowa, water is harder, with many values greater than 180 mg/L as $CaCO_3$.

To find the total hardness of a water, sum the contributions of all divalent cations after converting their concentrations to a common constituent. To convert the concentration of specific cations (from mg/L) to hardness (as mg/L $CaCO_3$), use the following expression, where $M^{2+}$ represents a divalent cation:

$$\frac{M^{2+} \text{ in mg}}{L} \times \frac{50}{\text{eqv wt of } M^{2+} \text{ in g/eqv}} = \frac{mg}{L} \text{ as } CaCO_3 \qquad (2.13)$$

The 50 in Equation 2.13 represents the equivalent weight of calcium carbonate (100 g $CaCO_3$/2 equivalents). The equivalent weights (in units of g/eqv) of other divalent cations are Mg, 24/2; Ca, 40/2; Mn, 55/2; Fe, 56/2; and Sr, 88/2.

The Kyoto Protocol regulates six major greenhouse gases. It was adopted in Kyoto, Japan, on December 11, 1997, and entered into force on February 16, 2005. It sets binding targets for 37 industrialized countries and the European Union to reduce greenhouse gas emissions. Each gas has a different ability to absorb heat in the atmosphere (the radiative forcing), so each differs in its global warming potential.

The **global warming potential (GWP)** is a multiplier used to compare the emissions of different greenhouse gases to a common constituent, in this case carbon dioxide. The GWP is determined over a set time period, typically 100 years, over which the radiative forcing of the specific gas would result. GWPs allow policy makers to compare emissions and reductions of specific gases.

**Carbon dioxide equivalents** are a metric measure used to compare the mass emissions of greenhouse gases to a common constituent, based on the specific gas's global warming potential. Units are mass based and typically a million metric tons of carbon dioxide equivalents. Table 2.3 provides global warming potentials for the six major greenhouse gases. Note that an equivalent mass release of two greenhouse gases does not have the same impact on global warming. For example, from Table 2.3, we can see that 1 ton of methane emissions equates to 25 tons of carbon dioxide emissions.

Greenhouse gas emissions are sometimes also reported as **carbon equivalents**. In this case, the mass of carbon dioxide equivalents is multiplied by 12/44 to obtain carbon equivalents. The multiplier 12/44 is

**United Nations Framework Convention on Climate Change**
http://unfccc.int

## Table / 2.3

**100-Year Global Warming Potentials (GWP) Used to Convert Mass Greenhouse Gas Emissions to Carbon Dioxide Equivalents ($CO_2e$)**

| Type of Emission | Multiplier for $CO_2$ Equivalents ($CO_2e$) |
|---|---|
| Carbon dioxide | 1 |
| Methane | 25 |
| Nitrous oxide | 298 |
| Hydrofluorocarbons (HFCs) | 124–14,800 (depends on specific HFC) |
| Perfluorocarbons (PFCs) | 7,390–12,200 (depends on specific PFC) |
| Sulfur hexafluoride ($SF_6$) | 22,800 |

SOURCE: Values from Intergovernmental Panel on Climate Change.

## example/2.14   Determination of a Water's Hardness

Water has the following chemical composition: $[Ca^{2+}] = 15$ mg/L; $[Mg^{2+}] = 10$ mg/L; $[SO_4^{2-}] = 30$ mg/L. What is the total hardness in units of mg/L as $CaCO_3$?

### solution

Find the contribution of hardness from each divalent cation. Anions and all nondivalent cations are not included in the calculation.

$$\frac{15 \text{ mg Ca}^{2+}}{L} \times \left( \frac{\dfrac{50 \text{ g CaCO}_3}{\text{eqv}}}{\dfrac{40 \text{ g CaCO}_3}{2 \text{ eqv}}} \right) = \frac{38 \text{ mg}}{L} \text{ as CaCO}_3$$

$$\frac{10 \text{ mg Mg}^{2+}}{L} \times \left( \frac{\dfrac{50 \text{ g CaCO}_3}{\text{eqv}}}{\dfrac{24 \text{ g CaCO}_3}{2 \text{ eqv}}} \right) = \frac{42 \text{ mg}}{L} \text{ as CaCO}_3$$

Therefore, the total hardness is $38 + 42 = 80$ mg/L as $CaCO_3$. This water is moderately hard.

Note that if reduced iron ($Fe^{2+}$) or manganese ($Mn^{2+}$) were present, they would be included in the hardness calculation.

the molecular weight of carbon (C) divided by the molecular weight of carbon dioxide ($CO_2$). Table 2.4 shows some relevant U.S. greenhouse gas emissions in units of $CO_2$ equivalents (abbreviated $CO_2e$). Note the expected large contribution associated with energy use from burning

## Table / 2.4

**U.S. Greenhouse Gas Emissions from Sources Relevant to Environmental and Civil Engineering**
Total greenhouse gas emissions in 2004 were 7,074.4 Tg $CO_2$ equivalents.

| Source (Gas) | $CO_2$ equivalents (Tg) | Source (Gas) | $CO_2$ equivalents (Tg) |
|---|---|---|---|
| Fossil fuel combustion ($CO_2$) | 5,656.6 | Agricultural soil management ($N_2O$) | 261.5 |
| Iron and steel production ($CO_2$) | 51.3 | Manure management ($N_2O$) | 17.7 |
| Cement manufacture ($CO_2$) | 45.6 | Human sewage ($N_2O$) | 16.0 |
| Municipal solid waste combustion ($CO_2$) | 19.4 | Municipal solid waste combustion ($N_2O$) | 0.5 |
| Soda ash manufacture and consumption ($CO_2$) | 4.2 | Mobile combustion ($N_2O$) | 42.8 |
| Landfills ($CH_4$) | 140.9 | Stationary combustion ($N_2O$) | 13.7 |
| Manure management ($CH_4$) | 39.4 | Substitution of ozone-depleting substances (HFCs, PFCs, $SF_6$) | 103.3 |
| Wastewater treatment ($CH_4$) | 36.9 | Electrical transmission and distribution (HFCs, PFCs, $SF_6$) | 13.8 |
| Rice cultivation ($CH_4$) | 7.6 | Semiconductor manufacture (HFCs, PFCs, $SF_6$) | 4.7 |

SOURCE: Data from EPA, 2006.

fossil fuels. But also note the amount of greenhouse gas emissions associated with other human activities. It is clear from this table that sustainable development will require that every engineering assignment consider how to reduce the overall emissions of greenhouse gases.

### 2.5.3  REPORTING PARTICLE CONCENTRATIONS IN AIR AND WATER

The concentration of particles in an air sample is determined by pulling a known volume (for instance, several thousand $m^3$) of air through a filter. The increase in weight of the filter due to collection of particles on it can be determined. Dividing this value by the volume of air passed through the filter gives the **total suspended particulate** (TSP) concentration in units of $g/m^3$ or $\mu g/m^3$.

In aquatic systems and in the analytical determination of metals, the solid phase is distinguished by filtration using a filter opening of 0.45 $\mu$m. This size typically determines the cutoff between the *dissolved* and *particulate* phases. In water quality, solids are divided into a *dissolved* or *suspended* fraction. This is done by a combination of filtration and evaporation procedures. Each of these two types of solids can be further broken down into a *fixed* and *volatile* fraction. Figure 2.2 shows the analytical differences between total solids, total suspended solids, total dissolved solids, and volatile suspended solids.

# example/2.15 Carbon Equivalents as a Common Constituent

The U.S. greenhouse gas emissions reported in 2004 were 5,988 teragrams (Tg) $CO_2$e of carbon dioxide ($CO_2$), 556.7 Tg $CO_2$e of methane ($CH_4$), and 386.7 Tg $CO_2$e of $N_2O$. How many gigagrams (Gg) of $CH_4$ and $N_2O$ were emitted in 2004? There are 1,000 gigagrams in one teragram.

## solution

The solution is a simple unit conversion:

$$\text{Tg } CO_2e = (\text{Gg of gas}) \times GWP \times \frac{\text{Tg}}{1,000 \text{ Gg}}$$

For methane:

$$556.7 \text{ Tg } CO_2e = (\text{Gg of methane gas}) \times 25 \times \frac{\text{Tg}}{1,000 \text{ Gg}}$$

$2.23 \times 10^4$ Gg of methane were emitted in 2004.

For $N_2O$:

$$386.7 \text{ Tg } CO_2e = (\text{Gg of } N_2O \text{ gas}) \times 298 \times \frac{\text{Tg}}{1,000 \text{ Gg}}$$

$1.30 \times 10^3$ Gg of nitrogen oxide were emitted in 2004.

If you go to the U.S. Environmental Protection Agency Web site (www.epa.gov), you can learn more about emissions and sinks of different greenhouse gases in the United States. The Intergovernmental Panel on Climate Change Web site (www.ipcc.ch) has updated information on the status of global climate change.

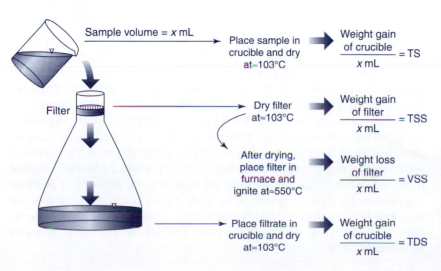

Place sample in crucible and dry at≈103°C → $\dfrac{\text{Weight gain of crucible}}{x \text{ mL}}$ = TS

Dry filter at≈103°C → $\dfrac{\text{Weight gain of filter}}{x \text{ mL}}$ = TSS

After drying, place filter in furnace and ignite at≈550°C → $\dfrac{\text{Weight loss of filter}}{x \text{ mL}}$ = VSS

Place filtrate in crucible and dry at≈103°C → $\dfrac{\text{Weight gain of crucible}}{x \text{ mL}}$ = TDS

**Figure 2.2** Analytical Differences between Total Solids (TS), Total Suspended Solids (TSS), Volatile Suspended Solids (VSS), and Total Dissolved Solids (TDS)

From Mihelcic (1999). Reprinted with permission of John Wiley & Sons, Inc.

**Total solids (TS)** are determined by placing a well-mixed water sample of known volume in a drying dish and evaporating the water at 103°C to 105°C. The increase in the weight of the drying dish is due to the total solids, so to determine total solids, divide the increase in weight gain of the drying dish by the sample volume. Concentrations typically are reported in mg/L.

To determine **total dissolved solids (TDS)** and **total suspended solids (TSS)**, first filter a well-mixed sample of known volume through a glass-fiber filter with a 2 μm opening. The suspended solids are the particles caught on the filter. To determine the concentration of TSS, dry the filter at 103°C to 105°C, determine the weight increase in the filter, and then divide this weight gain by the sample volume. Results are given in mg/L. Suspended solids collected on the filter may harm aquatic ecosystems by impairing light penetration or acting as a source of nutrients or oxygen-depleting organic matter. Also, a water high in suspended solids may be unsuited for human consumption or swimming.

The TDS are determined by collecting the sample that passes through the filter, drying this filtrate at 103°C to 105°C, and then determining the weight gain of the drying dish. This weight gain divided by the sample volume is the concentration of TDS, stated in mg/L. Dissolved solids tend to be less organic in composition and consist of dissolved cations and anions. For example, hard waters are also high in dissolved solids.

TS, TDS, and TSS can be further broken down into a fixed and volatile fraction. For example, the volatile portion of the TSS is termed the **volatile suspended solids (VSS)**, and the fixed portion is termed the **fixed suspended solids (FSS)**. The way to determine the volatile fraction of a sample is to take each sample just discussed and ignite it in a furnace at 500°C (±50°C). The weight loss due to this high-temperature ignition provides the volatile fraction, and the fixed fraction is what sample remains after ignition.

In wastewater treatment plants, the suspended solids or volatile fraction of suspended solids are used as a measure of the number of microorganisms in the biological treatment process. Figure 2.3 shows how to relate the various solid determinations.

| TS | = | TDS | + | TSS |
|----|---|-----|---|-----|
|    |   | =   |   | =   |
| TVS | = | VDS | + | VSS |
|    |   | +   |   | +   |
| TFS | = | FDS | + | FSS |

**Figure 2.3** **Relationships among the Various Measurements of Solids in Aqueous Samples** For example, if the TSS and VSS are measured, the FSS can be determined by difference.

From Mihelcic (1999). Reprinted with permission of John Wiley & Sons, Inc.

### 2.5.4 REPRESENTATION BY EFFECT

In some cases, the actual concentration of a specific substance is not used at all, especially in instances where mixtures of ill-defined chemicals are present (for example, in untreated sewage). Instead, **representation by effect** is used. With this approach, the strength of the solution or mixture is defined by some common factor on which all the chemicals within the mixture depend. An example is oxygen depletion from biological and chemical decomposition of the chemical mixture. For many organic-bearing wastes, instead of identifying the hundreds of individual compounds that may be present, it is more convenient to report the effect, in units of the milligrams of oxygen that can be consumed per liter of water. This unit is referred to as either **biochemical oxygen demand (BOD)** or **chemical oxygen demand (COD)**.

## example/2.16  Determining Concentrations of Solids in a Water Sample

A laboratory provides the following analysis obtained from a 50 mL sample of wastewater: total solids = 200 mg/L, total suspended solids = 160 mg/L, fixed suspended solids = 40 mg/L, and volatile suspended solids = 120 mg/L.

1. What is the concentration of total dissolved solids of this sample?

2. Suppose this sample was filtered through a glass-fiber filter, and the filter was then placed in a muffle furnace at 550°C overnight. What would be the weight of the solids (in mg) remaining on the filter after the night in the furnace?

3. Is this sample turbid? Estimate the percent of the solids that are organic matter.

## solution

1. Refer to Figure 2.3 to see the relationship between the various forms of solids. TDS equals TS minus TSS; thus,

$$TDS = \frac{200 \text{ mg}}{L} - \frac{160 \text{ mg}}{L} = \frac{40 \text{ mg}}{L}$$

2. The solids remaining on the filter are suspended solids. (Dissolved solids would pass through the filter.) Because the filter was subjected to a temperature of 550°C, the measurement was made for the volatile and fixed fraction of the suspended solids, that is, the VSS and FSS. However, during the ignition phase, the volatile fraction was burned off, while what remained on the filter was the inert or fixed fraction of the suspended solids. Thus, this problem is requesting the fixed fraction of the suspended solids. Accordingly, the 50 mL sample had FSS of 40 mg/L. Therefore,

$$FSS = \frac{40 \text{ mg}}{L} = \frac{\text{wt of suspended solids remaining on filter after ignition}}{\text{mL sample}}$$

$$= \frac{x}{50 \text{ mL}}$$

The unknown, $x$, can be solved for and equals 2 mg.

3. The sample is turbid because of the suspended particles, measured as TSS. If the sample was allowed to sit for some time period, the suspended solids would settle, and the overlaying water might not appear turbid. The solids found in this sample contain at least 60 percent organic matter. The total solids concentration is 200 mg/L, and of this, 120 mg/L are volatile suspended solids. Because volatile solids consist primarily of organic matter, we can conclude that approximately 60 percent (120/200) of the solids are organic.

## Key Terms

- biochemical oxygen demand (BOD)
- carbon dioxide equivalents
- carbon equivalents
- chemical oxygen demand (COD)
- common constituent
- equivalent basis
- equivalent weight
- fixed suspended solids (FSS)
- global warming potential (GWP)
- hardness
- ideal gas law
- Intergovernmental Panel on Climate Change

- mass/mass concentrations
- molality
- molar
- molarity
- mole fraction
- mole ratio
- moles per liter
- normality
- partial pressure
- parts per billion by mass ($ppb_m$)
- parts per billion by volume ($ppb_v$)
- parts per million by mass (ppm or $ppm_m$)

- parts per million by volume ($ppm_v$)
- parts per trillion by mass ($ppt_m$)
- percent by mass
- representation by effect
- total dissolved solids (TDS)
- total solids (TS)
- total suspended particulates (TSP)
- total suspended solids (TSS)
- units of mg $CaCO_3$/L
- universal gas constant
- volatile suspended solids (VSS)

# chapter/Two Problems

**2.1** (a) During drinking water treatment, 17 lb. of chlorine are added daily to disinfect 5 million gallons of water. What is the aqueous concentration of chlorine in mg/L? (b) The *chlorine demand* is the concentration of chlorine used during disinfection. The *chlorine residual* is the concentration of chlorine that remains after treatment so the water maintains its disinfecting power in the distribution system. If the chlorine residual is 0.20 mg/L, what is the chlorine demand in mg/L?

**2.2** A water sample contains 10 mg $NO_3^-$/L. What is the concentration in (a) $ppm_m$, (b) moles/L, (c) mg $NO_3^-$-N, and (d) $ppb_m$?

**2.3** A liquid sample has a concentration of iron (Fe) of 5.6 mg/L. The density of the liquid is 2,000 gm/L. What is the Fe concentration in $ppm_m$?

**2.4** Coliform bacteria (for example, *E. coli*) are excreted in large numbers in human and animal feces. Water that meets a standard of less than one coliform per 100 mL is considered safe for human consumption. Is a 1 L water sample that contains 9 coliforms safe for human consumption?

**2.5** The treated effluent from a domestic wastewater treatment plant contains ammonia at 9 mg N/L and nitrite at 0.5 mg N/L. Convert these concentrations to mg $NH_3$/L and mg $NO_2^-$/L.

**2.6** Nitrate concentrations exceeding 44.3 mg $NO_3^-$/L are a concern in drinking water due to the infant disease known as methemoglobinemia. Nitrate concentrations near three rural wells were reported as 0.01 mg $NO_3^-$-N, 1.3 $NO_3^-$-N, and 20 $NO_3^-$-N. Do any of these wells exceed the 44.3 $ppm_m$ level?

**2.7** Mirex (MW = 540) is a fully chlorinated organic pesticide manufactured to control fire ants. Due to its structure, mirex is very unreactive, so it persists in the environment. Water samples from Lake Erie have had mirex measured as high as 0.002 μg/L, and lake trout samples have had 0.002 μg/g. (a) In the water samples, what is the aqueous concentration of mirex in units of (i) $ppb_m$, (ii) $ppt_m$, and (iii) μM? (b) In the fish samples, what is the concentration of mirex in (i) $ppm_m$ and (ii) $ppb_m$?

**2.8** Chlorophenols impart unpleasant taste and odor to drinking water at concentrations as low as

5 mg/m³. They are formed when the chlorine disinfection process is applied to phenol-containing waters. What is the threshold for unpleasant taste and odor in units of (a) mg/L, (b) μg/L, (c) $ppm_m$, and (d) $ppb_m$?

**2.9** The concentration of monochloroacetic acid in rainwater collected in Zurich was 7.8 nanomoles/L. Given that the formula for monochloroacetic acid is $CH_2ClCOOH$, calculate the concentration in μg/L.

**2.10** Assume that concentrations of Pb, Cu, and Mn in rainwater collected in Minneapolis were found to be 9.5, 2.0, and 8.6 μg/L, respectively. Express these concentrations as nmole/L, given that the atomic weights are 207, 63.5, and 55, respectively.

**2.11** The dissolved oxygen (DO) concentration is measured as 0.5 mg/L in the anoxic zone and 8 mg/L near the end of a 108-ft.-long aerated biological reactor. What are these two DO concentrations in units of (a) $ppm_m$, (b) moles/L?

**2.12** Assume that the average concentration of chlordane—a chlorinated pesticide now banned in the United States—in the atmosphere above the Arctic Circle in Norway is 0.6 pg/m³. In this measurement, approximately 90 percent of this compound is present in the gas phase; the remainder is adsorbed to particles. For this problem, assume that all the compound occurs in the gas phase, the humidity is negligibly low, and the average barometric pressure is 1 atm. Calculate the partial pressure of chlordane. The molecular formula for chlordane is $C_{10}Cl_8H_6$. The average air temperature through the period of measurement was −5°C.

**2.13** What is the concentration in (a) $ppm_v$ and (b) percent by volume of carbon monoxide (CO) with a concentration of 103 μg/m³? Assume a temperature of 25°C and pressure of 1 atm.

×**2.14** Ice-resurfacing machines use internal combustion engines that give off exhaust containing CO and $NO_x$. Average CO concentrations measured in local ice rinks have been reported as high as 107 $ppm_v$ and as low as 36 $ppm_v$. How do these concentrations compare with an outdoor air quality 1-h standard of 35 mg/m³? Assume the temperature equals 20°C.

**2.15** Formaldehyde is commonly found in the indoor air of improperly designed and constructed buildings. If the concentration of formaldehyde in a home is 0.7 ppm$_v$ and the inside volume is 800 m$^3$, what mass (in grams) of formaldehyde vapor is inside the home? Assume $T = 298$ K and $P = 1$ atm. The molecular weight of formaldehyde is 30.

**2.16** The concentration of ozone ($O_3$) in Beijing on a summer day ($T = 30°C, P = 1$ atm) is 125 ppb$_v$. What is the $O_3$ concentration in units of (a) $\mu g/m^3$ and (b) moles of $O_3$ per $10^6$ moles of air?

**2.17** A balloon is filled with exactly 10 g of nitrogen ($N_2$) and 2 g of oxygen ($O_2$). The pressure in the room is 1.0 atm, and the temperature is 25°C. (a) What is the oxygen concentration in the balloon, expressed as percent by volume? (b) What is the volume (in liters) of the balloon after it has been blown up?

**2.18** A gas mixture contains $1.5 \times 10^{-5}$ mole CO and has a total of 1 mole. What is the CO concentration in ppm$_v$?

**2.19** "Clean" air might have a sulfur dioxide ($SO_2$) concentration of 0.01 ppm$_v$, while "polluted" air might have a concentration of 2 ppm$_v$. Convert these two concentrations to $\mu g/m^3$. Assume a temperature of 298 K.

**2.20** Carbon monoxide (CO) affects the oxygen-carrying capacity of your lungs. Exposure to 50 ppm$_v$ CO for 90 minutes has been found to impair one's ability to discriminate stopping distance; therefore, motorists in heavily polluted areas may be more prone to accidents. Are motorists at a greater risk of accidents if the CO concentration is 65 mg/m$^3$? Assume a temperature of 298 K.

**2.21** The Department of Environmental Quality determined that toxaphene concentrations in soil exceeding 60 $\mu g/kg$ (regulatory action level) can pose a threat to underlying groundwater. (a) If a 100 g sample of soil contains $10^{-5}$ g of toxaphene, what are the (a) toxaphene soil and (b) regulatory action level concentrations reported in units of ppb$_m$?

**2.22** Polycyclic aromatic hydrocarbons (PAHs) are a class of organic chemicals associated with the combustion of fossil fuels. Undeveloped areas may have total PAH soil concentrations of 5 $\mu g/kg$, while urban areas may have soil concentrations that range from 600 $\mu g/kg$ to 3,000 $\mu g/kg$. What is the concentration of PAHs in undeveloped areas in units of ppm$_m$?

**2.23** The concentration of toluene ($C_7H_8$) in subsurface soil samples collected after an underground storage tank was removed indicated the toluene concentration was 5 mg/kg. What is the toluene concentration in ppm$_m$?

**2.24** While visiting Zagreb, Croatia, Arthur Van de Lay visits the Mimara Art Museum and then takes in the great architecture of the city. He stops at a café in the old town and orders a bottle of mineral water. The reported chemical concentration of this water is: $[Na^+] = 0.65$ mg/L, $[K^+] = 0.4$ mg/L, $[Mg^{2+}] = 19$ mg/L, $[Ca^{2+}] = 35$ mg/L, $[Cl^-] = 0.8$ mg/L, $[SO_4^{2-}] = 14.3$ mg/L, $[HCO_3^-] = 189$ mg/L, $[NO_3^-] = 3.8$ mg/L. The pH of the water is 7.3. (a) What is the hardness of the water in mg/L CaCO$_3$? (b) Is the chemical analysis correct?

**2.25** In 2004, U.S. landfills emitted approximately 6,709 Gg of methane emissions, and wastewater treatment plants emitted 1,758 Gg of methane. How many Tg of CO$_2$ equivalents did landfills and wastewater plants emit in 2004? What percent of the total 2004 methane emissions (and greenhouse gas emissions) do these two sources contribute? (Total methane emissions in 2004 were 556.7 Tg CO$_2$e, and total greenhouse gas emissions were 7,074.4 CO$_2$e.)

**2.26** Mobile combustion of $N_2O$ in 2004 emitted 42.8 Tg CO$_2$e. How many Gg of $N_2O$ was this?

**2.27** A laboratory provides the following solids analysis for a wastewater sample: TS $= 200$ mg/L; TDS $= 30$ mg/L; FSS $= 30$ mg/L. (a) What is the total suspended solids concentration of this sample? (b) Does this sample have appreciable organic matter? Why or why not?

**2.28** A 100 mL water sample is collected from the activated sludge process of municipal wastewater treatment. The sample is placed in a drying dish (weight $= 0.5000$ g before the sample is added) and then placed in an oven at 104°C until all moisture is evaporated. The weight of the dried dish is recorded as 0.5625 g. A similar 100 mL sample is filtered, and the 100 mL liquid sample that passes through the filter is collected and placed in another drying dish (weight $= 0.5000$ g before the sample is added). This sample is dried at 104°C, and the dried dish's weight is recorded as 0.5325 g. Determine the concentration (in mg/L) of (a) total solids, (b) total suspended solids, (c) total dissolved solids, and (d) volatile suspended solids. (Assume VSS $= 0.7 \times$ TSS.)

# References

Eaton, A. D., L. S. Clesceri, E. W. Rice, A. E. Greenberg (Eds). 2005. *Standard Methods for the Examination of Water and Wastewater*, 21st ed. Washington, D. C.: American Public Health Association, Water Environment Federation, American Waterworks Association, 1368 pp.

Environmental Protection Agency (EPA). 2006. Inventory of U.S. Greenhouse Gas Emissions and Sinks: 1990–2004. (April), 430-R-06-002.

Mihelcic, J. R. 1999. *Fundamentals of Environmental Engineering*. New York: John Wiley & Sons.

# chapter/Three Chemistry

James R. Mihelcic, Noel R. Urban, Judith A. Perlinger

*This chapter presents several important chemical processes that describe the behavior of chemicals in both engineered and natural systems. The chapter begins with a discussion of the difference between activity and concentration. It then covers reaction stoichiometry and thermodynamic laws, followed by application of these principles to a variety of equilibrium processes. The basis of chemical kinetics is then explained, as are the rate laws commonly encountered in environmental problems.*

## Major Sections

3.1 Approaches in Environmental Chemistry

3.2 Activity and Concentration

3.3 Reaction Stoichiometry

3.4 Thermodynamic Laws

3.5 Volatilization

3.6 Air–Water Equilibrium

3.7 Acid–Base Chemistry

3.8 Oxidation-Reduction

3.9 Precipitation-Dissolution

3.10 Adsorption, Absorption, and Sorption

3.11 Kinetics

## Learning Objectives

1. Use ionic strength to calculate activity coefficients for electrolytes and nonelectrolytes.

2. Write balanced chemical reactions.

3. Relate the first and second laws of thermodynamics to engineering practice.

4. Write and apply equilibrium expressions for volatilization, air–water, acid–base, oxidation-reduction, precipitation-dissolution, and sorption reactions.

5. Apply mass balance principles to predict the distribution of chemicals among different environmental media.

6. Estimate how concentrations will change during the course of reactions using kinetic rate expressions for zero-order, first-order, and pseudo first-order reactions.

7. Determine how temperature affects rate constants.

8. Discuss how chemical properties affect the chemicals' distribution on a local and global basis.

## 3.1 Approaches in Environmental Chemistry

Chemistry is the study of the composition, reactions, and characteristics of matter. It is important because the ultimate fate of many chemicals discharged to air, water, soil, and treatment facilities is controlled by their reactivity and chemical speciation. Design, construction, and operation of treatment processes thus depends on fundamental chemical processes. Furthermore, individuals who predict (model) how chemicals move through indoor environments, groundwater, surface water, soil, the atmosphere, or a reactor are interested in whether a chemical degrades over time and how to mathematically describe the rate of chemical disappearance or equilibrium conditions.

Two very different approaches are used in evaluating a chemical's fate and treatment: kinetics and equilibrium. **Kinetics** deals with the rates of reactions, and **equilibrium** deals with the final result or stopping place of reactions. The kinetic approach is appropriate when the reaction is slow relative to our time frame or when we are interested in the rate of change of concentration. The equilibrium approach is useful whenever reactions are very fast, whenever we want to know in which direction a reaction will go, or whenever we want to know the final, stable conditions that will exist at equilibrium. If reactions happen very rapidly relative to the time frame of our interest, the final conditions that result from the reaction are likely to be of more interest than the rates at which the reaction occurs. In this case, an equilibrium approach is used. Examples of rapid reactions in the aqueous phase include acid–base reactions, complexation reactions, and some phase-transfer reactions, such as volatilization.

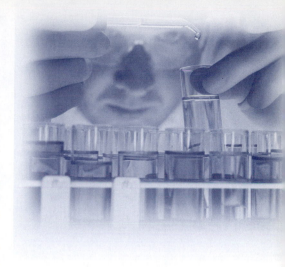

**Green Chemistry**
http://www.epa.gov/greenchemistry

## 3.2 Activity and Concentration

For a substance dissolved in a solvent, the **activity** can be thought of as the effective or apparent concentration, or that portion of the true mole-based concentration of a species that participates in a chemical reaction, normalized to the standard state concentration. In many environmental situations, *activity* and *concentration* are used interchangeably. Places where they begin to greatly differ include seawater, briny groundwaters, recycled and reused streams, and highly concentrated waste streams. Activity typically is designated by { } brackets, and concentration by [ ] brackets.

In an **ideal system**, the molar free energy of a solute in water depends on the mole fraction. However, this fraction does not reflect the effect of other dissolved species or the composition of the water, both of which also affect a solute's molar free energy. Chemical species interact by covalent bonding, van der Waals interactions, volume exclusion effects, and long-range electrostatic forces (repulsion and attraction between ions). In dilute aqueous systems, most interactions are caused by long-range electrostatic forces. On a molecular scale, these interactions can lead to local variations in the electron potential of the solution, resulting in a decrease in the total free energy of the system.

The use of activity instead of concentration accounts for these non-ideal effects. Activity is related to concentration by use of activity

**UNEP Chemicals Branch**
http://www.chem.unep.ch/

coefficients. **Activity coefficients** depend on the solution's ionic strength. Several equations (not described here in detail), developed specifically for either electrolytes (ions) or nonelectrolytes (uncharged species), express the activity coefficient of an individual species as a function of the ionic strength.

The **ionic strength** of a solution (referred to as $I$ or $\mu$) has units of moles/liter and is a measure of the long-range electrostatic interactions in that solution. Ionic strength can be calculated as follows:

$$\mu = 1/2 \; \Sigma_i C_i z_i^2 \tag{3.1}$$

where $C$ is the molar concentration of an ionic species $i$ in solution, and $z_i$ is the charge of the ion. In most natural waters, the ionic strength is derived primarily from the major background cations and anions. Freshwaters typically have an ionic strength of 0.001 to 0.01 M, and the ocean has an ionic strength of approximately 0.7 M. The ionic strength of aqueous systems rarely exceeds 0.7 M. Fortunately, it can be correlated to easily measured water-quality parameters such as total dissolved solids (TDS) or specific conductance:

$$\mu = 2.5 \times 10^{-5} \, (\text{TDS}) \tag{3.2}$$

where TDS is in mg/L, or

$$\mu = 1.6 \times 10^{-5} \, (\text{specific conductance}) \tag{3.3}$$

where specific conductance is in micromhos per centimeter ($\mu$mho/cm) and is measured with a conductivity meter.

The methods for calculating activity coefficients for electrolytes and nonelectrolytes are summarized in Figure 3.1. *Electrolytes* (for example, $Pb^{2+}$, $SO_4^{2-}$, $HCO_3^{2-}$) have a charge associated with them; *nonelectrolytes* (for example, $O_2$, $C_6H_6$) do not.

---

example/3.1 **Calculating Ionic Strength and Activity Coefficients for Electrolytes**

Calculate the ionic strength and all the individual activity coefficients for a 1 L solution in which 0.01 mole of $FeCl_3$ and 0.02 mole of $H_2SO_4$ are dissolved.

solution

After the two compounds are placed in water, they will completely dissociate to form 0.01 M $Fe^{3+}$, 0.04 M $H^+$, 0.03 M $Cl^-$, and 0.02 M $SO_4^{2-}$. The ionic strength is calculated by Equation 3.1:

$$\mu = 1/2 \, [0.01(3+)^2 + 0.04(1+) + 0.03(1-) + 0.02(2-)^2] = 0.12 \text{ M}$$

### example/3.1 Continued

This ionic strength is relatively high but still less than that of seawater.

The Davies approximation (see Figure 3.1) is useful for calculating activity coefficients for electrolytes when $\mu < 0.5$ M:

$$\gamma(H^+) = 0.74,\ \gamma(Cl^-) = 0.74,\ \gamma(SO_4^{2-}) = 0.30,\ \gamma(Fe^{3+}) = 0.065$$

The activity coefficients of ions with higher valence deviate much more from 1.0 for a given ionic strength; that is, for electrolytes, use of activity coefficients is much more important for ions with a higher valence, because they are strongly influenced by the presence of other ions. Thus, while at a particular ionic strength, it may not be important to calculate activity coefficients for monovalent ions, it may be very important for di-, tri-, and tetravalent ions.

**STEP 1**

After deciding whether ionic strength effects are important in a particular situation, calculate ionic strength from

$$\mu = \frac{1}{2}\Sigma_i C_i z_i^2 \text{ (Equation 3.1)}$$

or

estimate ionic strength after measuring the solution's total dissolved solids (Equation 3.2) or conductivity (Equation 3.3)

**STEP 2**

If species is an electrolyte, *$\gamma$ will always be $\leq 1$*

If species is a nonelectrolyte, *$\gamma$ will always be $\geq 1$*

For low ionic strengths, $\mu < 0.1$ M,

For high ionic strengths, $\mu < 0.5$ M,

For all ionic strengths, use

$$\log \gamma_i = k_s \times \mu$$

use the Güntelberg (or similar) approximation:

use the Davies (or similar) approximation:

$$\log \gamma_i = \frac{-A\,z_i^2\,\mu^{\frac{1}{2}}}{1 + \mu^{\frac{1}{2}}}$$

$$\log \gamma_i = \frac{-A\,z_i^2\,\mu^{\frac{1}{2}}}{[1 + \mu^{\frac{1}{2}}] - 0.3\mu}$$

**Figure 3.1**  **Two-Step Process to Determine Activity Coefficients for Electrolytes and Nonelectrolytes**

From Mihelcic (1999). Reprinted with permission of John Wiley & Sons, Inc.

example/3.2 **Calculating Activity Coefficients for Nonelectrolytes**

An air stripper is used to remove benzene ($C_6H_6$) from seawater and freshwater. Assume the ionic strength of seawater is 0.7 M and that of freshwater is 0.001 M. What is the activity coefficient for benzene in seawater and in freshwater?

**solution**

Because benzene is a nonelectrolyte, use the expression in Figure 3.1 to determine the activity coefficients. The value for $k_s$ (the salting-out coefficient) for benzene is 0.195.

$$\log \gamma = k_s \times \mu$$

$$\log \gamma = 0.195 \times (0.001 \text{ M}): \text{results in } \gamma \text{ (freshwater)} = 1$$

$$\log \gamma = 0.195 \times (0.7 \text{ M}): \text{results in } \gamma \text{ (seawater)} = 1.4$$

For freshwater, the activity coefficient does not deviate much from 1. It turns out there is little deviation for *nonelectrolytes* when $\mu < 0.1$ M. Therefore, determining activity coefficients for non-electrolytes becomes important for solutions with high ionic strengths. For most dilute environmental systems, activity coefficients for electrolytes and nonelectrolytes usually are assumed equal to 1. Places where they can gain importance are in the ocean, estuaries, briny groundwaters, and some recycled or reused waste streams.

## 3.3 Reaction Stoichiometry

**The law of conservation of mass** states that in a closed system, the mass of material present remains constant; the material may change form, but the total mass remains the same.

When this law is combined with our understanding that elements may combine with one another in numerous ways but are not converted from one to another (except for nuclear reactions), we arrive at the basis for reaction **stoichiometry**: in a closed system, the number of atoms of each element present remains constant. Therefore, in any single chemical reaction, the number of atoms of each element must be the same on both sides of the reaction equation.

A corollary of the law of conservation of mass is that electrical charges are also conserved; that is, the sum of charges on each side of a reaction equation must be equal. Electrical charges result from the balance between the numbers of protons and electrons present. Protons and electrons both have mass, and neither is converted into other subatomic particles during chemical reactions. Therefore, the total number of protons and electrons must remain constant in a closed system. It follows that the balance between the number of protons and electrons also must remain constant in a closed system. This means that reactions must be balanced in terms of mass and charge, and stoichiometry can be used not only for converting units of concentration but also for calculating chemical inputs and outputs.

## 3.4 Thermodynamic Laws

As the roots of the word imply (*thermo* equals heat; *dynamo* equals change), **thermodynamics** deals with conversions of energy from one form to another. Table 3.1 provides an overview of the **first law of thermodynamics** and **second law of thermodynamics**. Figure 3.2

### Table / 3.1

**Overview of the First and Second Laws of Thermodynamics**

| Law | What It Tells Us | Mathematical Expression | What It Means to Us |
|---|---|---|---|
| First law of thermodynamics | Energy is conserved; it may be converted from one form to another, but the total amount in a closed system is constant. In an open system, one must account for fluxes across the system boundaries. | For an open system: $dU = dQ - dW + dG$ where $U$ = internal energy content, $Q$ = heat content, $W$ = work done, and $G$ = energy of chemical inputs | This relationship demonstrates that the chemical potential (the energy within the chemical bonds of a molecule) constitutes a part of the total energy of the system. In a closed system (in which case, the third term on the right would be absent), reactions that change the chemical potential without changing the internal energy content must result in equivalent changes in heat content and in the pressure-volume work performed. |
| Second law of thermodynamics | All systems tend to lose useful energy and approach a state of minimum free energy or an equilibrium state. Thus, a process will proceed spontaneously (without energy put into the system from the outside) only if the process leads to a decrease in the free energy of the system (i.e., $\Delta G < 0$). | Formal definition of *Gibbs free energy:* $$G = \sum_i \mu_i \times N_i = H - T \times S$$ | The Gibbs free energy is related to the system's enthalpy ($H$), entropy ($S$), and temperature ($T$). The energy of inter- and intra-molecular bonds that bind various atoms and molecules together is termed *enthalpy*, while *entropy* refers to the disorder of the system. The chemical potential of all substances present, $\mu_i$, multiplied by the abundance of those substances, $N_i$, is equal to the combination of enthalpy and entropy present. |

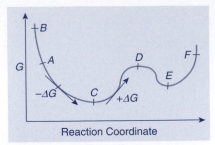

**Figure 3.2  Change in Free Energy (G) during a Reaction**  If the change leads to a decrease in free energy (i.e., for the forward reaction, if the slope of a tangent to the curve is negative), then the reaction can proceed spontaneously. Points C and E represent possible equilibrium points because the slopes of tangents at these points would be zero.

From Mihelcic (1999). Reprinted with permission of John Wiley & Sons, Inc.

© Elena Korenbaum/iStockphoto.

illustrates the change in free energy (G) during a reaction. In Figure 3.2 a process could proceed if it reduced the free energy from its value at point A in the direction of point C, but it could not proceed if it raised the energy in the direction of point B. The process could proceed from A as far as point C, but it could not go further toward point D. A reaction could also proceed from point D toward point C or point E. This is because moving in either direction results in a decrease in free energy.

Point E is called a **local equilibrium**. It is not the minimum possible energy point of the system (point C is), but to leave point E requires an input of energy. Hence, if the free energy of a system under all conditions could be quantified, we could then determine the changes that could occur spontaneously in that system (that is, any changes that would cause a decrease in the free energy).

$\Delta G$ is the free-energy change under *ambient conditions* (the prevailing environmental conditions). The value of $\Delta G$ is calculated according to the following relationship:

$$\Delta G = \Delta G^0 + RT \ln (Q) \tag{3.4}$$

where $\Delta G^0$ is the change in free energy determined under *standard conditions*, $R$ is the gas constant, $T$ is the ambient temperature in K, and $Q$ is the reaction quotient. $\Delta G^0$ is determined from reaction stoichiometry and tabulated values as described in most chemistry books. Other references provide detail on determination and application of this term (see, e.g., Mihelcic, 1999).

The reaction quotient $Q$ is defined as the product of the activities (apparent concentrations of the reaction products) raised to the power of their stoichiometric coefficients, divided by the product of the activities (or concentrations) of the reactants raised to the power of their stoichiometric coefficients. Thus, for the generalized reaction

$$aA + bB \leftrightarrow cC + dD \tag{3.5}$$

in which $a$ moles of compound $A$ react with $b$ moles of compound $B$ to form $c$ moles of compound $C$ and $d$ moles of compound $D$, $Q$ is given by

$$Q = \frac{\{C\}^c \{D\}^d}{\{A\}^a \{B\}^b} \tag{3.6}$$

As noted in Example 3.2, activity coefficients ($\gamma$) are usually assumed to equal 1; thus, $Q$ can be calculated based on concentrations. Table 3.2 describes the four rules used to determine what value to use for the activity (concentration) [$i$] in Equation 3.6. Following these rules is essential to make activities and reaction quotients dimensionless.

Only reactions that result in thermodynamically favorable changes in their energy state can occur. This change in energy state is called **Gibbs free energy** change and is denoted $\Delta G$. It is this change in energy state that defines the equilibrium condition. However, not all reactions that occur would result in a favorable

## Table / 3.2

**Rules for Determining Value of [i]**  These rules determine what value to use for the activity (i.e., concentration) termed [i] of a chemical species i. Following these rules is essential to make activities and reaction quotients dimensionless.

| | |
|---|---|
| Rule 1 | For liquids (e.g., water): [i] is equal to the mole fraction of the solvent. In aqueous solutions, the mole fraction of water can be assumed to equal 1. Thus, $[H_2O]$ always equals 1. |
| Rule 2 | For pure solids in equilibrium with a solution (e.g., $CaCO_{3(s)}$, $Fe(OH)_{3(s)}$): [i] always equals 1. |
| Rule 3 | For gases in equilibrium with a solution (e.g., $CO_{2(g)}$, $O_{2(g)}$): [i] equals the partial pressure of the gas (units of atm). |
| Rule 4 | For compounds dissolved in water: [i] is always reported in units of moles/L (not mg/L or $ppm_m$). |

change in Gibbs free energy, and the magnitude of this energy change seldom is related to the rate of the reaction. For a reaction to occur, it generally is necessary that atoms collide and that this collision have the right orientation and enough energy to overcome the **activation energy** required for the reaction. These energetic relationships are shown in Figure 3.3.

Equilibrium is defined as the state (or position) with the minimum possible free energy. This occurred at point C in Figure 3.2. The value of the slope at the point of equilibrium (point C) is 0. In other words, the change in free energy is zero at equilibrium. If the change in free energy ($\Delta G$) is equal to 0 at equilibrium, the reaction quotient at equilibrium (see Equation 3.6) is usually written with a special symbol, K, and is provided a special name, the **equilibrium constant.**

The equilibrium constant for the reaction written as Equation 3.5 is given by the equilibrium reaction quotient, $Q_{eqn}$:

$$Q_{eqn} = \frac{[C]^c[D]^d}{[A]^a[B]^b} = K \qquad (3.7)$$

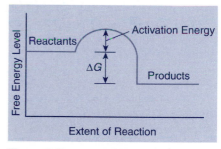

**Figure 3.3   Energetic Relationships Required for a Reaction to Occur** $\Delta G$ is called the Gibbs free energy change.

From Mihelcic (1999). Reprinted with permission of John Wiley & Sons, Inc.

## Box / 3.1    Effect of Temperature on the Equilibrium Constant

Most tabulated equilibrium constants are recorded at 25°C. The **van't Hoff relationship** (Equation 3.8) is used to convert equilibrium constants to temperatures other than those for which the tabulated values are provided. Van't Hoff discovered that the equilibrium constant (K) varied with absolute temperature and the enthalpy of a reaction ($\Delta H^0$). Van't Hoff proposed the following expression to describe this:

$$\frac{d \ln K}{dT} = \frac{\Delta H^0}{RT^2} \qquad (3.8)$$

Here $\Delta H^0$ is found from the heat of formation $(\Delta H^0_f)$ for the reaction of interest determined at standard conditions. Most temperatures encountered in environmental problems are relatively small (for example, 0°C to 40°C). Therefore, the temperature differences are not that large. If $\Delta H^0$ is assumed not to change over the temperature range investigated, Equation 3.8 can be integrated to yield

$$\ln\left[\frac{K_2}{K_1}\right] = \frac{\Delta H^0}{R} \times \left(\frac{1}{T_1} - \frac{1}{T_2}\right) \qquad (3.9)$$

Equation 3.9 can be used to calculate an equilibrium constant for any temperature (i.e., temperature 2, $T_2$) if the equilibrium constant is known at another absolute temperature ($T_1$, which is usually 20°C or 25°C).

The equilibrium constant is useful because it provides the ratio of the concentration (or activity) of individual reactants and products for any reaction at equilibrium. Remember, activity coefficients must be included if conditions are not ideal and these coefficients are raised to appropriate stoichiometric values.

Do not confuse the equilibrium constant, $K$, with the reaction rate constant, $k$, which we will discuss later in this chapter. $K$ is constant for a specific reaction (as long as temperature is constant). As reviewed in Figure 3.4, equilibrium constants and partition coefficients are defined for reactions that describe volatilization (saturation vapor pressure), air–water exchange (Henry's law constant, $K_H$), acid–base chemistry ($K_a$ and $K_b$), oxidation-reduction reactions ($K$), precipitation-dissolution reactions ($K_{sp}$), and sorptive partitioning ($K_d$, $K_p$, $K_{oc}$, $K$).

## 3.5   Volatilization

A key step in the transfer of pollutants between different environmental media is volatilization. All liquids and solids exist in equilibrium with a gas or vapor phase. **Volatilization** (synonymous with *evaporation* for the case of water) is the transformation of a compound from its liquid state to its gaseous state. *Sublimation* is the word used for transformation from the solid to gaseous state. The reverse reaction is termed *condensation*.

Most people have firsthand experience with the phenomenon of sublimation of water. The water vapor in the atmosphere (the humidity) is a function of temperature. Modern refrigerators prevent frost buildup by maintaining a low humidity inside the refrigerator; any ice that forms is sublimed, or vaporized. Similarly, the amount of snow on the ground decreases in periods between snowfalls, partially due to the sublimation of the snow. Many organic pollutants volatilize more readily than water. The fumes from gasoline, paint thinners, waxes, and glue attest to the volatility of organic chemicals contained in these commonly used products. Volatilization of chemicals can result in regional and long-range transport of the chemicals to places

| Section | Reaction | Equilibrium Constant Notation |
|---|---|---|
| 3.5 Volatilization | 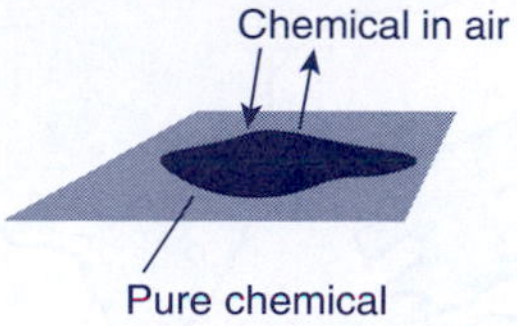 Chemical in air / Pure chemical | Partial pressure |
| 3.6 Air–Water Equilibrium | 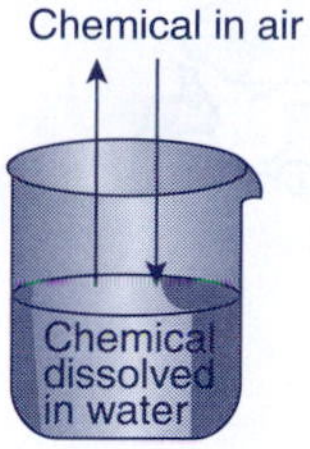 Chemical in air / Chemical dissolved in water | Henry's constant ($K_H$ or $H$) |
| 3.7 Acid–Base Chemistry | Acid $\leftrightarrow$ Conjugate base + H$^+$ or Base $\leftrightarrow$ Conjugate acid + OH$^-$ | Acidity ($K_a$) or basicity constants ($K_b$) |
| 3.8 Oxidation-Reduction | Electron acceptor (oxidant) + e$^- \longrightarrow$ Electron donor (reductant) | Equilibrium constant ($K$) |
| 3.9 Precipitation-Dissolution | Solid chemical $\leftrightarrow$ Dissolved chemical | Solubility Product ($K_{sp}$) |
| 3.10 Adsorption, Absorption, Sorption | 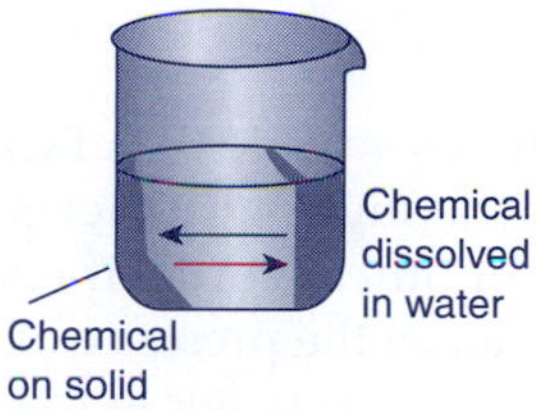 Chemical dissolved in water / Chemical on solid | Soil–water partition ($K_p$, $K_d$, $K_{oc}$); Freundlich parameters ($K$, $1/n$) |

**Figure 3.4** Important Equilibrium Processes for Environmental Engineering

Adapted from Mihelcic (1999). Reprinted with permission of John Wiley & Sons, Inc.

far away, where adverse environmental effects can be detected (see Figure 3.5).

The equilibrium between a gas and a pure liquid or solid phase is determined by the saturated vapor pressure of a compound. **Saturated vapor pressure** is defined as that partial pressure of the gas phase of a substance that exists in equilibrium with the liquid or solid phase of the substance at a given temperature.

The more volatile a compound, the higher its saturated vapor pressure. For example, the saturated vapor pressure of the solvent tetrachloroethylene (PCE) is 0.025 atm at 25°C, while the saturated vapor pressure of the pesticide lindane is $10^{-6}$ atm at the same temperature.

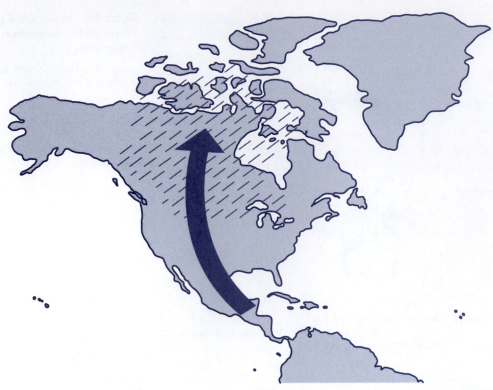

**Figure 3.5**  **Spread of Chemicals through Volatilization**  The process of exporting toxic chemicals to other countries that then return by atmospheric transport has been termed "the circle of poison." Persistent organic pollutants (POPs) become concentrated in the food chain, where they can cause toxic effects on animal reproduction, development, and immunological function. The U.S. State Department has termed POPs "one of the great environmental challenges the world faces." POPs include polychlorinated biphenyls (PCBs), polychlorinated dibenzo-*p*-dioxins, and furans, and pesticides such as DDT, toxaphene, chlordane, and heptachlor.

From Mihelcic (1999). Reprinted with permission of John Wiley & Sons, Inc.

**Class Discussion**

Though banned for use in many developed countries, persistent organic pollutants are still manufactured for export and/or remain widely used and unregulated in many parts of the world. These chemicals volatilize more easily in the warm surface temperatures found in the southern U.S. and subtropical and tropical regions of the world. They then condense and deposit in high latitudes where temperatures are cooler. Wildlife such as seals, killer whales, and polar bears along with human populations, such as the Inuit, that reside in the Arctic are unfairly burdened with the risk associated with production and use of these chemicals. Given this information, what equitable solutions that consider future generations can you think of?

Clearly, lindane is much less volatile than PCE. For the sake of comparison, water at 25°C has a slightly higher saturated vapor pressure (0.031 atm) than PCE. In other words, if containers or spills of PCE were left exposed to the air in the presence of containers of water, there would be about as much PCE as water in the atmosphere of an indoor-air environment.

The equilibrium between gas and liquid phases can be expressed in the usual form of a chemical reaction with an equilibrium constant:

$$H_2O_{(l)} \leftrightarrow H_2O_{(g)} \qquad \textbf{(3.10)}$$

Equation 3.10 indicates that liquid water is in equilibrium with gaseous water (water vapor). The equilibrium constant (called the *saturated vapor pressure*) for this reaction is:

$$K = \frac{[H_2O_{(g)}]}{[H_2O_{(l)}]} = P_{H_2O} \qquad \textbf{(3.11)}$$

where $P_{H_2O}$ is the partial pressure of water. Because the concentration (assumed to equal the activity, i.e., $\gamma = 1.0$) of a pure liquid is defined as 1.0 (remember Rule 1 in Table 3.2), the equilibrium constant is simply equal to the concentration in the vapor phase (called the saturated vapor pressure). One way of expressing gas-phase concentrations is as partial pressures; hence, the equilibrium constant for volatilization often is expressed in units of atmospheres.

If a mixture of miscible (mutually soluble) liquids—rather than a pure liquid—was present, the denominator in Equation 3.11 would be the concentration of the individual liquid ($A$) in mole fractions, $X_A$:

$$K = \frac{[A_{(g)}]}{[A_{(l)}]} = \frac{P_A}{X_A} \qquad (3.12)$$

Equation 3.12 is known as **Raoult's law**. The constant, $K$, equals the saturated vapor pressure. Raoult's law is useful whenever a mixture of chemicals (for example, gasoline, diesel fuel, or kerosene) is spilled.

The vapor pressure for all compounds increases with temperature, and at the boiling point of the compound, the vapor pressure equals atmospheric pressure. This statement has practical consequences. First, atmospheric concentrations of volatile substances tend to be higher in summer than in winter, in the day versus at night, and in warmer locations. Second, for any structurally similar group of liquid chemicals exposed to the air, the equilibrium gas-phase concentrations will decrease in order of increasing boiling points.

**Class Discussion**

Are there mercury advisories for fishing in your region? What are the sources of the mercury besides fossil fuel combustion, and what population segments are most at risk? Are the benefits and environmental risk equally distributed among all segments of society?

---

**Box / 3.2    The Complexity of Environmental Problems**

One feature of environmental problems is that they seldom are confined to just one medium. For example, a lot of the mercury discharged into the environment is first emitted as an air pollutant, but its most damaging effects occur in lakes after it moves through the atmosphere, is deposited into a lake, and then undergoes a biological transformation process called **methylation**. This allows mercury to bioaccumulate in fish, a process that has resulted in thousands of fishing advisories in U.S. lakes.

Much of the mercury is released from combustion of coal associated with electricity production. Readers are encouraged to turn back to Figure 1.7a (Chapter 1), which shows projected U.S. electricity production to the year 2030 coming primarily from combustion of coal, not sources of renewable energy. In addition, China is expected to double its coal consumption by 2020, and the migration of its population from rural to urban areas is resulting in increased energy use per capita. China also is expected to surpass the United States in greenhouse gas emissions by the year 2009, mainly because of its plan to consume its vast stores of coal.

The burning of fossil fuels such as coal not only leads to climate change but also is incurring future economic, social, and environmental costs that will be assumed by current and future generations—all from the environmental release of the neurotoxin mercury. Engineering systems and developing public policies that conserve energy, use renewable energy sources, and right size the buildings and other components of the built environment can have several mutually beneficial impacts to the economy, society, and the environment.

## example/3.3 Calculation of Gaseous Concentration in a Confined Area

On a Friday afternoon, a worker spills 1 L of tetrachloroethylene (PCE) on a laboratory floor. The worker immediately closes all the windows and doors and turns off the ventilation in order to avoid contaminating the rest of the building. The worker notifies the appropriate safety authority, but it is Monday morning before the safety official stops by with a crew to clean up the laboratory. Should the cleanup crew bring a mop or an air pump to clean up the room? The volume of the laboratory is 340 m$^3$, and the temperature in the room is 25°C. For PCE, the vapor pressure is 0.025 atm, the liquid density at 25°C is 1.62 g/cm$^3$, and the molecular weight is 166 g/mole.

## solution

PCE is a volatile chemical. The problem asks how much of the 1 L of spilled PCE remained on the floor versus how much volatilized into the air. If any PCE remained on the floor, the partial pressure of PCE in the air would be 0.025 atm. The ideal gas law can be used to solve for the number of moles present in the air (the term $n/V$ would provide the concentration):

$$n = \frac{PV}{RT} = \frac{(0.025 \text{ atm}) \times (340 \text{ m}^3) \times \left(\frac{1,000 \text{ L}}{\text{m}^3}\right)}{\frac{0.08205 \text{ L-atm}}{\text{mole-K}} \times (298 \text{ K})} = 348 \text{ moles}$$

www.epa.gov/gcc

The density of PCE can be used to determine that the 1 L spill weighs 1,620 g. Using the molecular weight of PCE, the 1 L spill would contain 9.8 moles of PCE. This is much less than the amount that could potentially volatilize into the air in the room (348 moles), assuming equilibrium has been attained. Thus, it can be concluded that no PCE would remain on the floor, and it would be entirely in the air. The cleanup crew should arrive at work equipped with air pumps and filters.

This problem demonstrates another important point: Chemistry and engineering need to become "green." If a green chemical were substituted for the solvent PCE in the required use, there would be no risk and, thus, no concern related to the spill. Better yet, perhaps the process that the PCE was used for could be changed so that no chemical is required at all. This type of thinking would result in reduced health costs, because workers would not be exposed to toxic chemicals. Other savings would result because there would be no requirement to pay the cleanup crew for remediation, no energy needed for the remediation phase, less paperwork associated with regulations that govern the handling and storage of the PCE, and no future liability associated with storage and use of PCE. The company might also be able to increase its market share by promoting that its facility is more socially and environmentally responsible.

## 3.6  Air–Water Equilibrium

The **Henry's law constant, $K_H$,** is used to describe a chemical's equilibrium between the air and water (often termed the dissolved or aqueous) phases. This situation is referred to as air–water equilibrium. **Henry's law** is just a special case of Raoult's law (Equation 3.12) applied to dilute systems (most environmental situations are dilute). Because the mole fraction of a dissolved substance in a dilute system is a very small number, concentrations such as moles/L typically are used rather than mole fractions. Equation 3.12 can also be used to estimate Henry's law constants in the absence of reliable experimental data. To determine a Henry's law constant for a particular chemical, divide the saturated vapor pressure of the chemical by its aqueous solubility.

The units of Henry's law constant vary depending on whether the air–water exchange reaction is written in the forward direction for transfer from the gas phase into aqueous phase or from the aqueous phase into the gas phase. In addition, Henry's law constants may also be unitless. Thus, it is important to use the proper units, understand why particular units are used, and be able to convert between different units.

### 3.6.1  HENRY'S LAW CONSTANT WITH UNITS FOR A GAS DISSOLVING IN A LIQUID

The air–water exchange of a gas (in this case, oxygen) from the atmosphere into water in the forward direction (depicted in Figure 3.4) can be written as

$$O_{2(g)} \leftrightarrow O_{2(aq)} \tag{3.13}$$

The equilibrium expression for this reaction is:

$$K_H = \frac{[O_{2(aq)}]}{[O_{2(g)}]} = \frac{[O_{2(aq)}]}{P_{O_2}} \tag{3.14}$$

The value of the Henry's law constant, $K_H$, at 25°C for oxygen is $1.29 \times 10^{-3}$ moles/L-atm. In this case, the units of $K_H$ are moles/L-atm. The reaction was written as oxygen gas transferring into the aqueous phase in the forward direction because in this case we are concerned with how the composition of the gas affects the composition of the aqueous solution. Thus, the equilibrated dissolved oxygen saturation concentration in surface waters is a function of the partial pressure of oxygen in the atmosphere and the Henry's law constant.

The concentration of **dissolved oxygen** in water equilibrated with the atmosphere is 14.4 mg/L at 0°C and 9.2 mg/L at 20°C. This value demonstrates that oxygen solubility in water depends on water temperature (one reason trout like colder waters). For the reaction

## example/3.4 Using Henry's Law Constant to Determine the Aqueous Solubility of Oxygen

Calculate the concentration of dissolved oxygen (units of moles/L and mg/L) in a water equilibrated with the atmosphere at 25°C. The Henry's law constant for oxygen at 25°C is $1.29 \times 10^{-3}$ mole/L-atm.

### solution

The partial pressure of oxygen in the atmosphere is 0.21 atm. Equation 3.14 can be rearranged to yield

$$K_H \times P_{O_2} = [O_{2(aq)}] = (1.29 \times 10^{-3}\frac{mole}{L\text{-atm}}) \times 0.21 \text{ atm}$$

$$= 2.7 \times 10^{-4} \frac{mole}{L}$$

Thus, the solubility of oxygen at this temperature is $2.7 \times 10^{-4}$ moles/L. If this value is multiplied by the molecular weight of oxygen (32 g/mole), the solubility can be reported as 8.7 mg/L.

described in Equation 3.13, the change in heat of formation ($\Delta H^0$) at standard conditions is −3.9 kcal. Because $\Delta H^0$ is negative, Equation 3.13 could be written as

$$O_{2(g)} \leftrightarrow O_{2(aq)} + \text{heat} \qquad (3.15)$$

An increase in the temperature (or adding heat to the system) will, according to Le Châtelier's principle, favor the reaction that tends to diminish the increase in temperature. The effect is to drive the reaction in Equation 3.15 to the left, which consumes heat, diminishing the temperature increase in the process. Therefore, at equilibrium, more oxygen will be present in the gas phase at an increased temperature; thus, the solubility of dissolved oxygen will be lower at the increased temperature.

### 3.6.2 DIMENSIONLESS HENRY'S LAW CONSTANT FOR A SPECIES TRANSFERRING FROM THE LIQUID PHASE INTO THE GAS PHASE

In the case for the transfer of a chemical dissolved in the aqueous phase into the atmosphere, the chemical equilibrium between the gas and liquid phase chemical is described by a reaction written in reverse of Equation 3.13. For example, for the chemical trichloroethylene (TCE),

$$\text{TCE}_{(aq)} \leftrightarrow \text{TCE}_{(g)} \qquad \textbf{(3.16)}$$

In this case, the equilibrium expression is written as

$$K_H = \frac{[\text{TCE}_{(g)}]}{[\text{TCE}_{(aq)}]} \qquad \textbf{(3.17)}$$

where the gas phase TCE is described by units of moles/liter of gas, not as partial pressure. Accordingly, the Henry's law constant, $K_H$, has units of moles per liter of gas divided by moles/liter of water, which cancel out. Therefore, the Henry's law constant in this case is termed *dimensionless* by some. In fact, it really has units of liters of water per liters of air. Other units of Henry's law constant include atm and L-atm/mole.

Henry's law constants that have units and those without units can be related using the ideal gas law. Several unit conversions for Henry's law constant are provided in Table 3.3.

## Table / 3.3

**Unit Conversion of Henry's Law Constants**

$$K_H\left(\frac{L_{H_2O}}{L_{Air}}\right) = \frac{K_H\left(\dfrac{\text{L-atm}}{\text{mole}}\right)}{RT}$$

$$K_H\left(\frac{\text{L-atm}}{\text{mole}}\right) = K_H\left(\frac{L_{H_2O}}{L_{Air}}\right) \times RT$$

$$K_H\left(\frac{L_{H_2O}}{L_{Air}}\right) = \frac{K_H(\text{atm})}{RT \times 55.6 \dfrac{\text{mole } H_2O}{L_{H_2O}}}$$

$$K_H\left(\frac{\text{L-atm}}{\text{mole}}\right) = \frac{K_H(\text{atm})}{55.6 \dfrac{\text{mole } H_2O}{L_{H_2O}}}$$

$$K_H(\text{atm}) = K_H\left(\frac{\text{L-atm}}{\text{mole}}\right) \times 55.6 \frac{\text{mole } H_2O}{L_{H_2O}}$$

$$K_H(\text{atm}) = K_H\left(\frac{L_{H_2O}}{L_{Air}}\right) \times RT \times 55.6 \frac{\text{mole } H_2O}{L_{H_2O}}$$

$$R = 0.08205 \frac{\text{atm-L}}{\text{mole-K}}$$

SOURCE: From Mihelcic (1999), reprinted with permission of John Wiley & Sons, Inc.

**example/3.5** Conversion between Dimensionless and Nondimensionless Henry's Law Constants

The Henry's law constant for the reaction transferring oxygen from air into water is $1.29 \times 10^{-3}$ moles/L-atm at 25°C. What is the dimensionless $K_H$ for the transfer of oxygen from water into air at 25°C?

**solution**

The problem is requesting a Henry's law constant for the reverse reaction. Therefore, the Henry's law constant provided equals the inverse of $1.29 \times 10^{-3}$ moles/L-atm, or 775 L-atm/mole for the transfer of aqueous oxygen into the gas phase. Solve using the ideal gas law:

$$K_H(\text{dimensionless}) = \frac{\dfrac{775 \text{ L-atm}}{\text{mole}}}{\left(\dfrac{0.08205 \text{ L-atm}}{\text{mole-K}}\right) \times (298 \text{ K})} = 32$$

## 3.7 Acid–Base Chemistry

Acid-base chemistry is important in treatment of pollution and in understanding the fate and toxicity of chemicals discharged to the environment.

### 3.7.1 pH

By definition, the **pH** of a solution is:

$$pH = -\log[H^+] \tag{3.18}$$

where $[H^+]$ is the concentration of the hydrogen ion. The pH scale in aqueous systems ranges from 0 to 14, with acidic solutions having a pH below 7, basic solutions having a pH above 7, and neutral solutions having a pH near 7. Ninety-five percent of all natural waters have a pH between 6 and 9. Rainwater not affected by anthropogenic acid-rain emissions has a pH of approximately 5.6 due to the presence of dissolved carbon dioxide that originates in the atmosphere.

The concentrations of $OH^-$ and $H^+$ are related to one another through the equilibrium reaction for the dissociation of water:

$$H_2O \leftrightarrow H^+ + OH^- \tag{3.19}$$

The equilibrium constant for the dissociation of water ($K_w$) for Equation 3.19 equals $10^{-14}$ at 25°C. Thus,

$$K_w = 10^{-14} = [H^+] \times [OH^-] \tag{3.20}$$

## Table / 3.4

**Dissociation Constant for Water at Various Temperatures and Resulting pH of a Neutral Solution**

| Temperature (°C) | $K_w$ | pH of Neutral Solution |
|---|---|---|
| 0 | $0.12 \times 10^{-14}$ | 7.47 |
| 15 | $0.45 \times 10^{-14}$ | 7.18 |
| 20 | $0.68 \times 10^{-14}$ | 7.08 |
| 25 | $1.01 \times 10^{-14}$ | 7.00 |
| 30 | $1.47 \times 10^{-14}$ | 6.92 |

SOURCE: From Mihelcic (1999), reprinted with permission of John Wiley & Sons, Inc.

Equation 3.20 allows the determination of the concentration of $H^+$ or $OH^-$ if the other is known. Table 3.4 shows the range of $K_w$ at temperatures of environmental significance. At 25°C in pure water, $[H^+]$ equals $[OH^-]$; thus $[H^+] = 10^{-7}$, and the pH of pure water is equal to 7.00. However, at 15°C, $[H^+]$ equals $10^{-7.18}$, so the pH of a neutral solution at this temperature is equal to 7.18.

### 3.7.2 DEFINITION OF ACIDS AND BASES AND THEIR EQUILIBRIUM CONSTANTS

Acids and bases are substances that react with hydrogen ions ($H^+$). An **acid** is defined as a species that can release or donate a hydrogen ion (also called a proton). A **base** is defined as a chemical species that can accept or combine with a proton. Equation 3.21 shows an example of an acid (HA) associated with a conjugate base ($A^-$):

$$HA \leftrightarrow H^+ + A^- \qquad (3.21)$$

Acids that have a strong tendency to dissociate (the Reaction in Equation 3.21 goes far to the right) are called *strong acids*, while acids that have less of a tendency to dissociate (the Reaction in Equation 3.21 goes just a little to the right) are called *weak acids*.

The strength of an acid is indicated by the magnitude of the equilibrium constant for the dissociation reaction. The equilibrium constant for the reaction depicted in Equation 3.21 is:

$$K_a = \frac{[H^+][A^-]}{[HA]} \qquad (3.22)$$

where $K_a =$ is the equilibrium constant for the reaction when an acid is added to water. At equilibrium, a strong acid will dissociate and show

## Table / 3.5

### Common Acids and Bases and Their Equilibrium Constants When Added to Water at 25°C

| Acids | | | Bases | | |
|---|---|---|---|---|---|
| | Name | $pK_a = -\log K_a$ | | Name | $pK_b = -\log K_b$ |
| $HCl$ | Hydrochloric | $-3$ | $Cl^-$ | Chloride ion | 17 |
| $H_2SO_4$ | Sulfuric | $-3$ | $HSO_4^-$ | Bisulfate ion | 17 |
| $HNO_3$ | Nitric | $-1$ | $NO_3^-$ | Nitrate ion | 15 |
| $HSO_4^-$ | Bisulfate | 1.9 | $SO_4^{2-}$ | Sulfate ion | 12.1 |
| $H_3PO_4$ | Phosphoric | 2.1 | $H_2PO_4^-$ | Dihydrogen phosphate | 11.9 |
| $CH_3COOH$ | Acetic | 4.7 | $CH_3COO^-$ | Acetate ion | 9.3 |
| $H_2CO_3^*$ | Carbon dioxide and carbonic acid | 6.3 | $HCO_3^-$ | Bicarbonate | 7.7 |
| $H_2S$ | Hydrogen sulfide | 7.1 | $HS^-$ | Bisulfide | 6.9 |
| $H_2PO_4^-$ | Dihydrogen phosphate | 7.2 | $HPO_4^{2-}$ | Monohydrogen phosphate | 6.8 |
| $HCN$ | Hydrocyanic | 9.2 | $CN^-$ | Cyanide ion | 4.8 |
| $NH_4^+$ | Ammonium ion | 9.3 | $NH_3$ | Ammonia | 4.7 |
| $HCO_3^-$ | Bicarbonate | 10.3 | $CO_3^{2-}$ | Carbonate | 3.7 |
| $HPO_4^{2-}$ | Monohydrogen phosphate | 12.3 | $PO_4^{3-}$ | Phosphate | 1.7 |
| $NH_3$ | Ammonia | 23 | $NH_2^-$ | Amide | $-9$ |

SOURCE: From Mihelcic (1999), reprinted with permission of John Wiley & Sons, Inc.

high concentrations of $H^+$ and $A^-$ and a smaller concentration of HA. This means that when a strong acid is added to water, the result is a much larger negative free-energy change than when adding a weaker acid. Thus, for strong acids, the equilibrium constant $K_a$ will be large (and $\Delta G$ will be very negative). Similarly, the $K_a$ for a weak acid will be small (and $\Delta G$ will be less negative).

Just as pH equals $-\log [H^+]$, **$pK_a$** is the negative logarithm of the acid dissociation constant (that is, $pK_a = -\log(K_a)$). Table 3.5 provides values of equilibrium constants for some acids and bases of environmental importance. The table shows that the $pK_a$ of a weak acid is larger than the $pK_a$ of a strong acid.

The $pK_a$ of an acid is related to the pH at which the acid will dissociate. Strong acids are those that have a $pK_a$ below 2. They can be assumed to dissociate almost completely in water in the pH range 3.5–14. HCl, $HNO_3$, $H_2SO_4$, and $HClO_4$ are four very strong acids commonly encountered in environmental situations. Likewise, their conjugate bases ($Cl^-$, $NO_3^-$, $SO_4^{2-}$ and $ClO_4^-$) are so weak that in the pH range of 3.5–14, they are assumed to never exist with protons.

## example/3.6 Acid–Base Equilibrium

What percentage of total ammonia (i.e., $NH_3 + NH_4^+$) is present as $NH_3$ at a pH of 7? The $pK_a$ for $NH_4^+$ is 9.3; therefore,

$$K_a = 10^{-9.3} = \frac{[NH_3][H^+]}{[NH_4^+]}$$

## solution

The problem is requesting

$$\frac{[NH_3]}{([NH_4^+] + [NH_3])} \times 100\%$$

Solving this problem requires another independent equation because the preceding expression has two unknowns. The equilibrium expression for the $NH_4^+/NH_3$ system provides the second required equation:

$$10^{-9.3} = \frac{[NH_3] \times [H^+]}{[NH_4^+]} = \frac{[NH_3] \times [10^{-7}]}{[NH_4^+]}$$

Thus, at pH = 7, $[NH_4^+] = 200 \times [NH_3]$. This expression can be substituted into the first expression, yielding

$$\frac{[NH_3]}{(200[NH_3] + [NH_3])} \times 100\% = 0.5\%$$

At this neutral pH, almost all of the total ammonia of a system exists as ammonium ion ($NH_4^+$). In fact, only 0.5 percent exists as $NH_3$!

The form of total ammonia most toxic to aquatic life is $NH_3$. It is toxic to several fish species at concentrations above 0.2 mg/L. Thus, wastewater discharges with a pH less than 9 have most of the total ammonia in the less toxic $NH_4^+$ form. This is one reason why some wastewater discharge permits for ammonia specify that the pH of the discharge must also be less than 9.

### 3.7.3 CARBONATE SYSTEM, ALKALINITY, AND BUFFERING CAPACITY

Figure 3.6 shows the important components of the **carbonate system**. The concentration of **dissolved carbon dioxide** in water equilibrated with the atmosphere (partial pressure of $CO_2$ is $10^{-3.5}$ atm) is $10^{-5}$ moles/L. This is a significant amount of carbon dioxide dissolved in water. This reaction can be written as follows:

$$CO_{2(g)} \leftrightarrow CO_{2(aq)} \qquad (3.23)$$

where $K_H = 10^{-1.5}$ moles/L-atm.

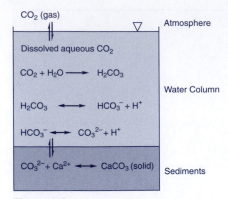

CO₂ (gas) — Atmosphere

Dissolved aqueous CO₂

$CO_2 + H_2O \longrightarrow H_2CO_3$

Water Column

$H_2CO_3 \longleftrightarrow HCO_3^- + H^+$

$HCO_3^- \longleftrightarrow CO_3^{2-} + H^+$

$CO_3^{2-} + Ca^{2+} \longleftrightarrow CaCO_3 \text{ (solid)}$ — Sediments

**Figure 3.6** **Important Components of the Carbonate System**

From Mihelcic (1999). Reprinted with permission of John Wiley & Sons, Inc.

### Class Discussion

Some scientists have suggested that we add large quantities of iron to the world's oceans to precipitate out carbonate, thus shifting carbonate system chemistry so the oceans take up more carbon dioxide from the atmosphere. Do you consider this geoengineering of the environment a sustainable solution for the current problem of the world emitting too many greenhouse gases?

Upon dissolving in water, dissolved $CO_2$ undergoes a hydration reaction by reacting with water to form carbonic acid:

$$CO_{2(aq)} + H_2O \leftrightarrow H_2CO_3 \qquad (3.24)$$

where $K = 10^{-2.8}$. This reaction has important implications for the chemistry of water in contact with the atmosphere. First, water in contact with the atmosphere (for example, rain) has the relatively strong acid, carbonic acid, dissolved in it. Thus, the pH of rainwater not impacted by anthropogenic emissions will be below 7. The pH of "unpolluted" rainwater is approximately 5.6. Thus, acid rain, which typically has measured pH values of 3.5 to 4.5, is approximately 10 to 100 times more acidic than natural rainwater, but not 10,000 times more acidic because natural rainwater is not neutral with a pH of 7.0.

In addition, because natural rainwater is slightly acidic and the partial pressure of carbon dioxide in soil may also be high from biological activity, water that contacts rocks and minerals can dissolve ions into solution. Inorganic constituents dissolved in freshwater and the dissolved salts in the oceans have their origin in minerals and the atmosphere. Carbon dioxide from the atmosphere provides an acid that can react with the bases of rocks, releasing the rock constituents into water, where they can either remain dissolved or precipitate into a solid phase.

It is difficult to distinguish analytically the difference between $CO_{2(aq)}$ and true $H_2CO_3$. Therefore, the term $H_2CO_3^*$ has been defined to equal the concentration of $CO_{2(aq)}$ plus the concentration of true $H_2CO_3$. However, $H_2CO_3^*$ can be approximated by $CO_{2(aq)}$ because true $H_2CO_3$ makes up only about 0.16 percent of $H_2CO_3^*$. Thus, the concentration of $H_2CO_3^*$ in waters equilibrated with the atmosphere is approximately $10^{-5}$ M.

$H_2CO_3^*$ is in equilibrium with bicarbonate ion as follows:

$$H_2CO_3^* \leftrightarrow HCO_3^- + H^+ \qquad (3.25)$$

where $K_{a1} = 10^{-6.3}$. Also, bicarbonate is in equilibrium with carbonate ion as follows:

$$HCO_3^- \leftrightarrow CO_3^{2-} + H^+ \qquad (3.26)$$

where $K_{a2} = 10^{-10.3}$.

According to our definition of an acid and base, bicarbonate can act as either an acid or a base. Bicarbonate and carbonate are also common bases in water. The *total inorganic carbon* content of a water sample is defined as follows:

$$\text{Total inorganic carbon} = [H_2CO_3^*] + [HCO_3^-] + [CO_3^{2-}] \qquad (3.27)$$

In the pH range of most natural waters (pH 6–9), $H_2CO_3^*$ and $CO_3^{2-}$ are small relative to $HCO_3^-$. Therefore, $HCO_3^-$ is the predominant component in Equation 3.27.

## Table / 3.6

**Explanation of Alkalinity and Buffering Capacity**

| Term | Definition |
|------|-----------|
| Alkalinity | Measure of a water's capacity to neutralize acids |
| | Alkalinity(moles/L) = $[HCO_3^-] + 2[CO_3^{2-}] + [OH^-] - [H^+]$ |
| | In most natural waters near pH = 6–8, the concentration of bicarbonate ($HCO_3^-$) is significantly greater than that of carbonate ($CO_3^{2-}$) or hydroxide ($OH^-$); therefore, the total alkalinity can be approximated by the bicarbonate concentration. |
| Buffering capacity | Ability of a water to resist changes in pH when either acidic or alkaline material is added |
| | In most freshwater systems, the buffering capacity is due primarily to the bases ($OH^-$, $CO_3^{2-}$, $HCO_3^-$) and acids ($H^+$, $H_2CO_3^*$, $HCO_3^-$). |

Table 3.6 provides definitions and descriptions of two important terms related to the carbonate system: **alkalinity** and **buffering capacity**. In the majority of natural freshwaters, alkalinity is caused primarily by $HCO_3^-$, $CO_3^{2-}$, and $OH^-$. In some natural waters and industrial waters, other salts of weak acids that may be important in determining a solution's alkalinity are borates, phosphates, ammonia, and organic acids. For example, anaerobic digester supernatant and municipal wastewaters contain large amounts of bases such as ammonia ($NH_3$), phosphates ($HPO_4^{2-}$ and $PO_4^{3-}$), and bases of various organic acids. The bases of silica ($H_3SiO_4^-$) and boric acid ($B(OH)_4^-$) can contribute to alkalinity in the oceans.

In most natural waters, the buffering capacity is due primarily to the bases ($OH^-$, $CO_3^{2-}$, $HCO_3^-$) and acids ($H^+$, $H_2CO_3^*$, $HCO_3^-$). Many lakes in the United States (for instance, in New England and the upper Midwest) have a low buffering capacity and consequently have been strongly influenced by acidic deposition (acid rain). This is because the geology of the basins that underlie these lakes is such that the slow dissolution of the underlying rocks and minerals does not result in the release of much alkalinity.

## 3.8 Oxidation-Reduction

Some chemical reactions occur because electrons are transferred between different chemical species. These reactions are called **oxidation-reduction** or **redox reactions**. Oxidation-reduction reactions control the fate and speciation of many metals and organic pollutants in natural environments, and numerous treatment processes employ redox chemistry. Also, many biological processes are just redox reactions mediated by microorganisms. The most commonly

## Table / 3.7

### Conventions for Assigning Oxidation State to Common Atoms (H, O, N, S) in Molecules

1. The overall charge on a molecule = $\sum$ charges (oxidation state) of its individual atoms.
2. The atoms in the molecule of concern have the following oxidation state; however, these numbers should be set equal to other numbers in reverse order (apply a different number to S before applying N, and so on) such that Convention 1 is always satisfied.

| Atom | Oxidation State |
|------|-----------------|
| $H^+$ | +1 |
| O | −2 |
| N | −3 |
| S | −2 |

**Redox Chemistry in Sediments**

www.wiley.com/college/mihelcic

used wastewater treatment processes involve redox reactions that oxidize organic carbon to $CO_2$ (while reducing oxygen to water) and oxidize and reduce various forms of nitrogen.

For molecules composed of single, charged atoms, the **oxidation state** is simply the charge on the atom; for example, the oxidation state of $Cu^{2+}$ is +2. In molecules containing multiple atoms, each atom is assigned an oxidation state according to the conventions provided in Table 3.7.

In a redox reaction, a molecule's oxidation state either goes up (in which case the molecule is *oxidized*) or down (in which case the molecule is *reduced*). Oxidized species can be depicted as reacting with free electrons ($e^-$) in half-reactions such as the following:

$$\text{electron acceptor (oxidant)} + e^- \rightarrow \text{electron donor (reductant)}$$

$$(3.28)$$

In this reaction, the species gaining the electron (the **electron acceptor** or *oxidant*) is reduced to form the corresponding reduced species; reduced molecules can donate electrons (the **electron donor**) and serve as *reductants*.

Consider two examples. In the first, ammonia nitrogen (oxidation state of −3) can be converted through nitrification and denitrification to $N_2$ gas (oxidation state of 0). In addition, important atmospheric pollutants include NO (oxidation state of +2) and $NO_2$ (oxidation state of +4). This conversion of nitrogen to different compounds occurs through many redox reactions. In the second example, acid rain is caused by emissions of $SO_2$ (sulfur oxidation state of +4), which is oxidized in the atmosphere to sulfate ion, $SO_4^{2-}$ (sulfur oxidation state of +6). Sulfate ion returns to Earth's surface in dry or wet deposition as sulfuric acid.

## example/3.7 Determining Oxidation States

Determine the oxidation states of sulfur in sulfate ($SO_4^{2-}$) and bisulfide ($HS^-$).

### solution

We expect the sulfur in sulfate to be more highly oxidized (due to the presence of oxygen in the molecule) than in bisulfide (due to the presence of hydrogen). The overall charge of $-2$ on sulfate must be maintained, and since the charge on each oxygen atom is $-2$ (see Table 3.7), the charge on the sulfur in sulfate must be $-2 - 4(-2) = +6$. To maintain the overall charge of $-1$ on bisulfide, the charge on sulfur must be $-1 - (+1) = -2$. Here the charge on $H^+$ was $+1$ (see Table 3.7). As expected, the sulfur found in sulfate is more oxidized than sulfur in bisulfide.

## 3.9 Precipitation-Dissolution

**Precipitation-dissolution** reactions involve the dissolution of a solid to form soluble species (or the reverse process whereby soluble species react to precipitate out of solution as a solid). Common precipitates include hydroxide, carbonate, and sulfide minerals.

A reaction that sometimes occurs in homes is the precipitation of $CaCO_3$. If waters are hard, this compound forms a scale in tea kettles, hot-water heaters, and pipes. Much effort is devoted to preventing excessive precipitation of $CaCO_3$ in municipal and industrial settings, and the process of removing divalent cations from water is referred to as water softening.

The reaction common to all of these situations is the conversion of a solid salt into dissolved components. In this example, the solid is calcium carbonate:

$$CaCO_{3(s)} \leftrightarrow Ca^{2+} + CO_3^{2-} \qquad (3.29)$$

Here, the subscript (s) denotes that the species is a solid. The equilibrium constant for such a reaction is referred to as the **solubility product,** $K_{sp}$. At equilibrium for the reaction in Equation 3.29, the $K_{sp}$ is equal to $Q$:

$$K_{sp} = \frac{\left[Ca^{2+}\right]\left[CO_3^{2-}\right]}{\left[CaCO_{3(s)}\right]} = \left[Ca^{2+}\right]\left[CO_3^{2-}\right] \qquad (3.30)$$

**Solubility** is defined as the maximum quantity (generally expressed as mass) of a substance (the solute) that can dissolve in a unit volume of solvent under specified conditions. Because the activity (which we assume equals concentration) of a solid is defined as equal to 1.0 (Rule 2 of Table 3.2), the equilibrium constant, $K_{sp}$, is equal to the

## Table / 3.8

**Common Precipitation-Dissolution Reactions, the Associated Solubility Product $K_{sp}$, and Significance**

| Equilibrium Equation | $K_{sp}$ at 25°C | Significance |
|---|---|---|
| $CaCO_3(s) \leftrightarrow Ca^{2+} + CO_3^{2-}$ | $3.3 \times 10^{-9}$ | Hardness removal, scaling |
| $MgCO_3(s) \leftrightarrow Mg^{2+} + CO_3^{2-}$ | $3.5 \times 10^{-5}$ | Hardness removal, scaling |
| $Ca(OH)_2(s) \leftrightarrow Ca^{2+} + 2OH^-$ | $6.3 \times 10^{-6}$ | Hardness removal |
| $Mg(OH)_2(s) \leftrightarrow Mg^{2+} + 2OH^-$ | $6.9 \times 10^{-12}$ | Hardness removal |
| $Cu(OH)_2(s) \leftrightarrow Cu^{2+} + 2OH^-$ | $7.8 \times 10^{-20}$ | Heavy-metal removal |
| $Zn(OH)_2(s) \leftrightarrow Zn^{2+} + 2OH^-$ | $3.2 \times 10^{-16}$ | Heavy-metal removal |
| $Al(OH)_3(s) \leftrightarrow Al^{3+} + 3OH^-$ | $6.3 \times 10^{-32}$ | Coagulation |
| $Fe(OH)_3(s) \leftrightarrow Fe^{3+} + 3OH^-$ | $6 \times 10^{-38}$ | Coagulation, iron removal |
| $CaSO_4(s) \leftrightarrow Ca^{2+} + SO_4^{2-}$ | $4.4 \times 10^{-5}$ | Flue gas desulfurization |

SOURCE: From Mihelcic (1999), reprinted with permission of John Wiley & Sons, Inc.

solubility product. Thus, if we know the equilibrium constant and the concentration of one of the species, we can determine the concentration of the other species.

No precipitate will form if the product of the concentrations of the ions is less than $K_{sp}$ (in Equation 3.30, $Ca^{2+}$ and $CO_3^{2-}$ are the species). This solution is described as *undersaturated*. Likewise, if the product of the concentrations of the ions exceeds $K_{sp}$, the solution is described as *supersaturated*, and the solid species will precipitate until the product of the ion concentrations equals $K_{sp}$. Table 3.8 provides some important solubility products and the associated reactions.

## 3.10   Adsorption, Absorption, and Sorption

**Sorption** is a nonspecific term that can refer to either or both process(es) of **adsorption** of a chemical at the solid surface and/or **absorption** (partitioning) of the chemical into the volume of the solid. In the case of organic pollutants, sorption is a key process determining fate, and the chemical is commonly absorbed into the organic fraction of the particle due to favorable energetics of this process. The *sorbate* (adsorbate or absorbate) is the substance transferred from the gas or liquid phase to the solid phase. The *sorbent* (adsorbent or absorbent) is the solid material onto or into which the sorbate accumulates. Solids that sorb chemicals may be either natural (for example, surface soil, harbor or river sediment, aquifer material) or anthropogenic (for example, activated carbon) materials.

Figure 3.7 shows a schematic of sorption processes for naphthalene sorbing to a natural solid such as a soil particle from the water phase.

## example/3.8 Precipitation-Dissolution Equilibrium

What pH is required to reduce a high concentration of dissolved $Mg^{2+}$ to 43 mg/L? $K_{sp}$ for the following reaction is $10^{-11.16}$.

$$Mg(OH)_{2(s)} \leftrightarrow Mg^{2+} + 2OH^-$$

## solution

In this situation, the dissolved magnesium is removed from solution as a hydroxide precipitate. First, the concentration of $Mg^{2+}$ is converted from mg/L to moles/L:

$$[Mg^{2+}] = \frac{43 \text{ mg}}{L} \times \frac{g}{1,000 \text{ mg}} \times \frac{1 \text{ mole}}{24 \text{ g}} = 0.0018 \text{ M}$$

Then, the equilibrium relationship is written as

$$10^{-11.16} = \frac{[Mg^{2+}] \times [OH^-]^2}{[Mg(OH)_{2(s)}]}$$

Substituting values for all the known parameters,

$$10^{-11.16} = \frac{[0.0018] \times [OH^-]^2}{1}$$

Solve for $[OH^-] = 6.2 \times 10^{-5}$ M. This results in $[H^+] = 10^{-9.79}$ M, so pH = 9.79. At this pH, any magnesium in excess of 0.0018 M will precipitate as $Mg(OH)_{2(s)}$ because the solubility of $Mg^{2+}$ will be exceeded.

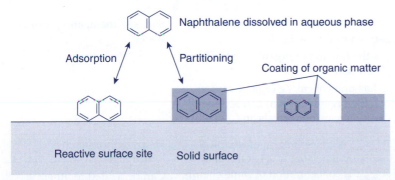

**Figure 3.7   Sorption of an Organic Chemical (Naphthalene) onto a Natural Material such as a Soil or Sediment Particle**   This typically occurs when the sorbate either sorbs onto reactive surface sites (adsorption) or absorbs or partitions into organic matter that coats the particle (the sorbent). The sorption process influences the mobility, natural degradation, and engineered remediation of pollutants.

From Mihelcic (1999). Reprinted with permission of John Wiley & Sons, Inc.

Why does this sorption occur? From a thermodynamic viewpoint, molecules always prefer to be in a lower energy state. A molecule adsorbed onto a surface has a lower energy state on a surface than in the aqueous phase. Therefore, during the process of equilibration, the molecule is attracted to the surface and a lower energy state. Attraction of a molecule to a surface can be caused by physical and/or chemical forces. Electrostatic forces govern the interactions between most adsorbates and adsorbents. These forces include dipole–dipole interactions, dispersion interactions or London–van der Waals force, and hydrogen bonding. During sorption to soils and sediments,

## Table / 3.9

### Common Terms Used to Describe Sorption Isotherms and Other Partitioning Phenomena

| Isotherm or Other Partitioning Term | Usually Presented as: | Symbols and Units | Common Application |
|---|---|---|---|
| **Freundlich isotherm** | $q = KC^{1/n}$ (Equation 3.31) | $q$ = mass of adsorbate adsorbed per unit mass of adsorbent after equilibrium (mg/g). $C$ = mass of adsorbate in the aqueous phase after equilibrium (mg/L). $K$ = Freundlich isotherm capacity parameter $((mg/g)(L/mg)^{1/n})$. $1/n$ = Freundlich isotherm intensity parameter (unitless). | Drinking water and air treatment where adsorbents such as activated carbon are used |
| **Linear isotherm** Special case of Freundlich isotherm where $1/n = 1$ (i.e., dilute systems) | $K = \dfrac{q}{C}$ (Equation 3.32) | $q$ and $C$ are same as Freundlich isotherm. $K$ = soil– or sediment–water partition (or distribution) coefficient, also written as $K_P$ or $K_d$ (units of $cm^3/g$ or L/kg). | Dilute systems, especially soil, sediment, and groundwater |
| **Normalizing $K$ for organic carbon*** | $K_{oc} = \dfrac{K}{f_{oc}}$ (Equation 3.33) | $K$ is the same as the linear isotherm (also referred to as $K_p$, $K_d$). $f_{oc}$ is the fraction of organic carbon for a specific soil. $K_{oc}$ has units of $cm^3/g$ organic carbon (or L/kg organic carbon) and sediment (1% organic carbon equals an $f_{oc}$ of 0.01). | Soil, sediment, and groundwater |
| **Octanol–water partition coefficient** | $K_{ow} = \dfrac{[A]_{octanol}}{[A]_{water}}$ (Equation 3.34) | $[A]_{octanol}$ is concentration of chemical dissolved in octanol ($C_8H_{17}OH$), and $[A]_{water}$ is concentration of same chemical dissolved in same volume of water. $K_{ow}$ is unitless and usually reported as $\log K_{ow}$. | Helps determine the hydrophobicity of a chemical. Can be related to other environmental properties such as $K_{oc}$ and bioconcentration factors |

*It has been shown that for soils and sediments with a fraction of organic carbon ($f_{oc}$) greater than 0.001 (0.1%) and low equilibrium solute concentrations ($< 10^{-5}$ molar or 1/2 the aqueous solubility), the soil–water partition coefficient ($K_P$) can be normalized to the soil's organic carbon content.

hydrophobic partitioning—a phenomenon driven by entropy changes—can also account for the interaction of a hydrophobic (water-fearing) organic chemical with a surface.

Table 3.9 provides examples of some common sorption isotherms and related partitioning phenomena. A *sorption isotherm* is a relationship that describes the affinity of a compound for a solid in water or gas at constant temperature (*iso-* means constant, and *therm* refers to temperature). The two sorption isotherms covered in Table 3.9 are the **Freundlich isotherm** and the **linear isotherm**. Figure 3.8 shows the relationship between the Freundlich and linear isotherms for various ranges of $1/n$. Here, $1/n$ is the Freundlich isotherm intensity parameter (unitless).

A problem with the value of the **soil–water partition coefficient**, $K$ (Equation 3.32, shown in Table 3.9), is that it is chemical- and sorbent-specific. Thus, although $K$ could be measured for every relevant system, this would be time-consuming and costly. Fortunately, when the solute is a neutral, nonpolar organic chemical, the soil–water partition coefficient can be normalized for organic carbon, in which case it remains chemical-specific but no longer sorbent-specific. $K_{oc}$ is called the **soil–water partition coefficient normalized to organic carbon**. $K_{oc}$ has units of $cm^3/g$ organic carbon or L/kg organic carbon (see Equation 3.33 in Table 3.9).

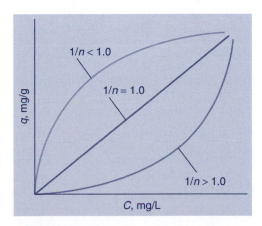

**Figure 3.8**  **Freundlich Isotherm Plotted for Different Values of $1/n$**

For values of $1/n$ less than 1, the isotherm is considered favorable for sorption because low values of the sorbate liquid-phase concentration yield large values of the solid-phase concentration. This means that it is energetically favorable for the sorbate to be sorbed. At higher aqueous concentrations, the ability of the solid to sorb the chemical decreases as the active sorption sites become saturated with sorbate molecules. For $1/n$ values greater than 1, the isotherm is considered unfavorable for sorption because high values of the liquid-phase sorbate concentration are required to get sorption to occur on the sorbent. However, as sorption occurs, the surface is modified by the sorbing chemical and made more favorable for additional sorption. If the $1/n$ value equals 1, the isotherm is termed a linear isotherm.

From Mihelcic (1999). Reprinted with permission of John Wiley & Sons, Inc.

## example/3.9 Adsorption Isotherm Data Analysis

A methyl tertiary-butyl ether (MTBE) adsorption isotherm was performed on a sample of activated carbon. The isotherm was performed at 15°C using 0.250 L amber bottles with an initial MTBE concentration, $C_0$, of 150 mg/L. The three left columns of Table 3.10 provide the isotherm data. Determine the Freundlich isotherm parameters ($K$ and $1/n$).

## solution

The values of the MTBE adsorbed for each isotherm point ($q$) and the logarithm values of $C$ and $q$ can be determined and inputted into Table 3.10 (3 left columns). To determine the Freundlich isotherm parameters, fit the logs of the isotherm data, log $q$ versus log $C$, using the linear form of Equation 3.31 (Table 3.9), expressed as

$$\log q = \log K + \left(\frac{1}{n}\right)\log C$$

Graph log $q$ versus log $C$, as shown in Figure 3.9, and use a linear regression to fit the data to determine $K$ and $1/n$.

## Table / 3.10

### Isotherm Data and Results Used in Example 3.9

| Isotherm Data | | | Results | | |
|---|---|---|---|---|---|
| Initial MTBE Concentration, $C_0$ (mg/L) | Mass of GAC, $M$ (g) | MTBE equilibrium liquid-phase concentration, $C$, mg/L | $q = (V/M) \times (C_0 - C)$ (mg/g) | log $q$ | log $C$ |
| 150 | 0.155 | 79.76 | 113.290 | 2.0542 | 1.9018 |
| 150 | 0.339 | 42.06 | 79.602 | 1.9009 | 1.6239 |
| 150 | 0.589 | 24.78 | 53.149 | 1.7255 | 1.3941 |
| 150 | 0.956 | 12.98 | 35.832 | 1.5543 | 1.1133 |
| 150 | 1.71 | 6.03 | 21.048 | 1.3232 | 0.7803 |
| 150 | 2.4 | 4.64 | 15.142 | 1.1802 | 0.6665 |
| 150 | 2.9 | 3.49 | 12.630 | 1.1014 | 0.5428 |
| 150 | 4.2 | 1.69 | 8.828 | 0.9459 | 0.2279 |

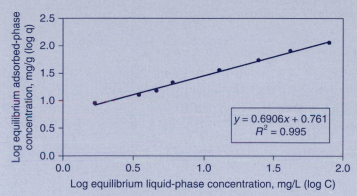

**Figure 3.9** Freundlich Isotherm Data Graphed for Example 3.9 to Determine $K$ and $1/n$

From Figure 3.9, the linear form of Equation 3.31 with values for $K$ and $1/n$ added is expressed as

$$\log q = 0.761 + (0.6906) \log C$$

Here, $\log K = 0.761$, so $K = 10^{0.761} = 5.77 (\text{mg/g})(\text{L/mg})^{1/n}$. Thus, $K = 5.77 (\text{mg/g})(\text{L/mg})^{1/n}$, and $1/n = 0.6906$.

For systems with a relatively high amount of organic carbon (greater than 0.1 percent), $K_{oc}$ can be directly correlated to a parameter called the **octanol–water partition coefficient, $K_{ow}$**, of a chemical. Values of $K_{ow}$ range over many orders of magnitude, so $K_{ow}$ usually is reported as $\log K_{ow}$. Table 3.11 lists some typical values of $\log K_{ow}$ for a wide variety of chemicals. Values of $K_{ow}$ for environmentally significant chemicals range from approximately $10^1$ to $10^7$ ($\log K_{ow}$ range of 1–7). The higher the value, the greater the tendency of the compound to partition from the water into an organic phase. Chemicals with high values of $K_{ow}$ are hydrophobic (water-fearing).

The magnitude of an organic chemical's $K_{ow}$ can tell a lot about the chemical's ultimate fate in the environment. For example, the values in Table 3.11 indicate that very hydrophobic chemicals such as 2,3,7,8-TCDD are more likely to bioaccumulate in the lipid portions of humans and animals. Conversely, chemicals such as benzene, trichloroethylene (TCE), tetrachloroethylene (PCE), and toluene are frequently identified as groundwater contaminants because they are relatively soluble and easily dissolve in groundwater recharge that is infiltrating vertically toward an underlying aquifer. This is in contrast to pyrene or 2,3,7,8-TCDD, which are both likely to be confined near the soil's surface in the location of the spill.

## Table / 3.11

### Examples of log $K_{ow}$ for Some Environmentally Significant Chemicals

| Chemical | log $K_{ow}$ |
|---|---|
| Phthalic acid | 0.73 |
| Benzene | 2.17 |
| Trichloroethylene | 2.42 |
| Tetrachloroethylene | 2.88 |
| Toluene | 2.69 |
| 2,4,-dichlorophenoxyacetic acid | 2.81 |
| Naphthalene | 3.33 |
| 1,2,4,5-Tetrachlorobenzene | 4.05 |
| Phenanthrene | 4.57 |
| Pyrene | 5.13 |
| 2,3,7,8-tetrachlorodibenzo-$p$-dioxin | 6.64 |

## Box / 3.3  Partitioning Behavior of Organic Chemicals and Their Environmental Transport and Fate

Many of the thousands of chemicals that enter the market each year are not tested for toxicity, and when such chemicals enter the environment, their fate depends to a great extent on their chemical structure and associated chemical properties, including $K_H$, $K_{ow}$, and $K_{oa}$.

$K_{oa}$ is the **octanol–air partition coefficient**, which equals the ratio of the octanol–water partition coefficient ($K_{ow}$) to the dimensionless Henry's law constant ($K_H$). Table 3.12 compiles these values for many chemicals of environmental importance.

## Table / 3.12

### Natural Logarithms of the Dimensionless Henry's Law Constant ($K_H$), Octanol–Water Partition Coefficient ($K_{ow}$), and Octanol–Air Coefficient ($K_{oa}$) of Organic Pollutants

| Compound | log $K_H$ | log $K_{ow}$ | log $K_{oa}$ | Number in Figure 3.11 |
|---|---|---|---|---|
| 1,1,1-trichloroethane | 0.16 | 2.49 | 2.33 | 3 |
| 1,1-dichloroethane | −0.61 | 1.79 | 2.40 | 6 |
| 1,1-dichloroethene (vinylidene chloride) | 0.10 | 1.48 | 1.38 | 2 |
| 1,2,4,5-tetrachlorobenzene | −1.30 | 4.72 | 6.02 | 10 |

## Table / 3.12

| Compound | log $K_H$ | log $K_{ow}$ | log $K_{oa}$ | Number in Figure 3.11 |
|---|---|---|---|---|
| 1,2-dichloroethane | −1.27 | 1.46 | 2.73 | 11 |
| 1,2-dichloropropane | −0.92 | 2.28 | 3.20 | 7 |
| 1,3,5-trichlorobenzene | −0.36 | 4.19 | 4.55 | 4 |
| 2,3,7,8-tetrachlorodibenzo-$p$-dioxin | −2.87 | 6.80 | 9.67 | 21 |
| 2,4-dichlorophenoxyacetic acid (2,4-D) | −0.10 | 2.68 | 2.78 | 3 |
| 2-chlorophenol | −3.24 | 2.19 | 5.43 | 23 |
| 3-chlorophenol | −4.16 | 2.48 | 6.64 | 27 |
| 4-chlorophenol | −4.43 | 2.42 | 6.85 | 27 |
| 4-$n$-nonylphenol | −2.89 | 5.76 | 8.65 | 22 |
| acenaphthene | −2.29 | 4.20 | 6.49 | 17 |
| acetic acid | −4.95 | −0.25 | 4.70 | 29 |
| alachlor | −6.04 | 2.95 | 8.99 | 31 |
| aldicarb | −7.23 | 0.89 | 8.12 | 33 |
| aldrin | −1.69 | 6.19 | 7.88 | 15 |
| anthracene | −2.80 | 4.68 | 7.48 | 20 |
| atrazine | −6.93 | 2.65 | 9.58 | 32 |
| benzene | −0.65 | 2.17 | 2.82 | 6 |
| benzo($a$)pyrene | −4.79 | 6.13 | 10.92 | 28 |
| bromoform | −1.62 | 2.67 | 4.29 | 12 |
| chloroform | −0.84 | 1.95 | 2.79 | 6 |
| $cis$-1,2-dichloroethene | −0.66 | 1.86 | 2.52 | 6 |
| $cis$-chlordane | −2.91 | 6.00 | 8.91 | 22 |
| DDT | −3.30 | 6.36 | 9.66 | 24 |
| dibenzo-$p$-dioxin | −2.32 | 4.30 | 6.62 | 17 |
| dichloromethane | −0.93 | 1.31 | 2.24 | 6 |
| ethyl benzene | −0.50 | 3.20 | 3.70 | 5 |

(Continued)

**Natural Logarithms of the Dimensionless Henry's Law Constant ($K_H$), Octanol–Water Partition Coefficient ($K_{ow}$), and Octanol–Air Coefficient ($K_{oa}$) of Organic Pollutants**

Table / 3.12

| Compound | log $K_H$ | log $K_{ow}$ | log $K_{oa}$ | Number in Figure 3.11 |
|---|---|---|---|---|
| heptachlor | −1.03 | 4.92 | 5.95 | 10 |
| hexachlorobenzene (HCB) | −1.44 | 5.80 | 7.24 | 12 |
| hexachloroethane | −1.01 | 3.93 | 4.94 | 9 |
| lindane | −3.94 | 3.78 | 7.72 | 26 |
| *m*-dichlorobenzene (1,3-DCB) | −0.86 | 3.47 | 4.33 | 8 |
| methyl ethyl ketone (butanone) | −2.60 | 0.29 | 2.89 | 19 |
| methyl tertiary-butyl ether (MTBE) | −1.54 | 0.94 | 2.48 | 11 |
| *m*-xylene | −0.53 | 3.30 | 3.83 | 5 |
| naphthalene | −1.74 | 3.33 | 5.07 | 14 |
| nitrobenzene | −3.12 | 1.85 | 4.97 | 23 |
| *o*-dichlorobenzene (1,2-DCB) | −1.04 | 3.40 | 4.44 | 8 |
| *o*-xylene | −0.69 | 3.16 | 3.85 | 5 |
| parathion | −5.31 | 3.81 | 9.12 | 30 |
| PCB-101 | −1.83 | 6.36 | 8.19 | 15 |
| PCB-180 | −2.36 | 7.36 | 9.72 | 18 |
| PCB-28 | −1.89 | 5.62 | 7.51 | 16 |
| *p*-dichlorobenzene (1,4-DCB) | −1.04 | 3.45 | 4.49 | 8 |
| phenanthrene | −2.85 | 4.57 | 7.42 | 20 |
| *p*-xylene | −0.55 | 3.27 | 3.82 | 5 |
| pyrene | −3.32 | 5.13 | 8.45 | 25 |
| tetrachloroethylene (perchloroethylene PCE) | 0.08 | 2.88 | 2.80 | 3 |
| tetrachloromethane (carbon tetrachloride) | 0.04 | 2.77 | 2.73 | 3 |
| toluene (methyl benzene) | −0.60 | 2.69 | 3.29 | 7 |
| toxaphene | 0.41 | 4.21 | 3.80 | 1 |
| *trans*-1,2-dichloroethene | −0.59 | 2.09 | 2.68 | 6 |
| *trans*-chlordane | −2.86 | 6.00 | 8.86 | 22 |

## Table / 3.12

| Compound | log $K_H$ | log $K_{ow}$ | log $K_{oa}$ | Number in Figure 3.11 |
|---|---|---|---|---|
| trichloroethylene (trichloroethene TCE) | −0.35 | 2.42 | 2.73 | 6 |
| vinyl chloride (chloroethene) | 0.04 | 1.27 | 1.23 | 2 |
| α-hexachlorocyclohexane (α-HCH) | −3.63 | 3.81 | 7.44 | 26 |

SOURCE: Values compiled from Beyer et al. (2002); Montgomery (1993); and Schwartzenbach et al. (2003).

Figure 3.10 shows how $K_{oc}$ and $K_{ow}$ are linearly correlated for a set of 72 chemicals that span many ranges of hydrophobicity. $K_{ow}$ has also been correlated to other environmental properties such as bioconcentration factors and aquatic toxicity. $K_{oc}$ can then be related to the site-specific soil–water partition coefficient ($K_p$) by knowledge of the system's organic carbon content, using Equation 3.33 (Table 3.9).

As demonstrated in Figure 3.11, we can predict where a chemical is likely to reside in the environment and the major modes of environmental transport based on the chemical's properties, assuming that the chemicals do not undergo degradation once they enter the environment (that is, they are persistent) and that they reach equilibrium between the atmosphere, surface waters (sea- and freshwaters), and soils and sediments in the global environment.

www.wiley.com/college/mihelcic
"Where does the POP Go?"

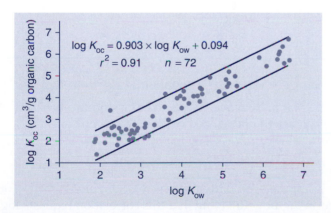

**Figure 3.10** **Scatter Plot of log $K_{oc}$ (cm³/g Organic Carbon) versus log $K_{ow}$ for 72 Chemicals** The relationship is given by the equation $\log(K_{oc}[\text{cm}^3/\text{g}]) = 0.903 \log(K_{ow}) + 0.094$ ($n = 72$, $r^2 = 0.91$) The heavy lines represent the 90 percent confidence intervals for the correlation. Individuals seeking values of $K_{oc}$ should consult a data set that has undergone a quality check or use an appropriate, statistically validated correlation to estimate the value of $K_{oc}$.

From Baker et al. (1997). Copyright WEF, reprinted with permission.

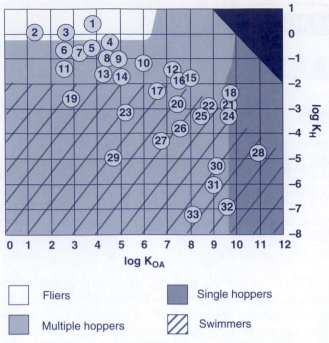

**Figure 3.11 Major Modes of Environmental Transport of Hypothetical Perfectly Persistent Organic Pollutants** Compounds are identified by the numbers shown in Table 3.12. Chemicals that have high dimensionless Henry's law constants ($K_H$) and low octanol–air partition coefficients ($K_{oa}$) (termed *fliers*) are transported globally in the atmosphere. Chemicals that have both low Henry's law constants and low octanol–air partition coefficients (termed *swimmers*) are transported globally in surface water. Chemicals that have intermediate Henry's law constants and octanol–air coefficients tend to be present in sediments and soils (called *single hoppers*) or in multiple phases (termed *multiple hoppers*). Because water does not travel as rapidly as air, *swimmers* are not transported as quickly as are *fliers*.

---

example/3.10 Determination of $K_{oc}$ from $K_{ow}$

The log $K_{ow}$ for anthracene is 4.68. What is anthracene's soil–water partition coefficient normalized to organic carbon?

solution

Use an appropriate correlation between log $K_{oc}$ and log $K_{ow}$ (such as provided in Figure 3.10). Note that this correlation requests log $K_{oc}$, not $K_{ow}$:

$$\log K_{oc} = 0.903(4.68) + 0.094 = 4.32$$

Therefore, $K_{oc} = 10^{4.32}$ cm$^3$/g organic carbon.

# example/3.11 Use of $K_{oc}$ to Predict Aqueous Concentration

Anthracene has contaminated harbor sediments, and the solid portion of sediments is in equilibrium with the pore water. If the organic carbon content of sediments is 5 percent and the solid sediment anthracene concentration is 50 µg/kg sediment, what is the pore water concentration of anthracene at equilibrium? In example 3.10, log $K_{oc}$ for anthracene was estimated to be 4.32.

## solution

An organic carbon (OC) content of 5 percent means that the fraction of organic carbon, $f_{oc}$, is 0.05. Use Equation 3.33 (from Table 3.9) to find the sediment-specific partition coefficient, $K$:

$$K\frac{cm^3}{g\ sediment} = \frac{10^{4.32}\ cm^3}{g\ OC} \times \frac{0.05\ g\ OC}{g\ sediment} = 1{,}045\ cm^3/g\ sediment$$

The equilibrium aqueous-phase concentration, $C$, is then derived from the equilibrium expression given in Equation 3.32 (Table 3.9):

$$C = \frac{q}{K} = \frac{\dfrac{50\ \mu g}{kg\ sediment} \times \dfrac{kg}{1{,}000\ g}}{\dfrac{1{,}045\ cm^3}{g\ sediment}} \times \frac{cm^3}{mL} \times \frac{1{,}000\ mL}{L} = \frac{0.048\ \mu g}{L}$$

Note that the aqueous-phase concentration of anthracene is relatively low compared with the sediment-phase concentration (50 $ppb_m$ in the sediments and 0.048 $ppb_m$ in the pore water). This is because anthracene is hydrophobic. Its aqueous solubility is low (and $K_{ow}$ is high), so it prefers to partition into the solid phase.

Also, the solid phase is high in organic carbon content. A sand-gravel aquifer would be much lower in organic carbon ($f_{oc}$ very low); therefore, less of the anthracene would partition from the aqueous into the solid phase.

# example/3.12 Partitioning of Chemical between Air, Water, and Soil Phases

A student uses a reactor to mimic the environment for a class demonstration. The sealed 1 L reactor contains 500 mL water, 200 mL soil (1 percent organic carbon and density of 2.1 g/cm$^3$), and 300 mL air. The temperature of the reactor is 25°C. After adding 100 µg TCE to the reactor, the student incubates the reactor until equilibrium is achieved between all three phases. The Henry's law constant for TCE is 10.7 L-atm/mole at 25°C, and TCE has a log $K_{ow}$ of 2.42. Assuming that no chemical or biological degradation of TCE occurs during the incubation, what is the aqueous-phase concentration of TCE at equilibrium? What is the mass of TCE in the aqueous, air, and sorbed phases after equilibrium is attained?

## solution

1. Set up a simplified mass balance that equates the total mass of TCE added to the mass of TCE in each phase at equilibrium:

   Total mass of TCE added = [mass of aqueous TCE]

   + [mass of gaseous TCE] + [mass of sorbed TCE]

   $$100 \ \mu g = [V_{aq} \times C_{aq}] + [V_{air} \times C_{air}] + [M_{soil} \times C_{sorbed}]$$

   The problem is requesting $C_{aq}$. The three known parameters are $V_{aq} = 500$ mL, $V_{air} = 300$ mL, and mass of soil $= M_{soil} = V_{soil} \times$ density of soil $= 200$ mL $\times$ cm$^3$/mL $\times$ 2.1 g/cm$^3 = 420$ g. The three unknowns are $C_{aq}$, $C_{air}$, and $C_{sorbed}$; however, $C_{air}$ can be related to $C_{aq}$ by a Henry's law constant, and $C_{sorbed}$ can be related to $C_{aq}$ by a soil–water partition coefficient.

2. Convert the Henry's law constant to dimensionless form. $K_H = 10.7$ L-atm/mole (by the units, we can tell this Henry's law constant is for the reaction written in the following direction: $C_{aq} \leftrightarrow C_{air}$). Convert to dimensionless form using the ideal gas law (see Table 3.3)

   $$\frac{\dfrac{10.7 \ \text{L-atm}}{\text{mole}}}{\dfrac{0.08205 \ \text{L-atm}}{\text{mole-K}}(298 \ \text{K})} = 0.44$$

   The Henry's law constant of 0.44 is equal to $C_{air}/C_{aq}$, so $C_{air} = 0.44 \ C_{aq}$.

3. Determine the soil–water partition coefficient. Remember, $K = K_{oc} \times f_{oc}$, and 1 percent organic carbon means $f_{oc} = 0.01$. Because $K_{oc}$ and $K$ are not provided, estimate $K_{oc}$ from $K_{ow}$: log $K_{oc} = 0.903 \times 2.42 + 0.094 = 2.28$. Therefore, $K_{oc} = 10^{2.28}$, and

   $$K = 10^{2.28} \times 0.01 = \frac{1.9 \ \text{cm}^3}{\text{g}} \quad C_{sorbed} = \frac{1.9 \ \text{cm}^3}{\text{g}} \times C_{aq}$$

   Accordingly, substitute into the mass balance so all concentrations are in terms of $C_{aq}$:

   $$100 \ \mu g = [500 \ \text{mL} \times C_{aq}] + [300 \ \text{mL} \times 0.44 \ C_{aq}]$$

   $$+ \left[420 \ \text{g} \times \frac{1.9 \ \text{cm}^3}{\text{g}} \times \frac{\text{mL}}{\text{cm}^3} \times C_{aq}\right]$$

$$100 \ \mu g = C_{aq}\Big\{500 \ \text{mL} + [300 \ \text{mL} \times 0.44]$$

$$+ \Big[420 \ g \times \frac{1.9 \ \text{cm}^3}{g} \times \frac{\text{mL}}{\text{cm}^3}\Big]\Big\}$$

$$100 \ \mu g = C_{aq}[500 \ \text{mL} + 132 \ \text{mL} + 798 \ \text{mL}]$$

$$C_{aq} = 0.070 \ \mu g/\text{mL} = 0.070 \ \text{mg}/\text{mL} = 70 \ \text{ppb}_m$$

The total mass of TCE in the aqueous phase is 35 μg; in the air phase, it is 9.2 μg; sorbed to soil, it is 55.8 μg.

The mass of chemical found in each of the three phases is a function of the combined effects of partitioning between each phase. The amount of chemical that partitions to each phase is based on the physical/chemical properties of the chemical (e.g., Henry's law constant, log $K_{ow}$) and soil/sediment properties ($f_{oc}$). This is very important when determining where a chemical migrates in the environment or an engineered system, as well as in determining what method of treatment should be selected.

Chemicals that have high Henry's law constants (dimensionless) and low octanol–air partition coefficients tend to be present primarily in the atmosphere at equilibrium. They are termed *fliers*, because according to their properties, they will be transported globally in the atmosphere. Chemicals that have low Henry's law constants and low octanol–air partition coefficients tend to exist primarily in surface waters. They are termed *swimmers*, because they are transported globally in surface water. Because water does not travel as rapidly as air (river velocities are typically less than 5 m/s, while wind speeds are in the range of 1–30 m/s), *swimmers* are not transported as quickly as are *fliers*. Chemicals with intermediate Henry's law constants and octanol–air coefficients tend to be present in sediments and soils (the *single hoppers*) or in multiple phases (the *multiple hoppers*).

The chemicals included in Figure 3.11 were selected to include compounds discussed in this and later chapters of this text. Return to this section to review the environmental transport and fate of these chemicals, as well as the type of treatment method best suited to a given pollutant. For example, if vinyl chloride (compound 2) enters the environment and comes into contact with the atmosphere, it is predicted be a flier and would need to be removed from the gas phase through treatment.

**Persistent Organic Pollutants**
http://www.chem.unep.ch/pops

**Class Discussion**
Should a chemical like DDT be banned globally or considered a viable solution to the unfair burden of malaria that inflicts many parts of the developing world, especially Africa. What equitable solutions that consider future generations of humans and wildlife can you think of?

## 3.11   Kinetics

The kinetic approach to environmental chemistry addresses the rate of reactions. Concepts include the rate law, zero-order and first-order reactions, half-life, and factors that affect the rate of reaction.

### 3.11.1 THE RATE LAW

The **rate law** expresses the dependence of the reaction rate on measurable, environmental parameters. Of particular interest is the dependence of the rate on the concentrations of the reactants. Other parameters that may influence the reaction rate include temperature and the presence of catalysts (including microorganisms).

The rate of an irreversible reaction and the exact form of the rate law depend on the mechanism of the reaction. Consider the hydrolysis of dichloromethane (DCM). In this reaction, one molecule of DCM reacts with a hydroxide ion ($OH^-$) to produce chloromethanol (CM) and chloride ion:

$$\begin{array}{ccc} & \text{Cl} & & & & \text{Cl} \\ & | & & & & | \\ \text{H}-\text{C}-\text{H} & + \text{ OH}^- & \longrightarrow & \text{H}-\text{C}-\text{H} & + & \text{Cl}^- \\ & | & & & & | \\ & \text{Cl} & & & & \text{OH} \end{array}$$

For the reaction depicted here to occur, one molecule of DCM must collide and react with one molecule of $OH^-$. The rate of an irreversible binary reaction is proportional to the concentration of each chemical species. For the hydrolysis of DCM, it can be written as

$$\mathbf{R = k[DCM][OH^-]} = -d[DCM]/dt$$
$$= -d[OH^-]/dt = d[CM]/dt = d[Cl^-]/dt \qquad \textbf{(3.35)}$$

where $R$ is the rate of reaction, $k$ is the rate constant for this particular reaction, [DCM] is the concentration of dichloromethane, $[OH^-]$ is the concentration of hydroxide ion, [CM] is the concentration of chloromethanol, $[Cl^-]$ is the concentration of chloride ion and, $t$ is time. The negative signs in Equation 3.35 indicate that the products' concentrations are decreasing over time.

The bold portion on the left side of Equation 3.35 is referred to as the reaction's *rate law*, which expresses the dependence of the reaction rate on the concentrations of the reactants. The rate law in this case would be called first order with respect to DCM and first order with respect to $OH^-$. The term **first order** indicates that each species is raised to the first power. The rate law is second order overall because it involves the product of two species, each raised to the first power. Because the reaction was depicted as irreversible, it was assumed that the concentration of products did not influence the rate of the forward reaction.

To generalize these terms, a hypothetical rate law can be constructed for a generic irreversible reaction of $a$ moles of species $A$ reacting with $b$ moles of species $B$ to yield products, $P$. The rate law is written as

$$R = k[A]^a[B]^b \qquad \textbf{(3.36)}$$

This reaction would be termed $a$th order with respect to $A$ and $b$th order with respect to $B$. The **overall order** of the reaction would be

$(a + b)$. This reaction is termed an **elementary reaction** because the reaction order is controlled by the stoichiometry of the reaction. That is, $a$ equals the molar stoichiometric coefficient of species $A$, and $b$ equals the molar stoichiometric coefficient for $B$.

The order of a reaction should be determined experimentally, because it often does not correspond to the reaction stoichiometry. This is because the mechanism or steps of the reaction do not always correspond to that shown in the reaction equation.

The collision-based reaction of the hydrolysis of dichloromethane can be contrasted with some biological transformations of organic chemicals that occur in treatment plants or natural environments where soils and sediments are present. In some of these situations, zero-order transformations are observed. A reaction is termed **zero-order** when it does not depend on the concentration of the compound involved in the reaction. Zero-order kinetics can be due to several items, including the rate-limiting diffusion of oxygen from the air into the aqueous phase, which may be slower than the demand for oxygen by the microorganism biodegrading the chemical. Another explanation for an observation of zero-order kinetics is the slow, rate-limiting movement of a chemical (required by the microorganisms for energy and growth) that has a low water solubility ($ppb_m$ and $ppm_m$ range) from an oil or soil/sediment phase into the aqueous phase, where the chemical is then available for the organism to utilize.

One chemical that has been observed to have zero-order kinetics of biodegradation is 2,4-D, an herbicide commonly used by farmers and households. 2,4-D can be transported into a river or lake by horizontal runoff or vertical migration to groundwater that is hydraulically connected to a lake or river. It has been found to disappear in lake water according to zero-order kinetics. The rate law for this type of reaction can be written as

$$R = -d[2,4\text{-D}]/dt = k \tag{3.37}$$

## 3.11.2  ZERO-ORDER AND FIRST-ORDER REACTIONS

Many environmental situations can be described by zero-order or first-order kinetics. Figure 3.12 compares the major differences between these two types of kinetics. In this section, we discuss these kinetic expressions in depth by first constructing a generic chemical reaction whereby a chemical, $C$, is converted to some unknown products:

$$C \rightarrow \text{products} \tag{3.38}$$

The rate law that describes the decrease in concentration of chemical $C$ with time can be written as:

$$d[C]/dt = -k[C]^n \tag{3.39}$$

Here, $[C]$ is the concentration of $C$, $t$ is time, $k$ is a rate constant that has units dependent on the order of the reaction, and the reaction order, $n$, typically is an integer (0, 1, 2).

| Reaction Order | Rate Law | Integrated Form of Rate Law | Plot of Concentration versus Time | Linearized Plot of Concentration versus Time | Half-Life, $t$ | Example Units of Rate Constant, $k$ |
|---|---|---|---|---|---|---|
| Zero | $\dfrac{d[C]}{dt} = -k$ | $[C] = [C_0] - kt$ | $[C]$ vs. Time (slope $= k$) | Same as $[C]$ vs. time | $\dfrac{0.5[C_0]}{k}$ | moles/L-s<br>mg/L-s |
| First | $\dfrac{d[C]}{dt} = -k[C]$ | $[C] = [C_0]e^{-kt}$ | $[C]$ vs. Time | $\ln[C]$ vs. Time (slope $= k$) | $\dfrac{0.693}{k}$ | $s^{-1}$, $min^{-1}$, $h^{-1}$, $day^{-1}$ |

**Figure 3.12** **Summary of Zero- and First-Order Rate Expressions**   Note the differences between each of these expressions.

From Mihelcic (1999). Reprinted with permission of John Wiley & Sons, Inc.

ZERO-ORDER REACTION If $n$ is 0, Equation 3.39 becomes

$$d[C]/dt = -k \tag{3.40}$$

*This is the rate law describing a zero-order reaction.* Here, the rate of disappearance of $C$ with time is zero-order with respect to $C$, and the overall order of the reaction is zero-order.

Equation 3.40 can be rearranged and integrated for the following conditions; at time 0, the concentration of $C$ equals $C_0$, and at some future time $t$, the concentration equals $C$:

$$\int_{C_0}^{C} d[C] = -k \int_{0}^{t} dt \tag{3.41}$$

Integration of Equation 3.41 yields

$$[C] = [C]_0 - kt \tag{3.42}$$

A reaction is zero-order if concentration data plotted versus time result in a straight line (illustrated in Figure 3.12). The slope of the resulting line is the zero-order rate constant $k$, which has units of concentration/time (for example, moles/liter-day).

FIRST-ORDER REACTION If $n = 1$, Equation 3.39 becomes:

$$d[C]/dt = -k[C] \tag{3.43}$$

This is the rate law for a first-order reaction. Here, the rate of disappearance of $C$ with time is first-order with respect to $[C]$, and the overall order of the reaction is first-order.

Equation 3.43 can be rearranged and integrated for the same two conditions used in Equation 3.40 to obtain an expression that describes the concentration of $C$ with time:

$$[C] = [C]_0\, e^{-kt} \qquad\qquad (3.44)$$

Here, $k$ is the first-order reaction rate constant and has units of time$^{-1}$ (for example, hr$^{-1}$, day$^{-1}$).

A reaction is first-order when the natural logarithm of concentration data plotted versus time results in a straight line. The slope of this straight line is the first-order rate constant, $k$, as illustrated in Figure 3.12.

There are some important things to note about first- and zero-order chemical reactions. First, when comparing the concentration over time in the two reactions (as shown in the figure), the rate of the first-order reaction (slope of concentration data versus time) decreases over time, while in the zero-order reaction, the slope remains constant over time. This suggests that the rate of a zero-order reaction is independent of chemical concentration (see Equation 3.42), while the rate of a first-order reaction is dependent on the concentration of the chemical (see Equation 3.44). Thus, a chemical whose disappearance follows concentration-dependent kinetics, like first-order, will disappear more slowly as its concentration decreases.

**Kinetics of Iron Hydrolysis**

### 3.11.3 PSEUDO FIRST-ORDER REACTIONS

There are many circumstances in which the concentration of one participant in a reaction remains constant during the reaction. For example, if the concentration of one reactant initially is much higher than the concentration of another, it is impossible for the reaction to cause a significant change in the concentration of the substance with the high

---

## example/3.13 Use of Rate Law

How long will it take the carbon monoxide (CO) concentration in a room to decrease by 99 percent after the source of carbon monoxide is removed and the windows are opened? Assume the first-order rate constant for CO removal (due to dilution by incoming clean air) is 1.2/hr. No chemical reaction is occurring.

## solution

This is a first-order reaction, so use Equation 3.44. Let $[CO]_0$ equal the initial CO concentration. When 99 percent of the CO goes away, $[CO] = 0.01 \times [CO]_0$. Therefore,

$$0.01 \times [CO]_0 = [CO]_0\, e^{-kt}$$

where $k = 1.2$/hr. Solve for $t$, which equals 3.8 hr.

initial concentration. Alternatively, if the concentration of one substance is buffered at a constant value (for example, pH in a lake does not change because it is buffered by the dissolution and precipitation of alkalinity-containing solid $CaCO_3$), then the concentration of the buffered species will not change, even if the substance participates in a reaction. A **pseudo first-order** reaction is used in these situations. It can be modeled as if it were a first-order reaction. Consider the following *irreversible elementary reaction*:

$$aA + bB \rightarrow cC + dD \tag{3.45}$$

The rate law for this reaction is:

$$R = k[A]^a[B]^b \tag{3.46}$$

If the concentration of A does not change significantly during the reaction for one of the reasons previously discussed (that is, $[A_0] \gg [B_0]$ or $[A] \cong [A_0]$), the concentration of A may be assumed to remain constant and can be incorporated into the rate constant, $k$. The rate law then becomes

$$R = k'[B]^b \tag{3.47}$$

where $k'$ is the pseudo first-order rate constant and equals $k[A_0]^a$. This manipulation greatly simplifies the rate law for the disappearance of substance B:

$$d[B]/dt = -k'[B]^b \tag{3.48}$$

If $b$ is equal to 1, then the solution of Equation 3.48 is identical to that for Equation 3.44. In this case, the pseudo first-order expression can be written as follows:

$$[B] = [B_0]e^{-k't} \tag{3.49}$$

example/3.14 Pseudo First-Order Reaction

Lake Silbersee is located in the German city of Nuremberg. The lake's water quality has been diminished because of high hydrogen sulfide concentrations (which have a rotten-egg smell) that originate from a nearby leaking landfill. To combat the problem, the city decided to aerate the lake in an attempt to oxidize the odorous $H_2S$ to nonodorous sulfate ion according to the following oxidation reaction:

$$H_2S + 2O_2 \rightarrow SO_4^{2-} + 2H^+$$

It has been determined experimentally that the reaction follows first-order kinetics with respect to both oxygen and hydrogen sulfide concentrations:

$$d[H_2S]/dt = -k[H_2S][O_2]$$

The present rate of aeration maintains the oxygen concentration in the lake at 2 mg/L. The rate constant $k$ for the reaction was determined experimentally to be 1,000 L/mole-day. If the aeration completely inhibited anaerobic respiration and thus stopped the production of sulfide, how long would it take to reduce the $H_2S$ concentration in the lake from 500 $\mu$M to 1 $\mu$M?

## solution

The dissolved oxygen of the lake is maintained at a constant value and therefore is a constant. It can be combined with the rate constant to make a pseudo first-order rate constant. Thus,

$$[H_2S] = [H_2S]_0\, e^{-k't}$$

where $k' = k[O_2]$

$$1\mu M = 500\ \mu M \times e^{\left\{-\frac{1,000\,L}{\text{mole-day}} \times \frac{2\,mg}{L} \times \frac{g}{1,000\,mg} \times \frac{mole}{32g} \times t\right\}}$$

Solve for the time: $t = 100$ days.

### 3.11.4 HALF-LIFE AND ITS RELATIONSHIP TO THE RATE CONSTANT

It often is useful to express a reaction in terms of the time required to react one-half of the concentration initially present. The **half-life**, $t_{1/2}$, is defined as the time required for the concentration of a chemical to decrease by one-half (for example, $[C] = 0.5 \times [C]_0$). The relationship between half-life and the reaction rate constant depends on the order of the reaction, as shown previously in Figure 3.12.

For zero-order reactions, the half-life can be related to the zero-order rate constant, $k$. To do this, substitute $[C] = 0.5 \times [C]_0$ into Equation 3.42:

$$0.5[C]_0 = [C]_0 - kt_{1/2} \tag{3.50}$$

Equation 3.50 can be solved for the half-life:

$$t_{1/2} = \frac{0.5 \times [C]_0}{k} \tag{3.51}$$

Likewise, for a first-order reaction, the half-life can be related to the first-order rate constant, $k$. In this case, substitute $[C] = 0.5 \times [C]_0$ into Equation 3.44:

$$0.5[C]_0 = [C]_0\, e^{-kt} \tag{3.52}$$

**Radioactive Decay and Popping Popcorn**

www.wiley.com/college/mihelcic

## example/3.15 Converting a Rate Constant to Half-Life

Subsurface half-lives for benzene, TCE, and toluene are listed as 69, 231, and 12 days, respectively. What are the first-order rate constants for all three chemicals?

### solution

The model only accepts concentration-dependent, first-order rate constants. Thus, to solve the problem, convert half-life to a first-order rate constant with the use of Equation 3.53:

For benzene,

$$t_{1/2} = \frac{0.693}{k} = \frac{0.693}{69 \text{ days}} = 0.01/\text{day}$$

Similarly, $k_{TCE} = 0.058/\text{day}$, and $k_{toluene} = 0.058/\text{day}$.

## example/3.16 Use of Half-Life in Determining First-Order Decay

After the Chernobyl nuclear accident, the concentration of $^{137}Cs$ in milk was proportional to the concentration of $^{137}Cs$ in the grass that cows consumed. The concentration in the grass was, in turn, proportional to the concentration in the soil. Assume that the only reaction by which $^{137}Cs$ was lost from the soil was through radioactive decay and the half-life for this isotope is 30 years. Calculate the concentration of $^{137}Cs$ in cow's milk after 5 years if the concentration in milk shortly after the accident was 12,000 bequerels (Bq) per liter. (*Note*: A bequerel is a measure of radioactivity; 1 bequerel equals 1 radioactive disintegration per second.)

### solution

Because the half-life equals 30 years, the rate constant $k$ can be determined from Equation 3.53:

$$k = \frac{0.693}{t_{1/2}} = \frac{0.693}{30 \text{ yr}} = 0.023/\text{yr}$$

Therefore:

$$\left[^{137}Cs\right]_{t=5} = \left[^{137}Cs\right]_{t=0} \exp(-kt) = 12{,}000 \text{ Bq/L} \times \exp\left(\frac{-0.023}{\text{yr}} \times 5 \text{ yr}\right) = 10{,}700 \text{ Bq/L}$$

The half-life for a first-order relationship then is given by

$$t_{1/2} = \frac{0.693}{k} \qquad (3.53)$$

## 3.11.5 EFFECT OF TEMPERATURE ON RATE CONSTANTS

Rate constants typically are determined and compiled for temperatures at 20°C or 25°C. However, groundwaters usually have temperatures around 8°C to 12°C, and surface waters, wastewaters, and soils generally have temperatures ranging from 0°C to 30°C. Thus, when a different temperature is encountered, you must first determine if the effect of temperature is important, and secondly, if important, determine how to convert the rate constant for the new temperature.

The **Arrhenius equation** is used to adjust rate constants for changes in temperature. It is written as

$$k = Ae^{-(Ea/RT)} \qquad (3.54)$$

where $k$ is the rate constant of a particular order, $A$ is termed the pre-exponential factor (same units as $k$), $Ea$ is the **activation energy** (kcal/mole), $R$ is the gas constant, and $T$ is temperature (K). The activation energy, $Ea$, is the energy required for the collision to result in a reaction. The pre-exponential factor is related to the number of collisions per time, so it is different for gas- and liquid-phase reactions. The pre-exponential factor, $A$, has a small dependence on temperature for many reactions; however, most environmental situations span a relatively small temperature range. Its value depends to a great extent on the number of molecules that collide in a reaction. For example, unimolecular reactions exhibit values of $A$ that can be several orders of magnitude greater than bimolecular reactions.

---

example/3.17 **Effect of Temperature on CBOD Rate Constant**

The rate constant for carbonaceous biochemical oxygen demand (CBOD) at 20°C is 0.1/day. What is the rate constant at 30°C? Assume $\Theta = 1.072$.

solution

Using Equation 3.55,

$$k_{30} = 0.1/\text{day}\left[1.072^{(30°C-20°C)}\right] = 0.2/\text{day}$$

This example demonstrates that, for biological systems used in wastewater treatment, we would often observe a doubling in the biological reaction with every 10°C increase in the temperature.

---

A plot of $\ln(k)$ versus $1/T$ can be used to determine $Ea$ and $A$. After $Ea$ and $A$ are known for a particular reaction, Equation 3.54 can be used to adjust a rate constant for changes in temperature.

The Arrhenius equation is the basis for another commonly used relationship between rate constants and temperature used for biological processes over narrow temperature ranges. The *carbonaceous biochemical oxygen demand (CBOD)* rate constant, $k$, known at a particular temperature, typically is converted to other temperatures using the following expression:

$$k_{T_2} = k_{T_1} \times \Theta^{(T_2 - T_1)} \tag{3.55}$$

where $\Theta$ is a **dimensionless temperature coefficient**. In fact, $\Theta$ equals $\exp\{Ea \div [R \times T_1 \times T_2]\}$, as can be seen from the Arrhenius equation. $\Theta$ is temperature dependent and has been found to range from 1.056 to 1.13 for biological decay of municipal sewage.

### 3.11.6 PHOTOCHEMICAL REACTION KINETICS

**Ozone Layer Protection and the Montreal Protocol**
http://ozone.unep.org/

Photochemical reactions usually are best described using a kinetic approach. These reactions can occur through direct or indirect reactions of molecules with light, and they may be catalyzed by chemicals occurring naturally in the environment or by chemicals emitted through human activities. An example of photochemistry in our daily lives is the fading of fabric dyes exposed to sunlight. Another example is perhaps the most important photochemical reaction in the world: photosynthesis.

Light is differentiated according to its wavelength. Table 3.13 shows the entire *electromagnetic spectrum*. Light can be envisioned to consist of small bundles of energy called *photons*, which can be absorbed or emitted by matter. The energy of a photon, $E$ (units of joules), is computed as follows:

$$E = \frac{hc}{\lambda} \tag{3.56}$$

## Table / 3.13

**Electromagnetic Spectrum** Photochemical reactions involving UV light are important in creation of the ozone hole and in formation of urban smog. Infrared light is important in understanding the greenhouse effect.

| Wavelength (nm) | Range |
| --- | --- |
| <50 | X-rays |
| 50–400 | Ultraviolet (UV) |
| 400–750 | Visible (400–450 = violet, and 620–750 = red) |
| >750 | Infrared |

where $h$ equals Planck's constant ($6.626 \times 10^{-34}$ J-s), $c$ is the speed of light ($3 \times 10^8$ m/s), and $\lambda$ is the light's specific wavelength. Equation 3.56 shows that greater energy is contained in photons with a shorter wavelength.

This light energy can be absorbed by a molecule. A molecule that absorbs light energy has its energy increased by rotational, vibrational, and electronic excitation. The molecule then typically has a very short time (a fraction of a second) to either use the energy in a photochemical reaction or lose it, most likely as heat.

All atoms and molecules have a favored wavelength at which they absorb light. That is, an atom or molecule will absorb light within a specific range of wavelengths. Greenhouse gases such as water vapor, $CO_2$, $N_2O$, and $CH_4$ absorb energy emitted by Earth as infrared light, while the major components of the atmosphere ($N_2$, $O_2$, Ar) are incapable of absorbing infrared light. It is this capture of energy released by Earth that partially contributes to the warming of the planet's surface. Anthropogenic emissions of greenhouse gases such as $CO_2$, $CH_4$, and CFCs have increased the amount of this "captured" energy.

Another example of molecules absorbing light energy is in the filtering of the ultraviolet light that enters Earth's atmosphere. $O_2$ molecules located above the stratosphere (the region of the atmosphere 15 to 50 km above Earth's surface) filter out (or absorb) most of the incoming UV light in the range of 120–220 nm, and other gases such as $N_2$ filter out the UV light with wavelengths smaller than 120 nm. This means no UV light with a wavelength below 220 nm reaches Earth's surface. All of the UV light in the range of 220–290 nm is filtered out by ozone ($O_3$) molecules in the stratosphere with a little help from $O_2$ molecules. However, $O_3$ alone filters a fraction of UV light in the range of 290–320 nm, and the remainder makes it to our planet's surface. Overexposure to this portion of the light spectrum can result in malignant and nonmalignant skin cancer, and damage the human immune system, and inhibit plant and animal growth. Most of the UV light in the range of 320–400 nm reaches Earth's surface, but fortunately this type of UV light is the least harmful to the planet's biological systems.

**Learn More About Ozone**

http://ozone.unep.org

**Class Discussion**

Stratospheric ozone depletion is a major public health issue in New Zealand. Why not in the U.S.? In your discussion consider differences in geography, demographics, culture, and governance.

---

### Box / 3.4   Ozone Cycling in the Stratosphere

Stratospheric $O_3$ is formed by the reaction of atomic oxygen (O·) with dimolecular oxygen ($O_2$) according to the following reaction:

$$O\cdot + O_2 \rightarrow O_3 \qquad (3.57)$$

The (O·) required in Equation 3.57 is derived from the reaction of $O_2$ with UV photons ($\lambda = 241$ nm) according to the following reaction:

$$O_2 + UV\ photon \rightarrow 2O\cdot \qquad (3.58)$$

However, in the stratosphere, the majority of oxygen exists as $O_2$, so only a little O· is available. Therefore, even though there is little O· in the stratosphere relative to $O_2$, the small amounts of O· created here will react with the abundant $O_2$ to form ozone ($O_3$), according to Equation 3.57.

Equations 3.57 and 3.58 explain the natural formation of ozone in the stratosphere. They also provide insight as to why the concentration of ozone is much higher in the stratosphere ($ppm_v$ levels) versus the troposphere ($ppb_v$ levels). This is because the stratosphere contains much more O· than the troposphere, where O· is not produced by natural mechanisms in large amounts except under human-induced conditions of smog formation (discussed later in this section).

Ozone is destroyed naturally in the stratosphere by the reaction of ozone with UV photons with $\lambda = 320$ nm:

$$O_3 + UV\ photon \rightarrow O_2 + O\cdot \qquad (3.59)$$

Single oxygen molecules can thus either react with $O_2$ to form more $O_3$, or else destroy $O_3$ to create $O_2$. Fortunately, the destruction reaction has a relatively high exponential factor, so this natural destruction reaction occurs at a slow rate.

Ozone is destroyed continuously in the stratosphere, and the amount of species that destroy ozone moving into the stratosphere has been augmented in recent years by human activities. Many of the processes that destroy ozone share a general mechanism, using X as the reactive species, as follows:

$$X + O_3 \rightarrow XO + O_2 \qquad (3.60)$$
$$XO + O\cdot \rightarrow X + O_2 \qquad (3.61)$$

Overall reaction:

$$O\cdot + O_3 \rightarrow 2O_2 \qquad (3.62)$$

The species X act as catalysts, speeding up the reaction between $O_3$ and O. The more common species of X have been identified in three categories: HO radicals (H·, OH·, OOH·), NO radicals (NO·, $NO_2$·), and ClO radicals (Cl·, ClO·). Near the stratosphere, HO radicals account for as much as 70 percent of the total ozone destruction.

*Free-radical chlorine atoms* are very efficient catalysts in the destruction of ozone. Thus, the greatest threat to stratospheric $O_3$ is from chlorine-containing chemicals. Fortunately, 99 percent of stratospheric Cl is stored in nonreactive forms such as HCl and chlorine nitrate ($ClONO_2$). The amount of stratospheric chlorine has increased in recent decades, however, and during the Antarctic spring, a lot of this stored chlorine is released into the active catalytic forms, Cl· and ClO·. A naturally occurring chlorine-containing chemical is chloromethane ($CH_3Cl$), which is formed over the world's oceans and may be transported up into the stratosphere. $CH_3Cl$ molecules can react with UV photons (wavelength of 200–280 nm) to produce chlorine-free radicals, Cl·.

The major anthropogenic source of chlorine is from the movement of chlorofluorocarbons (CFCs) into the stratosphere and subsequent release of chlorine-free radicals. CFCs (known commercially as Freon) were widely used in the northern hemisphere beginning in the 1930s. The three most commonly used CFCs were CFC-12 ($CF_2Cl_2$, used extensively as a coolant and refrigerant, and embedded in rigid plastic foam), CFC-11 ($CFCl_3$, used to blow holes in soft plastic such as cushions, carpet padding, and car seats), and CFC-13 ($CF_2Cl$-$CFCl_2$, used to clean circuit boards). CFCs are relatively stable in the troposphere, but after being transported up into the stratosphere, they can undergo photochemical reactions that release the catalytic chlorine free radical, Cl·. For example, the breakdown of CFC-12 occurs as follows:

$$CF_2Cl_2 + UV\ photon\ (200-280\ nm)$$
$$\rightarrow CF_2Cl\cdot + Cl\cdot \qquad (3.63)$$

Another Cl· can subsequently be released from the $CF_2Cl$·.

# Key Terms

- absorption
- acid
- acid–base chemistry
- activation energy
- activity

- activity coefficients
- adsorption
- air–water equilibrium
- alkalinity
- Arrhenius equation

- base
- binary reaction
- buffering capacity
- carbonate system
- concentration

- dimensionless Henry's law constant
- dimensionless temperature coefficient
- dissolved carbon dioxide
- dissolved oxygen
- electron acceptor
- electron donor
- elementary reaction
- equilibrium
- equilibrium constant ($K$)
- first law of thermodynamics
- first order
- Freundlich isotherm
- Gibbs free energy
- half-life
- Henry's law
- Henry's law constant ($K_H$)
- hydrophobic partitioning
- ideal system
- ionic strength
- kinetics
- $K_a$
- $K_b$
- $K_P$
- $K_w$
- law of conservation of mass
- linear isotherm
- local equilibrium
- methylation
- octanol–air partition coefficient ($K_{oa}$)
- octanol–water partition coefficient ($K_{ow}$)
- overall order
- oxidation-reduction
- oxidation state
- partition coefficient
- pH
- photosynthesis
- p$K_a$
- precipitation-dissolution
- pseudo first-order
- Raoult's law
- rate law
- redox reactions
- saturated vapor pressure
- second law of thermodynamics
- sediment–water partition coefficient
- soil–water partition coefficient
- soil–water partition coefficient normalized to organic carbon ($K_{oc}$)
- solubility product ($K_{sp}$)
- sorption
- stoichiometry
- thermodynamics
- van't Hoff relationship
- volatilization
- zero order

## chapter/Three Problems

**3.1** How many grams of NaCl would you need to add to a 1 L water sample (pH = 7) so the ionic strength equaled 0.1 M?

**3.2** The chemical 1,4-dichlorobenzene (1,4-DCB) is sometimes used as a disinfectant in public lavatories. At 20°C (68°F), the vapor pressure is $5.3 \times 10^{-4}$ atm. (a) What would be the concentration in the air in units of $g/m^3$? The molecular weight of 1,4-DCB is 147 g/mole. (b) An alternative disinfectant is 1-bromo-4-chlorobenzene (1,4-CB). The boiling point of 1,4-CB is 196°C, whereas the boiling point of 1,4-DCB is 180°C. Which compound would cause the highest concentrations in the air in lavatories? (Explain your answer.)

**3.3** The boiling temperatures of chloroform (an anesthetic), carbon tetrachloride (commonly used in the past for dry cleaning), and tetrachloroethylene (previously used as a degreasing agent) are 61.7°C, 76.5°C, and 121°C. The vapor pressure of a chemical is directly proportional to the inverse of the chemical's boiling point. If a large quantity of these compounds were spilled in the environment, which compound would you predict to have higher concentrations in the air above the site? (Explain your answer.)

**3.4** What would be the saturation concentration (mole/L) of oxygen ($O_2$) in a river in winter when the air temperature is 0°C if the Henry's law constant at this temperature is $2.28 \times 10^{-3}$ mole/L-atm? What would the answer be in units of mg/L?

**3.5** The log Henry's law constant (units of L-atm/mole and measured at 25°C) for trichloroethylene is 1.03; for tetrachloroethylene, 1.44; for 1,2-dimethyl-benzene, 0.71; and for parathion, −3.42. (a) What is the dimensionless Henry's law constant for each of these chemicals? (b) Rank the chemicals in order of ease of stripping from water to air.

**3.6** The dimensionless Henry's law constant for trichloroethylene (TCE) at 25°C is 0.4. A sealed glass vial is prepared that has an air volume of 4 mL overlying an aqueous volume of 36 mL. TCE is added to the aqueous phase so that initially it has an aqueous phase concentration of 100 ppb. After the system equilibrates, what will be the concentration (in units of $\mu g/L$) of TCE in the aqueous phase?

**3.7** The Henry's law constant for $H_2S$ is 0.1 mole/L-atm, and

$$H_2S_{(aq)} \rightleftharpoons HS^- + H^+$$

where $K_a = 10^{-7}$. If you bubble pure $H_2S$ gas into a beaker of water, what is the concentration of $HS^-$ at a pH of 5 in (a) moles/L, (b) mg/L, and (c) $ppm_m$?

**3.8** Determine the equilibrium pH of aqueous solutions of the following strong acids or bases: (a) 15 mg/L of $HSO_4^-$, (b) 10 mM NaOH, and (c) 2,500 $\mu g/L$ of $HNO_3$.

**3.9** What would be the pH if $10^{-2}$ moles of hydrofluoric acid (HF) were added to 1 L pure water? The $pK_a$ of HF is 3.2.

**3.10** When $Cl_2$ gas is added to water during the disinfection of drinking water, it hydrolyzes with the water to form HOCl. The disinfection power of the acid HOCl is 88 times better than its conjugate base, $OCl^-$. The $pK_a$ for HOCl is 7.5. (a) What percentage of the total disinfection power (HOCl + $OCl^-$) exists in the acid form at pH = 6? (b) At pH = 7?

**3.11** (a) What is the solubility (in moles/L) of $CaF_2$ in pure water at 25°C? (b) What is the solubility of $CaF_2$ if the temperature is raised 10°C? (c) Does the solubility of $CaF_2$ increase, decrease, or remain the same if the ionic strength is raised? (Explain your answer.)

**3.12** At a wastewater treatment plant, $FeCl_{3(s)}$ is added to remove excess phosphate from the effluent. Assume the following reactions occur:

$$FeCl_{3(s)} \rightleftharpoons Fe^{3+} + 3Cl^-$$
$$Fe^{3+} + PO_4^{3-} \rightleftharpoons FePO_{4(s)}$$

The equilibrium constant for the second reaction is $10^{26.4}$. $K_{sp} = 10^{-26.4}$ for the reverse reaction. What concentration of $Fe^{3+}$ is needed to maintain the phosphate concentration below the limit of 1 mg P/L?

**3.13** One method to remove metals from water is to raise the pH and cause them to precipitate as their metal hydroxides. (a) For the following reaction, compute the standard free energy of reaction:

$$Cd^{2+} + 2OH^- \rightleftharpoons Cd(OH)_{2(s)}$$

(b) The pH of water initially was 6.8 and then was raised to 8.0. Is the dissolved cadmium concentration reduced to below 100 mg/L at the final pH? Assume the temperature of the water is 25°C.

**3.14**   Naphthalene has a log $K_{ow}$ of 3.33. Estimate its soil–water partition coefficient normalized to organic carbon and the 95 percent confidence interval of your estimate.

**3.15**   Atrazine, an herbicide widely used for corn, is a common groundwater pollutant in the corn-producing regions of the United States. The log $K_{ow}$ for atrazine is 2.65. Calculate the fraction of total atrazine that will be adsorbed to the soil given that the soil has an organic carbon content of 2.5 percent. The bulk density of the soil is 1.25 g/cm$^3$; this means that each cubic centimeter of soil (soil plus water) contains 1.25 g soil particles. The porosity of the soil is 0.4.

**3.16**   A first-order reaction that results in the destruction of a pollutant has a rate constant of 0.1/day. (a) How many days will it take for 90 percent of the chemical to be destroyed? (b) How long will it take for 99 percent of the chemical to be destroyed? (c) How long will it take for 99.9 percent of the chemical to be destroyed?

**3.17**   A bacteria strain has been isolated that can cometabolize tetrachloroethane (TCA). This strain can be used for the bioremediation of hazardous-waste sites contaminated with TCA. Assume that the biodegradation rate is independent of TCA concentration (that is, the reaction is zero-order). In a bioreactor, the rate for TCA removal was 1 μg/L-min. What water retention time would be required to reduce the concentration from 1 mg/L in the influent to 1 μg/L in the effluent of a reactor? Assume the reactor is completely mixed.

**3.18**   Assume $PO_4^{3-}$ is removed from municipal wastewater through precipitation with $Fe^{3+}$ according to the following reaction: $PO_4^{3-} + Fe^{3+} \rightarrow$ $FePO_{4(s)}$. The rate law for this reaction is

$$\frac{d[PO_4^{3-}]}{dt} = -k[Fe^{3+}][PO_4^{3-}]$$

(a) What is the reaction order with respect to $PO_4^{3-}$?
(b) What order is this reaction overall?

**3.19**   Ammonia ($NH_3$) is a common constituent of many natural waters and wastewaters. When water containing ammonia is treated at a water treatment plant, the ammonia reacts with the disinfectant hypochlorous acid (HOCl) in solution to form mono-chloroamine ($NH_2Cl$) as follows:

$$NH_3 + HOCl \rightarrow NH_2Cl + H_2O$$

The rate law for this reaction is

$$\frac{d[NH_3]}{dt} = -k[HOCl][NH_3]$$

(a) What is the reaction order with respect to $NH_3$?
(b) What order is this reaction overall? (c) If the HOCl concentration is held constant and equals $10^{-4}$ M, and the rate constant equals $5.1 \times 10^6$ L/mole-s, calculate the time required to reduce the concentration of $NH_3$ to one-half its original value.

**3.20**   Nitrogen dioxide ($NO_2$) concentrations are measured in an air-quality study and decrease from 5 ppm$_v$ to 2 ppm$_v$ in four min with a particular light intensity. (a) What is the first-order rate constant for this reaction? (b) What is the half-life of $NO_2$ during this study? (c) What would the rate constant need to be changed to in order to decrease the time required to lower the $NO_2$ concentration from 5 ppm$_v$ to 2 ppm$_v$ in 1.5 min?

**3.21**   If the rate constant for the degradation of bio-chemical oxygen demand (BOD) at 20°C is 0.23/day, what is the value at 5°C and 25°C? Assume that $\Theta$ equals 1.1.

# References

Baker, J. R., J. R. Mihelcic, D. C. Luehrs, and J. P. Hickey. 1997. "Evaluation of Estimation Methods for Organic Carbon Normalized Sorption Coefficients." *Water Environment Research* 69:136–145.

Beyer, A., F. Wania, T. Gouin, D. Mackay, and M. Matthies. 2002. "Selecting Internally Consistent Physicochemical Properties of Organic Compounds." *Environ. Toxicol. Chem.* 21:941–953.

Mihelcic, J. R. 1999. *Fundamentals of Environmental Engineering.* New York: John Wiley & Sons.

Schwarzenbach, R. P., P. M. Gschwend, and D. M. Imboden. 2003. *Environmental Organic Chemistry.* New York: John Wiley & Sons.

Wania, F. 2006. "Potential of Degradable Organic Chemicals for Absolute and Relative Enrichment in the Arctic." *Environ. Sci. & Technol.* 40:569–577.

# chapter/Four Physical Processes

Richard E. Honrath Jr.,
James R. Mihelcic, Julie Beth
Zimmerman, Alex S. Mayer

*In this chapter, readers will learn about the physical processes that are important in the movement of pollutants through the environment and processes used to control and treat pollutant emissions. The chapter begins with a study of the use of material and energy balances and the processes of advection and dispersion. The final section of this chapter extends this description of transport processes with a look at the movement of particles in fluids, specifically the velocity of a settling particle and the discharge of groundwater.*

## Major Sections

## Learning Objectives

1. Use the law of conservation of mass to write a mass balance that includes rate of chemical production or disappearance.
2. Determine whether a situation is at steady or nonsteady state, and apply this information to the mass balance.
3. Differentiate batch reactors, completed mixed flow reactors, and plug flow reactors.
4. Relate a reactor's retention time to reactor volume and flow.
5. Differentiate forms of energy, and write an energy balance.
6. Relate an energy balance to the greenhouse effect.
7. Relate temperature change to sea level rise under different population, economic growth, and energy management scenarios.
8. Differentiate and employ the transport processes of advection, dispersion, and diffusion.
9. Apply Fick's law, Stokes' law, and Darcy's law to environmental problems.

## 4.1 Mass Balances

The **law of conservation of mass** states that mass can neither be produced nor destroyed. Conservation of mass and conservation of energy provide the basis for two commonly used tools: the **mass balance** and the energy balance. This section discusses mass balances, and energy balances are the topic of Section 4.2.

The principle of conservation of mass means that if the amount of a chemical increases somewhere (for example, in a lake), then that increase cannot be the result of some "magical" formation. The chemical must have been either carried into the lake from elsewhere or produced via chemical or biological reaction from other compounds that were already in the lake. Similarly, if reactions produced the mass increase of this chemical, they must also have caused a corresponding decrease in the mass of some other compound(s).

In terms of sustainability, this same principle of mass balance can be thought of in terms of the use of finite material and energy sources. For example, the consumption of fossil-based energy sources—oil, gas, and coal—must maintain a mass balance. As a result, as these resources are combusted for energy, the original source is depleted, and wastes are generated in the form of emissions to the air, land, and water. While the mass of carbon remains constant, much of it is removed from the energy-intensive form of oil, gas, or coal and is converted to carbon dioxide, a greenhouse gas.

Conservation of mass provides a basis for compiling a budget of the mass of any chemical. In the case of a lake, this budget keeps track of the amounts of chemical entering and leaving the lake and the amounts formed or destroyed by chemical reaction. This budget can be balanced over a given time period, much as a checkbook is balanced. Equation 4.1 describes the mass balance:

$$
\begin{aligned}
\text{mass at time } t + \Delta t = {} & \text{mass at time } t \\
& + \left( \begin{array}{c} \text{mass entering} \\ \text{from } t \text{ to } t + \Delta t \end{array} \right) - \left( \begin{array}{c} \text{mass exiting} \\ \text{from } t \text{ to } t + \Delta t \end{array} \right) \\
& + \left( \begin{array}{c} \text{net mass of chemical produced} \\ \text{from other compounds by} \\ \text{reactions between } t \text{ and } t + \Delta t \end{array} \right)
\end{aligned}
\tag{4.1}
$$

Each term of Equation 4.1 has units of mass. This form of balance is most useful when there is a clear beginning and end to the balance period ($\Delta t$), so that the change in mass over the balance period can be determined. Continuing our earlier analogy, when balancing a checkbook, a balance period of one month is often used.

In environmental problems, however, it usually is more convenient to work with values of **mass flux**—the rate at which mass enters or leaves a system. To develop an equation in terms of mass flux, the mass balance equation is divided by $\Delta t$ to produce an equation with units of mass per unit time. Dividing Equation 4.1 by $\Delta t$ and moving the first term on the right (mass at time $t$) to the left-hand side yields:

$$\frac{(\text{Mass at time } t + \Delta t) - (\text{mass at time } t)}{\Delta t} = \frac{\left(\begin{array}{c}\text{mass entering from} \\ t \text{ to } t + \Delta t\end{array}\right)}{\Delta t}$$

$$-\frac{\left(\begin{array}{c}\text{mass exiting from} \\ t \text{ to } t + \Delta t\end{array}\right)}{\Delta t} + \frac{\left(\begin{array}{c}\text{net chemical} \\ \text{production from} \\ t \text{ to } t + \Delta t\end{array}\right)}{\Delta t} \qquad (4.2)$$

Note that each term in Equation 4.2 has units of mass/time. The left side of Equation 4.2 is equal to $\Delta m / \Delta t$.

In the limit as $\Delta t \to 0$, the left side becomes $dm/dt$, the rate of change of chemical mass in the lake. As $\Delta t \to 0$, the first term on the right side of Equation 4.2 becomes the rate at which mass enters the lake (the mass flux into the lake), and the second term becomes the rate at which mass exits the lake (the mass flux out of the lake). The last term of Equation 4.2 is the *net rate* of chemical production or loss.

The symbol $\dot{m}$ refers to a mass flux with units of mass/time. Substituting mass flux, the equation for mass balances can be written as follows:

$$\left(\begin{array}{c}\text{mass} \\ \text{accumulation} \\ \text{rate}\end{array}\right) = (\text{mass flux in}) - (\text{mass flux out}) + \left(\begin{array}{c}\text{net rate of} \\ \text{chemical} \\ \text{production}\end{array}\right)$$

or

$$\boxed{\frac{dm}{dt} = \dot{m}_{\text{in}} - \dot{m}_{\text{out}} + \dot{m}_{\text{reaction}}} \qquad (4.3)$$

Equation 4.3 is the governing equation for mass balances used throughout environmental engineering and science.

## 4.1.1  CONTROL VOLUME

A mass balance is meaningful only in terms of a specific region of space, which has boundaries across which the terms $\dot{m}_{\text{in}}$ and $\dot{m}_{\text{out}}$ are determined. This region is called the **control volume**.

In the previous example, we used a lake as our control volume and included mass fluxes into and out of the lake. Theoretically, any volume of any shape and location can be used as a control volume. Realistically, however, certain control volumes are more useful than others. The most important attribute of a control volume is that it have boundaries over which $\dot{m}_{\text{in}}$ and $\dot{m}_{\text{out}}$ can be calculated.

$Q_{in}, C_{in} \rightarrow$

$\rightarrow Q, C$

**Figure 4.1** **Schematic Diagram of a Completely Mixed Flow Reactor (CMFR)** The stir bar is used as a symbol to indicate that the CMFR is well mixed.

From Mihelcic (1999). Reprinted with permission of John Wiley & Sons, Inc.

## 4.1.2 TERMS OF THE MASS BALANCE EQUATION FOR A CMFR

A well-mixed tank is an analogue for many control volumes used in environmental situations. For example, in the lake example, it might be reasonable to assume that the chemicals discharged into the lake are mixed throughout the entire lake. Such a system is called a **completely mixed flow reactor (CMFR)**. Other terms, most commonly *continuously stirred tank reactor (CSTR)*, are also used for such systems. A schematic diagram of a CMFR is shown in Figure 4.1.

The following discussion describes each term in a mass balance of a hypothetical compound within the CMFR.

### MASS ACCUMULATION RATE ($dm/dt$)
The rate of change of mass within the control volume, $dm/dt$, is referred to as the **mass accumulation rate**. To directly measure the mass accumulation rate would require determining the total mass within the control volume of the compound for which the mass balance is being conducted. This is usually difficult, but it is seldom necessary. If the control volume is well mixed, then the concentration of the compound is the same throughout the control volume, and the mass in the control volume is equal to the product of that concentration, $C$, and the volume, $V$. (To ensure that $C \times V$ has units of mass/time, express $C$ in units of mass/volume.) Expressing mass as $C \times V$, the mass accumulation rate is equal to

$$\frac{dm}{dt} = \frac{d(VC)}{dt} \qquad (4.4)$$

In most cases (and in all cases in this text), the volume is constant and can be moved outside the derivative, resulting in

$$\frac{dm}{dt} = V\frac{dC}{dt} \qquad (4.5)$$

In any mass balance situation, once a sufficient amount of time has passed, conditions will approach **steady state**, meaning that conditions no longer change with time. In steady-state conditions, the concentration—and hence the mass—within the control volume remains constant. In this case, $dm/dt = 0$. If, however, insufficient time has passed since a flow, inlet concentration, reaction term, or other problem condition has changed, the mass in the control volume will vary with time, and the mass balance will be **nonsteady state**.

The amount of time that must pass before steady state is reached depends on the conditions of the problem. To see why, consider the approach to steady state of the amount of water in two large, initially empty sinks. In the first sink, the faucet is opened halfway, and the drain is opened slightly. Initially, the mass of water in the sink increases over time, since the faucet flow exceeds the flow rate out of the drain. Conditions are changing, so this is a nonsteady-state situation. However, as the water level in the sink rises, the flow rate out of the drain will increase, and eventually the drain flow will equal the faucet flow. At this point, the water level will cease rising, and the situation will have reached steady state.

If this experiment is repeated with a second sink, but this time with the drain opened fully, the drain flow will increase more rapidly and will equal the faucet flow while the water level in the sink is still low. In this case, steady state will be reached more rapidly. In general, the speed at which steady state is approached depends on the magnitude of the mass flux terms, relative to the total mass in the control volume. Determining whether or not a mass balance problem is steady state is something of an art. However, if conditions of the problem have changed recently, then the problem is probably a nonsteady state. Conversely, if conditions have remained constant for a very long time, it is probably a steady-state problem. Treating a steady-state problem as nonsteady state will always result in the correct answer, while treating a nonsteady-state problem as steady state will not. This does not mean that all problems should be treated as nonsteady state, however. Nonsteady-state solutions generally are more difficult, so it is advantageous to identify steady state whenever present.

In terms of emissions to the environment, steady state often is equated with nature's ability to assimilate wastes at the rate at which they are released. For example, in the case of **carbon dioxide emissions** from burning fossil fuels, at steady state the rate of emissions would equal the total of all removal rates from the atmosphere. These include uptake by the oceans and the small fraction of uptake by plants for photosynthesis that is not balanced by respiration, which releases carbon dioxide. Eventually, as the carbon dioxide concentration in the atmosphere rises, the rate of uptake by the oceans will balance the rate of emissions from fossil fuel burning. However, for that to happen, the concentration in the atmosphere would have to increase significantly, and the dissolved carbon dioxide would have to become well mixed throughout the ocean. Since these processes take centuries to millennia, carbon dioxide emissions accumulate in the atmosphere, where they contribute to the greenhouse effect. A similar situation can occur for the release of industrial chemicals to the environment. Currently, the ease of assimilation by the environment often is ignored when chemicals are selected or designed and manufactured for uses that result in release to the environment. In many cases, the result is accumulation in the environment in a system that is not at steady state. This is of particular concern with chemicals that bioaccumulate (build up in organisms), becoming more concentrated in organisms further up the food chain.

**Indoor Air in Large Buildings**
http://www.epa.gov/iaq/largebldgs/

**Greeenhouse Gas Emissions from Transportation**
http://www.epa.gov/otaq/climate

example/4.1 Determining whether a Problem is Steady State

For each of the following mass balance problems, determine whether a steady-state or nonsteady-state mass balance would be appropriate.

1. Vision a mass balance on chloride ($Cl^-$) dissolved in a lake. Two rivers bring chloride into the lake, and one river removes chloride. No significant chemical reactions occur, as chloride is soluble and nonreactive. What is the annual average concentration of chloride in the lake?

2. A degradation reaction within a well-mixed tank is used to destroy a pollutant. Inlet concentration and flow are held constant, and the system has been operating for several days. What is the pollutant concentration in the effluent, given the inlet flow and concentration and the first-order decay rate constant?

3. The source of pollutant in problem 2 is removed, resulting in an instantaneous decline of the inlet concentration to zero. How long would it take until the outlet concentration reaches 10 percent of its initial value?

## solution

1. Over an annual period, river flows and concentrations can be assumed to be relatively constant. Since conditions are not changing, and since a single value independent of time is requested for chloride concentration, the problem is steady state.

2. Again, conditions in the problem are constant and have remained so for a long time, so the problem is steady state. Note that the presence or absence of a chemical reaction does not provide any information on whether the problem is steady state.

3. Two clues reveal that this problem is nonsteady state. First, conditions have changed recently: the inlet concentration dropped to zero. Second, the solution requires calculation of a time period, which means conditions must be varying with time.

**MASS FLUX IN** $(\dot{m}_{in})$ Often, the volumetric flow rate, $Q$, of each input stream entering the control volume is known. In Figure 4.1, the pipe has a flow rate of $Q_{in}$, with corresponding chemical concentration of $C_{in}$. The *mass flux into the CMFR* is then given by the following equation:

$$\dot{m}_{in} = Q_{in} \times C_{in} \qquad (4.6)$$

If it is not immediately clear how $Q \times C$ results in a mass flux, consider the units of each term:

$$\dot{m} = Q \times C$$

$$\frac{\text{mass}}{\text{time}} = \frac{\text{volume}}{\text{time}} \times \frac{\text{mass}}{\text{volume}}$$

Note that the concentration must be expressed in units of mass/volume.

If the volumetric flow rate is not known, it may be calculated from other parameters. For example, if the fluid velocity $v$ and the cross-sectional area $A$ of the pipe are known, then $Q = v \times A$.

In some situations, mass may enter the control volume through direct emission into the volume. In this case, the emissions frequently are specified in mass flux units mass/time, which can be used in a mass balance directly. For example, if a mass balance is performed on the air pollutant carbon monoxide (CO) over a city, we would use estimates of the total CO emissions (in units of tons/day) from automobiles and power plants in the city.

Another way to describe the flux is in terms of a flux density, $J$, times the area through which the flux occurs. $J$ has units of mass/area-time and is discussed further under the topic of diffusion. This type of flux notation is most useful at interfaces where there is no fluid flow, such as the interface between the air and water at the surface of a lake.

Often, the mass flux is composed of several terms. For example, a tank may have more than one inlet, or the air over a city may receive CO blowing from an upwind urban area in addition to its own emissions. In such cases, $\dot{m}_{in}$ is the sum of all individual contributions to mass input fluxes.

**Learn about the Chesapeake Bay**
http://www.chesapeakebay.net/

**Tampa Bay Estuary Program**
http://www.tbep.org

MASS FLUX OUT ($\dot{m}_{out}$) In most cases, there is only one effluent flow from a CMFR. Then the mass flux out may be calculated as $\dot{m}_{in}$ was calculated in Equation 4.6:

$$\dot{m}_{out} = Q_{out} \times C_{out} \tag{4.7}$$

In the case of a well-mixed control volume, the concentration is constant throughout. Therefore, the concentration in flow exiting the control volume is referred to simply as $C$, the concentration in the control volume, and

$$\dot{m}_{out} = Q_{out} \times C \tag{4.8}$$

NET RATE OF CHEMICAL REACTION ($\dot{m}_{reaction}$) The term $\dot{m}_{reaction}$ or $\dot{m}_{rxn}$ refers to the net rate of production of a compound from chemical or biological reactions. It has units of mass/time. Thus, if other compounds react to form the compound, $\dot{m}_{rxn}$ will be greater than zero; if the compound reacts to form some other compound(s), resulting in a loss, $\dot{m}_{rxn}$ will be negative.

Although the chemical reaction term in a mass balance has units of mass/time, chemical reaction rates usually are expressed in terms of concentration, not mass. Thus, to calculate $\dot{m}_{rxn}$, we multiply the rate of change of concentration by the CMFR volume to obtain the rate of change of mass within the control volume:

$$\dot{m}_{rxn} = V \times \left(\frac{dC}{dt}\right)_{\text{reaction only}} \tag{4.9}$$

where $(dC/dt)_{\text{reaction only}}$ is obtained from the rate law for the reaction and is equal to the rate of change in concentration that would occur if the reaction took place in isolation, with no influent or effluent flows.

Mass flux due to reaction may take various forms. The following are the most common:

- **Conservative compound.** Compounds with no chemical formation or loss within the control volume are termed **conservative compounds**. Conservative compounds are not affected by chemical or biological reactions, so $(dC/dt)_{\text{reaction only}} = \dot{m}_{\text{reaction}} = 0$. The term *conservative* is used for these compounds because their mass is truly conserved: what goes in equals what goes out.

- **Zero-order decay.** The rate of loss of the compound is constant. For a compound with **zero-order decay**, $(dC/dt)_{\text{reaction only}}$ equals $-k$, and $\dot{m}_{\text{rxn}}$ equals $-Vk$. Zero-order reactions are discussed in Chapter 3.

- **First-order decay.** For a compound with **first-order decay**, the rate of loss of the compound is directly proportional to its concentration: $(dC/dt)_{\text{reaction only}}$ equals $-kC$. For such a compound, $\dot{m}_{\text{rxn}}$ equals $-VkC$. First-order reactions are discussed in Chapter 3.

- **Production at a rate dependent on the concentrations of other compounds in the CMFR.** In this situation, the chemical is produced by reactions involving other compounds in the CMFR, and $(dC/dt)_{\text{reaction only}}$ is greater than zero.

**STEPS IN MASS BALANCE PROBLEMS** Solution of mass balance problems involving CMFRs generally will be straightforward if the problem is done carefully. Most difficulties in solving mass balance problems arise from uncertainty regarding the location of control volume boundaries or values of the individual terms in the mass balance. Therefore, the following steps will assist in solving each mass balance problem:

1. Draw a schematic diagram of the situation, and identify the control volume and all influent and effluent flows. All mass flows that are known or to be calculated must cross the control volume boundaries, and it should be reasonable to assume that the control volume is well mixed.

2. Write the mass balance equation in general form:

$$\frac{dm}{dt} = \dot{m}_{\text{in}} - \dot{m}_{\text{out}} + \dot{m}_{\text{rxn}}$$

3. Determine whether the problem is steady state ($dm/dt = 0$) or nonsteady state ($dm/dt = V \times dC/dt$).

4. Determine whether the compound being balanced is conservative $\dot{m}_{\text{rxn}} = 0$ or nonconservative ($\dot{m}_{\text{rxn}}$ must be determined based on the reaction kinetics and Equation 4.9).

5. Replace $\dot{m}_{\text{in}}$ and $\dot{m}_{\text{out}}$ with known or required values, as just described.

6. Finally, solve the problem. This will require solution of a differential equation in nonsteady-state problems and solution of an algebraic equation in steady-state problems.

## Table / 4.1

**Summary of CMFR Examples**

| Example Number | Form of $dm/dt$ | Form of $\dot{m}_{reaction}$ |
|---|---|---|
| Example 4.2 | Steady state | Conservative |
| Example 4.3 | Steady state | First-order decay |
| Example 4.4 | Nonsteady state | First-order decay |
| Example 4.5 | Nonsteady state | Conservative |

SOURCE: Mihelcic (1999). Reprinted with permission of John Wiley & Sons, Inc.

### 4.1.3 REACTOR ANALYSIS: THE CMFR

**Reactor analysis** refers to the use of mass balances to analyze pollutant concentrations in a control volume that is either a chemical reactor or a natural system modeled as a chemical reactor. Ideal reactors can be divided into two types: completely mixed flow reactors (CMFRs) and plug-flow reactors (PFRs). CMFRs are used to model well-mixed environmental reservoirs. PFRs, described in section 4.1.5, behave essentially like pipes and are used to model situations such as downstream transport in a river in which fluid is not mixed in the upstream–downstream direction.

This section presents several examples involving CMFRs in different combinations of steady-state or nonsteady-state conditions and conservative or nonconservative compounds, as summarized in Table 4.1. Example 4.2 demonstrates the use of CMFR analysis to determine the concentration of a substance resulting from the mixing of two or more influent flows. Examples 4.3 through 4.5 refer to the tank depicted in Figure 4.1 and demonstrate steady-state and nonsteady-state situations with and without first-order chemical decay. Calculations analogous to those in Examples 4.3 through 4.5 can be used to determine the concentration of pollutants exiting a treatment reactor, the rate of increase of pollutant concentrations within a lake resulting from a new pollutant source, or the period required for pollutant levels to decay from a lake or reactor once a source is removed.

**CMFR and PFR Simulators**

www.wiley.com/college/mihelcic

example/4.2 **Steady-State CMFR with Conservative Chemical: The Mixing Problem**

A pipe from a municipal wastewater treatment plant discharges 1.0 m³/s of poorly treated effluent containing 5.0 mg/L of phosphorus compounds (reported as mg P/L) into a river with an upstream flow rate of 25 m³/s and a background phosphorus concentration of 0.010 mg P/L (see Figure 4.2). What is the resulting concentration of phosphorus (in mg/L) in the river just downstream of the plant outflow?

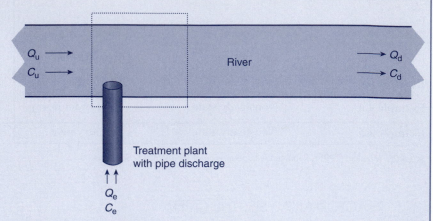

**Figure 4.2  Mixing Problem Used in Example 4.2**  The control volume is indicated by the area inside the dotted lines.

From Mihelcic (1999). Reprinted with permission of John Wiley & Sons, Inc.

## solution

To solve this problem, apply two mass balances: one to determine the downstream volumetric flow rate ($Q_d$) and a second to determine the downstream phosphorus concentration ($C_d$). First, a control volume must be selected. To ensure that the input and output fluxes cross the control volume boundaries, the control volume must cross the river upstream and downstream of the plant's outlet and must also cross the discharge pipe. The selected control volume is shown in Figure 4.2 within dotted lines. It is assumed to extend downriver far enough that the discharged wastewater and the river water become well mixed before leaving the control volume. As long as that assumption is met, it makes no difference to the analysis how far downstream the control volume extends.

Before beginning the analysis, determine whether this is a steady-state or nonsteady-state problem and whether the chemical reaction term will be nonzero. Because the problem statement does not refer to time, and it seems reasonable to assume that both the river and waste stream discharge have been flowing for some time and will continue to flow, this is a steady-state problem. In addition, this problem concerns the concentration resulting from rapid mixing of the river and effluent flows. Therefore, we can define our control volume to be small and can safely assume that chemical or biological degradation is insignificant during the time spent in the control volume, so we treat this as a steady-state problem.

1. Determine the downstream flow rate, $Q_d$. To find $Q_d$, conduct a mass balance on the total river water mass. In this case, the "concentration" of river water in (mass/volume) units is simply the density of the water, $\rho$:

$$\frac{dm}{dt} = \dot{m}_{in} - \dot{m}_{out} + \dot{m}_{rxn}$$

$$= \rho Q_{in} - \rho Q_{out} + 0$$

where the term $\dot{m}_{rxn}$ has been set to zero because the mass of water is conserved. Since this is a steady-state problem, $dm/dt = 0$. Therefore, as long as the density $\rho$ is constant, $Q_{in} = Q_{out}$, and $(Q_u + Q_e) = 26 \, m^3/s = Q_d$.

2. Determine the phosphorus concentration downstream of the discharge pipe, $C_d$. To find $C_d$, use the standard mass balance equation with steady-state conditions and with no chemical formation or decay:

$$\frac{dm}{dt} = \dot{m}_{in} - \dot{m}_{out} + \dot{m}_{rxn}$$

$$0 = (C_u Q_u + C_e Q_e) - C_d Q_d + 0$$

Solve for $C_d$:

$$C_d = \frac{C_u Q_u + C_e Q_e}{Q_d}$$

$$= \frac{(0.010 \, mg/L)(25 \, m^3/s) + (5.0 \, mg/L)(1.0 \, m^3/s)}{26 \, m^3/s}$$

$$= 0.20 \, mg/L$$

---

example/4.3 Steady-State CMFR with First-Order Decay

The CMFR shown in Figure 4.1 is used to treat an industrial waste, using a reaction that destroys the pollutant according to first-order kinetics, with $k = 0.216/day$. The reactor volume is 500 $m^3$, the volumetric flow rate of the single inlet and exit is 50 $m^3/day$, and the inlet pollutant concentration is 100 mg/L. What is the outlet concentration after treatment?

solution

An obvious control volume is the tank itself. The problem requests a single, constant outlet concentration, and all problem conditions are constant. Therefore, this is a steady-state problem ($dm/dt = 0$). The mass balance equation with a first-order decay term ($[dC/dt]_{reaction \; only} = -kC$ and $\dot{m}_{rxn} = -VkC$) is

## example/4.3 Continued

$$\frac{dm}{dt} = \dot{m}_{in} - \dot{m}_{out} + \dot{m}_{rxn}$$

$$0 = QC_{in} - QC - VkC$$

Solve for $C$:

$$C = C_{in} \times \frac{Q}{Q + kV}$$

$$= C_{in} \times \frac{1}{1 + \left(k \times \dfrac{V}{Q}\right)}$$

Substituting the given values, the numerical solution is

$$C = 100 \text{ mg/L} \times \frac{50 \text{ m}^3/\text{day}}{50 \text{ m}^3/\text{d} + (0.216/\text{day})(500 \text{ m}^3)}$$

$$= 32 \text{ mg/L}$$

## example/4.4 Nonsteady-State CMFR with First-Order Decay

The manufacturing process that generates the waste in Example 4.3 has to be shut down, and starting at $t = 0$, the concentration $C_{in}$ entering the CMFR is set to 0. What is the outlet concentration as a function of time after the concentration is set to 0? How long does it take the tank concentration to reach 10 percent of its initial, steady-state value?

## solution

The tank is again the control volume. In this case, the problem is clearly nonsteady-state, because conditions change as a function of time. The mass balance equation is

$$\frac{dm}{dt} = \dot{m}_{in} - \dot{m}_{out} + \dot{m}_{rxn}$$

$$V\frac{dC}{dt} = 0 - QC - kCV$$

Solve for $dC/dt$:

$$\frac{dC}{dt} = -\left(\frac{Q}{V} + k\right)C$$

To determine $C$ as a function of time, the preceding differential equation must be solved. Rearrange and integrate:

$$\int_{C_0}^{C_t} \frac{dC}{dt} = \int_0^t -\left(\frac{Q}{V} + k\right) dt$$

Integration yields

$$\ln C - \ln C_0 = -\left(\frac{Q}{V} + k\right) t$$

Because $\ln x - \ln y$ is equal to $\ln(x/y)$, we can rewrite this equation as

$$\ln\left(\frac{C}{C_0}\right) = -\left(\frac{Q}{V} + k\right) t$$

which yields

$$\frac{C_t}{C_0} = e^{-(Q/V + k)t}$$

We can verify that this solution is reasonable by considering what happens at $t = 0$ and $t = \infty$. At $t = 0$, the exponential term is equal to 1, and $C = C_0$, as expected. As $t \rightarrow \infty$, the exponential term approaches 0, and concentration declines to 0—again as expected—since $C_{in}$ is equal to 0.

We can now plug in values to determine the dependence of $C$ on time. Example 4.3 provides $Q$ and $V$. The initial concentration is equal to the concentration before $C_{in}$ was set to 0, which was found to be 32 mg/L in Example 4.3. Plugging in these values yields the outlet concentration as a function of time:

$$C_t = 32\,\text{mg/L} \times \exp\left[-\left(\frac{50\,\text{m}^3/\text{day}}{500\,\text{m}^3} + \frac{0.216}{\text{day}}\right) t\right]$$

$$= 32\,\text{mg/L} \times \exp\left(-\frac{0.316}{\text{day}} t\right)$$

This solution is plotted in Figure 4.3a.

How long will it take the concentration to reach 10 percent of its initial, steady-state value? That is, at what value of $t$ is $C_t/C_0 = 0.10$? At the time when $C_t/C_0 = 0.10$,

$$\frac{C}{C_0} = 0.10 = \exp\left(-\frac{0.316}{\text{day}} t\right)$$

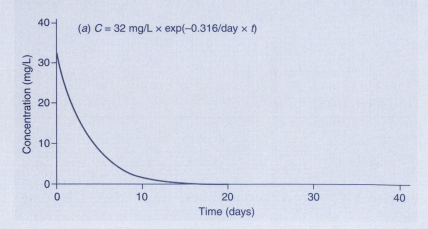

(a)

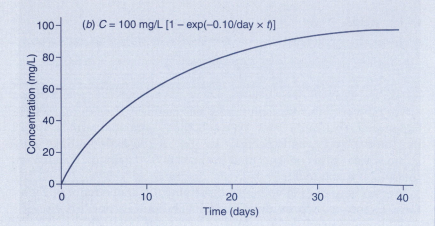

(b)

**Figure 4.3** **Concentration versus Time Profiles for the Solutions to Examples 4.4 and 4.5** (a) First-order decay in concentration resulting from the removal of min at time zero. The decay in concentration results from the sum of chemical reaction loss ($\dot{m}_{rxn}$) and the mass flux out term ($\dot{m}_{out}$). (b) Exponential approach to steady-state conditions when a reactor is started with initial concentration equal to zero. In the absence of a chemical reaction loss term, concentration in the reactor exponentially approaches the inlet concentration.

From Mihelcic (1999). Reprinted with permission of John Wiley & Sons, Inc.

Taking the natural logarithm of both sides,

$$\ln 0.10 = -2.303 = -\frac{0.316}{\text{day}}t$$

Therefore, $t = 7.3$ days.

# example/4.5 Nonsteady-State CMFR, Conservative Substance

The CMFR reactor depicted in Figure 4.1 is filled with clean water prior to being started. After start-up, a waste stream containing 100 mg/L of a conservative pollutant is added to the reactor at a flow rate of 50 m³/day. The volume of the reactor is 500 m³. What is the concentration exiting the reactor as a function of time after it is started?

## solution

Again, the tank will serve as a control volume. We are told that the pollutant is conservative, so $\dot{m}_{rxn} = \phi$. The problem asks for concentration as a function of time, so the mass balance must be nonsteady-state. The mass balance equation is

$$\frac{dm}{dt} = \dot{m}_{in} - \dot{m}_{out} + \dot{m}_{rxn}$$

$$V\frac{dC}{dt} = QC_{in} - QC + 0$$

Solve for $dC/dt$:

$$\frac{dC}{dt} = -\left(\frac{Q}{V}\right)(C - C_{in})$$

Because of the extra term on the right ($C_{in}$), this equation cannot be immediately solved. However, with a change of variables, we can transform the mass balance equation into a simpler form that can be integrated directly, using the same method as in Example 4.4. Let $y = (C - C_{in})$. Then $dy/dt = (dC/dt) - d(C_{in}/dt)$. Since $C_{in}$ is constant, $dC_{in}/dt = 0$, so $dy/dt = dC/dt$. Therefore, the last of the preceding equations is equivalent to

$$\frac{dy}{dt} = -\frac{Q}{V}y$$

Rearrange and integrate:

$$\int_{y(0)}^{y(t)}\frac{dy}{y} = \int_{0}^{t}-\frac{Q}{V}dt$$

Integration yields:

$$\ln\left(\frac{y(t)}{y(0)}\right) = -\frac{Q}{V}t$$

or

$$\frac{y(t)}{y(0)} = e^{-(Q/V)t}$$

Replacing $y$ with $(C - C_{in})$ results in the following equation:

$$\frac{C - C_{in}}{C_0 - C_{in}} = e^{-(Q/V)t}$$

Since clean water is present in the tank at start-up, $C_0 = 0$:

$$\frac{C - C_{in}}{-C_{in}} = e^{-(Q/V)t}$$

Rearrange to solve for $C$:

$$C - C_{in} = -C_{in}e^{-(Q/V)t}$$

$$C = C_{in} \times \left(1 - e^{-(Q/V)t}\right)$$

This is the solution to the question posed in the problem statement.
Note what happens as $t \rightarrow \infty$: $e^{-(Q/V)t} \rightarrow 0$, and $C \rightarrow C_{in}$. This is not surprising, since the substance is conservative. If the reactor is run long enough, the concentration in the reactor will eventually reach the inlet concentration. This final equation (plotted in Figure 4.3b) provides $C$ as a function of time. This can be used to determine how long it would take for the concentration to reach, say, 90 percent of the inlet value.

### 4.1.4 BATCH REACTOR

A reactor that has no inlet or outlet flows is termed a **batch reactor**. It is essentially a tank in which a reaction is allowed to occur. After one batch is treated, the reactor is emptied, and a second batch can be treated. Because there are no flows, $\dot{m}_{in} = 0$, and $\dot{m}_{out} = 0$. Therefore, the mass balance equation reduces to

$$\frac{dm}{dt} = \dot{m}_{rxn} \tag{4.10}$$

or

$$V\frac{dC}{dt} = V\left(\frac{dC}{dt}\right)_{reaction\ only} \tag{4.11}$$

Simplifying:

$$\frac{dC}{dt} = \left(\frac{dC}{dt}\right)_{reaction\ only} \tag{4.12}$$

Thus, in a batch reactor, the change in concentration with time is simply that which results from the chemical reaction. For example, for a first-order decay reaction, $r = -kC$. Thus,

$$\frac{dC}{dt} = -kC \tag{4.13}$$

or

$$\frac{C_t}{C_0} = e^{-kt} \qquad (4.14)$$

## 4.1.5 PLUG-FLOW REACTOR

The **plug-flow reactor (PFR)** is used to model the chemical transformation of compounds as they are transported in systems resembling pipes. A schematic diagram of a PFR is shown in Figure 4.4. PFR pipes may represent a river, a region between two mountain ranges through which air flows, or a variety of other engineered or natural conduits through which liquids or gases flow. Of course, a pipe in this model can even represent a pipe. Figure 4.5 illustrates examples of a PFR in an engineered system (Figure 4.5a) and a PFR in a natural system (Figure 4.5b).

As fluid flows down the PFR, the fluid is mixed in the radial direction, but mixing does not occur in the axial direction. That is, each plug of fluid is considered a separate entity as it flows down the pipe. However, time passes as the plug of fluid moves downstream (or downwind). Thus, there is an implicit time dependence, even in steady-state PFR problems. However, because the velocity of the fluid ($v$) in the PFR is constant, time and downstream distance ($x$) are interchangeable, and

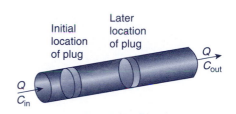

**Figure 4.4** **Schematic Diagram of a Plug-Flow Reactor**

From Mihelcic (1999). Reprinted with permission of John Wiley & Sons, Inc.

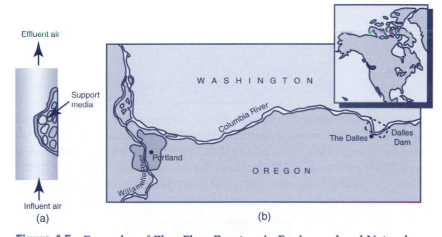

(a)

(b)

**Figure 4.5** **Examples of Plug-Flow Reactors in Engineered and Natural Systems** (a) Packed-tower biofilters are used to remove odorous air emissions, such as hydrogen sulfide ($H_2S$), from gas-phase emissions. Biofilters consist of a column packed with a support medium, such as rocks, plastic rings, or activated carbon, on which a biofilm is grown. Contaminated water or air is passed through the filter, and bacterial degradation results in the desired reduction of pollutant emissions. (b) The Columbia River flows 1,200 mi. from its source in Canada to the Pacific Ocean. Before reaching the ocean, the Columbia River flows southward into the United States and forms the border between Oregon and Washington. Shown is a stretch of the river near The Dalles, Washington, where the river once narrowed and spilled over a series of rapids, christened *les Dalles*, or the trough, by early French explorers. A large dam has since been constructed near The Dalles. The section of the river downstream of the dam could be modeled as a plug-flow reactor.

From Mihelcic (1999). Reprinted with permission of John Wiley & Sons, Inc.

$t = x/v$. That is, a fluid plug always takes an amount of time equal to $x/v$ to travel a distance $x$ down the reactor. This observation can be used with the mass balance formulations just given to determine how chemical concentrations vary during flow through a PFR.

To develop the equation governing concentration as a function of distance down a PFR, we will analyze the evolution of concentration with time within a single fluid plug. The plug is assumed to be well mixed in the radial direction but does not mix at all with the fluid ahead or behind it. As the plug flows downstream, chemical decay occurs, and concentration decreases. The mass balance for mass within this moving plug is the same as that for a batch reactor:

$$\frac{dm}{dt} = \dot{m}_{in} - \dot{m}_{out} + \dot{m}_{rxn} \tag{4.15}$$

$$V\frac{dC}{dt} = 0 - 0 + V\left(\frac{dC}{dt}\right)_{reaction\ only} \tag{4.16}$$

where $\dot{m}_{in}$ and $\dot{m}_{out}$ are set equal to zero because there is no mass exchange across the plug boundaries.

Equation 4.16 can be used to determine concentration as a function of flow time within the PFR for any reaction kinetics. In the case of first-order decay, $V(dC/dt)_{reaction\ only} = -VkC$, and

$$V\frac{dC}{dt} = -VkC \tag{4.17}$$

which results in:

$$\frac{C_t}{C_0} = \exp(-kt) \tag{4.18}$$

It is generally desirable to express the concentration at the outlet of PFR in terms of the inlet concentration and PFR length or volume, rather than time spent in the PFR. In a PFR of length $L$, each plug travels for a period $\theta = L/v = L \times A/Q$, where $A$ is the cross-sectional area of the PFR and $Q$ is the flow rate. The product of length and cross-sectional area is simply the PFR volume, so Equation 4.18 is equivalent to

$$\boxed{\frac{C_{out}}{C_{in}} = \exp\left(-\frac{kV}{Q}\right)} \tag{4.19}$$

Equation 4.19 has no time dependence. Although concentration within a given plug changes over time as that plug flows downstream, the concentration at a given fixed location within the PFR is constant with respect to time, since all plugs reaching that location have spent an identical period in the PFR.

COMPARISON OF THE PFR TO THE CMFR The ideal CMFR and the PFR are fundamentally different and thus behave differently. When a parcel of fluid enters the CMFR, it is immediately mixed throughout the entire volume of the CMFR. In contrast, each parcel of fluid entering the PFR remains separate during its passage through the reactor. To

**Mississippi River Basin**
http://www.epa.gov/msbasin/

**Gulf of Mexico Program**
http://www.epa.gov/gmpo/

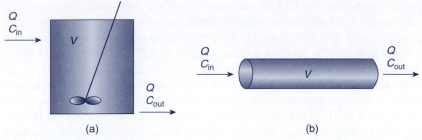

(a)            (b)

**Figure 4.6** Comparison of (a) Completely Mixed Flow Reactor and (b) Plug-Flow Reactor

From Mihelcic (1999). Reprinted with permission of John Wiley & Sons, Inc.

highlight these differences, consider an example involving the continuous addition of a pollutant to each reactor, with destruction of the pollutant within the reactor according to first-order kinetics. The two reactors are depicted in Figure 4.6. This example assumes the incoming concentration ($C_{in}$), the flow rate ($Q$), and the first-order reaction rate constant ($k$) are known and are the same for both reactors. Consider two common problems:

1. If the volume $V$ is known (the same for both reactors), what is the resulting outlet concentration ($C_{out}$) exiting the CMFR and PFR?

2. If an outlet concentration is specified, what volume of reactor is required for the CMFR and for the PFR? Table 4.2 summarizes the results of this comparison and lists the input variables.

The results shown in Table 4.2 indicate that, for equal reactor volumes, the PFR is more efficient than the CMFR and, for equal outlet concentrations, a smaller PFR is required. Why is this? The answer has to do with the fundamental difference between the two reactors—fluid parcels entering the PFR travel downstream without mixing, while fluid

## Table / 4.2

**Comparison of CMFR and PFR Performance***

Example 1. Determine $C_{out}$, given $V = 100$ L, $Q = 5.0$ L/s, $k = 0.05$/s.

| CMFR | PFR |
|---|---|
| $C_{out} = C_{in}/(1 + kV/Q)$ | $C_{out} = C_{in}\exp(-kV/Q)$ |
| $C_{out}/C_{in} = 0.50$ | $C_{out}/C_{in} = 0.37$ |

Example 2. Determine $V$, given $C_{out}/C_{in} = 0.5$, $Q = 5.0$ L/s, $k = 0.05$/s.

| CMFR | PFR |
|---|---|
| $V = (C_{in}/C_{out} - 1) \times (Q/k)$ | $V = -(Q/k)\ln(C_{out}/C_{in})$ |
| $V = 100$ L | $V = 69$ L |

*Example 1 compares the effluent concentration ($C_{out}$) for a PFR and CMFR of the same volume; Example 2 compares the volume required for each reactor type if a given percent removal is required.

SOURCE: Mihelcic (1999). Reprinted with permission of John Wiley & Sons, Inc.

parcels entering the CMFR are immediately mixed with the low-concentration fluid within the reactor. Since the rate of chemical reaction is proportional to concentration, the rate of chemical reaction within the CMFR is reduced, relative to that within the PFR. This effect is illustrated in Figure 4.7. The mass flux due to reaction is equal to $-VkC$ in both reactors. However, in the PFR, concentration decreases exponentially as each plug passes through the PFR, as shown by the solid curve in Figure 4.7. The average mass flux due to reaction in the PFR is simply the average value of this curve—the value indicated by the dashed line in Figure 4.7. In contrast, dilution as the incoming fluid is mixed into the CMFR immediately reduces the influent concentration to that within the CMFR, resulting in a reduced rate of destruction, indicated by the dotted line in Figure 4.7.

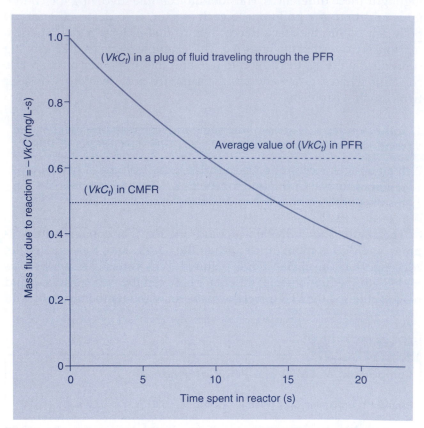

**Figure 4.7  Origin of the Higher Destruction Efficiency of a PFR under Conditions of First-Order Decay**   The rate of chemical destruction ($m_{rxn} = -VkC$) is shown as a function of time spent in the reactor for a PFR (solid line) and CMFR (dotted line) for the conditions given in Example 1 of Table 4.2. Since concentration changes as each plug passes through the PFR ($C = C_{in} \exp[-k\Delta t]$), the value of $m_{rxn}$ changes. The average rate of destruction in the PFR is shown by a dashed line. The rate of chemical destruction is constant throughout the well-mixed CMFR and is equal to $-VkC$. Since the high inlet concentration is diluted immediately on entering the CMFR, the rate of reaction is lower than that throughout most of the PFR and is lower than the average rate of reaction in the PFR.

From Mihelcic (1999). Reprinted with permission of John Wiley & Sons, Inc.

**Response to Inlet Spikes.** CMFRs and PFRs also differ in their response to spikes in the inlet concentration. In many pollution-control systems, inlet concentrations or flows are not constant. For example, flow into municipal wastewater treatment plants varies dramatically over the course of each day. It often is necessary to ensure that a temporary increase in inlet concentration does not result in excessive outlet concentrations. And as will be seen in Chapter 8, low-impact development technology such as bioretention cells are designed to first store storm water from a developed urban area and then release it back into the environment at a slow rate, reducing spikes in inlet concentrations and reducing the chance of overburdening wastewater treatment facilities.

When source reduction techniques are not in place, reducing or eliminating spikes in outlet concentration requires the use of CMFRs, as a result of the mixing that occurs within CMFRs but not within PFRs. Consider the effect of a temporary doubling of the concentration entering a PFR and CMFR: each is designed to reduce the influent concentration by the same amount with the flow, first-order decay rate constant, and required degree of destruction equal to the values given in Example 2 of Table 4.2.

The resulting changes in outlet concentration for the PFR and CMFR are shown in Figure 4.8. The concentration in fluid exiting the CMFR begins to rise immediately after the inlet concentration increases, as the more concentrated flow is mixed throughout the CMFR. The outlet concentration does not immediately double in response to the doubled inlet concentration, however, because the higher-concentration influent flow is diluted by the volume of low-concentration fluid within the CMFR. The CMFR outlet concentration rises exponentially and would eventually double, but the inlet spike does not last long enough for this to occur. In contrast, the outlet concentration exiting the PFR does not change until enough time has passed for the first plug of higher-concentration fluid to traverse the length of the PFR. At that time, the outlet concentration doubles, and it remains elevated for a period equal to the duration of the inlet spike.

**Selection of CMFR or PFR.** Selection of a CMFR or PFR in an engineered system is based on the considerations just described: control efficiency as a function of reactor size and response to changing inlet conditions. In many cases, the optimal choice is to use a CMFR to reduce sensitivity to spikes, followed by a PFR for efficient use of resources. Deciding between a CMFR and PFR has other environmental implications. If one reactor design is found to be more efficient than the other for a given set of operating conditions, using the more efficient design can cut energy requirements, waste production, and use of operating materials.

In natural systems, the choice is based on whether or not the system is mixed (in which case a CMFR would be used to model the system) or flows downstream without mixing (requiring use of a PFR). In some cases, it is necessary to use both the CMFR and PFR models. A common example of this involves effluent flow into a river. A CMFR is used to define a mixing problem, as was done in Example 4.2. This sets

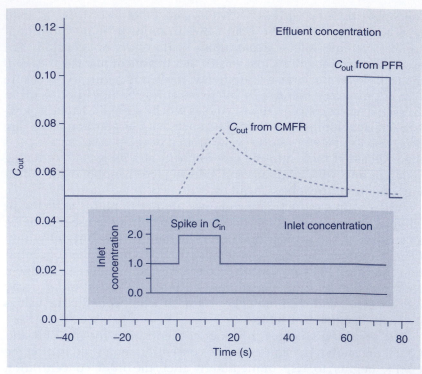

**Figure 4.8** **Response of a CMFR and PFR to a Temporary Increase in Inlet Concentration** The influent concentration, shown in the lower, inset figure, increases to 2.0 during the period $t = 0$–15 s. The resulting concentrations exiting the CMFR and PFR of Example 2 in Table 4.2 are shown as a function of time before, during, and after the temporary doubling of inlet concentration. The concentration exiting the CMFR is shown with a dashed line; the concentration exiting the PFR is shown with a solid line. The maximum concentration reached in the CMFR effluent is less than that reached in the PFR effluent because the increased inlet concentration is diluted by the volume of lower-concentration fluid within the CMFR.

From Mihelcic (1999). Reprinted with permission of John Wiley & Sons, Inc.

## example/4.6 Required Volume for a PFR

Determine the volume required for a PFR to obtain the same degree of pollutant reduction as the CMFR in Example 4.3. Assume that the flow rate and first-order decay rate constant are unchanged ($Q = 50 \, \text{m}^3/\text{day}$, and $k = 0.216/\text{day}$).

## solution

The CMFR in Example 4.3 achieved a pollutant decrease of $C_{out}/C_{in} = 32/100 = 0.32$. From Equation 4.19,

$$\frac{C_{out}}{C_{in}} = e^{-(kV/Q)}$$

example/4.6 Continued

or:

$$0.32 = \exp - \left( \frac{0.216/\text{day} \times V}{50\,\text{m}^3/\text{day}} \right)$$

Solve for $V$:

$$V = \ln 0.32 \times \frac{50\,\text{m}^3/\text{day}}{-0.216/\text{day}}$$

$$= 264\,\text{m}^3$$

As expected, this volume is smaller than the 500 m³ required for the CMFR in Example 4.3.

the inlet concentration for a PFR, which is used to model degradation of the pollutant as it flows further downstream. (This type of problem is investigated in Chapter 8 for dissolved oxygen in rivers.)

### 4.1.6  RETENTION TIME AND OTHER EXPRESSIONS FOR V/Q

A number of terms—including **retention time**, *detention time*, and *residence time*—refer to the average period spent in a given control volume, $\theta$. The retention time is given by

$$\theta = \frac{V}{Q} \tag{4.20}$$

where $V$ is the volume of the reactor and $Q$ is the total volumetric flow rate exiting the reactor. Examples 4.7 and 4.8 illustrate the calculation and application of retention time.

example/4.7 Retention Time in a CMFR and a PFR

Calculate the retention times in the CMFR of Example 4.3 and the PFR of Example 4.6.

solution

For the CMFR,

$$\theta = \frac{V}{Q} = \frac{500\,\text{m}^3}{50\,\text{m}^3/\text{day}} = 10\,\text{days}$$

For the PFR,

$$\theta = \frac{V}{Q} = \frac{264\,\text{m}^3}{50\,\text{m}^3/\text{day}} = 5.3\,\text{days}$$

## example/4.8 Retention Times for the Great Lakes

The Great Lakes region is shown in Figure 4.9. Calculate the retention times for Lake Michigan and Lake Ontario, using the data provided in Table 4.3.

## solution

For Lake Michigan,

$$\theta = \frac{4{,}900 \times 10^9\,\mathrm{m}^3}{36 \times 10^9\,\mathrm{m}^3/\mathrm{yr}} = 136\,\mathrm{yr}$$

### Table / 4.3

**Volume and Flows for the Great Lakes**

| Lake | Volume $10^9$ m$^3$ | Outflow $10^9$ m$^3$/yr |
|------|------|------|
| Superior | 12,000 | 67 |
| Michigan | 4,900 | 36 |
| Huron | 3,500 | 161 |
| Erie | 468 | 182 |
| Ontario | 1,634 | 211 |

SOURCE: Mihelcic (1999). Reprinted with permission of John Wiley & Sons, Inc.

For Lake Ontario,

$$\theta = \frac{1{,}634 \times 10^9\,\mathrm{m}^3}{212 \times 10^9\,\mathrm{m}^3/\mathrm{yr}} = 8\,\mathrm{yr}$$

These values mean that Lake Michigan changes its water volume completely once every 136 years and Lake Ontario once every 8 years. The higher flow and smaller volume of Lake Ontario result in a significantly shorter retention time. This means pollutant concentrations can increase in Lake Ontario much more quickly than they can in Lake Michigan and will drop much more quickly in Lake Ontario if a pollutant source is eliminated—provided that flow out of the lakes is the dominant pollutant sink.

These values of $\theta$ can be used to determine whether it would be appropriate to model the lakes as CMFRs in a mass balance problem. Temperate lakes generally are mixed twice per year. Therefore, over the period required for water to flush through Lakes Michigan and Ontario, the lakes would be mixed many times. It would therefore be appropriate to model the lakes as CMFRs in mass balances involving pollutants that do not decay significantly in less than approximately 1 yr.

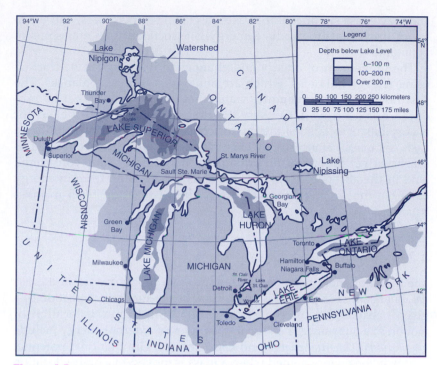

**Figure 4.9  The North American Great Lakes** The Great Lakes are an important part of the physical and cultural heritage of North America. The Great Lakes contain approximately 18 percent of the world's supply of freshwater, making them the largest system of available surface freshwater (only the polar ice caps contain more freshwater). The first humans arrived in the area approximately 10,000 years ago. Around 6,000 years ago, copper mining began along the south shore of Lake Superior, and hunting/fishing communities were established throughout the area. Population in the region in the 16th century is estimated between 60,000 and 117,000—a level that resulted in few human disturbances. Today, the combined Canadian and U.S. population in the region exceeds 33 million. Increases in human settlement and exploitation over the past 200 years have caused many disturbances to the ecosystem. Today, the outflow from the Great Lakes is less than 1 percent per year. Therefore, pollutants that enter the lakes by air, direct discharge, or from nonpoint pollution sources may remain in the system for a long period of time.

From Mihelcic (1999). Reprinted with permission of John Wiley & Sons, Inc.

## 4.2  Energy Balances

Modern society is dependent on the use of energy. Such use requires transformations in the form of energy and control of energy flows. For example, when coal is burned at a power plant, the chemical energy present in the coal is converted to heat, which is then converted in the plant's generators to electrical energy. Eventually, the electrical energy is converted back into heat for warmth or used to do work. However, energy flows and transformation can also cause environmental problems. For example, thermal heat energy from electrical power plants can result in increased temperatures in rivers used for cooling water, greenhouse pollutants in the atmosphere alter Earth's energy balance and may cause significant increases in global temperatures, and burning of fossil fuels to produce energy is associated with emissions of pollutants.

The movement of energy and changes in its form can be tracked using **energy balances**, which are analogous to mass balances. The **first law of thermodynamics** states that energy can neither be produced nor destroyed. Conservation of energy provides a basis for energy balances, just as the law of conservation of mass provides a basis for mass balances. However, all energy balances are treated as conservative; as long as all possible forms of energy are considered (and in the absence of nuclear reactions), there is no term in energy balances that is analogous to the chemical-reaction term in mass balances.

### 4.2.1 FORMS OF ENERGY

The forms of energy can be divided into two types: *internal* and *external*. Energy that is part of the molecular structure or organization of a given substance is internal. Energy resulting from the location or motion of a substance is external. Examples of external energy include *gravitational potential energy* and *kinetic energy*. Gravitational potential energy is the energy gained when a mass is moved to a higher location above the Earth. Kinetic energy is the energy that results from the movement of objects. When a rock thrown off a cliff accelerates toward the ground, the sum of kinetic and potential energy is conserved (neglecting friction); as the rock falls, it loses potential energy but increases in speed, gaining kinetic energy. Table 4.4 shows the mathematical representations of common forms of energy encountered in environmental engineering.

**Heat** is a form of internal energy—it results from the random motions of atoms. Heat is thus really a form of kinetic energy, although it is considered separately because the motion of the atoms cannot be seen. When a pot of water is heated, energy is added to the water. That

**Energy and the Environment**
http://www.epa.gov/energy/

### Table / 4.4

**Some Common Forms of Energy**

|  | Representation for Energy or Change in Energy |
| --- | --- |
| Heat internal energy | $\Delta E = \text{mass} \times c \times \Delta T$ |
| Chemical internal energy | $\Delta E = \Delta H_{rxn}$ at constant volume |
| Gravitational potential | $\Delta E = \text{mass} \times \Delta \text{ height}$ |
| Kinetic energy | $E = \dfrac{\text{mass} \times (\text{velocity})^2}{2}$ |
| Electromagnetic energy | $E = \text{Planck's constant} \times \text{photon frequency}$ |

SOURCE: Mihelcic (1999). Reprinted with permission of John Wiley & Sons, Inc.

energy is stored in the form of internal energy, and the change in internal energy of the water is expressed as follows:

$$\text{change in internal energy} = (\text{mass of } H_2O) \times c \times \Delta T \quad \textbf{(4.21)}$$

where $c$ is the heat capacity or specific heat of the water, with units of energy/mass-temperature. Heat capacity is a property of a given material. For water, the heat capacity is 4,184 J/kg-C (1 Btu/lb.-F).

---

### Box / 4.1    Another Energy Classification System: Renewable and Nonrenewable

Energy resources can be described as renewable and nonrenewable. **Renewable energy** sources can be replaced at a rate equal to or faster than the rate at which they are used (the source is continuously available). The sun, wind, and tides are examples of renewable energy feedstocks. **Nonrenewable energy** sources, on the other hand, are consumed faster than they can be replenished. Fossil-based feedstocks are considered nonrenewable because they cannot be replenished as fast as they are consumed (the source is finite). Each type of energy source, renewable and nonrenewable, falls within the categories listed in Table 4.4. For example, the energy contained in fossil fuels is present in the form of chemical internal energy, wind power comes from kinetic energy, and solar power uses electromagnetic energy.

---

*Chemical internal energy* reflects the energy in the chemical bonds of a substance. This form of energy is composed of two parts:

1. **The strengths of the atomic bonds in the substance.** When chemical reactions occur, if the sum of the internal energies of the products is less than that for the reactants, a reduction in chemical internal energy has occurred. As a result of the conservation of energy, this leftover energy must show up in a different form. Usually, the energy is released as heat. The most common example of this is the combustion of fuel, in which hydrocarbons and oxygen react to form carbon dioxide and water. The chemical bonds in carbon dioxide and water are much lower in energy than those in hydrocarbons, so combustion releases a significant amount of heat.

2. **The energy in the interactions between molecules.** Solids and liquids form as a result of interactions between adjacent molecules. These bonds are much weaker than the chemical bonds between atoms in molecules, but are still important in many energy balances. The energy required to break these bonds is referred to as *latent heat*. Values of latent heat are tabulated for various substances for the phase changes from solid to liquid and from liquid to gas. The latent heat of condensation for a given substance is equal to the heat released when a unit of mass of the substance condenses to form a liquid. (An equal amount of energy is required for evaporation.) The latent heat of fusion is equal to the heat released when a unit of mass solidifies. (Again, an equal amount of energy is required to melt the substance.)

**The Debate over Nuclear Power**
Con: http://www.nrdc.org/nuclear/plants/contents.asp
Pro: http://www.energy.gov/sciencetech/climatechange.htm

**Class Discussion**
Some advocate nuclear power as a source of energy to replace fossil fuels. Others see it as a security risk and having generational problems related to storing waste. Using a definition of sustainable development, does nuclear power have a role in our transformation towards a sustainable future?

## 4.2.2 CONDUCTING AN ENERGY BALANCE

In analogy with the mass balance equation (Equation 4.3), the following equation can be used to conduct energy balances:

$$\begin{pmatrix} \text{change in internal} \\ \text{plus external energy} \\ \text{per unit time} \end{pmatrix} = (\text{energy flux in}) - (\text{energy flux out})$$

or

$$\frac{dE}{dt} = \dot{E}_{\text{in}} - \dot{E}_{\text{out}} \tag{4.22}$$

The use of this relationship is illustrated in Examples 4.9 and 4.10. This same approach for calculating heat balances can be used to investigate the energy efficiency of different products, processes, and systems. In Chapter 14, a heat balance will be used for decision making about the energy efficiency of a building.

---

example/4.9 Heating Water: Scenario 1

A 40 gal. electric water heater heats tap water to a temperature of 10°C. The heating level is set to the maximum while several people take consecutive showers. If, at the maximum heating level, the heater uses 5 kW of electricity and the water use rate is a continuous 2 gal./min., what is the temperature of the water exiting the heater? Assume that the system is at steady state and the heater is 100 percent efficient; that is, it is perfectly insulated, and all of the energy used heats the water.

solution

The control volume is the water heater. Because the system is at steady state, $dE/dt$ is equal to zero. The energy flux added by the electric heater heats water entering the water heater to the temperature at the outlet. The energy balance is thus

$$\frac{dE}{dt} = 0 = \dot{E}_{\text{in}} - \dot{E}_{\text{out}}$$

The energy flux into the water heater comes from two sources: the heat content of the water entering the heater and the electrical heating element. The heat content of the water entering the heater is the product of the water-mass flux, the heat capacity, and the inlet temperature. The energy added by the heater is given as 5 kW.

The energy flux out of the water heater is just the internal energy of the water leaving the system $\left( \dot{m}_{\text{H}_2\text{O}} \times c \times T_{\text{out}} \right)$. There is no net

---

## example/4.9 Continued

conversion of other forms of energy. Therefore, the energy balance may be rewritten as follows:

$$0 = (\dot{m}_{H_2O}cT_{in} + 5\text{ kW}) - \dot{m}_{H_2O}cT_{out}$$

Each term of this equation is an *energy flux* and has the units of energy/time. To solve, place each term in the same units—in this case, watts (1 W equals 1 J/s, and 1,000 W = 1 kW). In addition, the water flow rate (gal/min) needs to be converted to units of mass of water per unit time using the density of water. Combining the first and third terms,

$$0 = \dot{m}_{H_2O}c(T_{out} - T_{in}) + 5\text{ kW} \qquad \textbf{(4.23)}$$

$$0 = \frac{2\text{ gal. }H_2O}{min} \times \frac{3.785\text{ L}}{gal} \times \frac{1.0\text{ kg}}{L} \times \frac{4,184\text{ J}}{kg \times °C} \times (T_{in} - T_{out}) + \frac{5,000\text{ J}}{s} \times \frac{60\text{ s}}{min}$$

$$= 3.16 \times 10^4 \frac{J}{min \times °C} \times (T_{in} - T_{out}) + 3.00 \times 10^5 \frac{J}{min}$$

Solve for $T_{out}$:

$$T_{out} = T_{in} + 9.5°C = (10 + 9.5) = 19.5°C$$

This is a cold shower! But it makes sense; many people have taken such a cold shower after the hot water in the tank was used up by previous showers.

## example/4.10 Heating Water: Scenario 2

Example 4.9 showed that it is necessary to wait until the water in the tank is reheated (hopefully, by passive solar energy!) before taking a hot shower. How long would it take the temperature to reach 54°C if no hot water were used during the heating period and the water temperature started at 20°C?

## solution

In this case, assuming the consumer is not taking advantage of solar energy, the only energy input is the electrical heat, and no energy is leaving the tank. Therefore, the rate of increase in internal energy is equal to the rate at which electrical energy is used:

$$\frac{dE}{dt} = \dot{E}_{in} - \dot{E}_{out} = \dot{E}_{in} - 0$$

From Table 4.4, $\Delta E = \text{mass} \times c \times \Delta T$, so we can express the relationship as follows:

$$\frac{dE}{dt} = \frac{(\text{mass of } H_2O) \times c \times \Delta T}{\Delta t}$$

and:

$$\frac{(\text{mass of } H_2O) \times c \times \Delta T}{\Delta t} = \dot{E}_{in} = 5{,}000 \text{ J/s}$$

This expression can be solved for the change in time, $\Delta t$, given that $\Delta T$ is equal to $54 - 20 = 34°C$:

$$\Delta t = \frac{(\text{mass of } H_2O) \times c \times \Delta T}{5{,}000 \text{ J/s}}$$

$$= \frac{\left(40 \text{ gal } H_2O \times \dfrac{3.785 \text{ L}}{\text{gal}} \times \dfrac{1.0 \text{ kg}}{\text{L}}\right)\left(4{,}184 \dfrac{\text{J}}{\text{kg} \times °C}\right)(54 - 20°C)}{5{,}000 \text{ J/s}}$$

$$= 4.3 \times 10^3 \text{ s} = 1.2 \text{ h}$$

M. Eric Honeycutt/iStockphoto.

Although we have neglected heat loss in Examples 4.9 and 4.10, in the real world heat loss often significantly affects the energy efficiency of processes, systems, and products. For example, heat loss through a poorly insulated hot-water heater tank requires additional energy to maintain the water at the desired temperature and can be a significant fraction of the total energy required. Similarly, windows and poor insulation in a house often act as conduits for heat loss, increasing the amount of energy required to maintain a comfortable temperature.

The additional energy use due to largely avoidable heat loss is significant: the energy used to offset unwanted heat losses and gains through windows in residential and commercial buildings costs the United States tens of billions of dollars every year. However, when properly selected and installed, windows can help minimize a home's heating, cooling, and lighting costs. This notion is further explored in Chapter 14 through a discussion of resistance values (R-values), heat balance, and energy efficiency.

## example/4.11 Thermal Pollution from Power Plants

The second law of thermodynamics states that heat energy cannot be converted to work with 100 percent efficiency. As a result, a significant fraction of the heat released in electrical power plants is lost as waste heat; in modern large power plants, this loss accounts for 65 to 70 percent of the total heat released from combustion.

A typical coal-fired electric power plant produces 1,000 MW of electricity by burning fuel with an energy content of 2,800 MW; 340 MW are lost as heat up the smokestack, leaving 2,460 MW to power turbines that drive a generator to produce electricity. However, the thermal efficiency of the turbines is only 42 percent. That means 42 percent of this power goes to drive the generator, but the rest (58 percent of 2,460 = 1,430 MW) is waste heat that must be removed by cooling water. Assume that cooling water from an adjacent river, which has a total flow rate of 100 $m^3/s$, is used to remove the waste heat. How much will the temperature of the river rise as a result of the addition of this heat?

## solution

This problem is similar to Example 4.9, because a specified amount of heat is added to a flow of water, and the resulting temperature rise must be determined. An energy balance can be written over the region of the river to which the heat is added. Here, $T_{in}$ represents the temperature of the water upstream, and $T_{out}$ represents the temperature after heating:

$$\frac{dE}{dt} = \dot{E}_{in} - \dot{E}_{out}$$

$$0 = \left(\begin{array}{c} 1{,}430 \text{ MW of heat} \\ \text{from power plant} \end{array}\right) + (\dot{m}_{H_2O} \times c \times T_{H_2O\ in})$$

$$-(\dot{m}_{H_2O} \times c \times T_{H_2O\ out})$$

Rearranging,

$$m_{H_2O} \times c \times (T_{out} - T_{in}) = 1{,}430 \text{ MW}$$

The remainder of this problem is essentially unit conversion. To obtain $\dot{m}_{H_2O}$ requires multiplication of the given river volumetric flow rate by the density of water (approximately 1,000 $kg/m^3$). The heat capacity of water, $c = 4{,}184$ J/kg-°C, also is required. Thus,

$$\left(100\frac{m^3}{s} \times 1{,}000\frac{kg}{m^3}\right) \times \left(4{,}184\frac{J}{kg \times °C}\right) \times \Delta T = 1{,}430 \times 10^6 \text{ J/s}$$

Solving for $\Delta T$:

$$\Delta T = 3.4°C$$

Consideration of this temperature increase is important, as the Henry's law constant for oxygen changes with temperature. This results in a reduced dissolved-oxygen concentration in the river in warmer water, which may be harmful to aquatic life.

© Mehmet Salih Guler/iStockphoto.

Earth's average surface temperature is determined by a balance between the energy provided by the sun and the energy radiated away by Earth to space. The energy radiated to space is emitted in the form of infrared radiation. As illustrated in Figure 4.10, some of this infrared radiation is absorbed in the atmosphere. The gases responsible for this absorption are called **greenhouse gases**; without them, Earth would not be habitable, as demonstrated in Box 4.2.

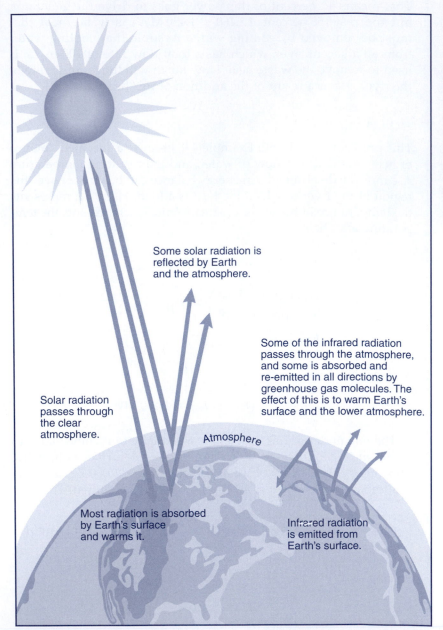

**Figure 4.10   The Greenhouse Effect**

Redrawn from *Our Changing Planet: The FY 1996 U.S. Global Change Research Program,* Report by the Subcommittee on Global Change Research, Committee on Environment and Natural Resources Research of the National Science and Technology Council, (supplement to the President's fiscal year 1996 Budget).

The energy balance of the Earth is being increasingly altered by human activities, mainly through the addition to the atmosphere of carbon dioxide from fossil fuel combustion. Calculate the global average temperature of Earth without greenhouse gases and show the effect greenhouse gases have on Earth's energy balance.

**Solution**

An energy balance can be written with the entire Earth as the control volume. For this system, the goal is to calculate Earth's annual average temperature. Over time periods of at least 1 yr, it is reasonable to assume that the system is at steady state. The energy balance is

$$\frac{dE}{dt} = 0 = \dot{E}_{in} - \dot{E}_{out}$$

The energy flux in is equal to the solar energy intercepted by Earth. At Earth's distance from the sun, the sun's radiation is 1,368 W/m², referred to as $S$. Earth intercepts an amount of energy equal to $S$ times the cross-sectional area of the Earth: $S \times \pi r_e^2$. However, because Earth reflects approximately 30 percent of this energy back to space, $\dot{E}_{in}$ equals only 70 percent of this value:

$$\dot{E}_{in} = 0.7\,S\,\pi R_e^2$$

The second term, $\dot{E}_{out}$, is equal to the energy radiated to space by Earth. The energy emitted per unit surface area of Earth is given by Boltzmann's law:

$$\left(\begin{array}{c}\text{Energy flux per}\\\text{unit area}\end{array}\right) = \sigma T^4$$

where $\sigma$ is Boltzmann's constant, equal to $5.67 \times 10^{-8}$ W/m²-K⁴. To obtain $\dot{E}_{out}$, this value is multiplied by Earth's total surface area, $4\pi R_e^2$. (The total surface area of the sphere is used here because energy is radiated away from Earth during both day and night.)

$$\dot{E}_{out} = 4\pi R_e^2 \sigma T^4$$

To solve the energy balance, set $\dot{E}_{in}$ equal to $\dot{E}_{out}$:

$$4\pi R_e^2 \sigma T^4 = 0.7 S \pi R_e^2$$

Simplify:

$$T^4 = \frac{0.7S}{4\sigma}$$

Plugging in the values for $S$ and $\sigma$ yields Earth's average annual temperature: $T = 255$ K, or $-18°C$.

This is too cold! In fact, the globally averaged temperature at Earth's surface is much warmer: 287 K. The reason for the difference is the presence of gases in the atmosphere that absorb the infrared radiation emitted by Earth and prevent it from reaching space. These gases, which include water vapor, $CO_2$, $CH_4$, and $N_2O$, were neglected in the initial energy balance. To include their influence, we can add a new term in the energy balance: the energy flux absorbed and retained by these gases. If the impact of greenhouse gas absorption is given by $E_{greenhouse}$, then the corrected $\dot{E}_{out}$ term is

$$\dot{E}_{out} = 4\pi R_e^2 \sigma T^4 - E_{greenhouse}$$

The reduction in $\dot{E}_{out}$ that results from greenhouse gas absorption is sufficient to cause the higher observed surface temperature. Clearly, this is largely a natural phenomenon, since surface temperatures were well above 255 K long before humans began burning fossil fuels. However, human activities—primarily the burning of fossil fuels—are changing the atmospheric composition to a significant extent and are increasing the magnitude of the greenhouse effect.

Atmospheric concentrations of carbon dioxide—the most important greenhouse gas—are shown in Figure 4.11. Increasing atmospheric concentrations of carbon dioxide—as well as those of methane, nitrous oxide, chlorofluorocarbons, and tropospheric ozone, which have occurred as a result of human activities—increase the value of $E_{greenhouse}$. This enhanced greenhouse effect, termed the **anthropogenic greenhouse effect**, currently is equivalent to an increase in the energy flux to Earth of approximately 2 W/m². Projections indicate that the increase could be as high as 5 W/m² over the next 50 years.

 **Global Energy Balance**

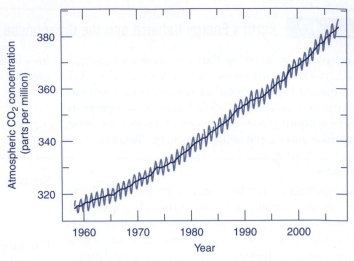

**Figure 4.11** **Global-Average Carbon Dioxide Concentration Trend** These measurements of $CO_2$ were made at the Mauna Loa, Hawaii, observatory by the National Oceanic and Atmospheric Administration. The annual increase of approximately 0.5 percent per year is attributed to fossil-fuel combustion and deforestation. The annual cycle is the result of photosynthesis and respiration, which result in a drawdown of $CO_2$ during the summer growing season and an increase during winter.

(Redrawn from data provided by NOAA; http://www.esrl.noaa.gov/gmd/ccgg/trends/).

As the energy absorbed by greenhouse gases increases, some other term in the energy balance must respond to maintain steady state. If the solar radiation absorbed by Earth remains constant, then Earth's average temperature must increase. The magnitude of the resulting temperature increase depends on the response of the complex global climate system, including changes in cloudiness and ocean circulation.

Current global climate models predict that the anthropogenic greenhouse effect will cause a global-average temperature increase (relative to 1990) of 1.1°C to 6.4°C by 2099 (IPCC, 2007). Resulting alterations to global and regional climate are predicted to include increased rainfall and increased frequency of severe storms, although some regions of the planet may experience increased frequency of drought or even regional cooling as a result of changes in atmospheric and oceanic circulation patterns. Several different scenarios for economic and population growth, material and energy efficiency technologies, and consumption patterns and the resulting predicted temperature changes and sea level rise are provided in Table 4.5.

### Class Discussion

Using the scenarios and outcomes listed in Table 4.5, how do population and continued use of fossil fuels affect the warming of the Earth and the rise in sea level?

## 4.3 Mass Transport Processes

Transport processes move chemicals from where they are generated, resulting in impacts that can be distant from the pollution source. In addition, transport processes are used in the design of treatment

## Table / 4.5

**Temperature Change and Sea Level Rise Resulting from Various Future Scenarios** Scenarios include economic and population growth, material and energy efficiency technology development, and consumption patterns for 2090–2099.

| Scenario | Temperature Change (°C at 2090–2099 relative to 1980–1999) | | Sea Level Rise (m at 2090–2099 relative to 1980–1999) |
|---|---|---|---|
| | *Best Estimate* | *Likely Range* | *Model-Based Range*[*] |
| **B1:** rapid economic growth toward a service and information economy; population peaks in midcentury and then declines; reductions in material intensity; clean/resource-efficient technologies; global solutions to sustainability, including improved equity | 1.8 | 1.1–2.9 | 0.18–0.38 |
| **A1T:** rapid economic growth; population peaks in midcentury and then declines; rapid introduction of new and efficient technologies; convergence among regions; non-fossil-fuel energy sources | 2.4 | 1.4–3.8 | 0.20–0.45 |
| **B2:** local solutions to sustainability; continuously increasing population; intermediate levels of economic development; less rapid and more diverse technological change | 2.4 | 1.4–3.8 | 0.20–0.43 |
| **A1B:** same as A1T except balance between fossil and non-fossil-fuel energy sources | 2.8 | 1.7–4.4 | 0.21–0.48 |
| **A2:** self-reliance and preservation of local identities; continuously increasing population; economic development that is primarily regionally oriented; slow and fragmented per capita economic growth and technological change | 3.4 | 2.0–5.4 | 0.23–0.51 |
| **A1FI:** same as A1T except fossil-intensive energy sources | 4.0 | 2.4–6.4 | 0.26–0.59 |

[*]Excluding future rapid dynamic changes in ice flow in the large glacial regions of Greenland and Antarctica.
SOURCE: IPCC (2007).

systems. Here, our discussion has two purposes: to provide an understanding of the processes that cause pollutant transport, and to present and apply the mathematical formulas used to calculate the resulting pollutant fluxes.

### 4.3.1 ADVECTION AND DISPERSION

Transport processes in the environment can be divided into two categories: advection and dispersion. **Advection** refers to transport with the mean fluid flow. For example, if the wind is blowing toward the east, advection will carry any pollutants present in the atmosphere toward the east. Similarly, if a bag of dye is emptied into the center of a river, advection will carry the resulting spot of dye downstream. In contrast, dispersion refers to the transport of compounds through the action of random motions. Dispersion works to eliminate sharp discontinuities in concentration and results in smoother, flatter concentration profiles. Advective and dispersive processes usually can be considered independently. For the spot of dye in a river, while advection moves the center of mass of the dye downstream, dispersion spreads out the concentrated spot of dye to a larger, less concentrated region.

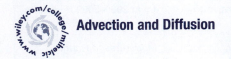

**Advection and Diffusion**

## DEFINITION OF THE MASS FLUX DENSITY Mass flux ($\dot{m}$, with units of mass/time), discussed earlier in this chapter, calculates the rates at which mass is transported into and out of a control volume in mass balances. Because mass balance calculations are always made with reference to a specific control volume, this value clearly refers to the rate at which mass is transported *across the boundary of the control volume*. However, in calculations of advective and dispersive fluxes, a specific, well-defined control volume will not be created. Instead, we determine the **flux density** across an imaginary plane oriented perpendicular to the direction of mass transfer.

The resulting mass flux density is defined as the rate of mass transferred across the plane per unit time per unit area. The symbol $J$ will be used to represent the flux density, expressed as the rate per unit area at which mass is transported across an imaginary plane. $J$ has units of (mass/time-length squared).

The total mass flux across a boundary ($\dot{m}$) can be calculated from the flux density. To do this, simply multiply $J$ by the area of the boundary:

$$\dot{m} = J \times A \qquad \text{(4.24)}$$

The mass transfer process that $J$ describes can result from advection, dispersion, or a combination of both processes.

## CALCULATION OF THE ADVECTIVE FLUX The **advective flux** refers to the movement of a compound along with flowing air or water. The advective-flux density depends simply on concentration and flow velocity:

$$J = C \times v \qquad \text{(4.25)}$$

The fluid velocity, $v$, is a vector quantity. It has both magnitude and direction, and the flux $J$ refers to the movement of pollutant mass in the same direction as the fluid flow. The coordinate system is generally defined so that the $x$-axis is oriented in the direction of fluid flow. In this case, the flux $J$ will reflect a flux in the $x$-direction, and the fact that $J$ is really a vector quantity will be ignored.

## DISPERSION Dispersion results from random motions of two types: the random motion of molecules and the random eddies that arise in turbulent flow. Dispersion from the random molecular motion is termed *molecular diffusion*; dispersion that results from turbulent eddies is called *turbulent dispersion* or *eddy dispersion*.

Calculate the average flux density $J$ of phosphorus downstream of the wastewater discharge of Example 4.2. The cross-sectional area of the river is 30 m$^2$.

### solution

In Example 4.2, the following conditions downstream of the spot where a pipe discharged to a river were determined: volumetric flow rate $Q = 26$ m$^3$/s and downstream concentration $C_d = 0.20$ mg/L. The average river velocity is $v = Q/A = (26$ m$^3$/s$)/(30$ m$^2) = 0.87$ m/s. Using the definition of flux density (Equation 4.25), we can solve for $J$:

$$J = \left[ (0.20 \text{ mg/L}) \times \frac{10^3 \text{ L}}{\text{m}^3} \right] \times (0.87 \text{ m/s})$$

$$= 174 \text{ mg/m}^2\text{-s or } 0.17 \text{ g/m}^2\text{-s}$$

**Fick's Law**  Fick's law is used to calculate the dispersive-flux density. It can be derived by analyzing the mass transfer that results from the random motion of gas molecules.[1] The purpose of this derivation is to provide a qualitative and intuitive understanding of why diffusion occurs, and the derivation is useful only for that purpose. In problems where it is necessary to calculate the diffusive flux, we will use Fick's law (Equation 4.35, derived in this section).

Consider a box that is initially divided into two parts, as shown in Figure 4.12. Each side of the box has a height and depth of one unit, and a width of length $\Delta x$. Initially, the left portion of the box is filled with 10 molecules of gas $x$, and the right side is filled with 20 molecules of gas $y$, as shown in the top half of Figure 4.12. What happens if the divider is removed?

Molecules are never stationary. All of the molecules in the box are constantly moving around, and at any moment, they have some probability of crossing the imaginary line at the center of the box. Assume that the molecules on each side are counted every $\Delta t$ seconds. The probability that a molecule crosses the central line during the period between observations can be defined as $k$, which is assumed to equal 20 percent (any value would do for the present purpose). The first time the box is checked, after a period $\Delta t$, 20 percent of the molecules originally on the left will have moved to the right, and 20 percent of the molecules originally on the right will have moved to the left. Counting the molecules on each side gives the situation shown in the bottom of Figure 4.12. Eight molecules of $x$ remain on the left, and two have moved to the right, while 16 molecules of $y$ remain on the right, four having moved to the left.

---

[1] This derivation is based closely on one presented by Fischer et al. (1979).

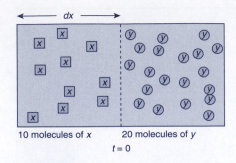

10 molecules of x          20 molecules of y

t = 0

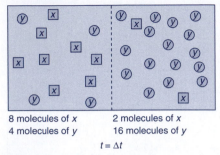

8 molecules of x      2 molecules of x
4 molecules of y     16 molecules of y

t = Δt

**Figure 4.12  Diffusion of Gas Molecules in a Box**   A box is divided into two regions of equal size. Ten gas molecules of one type (x) are added to the left side, while 20 gas molecules of another type (y) are added to the right side. Although they are distinguishable, the two types of molecules are identical in every physical respect. At time t = 0, the divider separating the two regions is removed. As a result of random motion, each molecule within the box has a 20 percent probability of moving to the opposite side of the box during each time interval of duration Δt. The result after one time interval is shown in the bottom figure.

From Fischer et al., *Mixing in Inland and Coastal Waters*, Copyright Elsevier (1979).

Because the boxes are equal in size, the concentration within each box is proportional to the number of molecules within it. Therefore, the random motion of molecules in the boxes has reduced the concentration difference between the boxes, with the difference falling from (20 − 0) to (16 − 4) for the molecules of y, and from (10 − 0) to (8 − 2) for the molecules of x. This result leads to a fundamental property of dispersive processes: dispersion moves mass from regions of high concentration to regions of low concentration and reduces concentration gradients.

The flux density J can also be derived for the two-box experiment. For this calculation, the situation shown in Figure 4.12 is used again, with the probability of any molecule crossing the central boundary during a period Δt equal to k. Since each molecule can be considered independently, the movement of a single molecule type—say, molecule y—can be analyzed.

Let $m_L$ be the total mass of molecule y in the left half of the box, and $m_R$ equal the mass in the right half. Since our box has unit height and depth, the area perpendicular to the direction of diffusion is one square unit. Thus, the flux density (the flux per unit area) is just equal to the rate of mass transfer across the boundary. The amount of mass transferred from left to right in a single time step is equal to $km_L$, since each molecule has a probability k of crossing the boundary, while the amount transferred from right to left during the same period is $km_R$. Thus, the net rate of mass flux from left to right across the boundary is equal to $(km_L − km_R)$ divided by Δt:

$$J = \frac{k}{\Delta t}(m_L − m_R) \qquad (4.26)$$

Since it is more convenient to work with concentrations than with total mass values, Equation 4.26 needs to be converted to concentration units. The concentration in each half of the box is given by

$$C_L = \frac{m_L}{\Delta x \times (\text{height}) \times (\text{depth})} \qquad (4.27)$$

Because the height and depth both equal 1, we can simplify:

$$= \frac{m_L}{\Delta x} \qquad (4.28)$$

For the right side of the box,

$$C_R = \frac{m_R}{\Delta x} \qquad (4.29)$$

Substituting CΔx for the mass in each half of the box, we can solve for the flux density:

$$J = \frac{k}{\Delta t}(C_L \Delta x − C_R \Delta x) \qquad (4.30)$$

$$= \frac{k}{\Delta t}(\Delta x)(C_L − C_R) \qquad (4.31)$$

Finally, note that as $\Delta x \to 0$, $(C_R - C_L)/\Delta x \to dC/dx$. Therefore, if we multiply Equation 4.31 by $(\Delta x/\Delta x)$,

$$J = \frac{k}{\Delta t}(\Delta x)(C_L - C_R)\frac{\Delta x}{\Delta x} \tag{4.32}$$

$$= \frac{k}{\Delta t}(\Delta x)^2 \frac{(C_L - C_R)}{\Delta x} \tag{4.33}$$

we obtain

$$J = -\frac{k}{\Delta t}(\Delta x)^2 \frac{dC}{dx} \tag{4.34}$$

The negative sign in this equation is simply a result of the convention that flux is positive when it flows from left to right, while the derivative is positive when concentration increases toward the right.

Equation 4.34 states that the flux of mass across an imaginary boundary is proportional to the concentration gradient at the boundary. Since the resulting flux cannot depend on arbitrary values of $\Delta t$ or $\Delta x$, the factor $k(\Delta x)^2/\Delta t$ must be constant. This product is the value called the **diffusion coefficient**, $D$. Rewriting Equation 4.34 results in Fick's law:

$$J = -D\frac{dC}{dx} \tag{4.35}$$

The units of the diffusion coefficient are clear from an analysis of the units of Equation 4.35 or from the units of the parameters in Equation 4.34; the diffusion coefficient has the same units as $k(\Delta x)^2/\Delta t$. Since $k$ is a probability and thus has no units, the units of $D$ are (length$^2$/time). Diffusion coefficients are commonly reported in units of cm$^2$/s.

Note the form of Equation 4.35:

$$\text{flux density} = (\text{constant}) \times (\text{concentration gradient}) \tag{4.36}$$

This form of equation will also appear later when Darcy's law is covered. Darcy's law governs the rate at which water flows through porous media, as in groundwater flow. The same equation also governs heat transfer, replacing the concentration gradient with a temperature gradient.

**Molecular Diffusion** The molecules-in-a-box analysis used earlier is essentially an analysis of molecular diffusion. Purely molecular diffusion is relatively slow. Table 4.6 lists typical values for the diffusion coefficient. These values are approximately $10^{-2}$ to $10^{-1}$ cm$^2$/s for gases and much lower, around $10^{-5}$ cm$^2$/s, for liquids. The difference in diffusion coefficient between gases and liquids is understandable because gas molecules are free to move much greater distances before being stopped by bumping into another molecule.

The diffusion coefficient also varies with temperature and the molecular weight of the diffusing molecule. This is because the average speed

## Table / 4.6

### Selected Molecular Diffusion Coefficients in Water and Air

| Compound | Temperature (°C) | Diffusion Coefficient (cm$^2$/s) |
|---|---|---|
| Methanol in $H_2O$ | 15 | $1.26 \times 10^{-5}$ |
| Ethanol* in $H_2O$ | 15 | $1.00 \times 10^{-5}$ |
| Acetic acid in $H_2O$ | 20 | $1.19 \times 10^{-5}$ |
| Ethylbenzene in $H_2O$ | 20 | $8.1 \times 10^{-6}$ |
| $CO_2$ in air | 20 | 0.151 |

*Of the two similar compounds methanol and ethanol, the less massive compound, methanol, has the higher diffusion coefficient.

SOURCE: Mihelcic (1999). Reprinted with permission of John Wiley & Sons, Inc.

of the random molecular motions depends on the kinetic energy of the molecules. As heat is added to a material and temperature increases, the thermal energy is converted to random kinetic energy of the molecules, and the molecules move faster. This results in an increase in the diffusion coefficient with increasing temperature. However, if the molecules have differing molecular weights, a heavier molecule moves more slowly at a given temperature, so the diffusion coefficient decreases with increasing molecular weight.

## example/4.13 Molecular Diffusion

The transport of polychlorinated biphenyls (PCBs) from the atmosphere into the Great Lakes is of concern because of health impacts on aquatic life and on people and wildlife that eat fish from the lakes. PCB transport is limited by molecular diffusion across a thin stagnant film at the surface of the lake, as shown in Figure 4.13. Calculate the flux density $J$ and the total annual amount of PCBs deposited into Lake Superior if the transport is by molecular diffusion, the PCB concentration in the air just above the lake's surface is $100 \times 10^{-12}$ g/m$^3$, and the concentration at a height of 2.0 cm above the water surface is $450 \times 10^{-12} =$ g/m$^3$. The diffusion coefficient for PCBs is equal to 0.044 cm$^2$/s, and the surface area of Lake Superior is $8.2 \times 10^{10}$ m$^2$. (The PCB concentration in the air at the air–water interface is determined by Henry's law equilibrium with dissolved PCBs.)

## solution

To calculate the flux density, first determine the concentration gradient. Assume that concentration changes linearly with height

example/4.13 Continued

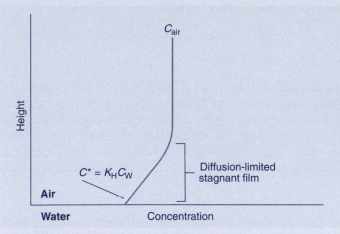

**Figure 4.13** **Variation of PCB Concentration with Height above Lake Superior** $C_{air}$ is the PCB concentration in the atmosphere above the lake, and $C^*$ is the concentration at the air–water interface, which is determined by Henry's law equilibrium with the dissolved PCB concentration. The flux of PCBs into the lake is determined by the rate of diffusion across a stagnant film above the lake.

From Mihelcic (1999). Reprinted with permission of John Wiley & Sons, Inc.

between the surface and 2.0 cm, as no concentration information was provided between those two heights. The gradient is then

$$\frac{dC}{dz} = \frac{450 \times 10^{-12}\,\text{g/m}^3 - 100 \times 10^{-12}\,\text{g/m}^3}{2.0\,\text{cm} - 0\,\text{cm}} \times \frac{10^2\,\text{cm}}{\text{m}}$$

$$= 1.8 \times 10^{-8}\,\text{g/m}^4$$

Fick's law (Equation 4.35) can be used to calculate the flux density:

$$J = -D\frac{dC}{dz}$$

$$= -(0.044\,\text{cm}^2/\text{s}) \times 1.8 \times 10^{-8}\,\text{g/m}^4 \times \frac{\text{m}^2}{10^4\,\text{cm}^2} \times \frac{3.15 \times 10^7\,\text{s}}{\text{yr}}$$

$$= -2.4 \times 10^{-6}\,\text{g/m}^2\text{-yr}$$

Here, the negative sign indicates the flux is downward, but it is not necessary to pay attention to the sign to determine that. Remember, diffusion always transports mass from higher concentration to lower concentration regions.

The total depositional flux is given by $\dot{m} = J \times A$:

$$\dot{m} = -2.4 \times 10^{-6} \text{g/m}^2\text{-yr} \times 8.2 \times 10^{10} \text{ m}^2$$
$$= -2.0 \times 10^5 \text{g/yr}$$

Thus, the total annual input of PCBs into Lake Superior from the atmosphere is approximately 200 kg. Although this is a small annual flux for such a large lake, PCBs do not readily degrade in the environment, and they bioaccumulate in the fish, resulting in unhealthy levels.

© Michael Braun/iStockphoto.

**Turbulent Dispersion** In **turbulent dispersion**, mass is transferred through the mixing of *turbulent eddies* within the fluid. This is fundamentally different from the processes that determine molecular diffusion. In turbulent dispersion, the random motion of the *fluid* does the mixing, while in molecular diffusion, the random motion of the pollutant molecules is important.

Random motions of the fluid generally are present in the form of whorls, or eddies. These are familiar in the form of eddies or whirlpools in rivers but occur in all forms of fluid flow. The size of turbulent eddies is several orders of magnitude larger than the mean free path of individual molecules, so turbulence moves mass much faster than does molecular diffusion. As a result, turbulent, or eddy dispersion coefficients used in Fick's law generally are several orders of magnitude larger than molecular-diffusion coefficients.

The value of the turbulent-dispersion coefficient depends on properties of the fluid flow. It does not depend on molecular properties of the compound being dispersed (as did the molecular-diffusion coefficient), because in turbulence, the molecules are simply carried along with the macroscale flow. For flow in pipes or streams, the most important flow property determining the turbulent dispersion coefficient is the flow velocity. Turbulence is present only at flow velocities above a critical level, and the degree of turbulence is correlated with velocity. More precisely, the presence or absence of turbulence depends on the *Reynolds number*, a dimensionless number that depends on velocity, width of the river or pipe, and viscosity of the fluid. In addition, the degree of turbulence depends on the material over which the flow occurs, so that flow over bumpy surfaces will be more turbulent than flow over smooth surfaces, and the increased turbulence will cause more rapid mixing. In lakes and in the atmosphere, buoyant mixing that results from temperature-induced density gradients also can cause turbulent mixing, even in the absence of currents.

Except in the case of transport across a boundary, such as at the air–water interface considered in Example 4.13, turbulent dispersion almost always entirely dominates molecular diffusion. This is because even an occasional amount of weak turbulence will cause more mixing than several days of molecular diffusion.

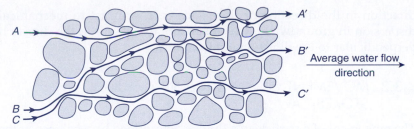

**Figure 4.14** **Process of Mechanical Dispersion in Groundwater Flow**
Two fluid parcels starting near each other at locations $B$ and $C$ are dispersed
to locations farther apart ($B'$ and $C'$) during transport through the soil pores,
while parcels from $A$ and $B$ are brought closer together, resulting in mixing of
water from the two regions.

From Hemond and Fechner, *Chemical Fate and Transport in the Environment.* Copyright Elsevier (1994).

Fick's law applies to turbulent dispersion just as it does to molecular diffusion. Thus, flux density calculations are the same for both processes; only the magnitude of the dispersion coefficient is different.

**Mechanical Dispersion**   The final dispersion process considered in this chapter is similar to turbulence, in that it is a result of variations in the movement of the fluid that carries a chemical. In **mechanical dispersion**, these variations are the result of (1) variations in the flow pathways taken by different fluid parcels that originate in nearby locations or (2) variations in the speed at which fluid travels in different regions.

Dispersion in groundwater flow provides a good example of the first process. Figure 4.14 shows a magnified depiction of the pores through which groundwater flows within a subsurface sample. (Note that, as shown in Figure 4.14, groundwater movement is not a result of underground rivers or creeks, but rather is caused by the flow of water through the pores of the soil, sand, or other material underground.) Because transport through the soil is limited to the pores between soil particles, each fluid particle takes a convoluted path through the soil, and as it is transported horizontally with the mean flow, it is displaced vertically a distance that depends on the exact flow path it took. The great variety of possible flow paths results in a random displacement in the directions perpendicular to the mean flow path. Thus, a spot of dye introduced into the groundwater flow between points $B$ and $C$ in the figure would be spread out, or dispersed, into the region between points $B'$ and $C'$ as it flows through the soil.

The second type of mechanical dispersion results from differences in flow speed. Anywhere that a flowing fluid contacts a stationary object, the speed at which the fluid moves will be slower near the object. For example, the speed of water flowing down a river is fastest in the center of a river and can be very slow near the edges. Thus, if a line of dye were somehow laid across the river at one point, it would be stretched out in the upstream/downstream direction as it flowed down the river, with the center part of the line moving faster than the edges. This type of dispersion spreads things out in the longitudinal

**Environmental Protection in Southern California**
http://www.epa.gov/region09/socal

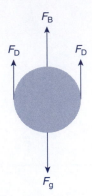

**Figure 4.15** **Forces Acting on a Particle Settling through Air or Water** The gravitational force $F_g$ is in the downward direction and is counteracted by the buoyancy force $F_B$ and the drag force $F_D$.

From Mihelcic (1999). Reprinted with permission of John Wiley & Sons, Inc.

direction in the direction of flow. This is in contrast to mechanical dispersion in groundwater, which spreads things out in the direction perpendicular to the direction of mean flow.

### 4.3.2 MOVEMENT OF A PARTICLE IN A FLUID: STOKES' LAW

The movement of a particle in a fluid is determined by a balance of the viscous drag forces resisting the particle movement with gravitational or other forces that cause the movement. In this section, a force balance on a particle is used to derive the relationship between particle size and settling velocity known as Stokes' law, and Stokes' law is used in examples involving particle-settling chambers.

**GRAVITATIONAL SETTLING** Consider the settling particle shown in Figure 4.15. To determine the velocity at which it falls (the settling velocity), a force balance will be conducted. Three forces act on the particle: the downward gravitational force, an upward buoyancy force, and an upward drag force.

The gravitational force $F_g$ is equal to the gravitational constant $g$ times the mass of the particle, $m_p$. In terms of particle density $\rho_p$, and diameter $D_p$, $m_p$ is equal to $(\rho_p \pi/6 D_p^3)$. Therefore,

$$F_g = \rho_p \frac{\pi}{6} D_p^3 g \tag{4.37}$$

The buoyancy force $F_B$ is a net upward force that results from the increase of pressure with depth within the fluid. The buoyancy force is equal to the gravitational constant times the mass of the fluid displaced by the particle:

$$F_B = \rho_f \frac{\pi}{6} D_p^3 g \tag{4.38}$$

where $\rho_f$ is equal to the fluid density.

The only remaining force to determine is the drag force, $F_D$. The drag force is the result of frictional resistance to the flow of fluid past the surface of the particle. This resistance depends on the speed at which the particle is falling through the fluid, the size of the particle, and the *viscosity*, or resistance to shear, of the fluid. Viscosity is essentially what one would qualitatively call the "thickness" of the fluid. Honey has a high viscosity, water has a relatively low viscosity, and the viscosity of air is much lower yet.

Over a wide range of conditions, the friction force can be correlated with the Reynolds number. Most particle-settling situations involve creeping flow conditions (Reynolds number less than 1). In this case, the Stokes' drag force can be used:

$$F_D = 3\pi\mu D_p v_r \tag{4.39}$$

where $\mu$ is the fluid viscosity (units of g/cm-s) and $v_r$ is the downward velocity of the particle relative to the fluid.

The net downward force acting on the particle is equal to the vector sum of all forces acting on the particle:

$$F_{\text{down}} = F_{\text{g}} - F_{\text{B}} - F_{\text{D}} \qquad (4.40)$$

$$= \rho_{\text{P}} \frac{\pi}{6} D_{\text{P}}^3 g - \rho_{\text{f}} \frac{\pi}{6} D_{\text{P}}^3 g - 3\pi\mu D_{\text{p}} v_{\text{r}} \qquad (4.41)$$

$$= (\rho_{\text{p}} - \rho_{\text{f}}) \frac{\pi}{6} D_{\text{P}}^3 g - 3\pi\mu D_{\text{p}} v_{\text{r}} \qquad (4.42)$$

The particle will respond to this force according to Newton's second law (force equals mass times acceleration). Thus,

$$F_{\text{down}} = m_{\text{p}} \times \text{acceleration} \qquad (4.43)$$

$$= m_{\text{p}} \times \frac{dv_{\text{r}}}{dt} \qquad (4.44)$$

This differential equation can be solved to determine the time-varying velocity of a particle that is initially at rest. The solution indicates that, in almost all cases of environmental interest, the period of time required before the particle reaches its final settling velocity is very short (much less than 1 s). For this reason, in this text, only the final (terminal) settling velocity is considered.

When the particle has reached terminal velocity, it is no longer accelerating, so $dv/dt = 0$. Thus, from Equation 4.44, $F_{\text{down}} = 0$. Setting $F_{\text{down}}$ equal to zero and noting that $v_{\text{r}}$ is equal to the settling velocity $v_{\text{s}}$ at terminal velocity, Equation 4.42 can be rearranged to yield

$$(\rho_{\text{p}} - \rho_{\text{f}}) \frac{\pi}{6} D_{\text{P}}^3 g = 3\pi\mu D_{\text{p}} v_{\text{s}} \qquad (4.45)$$

Solving for $v_{\text{s}}$:

$$v_{\text{s}} = \frac{g(\rho_{\text{p}} - \rho_{\text{f}})}{18\mu} D_{\text{P}}^2 \qquad (4.46)$$

**Sedimentation**

Equation 4.46 is referred to as **Stokes' law**. The resulting settling velocity is often called the Stokes' velocity. Stokes' law, so called because it is based on the Stokes' drag force, is the fundamental equation used to calculate terminal settling velocities of particles in both air and water. It is used in the design of treatment systems to remove particles from exhaust gases and from drinking water and wastewater as well as in analyses of settling particles in lakes and in the atmosphere.

An important implication of Stokes' law is that the settling velocity increases as the *square* of the particle diameter, so larger particles settle much faster than smaller particles. This result is used in drinking-water treatment: coagulation and flocculation are used to get small particles to come together and form larger particles, which can then be removed by gravitational settling in a reasonable amount of time. This process results in reduction in the turbidity (increase in the clarity) of the water. In contrast, particles with very small diameters settle extremely slowly. As a result, atmospheric particles with diameters less than 1 to 10 μm generally fall more slowly than the speed of turbulent eddies of air, with the result that they are not removed by gravitational settling.

Note that particle–particle interactions have been ignored in this derivation. Thus, Stokes' law is valid for *discrete particle settling*. In situations where particle concentration is extremely high, particles form agglomerations or mats, and Stokes' law may no longer be valid.

### 4.3.3  GROUNDWATER FLOW

**Groundwater Information**

http://water.usgs.gov/ogw

The driving force for **groundwater** flow is the difference between hydraulic head between two points, commonly referred to as the **hydraulic gradient**. The rate of groundwater flow is controlled by the magnitude of the gradient and permeability of the aquifer materials. This relationship is captured by **Darcy's law** (in its one-dimensional form):

$$q = -K\frac{dh}{dx} \tag{4.47}$$

where $q$ is the **specific discharge**, or flow per unit of cross-sectional area, $K$ is the hydraulic conductivity, and $h$ is the hydraulic head. The **hydraulic conductivity** is the ability of the porous medium to conduct a given fluid and is the formal term used to quantify the permeability of aquifer materials. The derivative $dh/dx$ is the hydraulic head gradient, or the change in hydraulic head with respect to distance.

**Hydraulic head** is a measure of the energy contained by water at a specific point per unit volume of water. It is expressed as follows:

$$h = \frac{p}{\rho g} + z = h_{\mathrm{p}} + z \tag{4.48}$$

**Groundwater Vulnerability in Africa**

http://www.unep.org/groundwaterproject

where $p$ is the pore water pressure (defined here as relative to atmospheric pressure), $\rho$ is the density of water, $g$ is the gravitational constant, $h_{\mathrm{p}} = p/\rho g$ is the pressure head, and $z$ is the elevation head, or the elevation of the point where the hydraulic head is being monitored. The elevation is measured as positive upward and relative to a common reference point, or datum. The definition of hydraulic head implies that groundwater flow is driven by differences in pore water pressure, elevation, or both.

In Figure 4.16, three wells are used to measure the pressure head—the distance from the measurement point to the water surface in the well. The wells are a type known as piezometers, which consist essentially of tubes that extend from the ground surface to an open measurement point. The elevation head in this case is the distance from the

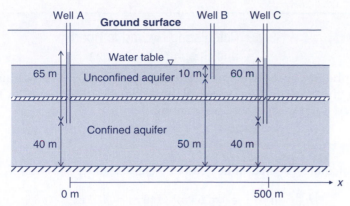

**Figure 4.16** **Hydraulic Head and Gradients in an Unconfined and Confined Aquifer** Vertical scale is exaggerated. Well B is in the unconfined aquifer. Wells A and C are in the confined aquifer.

datum to the measurement point. In the **unconfined aquifer**, the water level in the piezometer (well B) corresponds to the position of the water table. In the **confined aquifer**, the water levels in the piezometers (wells A and C) are above the top of the aquifer, indicating that the aquifer is under pressure.

---

## Box / 4.3 — Using Ceramic Porous Media to Treat Water at the Household Level in Developing-World Communities

The principles and equations that govern fluid flow through porous media can be applied to a wide variety of challenges related to the environment and sustainability. One example is the design and development of a popular "point of use" water treatment technology for the developing world—the ceramic or clay filter (see Figure 4.17). These filters are made with local materials, typically a mix of clay, water, and combustible organic material such as sawdust, flour, or rice husks. This mixture is then formed into a pot, air-dried, and fired in a kiln. Firing in the kiln forms the ceramic materials and combusts the organic material, making the filter porous and permeable to water. The pores are sized to remove larger water-borne particles and associated microorganisms to improve water quality and, subsequently, health.

The filter pots are suspended in a larger container so that when water is poured into the filter pot, it flows by gravity through the filter and into the lower container, where treated water can be accessed. The desired rate of water flow through the pots is 1 to 2 L/h.

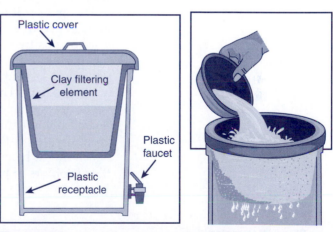

**Figure 4.17** **Clay Filter (Porous Medium) Used to Remove Turbidity and Associated Microorganisms from Water**

For more information, see Potters for Peace, www.pottersforpeace.org.

Using the equations to model water flow through the ceramic filter, we can determine the porosity necessary to yield the desired flow rate. This information gives the designer insight into the appropriate ratio of clay to combustible materials.

**example/4.14** Calculating Hydraulic Heads and Specific Discharges

Using the measurements provided in Figure 4.16 and a hydraulic conductivity of $10^{-3}$ cm/s, calculate the specific discharge in the confined aquifer shown in the same figure.

**solution**

First calculate, the hydraulic head at wells A and C:

$$h_A = h_{pA} + z_A = 65\,\text{m} + 40\,\text{m} = 105\,\text{m}$$
$$h_C = h_{pC} + z_C = 60\,\text{m} + 40\,\text{m} = 100\,\text{m}$$

Next, determine the hydraulic gradient between wells A and C, assuming that the derivative $dh/dx$ can be approximated with differences:

$$\frac{dh}{dx} \approx \frac{\Delta h}{\Delta x} = \frac{h_A - h_C}{x_A - x_C} = \frac{105\,\text{m} - 100\,\text{m}}{0\,\text{m} - 500\,\text{m}} = -0.01$$

To calculate the specific discharge, use Darcy's law (Equation 4.47):

$$q = -K\frac{dh}{dx} = -(10^{-3}\,\text{cm/s})(-0.01) = 10^{-5}\,\text{cm/s} = 10^{-7}\,\text{m/s}$$

Remember, the specific discharge has units of flow per unit cross-sectional area.

The answer illustrates two other important concepts. First, the implied positive sign on the specific discharge means that the flow is in the positive $x$ direction (left to right in Figure 4.16), although the gradient is negative. The negative sign in Darcy's law (Equation 4.47) indicates that the flow is in the opposite direction of the gradient. This makes sense if we recognize that the groundwater flows from a region of higher head to a region of lower head. Second, the magnitude of the answer, $10^{-7}$ m/s, indicates that the groundwater is flowing slowly relative to, say, flow in a river, which might be on the order of 0.1 m/s. In general, groundwater flow is orders of magnitude slower than surface water flow.

**Class Discussion**

The Ogallala aquifer stretches across the High Plains from South Dakota to western Texas. After researching its water use and natural recharge characteristics, discuss whether groundwater from this aquifer is a renewable resource as it is currently managed. How might future changes in climate and/or regional or global increases in population influence your discussion? If you conclude the Ogallala aquifer is not being managed as a renewable resource, describe technological and policy solutions to ensure groundwater for future generations.

## Key Terms

- advection
- advective flux
- anthropogenic greenhouse effect
- batch reactor
- carbon dioxide emissions
- completely mixed flow reactor (CMFR)
- confined aquifer
- conservative compound
- control volume
- Darcy's law
- diffusion coefficient
- dispersion
- energy balance
- Fick's law
- first law of thermodynamics
- first-order decay
- flux density
- greenhouse gases
- groundwater
- heat
- hydraulic conductivity
- hydraulic gradient
- hydraulic head
- law of conservation of mass
- mass accumulation rate
- mass balance
- mass flux
- mechanical dispersion
- nonrenewable energy
- nonsteady state
- plug-flow reactor (PFR)
- reactor analysis
- renewable energy
- retention time
- specific discharge
- steady state
- Stokes' law
- turbulent dispersion
- unconfined aquifer
- zero-order decay

# chapter/Four Problems

**4.1** A pond is used to treat a dilute municipal wastewater before the liquid is discharged into a river. The inflow to the pond has a flow rate of $Q = 4,000\ \text{m}^3/\text{day}$ and a BOD concentration of $C_{in} = 25\ \text{mg/L}$. The volume of the pond is $20,000\ \text{m}^3$. The purpose of the pond is to allow time for the decay of BOD to occur before discharge into the environment. BOD decays in the pond with a first-order rate constant equal to $0.25/\text{day}$. What is the BOD concentration at the outflow of the pond, in units of mg/L?

**4.2** A mixture of two gas flows is used to calibrate an air pollution measurement instrument. The calibration system is shown in Figure 4.18. If the calibration gas concentration $C_{cal}$ is $4.90\ \text{ppm}_v$, the calibration gas flow rate $Q_{cal}$ is $0.010\ \text{L/min}$, and the total gas flow rate $Q_{total}$ is $1.000\ \text{L/min}$, what is the concentration of calibration gas after mixing ($C_d$)? Assume the concentration upstream of the mixing point is zero.

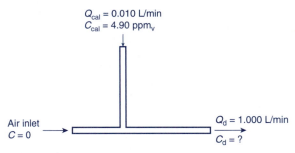

**Figure 4.18   Gas Calibration System**

From Mihelcic (1999). Reprinted with permission of John Wiley & Sons, Inc.

**4.3** Consider a house into which radon is emitted through cracks in the basement. The total volume of the house is $650\ \text{m}^3$ (assume the volume is well mixed throughout). The radon source emits 250 pCi/s. (A picoCurie [pCi] is a unit proportional to the amount of radon gas and indicates the amount of radioactivity of the gas.) Air inflow and outflow can be modeled as a flow of clean air into the house of $722\ \text{m}^3/\text{h}$ and an equal air flow out. Radon can be considered conservative in this problem. (a) What is the retention time of the house? (b) What is the steady-state concentration of radon in the house (units of pCi/L)?

**4.4** You are in an old spy movie and have been locked into a small room (volume $1,000\ \text{ft}^3$). You suddenly realize a poison gas has just started entering the room through a ventilation duct. You are safe as long as the concentration is less than $100\ \text{mg/m}^3$. If the ventilation air flow rate in the room is 100 cu. ft./min. and the incoming gas concentration is $200\ \text{mg/m}^3$, how long do you have to escape?

**4.5** In the simplified depiction of an ice rink with an ice-resurfacing machine operating (shown in Figure 4.19), points 1 and 3 represent the ventilation air intake and exhaust for the entire ice rink, and point 2 is the resurfacing machine's exhaust. Given that C indicates the concentration of carbon monoxide (CO), conditions at each point are as follows: point 1: $Q_1 = 3.0\ \text{m}^3/\text{s}, C_1 = 10\ \text{mg/m}^3$; point 2: emission rate $= 8\ \text{mg/s}$ of nonreactive CO; point 3: $Q_3, C_3$ unknown. The ice rink's volume ($V$) is $5.0 \times 10^4\ \text{m}^3$. (a) Define a control volume as the interior of the ice rink. What is the mass flux of CO into the control volume, in units of mg/s? (b) Assume that the resurfacing machine has been operating for a very long time and that the air within the ice rink is well mixed. What is the concentration of CO within the ice rink, in units of $\text{mg/m}^3$?

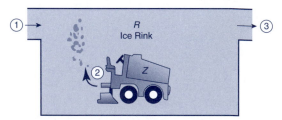

**Figure 4.19   Schematic Diagram of an Ice-Resurfacing Machine in an Ice Rink**

From Mihelcic (1999). Reprinted with permission of John Wiley & Sons, Inc.

**4.6** Poorly treated municipal wastewater is discharged to a stream. The river flow rate upstream of the discharge point is $Q_u = 8.7\ \text{m}^3/\text{s}$. The discharge occurs at a flow of $Q_d = 0.9\ \text{m}^3/\text{s}$ and has a BOD concentration of $50.0\ \text{mg/L}$. Assume the upstream BOD concentration is negligible. (a) What is the BOD concentration just downstream of the discharge point? (b) If the stream has a cross-sectional area of $10\ \text{m}^2$, what is the BOD concentration 50 km downstream? (BOD is removed with a first-order decay rate constant equal to 0.20/day.)

**4.7** Two towns, located directly across from each other, operate municipal wastewater treatment plants situated along a river. The river flow is 50 million gal. per day (50 MGD). Coliform counts are used as a measure to determine a water's ability to transmit disease to humans. The coliform count in the river upstream of the two treatment plants is 3 coliforms/100 mL. Town 1 discharges 3 MGD of wastewater with a coliform count of 50 coliforms/100 mL, and town 2 discharges 10 MGD of wastewater with a coliform count of 20 coliforms/100 mL. Assume the state requires the downstream coliform count not exceed 5 coliforms/100 mL. (a) Is the state water-quality standard being met downstream? (Assume coliforms do not die by the time they are measured downstream.) (b) If the state standard downstream is not met, the state has informed town 1 that it must treat its sewage further so the downstream standard is met. Use a mass balance approach to show that the state's request is unfeasible.

**4.8** In the winter, a stream flows at 10 m³/s and receives discharge from a pipe that contains road runoff. The pipe has a flow of 5 m³/s. The stream's chloride concentration just upstream of the pipe's discharge is 12 mg/L, and the runoff pipe's discharge has a chloride concentration of 40 mg/L. Chloride is a conservative substance. (a) Does wintertime salt usage on the road elevate the downstream chloride concentration above 20 mg/L? (b) What is the maximum daily mass of chloride (metric tons/day) that can be discharged through the road runoff pipe without exceeding the water quality standard?

**4.9** Calculate the hydraulic residence times (the retention time) for Lake Superior and for Lake Erie using data in Table 4.3.

**4.10** The total flow at a wastewater treatment plant is 600 m³/day. Two biological aeration basins are used to remove BOD from the wastewater and are operated in parallel. They each have a volume of 25,000 L. In hours, what is the aeration period of each tank?

**4.11** You are designing a reactor that uses chlorine in a PFR or CMFR to destroy pathogens in water. A minimum contact time of 30 min is required to reduce the pathogen concentration from 100 pathogens/L to below 1 pathogen/L through a first-order decay process. You plan on treating water at a

rate of 1,000 gal/min. (a) What is the first-order decay rate constant? (b) What is the minimum size (in gallons) of the reactor required for a plug flow reactor? (c) What size (in gallons) of CMFR would be required to reach the same outlet concentration? (d) Which type of reactor would you select if your treatment objective stated that "no discharge can ever be greater than 1 pathogen/L"? Explain your reasoning. (e) If the desired chlorine residual in the treated water after it leaves the reactor is 0.20 mg/L and the chlorine demand used during treatment is 0.15 mg/L, what must be the daily mass of chlorine added to the reactor (in grams)?

**4.12** The concentration of BOD in a river just downstream of a sewage treatment plant's effluent pipe is 75 mg/L. If the BOD is destroyed through a first-order reaction with a rate constant equal to 0.05/day, what is the BOD concentration 50 km downstream? The velocity of the river is 15 km/day.

**4.13** A $1.0 \times 10^6$ gallon reactor is used in a sewage treatment plant. The influent concentration is 100 mg/L, the effluent concentration is 25 mg/L, and the flow rate through the reactor is 500 gal/min. (a) What is the first-order rate constant for decay of BOD in the reactor? Assume the reactor can be modeled as a CMFR. Report your answer in units per hour. (b) Assume the reactor should be modeled as a PFR with first-order decay, *not* as a CMFR. In that case, what must be the first-order decay rate constant within the PFR reactor? (c) It has been determined that the outlet concentration is too high, so the residence time in the reactor must be doubled. Assuming all other variables remain constant, what must be the volume of the new CMFR?

**4.14** How many watts of power would it take to heat 1 L of water (weighing 1.0 kg) by 10°C in 1.0 h? Assume no heat losses occur, so all of the energy expended goes into heating the water.

**4.15** The concentration of a pollutant along a quiescent water-containing tube is shown in Figure 4.20. The diffusion coefficient for this pollutant in water is equal to $10^{-5}$ cm²/s. (a) What is the initial pollutant flux density in the x-direction at the following locations: $x = 0.5, 1.5, 2.5, 3.5,$ and 4.5? (b) If the diameter of the tube is 3 cm, what is the initial flux of pollutant mass in the x-direction at the same locations? (c) As time passes, this diffusive flux will change the shape of the concentration profile. Draw a sketch of concentration in the tube versus x-axis

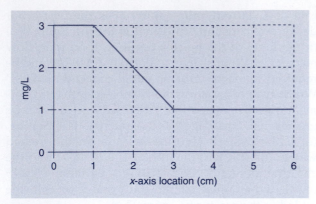

**Figure 4.20** **Hypothetical Concentration Profile in a Closed Pipe**

From Mihelcic (1999). Reprinted with permission of John Wiley & Sons, Inc.

location showing what the shape at a later time might look like. (It is not necessary to do any calculations to draw this sketch.) Assume that the concentration at $x = 0$ is held at 3 mg/L and the concentration at $x = 6$ is held at 1 mg/L. (d) Describe, in one paragraph, why the concentration profile changed in the way that you sketched in your solution to part (c).

**4.16** The tube in problem 4.15 is connected to a source of flowing water, and water is passed through the tube at a rate of 100 cm$^3$/s. If the pollutant

concentration in the water is constant at 2 mg/L, find: (a) the mass flux density of the pollutant through the tube due to advection; (b) the total mass flux through the tube due to advection.

**4.17** Two groundwater wells are located 1,000 m apart. The water level in well 1 is 50 m below the surface; in well 2, the water level is 75 m below the surface. The hydraulic conductivity is 1 m/day. Use Darcy's law to determine the specific discharge.

**4.18** Using the measurements provided in Figure 4.16 and a hydraulic conductivity of 10$^{-4}$ cm/s, calculate the specific discharge in the confined aquifer.

**4.19** Go to the Groundwater Network on the USGS web page (http://groundwaterwatch.usgs.gov). Research the impact of climate in your state discussing seasonal impacts on groundwater levels. Provide details about your sample location.

**4.20** Go to the World Health Organization web page (http://www.who.int/en). Write a two page report on the arsenic crisis addressing economic, social, and environmental issues.

**4.21** Go to the U. S. Geological Survey Web site (http://www.usgs.gov/ogw/aquiferbasics). Research a "principal aquifer" close to your home describing the extent and use at the aquifer.

# References

Fischer, H. B., E. J. List, J. Imberger, and N. H. Brooks. 1979. *Mixing in Inland and Coastal Waters*. New York: Academic Press.

Hemond, H. F., and E. J. Fechner. 1994. *Chemical Fate and Transport in the Environment*. San Diego: Academic Press.

Intergovernmental Panel on Climate Change (IPCC). 2007. "Summary for Policymakers." In *Climate Change 2007: The Physical Science Basis*. Contribution of Working Group I to the Fourth Assessment Report of the Intergovernmental Panel on Climate Change, ed. S. Solomon, D. Qin, M. Manning, Z. Chen, M. Marquis, K. B. Averyt, M. Tignor, and H. L. Miller. New York: Cambridge University Press.

Mihelcic, J. R. 1999. *Fundamentals of Environmental Engineering*. New York: John Wiley & Sons.

# chapter/Five Biology

Martin T. Auer, James R. Mihelcic, Julie Beth Zimmerman, Michael R. Penn

*In this chapter, readers are introduced to the fundamental biological principles governing ecosystems, with special attention to processes that mediate the fate of chemical substances in natural and engineered environments. The chapter begins with a discussion of ecosystem structure and function, including a description of population dynamics, that is, organism growth and attendant demand on resources. Production and consumption are then examined, leading to consideration of ecosystem trophic structure and energy flow. The chapter also introduces material flow in ecosystems, focusing on key biogeochemical cycles (e.g., oxygen, carbon, nitrogen, sulfur, and phosphorus) and effects of human activity on these flows. Finally, concepts relating to human and ecosystem health are explored, including biomagnification, biodiversity, and ecosystem health.*

## Major Sections

5.1 Ecosystem Structure and Function

5.2 Population Dynamics

5.3 Energy Flow in Ecosystems

5.4 Oxygen Demand: Biochemical, Chemical, and Theoretical

5.5 Material Flow in Ecosystems

5.6 Ecosystem Health and the Public Welfare

## Learning Objectives

1. Describe the relationships among individual organisms, species, and populations in ecosystem functions and structure.

2. Distinguish the exponential, logistic, and Monod models for population growth over time.

3. Identify and use the appropriate model to calculate changes in population over time.

4. Determine the carrying capacity of a population, and articulate how carrying capacity is affected by environmental conditions.

5. Use the Monod growth limitation model to calculate yield, substrate utilization, or biomass growth, and relate these terms to carrying capacity.

6. Discuss how human population growth, consumption, technology, and carrying capacity are related to the IPAT equation and ecological footprint.

7. Describe interconnections and energy/material transfer within a food web or ecosystem.

8. Define the following terms: biochemical oxygen demand (BOD), 5-day biochemical oxygen demand ($BOD_5$), ultimate biochemical oxygen demand (BODU), carbonaceous oxygen demand (CBOD), nitrogenous biochemical oxygen demand (NBOD), and theoretical oxygen demand (ThOD).

9. Describe the approach for calculating ThOD and the laboratory procedures for determining BOD.

10. Summarize the roles of photosynthesis and respiration in capturing and efficiently transferring energy in ecosystems.

11. Describe the flow of oxygen, carbon, nitrogen, sulfur, and phosphorus through ecosystems and the impact of anthropogenic activities on these flows.

12. Demonstrate how biological processes are related to issues of energy production and global carbon cycling.

13. Discuss the significance and application of bioaccumulation factors and bioconcentration factors.

14. Describe the benefits, threats, and indicators of biodiversity in relation to society, the economy, and ecosystem health and function.

**Biology** is defined as the scientific study of life and living things, often taken to include their origin, diversity, structure, activities, and distribution.

Biology includes the study of biotic effects. **Biotic** effects—those produced by or involving organisms—are important in many phases of environmental engineering. This chapter's exploration of environmental biology will focus on those activities—the ways organisms are affected by and have an effect on the environment. These include: (1) effects on humans (e.g., infectious disease); (2) impacts on the environment (e.g., species introductions); (3) impacts by humans (e.g., endangered species); (4) mediation of environmental transformation (e.g., breakdown of toxic chemicals); and (5) utilization in the treatment of contaminated air, water, and soil.

## 5.1 Ecosystem Structure and Function

In Figure 5.1, the Earth is conceptualized as comprising "great spheres" of living and nonliving material. The *atmosphere* (air), *hydrosphere* (water), and *lithosphere* (soil) constitute the **abiotic**, or nonliving, component. The **biosphere** contains all of the living things on Earth. Any intersection of the biosphere with the nonliving spheres—living things and their attendant abiotic environment—constitutes an **ecosystem**. Examples include natural (lake, grassland, forest, and desert) and

 **Global Ecosystems**

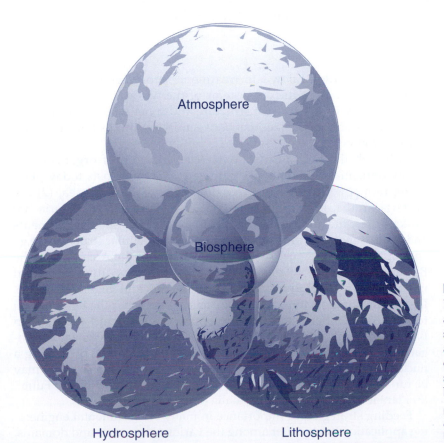

Atmosphere

Biosphere

Hydrosphere          Lithosphere

**Figure 5.1** **Earth's Great Spheres of Living and Nonliving Material** The atmosphere, hydrosphere, and lithosphere are the nonliving components, and the biosphere contains all the living components. The ecosphere is the intersection of the abiotic spheres and the biotic component.

Kupchella and Hyland, *Environmental Science*, 1st edn, © 1986, p. 5. Adapted by permission of Pearson Education, Inc., Upper Saddle River, NJ.

| Lake | Grassland | Biological waste treatment |
|---|---|---|
|  |  |  |
| **Physical–chemical environment:** water, coupled to the atmosphere and lake sediments and influenced by the meteorology characteristic of a specific latitude and altitude. | **Physical–chemical environment:** soil, coupled to the atmosphere and soil-water reserves and influenced by the meteorology characteristic of a specific latitude and altitude. | **Physical–chemical environment:** wastewater, largely uninfluenced by the meteorology characteristic of a specific latitude and altitude. |
| **Energy source: the sun** | **Energy source: the sun** | **Energy source: organic wastes (originally from the sun)** |
| **Primary production: algae, aquatic plants, and certain bacteria** | **Primary producers: grasses and flowers** | **Primary producers: none** |
| **Energy transfer: zooplankton, fish** | **Energy transfer: grasshoppers, ground squirrel, coyote** | **Energy transfer: bacteria, protozoans** |

**Figure 5.2**   An Ecosystem: Plants, Animals, Microorganisms, and their Physical–Chemical Environment

engineered (biological waste treatment) ecosystems (Figure 5.2). Taken together, all of the ecosystems of the world make up the *ecosphere*. **Ecology** is the study of structure and function of the ecosphere and its ecosystems: interactions between living things and their abiotic environment.

Although the field of taxonomy (classification of organisms) is highly dynamic and home to vigorous debate, biologists today place living things within one of three domains: (1) the Archaea, (2) the Bacteria, and (3) the Eukarya. The Archaea and the Bacteria are **prokaryotes**, meaning that their cellular contents, such as pigments and nuclear material, are not segregated within cellular structures (e.g., chloroplasts and the nucleus). While members of the Archaea and the Bacteria are similar in physical appearance, they differ in several important ways, including cellular composition and genetic structure. In our functional treatment of organisms, we consider the term *bacteria* to include members of both the domain Archaea and the domain Bacteria. The third domain, Eukarya, consists of organisms with segregated or compartmentalized organization—**eukaryotes**—possessing a nucleus and organelles such as chloroplasts. The domain Eukarya may be further divided into four kingdoms: (1) Protista (protists), (2) Fungi, (3) Plantae (plants), and (4) Animalia (animals).

Feeding strategies, of importance in many environmental engineering applications, also differ among the various kingdoms and domains.

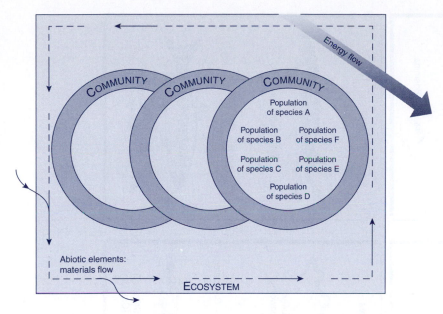

**Figure 5.3** Biotic Component of an Ecosystem, Organized According to Species, Populations, and Communities  In this schematic, energy flows through and chemicals cycle largely within the ecosystem. Natural, engineered, and industrial environments may be considered as ecosystems. For example, various biological processes employed for wastewater treatment (activated sludge, wetland, lagoon) have communities composed of a variety of microorganism populations. The nature of the ecosystem is determined by the physical design of the unit processes and by the chemical and biological character of the wastewater entering the system.

From Mihelcic (1999). Reprinted with permission of John Wiley & Sons, Inc.

Some organisms obtain their food by **absorption** (uptake of dissolved nutrients, as in kingdom Fungi), some through **photosynthesis** (fixation of light energy into simple organic molecules, as in kingdom Plantae), and some by *ingestion* (intake of particulate nutrients, as in kingdom Animalia). Some members of the kingdoms Plantae and Protista combine phototrophy and heterotrophy in mixotrophy, a practice where nutrition comes from photosynthesis and the uptake of dissolved and/or particulate organic carbon.

Domains may be subdivided into kingdoms, phyla, classes, orders, families, genera, and species. A **species** is a group of individuals that possesses a common gene pool and that can successfully interbreed. Each species is assigned a scientific name (genus plus species) in Latin, to avoid confusion associated with common names. Under this system of binomial nomenclature, *Sander vitreus* is the scientific name for the fish species commonly referred to as walleye, walleye pike, pike, pike perch, pickerel, yellow pike, yellow pickerel, yellow pike perch, or yellow walleye.

All of the members of a species in a given area make up a **population**—for example, the walleye population of a lake. All of the populations (of different species) that interact in a given system make up the **community**—for example, the fish community of a lake. Finally, as shown in Figure 5.3, all of the communities plus the abiotic factors make up the ecosystem (here, a lake) and the ecosystems, the ecosphere.

### 5.1.1 MAJOR ORGANISM GROUPS

A wide variety of organisms are encountered in **natural systems** (for example, lakes and rivers, wetlands, soil) and **engineered systems** (for example, wastewater treatment plants, landfills, constructed wetlands, and bioretention cells). Features of major organism groups especially important in environmental engineering are illustrated in Figure 5.4.

**Learn More on Ecosystems**
http://www.epa.gov/ebtpages/ecosystems.html

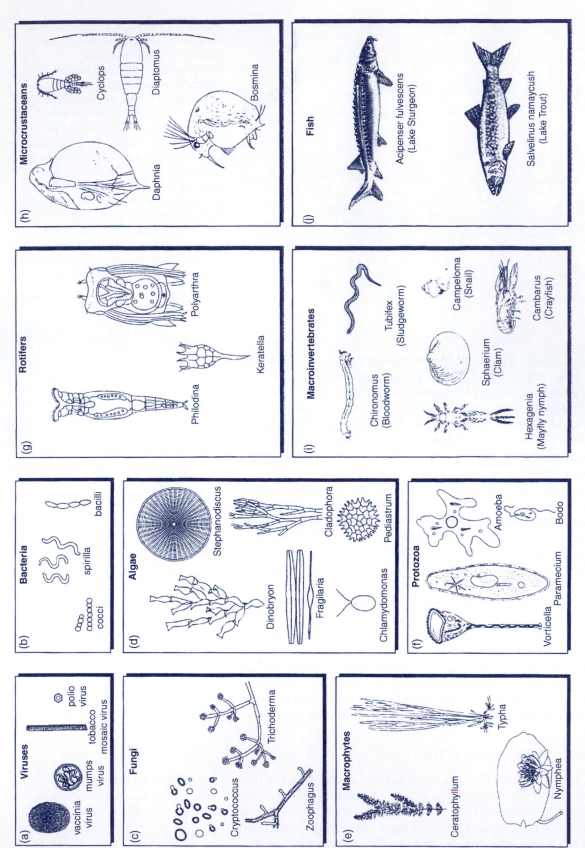

**Figure 5.4** **Major Organism Groups with Representative Members** The groups are (a) viruses, (b) bacteria, (c) fungi, (d) algae, (e) macrophytes, (f) protozoa, (g) rotifers, (h) microcrustaceans, (i) macroinvertebrates, and (j) fish.

From Mihelcic (1999). Reprinted with permission of John Wiley & Sons, Inc.

More than half of the endangered species in the United States are higher plants. The latest report from the **International Panel on Climate Change (IPCC)** indicates, with very high confidence, that recent warming is strongly affecting terrestrial biological systems, including poleward and upward shifts in species ranges, earlier spring greening, and greater potential for disturbances from pests and fire (IPCC, 2007). At the same time, higher plants significantly contribute to the removal of greenhouse gases from the atmosphere, removing as much $300 \times 10^{12}$ kg of carbon dioxide from the atmosphere every year—the equivalent of $CO_2$ emissions from approximately 40 billion automobiles.

As we move from consideration of individual organisms to populations and communities of organisms, we should not lose sight of the attributes of the species that make their roles in the ecosystem special. Furthermore, interaction among organism groups results in highly dynamic communities in both natural and engineered systems. Seasonal cycles and natural and human perturbations of the

**Climate Change and Ecosystems**
http://www.epa.gov/climatechange/effects/eco.html

---

**Box / 5.1    Producing Biodiesel from Algae**

Biodiesel is a renewable fuel that can be manufactured from vegetable oils, animal fats, waste restaurant grease, and algae. While a number of feedstocks are currently being explored for biodiesel production, algae have emerged as one of the most promising sources. Algae are fast-growing (with doubling times for many species on the order of hours) and can thrive in virtually any climate, overcoming a limitation of producing biodiesel from crops. The yield of oil from algae (some species have a 50 percent oil content) are much higher than those from traditional oilseeds—

potentially producing 250 times the amount of oil per acre as soybeans. Algae grown in photo-bioreactors like those shown in Figure 5.5 are harvested and pressed. Oil collected by pressing can then be converted to biodiesel through traditional transesterification reactions—the same ones used for vegetable oil. Overall, algae are remarkable and efficient biological factories capable of taking waste carbon dioxide, such as from municipal wastewater or power plant emissions, and converting it into a high-density liquid form of energy: natural oil.

**Figure 5.5    Conceptual Rendering of Large-Scale Photo-Bioreactors Used to Grow Algae**   Provided with sunlight, water, carbon dioxide, and some added nutrients, algae can produce about 10,000 gallons of biodiesel per acre per year.

environment often lead to dramatic shifts in population size and community structure. For example, the transparency or clarity of lakes varies with the quantity of soil particles and fertilizers delivered from terrestrial sources by tributary streams. The abundance of algae may, in turn, fluctuate with the size of the microcrustacean populations that graze on them and with the availability of nutrients introduced from the watershed. In some lakes, water clarity can go from "crystal clear" to "pea soup" to "crystal clear" over a matter of days as algal and microcrustacean populations wax and wane.

## 5.2 Population Dynamics

Population dynamics play a role in the fate of fecal bacteria discharged to surface waters, the efficiency of microorganisms in biological treatment, and substrate–organism interactions in the cleanup of contaminated soils. Other applications include control of nuisance algae growth in lakes, biomanipulation as a management approach for surface water quality, and the transfer of toxic chemicals through the food chain. Our ability to manage and protect the environment can be enhanced through an understanding of **population dynamics**, for example, by simulating or modeling the response of populations to environmental stimuli. In studying population dynamics, it is important to remember that, like other organisms, humans represent a population, one that may grow exponentially and experience the stress of approaching its carrying capacity.

### 5.2.1 UNITS OF EXPRESSION FOR POPULATION SIZE

Although it is the individual that is born, the individual that reproduces, and the individual that dies, in an environmental context these events are best appreciated by examining entire populations. And while it is possible to characterize individual populations through direct enumeration (for example, the number of alligators), the populations making up natural or engineered ecosystems include organisms of widely differing sizes. Thus, a "head count" provides a poor representation of population size and function where an estimate of all living material or biomass is desired. An alternative approach is to use a common constituent such as dry weight (g DW), organic carbon content (g C), or for plants, chlorophyll content (g Chl). For example, we might report plant biomass as g $DW/m^2$ for grasslands, metric tons C/hectare for forests, and mg $chl/m^3$ for lakes. In the wastewater treatment process, microorganisms exist in a mixture with waste solids. Here, biomass is expressed as total suspended solids (TSS) or volatile suspended solids (VSS), measures comparable to dry weight.

### 5.2.2 MODELS OF POPULATION GROWTH

A mass balance can be applied to the study of population dynamics in living organisms. Consider the case of the algal or bacterial community of a lake or river or the community of microorganisms in a waste treatment system. The mass balance on biomass may be written as follows:

$$V\frac{dX}{dt} = QX_{in} - QX \pm \text{reaction} \tag{5.1}$$

$V$ is volume (L), $X$ is biomass (mg/L), $t$ is time (days), $Q$ is flow (L/day), and *reaction* refers to all the kinetic processes mediating the growth or death of the organisms. Each term in Equation 5.1 has units of mass per time (mg/day).

To simplify the conceptual development of the models that follow, the flow terms will be ignored here (thus, $Q$ is equal to 0 in a batch reactor). Assuming that first-order kinetics adequately describe the reaction term (in this case, population growth), Equation 5.1 can be rewritten as follows:

$$V\frac{dX}{dt} = VkX \tag{5.2}$$

To simplify, divide both sides of Equation 5.2 by $V$:

$$\frac{dX}{dt} = kX \tag{5.3}$$

**Spreadsheets for
Bacterial Growth**

where $k$ is the first-order rate coefficient (time$^{-1}$). Because the reaction term is describing growth, the right side of Equation 5.3 is positive.

We will use this equation to develop realistic, but not overly complex, models to simulate the rates of organism growth. Three models are introduced here, describing unlimited (exponential), space-limited (logistic), and resource-limited (Monod) growth.

EXPONENTIAL OR UNLIMITED GROWTH The population dynamics of many organisms, from bacteria to humans, can be described using a simple expression, the **exponential-growth model**:

$$\frac{dX}{dt} = \mu_{max}X \tag{5.4}$$

Equation 5.4 is identical to Equation 5.3, with $\mu_{max}$, the maximum specific growth rate coefficient (day$^{-1}$) being a special case of the first-order rate constant $k$. The coefficient $\mu_{max}$ describes the condition where a full complement of energy reserves may be directed to growth, unaffected by feedback from crowding or resource competition or limitations. In addition to directing energy reserves toward growth, organisms must pay a "cost of doing business." Here, energy reserves mobilized through respiration are used to support cell maintenance and reproduction. In the terminology of wastewater engineering, this is *endogenous* (derived within) decay. The organism's respiratory demand may be represented as in Equation 5.4, using a first-order respiration or decay coefficient:

$$\frac{dX}{dt} = -k_dX \tag{5.5}$$

where $k_d$ is the respiration rate coefficient ($day^{-1}$). Here, the right-side term is negative because it represents a loss of biomass. In some situations, the definition of $k_d$ is expanded to include other losses, such as settling and predation.

Equations 5.4 and 5.5 may be combined:

$$\frac{dX}{dt} = (\mu_{max} - k_d)X \qquad (5.6)$$

and integrated to yield

$$X_t = X_0 e^{(\mu_{max} - k_d)t} \qquad (5.7)$$

where $X_t$ is the biomass at some time $t$ and $X_0$ is the initial biomass, reported as numbers or as a surrogate concentration, such as mg DW/L. The term ($\mu_{max} - k_d$) may also be thought of as the net effect of energy applied to growth minus the energy applied to respiration and termed $\mu_{net}$:

$$X_t = X_0 e^{\mu_{net} t} \qquad (5.8)$$

The expression utilized in Equation 5.7 will be retained here for clarity.

example/5.1 Exponential Growth and the Effect of the Specific Growth Rate on the Rate of Growth

Consider a population or community with an initial biomass ($X_0$) of 2 mg DW/L, a maximum specific growth rate ($\mu_{max}$) of 1.1/day, and a respiration rate coefficient of 0.1/day. Determine the biomass concentration (mg DW/L) over a time period of 10 days.

solution

Assume exponential growth. The biomass at any time is given by Equation 5.7:

$$X_t = X_0 e^{(\mu_{max} - k_d)t}$$

and

$$X_t = 2 \times e^{(1.1/day - 0.1/day)t}$$

Table 5.1 and Figure 5.6a present the results.

The J-shaped form of Figure 5.6a is typical of exponential growth. The steepness of the curve is determined by the value of the net specific growth rate coefficient ($\mu_{net} = \mu_{max} - k_d$). The influence of the value of $\mu_{net}$ on the shape of the growth curve is shown in Figure 5.6b.

## Table / 5.1

**Results of Calculations in Example 5.1 for Biomass as a Function of Time Using the Exponential-Growth Model**

| Time (days) | Biomass (mg DW/L) | Time (days) | Biomass (mg DW/L) |
|---|---|---|---|
| 1 | 5 | 6 | 807 |
| 2 | 15 | 7 | 2,193 |
| 3 | 40 | 8 | 5,962 |
| 4 | 109 | 9 | 16,206 |
| 5 | 297 | 10 | 44,053 |

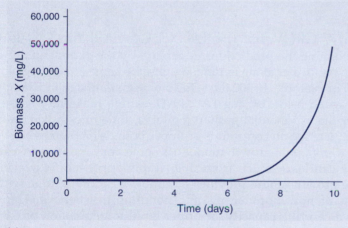

(a)

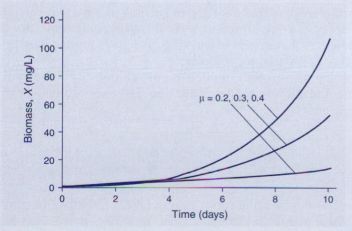

$\mu = 0.2, 0.3, 0.4$

(b)

**Figure 5.6** **Effect of Specific Growth Rate on Exponential Growth**
(a) Exponential population growth as determined in Example 5.1.
(b) Population growth according to the exponential model for three values of the specific growth rate coefficient. As $\mu$ increases, the rate of population growth ($dX/dt$) also increases.

From Mihelcic (1999). Reprinted with permission of John Wiley & Sons, Inc.

Note the similarity between the exponential growth model (Equation 5.4) applied here for living organisms,

$$\frac{dX}{dt} = \mu_{\max}X$$

with the expression for first-order decay introduced in Chapters 3 and 4 for application to chemical losses:

$$\frac{dC}{dt} = -kC$$

Both of these are first-order expressions; that is, both rates are a direct function of concentration (organism or chemical). However, organism concentrations typically increase exponentially (growth), while chemical concentrations decrease exponentially (decay).

**Modeling Growth of Endangered Yellowhead Birds in New Zealand**

## LOGISTIC GROWTH: THE EFFECT OF CARRYING CAPACITY

If we examine the predictions generated by the exponential-growth model a bit further along in time, we observe some interesting biomass levels. For example, in 100 days, the biomass simulated in Example 5.1 would reach $5.4 \times 10^{43}$ mg DW/L! Does this make any sense? No wonder the exponential-growth model is sometimes called *unlimited growth*: there are no constraints or upper bounds on biomass.

The exponential-growth model has some appropriate applications, and we can learn much from this simple approach. However, the **logistic-growth model** provides a framework more in tune with our concept of how populations and communities behave. Here, we invoke a **carrying capacity**, or upper limit, to population or community size (biomass) imposed by environmental conditions. Figure 5.7 illustrates the concept of carrying capacity and identifies space limitation and density-dependent losses such as disease and predation as components of environmental conditions. Food limitation is not addressed, as the concept of carrying capacity is limited here to space-related or nonrenewable resources.

The logistic-growth model is developed by modifying the exponential-growth model (Equation 5.6) to account for carrying-capacity effects:

$$\frac{dX}{dt} = (\mu_{\max} - k_{\mathrm{d}})\left(1 - \frac{X}{K}\right)X \qquad \text{(5.9)}$$

where $K$ is the carrying capacity (mg DW/L), that is, the maximum sustainable population biomass.

To appreciate the way in which carrying capacity mediates the rate of population growth, examine the behavior of the second term

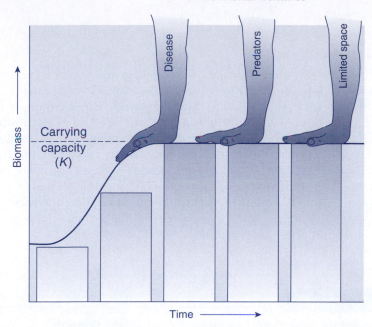

**Figure 5.7** **Effect on Biomass of Limitation by Nonrenewable (Space-Related) Resources as Manifested through Carrying Capacity** According to the logistic-growth model, environmental resistance (represented by the downward pressure of the hand) reduces the growth rate. At some time, the population reaches a carrying capacity that the population cannot exceed.

Adapted from Enger et al., 1983; figure from Mihelcic (1999). Reprinted with permission of John Wiley & Sons, Inc.

in parentheses in Equation 5.9. Note that when population size is small ($X \ll K$), Equation 5.9 reduces to the exponential-growth model, and as the carrying capacity is approached ($X \rightarrow K$), the population growth rate approaches zero. Equation 5.9 can be integrated to yield

$$X_t = \frac{K}{1 + \left[\left(\dfrac{K - X_0}{X_0}\right) e^{-(\mu_{max} - k_d)t}\right]} \qquad (5.10)$$

Equation 5.10 permits calculation of biomass as a function of time according to the logistic-growth model.

---

**example/5.2** Logistic Growth

Consider the population from Example 5.1 ($X_0 = 2\,\text{mg DW/L}$; $\mu_{max} = 1.1/\text{day}$; $k_d = 0.1/\text{day}$), but with a carrying capacity ($K$) of 5,000 mg DW/L. Determine the population biomass over a time period of 10 days.

example/5.2 Continued

## solution

Use the carrying-capacity term, $K$, to apply the logistic-growth model. Use Equation 5.10 to solve for the biomass concentration over the 10-day period. Table 5.2 and Figure 5.8a show the population biomass over time. In this example, the specific growth rate begins to decrease after several days and approaches zero. Also, the biomass concentration levels off over time as the carrying capacity (in this case, 5,000 mg DW/L) is approached.

Figure 5.8b compares the exponential and logistic growth models. Note that both models predict the same population behavior at low population numbers. This suggests that the exponential model may be appropriately applied under certain conditions.

## Table / 5.2

**Population over Time Determined for Logistic Growth Examined in Example 5.2**

| Time (days) | Biomass (mg DW/L) | $\mu_{max} - k_d$ (day$^{-1}$) |
|---|---|---|
| 0 | 2 | 1.000 |
| 1 | 7 | 0.999 |
| 2 | 22 | 0.996 |
| 3 | 72 | 0.986 |
| 4 | 232 | 0.954 |
| 5 | 695 | 0.861 |
| 6 | 1,745 | 0.651 |
| 7 | 3,201 | 0.360 |
| 8 | 4,276 | 0.145 |
| 9 | 4,757 | 0.049 |
| 10 | 4,924 | 0.015 |
| 11 | 4,977 | 0.005 |
| 12 | 4,993 | 0.001 |
| 13 | 4,998 | <0.001 |
| 14 | 4,999 | <0.001 |
| 15 | 5,000 | 0.000 |

## example/5.2 Continued

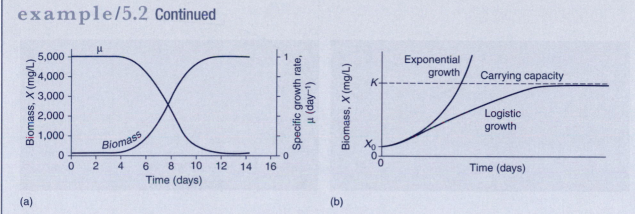

(a)                                              (b)

**Figure 5.8   Application of the Logistic-Growth Model**   (a) Population biomass and specific growth rate according to the logistic model as determined in Example 5.2. (b) A comparison of the exponential and logistic growth models. Both predict a similar population response in the early stages when population is small. However, the exponential model predicts that unlimited growth will continue, while the logistic model predicts an approach to the carrying capacity.

From Mihelcic (1999). Reprinted with permission of John Wiley & Sons, Inc.

**RESOURCE-LIMITED GROWTH: THE MONOD MODEL** In nature, it is more common for organisms to reach the limits imposed by reserves of **renewable resources**—for example, food—than to approach the limits established by carrying capacity. The relationship between nutrients and the population or community growth rate can be described using the **Monod model**. In this model, the maximum specific growth rate is modified to account for the effects of limitation, in this case by renewable resources:

$$\mu = \mu_{max} \frac{S}{K_s + S}$$   (5.11)

where $\mu$ is the specific growth rate (day$^{-1}$), $S$ is the nutrient or substrate concentration (mg S/L), and $K_s$ is the half-saturation constant (mg S/L).

The **substrate** or "food" in Equation 5.11 may be either a macronutrient (for instance, organic carbon in biological waste treatment) or a growth-limiting micronutrient (such as phosphorus in lakes). As illustrated in Figure 5.9a, the **half-saturation constant** ($K_s$) is defined as the substrate concentration ($S$) at which the growth rate is one-half of its maximum value, that is, $\mu = \mu_{max}/2$.

The magnitude of $K_s$ reflects the ability of an organism to consume renewable resources (substrate) at different substrate levels. Organisms with a low $K_s$ approach the maximum specific growth rate ($\mu_{max}$) at comparatively low substrate concentrations, while

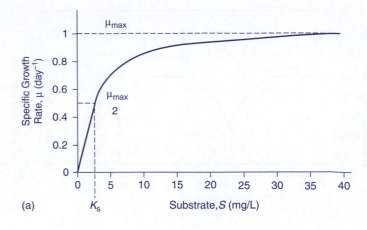

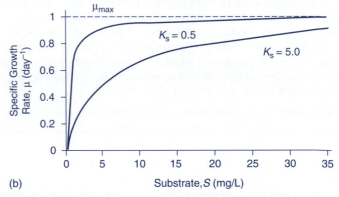

**Figure 5.9   The Monod Model**   (a) Basic Monod model, illustrating the relationship between the specific growth rate ($\mu$)  and substrate ($S$) concentration. At high substrate concentrations ($[S] \gg K_s$), $\mu$ approaches its maximum value $\mu_{max}$, and the growth is essentially independent of substrate concentration (i.e., zero-order kinetics). At low substrate concentrations ($[S] \ll K_s$), $\mu$ is directly proportional to substrate concentration (i.e., first-order kinetics). (b) Application of the Monod model illustrating the effect of variation in $K_s$. Organisms with a low $K_s$ approach their maximum specific growth rate at lower substrate concentrations and thus may have a competitive advantage.

From Mihelcic (1999). Reprinted with permission of John Wiley & Sons, Inc.

those with high $K_s$ values require higher levels of substrate to achieve the same level of growth. Figure 5.9b illustrates how variability in the half-saturation constant affects growth rate. The physiological basis for this phenomenon lies in the role of enzymes in catalyzing biochemical reactions; low half-saturation constants reflect a strong affinity of the enzyme for substrate.

The Monod model (Equation 5.11) can be substituted into Equation 5.6 (the exponential model) to yield

$$\frac{dX}{dt} = \left( \mu_{max} \frac{S}{K_s + S} - k_d \right) X \qquad (5.12)$$

## example/5.3 Resource-Limited Growth

Figure 5.10 shows population density as a function of time for two species of *Paramecium* (a protozoan) grown separately and in mixed culture. Grown separately, both species do well, acquiring substrate and achieving high biomass densities. In mixed culture, however, one species dominates, eliminating the other species. Organisms with a small $K_s$ have a competitive advantage because they can reach a high growth rate at lower substrate levels. This can be demonstrated by inspection of the Monod model.

A basic concept of ecology, the **principle of competitive exclusion**, states that two organisms cannot coexist if they depend on the same growth-limiting resource. How then do these two species of *Paramecium* manage to coexist in the natural world? Why isn't the poor competitor extinct?

## solution

The answer lies in another ecological principle, *niche separation*. The term *niche* refers to the unique functional role or "place" of an organism in the ecosystem. Organisms that are poorly competitive from a purely kinetic perspective (for example, $K_s$) can survive by exploiting a time or place where competition can be avoided.

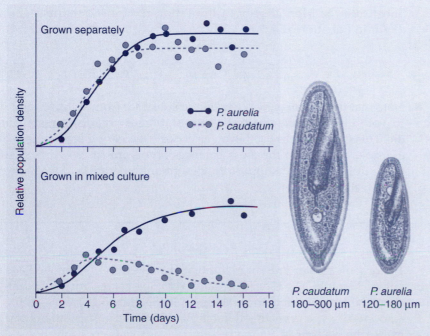

**Figure 5.10  Two Species of Paramecium Grown Separately and in Mixed Culture**  In separate culture, both species do well, acquiring substrate and achieving high biomass densities. In mixed culture, however, one species dominates and eliminates the other species

Adapted from Ricklefs (1983); figure from Mihelcic (1999). Reprinted with permission of John Wiley & Sons, Inc.

**YIELD COEFFICIENT: RELATING GROWTH AND SUBSTRATE UTILIZATION** While attention has been largely devoted here to tracking biomass, substrate fate may be of more interest in many engineering applications. To model substrate concentrations, or to relate substrate consumption to organism growth, we apply the **yield coefficient (Y)**, defined as the quantity of organisms produced per unit substrate consumed:

$$Y = \frac{\Delta X}{\Delta S} \tag{5.13}$$

$Y$ has units of biomass produced per mass of substrate consumed. A yield coefficient value of $Y = 0.2$ indicates that 20 mg of biomass are produced for every 100 mg of substrate consumed. Note that $Y$ for organic carbon is always less than 1, because organisms are not 100 percent efficient in converting substrate to biomass and because some energy must be expended for cell maintenance.

The yield coefficient is also commonly applied to relate the rate of substrate utilization $(dS/dt)$ to the rate of organism growth $(dX/dt)$:

$$\frac{dS}{dt} = -\frac{1}{Y}\left(\frac{dX}{dt}\right) \tag{5.14}$$

Substitute the Monod growth limitation model (Equation 5.12) for $dX/dt$ in Equation 5.14:

$$\frac{dS}{dt} = -\frac{1}{Y}\left(\mu_{max}\frac{S}{K_s + S} - k_d\right)X \tag{5.15}$$

Note that the role of $\mu_{max}$ is to remove substrate from solution, and the role of $k_d$ is to recycle substrate into solution as organisms respire and die. This expression is used in a variety of engineering applications, for example, to develop the mass balances on organism growth and substrate utilization that support the design and operation of waste treatment facilities.

example/5.4 Yield Coefficient

The organic matter present in municipal wastewater is removed at a rate of 25 mg $BOD_5$/L-hr in an aerated biological reactor. BOD (biochemical oxygen demand), defined in Section 5.4, refers to the amount of oxygen consumed in oxidizing a given amount of organic matter, here a representation by effect of substrate concentration.

Use the yield coefficient to compute the mass of microorganisms (measured as volatile suspended solids, or VSS) produced daily due to the consumption of organic matter by microorganisms in the aeration basin. Assume that the aeration basin has a volume of $1.5 \times 10^6$ L and the yield coefficient $Y$ equals 0.6 mg VSS/mg $BOD_5$.

## example/5.4 Continued

### solution

The yield coefficient $Y$ relates the rate of substrate (in this case, organic matter) disappearance to the rate of cell growth. This relationship (Equation 5.14) is written as follows:

$$\frac{dS}{dt} = -\frac{1}{Y}\frac{dX}{dt}$$

Therefore,

$$Y\frac{dS}{dt} = -\frac{dX}{dt}$$

Substitute the given values for $Y$ and the rate of substrate depletion:

$$\frac{0.6\ \text{mg VSS}}{\text{mg BOD}_5} \times \frac{25\ \text{mg BOD}_5}{\text{L-hr}} = \frac{15\ \text{mg VSS}}{\text{L-hr}}$$

Next, convert this value to a mass per day basic:

$$\frac{15\ \text{mg VSS}}{\text{L-hr}} \times 1.5 \times 10^6\ \text{L} \times \frac{24\ \text{hr}}{\text{day}} = \frac{5.4 \times 10^8\ \text{mg VSS}}{\text{day}}$$

$$= \frac{540\ \text{kg VSS}}{\text{day}}$$

Note that a lot of biological solids are produced at a wastewater treatment plant each day. This explains why engineers spend so much time designing and operating facilities to handle and dispose of the residual solids (sludge) generated at a wastewater treatment plant.

BIOKINETIC COEFFICIENTS The terms $\mu_{max}$, $K_s$, $Y$, and $k_d$ are commonly referred to as biokinetic coefficients because they provide information about the manner in which substrate and biomass change over time (kinetically). Values for these coefficients may be derived from thermodynamic calculations or through field observation and laboratory experimentation; literature compilations of coefficients derived in this fashion are available. Table 5.3 provides some representative values for the biokinetic coefficients as applied in municipal wastewater treatment.

BATCH GROWTH: PUTTING IT ALL TOGETHER Respiration and the growth-mediating mechanisms introduced earlier may be integrated into a single expression describing population growth in batch culture:

$$\frac{dX}{dt} = \left( \mu_{max}\frac{S}{K_s + S} - k_d \right)\left( 1 - \frac{X}{K} \right)X \tag{5.16}$$

## Table / 5.3

**Typical Values for Selected Biokinetic Coefficients for the Activated-Sludge Wastewater Treatment Process**

| Coefficient | Range of Values | Typical Value |
|---|---|---|
| $\mu_{max}$ | 0.1–0.5 hr$^{-1}$ | 0.12 hr$^{-1}$ |
| $K_s$ | 25–100 mg BOD$_5$/L | 60 mg BOD$_5$/L |
| $Y$ | 0.4–0.8 VSS/mg BOD$_5$ | 0.6 VSS/mg BOD$_5$ |
| $k_d$ | 0.0020–0.0030 hr$^{-1}$ | 0.0025 hr$^{-1}$ |

SOURCE: Tchobanoglous et al., 2003.

Then we can relate substrate utilization to Equation 5.16 through the yield coefficient:

$$\frac{dS}{dt} = -\frac{1}{Y}\left(\mu_{max}\frac{S}{K_s + S} - k_d\right)\left(1 - \frac{X}{K}\right)X \qquad (5.17)$$

Although of considerable importance in natural systems, the carrying-capacity term is not typically included in biokinetic models for municipal wastewater engineering, because these systems are designed to operate below their maximum sustainable biomass.

Figure 5.11 illustrates substrate utilization and the attendant phases of population growth in batch culture (no inflow or outflow), according to Equations 5.16 and 5.17. For simplicity, it is assumed that no substrate recycle occurs. Three phases of growth are described in Table 5.4: the exponential or log growth phase, the **stationary phase**, and the death phase. Certain simplifying assumptions regarding growth conditions during the exponential and death phases permit the calculation of substrate and biomass changes at those times.

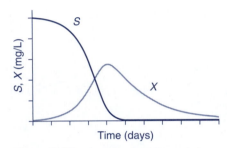

**Figure 5.11  Population Growth in Batch Culture** This graph illustrates an exponential-growth phase where initially substrate ($S$) is abundant and biomass ($X$) is low. This is followed by a stationary phase where substrate levels support a growth rate equal to the respiration rate and then a death phase where substrate is exhausted and the population is in decline due to unsupported respiratory demand.

## Table / 5.4

**Phases of Population Growth**

| Phase of Growth | Description | dX/dt |
|---|---|---|
| Exponential or log growth phase | Substrate uptake and growth are rapid. Growth pattern is well approximated by the exponential model (Equation 5.4). | $dX/dt > 0$ |
| Stationary phase | Growth slows due to substrate depletion (or crowding). For a brief period, gains through growth are exactly balanced by losses to respiration and death. | $dX/dt = 0$ |
| Death phase | Substrate is no longer available to support growth and losses to respiration and death. Approximated as an exponential decay (Equation 5.5). | $dX/dt < 0$ |

# example/5.5 Simplified Calculations of Substrate and Biomass

The differential equations that describe biomass and substrate dynamics (Equations 5.16 and 5.17) contain nonlinear terms that require specialized numerical methods for their solution. By applying certain simplifying assumptions, however, we can learn quite a bit about the dynamics of microbial populations and communities. Consider a population of microorganisms with the following characteristics growing in batch culture: initial biomass $X_0 = 10$ mg DW/L; maximum specific growth rate $\mu_{max} = 0.3$/day; half-saturation constant $K_s = 1$ mg/L; carrying capacity $K = 100,000$ mg DW/L; respiration rate coefficient $k_d = 0.05$/day; initial substrate concentration $S_0 = 2,000$ mg S/L; and yield coefficient $Y = 0.1$ mg DW/mg S.

1. Determine whether this population will ever approach its carrying capacity.

2. Calculate the population biomass after the first three days of growth.

3. Calculate the substrate concentration after the first three days of growth.

4. If the population peaks at 100 mg DW/L when the substrate runs out, calculate the biomass 10 days after the peak.

## solution

1. Changes in substrate and biomass concentrations over time are related by the yield coefficient as given by Equation 5.14. The maximum attainable biomass of this population, based on substrate availability, is given as the product of the maximum potential change in substrate concentration and the yield coefficient:

$$dX = dS \times Y = \frac{2,000 \text{ mg } S}{L} \times \frac{0.1 \text{ mg DW}}{\text{mg } S} = 200 \text{ mg DW}$$

   This is well below the carrying capacity of 100,000 mg DW/L; therefore, the population will not run out of substrate and never approach the carrying capacity.

2. Early in the growth phase when substrate concentrations are high (Monod term, $S/(K_s + S)$, approaches 1) and biomass concentrations are low (carrying-capacity term, $1 - X/K$, approaches 1 ), Equation 5.16 reduces to

$$\frac{dX}{dt} = (\mu_{max} - k_d)X$$

Integrate:

$$X_t = X_0 e^{(\mu_{max} - k_d)t}$$

$$X_3 = \frac{10 \text{ mg DW}}{L} \times e^{(0.3/\text{day} - 0.05/\text{day})3 \text{ day}} = \frac{21 \text{ mg DW}}{L}$$

3. The change in substrate concentration over the 3-day period is given by Equation 5.14:

$$\frac{dS}{dt} = -\frac{1}{Y}\left(\frac{dX}{dt}\right)$$

From the previous calculation, $dX/dt$ over three days is $X_3 - X_0 = 21 - 10 = 11 \text{ mg DW/L}$, and

$$\frac{dS}{dt} = -\left(\frac{0.1 \text{ mg DW}}{\text{mg } S}\right)^{-1} \times \frac{11 \text{ mg DW}}{L} = \frac{-110 \text{ mg } S}{L}$$

The substrate concentration after 3 days of growth is given by

$$S_3 = S_0 - \frac{dS}{dt} = 2{,}000 - 110 = \frac{1{,}890 \text{ mg } S}{L}$$

4. when substrate is exhausted, the Monod term equals 0, and Equation 5.16 reduces to Equation 5.5 and its analytical solution, $X_t = X_0 e^{-k_d t}$. In this case, the peak population decays according to first-order kinetics, so that 10 days after the peak,

$$X_t = X_0 e^{-k_d t} = 100 \times e^{(-0.05/\text{day} \times 10 \text{ day})} = \frac{61 \text{ mg DW}}{L}$$

**Class Discussion**

Having recently completed a study of population growth, we might ask ourselves which model best fits the data presented in Figure 5.12 and what that model suggests is likely to happen next.

GROWTH MODELS AND HUMAN POPULATION Despite the complexity of their reproduction, populations of humans can be simulated using the types of models described in this chapter. Human populations remained relatively unchanged for thousands of years, increasing much more rapidly in modern times (see Figure 5.12). In 2007, the average growth rate coefficient for the world's population was about 0.012 year$^{-1}$ (1.2 percent per year or 12 births per 1,000 people per year). At this rate, Earth's population will double every 60 years, with most of this growth occurring in urban areas.

The 18th-century British economist Thomas Malthus considered the questions of what model describes population data and what it predicts. Recognizing that population increased exponentially, Malthus concluded that such growth was checked only by "misery or vice," meaning war, pestilence, and famine. He asserted that while populations increased

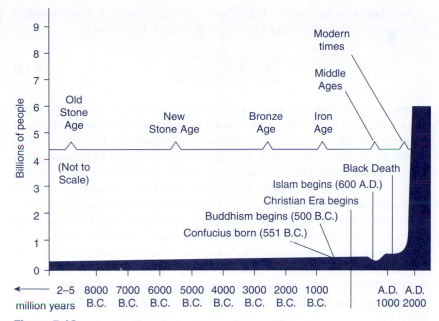

**Figure 5.12** Human Population Growth over Time

Adapted from Enger et al., *Environmental Science*, 1983, Wm. C. Brown Publishers with permission of the McGraw-Hill Companies. Figure reprinted from Mihelcic (1999), with permission of John Wiley & Sons, Inc.

exponentially, the "means of subsistence" (food) increased in a linear fashion. Therefore, it would be only a matter of time until demand outstripped supply, as shown in Figure 5.13, an event with catastrophic implications.

Two centuries following Malthus's predictions, we have not expended food resources, and the demand and supply curves shown in Figure 5.13 have not yet intersected. This is thanks to advances in agricultural production through the use of fertilizers and pesticides—which have their own environmental and health impacts, such as water scarcity, dependence on fossil fuels, soil erosion, loss of biodiversity, and chemical contamination of surface waters and groundwater.

In **The Limits to Growth**, the Club of Rome warned:

*If "present growth trends in world population, industrialization, pollution, food production, and resource depletion continue unchanged, the limits to growth will be reached sometime within the next one hundred years." They further predicted there would be an adverse impact on industrial capacity.*

This somewhat broader perspective on the threats that prescribe limits to growth is more consistent with our observations of the impacts of nutrient runoff, acid rain, contamination with heavy metals and toxic organic chemicals, depletion of the ozone layer, and carbon-driven global climate change. Malthus might have been surprised to see that we would "soil our nest" well before famine, the "last and most dreadful check to population," spread across our planet.

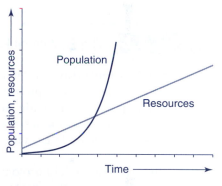

**Figure 5.13** Malthus's Predictions of Population and Resources Trends At some point, according to Malthus, demand will outstrip supply.

**Population Growth and Limiting Nutrients**

### Class Discussion

Engineers have a role to play in mitigating current threats and eliminating future insults with designs that improve the quality of life for Earth's population without the associated historical adverse environmental impacts. How do you see your role as a global citizen and as a professional?

## Class Discussion

While ecological footprint does provide general information about the impacts of consumption, what are some of the limitations to this approach? For example, is land area a good surrogate for all environmental impact? Is land contaminated during a process or at end of life considered differently than land used for growing organic crops? What challenges does this pose in using ecological footprint as an indicator of environmental sustainability?

**Calculate Your Personal Footprint**

www.myfootprint.org/en/

Another way to describe this phenomenon is through a relationship developed in the 1970s and known as the **IPAT equation**:

$$I = P \times A \times T \qquad\qquad (5.18)$$

where $I$ is the environmental impact, $P$ is population, $A$ is affluence, and $T$ is technology. Environmental impact ($I$) may be expressed in terms of resource depletion or waste accumulation. Population (P) refers to the size of the human population, affluence (A) to the level of consumption by that population, and technology (T) to the processes used to obtain resources and transform them into useful goods and wastes.

In addition to highlighting the contribution of population to environmental problems, the IPAT equation makes it clear that environmental problems involve more than pollution and are driven by multiple factors acting together to yield a compounding effect. The product of the affluence (A) and technology (T) terms in Equation 5.18 can be visualized as the per capita demand on ecosystem resources. This demand is sometimes quantified as an ecological footprint (see Box 5.2).

---

### Box / 5.2    The Ecological Footprint

An **ecological footprint** is a determination of the biologically productive area required to provide an individual's resource supplies and absorb the wastes their activities produce.

Another way to think of ecological footprint is as the ecological impact corresponding to the amount of nature an individual needs to occupy to keep intact his or her daily lifestyle (Wackernagel et al., 1997). It is generally assumed that, if the world's resources were allocated fairly, there would currently be 2.1 hectares of productive land allocated per person. This value is important because it serves as a benchmark for comparing the ecological footprint of the world's population. This benchmark number is derived from the fact that there is an amount of determined arable land, pasture, forest, ocean, and built-up environment. In addition, according to the World Commission on Environment and Development, a minimum of 12 percent of the world's ecological capacity should be preserved for biodiversity protection.

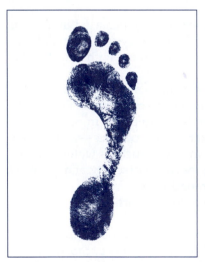

© Juri Samsonov/iStockphoto.

This value is incorporated into the calculation that determines there are only 2.1 hectares of productive land available for each person currently living on Earth.

Table 5.5 shows the ecological footprint and available ecological capacity (both as hectares per capita) of the world and selected nations. The table also shows the difference, or *ecological deficit*, which is determined by subtracting the footprint from the available ecological capacity of an individual country. Negative numbers indicate a deficit, and positive numbers indicate that some ecological capacity remains within the country's borders. The table shows that many of the world's nations are not currently sustainable if the rest of the world is to share their same level of current consumption of natural resources. This analysis assumes that the impact of an individual is felt within the physical land and ecosystems occupied by that person's home country. Of course, national borders do not bind many environmental emissions, so the impacts will most likely be felt beyond the home country.

## Table / 5.5

**Ecological Footprints around the World**

| | Ecological Footprint (ha per capita) | Biological Capacity (ha per capita) | Difference (ha per capita) |
|---|---|---|---|
| World | 2.7 | 2.1 | −0.6 |
| Bangladesh | 0.6 | 0.3 | −0.3 |
| Brazil | 2.4 | 7.3 | 4.9 |
| Canada | 7.1 | 20.0 | 13.0 |
| China | 2.1 | 0.9 | −1.2 |
| Germany | 4.2 | 1.9 | −2.3 |
| India | 0.9 | 0.4 | −0.5 |
| Japan | 4.9 | 0.6 | −4.3 |
| Jordan | 1.7 | 0.3 | −1.4 |
| Mexico | 3.4 | 1.7 | −1.7 |
| New Zealand | 7.7 | 14.1 | 6.4 |
| Nigeria | 1.3 | 1.0 | −0.4 |
| Russian Federation | 3.7 | 8.1 | 4.4 |
| South Africa | 2.1 | 2.2 | 0.1 |
| United Kingdom | 5.3 | 1.6 | −3.7 |
| United States | 9.4 | 5.0 | −4.4 |

SOURCE: Data from Living Planet Report, WWF, 2008.

On a global basis, the average footprint of the current world is 2.7 hectares per person, resulting in an ecological deficit of 0.6 hectare per person. It is clear that, to be ecologically sustainable, the world must either decrease its population or decrease the burden each person places on the environment (through more equitable sharing of the world's resources or wider use of green policies and technology), especially in developed countries that consume an unfair share of the world's resources. In fact, as the global population approaches 10 billion people in the upcoming decades, the footprint allocation of 2.1 hectares of productive land per person will approach 1 hectare, placing even more attention on issues of fairness and green policies and technologies.

In the United States, while the available ecological capacity is relatively large (5.0 hectares), there is an ecological footprint of 9.4 hectares, resulting in an overall deficit of 4.4 hectares per person. China's available ecological capacity is much lower than that of the United States (0.9 hectares per person), and its current footprint is much lower as well (2.1 hectares per person). China also has no additional "footprint" of its own to utilize. However, imagine what China's ecological deficit will be if its population of more than 1 billion attempts to emulate the current resource consumption patterns of the United States (remember the IPAT equation). In addition, you might ask yourself whether it is ethically responsible for citizens of the United States and other developed countries to consume the world's resources at current nonsustainable rates with the result that less-developed countries will lack resources to support future development.

**Class Discussion**

The IPAT equation also addresses the issue of how equity and sharing are critical factors in sustainable development. For example, as the 2 billion people currently living in the world who live in poverty (less than $2 per day) increase their affluence through increased personal income, other wealthy parts of the world will either have to share some of their affluence or provide green technology to those parts of the world to lessen the overall environmental impact of increases in population and affluence. What do you see as the global answer to this complex problem?

From Equation 5.18, it is apparent that only two options exist for reducing environmental impact:

1. Reduce population numbers ($P$), or

2. Reduce the magnitude of the per capita demand ($A \times T$).

The issues of population size ($P$), perhaps of paramount importance, and of affluence ($A$) are policy-oriented issues many engineers work on. Other efforts of the engineer in reducing environmental impact and advancing sustainability focus on ($T$), the design of greener technologies.

## 5.3   Energy Flow in Ecosystems

The character of Earth's many and varied ecosystems is determined to a large extent by their physical setting. Consider the changes in flora (plants) and fauna (animals) observed over the course of a long car trip, especially if traveling north–south or through dramatic changes in elevation. The physical setting includes climatic factors such as temperature (extreme values and duration of seasons), sunlight (day length and annual variation), precipitation (extremes and annual distribution), and wind. Other significant features of the physical setting include soil physics (particle size) and chemistry (pH, organic content, nutrients).

Given an appropriate physical setting, organisms require only two things from the environment: (1) energy to provide power and (2) chemicals to provide substance. Chemical elements are cycled within an ecosystem that could be regional or global, so continued function does not require that they be imported. Energy flows through and propels ecosystems; that is, it does not cycle but rather is converted to heat and lost for useful purposes forever.

### 5.3.1   ENERGY CAPTURE AND USE: PHOTOSYNTHESIS AND RESPIRATION

The sun is responsible, directly or indirectly, for virtually all of Earth's energy. Sunlight incident on an aquatic or terrestrial ecosystem is trapped by plant pigments, primarily chlorophyll, and that light energy is converted to chemical energy through a process termed **photosynthesis.**

**Artificial photosynthesis** is the term often used to describe engineered solar or photovoltaic cell systems designed to capture light energy and convert it into electrical energy. The use of solar energy to power engineered systems rather than the burning of fossil fuels can address many of the current environmental challenges including air pollution, climate change, and depletion of finite resources.

Chemical energy stored through photosynthesis is subsequently made available for use by organisms through respiration. Figure 5.14a provides a simplified representation of photosynthesis, represented as follows:

$$CO_2 + H_2O + \Delta \rightarrow C(H_2O) + O_2 \qquad (5.19)$$

where $\Delta$ is the sun's energy and $C(H_2O)$ is a general representation of organic carbon (for example, glucose, which is $C_6H_{12}O_6$ or $6\ C(H_2O)$). The free-energy change ($\Delta G$) for photosynthesis (Equation 5.19) is positive, so the reaction could not proceed without the input of energy from the sun.

Chlorophyll acts as an antenna, absorbing the light energy, which is then stored in the chemical bonds of the carbohydrates produced by this reaction. Oxygen is an important by-product of the process.

© Michael Krakowiak/iStockphoto.

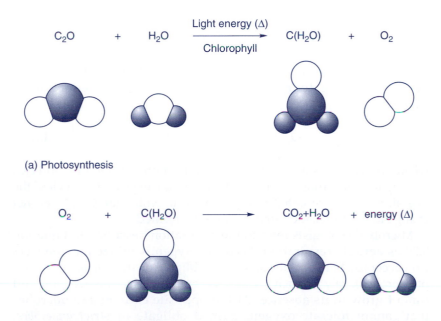

(a) Photosynthesis

(b) Respiration

**Figure 5.14** **Photosynthesis and Respiration** (a) Simplified version of photosynthesis, the process in which the sun's energy is captured by pigments such as chlorophyll and converted to chemical energy stored in the bonds of simple carbohydrates, e.g., $C(H_2O)$. More complex molecules (e.g., sugars, starches, cellulose) are then formed from simple carbohydrates. (b) Simplified version of respiration, the reverse of the photosynthetic process. The energy stored in chemical bonds (e.g., carbohydrates) is released to support metabolic needs.

Reprinted from Mihelcic (1999), with permission of John Wiley & Sons, Inc.

In relation to wastewater treatment, this **photosynthetic source** is considered to be a natural method for providing oxygen to wastewater, as happens in the aeration of lagoon-based treatment systems.

**Respiration** is the process by which the chemical energy stored through photosynthesis is ultimately released to do work in plants and other organisms (from bacteria to plants and animals):

$$C(H_2O) + O_2 \rightarrow CO_2 + H_2O + \Delta \qquad (5.20)$$

**Managing Ecosystems to Fight Poverty**
http://pdf.wri.org/wrro5_full_hires.pdf

Figure 5.14b provides a simplified representation of respiration. The reverse of photosynthesis, this reaction releases stored energy, making it available for cell maintenance, reproduction, and growth. The energy, denoted $\Delta$ in Equation 5.20, is equal to the free energy of reaction. Organisms are able to capture and utilize only a fraction (5 to 50 percent) of the total free energy of this reaction. Thus, all forms of life are by nature rather inefficient.

Respiration is what may be described chemically as an oxidation-reduction or **redox reaction**, which can be written in terms of the following two half-reactions. First, the oxidation of the organic carbon:

$$C(H_2O) + H_2O \rightarrow CO_2 + 4H^+ + 4e^- \qquad (5.21)$$

where the valence state of carbon goes from (0) in $C(H_2O)$ to (4+) in $CO_2$, yielding four electrons. And second, the reduction of oxygen:

$$O_2 + 4e^- + 4H^+ \rightarrow 2H_2O \qquad (5.22)$$

where the valence state of oxygen goes from (0) in $O_2$ to (2−) in $H_2O$, gaining four electrons. The two half-reactions can be added to yield the overall reaction presented in Equation 5.20. Note that there is no net change in electrons; they simply are redistributed.

Microbial ecologists refer to the respiration described in Equation 5.20 as **aerobic respiration**, because oxygen is utilized as the electron acceptor. Some bacteria, termed obligate or *strict aerobes* (or simply aerobes), rely exclusively on oxygen as an electron acceptor and cannot grow in its absence. At the opposite extreme are microbes that cannot tolerate oxygen, termed obligate or *strict anaerobes*. Facultative microbes can switch their metabolism between aerobic and anaerobic pathways, depending on the presence or absence of oxygen.

When oxygen is absent, **anaerobic respiration** takes place, utilizing a variety of other compounds as electron acceptors. Many bacteria can utilize oxygen as an electron acceptor but in its absence may utilize either nitrate or sulfate. Such bacteria are *facultative aerobes* and have a distinct ecological advantage over strict or obligate anaerobes or aerobes in environments that may be periodically devoid of oxygen. The

**Redox Reactions for Oxidation of Organic Matter Using Various Alternate Electron Acceptors**

| Electron Acceptor | Redox Reaction* | Equation No. |
|---|---|---|
| Nitrate | $C(H_2O) + NO_3^- \rightarrow N_2 + CO_2 + HCO_3^- + H_2O$ | (5.23) |
| Manganese | $C(H_2O) + Mn^{4+} \rightarrow Mn^{2+} + CO_2 + H_2O$ | (5.24) |
| Ferric iron | $C(H_2O) + Fe^{3+} \rightarrow Fe^{2+} + CO_2 + H_2O$ | (5.25) |
| Sulfate | $C(H_2O) + SO_4^{2-} \rightarrow H_2S + CO_2 + H_2O$ | (5.26) |
| Organic compounds | $C(H_2O) \rightarrow CH_4 + CO_2$ | (5.27) |

*Equations are not stoichiometrically balanced, so that participating species in the reactions may be more clearly emphasized.

terms *anaerobic* and *anoxic* are often used synonymously. In wastewater treatment and some natural-systems applications, **anoxic** refers to the case where oxygen is absent and respiration proceeds with nitrate as the electron acceptor.

Table 5.6 presents redox reactions for oxidation of organic matter using a variety of alternate electron acceptors, such as nitrate, manganese, and ferric iron. In the environment, these reactions are thought to take place in the sequence listed—the order of their favorability from a thermodynamic perspective. Thus, reduction of oxygen proceeds first, followed by nitrate, manganese, ferric iron, and sulfate, and finally fermentation occurs. This order is termed the **ecological redox sequence**, with each process carried out by different types of bacteria (for example, nitrate reducers and sulfate reducers).

Fermentation (Equation 5.27) is an anaerobic process mediated by yeasts and certain bacteria and differs from the other reactions in that organic matter is oxidized without an external electron acceptor. Here, organic compounds serve as both the electron donor and the electron acceptor, resulting in two end products, one of which is oxidized with respect to the substrate and the other of which is reduced. In the production of alcohol, for example, glucose ($C_6H_{12}O_6$, with C in the (0) valence state) is fermented to ethanol ($CH_2CH_3OH$, with C reduced to the (2−) valence state) and carbon dioxide ($CO_2$, with C oxidized to the (4+) valence state). **Methanogenesis** is a type of fermentation in which methane ($CH_4$) is an end product.

Organic matter produced in lakes and wetlands is broken down to carbon dioxide and stable, peat-like end products through aerobic and anaerobic respiration. Figure 5.15 illustrates the relative contribution of oxygen and various alternate electron acceptors to the oxidation of organic matter in the bottom waters and sediments of Onondaga Lake, located in New York. Approximately one-third of the organic matter decomposition during the summer period was

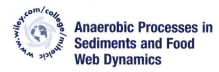

**Anaerobic Processes in Sediments and Food Web Dynamics**

www.wiley.com/college/mihelcic

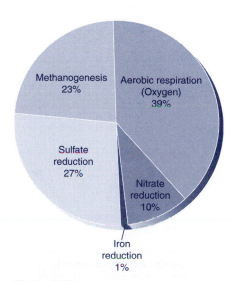

**Figure 5.15** Contribution of Various Terminal Electron Acceptors to the Oxidation of Organic Matter in the Bottom Waters of Onondaga Lake, New York

Data from Effler (1996); figure from Mihelcic (1999). Reprinted with permission of John Wiley & Sons, Inc.

accomplished aerobically, that is, having oxygen as the terminal electron acceptor, with the balance utilizing the alternate electron acceptors identified in Table 5.6.

## 5.3.2 TROPHIC STRUCTURE IN ECOSYSTEMS

In addition to energy, organisms require a source of carbon. Organisms that obtain their carbon from inorganic compounds (for example, $CO_2$ in Equation 5.19) are called **autotrophs**, loosely translated as self-feeders. This category includes photosynthetic organisms (green plants, including algae, and some bacteria) that use light as their energy source and nitrifying bacteria that use ammonia ($NH_3$) as their energy source. The simple carbohydrates produced through photosynthesis ($C(H_2O)$ in Equation 5.19) and the more complex organic chemicals synthesized later (for instance, starch, cellulose, fats, and protein) are collectively termed **organic matter**.

Organisms that depend on organic matter produced by others to obtain their carbon are termed **heterotrophs**, loosely translated as other feeders. This carbon source could be a simple molecule such as methane ($CH_4$) or a more complex chemical such as those listed previously. Animals and most bacteria derive both their carbon and energy from organic matter and thus are categorized as heterotrophs.

The amount of organic matter present at any point in time is the system's **biomass** (g C/L or DW/L, g C/m$^2$ or DW/m$^2$), and the rate of production of biomass is the system's **productivity** (g C/L-day or DW/L-day, g C/m$^2$-day or DW/m$^2$-day). *Primary production* refers to the photosynthetic generation of organic matter by plants and certain bacteria—for example, algae in lakes and field crops on land. *Secondary production* refers to the generation of organic matter by non-photosynthetic organisms—that is, those that consume the organic matter originating from primary producers to gain energy and materials and in turn generate more biomass through growth. Secondary producers include zooplankton in aquatic systems and cattle on land.

The trophic, or feeding structure, in ecosystems is composed of the abiotic environment and three biotic components: producers, consumers, and decomposers. Producers, most often plants, assimilate simple chemicals and utilize the sun's energy to produce and store complex, energy-rich compounds that provide an organism with substance and stored energy. Organisms that eat plants, extracting energy and chemical building blocks to make more complex substances, are primary consumers or **herbivores**. Those that consume herbivores are called secondary consumers or **carnivores**. Additional carnivorous trophic levels are possible (tertiary and quaternary consumers). Consumers that eat both plant and animal material are termed **omnivores**.

Figure 5.16 illustrates the various nutritional or trophic levels in a simple aquatic **food chain**. This is a linear subset of the more complex relationships and interactions that make up **food webs** such as the one shown in Figure 5.17. Likewise, the simple terrestrial food chain illustrated in Figure 5.18 is a linear subset of the corresponding food web in Figure 5.19.

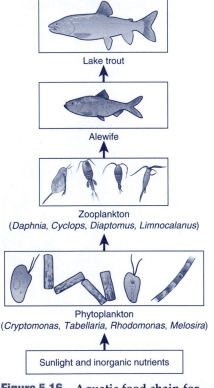

Lake trout

Alewife

Zooplankton
(*Daphnia, Cyclops, Diaptomus, Limnocalanus*)

Phytoplankton
(*Cryptomonas, Tabellaria, Rhodomonas, Melosira*)

Sunlight and inorganic nutrients

**Figure 5.16   Aquatic food chain for Lake Superior**   Phytoplankton are primary producers, zooplankton are primary consumers or herbivores, and the alewife and lake trout are secondary and tertiary consumers, respectively, both carnivores.

From Mihelcic (1999). Reprinted with permission of John Wiley & Sons, Inc.

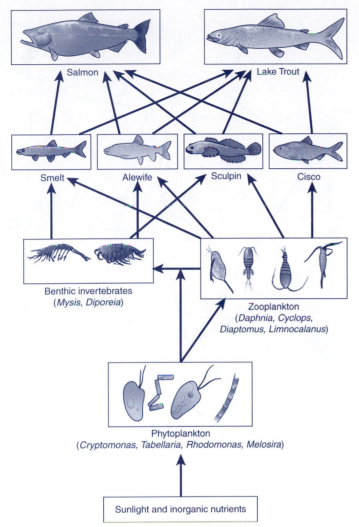

**Figure 5.17** **Food Web for Lake Superior** This illustration shows the more complex interrelationships commonly found in an ecosystem.

From Mihelcic (1999). Reprinted with permission of John Wiley & Sons, Inc.

## 5.3.3 THERMODYNAMICS AND ENERGY TRANSFER

The first law of thermodynamics states that energy cannot be created or destroyed, but it can be converted from one form to another. Applied to an ecosystem, this law suggests that no organism can create its own energy supply. For example, plants rely on the sun for energy, and grazing animals rely on plants (and thus indirectly on the sun). Organisms use the food energy they produce or assimilate to meet metabolic requirements for the performance of work, including cell maintenance, growth, and reproduction. Thus, ecosystems must import energy, and the needs of individual organisms must be met by transformations of that energy.

The second law of thermodynamics states that in every energy transformation, some energy is lost to heat and becomes unavailable to do work. In the food web, the inefficiency of energy transfer is reflected

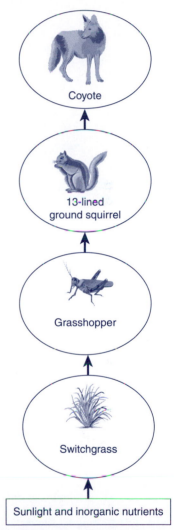

**Figure 5.18** **Simple Food Chain for a Prairie Ecosystem** This representation includes a primary producer (switchgrass) and primary (grasshopper), secondary (13-lined ground squirrel), and tertiary (coyote) consumers.

**Food, Energy, Climate, and Hunger**

http://www.fao.org/

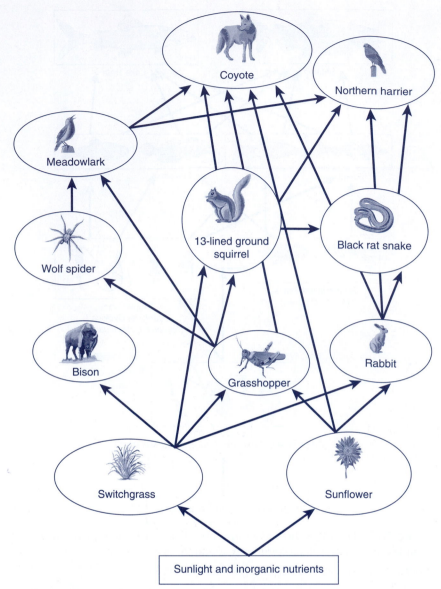

**Figure 5.19  Simplified Food Web for a Prairie Ecosystem**  The coyote, identified as a tertiary consumer and carnivore in the simple food chain (Figure 5.18), is seen within the context of the entire food web to be an omnivore. While the food web may seem complex, it is actually a simplification of the true ecosystem. That food web includes over 1,000 species of plants and animals, underscoring the true complexity of an ecosystem that seems, at first glance, quite simple.

Based on an above-ground food web developed for the Konza Prairie in Kansas by Dr. Anthony Joern of Kansas State University.

**More Information about Energy**

http://www.wri.org/topics/energy

in losses (Figure 5.20) (potentially recycled through the microbial loop) and respiration (heat). Because of this inefficiency, less energy is available at the higher levels of the energy pyramid (Figure 5.21). This explains why it takes a large amount of primary producers to support a single organism at the top of the food chain. (For example, a top predator may require a very large range or territory to support its energy

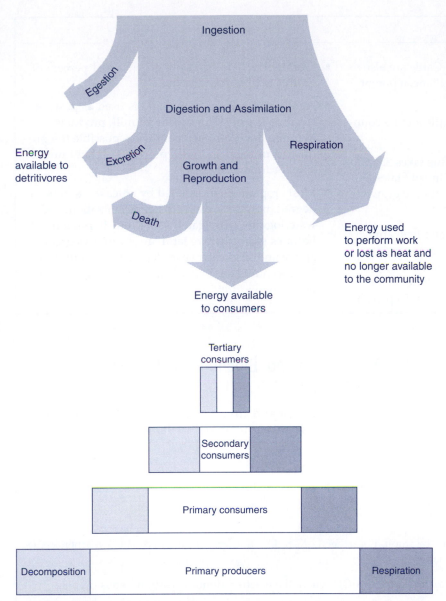

**Figure 5.20** **Energy Losses in a Food Web** A substantial part of the energy ingested by organisms is lost to egestion, excretion, death, and respiration. This inefficiency of energy transfer has a bearing on issues ranging from wastewater treatment microbiology to world population growth.

From Mihelcic (1999). Reprinted with permission of John Wiley & Sons, Inc.

**Figure 5.21** **Energy Pyramid Showing Loss of Energy to Detritivory and Respiration with Movement up the Food Chain** A surprisingly small fraction of the energy originally fixed remains available to transfer to higher trophic levels.

From Mihelcic (1999). Reprinted with permission of John Wiley & Sons, Inc.

needs.) The inefficiencies of energy transfer also have great bearing on our ability to feed an increasing global population that also is consuming more calories and meat because of increases in economic wealth.

---

**Box / 5.3** **The Food We Eat: Transfer of Energy up the Human Food Chain**

Each person in the United States annually consumes an average of 190 pounds of meat. More than 60 million people worldwide could be fed on the grain saved if Americans reduced their meat intake by just 10 percent. Where we live on the food chain has other impacts as well. Consider the following (Goodland, 1997):

- Seven pounds of cattle feed is required to produce a pound of beef, compared with 2 pounds of fish feed for some aquaculture species. Cattle are also one of the biggest producers of methane, a potent greenhouse gas.

- In the United States, 104 million cattle are the country's largest user of grain.

- Growing an acre of corn to feed cattle takes 535,000 gallons of water. Currently in the United States, 70 to 80 percent of the corn and soybeans are grown to produce meat.

- Agriculture consumes more freshwater than any other human activity (excluding electricity production). Worldwide, about 70 percent of freshwater is consumed (not recoverable) by agriculture. In the western United States, the figure is about 85 percent.

- Worldwide, food crops are grown on 11 percent of Earth's total fertile land area.

- Another 24 percent of the land is used as pasture to graze livestock for meat and milk products. Marginal land for pastures makes possible the production of meat and milk products on land unsuitable for food crops.

- Most cropland is threatened by at least one type of degradation (including erosion, salinization, and waterlogging of irrigated soils), and 10 million hectares of productive land are severely degraded and abandoned each year. Replacing agricultural land accounts for 60 percent of deforestation now occurring worldwide.

## 5.4  Oxygen Demand: Biochemical, Chemical, and Theoretical

The topic of **oxygen demand** involves a significant amount of new terminology and notation, summarized in Table 5.7.

## Table / 5.7

**Oxygen Demand: Definition and Notation**    All terms have units of mg $O_2$/L.

| | |
|---|---|
| BOD | *Biochemical oxygen demand*—the amount of oxygen utilized by microorganisms in oxidizing carbonaceous and nitrogenous organic matter. |
| CBOD | *Carbonaceous biochemical oxygen demand*—BOD where the electron donor is carbonaceous organic matter. |
| NBOD | *Nitrogenous biochemical oxygen demand*—BOD where the electron donor is nitrogenous organic matter. |
| ThOD | *Theoretical oxygen demand*—the amount of oxygen utilized by microorganisms in oxidizing carbonaceous and/or nitrogenous organic matter, assuming that all of the organic matter is subject to microbial breakdown, i.e., it is biodegradable. |
| $BOD_5$ | *5-day biochemical oxygen demand*—the amount of oxygen consumed (BOD exerted) over an incubation period of 5 days; the standard laboratory estimate of BOD. The 5-day BOD utilizes the notation $y_5$, referring to the BOD exerted ($y$) over 5 days of incubation. |
| $BOD_U$ | *Ultimate biochemical oxygen demand*—the amount of oxygen consumed (BOD exerted) when all of the biodegradable organic matter has been oxidized. The ultimate BOD utilizes the notation $L_o$, referring to its potential for oxygen consumption when proceeding to complete oxidation. |
| COD | *Chemical oxygen demand*—the amount of chemical oxidant, expressed in oxygen equivalents, required to completely oxidize a source of organic matter; COD and ThOD should be near equal. |

## 5.4.1 DEFINITION OF BOD, CBOD, AND NBOD

Organisms derive the energy required for maintenance of metabolic function, growth, and reproduction through the processes of fermentation and respiration. Both organic and inorganic matter may serve as sources of that energy. *Chemoheterotrophs* are organisms that utilize organic matter—$C(H_2O)$—as a carbon and energy source and, under aerobic conditions, consume oxygen in obtaining that energy:

$$C(H_2O) + O_2 \rightarrow CO_2 + H_2O + \Delta \qquad (5.28)$$

*Chemoautotrophs* are organisms that utilize $CO_2$ as a carbon source and inorganic matter as an energy source, and usually consume oxygen in obtaining that energy. An example of chemoautotrophy is nitrification, the microbial conversion of ammonia to nitrate (with bicarbonate ion contributing $CO_2$):

$$NH_4^+ + 2HCO_3^- + 2O_2 \rightarrow NO_3^- + 2CO_2 + 3H_2O + \Delta \qquad (5.29)$$

In these microbially mediated redox reactions, the electron donors are $C(H_2O)$ and $NH_4^+$, and the electron acceptor is $O_2$.

Oxygen is consumed in both reactions. **Biochemical oxygen demand (BOD)** can thus be defined as the amount of oxygen utilized by microorganisms in performing the oxidation. BOD is a measure of the "strength" of a water or wastewater: the greater the concentration of ammonia-nitrogen or degradable organic carbon, the higher the BOD.

The reactions described by Equations 5.28 and 5.29 are differentiated based on the source compound for the electron donor: *carbonaceous* and *nitrogenous*. Chemical strength (mg $C(H_2O)$/L or mg $NH_3$-N/L) is expressed here in terms of its impact on the environment (oxygen consumed, in mg BOD/L). This is representation by effect, as discussed in Chapter 2.

Dissolved oxygen is a critical requirement of the organism assemblage associated with a diverse and balanced aquatic ecosystem. Domestic and industrial wastes often contain high levels of BOD, which if discharged untreated, would seriously deplete oxygen reserves and reduce the diversity of aquatic life. To prevent degradation of receiving waters, systems are constructed where the supply of BOD and oxygen, the availability of the microbial populations that mediate the process, and the rate at which the oxidations themselves (Equations 5.28 and 5.29) proceed may be carefully controlled. The efficiency of BOD removal is a common performance characteristic of wastewater treatment plants, and BOD is a major feature of treatment plant discharge permits.

## 5.4.2 SOURCES OF BOD

The simple carbohydrates produced through photosynthesis are used by plants and animals to synthesize more complex carbon-based chemicals such as sugars and fats. These compounds are utilized by organisms as an energy source, exerting a **carbonaceous oxygen demand (CBOD)**, Equation 5.28. In addition, plants utilize ammonia

**Class Discussion**

Based on the information provided in Box 5.3, what type of diet would minimize the environmental impact associated with food consumption? What type of diet results in greater use of water and fossil-fuel based energy? Does feeding cattle via natural grasslands or corn consume less water or result in less soil erosion? Are these types of foods also healthier? Why or why not? Using a life cycle approach, what are the sustainability benefits of purchasing food from local sources?

**Biochemical Oxygen Demand Analysis**

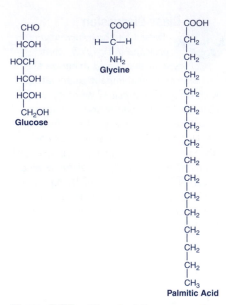

**Figure 5.22 Chemical Structure of Representative Carbonaceous and Nitrogenous Compounds** Municipal wastewater contains a vast number of different organic chemicals, including sugars (20–25%), amino acids (40–60%), and fatty acids (10%).

Adapted from Metcalf and Eddy (1989); illustration from Mihelcic (1999). Reprinted with permission of John Wiley & Sons, Inc.

| Table / 5.8 | |
|---|---|
| **Biochemical Oxygen Demand (BOD) of Selected Waste Streams** | |
| Origin | 5-day BOD (mg $O_2$/L) |
| River | 2 |
| Domestic wastewater | 200 |
| Pulp and paper mill | 400 |
| Commercial laundry | 2,000 |
| Sugar beet factory | 10,000 |
| Tannery | 15,000 |
| Brewery | 25,000 |
| Cherry-canning factory | 55,000 |

SOURCE: Nemerow, 1971.

to produce proteins, that is, complex, carbon-based chemicals with amino groups (-$NH_2$) as part of their structure. Proteins are ultimately broken down (proteolysis) to peptides and then amino acids. The process of deamination then further breaks down the amino acids, yielding a carbon skeleton (CBOD) and an amino group. Conversion of the amino group to ammonia (ammonification) completes the degradation process. The ammonia is then available to exert a **nitrogenous oxygen demand (NBOD)**, Equation 5.29, when utilized by microorganisms. Figure 5.22 illustrates the chemical structure of some representative carbonaceous and nitrogenous compounds.

Domestic wastewater and many industrial wastes are highly enriched in organic matter compared with natural waters. Proteins and carbohydrates constitute 90 percent of the organic matter in sewage. Sources include feces and urine from humans; food waste from sinks; soil and dirt from bathing, washing, and laundering; plus various soaps, detergents, and other cleaning products. Wastes from certain industries, such as breweries, canneries, and pulp and paper producers, also have elevated levels of organic matter.

Table 5.8 presents BOD values for some representative wastes. Even unpolluted natural waters contain some BOD, associated with the carbonaceous and nitrogenous organic matter derived from the watershed and from the waters themselves (for example, from decaying algae and macrophytes, leaf litter, and fecal matter from aquatic organisms). Dissolved-oxygen levels in surface waters—excluding those with excessive algal photosynthesis and attendant $O_2$ production—often are below the saturation level due to this "natural" BOD.

### 5.4.3 THEORETICAL OXYGEN DEMAND

**Theoretical oxygen demand (ThOD)**, given in mg $O_2$/L, is calculated from the stoichiometry of the oxidation reactions involved. A general approach for calculation of *carbonaceous ThOD* (Equation 5.28) is offered by a three-step process provided in Table 5.9.

The calculation is similar for the oxidation of ammonia ($NH_4^+$ in Equation 5.29) to nitrate: 2 moles (or 64 g) of oxygen are consumed for every mole (or 14 g) of ammonia-nitrogen oxidized. Note that ammonia is reported as mg N/L at 14 g/mole, not as ammonia (17 g/mole) or ammonium (18 g/mole). The stoichiometric coefficient for oxidation of nitrogenous wastes is thus 64/14 or 4.57. A waste containing 50 mg/L of $NH_3$-N would have a *nitrogenous ThOD* of 229 mg/L.

To determine the *total ThOD* of a waste stream containing multiple chemicals (for example, ammonia and organic matter), add the contributions of the component compounds (see Example 5.6).

The stoichiometry for oxidation of ammonia is invariant (and Equation 5.29 holds in all cases) because ammonia-nitrogen occurs in only one form with nitrogen in a single valence state. This is not the case for carbon-based compounds (Equation 5.28 does not hold in all cases), as they exist in a wide variety of forms with carbon in several valence states. For this reason, the reaction stoichiometry for each carbon-based compound must be inspected and the oxidation equations balanced individually.

## Table / 5.9

**Steps to Calculate the Carbonaceous Theoretical Oxygen Demand (ThOD)**

| Step | Description of Step | Example |
|---|---|---|
| Step 1 | Write the equation describing the reaction for oxidation of the carbon-based chemical of interest to carbon dioxide and water (e.g., for benzene, $C_6H_6$). | $C_6H_6 + O_2 \rightarrow CO_2 + H_2O$ |
| Step 2 | Balance the equation in the following sequence: (a) balance the number of carbon atoms; (b) balance the number of hydrogen atoms; and (c) balance the number of oxygen atoms. | For benzene, (a) place a 6 in front of $CO_2$ to balance the carbon; (b) place a 3 in front of $H_2O$ to balance the hydrogen; and (c) place a 7.5 in front of the oxygen to balance the oxygen:<br><br>$C_6H_6 + 7.5O_2 \rightarrow 6CO_2 + 3H_2O$ |
| Step 3 | Use the stoichiometry of the balanced chemical reaction, applying unit conversions, to determine the carbonaceous ThOD. | Assume the initial concentration of benzene = 156 mg/L:<br><br>$$\frac{156 \text{ mg benzene}}{L} \times \frac{1 \text{ mole benzene}}{78 \text{ g benzene}} \times \frac{7.5 \text{ mole } O_2}{\text{mole benzene}} \times \frac{32 \text{ g } O_2}{\text{mole } O_2} = \frac{480 \text{ mg } O_2}{L}$$ |

---

example/5.6 **Determination of Carbonaceous, Nitrogenous, and Total ThOD**

A waste contains 300 mg/L of $C(H_2O)$ and 50 mg/L of $NH_3$-N. Calculate the carbonaceous ThOD, the nitrogenous ThOD, and the total ThOD of the waste.

solution

Refer to Table 5.9 if you need a review of the process to write the balanced equation describing the oxidation of $C(H_2O)$ to $CO_2$ and water.

$$C(H_2O) + O_2 \rightarrow CO_2 + H_2O$$

The reaction shows that 1 mole of oxygen is required to oxidize each mole of $C(H_2O)$. The carbonaceous ThOD is determined from the stoichiometry:

$$300 \text{ mg } C(H_2O) \times \frac{g}{1,000 \text{ mg}} \times \frac{1 \text{ mole } C(H_2O)}{30 \text{ g } C(H_2O)} \times \frac{1 \text{ mole } O_2}{\text{mole } C(H_2O)}$$

$$\times \frac{32 \text{ g } O_2}{\text{mole } O_2} \times \frac{1,000 \text{ mg}}{g} = 320 \text{ mg/L}$$

Next, write the balanced equation describing oxidation of ammonia-nitrogen to nitrate:

$$NH_3 + 2O_2 \rightarrow NO_3^- + H^+ + H_2O$$

This reaction shows that 2 moles of oxygen are required to oxidize each mole of $NH_3$. Be aware that the ammonia concentration is reported as mg N/L, not mg $NH_3$/L. The nitrogenous ThOD is determined from the stoichiometry:

$$50 \text{ mg } NH_3 - N/L \times \frac{g}{1,000 \text{ mg}} \times \frac{1 \text{ mole } NH_3 - N}{14 \text{ g } NH_3 - N}$$

$$\times \frac{2 \text{ mole } O_2}{\text{mole } NH_3 - N} \times \frac{32 \text{ g } O_2}{\text{mole } O_2} \times \frac{1,000 \text{ mg}}{g} = 229 \text{ mg/L}$$

The total ThOD of the waste equals $320 + 229 = 549$ mg/L

## 5.4.4 BOD KINETICS

The ThOD calculation defines the oxygen requirement for complete oxidation of ammonia-nitrogen to nitrate-nitrogen or a carbon-based compound to carbon dioxide and water. ThOD does not, however, offer any information regarding the likelihood that the reaction will proceed to completion. For nitrogenous BOD, this is not an issue, because ammonia-nitrogen is readily oxidized and the theoretical and actual nitrogenous oxygen demand are identical.

Carbonaceous compounds, in contrast, are not all easily or completely oxidized by microorganisms (**biodegradation**), and the rate of that oxidation may vary widely among different sources of organic matter. For example, the carbon-containing compounds present in a Styrofoam cup are not as biodegradable as those found in tree leaves, and both are less biodegradable than the carbon compound that constitutes sugar. Thus, while a mass of carbon present as sugar or tree leaves or Styrofoam may have the same ThOD, their actual oxygen demand may be substantially different. Further, most wastes are a complex mixture of chemicals (say, Styrofoam + tree leaves + sugar), present in varying amounts, each with a different level of biodegradability. This characteristic of oxygen-demanding wastes is addressed through BOD kinetics.

Consider the oxidation of organic matter as a function of time, as shown in Figure 5.23. In Figure 5.23a $y_t$ is the CBOD exerted (oxygen consumed, in mg $O_2$/L), and $L_t$ is the CBOD remaining (potential to consume oxygen, in mg $O_2$/L) at any time, $t$. At $t = 0$, no CBOD has been exerted ($y_{t=0} = 0$), and all of the potential for oxygen consumption remains ($L_{t=0} = L_0$, the ultimate CBOD). As the oxidation process begins, oxygen is consumed (CBOD is exerted and $y_t$ increases), and the potential to consume oxygen is reduced (CBOD remaining, $L_t$,

iStockphoto.

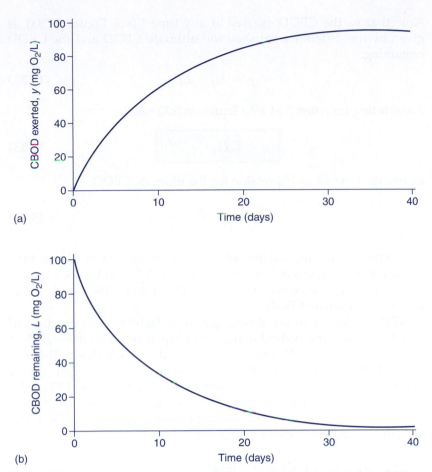

**Figure 5.23** **Biochemical Oxidation of Organic Matter as a Function of Time**
(a) CBOD exerted, $y$ and (b) CBOD remaining, $L$, as a function of time.

Mihelcic (1999). Reprinted with permission of John Wiley & Sons, Inc.

decreases). The rate at which CBOD is exerted is rapid at first, but later it slows and eventually approaches zero as all of the biodegradable organic matter has been oxidized. The total amount of oxygen consumed in oxidizing the waste is the **ultimate CBOD** ($L_0$).

The exponential decline in CBOD remaining ($L$, the potential to consume oxygen), illustrated in Figure 5.23b, can be modeled as a first-order decay:

$$\frac{dL}{dt} = -k_L L \tag{5.30}$$

where $k_L$ is the CBOD reaction rate coefficient (day$^{-1}$). Equation 5.30 can be integrated to yield the analytical expression depicted in Figure 5.23b:

$$L_t = L_0 e^{(-k_L t)} \tag{5.31}$$

Note that $y_t$, the CBOD exerted at any time $t$ (see Figure 5.23a), is given by the difference between the ultimate CBOD and the CBOD remaining:

$$y_t = L_0 - L_t \qquad \text{(5.32)}$$

Substituting Equation 5.31 into Equation 5.32 yields:

$$\boxed{y_t = L_0(1 - e^{-k_L t})} \qquad \text{(5.33)}$$

Rearrange to yield an expression for the ultimate CBOD:

$$L_0 = \frac{y_t}{(1 - e^{-k_L t})} \qquad \text{(5.34)}$$

Equation 5.31 is applied in predicting the change in CBOD over time in natural and engineered systems. Equation 5.34 can be used to convert laboratory measurements of CBOD ($CBOD_5$, discussed subsequently) to ultimate CBOD.

NBOD behaves in an almost identical fashion. The exertion of NBOD follows first-order kinetics, and Equations 5.30 through 5.34 apply, substituting $n$, $N$, and $k_N$ for $y$, $L$, and $k_L$ respectively. The ultimate NBOD ($N_0$) can be calculated from the ammonia-nitrogen content of the sample, based on the stoichiometry of Equation 5.28 (4.57 mg $O_2$ consumed per mg $NH_3$-N oxidized). As shown in Figure 5.24, NBOD exertion begins well after that of CBOD due to differences in the growth rates of the mediating organisms.

### 5.4.5 CBOD RATE COEFFICIENT

The reaction rate coefficient, $k_L$, utilized in CBOD calculations, is a measure of the biodegradability of a waste. The relationship between $k_L$ and the rates of CBOD exertion ($dy/dt$) and consumption ($dL/dt$) is illustrated in Figure 5.25a and Figure 5.25b, respectively. Typical ranges for this coefficient are presented in Table 5.10. Reductions in the magnitude of $k_L$ when moving from untreated sewage to treated sewage to unpolluted river water reflect progressive reductions in labile (biodegradable) organic carbon. As with other microbially mediated processes, values for the CBOD and NBOD reaction rate coefficients vary with temperature.

**Figure 5.24  First-Stage (CBOD) and Second-Stage (NBOD) Biochemical Oxygen Demand**  Exertion of nitrogenous demand lags that of carbonaceous demand because the nitrifying organisms grow more slowly than microorganisms that derive their energy from organic carbon.

Mihelcic (1999). Reprinted with permission of John Wiley & Sons, Inc.

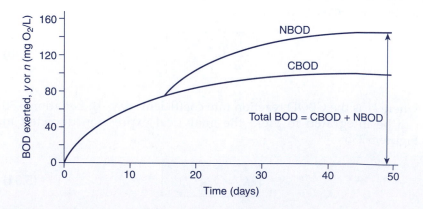

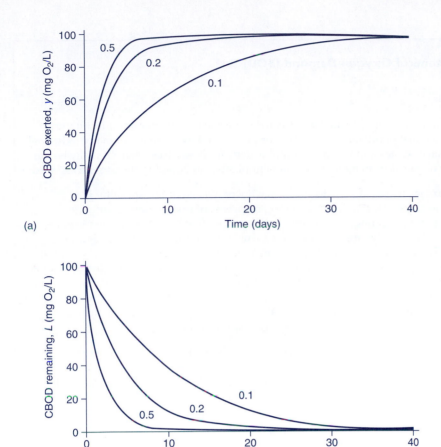

**Figure 5.25** **Variations in Rate at Which Organic Matter Is Stabilized, Reflected in the Reaction Rate Constant, $k_L$.** (a) CBOD exerted, $y$ (mg/L), showing effect of variation in $k_L$. (b) CBOD remaining, $L$ (mg/L), showing effect of variation in $k_L$.

Mihelcic (1999). Reprinted with permission of John Wiley & Sons, Inc.

Values for the CBOD reaction-rate coefficient can be determined experimentally in the laboratory using the *Thomas slope method*. Measurements of CBOD exerted ($y_t$) are made daily for 7 to 10 days. Values for the parameter $(t/y)^{1/3}$ are then calculated, and $k_L$ is determined from the slope and intercept of a plot of $(t/y)^{1/3}$ versus $t$, according to Equation 5.35:

$$k_L = 6.01 \frac{\text{slope}}{\text{intercept}} \qquad (5.35)$$

### 5.4.6 BOD: MEASUREMENT, APPLICATION, AND LIMITATIONS

No chemical method of analysis is available to measure BOD directly. It is necessary to call upon the same organisms that exert oxygen demand in nature to perform a bioassay. In the bioassay, water samples are incubated in a controlled environment for several days while

**Table / 5.10**

**Ranges of Values for the CBOD Rate Constant**

| Type of Sample | $k_L$ (day$^{-1}$) |
|---|---|
| Untreated municipal wastewater | 0.35–0.70 |
| Treated municipal wastewater | 0.10–0.35 |
| Unpolluted river water | <0.05 |

SOURCE: Values from Davis and Cornwell, 1991; Chapra, 1997.

## Table / 5.11

**Favorable Conditions for the Biochemical Oxygen Demand (BOD) Test**

| Condition | Description |
|-----------|-------------|
| Presence of appropriate microorganisms | Typically abundant in untreated and treated domestic wastewater and most natural waters, microbial populations are absent or present in low numbers in many industrial or disinfected wastes and in some natural waters. In these cases, microbes may be purchased or obtained from biological treatment plants and added to the sample as "seed." |
| Favorable and consistent incubation conditions | Samples are incubated at 20°C to encourage microbial activity (respiration is temperature dependent) and to facilitate comparison of results among sampling locations and the laboratories performing the assays. To ensure that oxygen is not completely depleted before the end of the incubation period, aerated dilution water is added. The dilution water may also contain inorganic nutrients (e.g., Fe, N, and P) required by the microbes. Some wastes may be toxic to microorganisms due to extreme pH or the presence of chemicals such as heavy metals. Sample pH can be adjusted and dilution water added to reduce or eliminate toxicity. The BOD bioassay is conducted in the dark to inhibit oxygen production through photosynthesis should algae be present, as is the case for many samples from natural waters. |
| Separating CBOD and NBOD | The standard BOD test measures both CBOD and NBOD. On occasion, it is necessary to separate the two processes to support plant design or operation. A chemical may be added to the water sample to inhibit nitrification, yielding CBOD as the sole result of the assay. NBOD may then be determined by difference from an analysis in which no inhibitor was added. In an inhibited assay, results are reported as CBOD; without inhibitor addition, results are reported as BOD. |

oxygen consumption is monitored. As described in Table 5.11, conditions for the test are established so the microorganisms experience an environment favorable for the oxidation of ammonia and organic carbon (exertion of BOD).

The cumulative exertion of BOD over the course of the assay is well described by the curve presented as Figure 5.24. The rate of BOD exertion gradually slows over time (as the organic matter is used up), and cumulative consumption asymptotically approaches a maximum value (the ultimate BOD, in mg $O_2$/L) after several days. As shown in Figure 5.24, it takes about 30 days to reach 95 percent of the ultimate BOD (with $k_L = 0.1$/day). Such a wait is impractical where results are required for wastewater plant operation, so a shorter 5-day assay has become standard. Here, only about 60 percent of the ultimate BOD is exerted, but an accurate estimate of the ultimate BOD can be obtained by applying Equation 5.33 and a value for $k_L$ to the 5-day measurement.

## example/5.7 Laboratory Determination of CBOD

A 15 mL wastewater sample is placed in a standard 300 mL BOD bottle, and the bottle is filled with dilution water. The bottle had an initial dissolved-oxygen concentration of 8 mg/L and a final dissolved-oxygen concentration of 2 mg/L. A blank (a BOD bottle filled

## example/5.7 Continued

with dilution water) run in parallel showed no change in dissolved oxygen over the 5-day incubation period. The BOD reaction rate coefficient for the waste is 0.4/day. Calculate the 5-day ($y_5$) and ultimate ($L_0$) BOD of the wastewater.

### solution

The 5-day BOD ($y_5$) is the amount of oxygen consumed over the 5-day period corrected for the dilution of the original sample. This can be written as follows:

$$y_5 = \frac{[DO_{initial} - DO_{final}]}{\left[\dfrac{mL\ sample}{total\ test\ volume}\right]} = \frac{\left[\dfrac{8\ mg\ O_2}{L} - \dfrac{2\ mg\ O_2}{L}\right]}{\left[\dfrac{15\ mL}{300\ mL}\right]} = \frac{120\ mg\ O_2}{L}$$

Equation 5.33, with $t = 5$ and $k_L = 0.4$/day, is then applied to determine the ultimate BOD:

$$L_0 = \frac{y_5}{(1 - e^{(-k_L t)})} = \frac{120}{(1 - e^{(-0.4/day \times 5\ day)})} = \frac{138\ mg\ O_2}{L}$$

## example/5.8 CSI for Environmental Engineers: Wastewater Forensics

A "midnight dumper" discharged a tank truck full of industrial waste in a gravel pit. The truck was spotted there 3 days ago, and a pool of pure waste remains. A laboratory technician determined that the waste has a 5-day BOD of 80 mg/L with a rate constant of 0.1/day. Three factories in the vicinity generate organic wastes: a winery (ultimate BOD = 275 mg/L), a vinegar manufacturer (ultimate BOD = 80 mg/L), and a pharmaceutical company (ultimate BOD = 200 mg/L). Determine the source of the waste.

### solution

The ultimate BOD of the waste may be calculated as follows:

$$L_0 = \frac{y_5}{1 - e^{(-K_L t)}} = \frac{80\ mg/L}{1 - e^{(-0.1/day \times 5\ day)}} = 203\frac{mg\ O_2}{L}$$

This value most closely matches the pharmaceutical manufacturer, but it fails to take into account the fact that the waste has been away from its source (and decaying) for 3 days. The original ultimate BOD may be calculated by

$$L_t = L_0 \times e^{-k_L t}$$

Rearrange and solve:

$$L_0 = \frac{L_t}{e^{-k_L t}} = \frac{203 \text{ mg/L}}{e^{(-0.1/\text{day} \times 3 \text{ day})}} = 274 \frac{\text{mg O}_2}{\text{L}}$$

The "midnight dumper" culprit appears to be the winery.

## 5.4.7 BOD TEST: LIMITATIONS AND ALTERNATIVES

While the 5-day BOD test remains a fundamental tool in waste treatment and water quality assessment, concerns regarding its logistics and accuracy have led to proposals for replacement by other measures. While relatively simple to perform, the test has three major shortcomings: (1) the time required to obtain results (5 days is almost unthinkable in today's world of real-time data acquisition); (2) the fact that it may not accurately assay waste streams that degrade over a time period longer than 5 days; and (3) the inherent inaccuracy of the procedure, largely due to variability in seed (bacteria).

Other analytes, such as total organic carbon (TOC), provide greater accuracy and precision but fail to readily distinguish biodegradable and nonbiodegradable organic carbon—precisely the objective of the BOD test in the first place. Further, the BOD test cannot be used to evaluate treatment efficiency for wastes that are poorly biodegradable or toxic. Here a different test is applied, one for **chemical oxygen demand (COD)**.

In this test, a sample containing an unknown amount of organic matter is added to a 250 mL flask. Also added to the flask are $AgSO_4$ (a catalyst to ensure complete oxidation of the organic matter); a strong acid ($H_2SO_4$), dichromate ($Cr_2O_7^{2-}$, a strong oxidizing agent); and, $HgCl_2$ (to provide a $Hg^{2+}$ ion that complexes the chloride ion, $Cl^-$). The chloride ion interferes with the test and is present in high amounts in many wastewater samples. This is because it can be oxidized to $Cl^0$ by dichromate as well as by organic matter. However, the complexed form of $Cl^-$ is not oxidized. Thus, if uncomplexed $Cl^-$ is allowed to be oxidized to $Cl^0$, it can result in a false-positive COD value.

The sample and all the reagents are combined, and the sample is refluxed for 3 hours. The organic matter is oxidized (donates electrons), and the chromium is reduced (accepts electrons) from the hexavalent form ($Cr^{6+}$) to the trivalent form ($Cr^{3+}$). Thus, the COD test determines how much of the hexavalent chromium is reduced during the COD test. After the sample is cooled to room temperature, the dichromate that remains in the system is determined by titration with ferrous ammonium sulfate. The amount of hexavalent chromium is then related to the amount of organic matter that was oxidized.

Results are expressed in oxygen equivalents (mg $O_2$/L), that is, the amount of oxygen required to completely oxidize the waste. The test is relatively quick (3 hours), and correlation to $BOD_5$ is easy to establish on a particular waste stream. For example, municipal wastewater has a ratio of 0.4 to 0.6.

Comparison of BOD and COD results can help identify the occurrence of toxic conditions in a waste stream or point to the presence of biologically resistant (refractory) wastes. For example, a $BOD_5/COD$ ratio approaching 1 may indicate a highly biodegradable waste, while a ratio approaching 0 suggests a poorly biodegradable material.

## 5.5 Material Flow in Ecosystems

The natural passage of chemicals, as mediated by organisms, occurs within **biogeochemical cycles**. Five chemicals are of particular importance in environmental engineering: C, O, N, P, and S. In addition, the **hydrologic cycle** (see Chapter 9) is of interest because it plays an important role in moving chemical elements through the ecosphere. This section considers each of these key cycles.

Note that humans use and cycle a much greater number of chemical elements in industrial applications than are found in living organisms. This can affect the environment through the mining of exotic elements, concentrating these once-dispersed chemicals, and exposing humans and natural systems to elevated concentrations, often damaging healthy function.

### 5.5.1 OXYGEN AND CARBON CYCLES

The **oxygen cycle** and **carbon cycle** are closely linked through the processes of photosynthesis (Equation 5.19) and respiration (Equation 5.20). Photosynthesis is the primary source term in the oxygen cycle and the origin of the organic carbon converted to carbon dioxide in the carbon cycle. Respiration is the major sink term in the oxygen cycle and is responsible for the conversion of organic carbon to carbon dioxide in the carbon cycle. Photosynthesis is carried out by plants and some bacteria, and respiration is carried out by all organisms, including those that photosynthesize. The interplay of photosynthesis and respiration plays a key role in regulating ecosystem energy balances and in maintaining the oxygen levels required by life in aquatic environments. Figure 5.26 depicts how the natural carbon cycle has been altered by humans.

### 5.5.2 NITROGEN CYCLE

The biogeochemical transformations embodied in the nitrogen cycle (Figure 5.27) are important in both natural and engineered systems. As a result of their association with bacteria and plants, many features of the nitrogen cycle are linked with the oxygen and carbon cycles. Plants take up and utilize nitrogen in the form of ammonia or nitrate, chemicals typically in short supply in agricultural soils, thus leading to requirements for fertilization. Certain bacteria and some plant species (such as legumes and clover) can also utilize atmospheric nitrogen ($N_2$), converting it to ammonia through a process termed **nitrogen fixation**. Plants incorporate ammonia and nitrate into a variety of organic compounds, such as proteins and nucleic acids, critical to metabolic

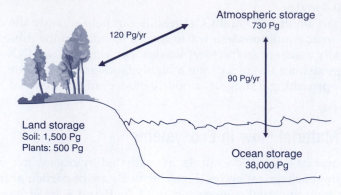

(a) Natural cycle

Atmospheric storage
730 Pg

120 Pg/yr

90 Pg/yr

Land storage
Soil: 1,500 Pg
Plants: 500 Pg

Ocean storage
38,000 Pg

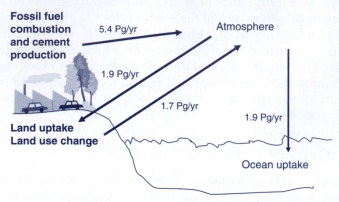

(b) Human perturbations

Fossil fuel
combustion
and cement
production

5.4 Pg/yr

Atmosphere

1.9 Pg/yr

1.7 Pg/yr

1.9 Pg/yr

Land uptake
Land use change

Ocean uptake

**Figure 5.26   Carbon Cycle under Natural Conditions and as Modified by Human Perturbation**   (a) The natural carbon cycle maintains relatively constant reservoirs of carbon in the air with balanced transfers among compartments. (b) Anthropogenic carbon emissions add to the atmospheric reservoir at rates that are not balanced through uptake by the land and oceans. The result is an increase in atmospheric $CO_2$ concentrations.

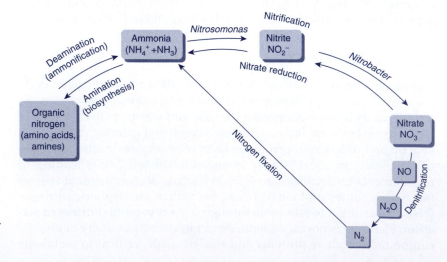

**Figure 5.27   The Nitrogen Cycle**

Snoeyink and Jenkins, Water Chemistry, 1980, reprinted with permission of John Wiley & Sons, Inc. (from Mihelcic 1999, reprinted with permission of John Wiley & Sons, Inc.).

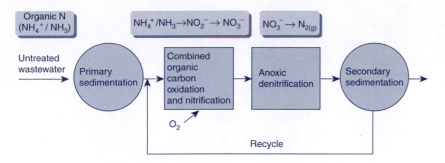

**Figure 5.28** Configuration of a Conventional Municipal Wastewater Treatment Plant for Removal of Nitrogen

Mihelcic (1999). Reprinted with permission of John Wiley & Sons, Inc.

function. Consumers (both herbivores and carnivores) transfer those nitrogen-rich compounds further up the food chain.

Nitrogenous compounds present in organisms are released to the environment through excretion and mortality, whether in nature or in the form of human waste. Nitrate is an important nutrient supporting plant growth and may be the limiting nutrient in some systems, especially coastal marine environments. Discharges to surface waters can create nuisance algal growth and attendant water-quality problems such as oxygen depletion.

Figure 5.28 describes the nitrogen cycle that is commonly observed in a municipal wastewater treatment plant. In the first step, **ammonification**, organic N (for example, protein) is converted to ammonia ($NH_4^+/NH_3$). The next step involves the transformation of ammonia to nitrite ($NO_2^-$) (by bacteria of the genus *Nitrosomonas*) and then to nitrate ($NO_3^-$) (by bacteria of the genus *Nitrobacter*), a process termed **nitrification**. This step requires the presence of oxygen. Nitrate may be transformed to nitrogen gas through the process of **denitrification**, where it is converted to nitrogen gas with subsequent release to the atmosphere. This microbially mediated transformation proceeds under anoxic conditions.

## 5.5.3 PHOSPHORUS CYCLE

Phosphorus-bearing minerals are poorly soluble, so most surface waters naturally contain very little of this important plant nutrient. When phosphorus is mined and incorporated into cleaning agents and fertilizers, biogeochemical cycling (the routing of the element through the environment) is vastly accelerated. Subsequent discharges to lakes and rivers, where phosphorus is the limiting nutrient, can stimulate nuisance algal growth and **eutrophication** (discussed further in Chapter 8), making lakes unpleasant and unavailable for a variety of uses. Thus, **phosphorus cycle** removal is an important feature of the wastewater treatment process stream.

## 5.5.4 SULFUR CYCLE

Like the oxygen and nitrogen cycles, the **sulfur cycle** (Figure 5.29) is to a large extent microbially mediated and thus linked to the carbon cycle. Sulfur reaches lakes and rivers as organic S, incorporated into materials such as proteins, and as inorganic S, primarily in the form of sulfate ($SO_4^{2-}$).

**Figure 5.29  Sulfur Cycle**  Reduction of sulfate ($SO_4^{2-}$) to hydrogen sulfide ($H_2S$) can lead to odor problems for wastewater collection systems and treatment plants. Oxidation of reduced sulfur can lead to acidification and discoloration of surface waters.

From Mihelcic (1999). Reprinted with permission of John Wiley & Sons, Inc.

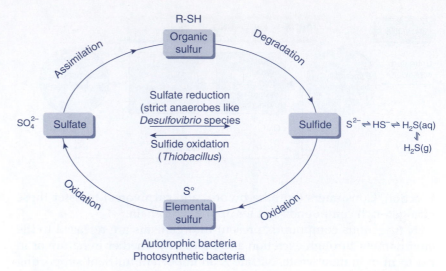

**Biogeochemical Cycles: Carbon, Nitrogen, Mercury, Toxics**

Hydrogen sulfide ($H_2S$) is malodorous and toxic to aquatic life at very low concentrations. Pyrite ($FeS_2$) is often found in and around geologic formations that are mined commercially, as in the case of coal or metals such as silver and zinc. Exposure of pyrite to the atmosphere initiates a three-step oxidation process catalyzed by bacteria including *Thiobacillus thiooxidans*, *Thiobacillus ferrooxidans*, and *Ferrobacillus ferrooxidans*:

$$4FeS_2 + 14O_2 + 4H_2O \rightarrow 4Fe^{2+} + 8SO_4^{2-} + 8H^+ \quad \textbf{(5.36)}$$

$$4Fe^{2+} + 8H^+ + O_2 \rightarrow 4Fe^{3+} + 2H_2O \quad \textbf{(5.37)}$$

$$4Fe^{3+} + 12H_2O \rightarrow 4Fe(OH)_{3(s)} + 12H^+ \quad \textbf{(5.38)}$$

**Acid Mine Drainage**

http://www.epa.gov/nps/acid_mine.html

This process yields acid mine drainage rich in sulfate, acidity, and ferric hydroxides (a yellowish-orange precipitate or floc termed "yellow boy"). While the sulfate is rather innocuous, acidity lowers the pH of the surface waters (often to levels severely impairing water quality), and the floc covers stream beds, eliminating macroinvertebrate habitat. In addition, the low pH of the water dissolves rocks and minerals, releasing hardness and total dissolved solids.

## 5.6  Ecosystem Health and the Public Welfare

All engineered projects should be designed, constructed, and operated in an environmentally benign manner that will ultimately serve society equitably and protect human and ecosystem health for future generations. This section introduces two topics that are important from this perspective: toxic substances and biodiversity.

### 5.6.1  TOXIC SUBSTANCES AND ECOSYSTEM AND HUMAN HEALTH

**Ecosystem Services for Tampa Bay, Coastal California, and Oregon's Willamette River Valley**

http://www.epa.gov/ord/esrp/index.htm

Toxic substances may influence ecosystem health directly through effects manifested at the population or community level and indirectly

Society derives many essential goods from natural ecosystems, including seafood, game animals, fodder, fuelwood, timber, and pharmaceutical products. These goods represent important and familiar parts of the economy. What has been less appreciated until recently is that natural ecosystems also perform fundamental life support services, including regulation of climate, water storage, flood control, buffering against extreme weather events, purification of air and water, regeneration of soil fertility, detoxification and decomposition of wastes, and production and maintenance of biodiversity.

Ecosystem services can be subdivided into five categories: (1) *provision*, such as the production of food and water; (2) *regulation*, such as the control of climate and disease; (3) *support*, including nutrient cycles and crop pollination; (4) *culture*, such as spiritual and recreational amenities; and (5) *preservation*, which includes the maintenance of diversity (Daily, 2000; Millennium Ecosystem Assessment, 2005). Such processes are worth many trillions of dollars annually. Yet because most of these benefits are not traded in economic markets, they carry no price tags that could alert society to

changes in ecosystem health influencing the supply of those benefits or deterioration of the systems that generate them.

To understand the magnitude of the economic implications of services provided by natural ecosystems, consider the following example. In New York City, where the quality of drinking water had fallen below standards required by the U.S. Environmental Protection Agency (EPA), authorities opted to restore the polluted Catskill Watershed, which had previously provided the city with the ecosystem service of water purification. Once the input of sewage and pesticides to the watershed area was reduced, natural abiotic processes such as soil adsorption and filtration of chemicals, together with biotic recycling via root systems and soil microorganisms, improved water quality to levels that met government standards. The cost of this investment in natural capital was estimated to be between $1 billion and $1.5 billion, contrasting dramatically with the estimated $6 billion to $8 billion cost of constructing a new water filtration plant with annual operating costs of $300 million (Chichilnisky and Heal, 1998).

by initiating an imbalance in ecosystem function (reducing or eliminating the role of a species or group of species). In addition, human health may be harmed through consumption of fish and wildlife contaminated with toxic substances.

**Bioconcentration** is the direct absorption of a chemical into an individual organism. Examples include mercury moving from water into phytoplankton across the cell wall or into fish via the gills.

**Bioaccumulation** (also termed biomagnification) refers to the accumulation of chemicals both by exposure to contaminated water (bioconcentration) and ingestion of contaminated food. For example, bioconcentration of a contaminant by plankton results in bioaccumulation in the next trophic level, fish.

Although there are significant losses in transferring energy and biomass up the food chain, as in the case of oxidation and excretion (see Figures 5.20 and 5.21), some chemicals (for example, mercury, PCBs, DDT, and some flame retardants) are retained by organisms. This retention, coupled with the loss of biomass, produces a concentrating effect in each successive level up the food chain.

Figure 5.30 depicts this concentrating effect for mercury in an aquatic system. Note that the seemingly small water concentration of 0.01 μg/L (ppb$_m$) increases by five orders of magnitude to 2,270 ppb$_m$ at the top predator level through bioconcentration and bioaccumulation. Is there a higher trophic level than fish? Yes, potentially humans, other mammals (for example, bears, seals, beluga whales) and birds

**Figure 5.30** **Bioaccumulation of Mercury in the Food Chain of Onondaga Lake, New York** Box size represents biomass (decreasing up the food chain through inefficiency of energy transfer), and shading represents the mercury concentration of the biomass (increasing up the food chain because it is retained as biomass is reduced). The concentration of mercury in the water column is ≈ 0.01 μg/L.

Data from Becker and Bigham (1995); figure from Mihelcic (1999). Reprinted with permission of John Wiley & Sons, Inc.

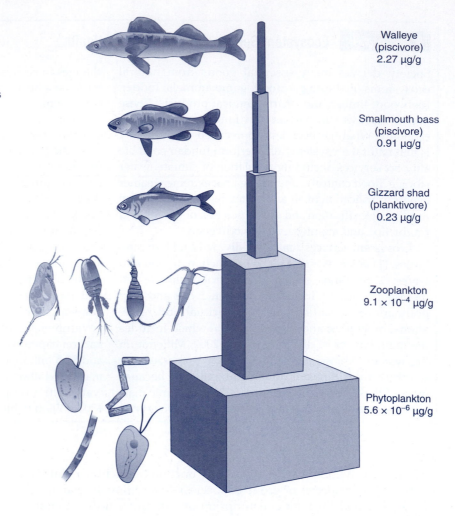

Walleye
(piscivore)
2.27 μg/g

Smallmouth bass
(piscivore)
0.91 μg/g

Gizzard shad
(planktivore)
0.23 μg/g

Zooplankton
$9.1 \times 10^{-4}$ μg/g

Phytoplankton
$5.6 \times 10^{-6}$ μg/g

such as gulls and eagles. Elevated levels of bioaccumulative substances are routinely observed in wildlife relying on fish as a significant portion of their diet.

Bioaccumulation of toxic substances has led to severe impacts on many species of wildlife. Table 5.12 illustrates the impact of chemicals such as DDT and lead bioaccumulating in bald eagles. In addition, bioaccumulation can contribute significantly to total human exposure, and thus environmental risk, for a particular chemical, as discussed in Chapter 6. Because of these effects and the threat posed to human populations, there is a pressing need for better understanding of the dynamics of bioaccumulation and its potential impact on humans and the environment.

The **bioconcentration factor (BCF)** is the ratio of the concentration of a chemical in an organism to that in the surrounding environmental medium (generally air or water) when uptake directly from that medium is the only mechanism considered.

The **bioaccumulation factor (BAF)** is the ratio of the concentration of a chemical in an organism to that in the surrounding medium when

## Table / 5.12

**Historical Numbers of Nesting Pairs of Bald Eagles in the Contiguous ("Lower 48") States** The ban on lead shot in hunting intended to reduce toxicity to waterfowl that often digest the pellets additionally reduced toxicity to eagles, which feed on dead or crippled waterfowl. Land use also has had a strong impact on successful nesting of eagles.

| Period | Number of Nesting Pairs | Explanation |
|--------|-------------------------|-------------|
| 1700s | 50,000 | Pre-settlement estimated population. |
| 1940 | Number unknown; "threatened with extinction" | Bald Eagle Protection Act prohibited killing, selling, or possessing the species. |
| 1960s | 400 | Habitat loss, lead poisoning, and reproductive failure due to bioaccumulation of the pesticide DDT; Rachel Carson's *Silent Spring* is published. |
| 1972 | <800 | DDT use banned in United States. |
| 1991 | 3,000 | Lead shot banned for waterfowl hunting. |
| 2007 | 10,000 | Removal from Endangered Species List by U.S. Fish and Wildlife Service |

SOURCE: U.S. Fish and Wildlife Service.

all potential uptake mechanisms (such as through food and water) are included.

Because the BCF addresses only passive uptake (absorption independent of organism-specific feeding patterns), it provides a means of comparing the potential risk of chemicals to organism and ecosystem health. Organisms with high lipid content tend to exhibit greater BCFs; for example, PCB concentrations are typically higher in fatty fish like trout and salmon than in largemouth and smallmouth bass, which are leaner. The phenomenon of chemical bioaccumulation has led to programs of fish consumption advisories in many states, from Florida to northern Minnesota. Specific recommendations are made for removal of fatty tissue when cleaning and preparing fish to minimize human consumption (and bioaccumulation) of contaminants. However, exposure to some contaminants, including mercury, is not reduced by selective trimming of fatty tissue, because this contaminant is uniformly distributed throughout the fish. In this case, the only way consumers can limit exposure is to control the amount of fish eaten per week.

**Class Discussion**

What fish advisories exist in your state or a neighboring state? What are the chemicals of concern? What is the origin of these chemicals (point discharge, non-point discharge, air)? What populations are most at risk? What sustainable solutions (policy and technological) can you implement to solve the problem?

**UNEP Biodiversity**
http://www.unep.org/themes/biodiversity

**Biodiversity in Africa**
http://www.eoearth.org/article/
Biodiversity_in_Africa

**Class Discussion**

We now live in one of the greatest mass extinctions in Earth's history. Discuss the role of engineers in this. Are they or aren't they partially responsible? How can they be involved in solutions to this problem?

## 5.6.2 BIODIVERSITY AND ECOSYSTEM HEALTH

The term, **biodiversity**, a contraction of *biological diversity*, refers to the great variety present in all forms of life. While the concept of biodiversity originally focused on individual species, many scientists now consider *genetic diversity* (that within a species) and *ecological diversity* of great importance. It is known that significant variation can exist in the genetic makeup of populations of the same species that are separated in time (for instance, by season) or in space. This variation may lead to differences in the response of populations to environmental stress or even the function of the species within an ecosystem. For this reason, fisheries biologists seeking to reestablish a species in a formerly degraded habitat seek young fish from local or regional populations to maintain the characteristics of the original gene pool. At the other end of this spectrum, perhaps it is wise to view ecosystems as representing a diversity of equal or greater value than that of the individual species, thus meriting the attention of those seeking to maintain biodiversity.

**THREATS TO BIODIVERSITY** Proceeding at a background rate of one species per million species per year, extinction has been outpaced by evolution, and Earth's biodiversity has experienced a steady increase in biodiversity over the past 600 million years. This increase has been punctuated by five episodes of mass extinction thought to be related to meteorite impacts, volcanic eruption, and climate change. But today, humankind has initiated its own meteorite impact, its own volcanic eruption:

*Because of the magnitude and speed with which the human species is altering the physical, chemical and biological world, biodiversity is being destroyed at a rate unprecedented in recent geologic time.*
—Thorne-Miller, 1999

While it is the species that becomes extinct, it is important to consider threats to biodiversity within the context of ecosystem structure and function. The Intergovernmental Panel on Climate Change (IPCC) reports:

*The resilience of many ecosystems is likely to be exceeded this century by an unprecedented combination of climate change, associated disturbances (e.g., flooding, drought, wildfire, insects, ocean acidification), and other global change drivers (e.g., land use change, pollution, overexploitation of resources).*
—IPCC, 2007

Certainly it is the species that is often targeted (for example, by hunting or poaching), but human stress may be manifested as well through negative effects on the ecosystem.

Threats to biodiversity can be organized into five categories: (1) overharvest; (2) habitat destruction; (3) species introductions; (4) chemical pollution; and (5) global atmospheric change. Of these, overharvest has perhaps the longest history. The hunting of waterfowl, harvest of wading birds for their feathers, and the relentless pursuit and killing of bison are well known from 19th-century America. More recently, overharvest from nonsustainable overfishing is threatening the future of marine fisheries.

Grassland and forest ecosystems that once covered thousands of square miles are now absent or so altered in composition as to be unrecognizable. Together with urban development, these land use changes have led to habitat fragmentation, with the formerly contiguous habitats required by some species now reduced to small patches. The process of species introduction is a natural feature of ecosystem evolution as conditions change and ranges expand. However, human acceleration (accidental or intentional) of this process can be critically damaging to native species and their ecosystem. (Consider, for example, the introduction of an assortment of mammals into New Zealand.)

The vulnerability of an ecosystem to introduction of new species increases through damage to its health (for instance, physical or chemical effects) or the loss of key biotic components (species). Alien **invasive species** are those that become established in a natural ecosystem and threaten native biological diversity. Invasive species have been introduced to aquatic environments intentionally (certain species of trout in Western rivers) and accidentally (sea lamprey through canal construction, zebra mussels in ballast water, and Asian carp through aquaculture escape).

In terrestrial ecosystems, land disturbance during construction of the built environment and the entry of motor vehicles into previously roadless areas have been documented to spread invasive species. The rapid spread and aggressive nature of invasive species is due, in part, to the absence of competition and predation pressure common to their home ecosystem. Invasive species contribute significantly to species extinction and loss of biodiversity, a problem that is increasing due to globalization.

In terms of chemical pollution, oil spills, nutrients input from wastewater treatment plants, agricultural and urban runoff, and heavy metals such as mercury and pesticide residues have had documented negative effects on ecosystems. Global climate change is also predicted to have a dramatic effect on ecosystem health. The IPCC reports, "Approximately 20-30% of plant and animal species assessed so far are likely to be at increased risk of extinction if increases in global average temperature exceed 1.5–2.5°C" (IPCC, 2007).

**SEEING VALUE IN BIODIVERSITY** Since passage of the Endangered Species Act in 1973, most Americans have become familiar with the concept and intrinsic value of biodiversity. However, the application of a philosophy supporting biodiversity may become complex in some management and regulatory arenas. In the case of the northern spotted owl in the old-growth forests of the Pacific Northwest, the timber industry took a strong stand against listing the species, arguing that ensuing joblessness and related socioeconomic impacts far outweighed the value of protecting the animal and biodiversity.

Three reasons are often supplied for encouraging programs supporting biodiversity. The first evolves from the concept that plants and animals may play a critical role in the development of more-productive food supplies and medicines that improve human health and prolong

## Invasive Species

http://www.epa.gov/owow/invasive_species

**Aldo Leopold** (1887–1948) was a U.S. ecologist, forester, and environmentalist. He was influential in the development of modern environmental ethics and in the movement for wilderness preservation. Aldo Leopold is considered to be a pioneer of wildlife management in the United States and was a lifelong fisher and hunter. In his *Sand County Almanac*, published shortly after his death, Leopold put forth the "Land Ethic," counseling, "A thing is right when it tends to preserve the integrity, stability, and beauty of the biotic community. It is wrong when it tends otherwise."

Leopold went on to state

*All ethics so far evolved rest upon a single premise: that the individual is a member of a community of interdependent parts. His instincts prompt him to compete for his place in that community, but*

Photo Courtesy of the Aldo Leopold Foundation Archives.

*his ethics prompt him also to co-operate (perhaps in order that there may be a place to compete for). The **land ethic** simply enlarges the boundaries of the community to include soils, waters, plants, and animals, or collectively: the land. In short, a land ethic changes the role of* Homo sapiens *from conqueror of the land-community to plain member and citizen of it. It implies respect for his fellow-members, and also respect for the community as such.\**

Leopold calls upon us to recognize our role as members of a diverse but fragile environmental community and to use our engineering skills and training in such a way as to protect the land and find value in biodiversity.

*\*quoted from "A Sand County Almanac: With Essays on Conservation" by Aldo Leopold (2001) with permission of Oxford University Press, Inc.*

life. For example, a species of maize discovered in Mexico several decades ago possesses disease resistance and growth patterns that may revolutionize the corn industry. In the field of medicine, examples include aspirin, a major component of which is derived from willow bark or the herb meadowsweet, and anticancer agents produced by the rosy periwinkle, a flowering shrub found only in Madagascar.

The second reason often given in support of biodiversity is that ecosystem structure is determined by the interactions of its components, so the loss of a single component (such as a species) could permanently and fatally disrupt ecosystem function. For example, invasive species can decimate native plants and animals, completely altering the face of an ecosystem. The United States is now experiencing such ecosystem disruption; for example, proliferation of the nonnative spotted knapweed is crowding out native species and reducing livestock forage. Purple loosestrife is becoming dominant in many wetlands, eliminating native plants that are more nutritive to wildlife.

The final reason is an ethical one related to the role of humans as stewards of vast and complicated ecosystems. Our society's environmental ethic has undergone an evolution of thought, moving from the apparently limitless resources of colonial America, through periods of expansion and industrialization to times when we find ourselves bumping up against the limits of the ecosphere and suffering the deterioration and loss of social, economic, and environmental amenities.

# Key Terms

- abiotic
- absorption
- aerobic respiration
- algae
- ammonification
- anaerobic respiration
- anoxic
- artificial photosynthesis
- autotrophs
- bacteria
- bioaccumulation
- bioaccumulation factor (BAF)
- biochemical oxygen demand (BOD)
- bioconcentration
- bioconcentration factor (BCF)
- biodegradation
- biodiversity
- biogeochemical cycles
- biology
- biomass
- biosphere
- biota
- biotic
- carbon
- carbonaceous oxygen demand (CBOD)
- carnivores
- carrying capacity
- chemical oxygen demand (COD)
- climate
- community
- death phase
- denitrification

- detritus
- ecological footprint
- ecological redox sequence
- ecology
- ecosystem
- ecosystem services
- endangered species
- energy
- eukaryotes
- eutrophication
- exponential-growth model
- exponential or log growth phase
- food chains
- food webs
- fungi
- greenhouse gas
- half-saturation constant
- herbivores
- heterotrophs
- higher animals
- higher plants
- hydrologic cycle
- International Panel on Climate Change (IPCC)
- invasive species
- IPAT equation
- land ethic
- land use
- Leopold, Aldo
- *The Limits to Growth*
- logistic-growth model
- macroinvertebrates
- methane
- methanogenesis

- microcrustaceans
- Monod model
- natural systems
- nitrification
- nitrogen fixation
- nitrogenous oxygen demand (NBOD)
- omnivores
- organic matter
- oxygen
- phosphorus
- photosynthesis
- photosynthetic source (of oxygen)
- population
- population dynamics
- principle of competitive exclusion
- productivity
- prokaryotes
- protozoa
- redox reaction
- renewable resources
- respiration
- rotifers
- species
- stationary phase
- stoichiometric
- substrate
- sulfur
- theoretical oxygen demand (ThOD)
- threatened species
- ultimate CBOD
- viruses
- yield coefficient ($Y$)

## chapter/Five Problems

**5.1** Mathematical models are used to predict the growth of a population, that is, population size at some future date. The simplest model is that for exponential growth. The calculation requires knowledge of the organism's maximum specific growth rate. A value for this coefficient can be obtained from field observations of population size or from laboratory experiments where population size is monitored as a function of time when growing at high substrate concentrations ($S \gg K_s$):

| Time (days) | Biomass (mg/L) |
|---|---|
| 0 | 50 |
| 1 | 136 |
| 2 | 369 |
| 3 | 1,004 |
| 4 | 2,730 |
| 5 | 7,421 |

Calculate $\mu_{max}$ for this population, assuming exponential growth; include appropriate units.

**5.2** Once a value for $\mu_{max}$ has been obtained, the model may be used to project population size at a future time. Assuming that exponential growth is sustained, what will the population size in Problem 5.1 be after 10 days?

**5.3** Exponential growth cannot be sustained forever because of constraints placed on the organism by its environment, that is, the system's carrying capacity. This phenomenon is described using the logistic-growth model. (a) Calculate the size of the population in Problem 5.1 after 10 days, assuming that logistic growth is followed and that the carrying capacity is 100,000 mg/L. (b) What percentage of the exponentially growing population size would this be?

**5.4** Food limitation of population growth is described using the Monod model. Population growth is characterized by the maximum specific growth rate ($\mu_{max}$) and the half-saturation constant for growth ($K_s$). (a) Calculate the specific growth rate ($\mu$) of the population in Problem 5.1 growing at a substrate concentration of 25 mg/L according to Monod kinetics if it has a $K_s$ of 50 mg/L. (b) What percentage of the maximum growth rate for the exponentially growing population size would this be?

**5.5** Laboratory studies have shown that microorganisms produce 10 mg/L of biomass in reducing the concentration of a pollutant by 50 mg/L. Calculate the yield coefficient, specifying the units of expression.

**5.6** When food supplies have been exhausted, populations die away. This exponential decay is described by a simple modification of the exponential-growth model. Engineers use this model to calculate the length of time that a swimming beach must remain closed following pollution with fecal material. For a population of bacteria with an initial biomass of 100 mg/L and a $k_d = 0.4$/day, calculate the time necessary to reduce the population size to 10 mg/L.

**5.7** A population having a biomass of 2 mg/L at $t = 0$ days reaches a biomass of 139 at $t = 10$ days. Assuming exponential growth, calculate the value of the specific growth coefficient.

**5.8** Fecal bacteria occupy the guts of warm-blooded animals and do not grow in the natural environment. Their population dynamics in lakes and rivers—that is, following a discharge of raw sewage—can be described as one of exponential decay or death. How many days would it take for a bacteria concentration of $10^6$ cell/mL to be reduced to the public health standard of $10^2$ cell/mL if the decay coefficient is 2/day?

**5.9** What is the ThOD of the following chemicals? Show the balanced stoichiometric equation with your work: (a) 5 mg/L $C_7H_3$; (b) 0.5 mg/L $C_6Cl_5OH$; (c) $C_{12}H_{10}$.

**5.10** A waste contains 100 mg/L ethylene glycol ($C_2H_6O_2$) and 50 mg/L $NH_3$-N. Determine the theoretical carbonaceous and the theoretical nitrogenous oxygen demand of the waste.

**5.11** Calculate the NBOD and ThOD of a waste containing 100 mg/L isopropanol ($C_3H_7OH$) and 100 mg/L $NH_3$-N.

**5.12** A waste contains 100 mg/L acetic acid ($CH_3COOH$) and 50 mg/L $NH_3$-N. Determine the theoretical carbonaceous oxygen demand, the theoretical nitrogenous oxygen demand, and the total theoretical oxygen demand of the waste.

**5.13** A waste has an ultimate CBOD of 1,000 mg/L and a $k_L$ of 0.1/day. What is its 5-day CBOD?

**5.14** Calculate the ultimate BOD of a waste that has a measured 5-day BOD of 20 mg/L, assuming a BOD rate coefficient of 0.15/day.

**5.15** According to the Web site maintained by Redefining Progress, its latest Footprint Analysis indicates that humans are exceeding their ecological limits by 39 percent. Go to the following Web site, and determine your own ecological footprint. Record your value, and compare it with those for your country and the world. Identify some changes you can make in your current lifestyle, and then rerun the footprint calculator to reflect those changes. Summarize the changes you make and how they affect your ecological footprint. The Web site is www.rprogress.org/ecological_footprint/about_ecological_footprint.htm.

**5.16** Go to the IPCC Web site (www.ipcc.ch). Choose a specific ecosystem to study, and use information from the Web site to research the impact of climate change on that ecosystem. Write a one-page essay (properly referenced) summarizing your findings.

**5.17** Biofuels are being suggested as a method to close the carbon loop. Do a library and Internet search on a particular type of biofuel. Write a one-page essay that addresses the link between biofuels and the global carbon cycle. Also address the impact that your particular biofuel may have on water quality, food supply, biodiversity, and air quality.

**5.18** According to the U.S. Environmental Protection Agency, mountain top coal mining consists of removal of mountaintops to expose coal seams and subsequent disposal of the associated mining overburden in adjacent valleys (see http://www.epa.gov/region3/mtntop/). In your own words, discuss the environmental impacts associated with mountain top mining. How does this relate to EPA's Healthy Waters Priority Program and does this method of providing energy fit into a sustainable future that considers social, environmental, and economic balance? How would you engineer the production of a secure, reliable, and sustainable energy supply?

**5.19** The U.S. Environmental Protection Agency defines a TMDL as the calculation of the maximum amount of a pollutant that a waterbody can receive and still meet water quality standards. The TMDL approach is a way to apply carrying capacity to a particular waterbody. Under section 303(d) of the Clean Water Act, states, territories, and authorized tribes are required to develop lists of impaired waters. Go the following web site (http://www.epa.gov/owow/tmdl/lawsuit.html) that lists the states which EPA is under court order or agreed in a consent decree to establish TMDLs for. Produce a clear and easy to read map of all 50 states that shows the locations of these states.

# References

Becker, D. S., and G. N. Bigham. 1995. "Distribution of Mercury in the Aquatic Food Web of Onondaga Lake, NY." *Water, Air and Soil Pollution* 80:563–571.

Chapra, S. C. 1997. *Surface Water-Quality Modeling*. New York: McGraw-Hill.

Chichilnisky, G., and G. Heal. 1998. "Economic Returns from the Biosphere," *Nature* 391:629–630.

Daily, G. C. 2000. "Management Objectives for the Protection of Ecosystem Services." *Environmental Science & Policy* 3:333–339.

Davis, M. L., and D. A. Cornwell. 1991. *Introduction to Environmental Engineering*. New York: McGraw-Hill.

Effler, S. W. 1996. *Limnological and Engineering Analysis of a Polluted Urban Lake: Prelude to Environmental Management of Onondaga Lake, New York*. New York: Springer.

Enger, E. D., I. R. Kormelink, B. F. Smith, and R. J. Smith. 1983. *Environmental Science*. Dubuque, Iowa: Wm. C. Brown Publishers.

Goodland, R. 1997. "Environmental Sustainability in Agriculture: Diet Matters." *Ecological Economics* 23 (3): 189–200.

Intergovernmental Panel on Climate Change (IPCC). 2007. "Summary for Policy Makers." In *Climate Change 2007: Impacts, Adaptation and Vulnerability*. Contribution of Working Group II to the Fourth Assessment Report of the Intergovernmental Panel on Climate Change, ed. M. L. Parry, O. F. Canziani, J. P. Palutikof, P. J. van der Linden, and C. E. Hanson, 7–22. Cambridge: Cambridge University Press.

Kupchella, C., and M. Hyland. 1986. *Environmental Science*. Needham Heights, Mass.: Allyn and Bacon.

Metcalf and Eddy, Inc. 1989. *Wastewater Engineering: Treatment, Disposal, Reuse*, 2nd ed. New York: McGraw-Hill.

Mihelcic, J. R. 1999. *Fundamentals of Environmental Engineering*. New York: John Wiley & Sons.

Millennium Ecosystem Assessment (MEA). 2005. *Ecosystems and Human Well-Being: Synthesis*. Washington, D.C.: Island Press.

Nemerow, N. L. 1971. *Liquid Waste of Industry: Theories, Practices, and Treatment*. Reading, Mass.: Addison-Wesley.

Ricklefs, R. E. 1983. *The Economy of Nature*, 2nd ed. New York: Chirun Press.

Snoeyink, V. L., and D. Jenkins. 1980. *Water Chemistry*. New York: John Wiley & Sons.

Tchobanoglous, G., F. L. Burton, and H. D. Stensel. 2003. *Wastewater Engineering*, 4th ed. Boston: Metcalf & Eddy; New York: McGraw-Hill.

Thorne-Miller, B. 1999. *The Living Ocean: Understanding and Protecting Marine Biodiversity*. Washington, D.C.: Island Press.

Wackernagel, M., L. Onisto, A. Callejas Linares, I. S. López Falfán, J. Méndez García, A.I. Suárez Guerrero, and M. Guadalupe Suárez Guerrero. 1997. *Ecological Footprints of Nations: How Much Nature Do They Use? How Much Nature Do They Have?* Centre de Estudios para la Sustentabilidad, Universidad Anáhuac de Xalapa, Mexico. Commissioned by the Earth Council for the Rio+5 Forum.

# chapter/Six Environmental Risk

James R. Mihelcic
and Julie Beth Zimmerman

*In this chapter, readers will learn about the distinction between a hazardous and a toxic chemical, the meaning of environmental risk, and methods for assessing environmental risk and incorporating it into engineering practice. Readers will also learn about the four components of a complete environmental risk assessment (hazard assessment, dose response assessment, exposure assessment, and risk characterization), the impact of site-specific conditions on exposure to chemicals, and ways that land use ultimately affects environmental risk. Finally, the chapter demonstrates the differences in developing a risk characterization for carcinogenic and noncarcinogenic compounds. The emphasis is on risk determination due to exposure to contaminated chemicals found in water, air, and food. The chapter also demonstrates the method used to determine allowable chemical concentrations in groundwater and contaminated soil.*

## Learning Objectives

1. Describe how to minimize or eliminate risk by designing for reduced hazard and/or reduced exposure.

2. Summarize the different types of hazard and their potential adverse impacts on human health and the environment.

3. Articulate the limitations of the risk assessment paradigm for the protection of human health and the environment and the factors that affect the toxicity of a given chemical, including uncertainty associated with data collection and interpretation.

4. Define the four components of a risk assessment, and discuss the difference between risk assessment and risk perception.

5. Calculate the acceptable concentration and acceptable risk associated with exposure to a carcinogenic and noncarcinogenic chemical for various and multiple exposure pathways, including chemical partitioning between soil, air, and water phases.

6. Understand the relationship of bioaccumulation/bioconcentration, food web cycles, and toxicity.

7. Define the terms *environmental justice* and *susceptible populations* in relation to risk assessment, and explain the roles engineers can play in addressing these topics.

8. Perform a basic risk assessment for carcinogens and noncarcinogens, given appropriate data including interpretation of a dose-response curve.

FISH CONTAMINATED
DO NOT EAT

215

## 6.1 Risk and the Engineer

For the last 60 years of the 20th century, global chemical production increased several hundred times, so that by the beginning of the 21st century, approximately 70,000 chemicals are commonly sold in commerce. Additionally, hundreds of new chemicals are introduced into the market each year. Individuals may be exposed to these chemicals at home, at school, in the workplace, while traveling, or simply while jogging in a large urban area. The indoor environment also is becoming an important place for exposure to chemicals, because Americans now spend 85 percent of their time indoors. Therefore, indoor environments, particularly those that are poorly ventilated and have chemical-emitting carpeting, coatings, and adhesives, can have a large impact on human health.

**Risk** is the likelihood of injury, disease, or death. In general terms,

$$\text{Risk} = f(\text{hazard, exposure}) \qquad (6.1)$$

**Environmental risk** is the risk resulting from exposure to a potential environmental hazard. Environmental hazards can be specific chemicals or chemical mixtures such as secondhand smoke and automobile exhaust. They can also be other hazards such as pathogens, stratospheric ozone depletion, climate change, and water scarcity. In this chapter, we will focus on environmental risk to humans derived from exposure to chemicals or substances. However, the concept of environmental risk can be applied to the health of plants, animals, and entire ecosystems, which support human livelihood and enhance our quality of life.

Table 6.1 summarizes many types of hazard, including physical hazards, toxicological hazards, and global hazards. It is important to note that the adjective *hazardous* does not only imply cancer-causing, but also includes any adverse impact to humans or to the environment as a result of exposure to a chemical or material.

The risk of a chemical may involve its toxic effects or the hazard it presents to workers or to a community—for example by causing an explosion. Risk has been historically managed by addressing the issue of **exposure**. For example, exposure may be limited by requiring that workers wear protective clothing or by developing warning signs for trucks that transport hazardous chemicals.

Because risk is the product of a function of hazard and of exposure, two implications become clear. As hazard approaches infinity, risk can only be reduced to near zero by reducing exposure to near zero. Conversely, as hazard approaches zero, exposure can approach infinity without significantly affecting risk. Green chemistry and engineering are methods to reduce hazard toward zero.

These relationships among risk, hazard, and exposure are extremely important because current methods to protect human health and the environment are closely tied to the risk paradigm and depend almost exclusively on controlling exposure. The resulting engineering efforts to lower the probability of exposure to a wide range of hazards including toxicants, reactive substances, flammables, and explosives have

## Table / 6.1

### Hazard Categories and Examples of Potential Hazard Manifestations

| Human Toxicity Hazards | Environmental Toxicity Hazards | Physical Hazards | Global Hazards |
|---|---|---|---|
| Carcinogenicity | Aquatic toxicity | Explosivity | Acid rain |
| Neurotoxicity | Avian toxicity | Corrosivity | Global warming |
| Hepatoxicity | Amphibian toxicity | Oxidizers | Ozone depletion |
| Nephrotoxicity | Phytotoxicity | Reducers | Security threat |
| Cardiotoxicity | Mammalian toxicity (nonhuman) | pH (acidic or basic) | Extreme weather events such as flooding |
| Pulmonary toxicity | | Violent reaction with water | Water scarcity |
| Hematological toxicity | | | Loss of biodiversity |
| Endocrine toxicity | | | Persistence |
| Immunotoxicity | | | Bioaccumulation |
| Reproductive toxicity | | | |
| Teratogenicity | | | |
| Mutagenicity (DNA toxicity) | | | |
| Dermal toxicity | | | |
| Ocular toxicity | | | |
| Enzyme interactions | | | |

been significant. However, this strategy is tremendously expensive. Worst of all, such an approach can and, as a probability function, will eventually fail. When exposure controls fail, risk is then equal to a function of hazard (see Equation 6.1). This relationship argues for the need to design molecules, products, processes, societal behavior, and systems so health and safety do not depend on controls or systems that can fail or be sabotaged (either intentionally or accidentally) but instead rely on the use of benign (minimally hazardous) chemicals and materials.

During the early stages of design of a manufacturing process, there is a great amount of flexibility in developing solutions that prevent or minimize risk by making decisions that eliminate the use and production of hazardous chemicals. An engineer would not intentionally design a manufacturing process if a direct result were that the process would cause plant workers and community members to contract cancer or that fish located in a local stream would die after exposure to wastewater discharge. Sadly, though, these adverse human and ecosystem impacts are the result of many of our current practices of engineering design. However, engineers are now embracing "green chemistry"

### Class Discussion

*Does the current method of managing risk by reducing exposure seem like a particularly proactive or innovative method? Wouldn't a better method be to eliminate or at least minimize the hazard? Note that in Equation 6.1 and the definition of environmental risk that followed, there is zero risk if there is zero exposure, which is difficult to achieve consistently and constantly.*

## example/6.1 Limitations of Controlling Exposure

Describe a past or current event where the failure of an engineered control system allowed for the exposure of humans and the environment to a hazardous release.

### answer

Many answers are possible. One example occurred in the early-morning hours of December 3, 1984, when a holding tank containing 43 tons of stored methyl isocyanate (MIC) from a Union Carbide factory in Bhopal, India, overheated and released a toxic, heavier-than-air MIC gas mixture. MIC is an extremely reactive chemical and is used in production of the insecticide carbaryl.

The post-accident analysis of the process showed the accident started when a tank containing MIC leaked. It is presumed that the scientific reason for the accident was that water entered the tank where about 40 m$^3$ of MIC was stored. When water and MIC mixed, an exothermic chemical reaction started, producing a lot of heat. As a result, the safety valve of the tank burst because of the increase in pressure. This reaction was so violent that the coating of concrete around the tank also broke. It is presumed that between 20 and 30 tons of MIC were released during the hour that the leak took place.

The gas leaked from a 30 m high chimney, and this height was not enough to reduce the effects of the discharge. The reason was that the high moisture content (aerosol) in the discharge evaporated and gave rise to a heavy gas that rapidly sank to the ground, where people had their residences. According to the Bhopal Medical Appeal, around 500,000 people were exposed. Approximately 20,000 are believed to have died as a result; on average, roughly one person dies every day from the effects. Over 120,000 continue to suffer from effects including breathing difficulties, cancer, serious birth defects, blindness, and other problems.

and "green engineering" as a means to develop chemicals, materials, processes, and services that reduce or eliminate the use and generation of hazardous substances, leading to reduced risk to human health and the environment by reducing hazard.

Take buildings, for example. Reflect for a few minutes on the large variety of materials used in building construction, the large list of materials and coatings used to decorate and furnish a building, and the large number of chemicals used during operation and maintenance of the building. How many of these structural materials, adhesives, sealants, floor and wall coverings, furniture components, and cleaning agents are selected based on the criteria to maximize the health and productivity of the building's inhabitants by minimizing potential adverse impact (the risk) to humans or the environment? Unfortunately the answer to this question is "very few." Green building design takes into consideration

## Table / 6.2

**Estimated Economic Benefits to U.S. Society if Architects and Engineers Design and Operate Buildings with Consideration of Health**

| |
|---|
| $6 billion–$14 billion from reduced respiratory disease |
| $1 billion–$4 billion from reduced allergies and asthma |
| $10 billion–$30 billion from reduced sick-building syndrome |
| $20 billion–$160 billion from increased worker productivity unrelated to health |

SOURCE: Fisk (2000).

© Skip O'Donnell/iStockphoto.

the health of building occupants along with the impact on the environment associated with material choices.

Poorly designed and managed indoor environments have a large adverse economic impact on society that is associated with increased health costs and lower worker productivity. As shown in Table 6.2, huge savings could result from reducing that impact in the United States. In high-mortality developing countries, indoor air pollution is now responsible for up to 3.7 percent of the burden of disease. Remember from Chapter 1 that 33 percent of the environmental risk that leads to loss of disability-free days in the world is due to indoor air pollution from burning solid fuels.

## 6.2    Risk Perception

**Risk perception** examines the judgments people make when they are asked to characterize and evaluate hazardous activities and technologies. People make qualitative or quantitative judgments about the current and desired riskiness of many different hazards through everyday choices and behaviors. These decisions are based on the likelihood (probability) of an injury by a specific hazard and the severity of consequences associated with that injury.

Our judgments about risks are based on several considerations. One important factor is how familiar we are with the hazard. If we believe we know a lot about a hazard because we are often exposed to it, we often underestimate the degree of risk. Another factor is whether or not we are voluntarily interacting with a hazard. When a person voluntarily takes on a risk, he or she also usually underestimates the chances of a resulting injury. This may have to do with how much control individuals feel they have over the situation. Examples of voluntary risk include smoking, driving a car faster than the speed limit, and participating in extreme activities such as mountaineering or skydiving. Also, individuals often feel it is more acceptable to choose a risk than to be put at risk by government or industry.

**Risk Perception**

www.wiley.com/college/mihelcic

iStockphoto.

**Identify TRI Facilities by Zip Code**
http://www.scorecard.org/

**Toxics Release Inventory**
http://www.epa.gov/tri

This attitude toward involuntary versus voluntary risk explains why there is usually a public outcry when a factory contaminates local drinking water or indoor air quality is found to be unsafe. In these cases, the added risk from exposure to contaminated water and air is not voluntary. Individuals feel as if they are being subjected to hazards beyond their control and without their knowledge. Examples of involuntary risk are inhaling secondhand smoke, having a highway or high-voltage power lines placed in your community, and having pesticide residue on the outside of your produce.

## 6.3 Hazardous Waste and Toxic Chemicals

Exposure to a toxic or hazardous chemical can result in death, disease, or some other adverse impact such as a birth defect, infertility, stunted growth, or a neurological disorder. For humans, this contact with a chemical is typically through ingestion, inhalation, or skin contact. Exposure to the chemical can be associated with drinking water, eating food, ingesting soil and dust, inhaling airborne contaminants that could be in a vapor or particulate form, and contacting chemicals that are transported through the skin.

The **Toxics Release Inventory (TRI)** provides information to the public about hazardous waste and toxic chemicals. This inventory was established under the **Emergency Planning and Community Right-to-Know Act (EPCRA)** of 1986 and expanded by the Pollution Prevention Act of 1990. The TRI is a publicly available database, published by the Environmental Protection Agency (EPA), that contains information on releases of nearly 650 chemicals and chemical categories, submitted by over 23,000 industrial and federal facilities. The TRI tracks disposal or other releases both on-site and off-site, including wastes directly to air, land, surface water, and groundwater. It also provides information on other waste management strategies, such as recycling, energy recovery, treatment, and discharge to wastewater treatment plants.

The TRI database can be searched by year, geographical location, chemical released, or industry type. Citizens and emergency-response personnel can look up emissions of toxic chemicals in their community. The TRI provides the public with unprecedented access to information about toxic chemical releases and other waste management activities on a local, state, regional, and national level. One goal of the TRI is to empower citizens, through information, to hold companies and local governments accountable for how toxic chemicals are managed.

Figure 6.1 shows the total mass of TRI emissions since 1988, along with the number of facilities reporting releases. The TRI data help the public, government officials, and industry meet three objectives: (1) identify potential concerns and gain a better understanding of potential risks; (2) identify priorities and opportunities to work with industry and government to reduce toxic-chemical disposal or other releases and potential risks associated with them; and (3) establish reduction targets and measure progress toward reduction goals.

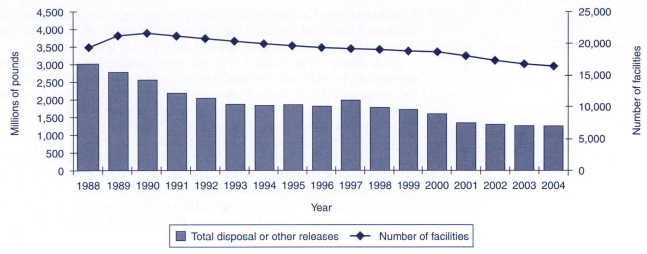

**Figure 6.1** Toxic Release Inventory (TRI) Emissions (in million pounds) since 1988 and Number of Facilities Reporting Emissions

Adapted from www.epa.gov/tri.

## 6.3.1 HAZARDOUS WASTE

In the United States, a **hazardous waste** is a regulatory subset of a solid waste. Solid wastes are defined under the **Resource Conservation and Recovery Act (RCRA)**. This regulatory definition says nothing about the waste's physical state, so some solid wastes are in liquid form. In the United States, **solid wastes** are legally defined as any discarded material not excluded by 40 C.F.R. 261.4(a). *Excluded wastes* include items such as domestic sewage, household hazardous waste, fly ash and bottom ash from coal combustion, and manure returned to soil. C.F.R. is the abbreviation for Code of Federal Regulations, the document in which federal regulations are published. The number 40 indicates the section of the C.F.R. related to the environment. The C.F.R. can be accessed via the Internet.

Thus, a hazardous waste denotes a regulated waste. Only certain waste streams are designated as hazardous under federal regulations. Wastes are classified as hazardous based on: (1) physical characteristics such as reactivity, corrosivity, and ignitability; (2) toxicity; (3) the quantity generated; and (4) the history of the chemical in terms of environmental damage it caused and the likely environmental fate. Hazardous wastes thus may or may not exhibit toxicity.

## 6.3.2 TOXICITY

Environmental toxicology, also known as environmental health sciences, is an interdisciplinary field dealing with the effects of chemicals on living organisms. Because energy and material are distributed and cycled through food webs, it is likely that an impact on one level will be reflected in other levels as well. For example, there is evidence that elevated PCB levels in fish results in adverse health effects to children born from mothers who included fish in their diet. While **bioaccumulation**

**Access the Code of Federal Regulations**
http://www.gpoaccess.gov/cfr/

**Construction and Demolition Debris**
http://www.epa.gov/epawaste/conserve/rrr/imr/cdm/index.htm

 **Toxicity Testing**

(concentration of a chemical builds up in an organism over time) of PCBs may have had no direct adverse effect on adult fish, there was an impact on some fish offspring and the next trophic level (humans).

Toxic effects can be divided into two types: **carcinogenic** and **noncarcinogenic**. A carcinogen promotes or induces tumors (cancer), that is, the uncontrolled or abnormal growth and division of cells. Carcinogens act by attacking or altering the structure and function of DNA within a cell. Many carcinogens seem to be site-specific; that is, a particular chemical tends to attack a specific organ. In addition, carcinogens may be categorized based on whether they cause direct or indirect effects: *primary carcinogens* directly initiate cancer; *pro-carcinogens* are not carcinogens but are metabolized to form carcinogens and thus indirectly initiate cancer; *co-carcinogens* are not carcinogens but enhance the carcinogenicity of other chemicals; and *promoters* enhance the growth of cancer cells.

---

### Box / 6.1    How Cancer Develops

*Cancer refers to a group of diseases involving abnormal, malignant tissue growth. Research has revealed that the development of cancer involves a complex series of steps, and carcinogens may operate in a number of different ways. Ultimately, cancer results from a series of defects in genes controlling cell growth, division, and differentiation. Genetic defects leading to cancer may occur because a chemical (or other carcinogenic agent) damages DNA directly. Alternatively, an agent may have indirect effects that increase the likelihood, or accelerate the onset, of cancer without directly interacting with DNA. For example, an agent might interfere with DNA repair mechanisms, thereby increasing the likelihood that cell division will give rise to cells with damaged DNA. An agent might also increase rates of cell division, thus increasing the potential for genetic errors to be introduced as cells replicate their DNA in preparation for division.*

(EPA, "Fact Sheet for Guidelines for Carcinogen Risk Assessment," March 2005)

---

Classification of a chemical as being carcinogenic to humans requires sufficient evidence that human exposure leads to a significantly higher incidence of cancer. Such evidence is often collected from workers in job environments where there is prolonged contact with a chemical. (This is called epidemiological data.) While there are few *known* human carcinogens (for example, benzene, vinyl chloride, arsenic, and hexavalent chromium), many chemicals are *probable* human carcinogens (for example, benzo(*a*)pyrene, carbon tetrachloride, cadmium, and PCBs), and hundreds of chemicals have *suggestive evidence* that they are carcinogens. As we will discuss later, chemicals are listed as **suspected carcinogens** when experimental evidence indicates increased cancer risk in test animals and insufficient information is available to show a direct cause–effect relationship for humans.

Noncarcinogenic effects include all toxicological responses other than carcinogenic, of which there are countless examples: organ damage (including kidney and liver), neurological damage, suppressed immunity, and birth and developmental (harming an organism's reproductive ability or intelligence) effects. For example, elevated lead levels in children have been shown to cause learning disorders and lower IQs. The toxic effects manifested following exposure to a

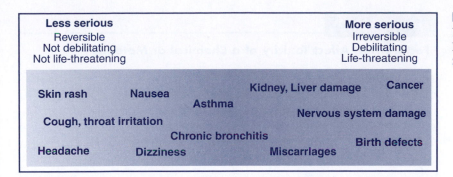

chemical often result from interference with enzyme (catalyst) systems that mediate the biochemical reactions critical for organ function. Figure 6.2 depicts the continuum of risks due to exposure to noncarcinogens as ranging from less serious to more serious. Risks that are reversible, not debilitating, and/or not life threatening are considered to be of less concern than those that are irreversible, debilitating, and/or life threatening.

Chemicals collectively known as **endocrine disruptors** exert their effects by mimicking or interfering with the actions of hormones, biochemical compounds that control basic physiological processes such as growth, metabolism, and reproduction. Endocrine disruptors may exert noncarcinogenic or carcinogenic effects. They are believed to contribute to breast cancer in women and prostate cancer

### Endocrine Disruptors

http://www.setac.org/node/100
http://www.who.int/ipcs/assessment/en/

---

**Box / 6.2    Endocrine-Disrupting Chemicals**

Endocrine-disrupting chemicals are chemicals that, when absorbed into the body, either mimic or block hormones and disrupt the body's normal functions. This disruption can happen through altering normal hormone levels, halting or stimulating the production of hormones, or changing the way hormones travel through the body, thus affecting the functions these hormones control.

These chemicals and substances are accumulating in fish and wildlife, and the number of warnings about eating of fish and wildlife due to endocrine disruptors is increasing and has reached over 30 percent of U.S. lakes and 15 percent of U.S. river miles. Studies document that these chemicals are accumulating in fish and wildlife to levels that are causing serious hormonal and reproductive effects in fish and wildlife at the top of the food chain, including wading birds, alligators, Florida panthers, minks, polar bears, seals, and beluga and orca whales. Many subpopulations with significant exposure are experiencing major reproductive effects, resulting in infertility and reproductive failures.

In humans, several health problems possibly linked to endocrine-disrupting chemicals have been recorded: (1) declines in sperm count in many countries; (2) a 55 percent increase in incidence of testicular cancer from 1979 to 1991 in England and Wales; (3) increases in prostate cancer; and (4) an increase in breast cancer in women, including an annual increase of 1 percent in the United States since the 1940s (Friends of the Earth, 2009).

In wildlife, the following are examples of effects that have been linked to endocrine-disrupting chemicals: (1) masculinization of female dog whelks (a type of shellfish); (2) eggs found in testes of roach fish in many rivers in the United Kingdom; (3) low egg viability, enlarged ovaries, and reduced penis size in Florida alligators; and (4) eggshell thinning and female–female pairing in birds (Friends of the Earth, 2009).

The risks associated with endocrine-disrupting chemicals are only just beginning to be discovered and quantified, because the doses that cause the effects are much lower than those traditionally tested in toxicity studies.

## Table / 6.3

### Factors That Affect Toxicity of a Chemical or Material

Form and innate chemical activity

Dosage, especially dose-time relationship

Exposure route

Species

Ability to be absorbed

Metabolism

Distribution within the body

Excretion

Presence of other chemicals

in men. Chemicals identified as endocrine disruptors include pesticides (such as DDT and its metabolites), industrial chemicals (such as some surfactants and PCBs), some prescription drugs, and other contaminants, such as dioxins (National Science and Technology Council, 1996).

The likelihood of a toxicological response is determined by the exposure to a chemical (one factor in Equation 6.1): a product of the chemical dose and the duration over which that dose is experienced. In humans, there are three major **exposure pathways**: ingestion (eating and/or drinking), inhalation (breathing), and dermal (skin) contact. Table 6.3 lists important factors that affect the toxicity of a chemical or material.

Some chemicals (for example, dioxin) can be lethal to test animals in very small doses, whereas others create problems only at much higher levels. Table 6.4 lists chemical compounds with widely varying

## example/6.2 Chromium Toxicity

Which form of chromium, Cr(III) or Cr(VI), is toxic?

### solution

The toxicity of chromium varies greatly depending on which oxidative state it is in. Cr(III), or $Cr^{+3}$, is relatively nontoxic, whereas Cr(VI), or $Cr^{+6}$, causes skin or nasal damage and lung cancer. Of course, chemicals can undergo oxidation and reduction reactions in environmental conditions, so release of the lower-toxicity form does not mean that the chromium will pose no risk to human health or the environment.

## Table / 6.4

**Oral Median Lethal Dose for Various Organisms and Chemicals**

| Chemical | Organism | LD$_{50}$ (mg chemical/kg body weight) |
|---|---|---|
| Methyl ethyl ketone | Rat | 5,500 |
| Fluoranthene | Rat | 2,000 |
| Pyrene | Rat | 800 |
| Pentachlorophenol | Mouse | 117 |
| Lindane | Mouse | 86 |
| Dieldrin | Mouse | 38 |
| Sarin (nerve gas) | Rat | 0.5 |

SOURCE: Values from Patnaik, 1992. Reprinted from Mihelcic, 1999, with permission of John Wiley & Sons, Inc.

toxicities. Here, **toxicity** is defined as causing death, an experimental endpoint that (for test animals) is more readily determined than, for example, lung cancer.

A common method of expressing toxicity is in terms of the **median lethal dose (LD$_{50}$)**, which is the dose that results in the death of 50 percent of a test organism population. The LD$_{50}$ is typically presented as the mass of contaminant dosed per mass (body weight) of the test organism, using units of mg/kg. Thus, a rodenticide with an LD$_{50}$ of 100 mg/kg would result in the death of 50 percent of a population of rats, each weighing 0.1 kg, if applied at a dose of 10 mg per rat. A dose of 20 mg per 0.1 kg rat should result in the death of more than 50 percent of the population, and a dose of 5 mg per 0.1 kg rat would result in the death of less than 50 percent.

A similar term, the **median lethal concentration (LC$_{50}$)** is typically used in studies of aquatic organisms and represents the ambient aqueous contaminant concentration (as opposed to injected or ingested dose) at which 50 percent of the test organisms die.

To identify LD$_{50}$ or LC$_{50}$, a series of experiments at various concentrations yields a **dose-response curve** as depicted in Figure 6.3. More subtle (behavioral or developmental) changes can also reflect a toxic response but are difficult to assess. These nonlethal end points are measured as an effective concentration that affects 50 percent of the population (EC$_{50}$).

Recall that what determines toxicity is not only the dose, but also duration of exposure to a chemical or substance. **Acute toxicity** refers to death (or some other adverse response) resulting from short-term (hours to days) exposure to a chemical. **Chronic toxicity** refers to a response resulting from long-term (weeks to years) exposure to a chemical.

Acute effects are typically experienced at higher contaminant concentrations than are chronic effects. For example, the EPA has established acute (1.7 µg/L) and chronic (0.91 µg/L) water-quality criteria

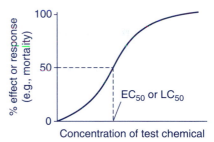

**Figure 6.3** Typical Form of Dose-Response Curve Used in Identifying EC$_{50}$ and LC$_{50}$ for Chemicals and Test Organisms

## Table / 6.5

**48-Hour LC$_{50}$ Values for 2,4-D for Selected Organisms**

| Species | LC$_{50}$ (mg/L) |
|---|---|
| *Daphnia magna* (zooplankton) | 25 |
| Fathead minnow | 325 |
| Rainbow trout | 358 |

SOURCE: Patnaik, 1992.

**Mercury Report to Congress**

http://www.epa.gov/mercury/reportover.htm

for mercury (II) to protect aquatic life in the Great Lakes from toxic effects. Here, the acute criterion is higher than the chronic value. As duration increases, the concentrations that can be tolerated without adverse effect are lower. Acute copper toxicity for rainbow trout has been shown to decrease from an LC$_{50}$ of 0.39 mg/L at a 12 hr duration to 0.13 mg/L at 24 hr to 0.08 mg/L at 96 hr. The toxicity of a specific chemical may also vary among species. Table 6.5 demonstrates this effect, comparing 48 hr LC$_{50}$ values for 2,4-dichlorophenoxyacetic acid (2,4-D), a common herbicide used on farms and household lawns, for various aquatic organisms.

While the concentrations of 2,4-D listed in Table 6.5 are not likely to be encountered in surface waters (although levels of agricultural chemicals in runoff do increase following spring rains and snowmelt), the observed variation in LC$_{50}$ values suggests a scenario in which sediment-living microcrustacean populations would be affected while fish populations would not. Such a scenario could potentially alter and disrupt the food web, with ecosystem-wide impacts. An understanding of food web function and the bioaccumulation and toxicity of contaminants (at each trophic level) is necessary to adequately assess the risk posed by the myriad chemical contaminants introduced into our environment (see Chapter 5).

The species-specific nature of toxicity presents a fundamental shortcoming in procedures commonly applied to estimate effects on humans based on experiments with test animals. Individual humans may be substantially more or less susceptible to the toxic effects of a specific compound at a given dose than are laboratory surrogate organisms. When animal studies are used to determine standards for human exposure, the uncertainties involved in utilizing the results are accounted for through use of conservative assumptions and application of safety factors that may result in an estimate that is conservative by several orders of magnitude—an approach based on a "better safe than sorry" philosophy. In addition, the fact that some wildlife may be more sensitive to toxic chemicals than humans has led to the promulgation of water-quality criteria in which the more stringent of wildlife- or human-health-based standards govern discharge limits. For example, the maximum contaminant level (MCL) for allowable chromium in drinking water is 0.1 mg/L, while the acute criterion for freshwater aquatic life is 21 µg/L. In this case, the wildlife standard is approximately one-fifth of the human-health-based value.

Sensitive segments of a population, known as **susceptible populations**, must receive distinct consideration in determining toxic effects of chemicals or substances. The embryonic, juvenile, elderly, and/or ill segments of any population (human or environmental) are likely to be more susceptible to adverse effects from chemical exposure than are healthy young adults. In some cases, the sex of an individual may also influence its susceptibility.

*Synergistic toxicity*, resulting from the exposure to multiple chemicals, is a phenomenon that is receiving increased attention. For example, consider two compounds with LC$_{50}$ values of 5 and 20 mg/L, respectively. When present together, their individual LC$_{50}$ values might drop to 3 and 10 mg/L, levels that are lower than the individual LC$_{50}$ values. In some cases, chemicals may have the opposite (antagonistic)

## Table / 6.6

**Potential Combined Toxicity Resulting from Exposure to a Mixture of Chemicals A and B**

| Type of Interaction | Toxic Effect, Chemical A | Toxic Effect, Chemical B | Combined Effect, Chemicals A + B |
|---|---|---|---|
| Additivity | 20% | 30% | 50% |
| Antagonism | 20% | 30% | 5% |
| Potentiation | 0% | 20% | 50% |
| Synergism | 5% | 10% | 100% |

effect, resulting in a combination that is less toxic than when present separately.

Table 6.6 provides an example of the possible effects chemical mixtures could have on combined toxicity. Note that the combined effect of the two chemicals (A and B) can be greater or less. This is an example of the difficulty in assessing the risks of chemical mixtures. Unfortunately, scientific studies of chronic synergistic effects are lacking, largely because countless numbers of chemicals and combinations of chemicals exist and because long-term experiments involve inherent difficulties.

## 6.4 Engineering Ethics and Risk

Engineers must understand environmental risk in order to protect all segments of society and all inhabitants of ecosystems. These individuals include the residents of communities in which the engineer resides, aquatic life residing in a river downstream of a construction site or treatment plant, and the global community of more than 6 billion people.

All too often, engineers work to minimize or eliminate the risk of an *average* member of society or an ecosystem inhabitant that is valued for recreational sport or commercial profit. Given the limitations of risk assessment and the multitude of uncertainties, engineers need to carefully consider all segments of society as well as ecosystem health (for example, biodiversity and endangered species). It is also important to recognize the susceptible segments of any population that may be significantly more sensitive to environmental exposure to a chemical or substance. For example, the impact of a chemical will vary with a person's age, gender, health status, occupation, and lifestyle.

In the phenomenon of **environmental justice**, certain segments of society that are socially economically disadvantaged may be burdened with a greater amount of environmental risk. An environmental justice issue is apparent in Santa Clara County, California, where facilities that are required to list their toxic emissions in the EPA's Toxic Release

**Global Burden of Disease**
http://www.who.int/quantifying_ehimpacts/en/

**Environmental Justice**
http://www.epa.gov/compliance/environmentaljustice/

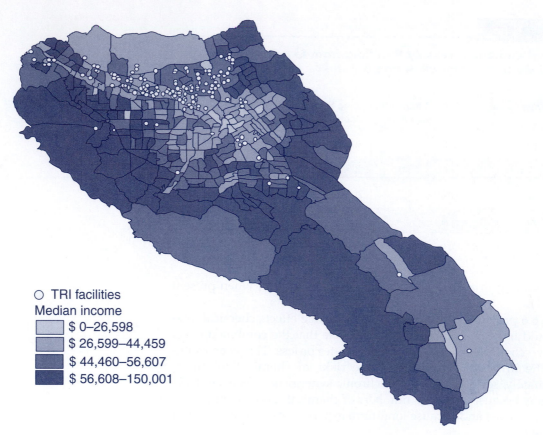

○ TRI facilities
Median income
▢ $ 0–26,598
▢ $ 26,599–44,459
▢ $ 44,460–56,607
▢ $ 56,608–150,001

**Figure 6.4**  Facilities Required to List their Toxic Emissions in the TRI, by Median Income   Note that the facilities that emit toxic chemicals are located primarily in areas with lower median incomes.

Meuser and Szacz, *American Behavioral Scientist* 43 (4), p. 602, copyright © 2000. Reprinted by Permission of SAGE Publications, Inc.

## Indigenous People and the Environment

http://www.unep.org/indigenous

### Class Discussion

Economically disadvantaged individuals could also be living in a location in the developing world where they are exposed to disease-causing pathogens in unsafe drinking water and bear the added burdens of large-scale impact of HIV/AIDs and chronic effects of malaria. Climate change melting Arctic ice is disturbing the subsistence lifestyles of the Inuit people, who live in Arctic regions, as well as polar bears that hunt in these areas. Is it fair that a greater risk is assumed by these segments of the global community?

Inventory are located in communities with lower median incomes (Figure 6.4). Economically disadvantaged people tend to inhabit places that expose them to a greater number or higher concentrations of toxic chemicals (for example, next to highways that contribute air pollutants, next to industry that emits chemicals of concern, downwind from incinerators); they live in buildings that have hazardous materials associated with older construction or are served by aging infrastructure; or they have employment that results in increased exposure to hazardous materials.

These higher-risk individuals may live in urban or rural areas, typically have little political clout, and often are members of economically disadvantaged minority groups. They could be segments of a population that have elevated exposure levels because of their hunting or fishing habits due to a more subsistence lifestyle (as in the case of Native Americans who eat more fish or parts of fish that contain greater concentrations of toxics than in the average American's diet). They include the African American communities located near the intensive number of oil- and chemical-processing facilities located along the lower Mississippi River.

**John Muir** at Merced River with Royal Arches and Washington Column in background, Yosemite National Park, California. John Muir Papers, Holt-Atherton Special Collections, University of the Pacific Library. Copyright 1984 Muir-Hanna Trust.

**John Muir**, the founder of the Sierra Club, stated, "Everybody needs beauty as well as bread, places to play in and pray in, where nature may heal and give strength to body and soul alike."

The EPA defines environmental justice as the "fair treatment and meaningful involvement of all people regardless of race, color, national origin, or income with respect to the development, implementation, and en-

http://www.who.int/en/

forcement of environmental laws, regulations, and policies."

The World Health Organization's constitution (first written in 1946) states, "The enjoyment of the highest attainable standard of health is one of the fundamental rights of every human being." In 2002, water was recognized as a basic right when the United Nations Committee on Economic, Social, and Cultural Rights agreed, "The right to water clearly falls within the category of guarantees essential for securing an adequate standard of living, particularly since it is one of the most fundamental conditions for survival."

Engineers have a responsibility to consider these at-risk individuals and the communities they inhabit and to minimize, or eliminate, the likelihood that they bear a greater proportion of environmental risk than wealthier, better educated, or more politically powerful segments of society. With our new and increasing knowledge of sustainable design and green engineering, we have the ability to meet these challenges while continuing to improve quality of life for *all* segments of society in both the developed and developing world by employing more benign chemicals and materials, reducing energy and material consumption, and taking a systems perspective.

## 6.5    Risk Assessment

Risk assessments address questions such as these: What health problems are caused by chemicals and substances released into the home, workplace, and environment? What is the probability that humans will experience an adverse health effect when exposed to a specific concentration of chemical? How severe will the adverse response be? The remainder of this chapter focuses primarily on how to quantify the risks associated with exposure to chemicals and other environmental agents and the subsequent impacts on human health.

The four components of a complete **risk assessment** are (1) hazard assessment; (2) dose-response assessment; (3) exposure assessment; and (4) risk characterization. Figure 6.5 depicts how these four items

**Risk Assessment**
http://www.epa.gov/risk/

**Class Discussion**

The three statements provided in Box 6.3 suggest that humans need wild places, the economically disadvantaged should not be burdened with a disproportionate percentage of environmental risk, and access to a basic level of freshwater and sanitation, a healthy workplace, and healthy environment are legal entitlements, instead of commodities or services that should be marginalized or privatized. Do you agree with these three statements? Should all people in the world be guaranteed some basic access to water that would guarantee some specified level of health? How about ecosystems?

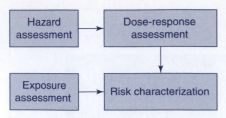

**Figure 6.5** **Components of a Complete Risk Assessment** The dose-response and exposure assessments are combined to yield a risk characterization.

are integrated. A risk assessment organizes and analyzes a large set of information embedded in the four components to determine whether some environmental hazard will result in an adverse impact on humans or the environment. The environmental hazard could be exposure to a specific chemical or a broader issue such as climate change.

## 6.5.1 HAZARD ASSESSMENT

A hazard assessment is not a risk assessment. A **hazard assessment** consists of a review and analysis of toxicity data, weighing evidence that a substance causes various toxic effects, and evaluating whether toxic effects in one setting will occur in other settings. The hazard assessment determines whether a chemical or substance is, or is not, linked to a particular health concern, whereas a risk assessment will take into account the hazard assessment as well as the exposure assessment.

Sources of toxicity data include test tube studies, animal studies, and human studies. Test tube studies are fast and relatively easy, so they are commonly used to screen chemicals. Animal studies may measure acute or chronic effects. They could investigate a general end point (for instance, death) or a more specialized end point (say, a birth defect). Controlled laboratory studies are commonly employed to determine the toxicity of specific chemicals to aquatic life. Human studies typically consist of case studies that alert society to a problem and more extensive controlled epidemiologic studies.

The best study to determine the impact on humans is the epidemiology study. **Epidemiology** is the study of diseases in populations of humans or other animals, specifically how, when, and where they occur. Epidemiologists attempt to determine what factors are associated with diseases (risk factors), and what factors may protect people or animals against disease (protective factors).

---

| Box / 6.4 | Human and Ecosystem Toxicological Databases |

**Integrated Risk Information System (IRIS)**

The Integrated Risk Information System (IRIS) is an electronic database that contains human health effect information for hundreds of chemicals (see www .epa.gov/iris/). IRIS provides information for two components of a risk assessment: the hazard assessment and the dose-response assessment. IRIS was developed by the EPA and is written for professionals involved in risk assessments, decision making, and regulatory activities. Thus, it is intended for use by individuals who do not have training in toxicology.

IRIS contains descriptive and quantitative information on hazard identification, oral slope factors, oral and inhalation unit risks for carcinogenic effects, as well as oral reference doses (RfDs) and inhalation reference concentrations (RfCs) for chronic noncarcinogenic health effects. These topics are discussed later in this chapter.

**Ecotoxicology**

The EPA maintains the ECOTOX database as a source for locating single-chemical toxicity data for aquatic life and terrestrial plants and wildlife (see www .epa.gov/ecotox/). This database can be used to assist ecological-hazard assessments and evaluate the potential hazard associated with wastewater effluent and/or leachate.

## Table / 6.7

### Difficulties of Epidemiology Studies

Matching control groups is difficult, because factors that lead to exposure to a chemical may be associated with other factors that affect health.

Society has become more mobile, so individuals may no longer live in the same community all of their life.

Death certificates typically measure only the cause of death, so they miss health conditions that individuals had over the course of their life.

Other toxicity end points besides death (e.g., miscarriages, infertility, learning disorders) might not be measured with use of death certificates.

Accurate exposure data can be difficult to obtain for a large group of individuals.

Large populations are required for these studies so that rigorous statistical analysis can be applied to the data.

Many diseases can take years to develop.

Epidemiological studies can be divided into two basic types depending on whether the events have already happened (retrospective) or whether the events may happen in the future (prospective). The most common studies are retrospective studies that are also called case-control studies. A case-control study may begin when an outbreak of disease is noted and the causes of the disease are not known, or when the disease is unusual within the population studied.

These types of studies have difficulties however, as summarized in Table 6.7. With epidemiological studies, it is extremely difficult to prove causation, meaning proof that a specific risk factor actually caused the disease being studied. Epidemiological evidence can, however, readily show that this risk factor is associated (correlated) with a higher incidence of disease in the population exposed to that risk factor. The higher the correlation is, the more certain the association.

**Weight of evidence** is a brief narrative that suggests the potential for whether a chemical or substance can act as a carcinogen to humans. Currently, weight of evidence is categorized by one of five descriptors listed in the left column of Table 6.8. Scientists who analyze the available data obtained from animal or human studies develop these descriptors for carcinogens. The right column of Table 6.8 summarizes how these descriptors are related to the quality and quantity of available data. IRIS (described in Box 6.4) provides information on the descriptor associated with particular chemicals.

### 6.5.2 DOSE-RESPONSE ASSESSMENT

DOSE A **dose** is the amount of a chemical received by a subject that can interact with the subject's metabolic process or other biological receptors after it crosses an outer boundary. Depending on the context,

**Dosage**

## Table / 6.8

**Explanation of Weight of Evidence Descriptors**

| Weight of Evidence Descriptor | Relationship of Descriptor to Scientific Evidence |
|---|---|
| Carcinogenic to humans | Convincing epidemiologic evidence demonstrates causality between human exposure and cancer, or evidence demonstrates exceptionally when there is strong epidemiological evidence, extensive animal evidence, knowledge of the mode of action, and information that the mode of action is anticipated to occur in humans and progress to tumors. |
| Likely to be carcinogenic to humans | Tumor effects and other key data are adequate to demonstrate carcinogenic potential to humans but do not reach the weight of evidence for being carcinogenic to humans. |
| Suggestive evidence of carcinogenic potential | Human or animal data are suggestive of carcinogenicity, which raises a concern for carcinogenic effects but is judged not sufficient for a stronger conclusion. |
| Inadequate information to assess carcinogenic potential | Data are judged inadequate to perform an assessment. |
| Not likely to be carcinogenic to humans | Available data are considered robust for deciding that there is no basis for carcinogenic human hazard concern. |

SOURCE: EPA, *Guidelines for Carcinogenic Risk Assessment*, March 29, 2005.

the dose may be (1) the amount of the chemical that is administered to the subject; (2) the amount administered to the subject that reaches a specific location in the organism (for example, liver); or (3) the amount available for interaction within the test organism after the chemical crosses a barrier such as a stomach wall or skin.

To calculate the dose associated with a chemical, determine the mass of the chemical administered per unit time, and divide that by the weight of the individual. In the case of an adult or child who is drinking water that contains the chemical of concern, the dose can be determined as shown in Example 6.3.

### example/6.3 Determining Dose

Assume that the chemical of concern has a concentration of 10 mg/L in drinking water, and that adults drink 2 liters of water per day, and children drink 1 liter of water per day. Assume also that an adult male weighs 70 kg, a female weighs 50 kg, and a child weighs 10 kg. What is the dose for each of these three members of society?

### solution

To find the dose associated with a chemical, determine the mass of the chemical taken in per unit time, and divide this by the weight of the individual. In this situation, the only route of exposure is from drinking contaminated water. The dose for the three segments of society can be determined as follows:

$$\text{adult female dose:} \quad \frac{10\,\dfrac{mg}{L} \times 2\,\dfrac{L}{day}}{50\,kg} = 0.40\,\frac{mg}{kg\text{-}day}$$

$$\text{adult male dose:} \quad \frac{10\,\dfrac{mg}{L} \times 2\,\dfrac{L}{day}}{70\,kg} = 0.29\,\frac{mg}{kg\text{-}day}$$

$$\text{child dose:} \quad \frac{10\,\dfrac{mg}{L} \times 1\,\dfrac{L}{day}}{10\,kg} = 1.0\,\frac{mg}{kg\text{-}day}$$

Note that, in this example, the dose received by the child and adult female is greater than that received by the adult male. This is one reason certain segments of society may be at greater risk when exposed to a specific chemical. Another reason is that, in most situations, children, older adults, and the sick are harmed to a greater extent by exposure to toxic chemicals and pathogens than are young, healthy adults. The same would hold true for plants and animals that inhabit ecosystems.

When determining the dose, scientists can account for absorption of the chemical. For example, the stomach wall may act as a barrier to the absorption of some chemicals that are ingested, while the skin may act as a barrier to chemicals that contact the hands. To account for the fact that 100 percent uptake (or absorption) of some chemicals does not occur, multiply the dose by the percent absorbed (termed $f$ in Example 6.4). The value of $f$ is typically 0 to 0.1 (1 to 10 percent) for metals and 0 to 1.0 (0 to 100 percent) for many organics.

In many states, the absorption efficiency applicable to dermal contact is considered to be 10 percent ($f = 0.10$) for contact with volatile or semivolatile organic chemicals and 1 percent ($f = 0.01$) for inorganic chemicals. For ingestion of contaminated soil and dust, it is assumed to be 100 percent ($f = 1.0$) for volatile organic chemicals and 100 percent for chemicals that sorb more strongly to soil (for example, PCBs and pesticides). However, the medium in which the chemical is present (water versus lipid, air versus water) may determine the extent of absorption.

DOSE RESPONSE In animal laboratory studies, health damage is typically measured over a range of doses (minimum of three). Because sample populations are kept low during these studies to save time and money, applied doses must be at relatively high concentrations,

## example/6.4 Accounting for Absorption Efficiency when Determining Dose

Assume scientists know that only 10 percent of the chemical discussed in Example 6.3 is absorbed through the stomach wall. In this case, the exposure to the chemical was only through drinking contaminated water. What is the dose for the three target populations?

## solution

Since 10 percent of the chemical is transported through the stomach wall, $f = 0.10$. The doses that account for incomplete transport of the chemical of concern for our three segments of society are as follows:

$$\text{adult female dose:} \quad \frac{10\,\dfrac{mg}{L} \times 0.10 \times 2\,\dfrac{L}{day}}{50\,kg} = 0.04\,\frac{mg}{kg\text{-}day}$$

$$\text{adult male dose:} \quad \frac{10\,\dfrac{mg}{L} \times 0.10 \times 2\,\dfrac{L}{day}}{70\,kg} = 0.029\,\frac{mg}{kg\text{-}day}$$

$$\text{child dose:} \quad \frac{10\,\dfrac{mg}{L} \times 0.10 \times 1\,\dfrac{L}{day}}{10\,kg} = 0.10\,\frac{mg}{kg\text{-}day}$$

Here the dose is much lower than when the effect of absorption was neglected; however, children and adult females still receive a greater dose than adult males.

that is, at concentrations higher than what are typically observed in the environment (for example, in the workplace and home). Accordingly, a **dose-response assessment** is performed to allow extrapolation of data obtained from laboratory studies performed at higher doses to lower doses that are more representative of everyday life. Because of this extrapolation process, the assessment may overlook hazards such as endocrine-disrupting effects, which can occur at extremely low doses. However, as described in this section, dose-response assessments are performed differently for carcinogens and noncarcinogens.

**Carcinogens** Scientists have knowledge of the carcinogenic effect of chemicals primarily through laboratory test animal studies. Laboratory studies are conducted at higher doses so scientists can observe statistical changes in response with dose. In this case, the adverse response is formation of a tumor or some other sign of cancer. Carcinogens are treated as having **no threshold effect**, that is, under the assumption that any exposure to a cancer-causing substance will, with some degree of uncertainty, result in the initiation of cancer.

Figure 6.6 shows an example of a dose-response assessment for a carcinogenic chemical. Because a conservative scientific approach is used and, as stated previously, evidence suggests there is no threshold effect, the intersection of the dose-response curve at low doses is through the zero intercept. Application of such a dose-response model implies that the probability of contracting cancer is zero only if the exposure to the carcinogen is zero. The slope of the dose-response curve at very low doses is called the **potency factor** or **slope factor**. The slope factor is used in risk assessments for carcinogens, as we will show later in this chapter. The slope factor is an upper-bound estimate of risk per increment of dose that can be used to estimate carcinogenic risk probabilities for different exposure levels.

As shown in Figure 6.6, the slope factor has units of inverse mg/kg-day, or $(mg/kg\text{-}day)^{-1}$. It equals the unit risk for a chronic daily intake of 1 mg/kg-day. Values of the slope factor for many carcinogenic chemicals are available in the IRIS database (see Box 6.4). As we will see later in several examples, to obtain the overall risk, we multiply the slope factor by the calculated dose.

For most risk assessments, we determine the average daily dose by assuming that an individual is exposed to the maximum concentration of the carcinogen over his or her lifetime. In this case, the individual adult is assumed to live 70 yr and weigh 70 kg. This 70 yr averaging time may be different than the actual time of exposure, as we will discuss later when we put everything together to conduct a risk assessment to determine a risk-based cleanup level of contaminants found in drinking water and soil.

**Noncarcinogens**    Noncarcinogenic chemicals do not induce tumors. In this case, the adverse end point would be a health impact such as liver disease, learning disorder, weight loss, or infertility. Obviously many end points do not result in cancer or death. An important point to understand is that, compared with carcinogens, noncarcinogens are assumed to have a **threshold effect**. That is, there is a **dose limit** below which it is believed there is no adverse impact. Figure 6.7 shows the dose response assessment for a noncarcinogen.

Several new terms are defined in Figure 6.7. First, a **no observable adverse effect level (NOAEL)** is present. The NOAEL is the dose (units of mg/kg-day) at which no adverse health effect is observed. A dose less than or equal to this level is considered safe. However, because there is uncertainty in this safe dose of a noncarcinogen, scientists apply safety factors to the NOAEL to determine the **reference dose (RfD)**.

Reference dose is defined by EPA as an estimate, with uncertainty spanning perhaps an order of magnitude, of a daily oral exposure to the human population (including sensitive subgroups) that is likely to be without an appreciable risk of deleterious effects during a lifetime.

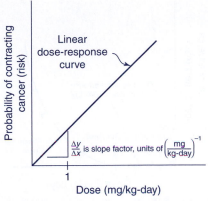

**Figure 6.6    Linear Dose-Response Relationship for a Carcinogenic Chemical or Substance**    The zero intercept indicates that, according to this model, there is no threshold effect, so the probability of contracting cancer is zero only if the exposure to a carcinogen is zero. The $y$-axis can be thought of as the probability of contracting cancer at a given dose. The $x$-axis is the dose (mg chemical per kg body weight per day). The slope of the dose-response curve near the intercept for a dose of 1 mg/kg-day is referred to as the potency factor or slope factor, with units of inverse mg/kg-day.

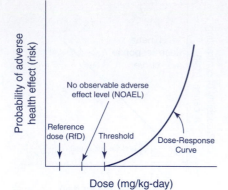

**Figure 6.7 Dose-Response Relationship for a Noncarcinogenic Chemical or Substance** Note the presence of a threshold dose below which no adverse response is observed for the response being assessed. The *y*-axis can be thought of as the probability of contracting an adverse effect at a given dose. The *x*-axis is the dose (mg chemical per kg body weight per day). The NOAEL is the dose at which no adverse health effect is observed. Doses less than or equal to this level can be considered safe. The reference dose (RfD) is an estimate of a lifetime dose that is likely to be without significant risk. RfDs are used for oral ingestion through exposure routes such as drinking water or eating food.

The RfD can be expressed mathematically:

$$RfD = \frac{NOAEL}{UF} \qquad (6.2)$$

Note that Equation 6.2 and Figure 6.7 both show that the RfD is lower than the NOAEL. The **uncertainty factor (UF)** typically ranges from 10 to 1,000. Application of the UF accounts for numerous uncertainties in applying NOAEL values to estimate RfD values. (Later, Box 6.6 discusses these uncertainties and how the UF values account for these uncertainties in risk characterization.)

Inclusion of the UF in determination of a *safe* dose shows that the reference dose (units of mg/kg-day) was developed to account for the uncertainty associated with conducting dose-response studies on small homogenous test animal populations for application to humans. The reference dose should also account for societal groups (such as children) who may be more sensitive to a chemical. As we will see later, reference doses are used in risk assessments for oral intake of noncarcinogens through drinking water or eating food.

The **reference concentration (RfC)** was developed as an estimate of an inhalation exposure (from breathing) for a given duration that is likely to be without an appreciable risk of adverse health effects over a lifetime. The reference concentration can be thought of as an estimate (with uncertainty of one order of magnitude or greater) of a continuous inhalation exposure to a noncarcinogen that is likely to be without significant risk to human populations. The IRIS database provides separate values for RfD and RfC.

### 6.5.3 EXPOSURE ASSESSMENT

The purpose of the **exposure assessment** is to determine the extent and frequency of human exposure to target chemicals. Some of the questions that are answered during the exposure assessment are listed in Table 6.9.

**Table / 6.9**

**Some Questions Answered during the Exposure Assessment**

What are the important sources of chemicals (e.g., pesticide application)?

What are the pathways (e.g., water, air, food) and routes of exposure (e.g., ingestion, inhalation, dermal contact)?

What amount of the chemical are people exposed to?

How often are people exposed?

How much uncertainty is associated with the estimates?

What segments of society (or ecosystem) are more at risk?

The exposure assessment can also determine the number of people exposed and the degree of absorption by various routes of exposure. Remember that the exposure assessment study should also determine the exposure of average individuals in society and high-risk groups (for example, workers, children, women, economically disadvantaged groups, older adults, area residents). Children typically have a more limited diet that may lead to relatively high but intermittent exposures. They also engage in behaviors such as crawling and mouthing (placing hands and objects in their mouth), which result in an increased exposure of chemicals via oral ingestion. The elderly and disabled may have sedentary lifestyles, which change their exposure. Pregnant and lactating women typically consume more water, which may lead to a different exposure assessment. Lastly, the many physiological differences between men and women, such as body weight and inhalation rates, could lead to important differences in exposures.

Exposure assessment can also be applied to a specific location. As mentioned earlier in this chapter, additional exposure could be associated with living next to a highway, incinerator, landfill, or factory. It could also be associated with living or working in a particular type of building, drinking a particular water supply, or eating a particular type and amount of food. Many details are considered, and a scientific study goes along with each of these scenarios.

| Table / 6.10 |
| --- |
| **Some Key Barriers to Brownfield Redevelopment** |
| Issues of liability |
| Differences in cleanup standards (can vary between state and federal governments) |
| Cost uncertainty associated with assessing contamination and cleanup |
| Obtaining financing, because lenders may want the government to waive liability associated with these sites |
| Community concerns |

---

**Box / 6.5    Brownfields**

According to the EPA, brownfields are "abandoned, idled, or under-used industrial and commercial sites where expansion or redevelopment is complicated by real or perceived environmental contamination." In the United States, there are an estimated half million brownfields, primarily located in urban areas. Concerns of environmental and economic justice are associated with these sites, because many are located in poorer communities.

Unfortunately, brownfields typically lie idle because purchasers, lenders, and developers stay away for liability reasons and seek out greenfield sites, that is, open spaces typically located on the edge of towns and cities. However, development of greenfield space is not desirable to society and the environment because of issues such as loss of farmland and its associated way of life, loss of open space and wildlife habitat, and problems of flooding associated with stormwater management and paving, which causes increased runoff.

Another set of undesirable impacts associated with developing green spaces is that no infrastructure is present (unlike in an urban area), so it must be built and paid for (which consumes energy and raw materials). In addition, the location of employment away from the urban core can isolate employers from workers who cannot afford ownership of an vehicle or might have to take several bus transfers to reach a job site. Also, development of green space usually results in future problems of sprawl and congestion.

In contrast, brownfields are typically located near existing built environment infrastructure, mass transportation, and labor. Thus, there are clear economic, social, and environmental benefits of redeveloping a brownfield instead of developing a greenfield site.

Table 6.10 lists some key barriers to brownfield redevelopment. Listening to the concerns of the community, engaging stakeholders, and working with local units of government in the planning process are critical components of the engineer's job in successful brownfield redevelopment.

You can learn more about brownfields by visiting the EPA's Web site, www.epa.gov/brownfields.

Due to space limitations, we will focus our discussion on how exposure assessment is related to usage of land for residential, commercial, and industrial purposes. Much of this activity is associated with decision making related to engineering abandoned or idle properties (termed **brownfields**, discussed in Box 6.5) into something that is beneficial to society and the environment. Many times, brownfields are contaminated from past activities at the site. Brownfield redevelopment requires that an engineer work with a diverse group of stakeholders—community members, nongovernmental organizations, government officials, financial lenders, real estate agents, and developers—in order to achieve a value-added use for the site.

Table 6.11 shows three types of land usage and associated parameters that might be used in an exposure assessment. The three types of land usage considered in Table 6.11 are residential, industrial, and commercial. Examples of specific activities that constitute each usage also are provided in the table. The commercial land use category is extremely varied, so it can be split into several subtypes. For example, commercial usage may encompass day-care centers, schools, gas stations, lumberyards, government buildings, professional offices, and commercial businesses that serve food. In all of these cases, there are different levels of restrictions on public access and different exposure levels to workers and customers.

Table 6.11 also estimates human intake of chemicals through mechanisms such as drinking water, breathing air, and ingesting soil (dust). There may also be dermal contact by direct contact with chemicals or contaminated soils. The estimates of these intake rates (IR) are based on scientific studies, the type of individual, and the activity that takes place at the site.

As you can imagine, site specificity is related to exposure assessment. For example, Table 6.12 shows the amount of soil that adheres to the body and is ingested daily for specific human populations based on land use and associated employment. To determine soil adherence, scientists need to know how much skin is exposed for potential dermal (skin) contact with chemicals and contaminated soil. As an example, scientists assume that an adult worker wears a short-sleeve shirt, long pants, and shoes. The amount of skin surface area exposed to dust and dirt for these assumptions of clothing is also a function of the worker's body weight. The total dermal area available for contact is thus assumed to be 3,300 cm$^2$. This assumes exposed skin consists of the head (1,200 cm$^2$), hands (900 cm$^2$), and forearms (1,200 cm$^2$).

Climatic conditions such as snow cover and freezing conditions are not assumed to affect the amount of soil ingested by humans because studies suggest that up to 80 percent of indoor air dust is from outdoor soils. It is believed that outdoor soil is transported inside buildings by air deposition, heating, ventilation and air-conditioning systems, and foot traffic. The indoor air environment clearly affects health, especially because, as stated earlier, Americans now spend 85 percent of their day inside some type of building. However, assessment of dermal exposure to contaminated soil would also consider

**Land Uses and Assumptions of Exposure Assessment Associated with Each Use**   The EPA publishes an *Exposure Factors Handbook* that provides more detail on specific values used in exposure assessment (National Center for Environmental Assessment, EPA/600/P-95/002F, 1997).

| Land Use | Examples of This Land Use | Intake Rates (IR) for Drinking Water; Air Inhalation, and Soil Ingestion* | Exposure Frequency (EF) (days per year) and Exposure Duration (ED) (years) |
|---|---|---|---|
| Residential (primary activity is residential) | Single-family dwellings, condominiums, apartment buildings | **Children drink** 1 L/day **Adults drink** 2 L/day **Adults inhale** 20 m$^3$/day **Children age 1–6 consume** 200 mg soil/day **Adults consume** 100 mg soil/day | **For drinking water** EF: 350 days/yr ED: 30 yr **For air inhalation** EF: 350 days/yr ED: 30 yr **For soil ingestion** ED: 6 yr for children 1–6 ED: 24 yr for adults EF: 350 days for children and adults |
| Industrial (primary activity is industrial, or zoning is industrial) | Manufacturing, utilities, industrial research and development, petroleum bulk storage | **Adults drink** 1 L/day **Adults inhale** 10 m$^3$/day | **For drinking water** EF: 245 days/yr ED: 21 yr **For air inhalation** EF: 245 days/yr ED: 21 yr **For soil ingestion** ED: 21 yr for adults EF: 245 days for children and adults |
| Commercial (use is a business or is intended to house, educate, or provide care for children, the elderly, the infirm, or other sensitive subpopulations) | Day-care centers, educational facilities, hospitals, elder-care facilities and nursing homes, retail stores, professional offices, warehouses, gas stations, auto services, financial institutions, government buildings | **Adults drink** 1 L/day **Adults inhale** 10 m$^3$/day | **For drinking water** EF: 245 days/yr ED: 21 yr **For air inhalation** EF: 245 days/yr ED: 21 yr **For soil ingestion** ED: 21 yr for adults EF: 245 days for children and adults |

*Recall that the average weight for a male, female, and child are 70 kg, 50 kg, and 10 kg, respectively.

## Table / 6.12

### Amount of Soil Assumed to Adhere to Surface of Skin and Taken In Daily for Specific Populations Based on Land Use and Associated Employment

| Target Population | Soil Adherence (mg soil/cm$^2$ skin) | Mass of Soil Taken In Daily (mg/day) |
|---|---|---|
| Adult living in residential area | 0.07 | 50 |
| Child living in residential area | 0.2 | 200 for ages 1-6<br>100 for all others |
| Adult worker at commercial III | 0.01 | 50 |
| Adult worker at commercial IV | 0.1 | 50 |
| Industrial worker | 0.2 | 50 |

Commercial III refers to gas stations, auto dealerships, retail warehouses. The worker population is engaged in activities at the property that are of a low soil-intensive nature.

Commercial IV refers to hotels, professional offices, banks. A groundskeeper worker population has been identified as an appropriate receptor population. They engage in activities at the property that are of a high soil-intensive nature.

the climatic conditions in northern areas (snow cover and frozen soil for a particular period of the year) that limits direct contact between soil and skin.

For engineers, knowledge of risk and exposure assessment provides information for determining whether a contaminated site needs to be remediated, and if so, to what level. Contaminated soil and groundwater at brownfield sites can be remediated using engineering technology. Alternatively, technological and institutional barriers can be used to minimize or prevent exposure. For example, paving a parking lot may prevent direct dermal contact with contaminated soils that lie underneath. Another example, in this case to prevent exposure to contaminated groundwater, would be for the property deed to place a restriction on placement of wells if a property is served by a municipal water supply.

### 6.5.4  RISK CHARACTERIZATION

As shown previously in Figure 6.5, the **risk characterization** takes into account the first three steps in risk assessment (hazard assessment, dose-response assessment, and exposure assessment). The risk characterization is specifically determined by integrating information from the dose-response and exposure assessments. The process is performed differently for carcinogens and noncarcinogens.

An important question is, What is an acceptable level of risk? Policy makers and scientists have determined that an acceptable environmental risk is a lifetime risk of 1 chance in a million ($10^{-6}$) of an adverse effect, and an unacceptable risk is 1 chance in 1,000 ($10^{-3}$) of an adverse effect.

A $10^{-6}$ risk means that if 1 million individuals were exposed to a toxic chemical at the same level and exposure, then 1 individual would have an adverse effect from this exposure. A $10^{-3}$ risk means that 1 individual would suffer an adverse effect if 1,000 people were exposed under the same conditions. Typically, state and federal governments have set the acceptable risk between $10^{-4}$ and $10^{-6}$, with $10^{-5}$ and $10^{-6}$ being the most commonly used values in policies set by states and the federal government. These values represent the increased risk due to exposure to the hazard over the background risk.

In the examples that follow in the next two subsections, a risk characterization can be used to determine an allowable concentration of a chemical in air, water, or soil for an acceptable risk. It can also be used to determine the resulting environmental risk for a particular chemical at a given concentration and the exposure scenario for that chemical in a particular environmental medium.

In the first scenario, policy makers would fix the acceptable risk at a predetermined level (say, $10^{-4}$ to $10^{-6}$), and the allowable concentration of the chemical in a particular medium that would result in that risk would be estimated. In the second situation, the concentration of the chemical in a particular medium is known, and the risk is determined.

CARCINOGENS When developing a risk characterization for carcinogens, an important point is that the dose is assumed to be an average daily dose received by a subject over a lifetime of exposure. For carcinogens, this lifetime of exposure is assumed to be 70 years. Later in this section, we will describe how this lifetime exposure is accounted for.

In simple terms, the risk associated with carcinogenic chemicals is equal to the dose multiplied by the unit risk associated with a dose of 1 mg/kg-day:

$$\text{risk} = \text{dose} \times \text{risk per unit dose} \qquad (6.3)$$

Remember that for carcinogens, the unit risk associated with a dose of 1 mg/kg-day is called the slope factor.

**Class Discussion**

When determining an acceptable level of risk, keep in mind that many individuals who are associated with an individual whose health is harmed also are indirectly harmed. The death or illness of an individual takes an emotional and financial toll on the individual's family members, friends, and coworkers. Also, broader economic and societal costs are associated with death and illness of an individual. Unfortunately, a typical risk characterization does not capture these broader societal and economic impacts. What are your personal and professional feelings about this issue? Are they the same, or do they differ?

example/6.5 Determining Risk

In Example 6.3, we determined that the dose for an adult male exposed to a chemical found in drinking water at 10 mg/L was 0.29 mg/kg-day. What is the risk associated with this exposure? Is this risk within acceptable guidelines?

Assume this dose is applied over a 70 yr lifetime and the chemical found in the water is benzene, a known carcinogen. The IRIS database provides an oral slope factor of 0.055 (mg/kg-day)$^{-1}$ for oral ingestion of benzene.

solution

Remember that previously we learned that the slope factor equals the unit risk for a chronic daily intake of 1 mg/kg-day. To determine the risk characterization, multiply the dose by the slope factor:

$$\text{risk} = 0.29 \, \frac{\text{mg}}{\text{kg-day}} \times 0.055 \, \frac{\text{kg-day}}{\text{mg}} = 1.59 \times 10^{-2}$$

This solution means that if 100 individuals were exposed to benzene at a concentration of 10 mg/L over their lifetime, 1.59 individuals would develop cancer. Extrapolated to a population of 10,000, this means that if all of them had similar exposure to benzene as this adult, 159 individuals would develop cancer. Extrapolated to a population of 1 million, we would expect 15,950 individuals to develop cancer.

This is well above acceptable risks of 1 in 10,000 ($10^{-4}$) and 1 in 1 million ($10^{-6}$). This is one reason the maximum contaminant level (MCL) for benzene in drinking water is 0.005 mg/L (or 5 µg/L), much lower than the 10 mg/L value stated in this example.

Example 6.5 assumed that individuals were exposed to the chemical carcinogen for their 70 yr lifetime. What happens in the case in which exposure is actually less than an individual's entire lifetime? For example, assume that exposure occurred over a 30 yr period of employment when a worker was exposed to the chemical only at work. In this case,

$$\text{risk} = \text{dose} \times \frac{\text{risk}}{\text{unit dose}} \times \frac{\text{time of exposure}}{\text{lifetime length}} \qquad (6.4)$$

Remember that for carcinogens, we assume the exposure takes place over a lifetime (70 yr); thus, the lifetime length is set at 70 yr. The *lifetime length* term in Equation 6.4 is referred to as the averaging time (AT) and typically has units of days. The AT for carcinogens is assumed to be 25,550 days (70 yr × 365 days/yr).

The *time of exposure* in Equation 6.4 is the exposure frequency (EF) multiplied by the exposure duration (ED). EF is the number of days an individual is exposed to the chemical per year. ED is the number of years an individual is exposed to the chemical.

Table 6.11 provides examples of exposure durations and exposure frequencies for different land use situations. For example, in the case of residential use and an exposure route of drinking water, the EF is assumed to be 350 days/yr (50 weeks), and the ED is assumed to be 30 yr. Application of this EF value assumes that an individual spends two weeks away from his or her house every year for vacation or other professional or family activities. Application of this ED value assumes that an individual resides in a house for only 30 yr of his or her life. Note the differences for other land uses. For example, Table 6.11 shows that, in an industrial setting, an average worker is on the job site only 245 days per year (so EF = 245 day/yr) and has an average employment history of 21 years (ED = 21 yr).

The assumed values of EF and ED can be used to develop an expression to determine the acceptable concentration of a chemical in drinking water for a stated acceptable risk:

$$\text{acceptable concentration} = \frac{\text{acceptable risk} \times \text{BW} \times \text{AT}}{\text{SF} \times \text{IR} \times \text{EF} \times \text{ED}} \qquad (6.5)$$

In Equation 6.5, BW is the average body weight of the target population, and IR is the ingestion rate, in this case, 2 liters of water per day (2 L/day).

Careful examination of Equation 6.5 shows that it is similar to the simpler-looking Equation 6.4. Some parameters were added to define the terms in Equation 6.4, and the equation was rearranged to set up the problem to compute the acceptable drinking-water concentration rather than the risk. The dose also is hidden in Equation 6.5. Here dose equals the IR multiplied by the acceptable concentration divided by the BW.

How would the acceptable risk in Example 6.6 change if the exposure assessment also showed that the target population consumed

**Class Discussion**

If you wanted to design a remediation system to eliminate exposure by residents to this toxaphene-contaminated groundwater in Example 6.6, what are some methods you could use? You could design, construct, and operate a groundwater remediation system that pumps the groundwater to the surface and treats the contaminated water in an aboveground reactor. Another approach could involve treating the chemical in place using in situ technology. This technology could utilize some biological, chemical, or physical process (or some combination of the three). What are some other nontreatment methods to eliminate exposure?

example/6.6 **Determining an Allowable Concentration of a Carcinogenic Chemical in Drinking Water**

Calculate an acceptable groundwater concentration for the chemical toxaphene if a residential development is placed above a groundwater aquifer contaminated with toxaphene. Assume you determine the risk for an adult who weighs 70 kg and consumes 2 L water per day from the contaminated aquifer. The state you work in has determined that an acceptable risk is 1 cancer occurrence per $10^5$ people. Use values from Table 6.11 for exposure frequency and exposure duration provided for residential land use.

solution

The IRIS database provides an oral slope factor for toxaphene of 1.1 per mg/kg-day. Remember that for carcinogens, AT is assumed to be 70 yr. Using Equation 6.5 and exposure data from Table 6.11, solve

example/6.6 Continued

for the acceptable concentration of toxaphene in the groundwater (assuming the only route of exposure is from drinking contaminated water):

$$\text{Concentration} = \frac{70 \text{ kg} \times 10^{-5} \times 70 \text{ yr} \times \dfrac{365 \text{ days}}{\text{yr}} \times \dfrac{1{,}000 \text{ μg}}{1 \text{ mg}}}{\dfrac{1.1 \text{ kg-day}}{\text{mg}} \times \dfrac{350 \text{ days}}{\text{yr}} \times 30 \text{ yr} \times \dfrac{2 \text{ L}}{\text{day}}}$$

$$= 0.77 \frac{\text{μg}}{\text{L}} \text{ or } 0.77 \text{ ppb}_{\text{m}}$$

Note that if the acceptable risk were 1 in 1 million ($10^{-6}$), the allowable toxaphene concentration would decrease to 0.077 μg/L (or 0.077 ppb$_{\text{m}}$).

30 g fish per day? The answer is simple. There would be no change in the acceptable risk unless for some reason you had knowledge that toxaphene was found in the fish. In this case, the toxaphene is found in groundwater below this residential neighborhood. We have no information to suggest that the fish these individuals consume came into contact with the contaminated groundwater. If the fish did contain the chemical, the ingestion of toxaphene in the fish would be

example/6.7 Determining an Acceptable Concentration of a Chemical in Indoor Air

Assume that the chemical benzene, a known carcinogen, is found in air at a constant concentration of 1 μg/m³. Calculate the risk for exposure to this benzene for an average adult who weighs 70 kg and inhales 20 m³ air/day with 50 percent absorption. The chemical's inhalation slope factor is 0.015 mg/kg-day.

solution

The risk is

$$\text{risk} = \frac{\dfrac{1 \text{ μg}}{\text{m}^3} \times 20 \dfrac{\text{m}^3}{\text{day}} \times 0.5 \times \dfrac{1 \text{ mg}}{1{,}000 \text{ μg}}}{70 \text{ kg} \times \left(\dfrac{0.015 \text{ mg}}{\text{kg-day}}\right)}$$

$$= 9.5 \times 10^{-3}$$

This value means that, for every 1,000 people continuously exposed to this chemical, 9.5 would develop cancer over their lifetime.

added to the calculation. That is, the exposure would be from drinking 2 L water per day and eating 30 g fish per day. Section 6.6 will provide an example problem that includes exposure from both water and fish.

Another solution may be to investigate whether there is a municipal water supply close to this community that could serve as their source of drinking water. In this case, a deed restriction would be placed on the property so individual property owners could not install a drinking-water well. Furthermore, a hydrogeological study may have to be conducted to assess whether the contaminated groundwater recharges into a stream or river, where the chemical could exert toxicity to aquatic life or perhaps contaminate a downstream drinking-water intake. This option is likely to be costly and will clean up only the contamination currently on site. It will not prevent new quantities of this or other toxic chemicals from being introduced to the site.

NONCARCINOGENS As stated previously, risk characterizations performed for noncarcinogens are handled differently than for carcinogens. Recall that noncarcinogens have a threshold dose, below which no adverse effect is estimated to occur. A safe dose referred to as the reference dose (RfD) estimates (with an uncertainty of one order of magnitude or greater) a lifetime dose of a noncarcinogen that is likely to be without significant risk to human populations. The IRIS database, discussed previously, provides values for RfDs.

The acceptable risk from exposure to a noncarcinogenic chemical is determined by calculating a **hazard quotient (HQ)**. For the exposure to a carcinogen, the dose is assumed to apply over a 70 yr lifetime. Noncarcinogenic effects are evaluated over the actual exposure duration, which can be quite short. Thus, the HQ is the average daily dose of a chemical received by a subject over the exposure period divided by the reference dose (RfD):

$$HQ = \frac{\text{average daily dose over exposure period}}{\text{RfD}} \qquad (6.6)$$

HQs that are less than or equal to 1 are considered safe; HQs greater than 1 are considered unsafe. This should make sense from careful study of Equation 6.6. If the HQ equals 1, the average daily dose to which an individual or community is exposed equals the reference dose (the safe dose).

---

**Box / 6.6**  **Accounting for Uncertainties in Dose-Response Assessment to Provide Conservative Estimates of Risk per Unit Dose**

In the dose-response assessment process, numerous sources of uncertainty cause uncertainty in estimates of the risk per unit dose used in Equation 6.3 for carcinogens (the slope factor) and noncarcinogens (the hazard quotient). However, public policy has been developed that accounts for these uncertainties and leads to conservative estimates of the values used in risk characterization.

In the case of carcinogens, the linear dose-response model that is commonly applied (Figure 6.6) is conservative because this model leads to a higher response (risk) estimate at low doses than other models, such as the S-shaped multi-hit model and the U-shaped dose-response curve. It is usually the case that the dose-response data generated using test animal studies must be extrapolated to significantly lower doses in human risk assessments. Furthermore, lack of a threshold level in the dose-response relationship for carcinogens provides a conservative estimate of risk. Even if the dose were one *molecule* per kilogram per day, some risk of getting cancer is estimated when no threshold level is assumed.

For noncarcinogens, application of uncertainty factors (UFs) in determining RfD and RfC (and thus HQ) values provides conservative estimates of HQ values. These uncertainty factors account for variation in susceptibility among the members of the human population (inter-individual or intraspecies variability), uncertainty in extrapolating animal data to humans (interspecies uncertainty), uncertainty in extrapolating from data obtained in a study with less-than-lifetime exposure (extrapolating from subchronic to chronic exposure), uncertainty in extrapolating from a lowest observative adverse effect level (LOAEL), rather than from a NOAEL value, and uncertainty associated with extrapolation when the database is incomplete.

The EPA has begun to recommend use of a *benchmark dose* concept to improve the quality of the RfD and RfC values and to reduce the number of uncertainty factors used. This approach uses all available data to estimate the NOAEL, rather than relying on a single point. This development provides an example of means by which public policy is continuously being improved in estimation of risk per unit dose values.

## example/6.8 Determining the Risk of Noncarcinogenic Chemicals

Is there an unsafe risk associated with daily inhalation of air with a concentration 0.010 $\mu g/m^3$ of the chemical toluene? In this problem, we will look at noncarcinogenic effects. The IRIS database states the inhalation RfD for toluene is 0.114 mg/kg-day. Assume an adult weighs 70 kg and inhales 20 $m^3$ of contaminated air per day.

### solution

Determine the average daily dose to which the individual is exposed, and divide this value by the RfD to determine the hazard quotient. The average daily dose is

$$\frac{0.010 \, \frac{\mu g}{m^3} \times 20 \, \frac{m^3}{day} \times \frac{mg}{1,000 \, \mu g}}{70 \, kg} = 2.8 \times 10^{-6} \, \frac{mg}{kg\text{-}day}$$

Use this value and the RfD to determine the HQ:

example/6.8 **Continued**

$$HQ = \frac{2.8 \times 10^{-6} \dfrac{mg}{kg\text{-}day}}{0.114 \dfrac{mg}{kg\text{-}day}} = 2.5 \times 10^{-5}$$

Because the HQ is much less than 1, the risk is acceptable. Individuals exposed to the chemical toluene should be free of adverse health effects in this instance.

Equation 6.6 and the information discussed previously can be used to develop an expression to determine the acceptable concentration of a noncarcinogenic chemical in drinking water:

$$\text{acceptable concentration} = \frac{HQ \times RfD \times BW \times AT}{IR \times EF \times ED} \qquad (6.7)$$

All the terms in Equation 6.7 have been defined previously: HQ is the hazard coefficient, RfD is the reference dose, BW is the body weight of the target population, AT is the averaging time, IR is the intake rate (in this case, the water ingestion rate), EF is the exposure frequency, and ED is the exposure duration.

example/6.9 **Determining Acceptable Concentrations of a Noncarcinogenic Chemical in Drinking Water**

Calculate the acceptable groundwater protection standard for a non-carcinogenic chemical. A commercial development is placed above a groundwater aquifer contaminated with the chemical and commercial establishments will use the groundwater as a source of drinking water. Assume you determine risk for an average adult who weighs 70 kg and consumes 1 L water per day. Use the values in Table 6.11 values for exposure frequency and exposure duration provided for commercial use.

**solution**

The IRIS database states that the chemical's RfD for oral ingestion is 0.01 mg/kg-day. Table 6.12 provided an EF of 245 days/yr and an ED of 21 yr. The AT for noncarcinogens is typically 30 yr. The accept-able concentration of the chemical in the groundwater (assuming the only route of exposure is from drinking contaminated water) is found from Equation 6.7:

$$\text{acceptable concentration} = \frac{HQ \times RfD \times BW \times AT}{IR \times EF \times ED}$$

$$= \frac{1 \times 0.01 \dfrac{mg}{kg\text{-}day} \times 70\,kg \times 30\,yr \times \dfrac{365\,days}{yr} \times \dfrac{1{,}000\,\mu g}{mg}}{1 \dfrac{L}{day} \times 245 \dfrac{days}{yr} \times 21\,yr}$$

$$= 1{,}490\,\mu g/L \approx 1{,}500\,\mu g/L = 1.5\,mg/L\;(\text{or } 1.5\,ppm_m)$$

Note the similarities and differences between Equation 6.7 and Equation 6.5. If the HQ in Equation 6.7 is set to 1 (the upper safe dose) and is multiplied by the RfD, the resulting value is the safe dose of the chemical. This computation is different from, yet similar to, the method to compute acceptable concentration for carcinogens in which an acceptable risk was first stated. The terms in Equation 6.7 for acceptable concentration, IR, and BW are all related to the actual dose received by the target population. For noncarcinogens, the averaging time (AT) is typically assumed to be 30 yr (not 70 yr, as with carcinogens).

## 6.6 More Complicated Problems with at Least Two Exposure Routes

Exposure assessments can become more complicated than shown in the examples presented previously. Up to this point, the dose determination assumed only one route of exposure. This final section investigates how environmental risk can consider several routes of exposure at one time and how some of the environmental partitioning processes studied in Chapters 3 and 5 are incorporated into more complex, multimedia problems in which a chemical partitions between the water, air, and/or solid phases.

The first example will use a risk characterization to determine a surface water quality standard for which the exposure assessment identified ingestion from eating contaminated fish that live in those waters in addition to drinking the contaminated water. The second example will use a risk characterization to determine acceptable cleanup standards for contaminated soil that can potentially leach vertically into the subsurface and contaminate underlying groundwater. In this situation, the chemical is not only dissolved in the pore water of the soil, but also may be sorbed to organic coatings on the soil particles and/or partitioned into air space voids found in the soil structure. Both examples

## Table / 6.13

**Exposure Assessment Assumptions Used by 3 Hypothetical States in Setting a Water Quality Standard for a Chemical That Bioaccumulates in Fish**

| | State 1 | State 2 | State 3 |
|---|---|---|---|
| Ingestion rate of water (L/day) | 2 | 2 | 2 |
| Body weight of adult (kg) | 70 | 70 | 70 |
| Ingestion rate of fish (g/day) | 6.5 | 30 | 15 |
| Bioaccumulation factors, i.e., measured concentration in fish divided by measured concentration in water (L water/kg fish) | 51,500 | 336,000 (cold water) 84,086 (warm water) | 7,310 |

will require us to use our knowledge of how a chemical partitions in the environment.

### 6.6.1 SETTING WATER QUALITY STANDARDS BASED ON EXPOSURE FROM DRINKING WATER AND EATING FISH

Table 6.13 provides information from three different hypothetical states we could independently use to set a surface water quality standard for a hypothetical chemical. Careful examination of Table 6.13 shows that the three states would obtain different water quality standards for the same chemical. How can that be? A further look at the information in Table 6.13 indicates that the states use different assumptions in their exposure assessment.

Each hypothetical state assumes the same weight of an adult (70 kg), the same ingestion rate of water (2 L/day), and the same cancer slope factor. However, each state assumes different rates of fish consumption by the adult population and a different magnitude of partitioning behavior of the chemical from water into the fish (the bioaccumulation factor). Note that state 2 even assumes that the type of fish found in cold water has a greater fat (lipid) content, so the chemical of concern (which is very hydrophobic) partitions to a greater extent into cold-water fish than warm-water fish.

The total dose in this case comes from drinking water and eating fish, and in this case, most of the dose results from eating contaminated fish. The high bioaccumulation factors in Table 6.13 indicate that the concentration of the chemical in fish is much greater than in the surface water to which the fish is exposed. When multiplied by the daily ingestion rate for fish, quite a large value results. This is because, in this case, the chemical is very hydrophobic, so it does not tend to dissolve to a great extent in water but instead partitions to a great extent into the fat (lipid) of the cold-water fish.

**Class Discussion**

*What are some segments of society that are exposed to higher risk as a result of eating contaminated fish?*

Related to the discussion topic, the segments of society exposed to higher risk from fish consumption are probably children (higher dose given low body weight), pregnant women (perhaps the chemical is even suspected to impair fetal development), or recreational or subsistence anglers who consume more fish or fish parts that contain higher lipid content than the general population (increased exposure due to increased consumption). This last group could be vacationers who are consuming a large amount of fish for a short duration during a fishing trip. More likely it is individuals who depend on these fish for subsistence. For example, in many parts of the United States, Native American and immigrant populations are known to consume more fish or to eat fish parts that contain higher contaminant concentrations than others consume.

### 6.6.2 HOW TO DETERMINE ALLOWABLE SOIL CLEANUP STANDARDS THAT PROTECT GROUNDWATER

Now let us turn our attention to determining an appropriate cleanup level to which an engineer should remediate at a site containing contaminated soil. For this particular problem, assume an exposure assessment indicates that the contaminated soil does not pose a direct threat to adults or children who ingest the contaminated soil or breathe vapors that may emit from the contaminated soil. This could be the case for leaking underground storage tanks where the contamination is below the soil surface and the resulting soil vapors do not reach the soil surface. In this case, the problem is that the contaminated soil acts as a source of pollution, and the chemical of concern may leach from the soil and contaminate underlying groundwater. The groundwater may serve as a source of drinking water for a home or municipality, or perhaps the groundwater recharges a stream, where the chemical can then exert toxicity to aquatic life. Figure 6.8 shows the complexity of this problem.

Solving this type of problem requires several steps, detailed in Table 6.14. We will focus our efforts on only one of several hundred chemicals found in petroleum products, benzene. We will also assume that the groundwater does not recharge a stream, so there is no concern about risk to an aquatic ecosystem.

We assumed residential use of the groundwater, and Example 6.6 provided us information on the exposure frequency, exposure duration, and averaging time for this particular type of land use. The allowable benzene concentration is determined as follows (using the appropriate slope factor of 0.055 $(mg/kg\text{-}day)^{-1}$ from the IRIS database).

$$\text{concentration} = \cfrac{70\,kg \times 10^{-5} \times 70\,yr \times \cfrac{365\,days}{yr} \times \cfrac{1{,}000\,\mu g}{mg}}{0.055\,\cfrac{kg\text{-}day}{mg} \times \cfrac{350\,days}{yr} \times 30\,yr \times \cfrac{2\,L}{day}} \qquad (6.8)$$

The allowable concentration is determined to be 15 $\mu g/L$ (15 $ppb_m$).

This value is a factor of three higher than the maximum contaminant level (MCL) for benzene, which is 5 $\mu g/L$. The MCL is the enforceable

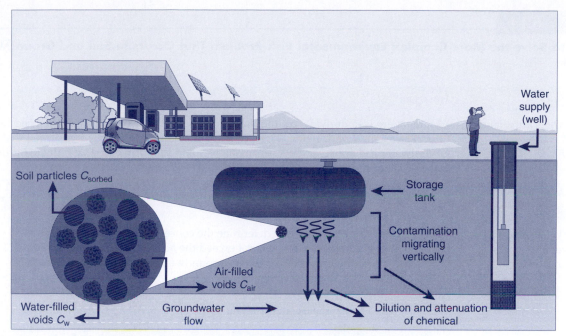

**Figure 6.8** **Complexity of a Situation in which a Leaking Underground Storage Tank Has Discharged a Chemical That Is Contaminating Subsurface Groundwater** The chemical may partition between air voids in the soil, water-filled voids in the soil, and organic coatings on soil. Henry's constant is used to relate the gaseous and aqueous concentrations of the chemical at equilibrium. A soil–water partition coefficient is used to relate the concentrations of the aqueous and sorbed phases for sorptive equilibrium. The chemical dissolved in water can leach vertically to the groundwater. In the process, it may chemically or biologically attenuate or be diluted by clean, upgradient groundwater.

standard set by the EPA, so this default value will be used to compute the soil cleanup standard in subsequent steps. MCLs are based upon treatment technologies, affordability, and other feasibility factors, such as availability of analytical methods, treatment technology, and costs for achieving various levels of removal. The EPA guidance for establishing an MCL states that MCLs are enforceable standards and are to be set as close to the maximum contaminant level goals (MCLGs) (health goals) as is feasible.

Step 4 from Table 6.14 is to determine how the concentration of the chemical is changed as it moves vertically through the unsaturated zone to the groundwater. An engineer could perform sophisticated hydrological modeling to determine the vertical movement of the benzene from the contaminated soil found in the unsaturated zone to the underlying saturated zone. For simplicity, let us assume that the migration of a contaminant from soil to underlying groundwater has two stages. First, the chemical must partition (from sorbed or gaseous phases) into the pore water that surrounds the contaminated area. Then the resulting leachate must be transported vertically to the underlying groundwater. As the dissolved chemical is transported downward with the infiltrating water, it may be transformed to a lower concentration by naturally occurring chemical or biological processes. It may also be

## Table / 6.14

### Steps to Solve the More Complex Environmental Risk Problem That Occurs in Soil and Groundwater Media

| Step | Procedure |
|------|-----------|
| **Step 1:** Determine land use and routes of exposure. | Assume there is underlying groundwater that is currently used (or might be used in the future) as a source of drinking water for a home or community. It was assumed that the groundwater did not recharge a stream or river and that drinking contaminated groundwater is the only way people may be exposed to the contamination. |
| **Step 2:** What is the acceptable risk? | Assume that a regulatory body has set the acceptable risk at $10^{-5}$. |
| **Step 3:** What is acceptable level of benzene in the groundwater? | Perform a risk characterization to determine the concentration of benzene in the contaminated groundwater that will not exceed the acceptable risk stated in step 2. The acceptable concentration of benzene in the groundwater is found from Equation 6.5 (repeated here): $$\text{acceptable concentration} = \frac{\text{acceptable risk} \times BW \times AT}{SF \times IR \times EF \times ED}$$ |
| **Step 4:** Determine how the concentration of the chemical is changed as it moves vertically through the unsaturated zone to the groundwater. | This is determined either by a detailed hydrogeological study or by making an assumption that provides insight into the dilution and attenuation of the chemical. |
| **Step 4a:** Determine the allowable concentration in the soil pore water that surrounds the contamination site from the acceptable concentration in the groundwater. | Use results from the hydrogeological study or a dilution attenuation factor (DAF). |
| **Step 4b:** Estimate an allowable soil concentration (total benzene per total mass of wet soil) from the allowable pore water concentration. | Apply knowledge of mass balance and chemical partitioning between air, soil, and aqueous phases. |

diluted with the uncontaminated water it meets that is found upgradient in the groundwater.

For simplicity, assume that a value of 16 accounts for all dilution and attenuation processes. (This value is equal to that actually used by some states to determine some risk-based soil cleanup regulations for benzene, for example.) That is, the chemical concentration in the pore water that surrounds the contamination divided by 16 will equal the concentration in the groundwater that someone might eventually be exposed to.

Step 4a is to determine the allowable concentration in the soil pore water that surrounds the contamination site from the acceptable concentration in the groundwater. Moving from the allowable concentration in the groundwater to the pore water, the allowable benzene concentration in the pore water within the contaminated soil zone that will leach vertically to not adversely affect the groundwater is

$$0.005\,\text{mg/L} \times 16 = 0.080\,\text{mg/L}$$

Remember that this value of 0.080 mg/L is not the concentration that someone would be exposed to if he or she drank the contaminated water. The value of 0.080 mg/L is the allowable concentration in the pore water of the contaminated soil. It will be decreased by dilution and attenuation so that the concentration that is eventually found in the groundwater is 0.005 mg/L.

Step 4b is to estimate an allowable soil concentration (total benzene per total mass of wet soil) from the allowable pore water concentration. The final step is to take the allowable aqueous phase concentration in the pore water (mg benzene per L water) and convert it to a soil concentration (mg benzene per kg wet soil). To make the problem easier, assume the benzene-air-water-soil system is at equilibrium. The amount of benzene that is partitioned at equilibrium among the pore water, air voids, and the sorbed soil phase is determined.

This total mass of the benzene (from the aqueous pore water, air, and sorbed phases) that is contained in a unit mass of soil is what is measured if you collect a subsurface soil sample and request a laboratory analysis of the soil. This value can also be thought of as the allowable concentration of benzene that can left in the soil (units of mg benzene per kg soil) that, if it leached vertically, would not result in a subsequent groundwater concentration that would harm human health at a stated acceptable risk.

To carry out this computation, we need subsurface property information. Assume the subsurface system is homogenous, the soil porosity is 0.3 percent, soil organic carbon content is 1 percent, and soil bulk density is 2.1 gm/cm$^3$. Assume also that one-third of the void spaces of the soil are filled with air and the remaining two-thirds of the void spaces are filled with water.

The total mass of chemical is equal to the mass of chemical sorbed to the soil plus the mass of chemical in the water-filled voids plus the mass of chemical in the air-filled voids. Mathematically, this statement can be written as follows, using information about equilibrium partitioning (Chapters 3 and 5):

allowable soil concentration

$$= (C_{\text{water}} \times K) + \left(\frac{C_{\text{water}} \times \theta_{\text{water}}}{\rho_b}\right) + \left(\frac{C_{\text{water}} \times \theta_{\text{air}} \times K_{\text{H}}}{\rho_b}\right) \quad \textbf{(6.9)}$$

Here, $C_{\text{water}}$ is the pore water concentration of the chemical determined in step 4a (0.08 mg/L), $K$ is the soil–water partition coefficient for our chemical of concern, $K_{\text{H}}$ is the dimensionless Henry's law constant for the chemical of concern, $\theta_{\text{water}}$ is the soil porosity filled with water, $\theta_{\text{air}}$

is soil porosity filled with air, and $\rho_b$ is the soil bulk density (assumed to be 2.1 g/cm$^3$).

In this case, we have the following values for these variables:

- $C_{water}$ was determined in step 4a to be 0.080 mg/L.

- $\theta_{water}$, the soil porosity filled with water, is equal to the porosity (0.3) multiplied by the fraction of voids that are filled with water (2/3 in this problem). In this case, $\theta_{water}$ is equal to 0.2 (expressed in units of $L_{water}/L_{total}$).

- $\theta_{air}$, the soil porosity filled with air voids, is equal to the porosity (0.3) multiplied by the fraction of voids that are filled with air (1/3 in this problem). In this case, $\theta_{air}$ equals 0.1 (expressed in units of $L_{air}/L_{total}$).

- $K$, the soil-water partition coefficient, can be estimated from the octanol–water partition coefficient, as discussed in Chapter 3. Assume benzene has a log $K_{ow}$ of 2.13. For benzene, the correlation from Figure 3.10 can be used to determine the soil–water partition coefficient normalized to organic carbon (log $K_{oc}$) as 2.02 ($K_{oc}$ equals $10^{2.02}$ L/kg organic carbon). Because the organic carbon content of the soil was stated to be 1 percent, Equation 3.32 can be used to determine that $K$ is 1.05 L/kg soil.

- The $K_H$ for benzene is 0.18 at 20°C. This value is for reaction of a chemical in the air phase going to the water phase. The units of $K_H$ in this situation are thus $L_{water}/L_{air}$. To simplify the problem, we will assume that the cooler temperature of the subsurface does not greatly affect the air–water partitioning behavior of benzene.

Plugging in these values for the unknowns in Equation 6.9 results in

allowable soil concentration

$$= \left( 0.08\, \frac{mg}{L} \times 1.05\, \frac{L}{kg\ soil} \right) + \left( \frac{0.08\, \frac{mg}{L} \times 0.2\, \frac{L_{water}}{L_{total}}}{2.1\, \frac{g_{soil}}{cm^3} \times 1{,}000\, \frac{cm^3}{L} \times \frac{kg}{1{,}000\ g_{soil}}} \right)$$

$$+ \left( \frac{0.08\, \frac{mg}{L} \times 0.18\, \frac{L_{water}}{L_{air}} \times 0.1\, \frac{L_{air}}{L_{total}}}{2.1\, \frac{g_{soil}}{cm^3} \times 1{,}000\, \frac{cm^3}{L} \times \frac{kg}{1{,}000\ g_{soil}}} \right) \qquad \textbf{(6.10)}$$

Solving for the concentrations of benzene in the sorbed phase, pore water phase, and air voids, respectively, results in

allowable soil concentration

$$= 0.084\ mg/kg\ +\ 0.0076\ mg/kg\ +\ 0.00069\ mg/kg \qquad \textbf{(6.11)}$$

$$(sorbed) \qquad\qquad (water) \qquad\qquad (air)$$

$$= 0.092\ mg\ benzene/kg\ soil$$

This allowable soil concentration of 0.092 mg/kg (0.092 ppm$_m$) is the concentration of benzene that can remain in the soil and still protect the groundwater resource to drinking-water standards. Any areas of the contaminated soil greater than this value would have to be removed or remediated to this lower cleanup level.

## Key Terms

- acute toxicity
- bioaccumulation
- brownfield
- carcinogenic
- chronic toxicity
- dose
- dose limit
- dose-response assessment
- dose-response curve
- Emergency Planning and Community Right-to-Know Act (EPCRA)
- endocrine disruptors
- environmental justice
- environmental risk
- epidemiology

- exposure
- exposure assessment
- exposure pathways
- hazard assessment
- hazard quotient (HQ)
- hazardous waste
- median lethal dose (LD$_{50}$)
- Muir, John
- no observable adverse effect level (NOAEL)
- no threshold effect
- noncarcinogenic
- potency factor
- reference concentration (RfC)
- reference dose (RfD)

- Resource Conservation and Recovery Act (RCRA)
- risk
- risk assessment
- risk characterization
- risk perception
- slope factor
- solid waste
- susceptible populations
- suspected carcinogens
- threshold effect
- toxicity
- Toxics Release Inventory (TRI)
- uncertainty factor (UF)
- weight of evidence

WARNING
FISH CONTAMINATED
DO NOT EAT

**6.1** What is the regulatory difference in the RCRA between a hazardous and toxic substance?

**6.2** Identify several types of risk during your commute to school. Which of these risks would classify as environmental risk?

**6.3** Identify the top three chemical releases in your hometown or university community using the EPA's Toxic Release Inventory database. What information can you find about the toxicity of these chemicals? Is it easy or difficult to find this information? Is the information consistent, or is it conflicting? Does it vary with source (government versus industry)?

**6.4** (a) List a different environmental risk associated with an indoor environment in the developed world and developing world. (b) What particular building occupants are at the greatest risk for the items you have identified?

**6.5** Return to Chapter 1 and go to the Web site of the World Health Organization (www.who.org). Based on the information there, write a referenced two-page essay on global environmental risk. How much environmental risk is from factors such as unsafe water and sanitation, indoor air, urban air, and climate?

**6.6** Recalling that the Environmental Protection Agency defines environmental justice as the "fair treatment and meaningful involvement of all people regardless of race, color, national origin, or income with respect to the development, implementation, and enforcement of environmental laws, regulations, and policies," research an issue of environmental justice in your hometown or state. What is the environmental issue? What groups of society are being harmed by the environmental issue? What injustice is taking place?

**6.7** Online at www.scorecard.org, you can search for the location and number of hazardous-waste sites by location. Use it to search for hazardous-waste sites in your hometown or a city close to your university. Comment on whether the number and location of hazardous-waste sites pose any environmental injustice to residents of the community you are investigating.

**6.8** List the four components of a complete risk assessment.

**6.9** The Integrated Risk Information System (IRIS), an electronic database that identifies human health effects related to exposure to hundreds of chemicals, is available at www.epa.gov/iris. Go to IRIS and determine (a) the weight of evidence descriptor; (b) the reference dose (RfD); and (c) the slope factor if available for the following six chemicals/substances: arsenic, methylmercury, ethylbenzene, methyl ethyl ketone, naphthalene, and diesel engine exhaust.

**6.10** Assume an adult inhales 23 m$^3$ of air every day and the inhalation absorption factor for the chemical of concern is 75 percent (so 25 percent of the chemical is exhaled). A wind analysis shows exposure is 30 percent of the time on an average annual basis. Assume that an adult lives in this area over his or her entire lifetime, and that the average air concentration at the property boundary of the source is 0.044 mg/m$^3$. Determine the dose for an adult male in mg/kg-day.

**6.11** (a) Determine the dose (in mg/kg-day) for a bioaccumulative chemical with BCF = $10^3$ that is found in water at a concentration of 0.1 mg/L. Calculate your dose for a 50 kg adult female who drinks 2 L lake water per day and consumes 30 g fish per day that is caught from the lake. (b) What percent of the total dose is from exposure to the water, and what percent is from exposure to the fish?

**6.12** Calculate a risk-based groundwater protection standard (in ppb) for the chemical 1,2-dichloroethane for a residential homeowner where the person's well used for drinking water is contaminated with 1,2-dichloroethane. Assume you are determining risk for an average adult who weighs 70 kg. The state where you work has determined that an acceptable risk is 1 cancer occurrence per $10^6$ people. Use the values for route of intake, exposure frequency, exposure duration, and averaging time provided for residential use in Table 6.11. Assume an oral slope factor for 1,2-dichloroethane of $9.2 \times 10^{-2}$ per (mg/kg)/day.

**6.13** (a) Calculate a risk-based groundwater protection standard for the chemical benzo(a)pyrene. Assume you are determining risk for an average adult female who weighs 50 kg and consumes 2 L water and eats 30 g fish per day. The state has determined that an acceptable risk is 1 cancer occurrence per $10^5$ people. Use the values for exposure frequency, exposure duration, and averaging time provided for residential land use. (b) According to EPA's Integrated Risk Information System (IRIS), what type of carcinogen is benzo(*a*)pyrene, using the weight of evidence of human and animal studies? (c) Assuming the chemical is leaching from some contaminated soil, estimate the allowable concentration of benzo(*a*)pyrene in the pore water of contaminated soil.

**6.14** Is there an unsafe risk associated with a 70 kg adult eating 15 g fish every day that contains 1 mg/kg of methylmercury? Methylmercury has been shown to cause developmental neuropsychological impairment in human beings. The RfD for methylmercury is $1 \times 10^{-4}$ mg/kg-day.

**6.15** Is there an unsafe risk associated with a 70 kg adult eating 15 g fish every day that contains 9.8 μg/kg of Arochlor 1254? Arochlor 1254 can exhibit non-carcinogenic effects in humans. Use the IRIS database to find any other information required to solve this problem.

**6.16** Concentrations of toxaphene in fish may impair human health and fish-eating birds (such as bald eagles) that feed on the fish. (a) If the log of the octanol–water partition coefficient (log $K_{ow}$) for toxaphene is assumed to equal 4.21, what is the expected concentration of toxaphene in fish? (Assume that the equilibrium aqueous phase toxaphene concentration is 100 ng/L.) (b) If it is assumed that an average person drinks 2 L untreated water daily and consumes 30 g contaminated fish, what route of exposure (drinking water or eating fish) results in the greatest risk from toxaphene in 1 year? (c) What route of exposure is greatest for a higher-risk group that is assumed to consume 100 g fish per day? Support all of your answers with calculations. Assume the following correlation applies to toxaphene and our problem's specific fish:

$$\log \text{BCF} = 0.85 \log K_{ow} - 0.07$$

**6.17** Identify a brownfield in your local community, hometown, or a nearby city. What was specifically done at the site? What are several social, economic, and environmental issues associated with restoring the brownfield site?

**6.18** The Code of Federal Regulations (CFR) is the codification of the general and permanent rules published in the Federal Register by the executive departments and agencies of the Federal Government. It can be accessed at http://www.gpoaccess.gov/cfr/. What CFR number is associated with the following sections? (for example, 50 CFR for Wildlife and Fisheries). (a) Protection of Environment; (b) Transportation, (c) Conservation of Power and Water Resources, (d) Public Health, and (e) Highways.

**6.19** Nairobi is a city in Kenya, Africa. Use the following web site to obtain the Report "City of Nairobi Environment Outlook." http://www.unep .org/DEWA/Africa/docs/en/NCEO_Report_FF_ New_Text.pdf. Develop a pie chart showing the top ten major causes of mortality in Nairobi in 2000.

**6.20** Research the safety of your personal care and household cleaning products using a web site such as http://lesstoxicguide.ca/index.asp?fetch=personal. Develop a table that lists 7 current personal care or household cleaning products used in your apartment, home, or dormitory. Add a second column that lists a less hazardous alternative for each of the 7 products.

# References

Fisk, W. J. 2000. "Health and Productivity Gains from Better Indoor Environments and Their Relationship with Building Energy Efficiency." *Annual Review of Energy and the Environment* 25:537–566.

Friends of the Earth, 2009. "Endocrine Disrupting Pesticides," http://www. foe.co.uk/index.html, accessed February 21, 2009.

Meuser, M., and A. Szacz. 2000. "Unintended, Inexorable: The Production of Environmental Inequalities in Santa Clara County, California." *American Behavioral Scientist* 43 (4): 602.

Mihelcic, J. R. 1999. *Fundamentals of Environmental Engineering.* New York: John Wiley & Sons.

National Science and Technology Council. 1996. The Health and Ecological Effects of Endocrine Disrupting Chemicals: A Framework for Planning. Committee on Environment and Natural Resources. http://www.epa.gov/endocrine/pubs/framework.pdf, accessed November 18, 2008.

Patnaik, P. 1992. *A Comprehensive Guide to the Hazardous Properties of Chemical Substances.* New York: Van Nostrand Reinhold.

# chapter/Seven  Green Engineering

Julie Beth Zimmerman
and James R. Mihelcic

*This chapter focuses on a design framework for green engineering—called the 12 principles of green engineering—and highlights the key approaches to advancing sustainability through engineering design. The chapter builds on earlier discussions of sustainability, metrics, general design processes, and challenges to sustainability. The current approach to design, manufacturing, and disposal is discussed in the context of examples and case studies from various sectors. This will provide a basis for identifying what to consider and how to address those considerations when designing products, processes, and systems that contribute to furthering sustainability. The fundamental engineering design topics that will be addressed include toxicity and benign alternatives, pollution prevention and source reduction, separations and disassembly, material and energy efficiencies and flows, systems analysis, biomimicry, and life cycle design, management, and analysis.*

## Learning Objectives

1. Define green engineering, and articulate the goals of this approach.
2. Describe the importance and benefits of approaching green engineering from a design perspective.
3. Summarize the differences and benefits of the levels in the pollution prevention hierarchy.
4. Discuss key strategies for sustainable design, including inherency, resiliency, systems thinking, and design of services rather than physical entities.
5. Diagram a complex system with feedback loops.
6. Apply life cycle costing to determine the time value of money for single payments and annuities.
7. Describe the components of a life cycle assessment (LCA), and perform a comparative LCA, including a goal statement, function/functional unit definition, and inventory, analysis, and assessment steps.
8. Develop sustainability indicators for a given product, process, or system.
9. Articulate the various policy mechanisms to encourage the implementation of green engineering.

## 7.1 What Is "Green Engineering?"

Engineers play a significant and vital role in nearly all aspects of our lives. They provide basic services such as water, sanitation, mobility, energy, food, health care, and shelter in addition to such advances as real-time communications and space exploration. The implementation of all of these engineering achievements can lead to benefits as well as problems in terms of the economy, society (particularly human health), and the environment—the three pillars of sustainability.

**Green engineering** is the design, discovery, and implementation of engineering solutions with an awareness of potential benefits and problems in terms of the environment, the economy, and society throughout the lifetime of the design. The green engineering approach is scalable and applies across molecular, product, process, and system design. This approach is as broad as the disciplines of engineering themselves and relies on the traditions of innovation, creativity, and brilliance that engineers use to find new solutions to any challenge. The goal of green engineering is to minimize adverse impacts while simultaneously maximizing benefits to the economy, society, and the environment.

The adverse impacts of conventional engineering design, often implemented without a sustainability perspective, can be found all around us in the form of water use inefficiencies, depletion of finite materials and energy resources, urban congestion, and degradation of natural systems as a result of human activity. Mutual benefits resulting from green engineering design include a competitive and growing economy in the global marketplace, improved quality of life for all species, and enhanced protection and restoration of natural systems.

The **principles of green engineering** (Anastas and Zimmerman, 2003), listed in Box 7.1, provide a framework for understanding green

---

**Box / 7.1    The 12 Principles of Green Engineering**

1. Designers need to strive to ensure that all material and energy inputs and outputs are as inherently non-hazardous as possible.

2. It is better to prevent waste than to treat or clean up waste after it is formed.

3. Separation and purification operations should be a component of the design framework.

4. System components should be designed to maximize mass, energy, and temporal efficiency.

5. System components should be output pulled rather than input pushed through the use of energy and materials.

6. Embedded entropy and complexity must be viewed as an investment when making design choices on recycle, reuse, or beneficial disposition.

7. Targeted durability, not immortality, should be a design goal.

8. Design for unnecessary capacity or capability should be considered a design flaw. This includes engineering "one size fits all" solutions.

9. Multi-component products should strive for material unification to promote disassembly and value retention. (minimize material diversity)

10. Design of processes and systems must include integration of interconnectivity with available energy and materials flows.

11. Performance metrics include designing for performance in commercial "after-life".

12. Design should be based on renewable and readily available inputs throughout the life-cycle.

(Anastas and Zimmerman, 2003)

engineering. They also represent techniques that can be used to make engineered solutions more sustainable. Students should carefully review them and see how each principle fits into their particular engineering discipline. The 12 principles should be thought of not as rules, laws, or inviolable standards. They are instead a set of guidelines for thinking in terms of sustainable design criteria that, if followed, can lead to useful advances in terms of sustainability challenges and improved design for a wide range of engineering problems.

In a complex system, there may be synergies in which progress toward achieving the goal of one principle will enhance progress toward several other principles. Other cases may require trade-offs between the application of two principles that can be resolved only by the specific choices and values of the practitioners within the context of society. In the end, the framework of the 12 principles is a tool to aid in *consciously* and *transparently* addressing design choices relevant to sustainability challenges.

## 7.2 Design

Embedded in the discussion of green engineering is the word *design*. **Design** is the engineering stage where the greatest influence can be achieved in terms of sustainable outcomes. At the design stage, engineers are able to select and evaluate the properties of the final outcome. This can include material, chemical, and energy inputs; effectiveness and efficiency; aesthetics and form; and intended specifications such as quality, safety, and performance.

The design stage also represents the time for innovation, brainstorming, and creativity, offering an occasion to integrate sustainability goals into the specifications of the product, process, or system. Sustainability should *not* be viewed as a design constraint. It should be utilized as an *opportunity* to leapfrog existing ideas or designs and drive innovative solutions that consider systematic benefits and impacts over the lifetime of the design. This potential is shown in Figure 7.1. Figure 7.1 demonstrates that allowing an increased number of degrees of freedom

**Green Vehicle Guide**
http://www.epa.gov/greenvehicles/Index.do

**Clean Vehicles 101**
http://www.ucsusa.org/clean_vehicles/clean_vehicles_101/

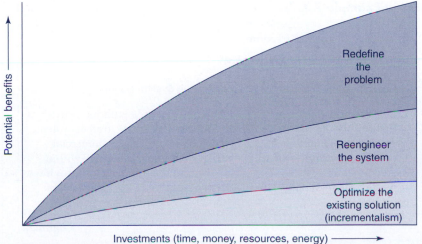

**Figure 7.1** **Increasing Potential Benefits with Increasing Degrees of Design Freedom for a Given Investment** Note that allowing an increased number of degrees of freedom to solve a problem frees up more design space to innovate and generate sustainable solutions.

to solve a challenge, address a need, or provide a service creates more design space to generate sustainable solutions.

For a given investment (time, energy, resources, capital), potential benefits can be realized. These benefits include increased market share, reduced environmental impact, minimized harm to human health, and improved quality of life. In the case where constraints require merely optimizing the existing solution or making incremental improvements, some modest gains can be achieved. However, if the degrees of freedom within the design space can be increased, more benefits can be realized. This is because the engineer has an opportunity to design a new solution that may appear very different in form but provides the same service. This may pose challenges if the new design is too embedded into an existing and constrained system. Ultimately, the most benefits can be achieved when the engineer designs with the most degrees of freedom—at the highest system scale—to ensure that each component within the system is sustainable, performs with the other system components, and meets the overall intended purpose.

example/7.1 Degrees of Freedom and Sustainable Design

In 2004 the average miles per gallon for a car on the road in the United States was 22. In response to concerns about global climate change, engineers are working toward a more innovative design to improve gas mileage and lower carbon dioxide emissions. What are the design opportunities for improvement scaled with increasing degrees of freedom, and what are the potential benefits?

solution

Table 7.1 shows three design solutions. As the degrees of freedom in the design increase, engineers in this example have more flexibility to innovate a solution to the problem.

## Table / 7.1

**Three Design Solutions Investigated in Example 7.1**

Increasing degrees of freedom

————————————————————▶

| | Incremental Improvement | Reengineer the System | Redefine the System Boundary |
|---|---|---|---|
| Design solution | Improve the efficiency of the Carnot engine; use lighter-weight materials (composites instead of metals) | Use a hybrid electric or fuel cell system for energy; change the shape of the car for improved aerodynamics; capture waste heat and energy for reuse | Meet mobility needs without individual car; implement a public transit system; design communities so commercial districts and employment are within walking and cycling distance; provide access to desired goods and services without vehicular transportation |

(Continued)

## Table / 7.1

|  | Incremental Improvement | Reengineer the System | Redefine the System Boundary |
|---|---|---|---|
| Potential realized benefits | Moderate fuel savings; moderate reductions in $CO_2$ emissions | Improved fuel savings; improved reductions in $CO_2$ emissions; improved material and energy efficiency | Elimination of the environmental impacts associated with the entire automobile life cycle; maximized fuel savings and $CO_2$ reductions; improved infrastructure; denser development (smart growth); improved health of society from walking and less air pollution |

The design phase offers unique opportunities in the life cycle of an engineered product, process, or system. As shown in Figure 7.2, it is at the design phase of a typical product that 70 to 75 percent of the cost is set, even though these costs will not be realized until much later in the product life cycle. The environmental costs are analogous to the economic ones. For example, it is also at the design phase that materials are specified. This often dictates the production process as well as operation and maintenance procedures (for example, painting, coating, rust inhibiting, cleaning, and lubricating).

As soon as a material is specified as a design decision, the entire life cycle of that material from acquisition through processing and the end of life is now included as a part of the environmental impacts of the

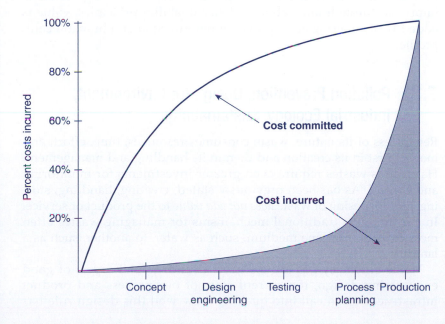

**Figure 7.2 Percent Costs Incurred versus Design Time Line** The costs can be thought of as economic or environmental. During the design phase, approximately 70 percent of the cost becomes fixed for development, manufacture, and use.

designed product, process, or system. Therefore, it is at the design phase that the engineer has the greatest ability to affect the environmental impacts associated with the final outcome. This process is similar for design of infrastructure such as water supply systems, wastewater treatment plants, buildings, transportation systems, and residential or commercial development.

As an example, think of all the materials and products that go into constructing and furnishing a building. At this point, the engineer needs to vision the future in regard to how these materials will be maintained, what cleaning agents will be used, what the water and energy demands of the building will be, what will happen to the building after its useful life is over, and what the fate of these materials will be at the end of the building's life. In terms of transportation systems, an engineer can think beyond the design of a new highway intended to relieve urban congestion, because in many areas, increased road volume potential leads to increased growth, which in turn can lead to more traffic and congested roadways.

It is also important to note that the design phase is when the engineer has the opportunity to incorporate increased efficiency; reduce waste of water, materials, and energy; reduce costs; and most important, impart new performance and capabilities. While many of the other attributes listed can be achieved through end-of-the-pipe control technologies, only at the design phase can the actual product, process, or system characteristics be changed. For example, in Chapter 6 we discussed how simple materials substitutions can eliminate exposure to chemicals altogether.

Adding new performance and/or capabilities often has improved environmental characteristics while offering the opportunity for improved competitiveness and market share, making the design better for many reasons. Simply controlling or minimizing waste through manufacturing or even end of life cannot alter or improve the fundamental nature of the design. When the design itself is improved through innovation and sustainability principles, value is added while offering an improved environmental and human health profile.

## 7.3 Pollution Prevention, Design for Environment, Industrial Ecology, Sustainability

Regardless of its nature, waste consumes resources, time, effort, and money, first in its creation and then in its handling and management. Hazardous wastes require even greater investments for monitoring and control. As has been previously stated, creating, handling, storing, and disposing of waste *does not add value* to the product or service. In addition, the traditional mechanisms for managing wastes often move wastes from one medium, such as water, to another, such as a landfill.

While efficiency has always been a fundamental aspect of good engineering design, the current state of our process and product infrastructure can call into question how well this design rule has been historically or systematically applied. One could argue that during the 20th century, engineering focused much more on design strategies for dealing with waste, such as treatment or disposal, rather than on innovative, disruptive technologies based on efficiencies. In response to regulations, subsidies, laws, and invested capital, it became far more important to make existing inefficient and unsustainable processes continue through the use of elegant and expensive technological interventions than to engage in fundamentally efficient and sustainable design. The result of this skewed focus is an extensive engineering portfolio of ways to monitor, control, and remediate waste.

Green engineering aims to refocus these efforts on sustainable, efficient design. This means avoiding waste in the first place wherever practicable and eliminating the concept of waste wherever possible.

**Pollution prevention** is focused on increasing the efficiency of a process to reduce the amount of pollution generated. This is the idea of incrementalism or **eco-efficiency**, where the current system is tweaked to be better than before. This does not take into account that the current design may not be the best or most appropriate for the current application. That is, the current product, process, or system was not designed with the intent of reducing waste and/or environmental impact. Instead, it is being improved within its current constraints, taking these considerations into account after the fact, after the design has been completed and often has been implemented.

The Pollution Prevention Act of 1990 (see Box 7.2) was passed to *encourage* (not regulate) pollution prevention in the United States. It establishes a **pollution prevention hierarchy** (Figure 7.3) as follows:

- **Source reduction**—Waste (hazardous substance, pollutant, or contaminants) should be prevented at the source (prior to recycling, treatment, or disposal).

**P2 at EPA**

http://www.epa.gov/p2

**Class Discussion**

*How does waste minimization and pollution prevention differ from green engineering?*

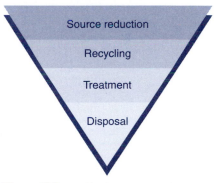

**Figure 7.3** Pollution Prevention Hierarchy

---

**Box / 7.2    Pollution Prevention Act of 1990**

The Pollution Prevention Act focused industry, government, and public attention on reducing the amount of pollution through cost-effective changes in production, operation, and raw-materials use. Opportunities for source reduction are often not realized because of existing regulations, and the industrial resources required for compliance focus on treatment and disposal. Source reduction is fundamentally different from and more desirable than waste management or pollution control (2 U.S.C. 13,101 and 13,102, s/s et seq., 1990).

*The Congress hereby declares it to be the national policy of the United States that pollution should be*

*prevented or reduced at the source whenever feasible; pollution that cannot be prevented should be recycled in an environmentally safe manner, whenever feasible; pollution that cannot be prevented or recycled should be treated in an environmentally safe manner whenever feasible; and disposal or other release into the environment should be employed only as a last resort and should be conducted in an environmentally safe manner.*

(2 U.S.C. 13,101b)

### Class Discussion

The pollution prevention hierarchy clearly shows the source reduction is favored over the other three aspects of pollution prevention. Disposal is the least preferred alternative. How does the pollution prevention hierarchy relate to industrial wastewater treatment and solid waste management?

### Where Can I Recycle My Stuff?

http://earth911.com

### Zero Waste

http://zerowaste.ca.gov/

- **Recycling**—Waste generated should be reused either in the process that created it or in another process.

- **Treatment**—Waste that cannot be recycled should be treated to reduce its hazard.

- **Disposal**—Waste that is not treated should be disposed of in an environmentally safe manner.

In the case of wastewater treatment, the pollution prevention hierarchy suggests that we should focus efforts on identifying ways to eliminate waste materials from being sewered and transported to a treatment plant, rather than devoting all of our efforts to improving the design of treatment facilities. In terms of solid-waste management, it is clear that landfill disposal is not the recommended alternative for managing a waste stream. In this case, an engineer would think beyond design of a landfill and focus on broader initiatives to reduce the amount of waste that is generated and discarded.

Allowing more degrees of design freedom and moving upstream for opportunities to redesign the product, process, or system offer greater opportunity for **waste minimization** or even waste elimination. While there may be current barriers, including scientific, technical, or economic, to zero-waste design, it is important to note that the concept of waste is human. In other words, there is nothing inherent about materials, energy, space, or time that makes it waste. It is waste only because no one has yet imagined or implemented a defined use for it.

If the creation of waste cannot be avoided under given conditions or circumstances, designers and engineers can consider alternative mechanisms to effectively exploit these resources for value-added purposes. For example, the waste could be beneficially used as a feedstock by capturing it and recycling/reusing it within the process, the organization, or beyond. This turns a cost and liability into a savings and benefit. Or perhaps construction waste could be captured on site, rather than discarding it to a landfill, so that it can be repurposed for other building applications.

It is important to consider that materials and energy that were utilized and are now "waste" have embedded entropy and complexity, representing an investment in cost and resources. This indicates that the recovery of waste as a feedstock represents both potential environmental and economic benefits.

**Industrial ecology** is the shifting of industrial processes from linear (open-loop) systems (type 1 systems), in which resource and capital investments move through the system to become waste, to ecological closed-loop systems where wastes become inputs for new processes (type 3 systems) (Graedel and Allenby, 1995). Figure 7.4 shows the differences between types 1, 2, and 3 systems. Note that type 2 systems have some cycling and generate some waste. Type 3 systems are based on natural systems where there is no waste. Instead, any waste produced in nature becomes a food or nutrient source for another species or natural process.

Think about whether processes you are familiar with are types 1, 2, or 3 systems. For example, conventional wastewater treatment is a type

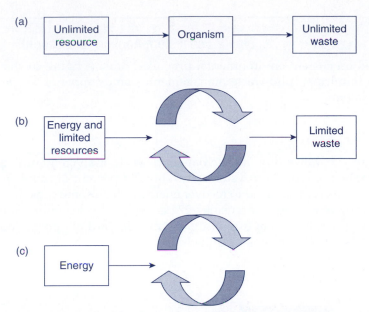

**Figure 7.4   Schematics of Types 1, 2, and 3 Ecosystems**   In this ecosystem, wastes are recycled back into the system. (a) The type 1 ecosystem is a traditional industrial system where resource and capital investments move through a system to become waste. (b) The type 2 ecosystem has some cycling and generates some waste. (c) The type 3 ecosystem is a closed-loop system where wastes become inputs for new processes.

2 ecosystem in the sense that some of the waste materials are recycled (for example, land application of biosolids and use of methane generated by anaerobic digestion to provide heat or electricity). However, these systems may still disperse nitrogen and phosphorus nutrients (once concentrated in food inputs) into aquatic environments. In this situation, added nutrients not only hurt water quality but become difficult to remediate once they accumulate in sediments. For a conventional wastewater treatment plant to move toward a type 3 ecosystem, the engineer could consider modifying the plant to transform aqueous-based ammonia nitrogen to gaseous nitrogen via nitrification and denitrification reactions. There could also be reuse of treated water for human consumption, agriculture, landscaping needs, and support of ecosystems.

Waste can be eliminated through the design of disruptive technologies that move toward sustainable products, processes, or systems that inherently reduce unnecessary inputs or the hazard associated with those inputs while still attaining the desired outcome. Another strategy to eliminate the concept of waste is to design molecules, products, processes, and systems to incorporate all of the inputs. These strategies suggest that the engineer design with the intention that all of the resources used be in desirable, value-added investments. This approach has both environmental and economic benefits in terms of reducing the life cycle impacts by eliminating the costs and impacts of resources and the associated costs of procurement and disposal for inputs that are no longer necessary.

**Class Discussion**

Can you think of other municipal or industrial systems that are either type 1 or type 2 ecosystems?

## example/7.2 Type 2 Ecosystem Applied to a Landfill

Based on current environmental practices, draw a figure to show how municipal solid-waste management is an example of a type 2 ecosystem.

## solution

Figure 7.5 shows how municipal solid-waste management is an example of a type 2 ecosystem. Currently, 17 percent of municipal solid waste is incinerated to recover energy (which some consider to be a form of recycling or reusing), 27 percent is recycled (aluminum, glass, paper) and composted (organic matter, yard clippings, food waste), and 56 percent is landfilled.

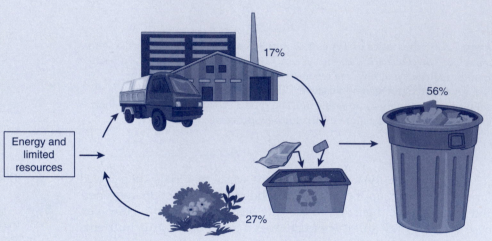

**Figure 7.5** Example of How Municipal Solid-Waste Management Is a Type 2 Ecosystem

**EPA's Design for Environment Program**
http://www.epa.gov/oppt/dfe/

**Design for environment (DFE)** or **eco-design**, is an approach to design in which environmental burdens are intentionally considered and eliminated where possible at the design phase. DFE strategies include source reduction, material recovery, and when these fail, the use of treatable as opposed to untreatable materials. Table 7.2 provides several methods for introducing these strategies.

When social, economic, and environmental impacts are considered for the long term at the design phase, the effort can be considered sustainable design. Remember that, as we discussed in Chapter 1, the goal of sustainability is to simultaneously align and advance benefits to society, the economy, and the environment. This strategy goes beyond pollution prevention and DfE to realize even greater benefits for current and future generations.

**Design for Environment (DFE) Strategies for Eliminating Environmental Burdens**

- Changes in material selection

- Changes in equipment selection; improved purchasing choices

- Improved operating practices

- Improved recovery and disposition practices

- Improved logistics

SOURCE: Cooper, 1999.

## 7.4    Fundamental Concepts

Green engineering and sustainable design involve several fundamental concepts, including inherency, life cycle, systems thinking, resilience, performance criteria, providing services, and financial analysis. This section discusses each of these concepts in greater detail.

### 7.4.1    INHERENCY

As was shown in the risk equation from the previous chapter, **risk** is a function of hazard and exposure:

$$\text{Risk} = f(\text{hazard}, \text{exposure}) \qquad (7.1)$$

In green chemistry, risk is minimized by reducing or eliminating the hazard. As the intrinsic hazard is decreased, there is less reliance on exposure controls and therefore less likelihood for failure. The ultimate goal would be completely benign materials or chemicals such that there is no need to control exposure. That is, the chemicals and materials would not cause harm if they are released to the environment or humans are exposed to them. The advances being made in moving toward inherently benign chemicals through green chemistry is significant and dramatic.

**Green chemistry**, which emerged as a cohesive area of research in 1991, is defined as the design of products and processes that reduce or eliminate the use and generation of hazardous substances. The green chemistry approach was outlined in the framework of the *12 Principles of Green Chemistry* (Anastas and Warner, 1998) and has served as the guiding document for the field. Green chemistry is one of the most fundamental fields related to science and technology for sustainability in that it focuses at the molecular level to design chemicals and materials to be inherently nonhazardous.

The fundamental research of green chemistry has been brought to bear on a diverse set of challenges, including energy, agriculture, pharmaceuticals and health care, biotechnology, nanotechnology, consumer

**Green Chemistry**
http://www.epa.gov/greenchemistry

**Table / 7.3**

**Examples of Green Chemistry**

- A dramatically more effective fire-extinguishing agent that eliminates halon and utilizes water in combination with an advanced surfactant

- Production of large-scale pharmaceutical active ingredients without the typical generation of thousands of pounds of toxic waste per pound of product

- Elimination of arsenic from wood preservatives that are used in lumber applied to household decks and playground equipment

- Introduction of the first commodity bio-based plastic that has the performance qualities needed for a multi-million-pound application, as a food packaging

- A new solvent system that eliminates large-scale ultra-pure water usage in computer chip manufacture, replacing it with liquid carbon dioxide, which allows for the production of next-generation chips

products, and materials. In each case, green chemistry has been successfully demonstrated to reduce intrinsic hazard, to improve material and energy efficiency, and to ingrain a life cycle perspective. Table 7.3 provides examples of green chemistry that illustrate its breadth of applicability.

It is clear from Chapter 6 that toxicity is a complex topic with many contributing factors. This reinforces the difficulty of performing a precise risk assessment. It further demonstrates that one of the best strategies for mitigating risk is to reduce or eliminate the intrinsic hazard of the chemicals or materials that are designed into a given product, process, or system.

It is also critical to note that other characteristics of a product, process, or system (besides toxicity) can be designed to be inherent. A design can be inherently more reliable, more durable, more resilient, and more efficient. The intention is to design the desired properties intrinsically, rather than controlling or maintaining them through external circumstances.

### 7.4.2 LIFE CYCLE

**The Life Cycle Initiative**

http://lcinitiative.unep.fr/

**Life cycle** considerations take into account the environmental performance of a product, process, or system through all phases from acquisition of raw materials to refining those materials, manufacturing, use, and end-of-life management (see Figure 7.6a). In the case of engineering infrastructure, the life stages would be site development, materials and product delivery, infrastructure manufacture, infrastructure use, and refurbishment, recycling, and disposal (see Figure 7.6b). In some cases, the transportation impacts of moving between these life cycle stages also are considered.

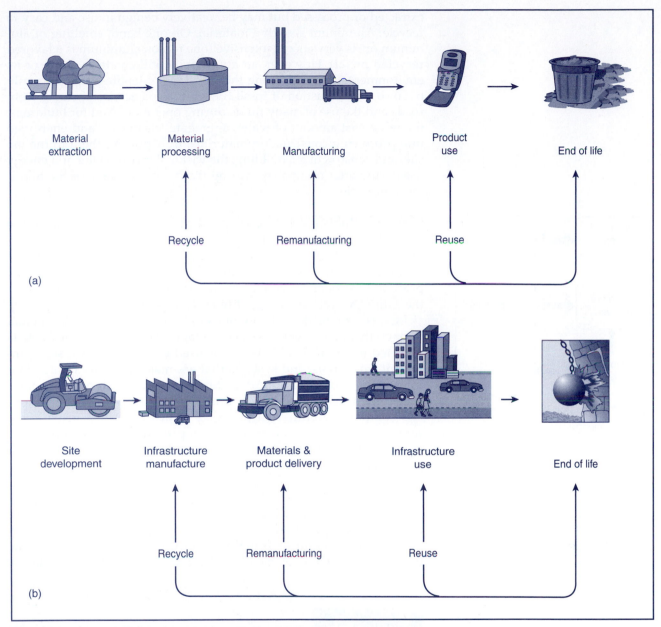

(a)

(b)

**Figure 7.6  Common Life Cycle Stages**  The most common stages for (a) a manufactured product and (b) engineered infrastructure that makes up the built environment.

Figure 7.6a and 7.6b also show, as feedback loops, the potential for recycling, remanufacturing, and reuse. While benefits are often associated with these various end-of-life handling strategies, they can also carry negative environmental impacts and should be included when making design or improvement designs and in life cycle calculations.

The entire life cycle must be considered, because different environmental impacts can occur during different stages. For example, some materials may have an adverse environmental consequence when extracted or processed but may be relatively benign in use and easy to recycle. Aluminum is such a material. On one hand, smelting of aluminum ore is very energy intensive (one reason aluminum is a favored recycled metal). However, an automobile will create the bulk of its environmental impact during the use stage of its life cycle, primarily because of combustion of fossil fuels but also because of runoff from roads and the use of many fluids during operation. And for buildings, though a vast amount of water, aggregate, chemicals, and energy go into producing construction materials, transporting them to the job site, and constructing a building, the vast majority of water and energy use occurs after occupancy, during the operation stage of the building's life cycle.

## ENVIRONMENTAL LIFE CYCLE ASSESSMENT (LCA)
The idea of life cycle considerations is to protect against applying green engineering and sustainable design on just one life cycle stage. To effectively capture these impacts across the entire life cycle of the product, process, or system, we must consider the environmental impacts for the entire life cycle, by using a **life cycle assessment (LCA)**.

LCA is a sophisticated way of examining the total environmental impact through every life cycle stage. The LCA framework is depicted in Figure 7.7. LCAs can be used to identify processes, components, materials, and systems that are major contributors to environmental impacts, compare different options within a particular process with the objective of minimizing environmental impacts, and compare two different products or processes that provide the same service. This analysis can then provide guidance on where opportunities exist for design decisions to improve environmental performance.

As shown in Figure 7.7, the first step in performing an LCA is to define the goal and scope. This can be accomplished by answering the questions listed in Table 7.4. In the first step of the LCA, the goal states the intended application and audience, while the scope of the study defines the **function** or **functional unit**, defined in Table 7.5. The functional unit serves as the basis of the LCA, the system boundaries,

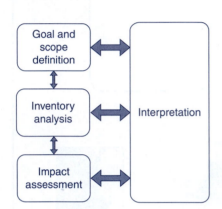

**Figure 7.7** **Components of the Life Cycle Assessment Framework**

**LCA 101**

http://www.epa.gov/nrmrl/lcaccess/lca101.html

## Table / 7.4

**Questions for the First Step of a Life Cycle Assessment (LCA): Defining the Goal and Scope**

- What is the purpose of the LCA? Why is the assessment being conducted?
- How will the results be used, and by whom?
- What materials, processes, or products are to be considered?
- Do specific issues need to be addressed?
- How broadly will alternative options be defined?
- What issues or concerns will the study address?

example/7.3 **Life Cycle Assessment Framework**

If you were asked to conduct an LCA on a new paper product, what would be a possible description of the goal and scope?

solution

This problem has several possible solutions.

**Solution 1**

*Possible goal:* To identify which stage in the life cycle of paper production emits significant greenhouse gases

*Possible scope:*

- Analysis of paper fibers that are chemically pulped

- Limit to one pulp and paper mill

- Conduct analysis on per-weight basis

**Solution 2**

*Possible goal:* To decide between this new paper product and the currently used paper product

*Possible scope:*

- Restrict the analysis to the manufacturing stage or each paper type (assume the same logging practices and that recycling rates at the end of life stage are identical for the two paper types)

- Neglect any material or component that composes less than 5 percent by weight of the final paper product

- Conduct analysis on the production of 10,000 sheets of paper

and the data requirements and assumptions (for example, that impacts from transportation between life cycle stages will be neglected).

Once the goal, scope, and functional unit have been defined, the next step of an LCA is to develop a flow diagram for the processes

**Table / 7.5**

**Definitions of Function and Functional Unit**

| Function | Functional Unit |
|---|---|
| Service provided by the system | Means of quantifying production function |
| Performance characteristic of a product | Basis for an LCA |
| | Reference for normalization of input and output data |

### example/7.4 Determining Function and Functional Unit in Terms of LCA

### example 1

If you are asked to conduct an LCA on two different laundry detergents, what could you use as the functional unit for the analysis?

### solution 1

The basis of the LCA could be the weight or volume of each laundry detergent necessary to run 1,000 washing machine cycles. (This says nothing about the performance of the laundry detergents—how clean the clothes are after washing—as that is assumed to be identical for the purpose of the LCA.)

### example 2

If you are asked to conduct an LCA on paper versus plastic grocery bags, what could you use as the functional unit for the analysis?

### solution 2

The basis of the LCA could be a set volume of groceries to be carried, in which case two plastic bags might be equivalent to one paper bag. Or the functional unit could be related to the weight of groceries carried, in which case you would need to determine whether paper or plastic bags are stronger and how many of each would be needed to carry the specified weight.

being evaluated and conduct an **inventory analysis**. The inventory analysis was shown as part of the LCA framework in Figure 7.7. It involves describing all of the inputs and outputs in a product's life cycle, beginning with what the product is composed of, where those materials came from, where they go, and the inputs and outputs related to those component materials during their lifetime. It is also necessary to include the inputs and outputs during the product's use, such as the use of electricity or batteries.

The purpose of the inventory analysis is to quantify what comes in and what goes out, including the energy and material associated with each stage in the life cycle. Inputs include all materials, both renewable and nonrenewable, and energy. It is important to remember that outputs include the desired products as well as by-products and wastes such as emissions to air, water, and land. While some of these data sources are public and can be found on the Internet, others require a

**Mercury and Lighting**

paid subscription for access to the data or are included in the purchase of life cycle assessment software.

Life cycle inventory data can come from many sources, including peer-reviewed literature, manufacturers or trade associations, and generic databases. The data may be averaged for a product (such as the average for all methods to produce 1 kg glass) or sector (average for all producers to produce 1 gal. drinking water) to maintain the proprietary nature of this information for individual companies.

When conducting a life cycle inventory analysis, it is important to consider the quality of data for inputs and outputs to the system. Data quality indicators include precision, completeness, representativeness, consistency, and reproducibility. Data quality is especially important to consider given that much of the available life cycle data originates in Europe and Japan, where the manufacturing processes and the grid energy mix to produce electricity may be different than in North America.

The third step in the LCA (as was shown in Figure 7.7) is to conduct an **impact assessment**. This step involves identifying all the environmental impacts associated with the inputs and outputs detailed in the inventory analysis. In this case, the environmental impacts from across the life cycle are grouped together in broad topics. Environmental impacts can include stressors such as resource depletion, water use, global-warming potential, ozone hole depletion, human toxicity, smog formation, and land use. This step often involves some assumptions about what human health and environmental impacts will result from a given emission scenario, which is similar to a risk assessment.

The final step in the impact assessment can be controversial, as it often involves weighting these broad environmental impact categories to yield a single score for the overall environmental performance of the product, process, or system being analyzed. This is often a societal consideration, which can vary among cultures. For example, South Pacific island nations may give greater weight to climate change, given their vulnerability to a rise in sea level, while other countries may give greater weight to impact on human health.

Consequently, the total impact score may be distorted by weighting factors. Also, for an identical life cycle inventory, the resulting decisions from the impact assessment may vary from country to country or from organization to organization. Impact assessments can also be performed using a score card with *subjective grades* assigned such as numbers 1–5, with 5 being the best, or a more visual grading system such as: ✓+, ✓, and ✓−.

Ultimately, LCA can provide insight into opportunities for improving the environmental impact of given product, process, or system. This can include selecting one of two options or identifying areas for improvement for a single option. LCA is extremely valuable in ensuring that environmental impact is being minimized across the entire life cycle and that impacts are not being shifted from one life cycle stage to another. This leads to a system that is globally optimized to reduce adverse effects of the specified product, process, or system.

**Applying Life Cycle Thinking to International Water and Sanitation Development Projects**
*http://cee.eng.usf.edu/peacecorps/ Resources.htm*

example/7.5 **Performing a Life Cycle Analysis on Pipe Materials**

The purpose of this extended example is to show the complete process and complexity of performing an LCA. We will examine four pipe products for this example.

solution

Because the solution to this problem is long, we have structured the solution into ten steps.

1. **Specify the study's goal:** Assess the life cycle impacts of four different pipe products for use in water distribution for a new community planned for sustainability. The four piping materials are polyvinyl chloride (PVC), polyvinylidene chloride (PVDC), polyethylene (PE), and aluminum (Al).

2. **Specify the study's scope:** For this example, the life cycle assessment will have a system boundary around the entire life cycle from raw-material extraction to manufacturing, use, and end of life (landfill, incineration, and recycling) (see Figure 7.8).

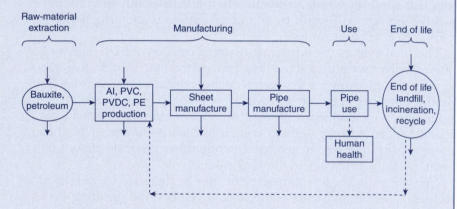

**Figure 7.8  System Boundary Used in Example 7.5**  The boundary encompasses the life cycle from raw-material extraction to manufacturing, use, and end of life.

3. **Determine the function:** Piping provides a physical product to transport water throughout the built environment. Desirable properties include ability to contain the media being transported, minimization of heat transfer, long life, and resistance to degradation. Such concerns as aesthetics, adhesives, and joining methods are outside the scope of this life cycle analysis.

4. **Determine the functional unit:** Assume 1,000 ft. of pipe material. The packaging life cycles are identical for each of the four products and therefore are not included in this study. Transportation from each of the four manufacturing facilities to the construction site is similar and negligible, and therefore is not included in this study.

5. **Develop process flow diagrams for production of the four different pipe materials:**

   Figure 7.9 shows the process flow diagram for polyvinyl chloride (PVC) pipe production.

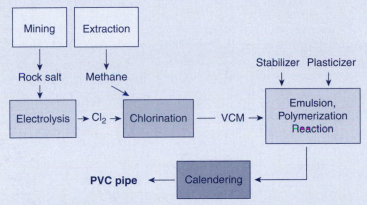

**Figure 7.9**   Process Flow Diagram for Polyvinyl Chloride (PVC) Pipe Production

Figure 7.10 shows the process flow diagram for polyvinylidene chloride (PVDC) pipe production.

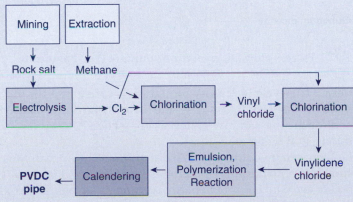

**Figure 7.10**   Process Flow Diagram for Polyvinylidene Chloride (PVDC) Pipe Production

Figure 7.11 shows the process flow diagram for polyethylene (PE) pipe production.

**Figure 7.11** Process Flow Diagram for Polyethylene (PE) Pipe Production

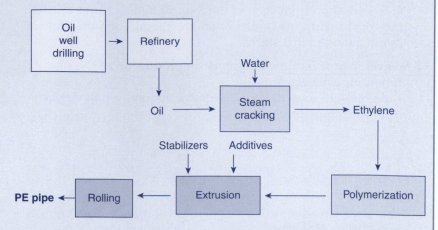

Figure 7.12 shows the process flow diagram for aluminum (Al) pipe production.

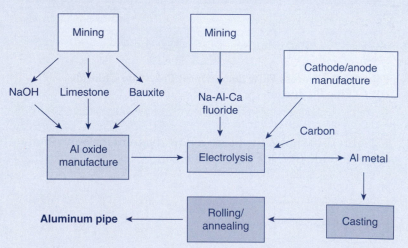

**Figure 7.12** Process Flow Diagram for Aluminum Pipe Production

6. **Perform a life cycle inventory of materials consumed for pipe material production:** In this case, the life cycle inventory is performed for the mass of raw materials consumed, energy, air emissions, and water emissions.

The life cycle inventory of the mass of specific raw materials consumed for pipe production is shown in Figure 7.13a.

The life cycle inventory of energy consumed for pipe material production is shown in Figure 7.13b. Note that the energy consumed for these products includes the energy embodied in the product (that is, if the oil used to produce PVC, PVDC, and

example/7.5 Continued

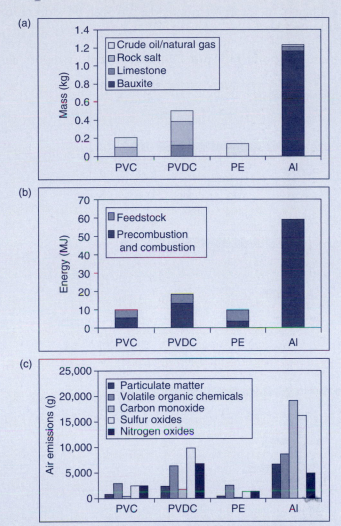

**Figure 7.13** **Life Cycle Inventories** (a) Life cycle inventory of mass of raw materials consumed for pipe production; (b) life cycle inventory of energy consumed for pipe material production; and (c) life cycle inventory of air emissions of criteria air pollutants (per functional unit).

PE were used for energy generation rather than for product production) as well as pre-combustion and combustion energy for all of the life cycle stages. Pre-combustion energy is the energy required to acquire the resource, and combustion energy is the energy consumed in the acquisition and manufacturing processes themselves.

The life cycle inventory of air emissions of criteria air pollutants is shown in Figure 7.13c. As we will discuss in Chapter 12, criteria air pollutants are those the Clean Air Act requires the Environmental Protection Agency (EPA) to regulate through National Ambient Air Quality Standards. They include ozone, particulate matter, carbon monoxide, nitrogen dioxide, sulfur dioxide, and lead.

The life cycle inventory of water emissions for several key emissions is shown in Figure 7.14. This inventory reports the life

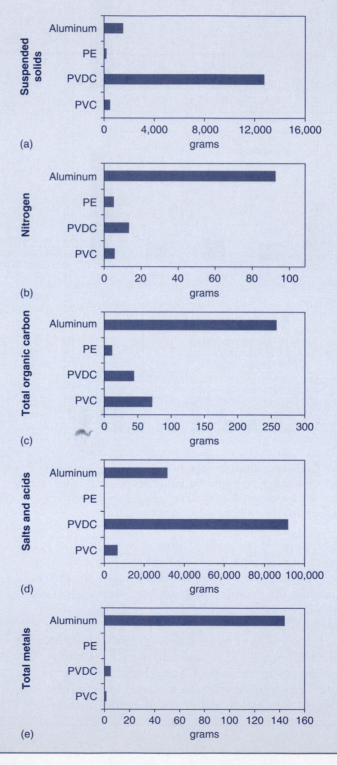

**Figure 7.14** **Life Cycle Inventory of Key Water Emissions** This inventory measures (a) suspended solids, (b) nitrogen, (c) total organic carbon, (d) salts and acids, and (e) total metals (per functional unit).

cycle water emissions for each pipe product based on the flow diagrams established in step 5, including suspended solids, nitrogen, salts and acids, total metals, and total organic carbon.

7. **Determine the life cycle inventory of the final life stage (end-of-life handling):** At the final life stage, the product can be landfilled, incinerated, or recycled (Figure 7.15).

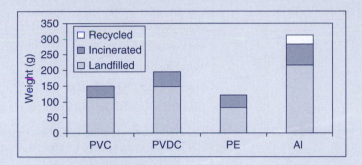

**Figure 7.15** Life Cycle Inventory of the Final Life Stage (End-of-Life Handling) (per functional unit)

8. **Determine the life cycle impact assessment of life cycle inventory data:** In this step, all of the inventory data, including raw-material and energy inputs and all of the outputs in terms of air, water, and solid waste, are grouped together into broad environmental-impact categories such as global-warming potential (Figure 7.16) and human health concerns (Table 7.6). In Figure 7.16, $GWP_{100}$ is global-warming potential in kilograms of $CO_2$ equivalents with a 100-year perspective (see Chapter 2). For categories such as human health effects, it is often necessary to include qualitative assessments, because—as we discussed in Chapter 6—performing risk assessments is difficult and complex.

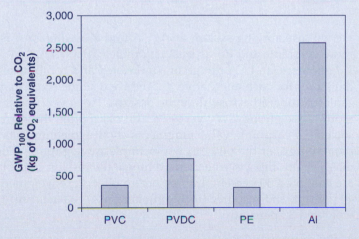

**Figure 7.16** Life Cycle Impact Assessment for Environmental Impact of Global-Warming Potential ($GWP_{100}$) (per functional unit)

9. **Interpret life cycle inventory and assessment:** As this point in the LCA, subjective evaluations are made of the impacts evaluated in step 8, comparing the impacts for each of the alternative designs to determine which one has the best environmental profile and to identify potential design improvements. In this case,

## Table / 7.6

**Life Cycle Impact Assessment for Human Health Concerns in Example 7.5**

| Material | Health Concerns |
|---|---|
| Dioxin (PVC/PVDC) | Carcinogen—potential endocrine disruptor and teratogen |
| | Exposure linked to diabetes |
| | Vinyl institute: 24.3 g/yr |
| | Greenpeace: 283–565 g/yr |
| | EPA: Insufficient data to evaluate claims |
| Plasticizers (PVC/PVDC) | Diethylhydroxylamine (DEHA): potential human carcinogen, teratogen, mutagen, and endocrine disruptor |
| | di-2-ethyl hexyl phthalate (DEHP): potential human carcinogen, teratogen, and endocrine disruptor |
| Migration of plasticizers (PVC/PVDC) | Concern for DEHA and DEHP as an additional exposure pathway |
| Aluminum | Possible link to Alzheimer's disease; negligible exposure from this product pathway |
| | Production of aluminum by smelting linked to asthma and chronic airway disease; causes acid mine drainage |

we will use a rating system of ✓+, ✓, and ✓− where ✓+ is more favorable and ✓− is least favorable (Table 7.7).

From here, an LCA can go one more step and combine all of the scores for each alternative to produce a single score that can be compared across all of the designs. However, this step requires weighting GWP versus health risks; often these are societal or corporate value judgments about which impacts are more important than others. For example, a weighting scheme may weight impacts where environmental emission impacts are weighted 30 percent, human health effects are weighted 40 percent, and resource (including land use) impacts are weighted 30 percent.

10. **Make recommendations:** Based on the life cycle assessment, the following recommendations could be made for environmental and human health improvements to produce more environmentally benign alternatives designs than those currently available or under consideration.

## Table / 7.7

**Life Cycle Interpretation of Inventory and Assessment for Example 7.5**

|  | PVC | PVDC | PE | Al |
|---|---|---|---|---|
| Material intensiveness | ✓ + | ✓ | ✓ + | ✓ − |
| Energy | ✓ + | ✓ | ✓ + | ✓ − |
| Health | ✓ − | ✓ − | ✓ + | ✓ |
| Air pollutants | ✓ | ✓ | ✓ + | ✓ − |
| Water pollutants | ✓ | ✓ − | ✓ + | ✓ − |
| Global-warming potential | ✓ + | ✓ | ✓ + | ✓ − |
| End of life | ✓ − | ✓ − | ✓ − | ✓ |

- Examine changes in manufacturing techniques to minimize or eliminate plasticizer migration. This will address the health concerns associated with human exposure to the plasticizers used in PVC and PVDC, improving their overall score.

- Develop alternative benign plasticizers through green chemistry. If alternative plasticizers are developed that are not of concern when humans are exposed to them, this would also improve the health scores of PVC and PVDC, minimizing their overall life cycle impacts.

- Minimize contact surface of pipe with potable water. Based on the health concerns associated with PVC, PVDC, and aluminum products, minimizing these impacts is important to addressing the overall life cycle burden. In this case, we would be controlling the circumstance in which the product is used, rather than designing an inherently more benign product. This strategy is not as effective and does not allow for as many potential realized benefits, because the chemicals of concern during use will continue to be an issue for human exposure at end of life. While minimizing contact would improve environmental performance, it is less beneficial than eliminating the plasticizer altogether or developing benign plasticizer alternatives.

- Reuse or recycle all aluminum pipe. Given the tremendous environmental impacts associated with the manufacturing of aluminum pipe, particularly the energy demand, a strategy to improve the life cycle performance of aluminum pipe would be to recycle or reuse the product rather than having to produce virgin product for each use.

# ECONOMIC LIFE CYCLE ASSESSMENT: LIFE CYCLE COSTING

Similar to a life cycle assessment for environmental impacts, a financial analysis can look at costs across the life cycle of a product, process, or system. In this case, the upfront purchase costs are included, in addition to any costs associated with use, maintenance, repair, operation, replacement (if the expected lifetime is shorter than needed), and end-of-life handling such as disposal or salvage. This provides a more realistic picture of the economics associated with a given design and allows for more informed decision making. The **life cycle cost (LCC)** is known as the total cost of ownership.

To gain an understanding of what is, and is not, a good design from a financial point of view through the life cycle, it is important to account for all present and future costs as a single metric. This allows for comparison on an equivalent basis. This raises the issue of the *time value of money*. Money today is worth more than money tomorrow, so money from two different time periods cannot be compared without adjustment. When performing analysis on LCC, it is important to put all of the costs and/or revenues in a common time period and to approximate costs or revenues that occur over the year as single year-end amounts.

The value of money from different time periods can be adjusted using an interest rate. Some standard calculations based on the time value of money are described in Table 7.8. They include **present value (PV), future value (FV), present value of an annuity (PVA),** and **future value of an annuity (FVA).**

**Interest** is the fee paid to borrow money, since the money being used in the present is not available for other expenditures or investments. The original amount lent or spent is called the *principal*. If interest is paid annually and the time period is greater than a year, the interest is said to be *compounded*. That is, the money in year 2 is the principal plus the interest from year 1. The money in year 3 is the principal plus the interest for years 1 and 2. In all of the following equations, the units for the interest rate ($i$) and the number of time periods ($n$) must match. For example, if an annual interest rate is reported, then the number of

## Table / 7.8

### Standard Calculations Based on the Time Value of Money

| | |
|---|---|
| Present value (PV) | Present value (PV) of an amount that will be received in the future |
| Future value (FV) | Future value (FV) of an amount invested (such as in a deposit account) now at a given rate of interest |
| Present value of an annuity (PVA) | Present value of a stream of (equally sized) future payments, such as a mortgage |
| Future value of an annuity (FVA) | Future value of a stream of payments (annuity), assuming the payments are invested at a given rate of interest |

periods must be in years. If the interest is provided as monthly, then the time periods must be monthly. Interest rates are assumed to be annual unless otherwise stated.

Based on these principles, the future value of money can be calculated as follows:

$$FV = PV \times (1 + i)^n \qquad \textbf{(7.2)}$$

where PV is the present value of money, FV is the future value of money, $i$ is the annual interest rate, and $n$ is the number of years.

---

example/7.6 **Determining the Future Value of Money**

If you invested $2,000 today in an account that paid 6 percent interest, with interest compounded annually, how much will be in the account at the end of 2 years? (Assume no withdrawals.)

solution

The future value of the initial $2,000 investment can be determined from Equation 7.2:

$$FV = \$2,000.00 \times (1 + 0.06)^2 = \$2,247.20$$

---

Equation 7.2 can be rearranged to solve for the present value of money:

$$PV = \frac{FV}{(1 + i)^n} \qquad \textbf{(7.3)}$$

This equation enables the discounting process of translating a future value or a set of future cash flows into a present value.

---

example/7.7 **Determining the Present Value of Money**

If you wanted to have $1 million 50 years in the future and could find a green investment with a guaranteed interest rate of 15 percent per year, how much would you need to invest today?

solution

This problem requests that Equation 7.3 be solved for the present value of $1 million.

$$PV = \frac{\$1,000,000}{(1 + 0.15)^{50}} = \$922.80$$

---

Equations 7.2 and 7.3 can be used to perform cost comparisons for investments in greener materials, equipment, and infrastructure, which perhaps have an initial higher price tag. These equations can also be used to determine whether an engineer should recommend purchase of low-cost equipment that has higher costs for maintenance, treatment, energy, permitting, and replacement. Decisions can then be determined on both an environmental basis (using LCA) and on an economic basis (using LCC analysis).

The present value of an annuity (PVA), which is the value of an annuity at time = 0, can then be calculated based on the sum of a series of present value calculations:

$$PVA = \sum_{j=1}^{n} A \frac{1}{(1+i)^j} = A\left[ \frac{1}{1+i} + \frac{1}{(1+i)^2} + \cdots + \frac{1}{(1+i)^n} \right] \quad (7.4)$$

where $A$ is the value of the individual payments in each year, $i$ is the interest rate, and $n$ is the number of years.

Therefore:

$$PVA = A \times \left( \frac{(1+i)^{n-1}}{i(1+i)^n} \right) = A \times \left[ \frac{1-(1+i)^{-n}}{i} \right] \quad (7.5)$$

## example/7.8 Life Cycle Cost Applied to Water Treatment System

Your company just received a 4-year contract to provide drinking water to a community. The current treatment system you are running has high operating costs, but the purchase price of the new system is expensive, as shown in the following table. Based on an LCC analysis, which system would you select?

| Options | Continue with Old | Buy New |
|---|---|---|
| Purchase Price ($) | 0 | 4,000 |
| Operating Cost ($/yr) | 2,000 | 500 |
| Lifetime (yr) | 4 | 4 |

## solution

Use Equation 7.5 to determine the present value of an annuity (PVA) for the old and the new system. This allows us to consider the higher operating costs, which occur every year over the 4-year contract.

$$PVA_{old} = \$2,000 \times \left[ \frac{1-(1+i)^{-4}}{i} \right]$$

$$PVA_{new} = \$4,000 + \$500 \times \left[ \frac{1-(1+i)^{-n}}{i} \right]$$

## example/7.8 Continued

The final outcome depends, of course, on the interest rate, $i$. If the bank provides an interest rate of 10 percent per year, then

$$PVA_{old} = \$6,340$$
$$PVA_{new} = \$5,585$$

Purchasing the new system would be better. If the bank provides an interest rate of 20 percent per year, then

$$PVA_{old} = \$5,180$$
$$PVA_{new} = \$5,295$$

Therefore, continuing with the older system would be better.

Other decisions to consider are material-handling costs, energy costs, ease of maintenance, permitting costs, and overall contribution to the community. Note that, in a replacement problem such as this, revenues can be added to analysis, but there is no need to do so if they are the same for both options. In this case, the profits from providing drinking water would be the same, so revenues were neglected.

In some situations, it will be necessary to make decisions between two different designs that do not have the same lifetime. For example, one system (system A) may be very poorly made and last only 2 years, while another system (system B) may provide the same function, be better made, and last 20 years.

Just as we cannot accurately compare money from two different time periods, it is not appropriate to compare systems with two different lifetimes. In this case, assume you will purchase and dispose of the design with the shorter lifetime often enough to equal the lifetime of the longer lasting design. In this case, you need to purchase and dispose of system A ten times to equal one system B. To perform an LCC analysis to make the purchase decision between system A and system B, convert the investment, operating, and end-of-life costs into a single annual payment.

## example/7.9 Annualized Costs with Different Service Lives

Based on the following costs for two indoor air-filtering systems and an interest rate of 15 percent per year, which alternative would you recommend to your client, based on the annual cost of each alternative?

## example/7.9 Continued

|  | Alternative 1 | Alternative 2 |
|---|---|---|
| Purchase price ($) | 10,000 | 20,000 |
| Annual operating cost ($/yr) | 1,500 | 1,000 |
| Salvage value ($) | 500 | 1,000 |
| Service life (yr) | 2 | 3 |

## solution

The problem requests you solve for A (using Equations 7.3 and 7.5), which is the value of individual payments in each year. This amount represents the annual energy costs that would need to be realized to justify installation of one air-filtering system over another on a purely economic basis. The annualized cost of alternative 1 can be represented schematically for each year by the following diagram, where costs (in dollars) are on the bottom, and revenues (in dollars) are on the top for each of the two years:

$0                         $500

                Year 1       Year 2

$10,000         $1,500       $1,500

$$PVA = \$10{,}000 + \$1{,}500 \times \left[ \frac{1 - (1 + 0.15)^{-2}}{0.15} \right] - \frac{\$500}{(1 + 0.15)^2}$$

$$= \$12{,}061$$

Now the present value of an annuity (PVA) needs to be translated into an annual cost by rearranging Equation 7.5 to solve for $A$:

$$A = PVA \times \left[ \frac{i}{1 - (1 + i)^{-n}} \right] = \$12{,}061 \times \left[ \frac{0.15}{1 - (1 + 0.15)^{-2}} \right]$$

$$= \$7{,}418$$

Schematically, this can be represented (costs on the bottom, revenues on the top) as follows:

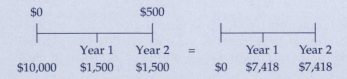

To calculate the annualized cost of alternative 2, convert all of the costs to the present value, and then calculate the annual cost:

$$PVA = \$20,000 + \$1,000 \times \left[\frac{1 - (1 + 0.15)^{-3}}{0.15}\right] - \frac{\$1,000}{(1 + 0.15)^3}$$

$$= \$21,626$$

To translate the PVA into an annual cost, rearrange the PVA equation to solve for $A$:

$$A = PVA \times \left[\frac{i}{1 - (1 + i)^{-n}}\right] = \$21,626 \times \left[\frac{0.15}{1 - (1 + 0.15)^{-3}}\right]$$

$$= \$9,472$$

Schematically, this can be represented as follows:

| $0 | | | $1,000 | | | | | |
|---|---|---|---|---|---|---|---|---|
| | Year 1 | Year 2 | Year 3 | = | | Year 1 | Year 2 | Year 3 |
| $20,000 | $1,000 | $1,000 | $1,000 | | $0 | $9,472 | $9,472 | $9,472 |

Based only on cost, you would select indoor air-filtering system 1, because it will be less expensive to operate on an annual basis ($7,418 versus $9,472).

## example/7.10 Annualized Costs Applied to Renewable Energy

Assume that installing a solar array for on-site energy generation costs $50,000. How much needs to be saved on annual energy bills to recover the cost of the system in 5 years? Assume an annual interest rate of 7 percent.

© olaf loose/iStockphoto.

### solution

In this case, we are provided the following values: PV = $50,000, $i$ = 7 percent per year, and $n$ = 5 years. We need to solve for the value of the individual payments in each year ($A$).
Rearrange Equation 7.5, and solve for $A$:

$$A = PVA \times \left[\frac{i}{1 - (1 + i)^{-n}}\right] = \$50,000 \times \left[\frac{0.07}{1 - (1 + 0.07)^{-5}}\right]$$

$$= \$12,195$$

example/7.10 **Continued**

On a purely economic basis, the annual energy bills would have to decrease by $12,195 to recover the costs of installing a $50,000 solar array over 5 years at a 7 percent annual interest rate. Using the same formula, this problem can be solved for different payback periods and interest rates.

**BEYOND ENVIRONMENTAL AND ECONOMIC LIFE CYCLE CONSIDERATIONS** While the previous two sections have provided information and skills for decision making based on environmental and economic life cycle considerations, there are other factors to be considered in terms of **intangibles**. Intangible costs and benefits include reputation, brand value, potential liability, stock performance, societal and quality-of-life benefits, stakeholder support, client and employee loyalty, innovation, and leadership. These characteristics, while currently difficult to quantify in terms of environmental or economic factors, can add significant value to an individual, project, community, or company. These intangibles are integral to understanding and communicating the long-term value of a project.

### 7.4.3 SYSTEMS THINKING

**Systems thinking** considers component parts of a system as having added characteristics or features when functioning within a system rather than isolated alone. This suggests that systems should be viewed in a holistic manner. Systems as a whole can be better understood when the linkages and interactions between components are considered in addition to the individual components.

The nature of systems thinking makes it extremely effective for solving the most difficult types of problems. For example, sustainability challenges are quite complex, depend on interactions and interdependence, and are currently managed or mitigated through disparate mechanisms.

One way to begin a systems analysis is through a **causal loop diagram (CLD)**. Causal loop diagrams provide a means to articulate the dynamic, interconnected nature of complex systems. These diagrams consist of arrows connecting variables (things that change over time) in a way that shows how one variable affects another. Each arrow in a causal loop diagram is labeled with an *s* or an *o*. *s* means that when the first variable changes, the second one changes in the *same* direction. (For example, increased profits lead to increased investments in research and development.) *o* means that the first variable causes a change in the *opposite* direction of the second variables. (For example, more green engineering innovations can lead to reduced environmental and human health liabilities.)

In CLDs, the arrows come together to form loops, and each loop is labeled with an *R* or a *B* (Figure 7.17). R means *reinforcing*; that is, the causal relationships within the loop create exponential growth or collapse. For instance, the more fossil-fuel-based energy consumed,

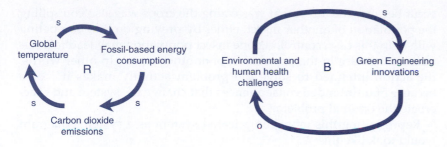

**Figure 7.17** **Examples of Reinforcing and Balancing Causal Loop Diagrams** Each arrow in a causal loop diagram is labeled with an *s* or an *o*. *s* means that when the first variable changes, the second one changes in the same direction. *o* means that the first variable causes a change in the opposite direction of the second variables. R means reinforcing, that is, the causal relationship within the loop to create exponential growth or collapse. B means balancing, that is, the casual influences in the loop keep variables in equilibrium. A shortcut to determining whether a loop is balancing or reinforcing is to count the number of "o's" in the loop. An odd number of "o's" indicates a balancing loop (i.e., an odd number of U-turns keeps you headed in the opposite direction); an even number or no "o's" means it is a reinforcing loop. (redrawn with permission of Daniel Aronson).

the more carbon dioxide that is emitted, the more global temperatures increase, and the more energy that needs to be consumed. B means *balancing*; that is, the causal influences in the loop keep variables in equilibrium. For example, the more profits generated by a company, the more research and development investments that can be made, which will lead to more green engineering innovations, reducing the number of environmental and human health liabilities, which leads to greater potential for profits.

CLDs can contain many different R and B loops, all connected with arrows. Drawing these diagrams can develop a deep understanding of the system dynamics. Through this process, opportunities for improvements will be highlighted. For example, the links between energy use, carbon emissions, and global temperatures may lead us to find ways to reduce energy consumption.

An example that illustrates the difference between systems thinking and the perspective taken by traditional forms of analysis is the action taken to reduce crop damage by insects (described by Aronson, 2003). When an insect is eating a crop, the conventional response is to spray the crop with a pesticide designed to kill that insect. Assume this pesticide was designed using green chemistry, so it is hazardous only to the target pest and is not hazardous to human health or the environment. This can be represented by a systems diagram such as the following:

$$\text{Pesticide application} \xrightarrow{\;\;O\;\;} \text{Insects damaging crops}$$

In this systems diagram, the arrow indicates the direction of causation—in this case, a change in the amount of pesticide applied causes a change in the numbers of insects damaging the crops. The letter after the arrow indicates how the two variables are related. In this case, the *o* means they change in the opposite direction—pesticide use goes up, and insect numbers go down.

The systems diagram is read, "A change in the amount of pesticide applied causes the number of insects damaging crops to change in the opposite direction." The more pesticide applied, the fewer insects there will be damaging crops (and the less total crop damage).

Taking a systems perspective of the situation, one may realize that while adding increasing amounts of pesticide does limit crop damage by insects, this is often not a long-term solution. What frequently happens instead is that, in subsequent years, the problem of crop damage increases, and the pesticide no longer appears effective. This can often

**Pesticides in U.S. Waters**
http://water.usgs.gov/nawqa/

occur because the insect that was eating the crops was also controlling the population of another insect, either by preying on it or competing with it. In this case, controlling one insect population may lead to a significant increase in the population of another insect. In other words, the action intended to solve the problem actually makes it worse because of **unintended consequences** that change the system and exacerbate the original problem.

Representing this more complicated system as a systems diagram would look like this:

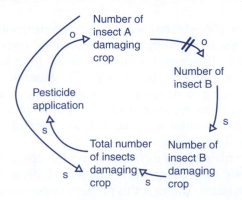

According to this diagram, the greater the pesticide application, the smaller the numbers of insect A (the original pest) that will eat the crop. This leads to an immediate decrease in the numbers of insects eating the crop (the intended effect). However, the smaller numbers of insect A eventually lead to greater numbers of insect B (the hash marks on the arrow indicate a delay), because insect A is no longer controlling the number of insect B to the same extent. This leads to a population explosion of insect B, to greater numbers of insect B damaging crops, and to greater numbers of insects damaging the crop—exactly the opposite of what was intended. Thus, although the short-term effects of applying the pesticide were exactly what were intended, the long-term effects are quite different.

With this picture of the system in mind, other actions with better long-term results have been developed, such as integrated pest management, which includes controlling the insect eating the crops by introducing more of its predators into the area.

© Thomasz Pietryszck/iStockphoto.

### 7.4.4 RESILIENCE

Another fundamental concept in designing for sustainability is the concept of **resilience**—the capacity of a system to survive, adapt, and grow in the face of unforeseen changes, even catastrophic incidents (Fiksel, 2003). Resilience is a common feature of complex systems, such as companies, cities, or ecosystems. These systems perpetually evolve through cycles of growth, accumulation, crisis, and renewal, and often self-organize into unexpected new configurations.

By the laws of thermodynamics, *closed systems* will gradually decay from order into chaos, tending toward maximum entropy. However, living systems are *open* in the sense that they continually draw upon

Provide an example of a distributed system composed of independent yet interactive elements that may deliver improved functionality and greater resilience. What are the potential benefits in terms of sustainability?

solution

A collection of distributed electric generators (for instance, fuel cells) connected to a power grid may be more reliable and fault-tolerant than centralized power generation (Fiksel, 2003). The sustainability benefits may include the following:

- Reduced resources necessary for transmission and distribution

- Reduced losses due to long-distance transmission and distribution, so less total energy needs to be generated to provide the same amount to the end user

- Possible credit given to owner for net reductions in area emissions

- Lower overall emissions if distributed energy source is cleaner than alternative (for example, fuel cells, landfill gas recovery, biomass)

- Potential for reduced emissions by producing energy only to meet current demand (much more flexibility in production levels with distributed systems)

external sources of energy and maintain a stable state of low entropy (Schrödinger, 1943). This enables resilient systems to withstand large perturbations without failure or collapse. That is, these systems are sustainable in terms of long-term survival and can adapt and evolve to a new equilibrium state. Given the uncertainty and vulnerability around sustainability challenges such as climate change, water scarcity, and energy demands, sustainable designs likely will need to incorporate resilience as a fundamental concept.

The idea of designing engineered systems for resilience would be to introduce more distributed and/or smaller systems that can continue to effectively function in uncertain situations with greater resilience. Examples include power generation and rainwater collection at the household and community level and decentralized wastewater treatment. Again, it is necessary to consider the life cycle impacts of the entire system when designing a new, distributed system with more redundancy to replace a more centralized system to understand the potential trade-offs between environmental and human health impacts for resilience. This is where the lifetime of the system becomes a crucial factor in the life cycle assessment.

### 7.4.5 DESIGN CRITERIA

Performance criteria are explicit goals that a design must achieve in order to be successful. These are often the minimum criteria used to define the design standards. For engineered designs, these criteria often include discussions of performance, safety, quality, and cost. For example, if a design does not perform its intended function, it does not matter if it is safe, of high quality, or inexpensive. The design has failed because it has not met one of the explicit minimum standards—that is, performance.

The success of engineering design strategies such as **total quality management** and **Six Sigma** that have explicit criteria for success related to zero defects or high efficiency suggests that sustainability goals can be treated similarly. For example, the design criteria for any new design may include "reduce or eliminate hazard to human health and the environment" or "maximize resilience." In this way, sustainability goals can be incorporated into design evaluation as a minimum standard for bringing the design to production or implementation. Treating sustainability goals as design criteria ensures that only the design solution(s) that successfully address these issues will be realized.

This ensures that instead of recognizing and addressing sustainability objectives after the design has been completed, engineers and managers make them an inherent part of the design conception process. A viable design is defined as one that considers sustainability goals. This prevents designs that are inherently unsustainable at the design stage from moving forward, since they do not meet the minimum standards, making sustainability analogous to traditional criteria for engineering design.

It is critical that green designs perform. That is, they need to serve their intended use as well as, if not better than, the conventional design to meet the intended function. If a design cannot compete in the marketplace based on performance or cost, it cannot provide any environmental or human health benefits because its potential will not be realized.

### 7.4.6 PRODUCTS VERSUS SERVICES

Creating physical entities to perform intended functions necessarily has an environmental and economic burden. One significant mechanism to reduce these burdens is to provide the same service or function without the creation of that physical entity. This implies that designs must be defined in terms of their function rather than the form they provide. This concept is closely related to the idea of maximizing the degrees of design freedom.

By designing for intended function rather than a prescribed physical form, organizations can realize life cycle benefits by eliminating the need to acquire and manufacture raw materials, produce a final product or process, and then manage it at the end-of-life stage. For example, rather than designing the physical infrastructure for telecommunications, including poles and wires that must be established to connect throughout the built environment, the design goal could be to provide high-quality, resilient telecommunications. This allows the designer to develop solutions such as cellular phones, which require many fewer resources (natural and economic) in terms of physical infrastructure

## example/7.12 Maintaining Function while Reducing Resource Consumption

Provide an example of a product being replaced by a service that reduces or eliminates consumption of resources (natural and economic) while providing the same function (Beckman, 2006).

## solution

Our example will investigate the decaffeination of coffee. Three processes are examined, as shown in Table 7.9. These three processes have different environmental impacts, detailed in the table.

### Table / 7.9

**Three Processes Examined to Decaffeinate Coffee**

| | | |
|---|---|---|
| Process 1 |  | Producing decaffeinated coffee by using methylene chloride as a solvent in the process: Methylene choride may be carcinogenic, as it has been linked to cancer of the lungs, liver, and pancreas in laboratory animals. Methylene chloride is also a mutagen/teratogen and crosses the placenta, causing fetal toxicity in women who are exposed to it during pregnancy (Bell et al., 1991). |
| Process 2 |  | Producing decaffeinated coffee by using supercritical carbon dioxide as the solvent: Supercritical carbon dioxide ($sCO_2$) is not regulated as a solvent by the Food and Drug Administration and is thought to be nontoxic. However, $sCO_2$ still requires significant amounts of energy to raise the temperature and pressure of the gas to where it is in a supercritical state. |
| Process 3 |  | Producing decaffeinated coffee by growing coffee plants with reduced caffeine content: This eliminates the need to transport and process coffee beans to achieve the desired function (decaffeination). Specifying the goal as "producing coffee without caffeine," rather than "designing a safer solvent system for decaffeination," allows more design degrees of freedom, and the maximum environmental and economic benefits can be realized by eliminating the use of solvents in achieving the desired function. |

while meeting the same intended goal. Of course, it is possible to evaluate the environmental and economic costs, benefits, and trade-offs of developing a telecommunications system based on poles and wires versus cellular telephones, including manufacturing and handling each individual handset at end of life, using life cycle assessment and life cycle costing.

### 7.4.7 INHERENTLY BENIGN MATERIALS AND ENERGY THROUGH GREEN CHEMISTRY

Green chemistry is devoted to the design of chemical products and processes that reduce or eliminate the use and generation of hazardous

## Nature Inspired Design

www.asknature.org

**Biomimicry**

(L) iStockphoto. (R) © Chanyut Sribua-rawd/ iStockphoto.

materials (Anastas and Warner, 1998). Green chemistry is devoted to addressing hazard through molecular design and the processes used to synthesize those molecules. Remember that hazard is defined extremely broadly (refer to Table 6.1 in Chapter 6) to include not only toxicity, but also impacts related to resource depletion and global-warming potential. Therefore, green chemistry would encourage the use of renewable resources and locally available materials (to reduce carbon dioxide emissions associated with transportation).

This emerging field also uses the lessons and processes of nature to inspire design through biomimicry (Benyus, 2002). **Biomimicry** (from bios, meaning life, and mimesis, meaning to imitate) is a design discipline that studies nature's best ideas and then imitates these designs and processes to solve human problems (see Box 7.3). Studying a leaf to invent a better solar cell is an example of this "innovation inspired by nature" (Benyus, 2002).

---

### Box / 7.3   Biomimicry

Biomimicry uses designs from nature to solve human problems. Roughly, we can distinguish three biological levels in biology after which technology can be modeled:

- Mimicking natural methods of manufacture of chemical compounds to create new ones

- Imitating mechanisms found in nature (for example, Velcro)

- Studying organizational principles from social behavior of organisms, such as the flocking behavior of birds or the emergent behavior of bees and ants

The examples are inspiring (from *Biomimicry: Innovation Inspired by Nature*, Janine M. Benyus, with permission of HarperCollins Publishers):

*Color without paint:* Packaging's final coat of paint or ink can have significant environmental impacts. Organisms use two methods to create color without paint: internal pigments and the structural color that makes tropical butterflies, peacocks, and hummingbirds so gorgeous. A peacock is a completely brown bird. Its "colors" result from light scattering off regularly spaced melanin rods, and interference effects through thin layers of keratin (the same stuff as your fingernails). What if packages could be dipped in something clear and nontoxic that played with light to create the illusion of color?

*Iridigm,* in San Francisco, is using structural color ideas from tropical butterflies to create a PDA screen that can be easily read in sunlight. In Japan, researchers are developing a liquid crystal display sign whose structure can be set using UV light, then reset for a different message, all without ink.

*Protection from microbes:* To protect against microbial squatters, a biomimic would look for clues in the skins of organisms that manage to keep themselves slime-free. Red and green algae (kelp) are able to stabilize a normally reactive compound called bromine to fend off microbes without harming the alga. Nalco engineer William McCoy borrowed this stabilization recipe to create Stabrex™, a chlorine alternative that keeps industrial cooling systems microbe-free.

*Staying clean without detergent:* If the goal is to keep packaging clean, inspiration is hiding in the microscopic surfaces of leaves. Plants like the swampland lotus can't afford to have dirt interfere with its interaction with sunlight. Using powerful scopes, German scientists found that lotus leaves have mountainous surfaces that keep dirt particles teetering on the peaks rather than adhering. Rainwater balls up and rolls the loose particles away. A number of new products are available in self-cleaning lotus-effect surfaces, including roof shingles, car paint, and a building façade paint, Lotusan, made by ispo. The paint dries with lotus-like bumps, and rainwater cleans the surface.

---

A few of the promising examples of inherently benign substances in the marketplace or under development include ethyl lactate solvents and plant-based coagulants for drinking-water treatment. Millions of pounds of toxic industrial solvents may be able to be replaced by environmentally friendly solvents made with ethyl lactate, an ester of lactic acid. Unlike other solvents, which may raise human health concerns, damage the ozone layer, or pollute groundwater, ethyl lactate is considered significantly more benign. Lactate ester solvents have numerous attractive environmental advantages, including being biodegradable, easy to recycle, noncorrosive, noncarcinogenic, and non–ozone depleting (Henneberry, 2002). As further evidence of the benign characteristics of this chemical, ester lactate is approved by the U.S. Food and Drug Administration for use in food products. Most importantly, ethyl lactate solvents are cost- and performance-competitive with traditional solvents in a wide range of applications.

In the case of water treatment for the developing world, technologies from the developed world often are not appropriate, low-cost, or based on local resources—in other words, they are unsustainable. For example, these technologies often depend on chlorination and metal salt coagulants (such as alum), which pose challenges throughout their life cycles, including ecotoxicological impacts when introduced into the environment as post-treatment sludge and damage to human health as a result of consumption in treated water. Plant-based coagulants, often traditional food crops, are bio-based and renewable, require only local materials and resources, are relatively low-cost, and can perform as well as, if not better than, conventional coagulants at removing turbidity and microorganisms, including pathogens, from drinking water. However, this approach does not address the need for residual protection against microbial growth during storage.

## 7.4.8 EFFICIENCY

While efficiencies of all types have always been a component of good design, the understanding of what can be accomplished continues to evolve. Designers no longer consider mass and energy to be the only goals for efficient use; now they also consider space and time. This can be achieved through intensification of products and processes into smaller, more distributed components.

Perhaps the most visible and successful embracing of this principle through the years has been in electronics products. A few decades ago, computers were the size of large rooms. Now computers with as much processing power fit into devices the size of a deck of cards or smaller. However, it is important to remember the overarching caveat and fundamental concept of a life cycle perspective.

Waste generation can be considered a design flaw, and one way to measure it for materials is in terms of an E factor. **E factor** is defined (Sheldon, 1992) to measure the efficiency of various chemical industries in terms of kilograms of material inputs relative to the kilograms of final product. E factor does not consider chemicals and materials that are not directly involved in the reaction, such as

## example/7.13 Determining the E Factor

Calculate the E factor for the desired product, given the following chemical production process:

$$CH_3CH_2CH_2CH_2OH + NaBr + H_2SO_4 \rightarrow$$
$$CH_3CH_2CH_2CH_2Br + NaHSO_4 + H_2O$$

Table 7.10 provides details about the molecules involved.

### Table / 7.10

**Information Needed for Example 7.13**

| Type | Molecular Formula | Molecular Weight | Weight (g) | Moles |
|---|---|---|---|---|
| Reactant | $CH_3CH_2CH_2CH_2OH$ | 74.12 | 0.8 (added) | 0.80 (added) |
| Reactant | NaBr | 102.91 | 1.33 (added) | 1.33 (added) |
| Reactant | $H_2SO_4$ | 98.08 | 2.0 (added) | 2.0 (added) |
| Desired product | $CH_3CH_2CH_2CH_2Br$ | 137.03 | 1.48 | 0.011 |
| Auxiliary | $NaHSO_4$ | | | |
| Auxiliary | $H_2O$ | | | |

## solution

$$E \text{ factor} = \frac{\Sigma \text{ kg inputs}}{\Sigma \text{ kg product}}$$

$$E \text{ factor} = \frac{0.0008 + 0.00133 + 0.002}{0.00148} = 2.8$$

In this example, 2.8 times more mass of material inputs are required than are obtained in the final product.

Be aware that this type of calculation is only a measure of mass efficiency and does not consider the toxicity of the materials used or generated.

solvents and rinse water. An E factor is determined by the following equation:

$$E \text{ factor} = \frac{\Sigma \text{ kg inputs}}{\Sigma \text{ kg product}} \tag{7.6}$$

According to Sheldon, bulk chemicals have an E factor of less than 1 to 5, compared with 5 to greater than 50 for fine chemicals, and 25 to more than 100 for pharmaceuticals.

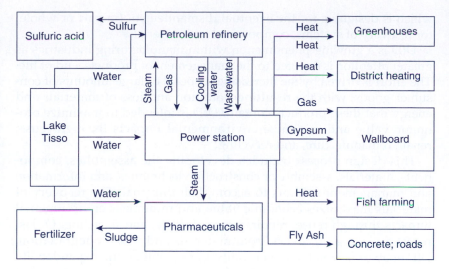

**Figure 7.18** Schematic Diagram of Eco-industrial Park in Kalundborg, Denmark, Depicting Material and Energy Flows  Here industries and commercial applications are interconnected so that local materials that are the waste of one process become a value-added feedstock for a nearby process.

From Wernick (1997).

## 7.4.9 INTEGRATION OF LOCAL MATERIAL AND ENERGY FLOWS

The green engineering principle of integrating material and energy flows is related to earlier discussions of industrial ecology and ecosystem types. While it is important to consider closed-loop systems where waste equals food (type 3), it is not necessary to always replicate this concept within a single process or even a single facility. Any design is implemented within a context of material and energy flows, some of which may exist beyond the facility but still within the local or regional community.

There is, perhaps, no better current example of a large-scale cross-process design for integration of material and energy flows than the eco-industrial park located in Kalundborg, Denmark (Figure 7.18). This arrangement represents the manifestation of industrial ecology whereby entire industries and commercial applications are interconnected so that local materials that are the waste of one process become a value-added feedstock for a nearby process. In this case, notice that what would normally be considered an environmental emission and liability—$SO_2$—is now a value-added feedstock for manufacture of gypsum ($CaSO_4$) wallboard. The same is true for fly ash generated by the power station, which is sold to a cement manufacturer. Similarly, sludge produced from pharmaceutical manufacturing, normally a disposal cost and environmental burden, is sold to a fertilizer company as value-added feedstock. The same is true with energy flows that are shared beneficially between various industrial and residential sectors of the community.

## 7.4.10 DESIGN FOR END-OF-LIFE HANDLING

It is imperative to consider the end of life of a product or system, including buildings, at the design phase, since choices are made about assembly structures, fasteners, and the number and type of materials. One strategy to address this issue is **design for disassembly (DfD)**,

**Class Discussion**

By utilizing readily available materials and energy and integrating them into the process or system, a designer can increase overall system efficiency, reduce costs by using waste as a feedstock rather than virgin material, and reduce impacts on human health and the environment. Does this type of waste–feedstock sharing arrangement have environmental benefits? What about economic benefits? The answer to both questions is yes. Think of an example, and discuss the environmental, social, and economic benefits.

which is designing for the eventual dismantlement (in part or whole) for recovery of systems, components, and materials.

DfD is a growing phenomenon within manufacturing industries as greater attention is devoted to the management of products' end of life. The effort is driven by the increasing disposal of large amounts of consumer goods, with the resulting pollution and loss of materials and energy that these products contain. DfD is intended to maximize economic value and minimize environmental impacts through reuse, repair, remanufacture, and recycling.

This design process includes developing the assemblies, components, materials, assembly or construction techniques, and information and management systems to accomplish this goal. The recovery of materials maximizes economic value and minimizes environmental impacts through reuse, repair, remanufacture, and recycling. Of last resort are energy recovery from materials and safe biodegradation. DfD enables flexibility, convertibility, upgradability, and expandability supporting efforts to avoid the disposal of products or removal of buildings.

## 7.5  Measuring Sustainability

An **indicator**, in general, is something that points to an issue or condition. Its purpose is to show you how well a system is working. If there is a problem, an indicator can help you determine what direction to take to address the issue. Indicators are as varied as the types of systems they monitor. However, as listed in Table 7.11, effective indicators have certain characteristics in common.

An example of an indicator is the gas gauge in your car. The gas gauge shows you how much gasoline is left in your car. If the gauge shows the tank is almost empty, you know it's time to fill up. Another example of an indicator is a midterm report card. It shows a student and instructor whether they are doing well enough to go to the next

### Table / 7.11

**Characteristics of Effective Indicators**

- Effective indicators are relevant; they show you something about the system that you need to know.

- Effective indicators are easy to understand, even by people who are not experts.

- Effective indicators are reliable; you can trust the information that the indicator is providing.

- Effective indicators are quantifiable; you can numerically measure the information the indicator is tracking.

- Effective indicators are based on accessible data; the information is available or can be gathered while there is still time to act.

SOURCE: With permission of Community Indicators Consortium (www.communityindicators.net).

grade or if extra help is needed. Both of these indicators provide information to help prevent or solve problems, hopefully before they become too severe.

An example of a common one-dimensional indicator of economic progress is gross domestic product (GDP). Note, however that many argue that GDP is insufficient to be used as a sustainability indicator because it measures economic productivity in areas that would not be considered in a vision of a more sustainable world—for example, prisons, pollution control, and cancer treatment.

While the principles of green engineering provide a framework for designers, many engaged in sustainability efforts also develop metrics or indicators to monitor their progress in meeting sustainability goals. A sustainability indicator (SI) measures the progress toward achieving a goal of sustainability. Sustainability indicators should be a collection of indictors that represent the multidimensional nature of sustainability, considering environmental, social, and economic facets. In terms of campus sustainability indicators, the University Leaders for a Sustainable Future states:

> Sustainability implies that the critical activities of a higher education institution are (at a minimum) ecologically sound, socially just and economically viable, and that they will continue to be so for future generations (ULSF, 2008).

Table 7.12 compares traditional versus sustainability indicators for a community, including what new information sustainability indicators provide about progress toward sustainability that is not captured by more traditional indicators.

**National Transportation Statistics: Transportation, Energy, Environment**
http://www.bts.gov/

**Sustainable Seattle**
http://www.sustainableseattle.org

**University Leaders for a Sustainable Future**
http://www.ulsf.org/

## 7.6 Policies Driving Green Engineering and Sustainability

There is a close, albeit often unrecognized, link between policy and engineering design. Policies are often aimed at protecting the public good in much the same way that green chemistry and green engineering are aimed at protecting human health and the environment. Policy can be a powerful driver influencing engineering design in terms of which material and energy sources are used through subsidies and/or strict regulations on emissions. In this way, policy can play a significant role in supporting engineering design for sustainability. Two main types of policies can affect design at this scale: regulations and voluntary programs.

### 7.6.1 REGULATIONS

A **regulation** is a legal restriction promulgated by government administrative agencies through rulemaking supported by a threat of sanction or a fine. While traditional environmental regulations focused on end-of-pipe releases, there is an emerging policy area focused on sustainable design. Two of the most established examples include **extended product responsibility (EPR)** initiatives and banning of specific substances.

## Table / 7.12

**Traditional Indicators versus Sustainability Indicators for a Community and What They Say about Sustainability**

| | | |
|---|---|---|
| **Economic Indicators** | **Traditional** | Median income<br>Per capita income relative to the U.S. average<br>Size of the economy as measured by GNP and GDP |
| | **Sustainable** | Number of hours of paid employment at the average wage required to support basic needs<br>Wages paid in the local economy that are spent in the local economy<br>Dollars spent in the local economy that pay for local labor and local natural resources<br>Percent of local economy based on renewable local resources |
| | **Emphasis of Sustainability Indicator** | What wage can buy<br>Defines basic needs in terms of sustainable consumption<br>Local financial resilience |
| **Environmental Indicators** | **Traditional** | Ambient levels of pollution in air and water<br>Tons of solid waste generated<br>Cost of fuel |
| | **Sustainable** | Use and generation of toxic materials (both in production and by end user)<br>Vehicle miles traveled<br>Percent of products produced that are durable, repairable, or readily recyclable or compostable<br>Total energy used from all sources<br>Ratio of renewable energy used at renewable rate to nonrenewable energy |
| | **Emphasis of Sustainability Indicator** | Measuring activities causing pollution<br>Conservative and cyclical use of materials<br>Use of resources at sustainable rate |
| **Social Indicators** | **Traditional** | Number of registered voters<br>SAT and other standardized-test scores |
| | **Sustainable** | Number of voters who vote in elections<br>Number of voters who attend town meetings<br>Number of students trained for jobs that are available in the local economy<br>Number of students who go to college and come back to the community |
| | **Emphasis of Sustainability Indicator** | Participation in democratic process<br>Ability to participate in the democratic process<br>Matching job skills and training to needs of the local economy |

SOURCE: Hart, 2007.

EPR, such as the European Union's (EU) Waste Electrical and Electronic Equipment (WEEE) directive, holds the original manufacturers responsible for their products throughout the life cycle. The WEEE directive aims to minimize the impact of electrical and electronic goods on the environment by increasing reuse and recycling and by reducing the amount of electrical and electronic equipment going to landfills. It seeks to achieve this by making producers responsible for financing the collection, treatment, and recovery of waste electrical equipment, and by obliging distributors (sellers) to allow consumers to return their waste equipment free of charge. This drives engineers to design electrical and electronic equipment with the principles of green engineering in mind. For example, these designs aim for ease of disassembly, recovery of complex components, and minimized material diversity. One positive impact of this approach from a company's perspective is that it reconnects the consumer with the manufacturer at the end-of-life stage of the life cycle.

Another policy approach to driving engineering design toward sustainability goals is to ban specific substances of concern. An example closely tied to the WEEE directive is the EU's **Restriction of Hazardous Substances (RoHS).** RoHS is focused on "the restriction of the use of certain hazardous substances in electrical and electronic equipment." This directive bans the placing on the EU market of new electrical and electronic equipment containing more than agreed levels of lead, cadmium, mercury, hexavalent chromium, polybrominated biphenyl (PBB), and polybrominated diphenyl ether (PBDE) flame retardants. By banning these chemicals of concern in significant levels, this directive is driving the implementation of green chemistry and green engineering principles in terms of designing alternative chemicals and materials that reduce or eliminate the use and generation of hazardous substances and prevent pollution.

**European Commission Initiatives**
http://ec.europa.eu/environment/

### 7.6.2 VOLUNTARY PROGRAMS

Another policy strategy for encouraging green engineering design is the establishment of **voluntary programs**. These programs are not mandated by law or enforceable but are meant to encourage and motivate desirable behaviors. The government, industry, or third-party nongovernmental organizations can sponsor these programs. While there are many different varieties of voluntary programs, two types that have been established with success are ecolabeling and preferential purchasing.

Environmental standards allow for an environmental assessment of a product's impact on such factors as air pollution, wildlife habitat, energy, natural resources, ozone depletion and global warming, and toxic contamination. Companies that meet environmental standards for their specific product or service can apply an **ecolabel**. Ecolabels attempt to provide an indicator to consumers of the product's environmental performance (for example, "recycled packaging" or "no toxic emissions"). Independent third parties, such as Green Seal, United States Green Buildings Council, and EnergyStar, provide unbiased verification of environmental labels and certifications, so these labels are

**EnergyStar**
http://www.energystar.gov/

**Green Seal**
http://www.greenseal.org/

**Green Buildings**
http://www.usgbc.org

the most reliable. First-party ecolabels are self-awarded, so they aren't independently verified. In the United States, these sorts of labels are governed by the Federal Trade Commission's (FTC's) guide for the use of environmental marketing claims and must be accurate. The FTC has brought action against several manufacturers for violating truth-in-advertising laws.

To further support these programs, many organizations are implementing **environmentally preferable or preferential purchasing** policies. These policies can be implemented by any organization (even your college or university) and mandate a preference to purchase products from office supplies to computers to industrial chemicals with improved environmental and human health profiles. By specifying purchases of this type, organizations are creating a demand in the marketplace for products and services with reduced impacts on human health and the environment, a very powerful tool to drive innovation in this area and to reduce costs of these products through economies of scale.

Table 7.13 provides reasons why companies adopt environmental purchasing policies and the types of savings they achieve as a result. Many companies adopted environmental purchasing policies for traditional business reasons, as described in Table 7.13. Although these reasons result in intangible benefits, there are specific examples of measurable reduced costs associated with environmentally preferable products. These include a lower purchase price (as in the case of remanufactured products), reduced operational costs (for example, through energy efficiency), reduced disposal costs (for example, more durable products), and reduced hazardous management costs (for example, less toxic products). In addition, purchasing environmentally preferable products may reduce an organization's potential future liability, improve the work environment, and minimize risks to workers.

## Table / 7.13

### Why Companies Adopt Environmental Purchasing Policies and What They Save

| Business Reasons for Adopting Environmental Purchasing Policies | Types of Savings Achieved through Environmentally Preferable Purchasing |
| --- | --- |
| Distinguishing a company and its products from competitors | Reduced repair and replacement costs when using more durable and repairable equipment |
| Avoiding hidden costs and pursuing cost savings | Reduced disposal costs from generating less waste |
| Increasing operating efficiency | Improved product design and performance of the product(s) |
| Joining an industry or international market trend | Increased employee safety and health at the facility |
| Recognizing market preferences; serving customers who have a stated interest in "environmentally friendly" products and practices | Reduced material costs for manufacturers |

SOURCE: EPA, 1999.

## 7.7 Designing a Sustainable Future

By applying the principles of green engineering and considering the fundamental concepts of sustainability, engineers can contribute to addressing the challenges traditionally associated with economic growth and development. This new awareness provides the potential to design a better tomorrow—one where our products, processes, and systems are more sustainable, including being inherently benign to human health and the environment, minimizing material and energy use, and considering the entire life cycle.

## Key Terms

- biomimicry
- causal loop diagram (CLD) design
- design for disassembly (DfD)
- design for environment (DFE)
- disposal
- eco-design
- eco-efficiency
- ecolabel
- E factor
- environmentally preferable or preferential purchasing
- extended product responsibility
- function
- functional unit
- future value (FV)
- future value of an annuity (FVA)

- green chemistry
- green engineering
- impact assessment
- indicator
- industrial ecology
- inherency
- intangibles
- interest
- inventory analysis
- life cycle
- life cycle assessment (LCA)
- life cycle cost (LCC)
- pollution prevention
- pollution prevention hierarchy
- present value (PV)

- present value of an annuity (PVA)
- principles of green engineering
- recycling
- regulation
- resilience
- Restriction of Hazardous Substances
- risk
- Six Sigma
- source reduction
- systems thinking
- total quality management
- treatment
- unintended consequences
- voluntary program
- waste minimization

**7.1** The expenses of owning and operating a traditional wastewater treatment facility are too high for the local community. The community is seeking ideas to address this issue. What are the design opportunities for improvement scaled with increasing degrees of freedom, and what are the potential benefits?

**7.2** The teaching and research laboratories at your school are found to be out of compliance with EPA regulations. The EPA would like a plan stating how the labs can come within the regulations. Based on the pollution prevention hierarchy, provide at least one action at each level that your university can take with regard to waste in the research labs. What are the advantages and disadvantages of each action?

**7.3** Visit EPA's Presidential Green Chemistry Challenge Award Web site at www.epa.gov/greenchemistry/pubs/pgcc/past.html. Select a past award-winning project. Based on the description of this project, what are the environmental, economic, and social benefits of this green chemistry advance?

**7.4** Discuss whether shoe A (leather) or shoe B (synthetic) is better for the environment, based on data in Table 7.14. Is it possible to weight one aspect (air, water, land pollution, or solid waste) as being more important than another? How? Why? Who makes these decisions in our society?

**7.5** You are preparing a life cycle assessment of different transportation options for getting from your house to work (10 miles each way). The options include bicycling, one person in a car, carpooling with three or more people, or taking the bus. Write a possible goal, scope, function, and functional unit for this LCA.

**7.6** Draw a causal loop diagram that links energy consumption, air emissions, climate change, water availability, Earth's temperature, and human health/quality of life.

**7.7** Draw and then explain in words a causal loop diagram for the following system elements: GDP, population, resource consumption, environmental quality, and health. Then redraw the diagram to include green engineering.

**7.8** Is centralized drinking-water treatment and distribution more or less resilient than point of use water treatment technologies? Why or why not? Does it matter whether these water treatment systems are implemented in the developed or developing world?

**7.9** Provide an example of a product's replacement by a service that reduces or eliminates consumption of (natural and economic) resources while providing the same function. What benefits are associated with the evolution from product to service for this example?

**7.10** You are about to buy a car that will last 7 years before you have to buy a new one, and Congress has just passed a new tax on greenhouse gases. Assume a 5% annual interest rate. You have two options: (a) Purchase a used car for $12,000, upgrade the catalytic converter at a cost of $1,000, and pay a $500 annual carbon tax. This car has a salvage value of $2,000. (b) Purchase a new car for $16,500 and pay only $100 annually in carbon tax. This car has a salvage value of $4,500. Based on the annualized cost of these two options, which car would you buy?

**7.11** Provide an example of a product either commercially available or currently under development that uses biomimicry as the basis for its design. Explain how the design

## Table / 7.14

**Hypothetical Example of Life Cycle Environmental Impacts of Shoes per 100 Pairs of Shoes Produced**

| Product | Energy Use (Btu) | Raw Material Consumption | Water Use (gal.) | Air Pollution (lb.) | Water Pollution | Hazardous and Solid Waste |
|---|---|---|---|---|---|---|
| Shoe A (leather) | 1 | Limited supply, some renewable | 2 | 4 | 2 lb. organic chemicals | 2 lb. hazardous sludge |
| Shoe B (synthetic) | 2 | Large supply, nonrenewable | 4 | 1 | 8 lb. inert inorganic chemicals | 1 lb. hazardous sludge<br>3 lb. nonhazardous solid waste |

is mimicking a product, process, or system found in nature.

**7.12**  Two reactants, benzyl alcohol and tosyl chloride, react in the presence of an auxiliary, triethylamine, and the solvent toluene to produce the product sulfonate ester (see Table 7.15). Calculate the E factor for the reaction. What would happen to the E factor if the solvents and auxiliary chemicals were included in the calculation? Should these types of materials and chemicals be included in an efficiency measure? Why or why not?

## Table / 7.15

### Information Useful for Problem 7.12

| Reactant | Benzyl alcohol | 10.81 g | 0.10 mole | MW 108.1 g/mole |
|---|---|---|---|---|
| Reactant | Tosyl chloride | 21.9 g | 0.115 mole | MW 190.65 g/mole |
| Solvent | Toluene | 500 g | | |
| Auxiliary | Triethylamine | 15 g | | MW 101 g/mole |
| Product | Sulfonate ester | 23.6 g | 0.09 mole | MW 262.29 g/mole |

**7.13**  Choose three of the principles of green engineering (Box 7.1). For each one, (a) explain the principle in your own words; (b) find an example (commercially available or under development), and explain how it demonstrates the principle; and (c) describe the associated environmental, economic, and societal benefits, identifying which ones are tangible and which ones are intangible.

**7.14**  Develop five sustainability metrics or indicators for a corporation or industrial sector analagous to those presented for communities in Table 7.12. Compare them with traditional business metrics or indicators. Describe what new information can be determined from the new sustainability metrics or indicators.

**7.15**  A car company has developed a new car, ecoCar, that gets 100 miles per gallon (mpg), but the cost is slightly higher than cars currently available on the market. What type of incentives could the manufacturer offer or ask Congress to implement to encourage customers to buy the new ecoCar?

**7.16**  The design team for a building project was formed at your company last week and they have already had 2 meetings. Why is it so important for you to get involved immediately in the design process?

**7.17**  EcoStar needs to determine how much to charge for the new greenCar model. Developing, producing, and marketing the greenCar costs EcoStar $100,000,000 per year for three years befor it goes to market. Once it goes on the market, EcoStar will not spend any additional money on the greenCar. EcoStar would like to make their investment in greenCar back 3 years after it goes on the market. If EcoStar anticipates selling 25,000 greenCars per year for those 3 years, how much do they need to change per greenCar to breakeven? Assume a discount rate of 10%.

**7.18**  To compare plastic and paper bags in terms of acquisition of raw materials, manufacturing and processing, use and disposal, we'll use data provided by Franklin Associates, a nationally known consulting firm whose clients include the U.S. Environmental Protection Agency as wll as many companies and industry groups. In 1990, Franklin Associates compared plastic bags to paper bags in terms of their energy and air/water emissions in manufacture, use, and disposal. The following chart is a result of their study:

| Life Cycle Stages | Air Emissions (pollutants) oz/bag | | Energy Required BTU/bag | |
|---|---|---|---|---|
| | Paper | Plastic | Paper | Plastic |
| Materials manufacture, product manufacture, product use | 0.0516 | 0.0146 | 905 | 464 |
| Raw materials acquisition, product disposal | 0.0510 | 0.0045 | 724 | 185 |

(a) Which bag would you choose if you were most concerned about air pollution? (Notice that the information does not tell you if these are toxic air emissions or greenhouse gases.) (b) If we assume that two plastic bags equal one paper bag, does the choice change? (c) Compare the energy required to produce each bag. Which bag takes less energy to produce?

# References

Anastas, P. T., and J. C. Warner. 1998. *Green Chemistry: Theory and Practice*. Oxford: Oxford University Press.

Anastas, P. T., and J. B. Zimmerman. 2003. "Design through the Twelve Principles of Green Engineering." *Environmental Science and Technology* 37 (5): 94A–101A.

Anastas, P. T., and J. B. Zimmerman. 2006. "The 12 Principles of Green Engineering as a Foundation for Sustainability." In *Sustainability Science and Engineering: Principles Book*, ed. M. Abraham. New York: Elsevier Science.

Aronson, D. 2003. "Targeted Innovation: Using Systems Thinking to Increase the Benefits of Innovation Efforts." *R&D Innovator* (now *Innovative Leader*) 6 (2).

Beckman, E. 2006. "Using Principles of Sustainability to Design 'Leap-Frog' Chemical Products." Proceedings of the 10th Annual Green Chemistry and Engineering Conference, June 2006, Washington, D.C.

Bell, B., P. Franks, N. Hildreth, and J. Melius. 1991. "Methylene Chloride Exposure and Birthweight in Monroe County, New York." *Environ Res* 55 (1): 31–39.

Benyus, J. M. 2002. *Biomimicry: Innovation Inspired Design*. New York: Harper Perennial.

Cooper, J. S., and B. Vigon. 1999. "Life Cycle Engineering Guidelines." Prepared for U.S. Environmental Protection Agency Office of Research and Development, National Risk Management Research Laboratory, by Battelle Memorial Institute, Contract No. CR822956.

Environmental Protection Agency (EPA). 1999. "Private Sector Pioneers: How Companies Are Incorporating Environmentally Preferable Purchasing." Report No. EPA742-R-99-001.

Fiksel, J. 2003. "Designing Resilient, Sustainable Systems." *Environmental Science and Technology* 37: 5330–5339.

Graedel, T., and B. Allenby. 1995. *Industrial Ecology*. Upper Saddle River, N.J.: Prentice Hall.

Hart, M. 2007. Sustainable Measures Web site, www.sustainablemeasures.com.

Henneberry, M. 2002. *Paint and Coatings Industrial Magazine* 6.

Schrödinger, E. 1943. *What Is Life?* Dublin: Dublin Institute for Advanced Studies.

Sheldon, R. A. 1992. *Chemistry & Industry* 903.

University Leaders for a Sustainable Future (ULSF). 2008. "Sustainability Assessment Questionnaire." ULSF Web site, www.ulsf.org/programs_saq.html.

Wernick, I., and J. Ausubel. 1997. *Industrial Ecology: Some Directions for Research*. Prepared for the Office of Energy and Environmental Systems, Lawrence Livermore National Laboratory.

# chapter/Eight Water Quality

Martin T. Auer, James R. Mihelcic, Noel R. Urban, Alex S. Mayer, Michael R. Penn

*Two closely allied disciplines are associated with water: water resources engineering deals with water quantity (for example, its storage and transport), and water quality engineering is concerned with the biological, chemical, and physical nature of water. This chapter provides key concepts, principles, and calculations that support an engineered approach to water quality management. Five systems will be considered: rivers, lakes, stormwater, wetlands, and groundwater. The chapter describes a relatively new approach for managing stormwater, called low-impact development. The presentation will focus on the management issues, characteristic to a particular system, commonly encountered in engineering practice.*

## Learning Objectives

1. Apply mass balance concepts and knowledge of plug flow reactors to investigate issues related to surface water quality.

2. Determine the oxygen deficit in a river.

3. Describe features of the dissolved oxygen (DO) sag curve, and determine the location of the critical point and the oxygen concentration at the critical point for a given flow and discharge scenario.

4. Describe the process of lake and reservoir stratification, and relate it to issues of water quality.

5. Describe eight methods of engineered lake management.

6. Relate issues of population growth, urbanization, and land use to surface water and groundwater quality.

7. Define the four types of wetlands, and describe functions of wetlands.

8. Discuss the social, economic, and environmental benefits provided by wetlands.

9. Summarize how protection of biodiversity can be integrated into engineered management of urban and rural waters.

10. Describe best management practices (BMPs) for controlling urban stormwater, such as rain gardens, permeable pavements, green roofs, and bioswales.

11. Design a bioretention cell for given precipitation and design scenarios.

12. Describe how advection and dispersion influence movement of groundwater contaminants.

13. Calculate travel time of a subsurface chemical with and without chemical retardation.

14. Describe a regional, national, and global water quality challenge, and present a solution that moves society toward sustainable management of the water resource.

## 8.1 Introduction

*Pollution* may be defined as the introduction of a substance to the environment at levels leading to lost beneficial use of a resource or degradation of the health of humans, wildlife, or ecosystems. Pollutants are discharged to aquatic systems from **point sources** (stationary locations such as an effluent pipe) and from **nonpoint sources** (also called *diffuse*) such as land runoff and the atmosphere. The mass flux of a pollutant is termed its **load** and is expressed in units of mass per unit time.

Engineered approaches to pollution management vary with the type of material in question. **Macropollutants** such as nitrogen, phosphorus, organic matter, and suspended solids are discharged to the rivers of the world by the tens of millions of tons per year. Figure 8.1a shows the BOD loading (in megatons per year) to global waterways for agricultural, domestic, and industrial sectors in 1995, as well as discharges expected in 2010 and 2020 for countries that are members and nonmembers of the Organization for Economic Cooperation and Development (OECD). Figure 8.1b shows the agricultural nitrogen loading for OECD and non-OECD countries for the same time periods. Note the large contribution of the agricultural sector to global BOD and nitrogen loadings.

The **Clean Water Act** calls for the maintenance of fishable–swimmable conditions in U.S. waters. The Environmental Protection Agency (EPA)

**Visualize Florida Water Issues**

*www.wateratlas.org*

### Class Discussion

What methods will protect water ecosystems for future generations as population and urbanization increase, changes occur in land use, and population and increased affluence drive increases in demand for food and biofuels.

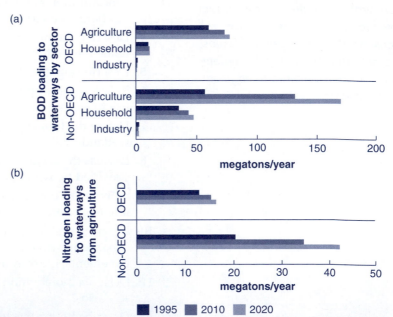

**Figure 8.1 Macropollutants in Global Waterways** (a) Annual BOD loading (megatons) into global waterways (in OECD and non-OECD countries) for agricultural, domestic, and industrial sectors for 1995 and estimated for 2010 and 2020. (b) Nitrogen loading (megatons) for OECD and non-OECD countries for agricultural sectors into global waterways for 1995 and estimated for 2010 and 2020.

Data from UNESCO, 2003.

has set standards to achieve this goal, retain beneficial uses, and protect human and ecosystem health. Some standards are technology based, requiring a particular level of treatment regardless of the condition of the receiving water. Other standards are water quality based, calling for additional treatment where conditions remain degraded following implementation of standard technologies. Under the **National Pollutant Discharge Elimination System (NPDES)**, permits are required for all those seeking to discharge effluents to surface water or groundwater. Standards may then be met by regulating the conditions of the permit, that is, the load that may be discharged. In cases where controls are not stringent enough to maintain the desired water quality, an analysis is conducted to establish the *total maximum daily load (TMDL)* that may be discharged to a water body, and permits are set accordingly.

## 8.2  River Water Quality

The treatment of river water quality in this section focuses on the management of **dissolved oxygen (DO)** in relation to the discharge of oxygen-demanding wastes. This is a classic issue in surface water quality that remains of interest today with respect to the issuance of discharge permits and the setting of total maximum daily loads for receiving waters.

### 8.2.1  DISSOLVED OXYGEN AND BOD

Dissolved oxygen is required to maintain a balanced community of organisms in lakes, rivers, and the ocean. When an oxygen-demanding waste (measured as BOD) is added to water, the rate at which oxygen is consumed in oxidizing that waste (**deoxygenation**) may exceed the rate at which oxygen is resupplied from the atmosphere (**reaeration**). This can lead to depletion of oxygen resources, with concentrations falling far below saturation levels (Figure 8.2). When oxygen levels drop below 4 to 5 mg $O_2/L$, reproduction by fish and macroinvertebrates is impaired. Oxygen depletion is often severe enough that anaerobic conditions develop, with an attendant loss of biodiversity and poor aesthetics (turbidity and odor problems). Figure 8.2 also illustrates the response of stream biota to BOD discharges.

Consideration of the fate of BOD following discharge to a river is a useful starting point for examining the impact of oxygen-demanding wastes on oxygen resources. Example 8.1 applies the concepts of the mixing basin (Chapter 4) and BOD kinetics (Chapter 5) to examining the oxidation of an organic waste following discharge to and mixing with a river.

In Example 8.1, more than 23 mg $O_2/L$ of ultimate CBOD is exerted over the 50 km stretch downstream of the discharge. To appreciate the impact of this demand on a river's oxygen resources, it is necessary to understand the capacity of water to hold oxygen (saturation) and the rate at which oxygen can be re-supplied from the atmosphere (reaeration).

**Clean Water Act**
http://www.epa.gov/regulations/laws/cwa.html

**Surf Your Watershed**
http://cfpub.epa.gov/surf/locate/index.cfm

© Galyna Andrushko/iStockphoto.

**Figure 8.2** Dissolved-Oxygen Sag Curve (a) and Associated Water Quality Zones ((b)–(d)) Reflecting Impacts on Physical Conditions and the Diversity and Abundance of Organisms.

From Mihelcic (1999). Reprinted with permission of John Wiley & Sons, Inc.

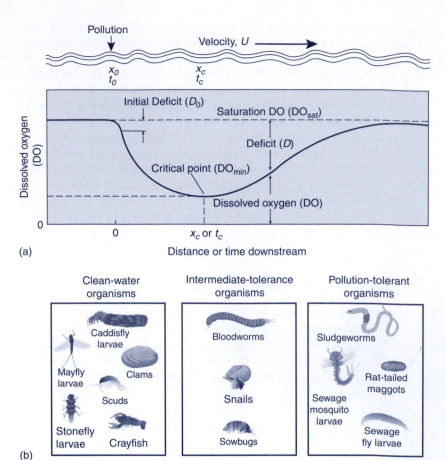

(a)

Clean-water organisms | Intermediate-tolerance organisms | Pollution-tolerant organisms

Caddisfly larvae, Mayfly larvae, Clams, Scuds, Stonefly larvae, Crayfish

Bloodworms, Snails, Sowbugs

Sludgeworms, Rat-tailed maggots, Sewage mosquito larvae, Sewage fly larvae

(b)

| | | Stream Zones | | | |
|---|---|---|---|---|---|
| | Clean water | Degradation | Damage | Recovery | Clean Water |
| Physical condition | Clear water; no bottom sludge | Floating solids; bottom sludge | Turbid water; malodorous gases; bottom sludge | Turbid water; bottom sludge | Clear water; no bottom sludge |
| Fish species | Cold or warm water game and forage fish; trout, bass, | Pollution-tolerant fish; carp, gar, buffalo | None | Pollution-tolerant fish; carp, gar, buffalo | Cold or warm water game and forage fish; trout, bass, |
| Benthic invertebrate | Clean water | Intermediate tolerance | Pollution-tolerant | Intermediate tolerance | Clean water |

(c)

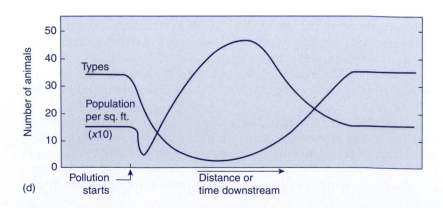

(d)

## example/8.1 Mixing Basin Calculation for CBOD

A waste with a 5-day CBOD ($y_5$) of 200 mg $O_2$/L and a $k_L$ of 0.1/day is discharged to a river at a rate of 1 $m^3$/s. Calculate the ultimate CBOD ($L_0$) of the waste before discharge to the river. Assuming instantaneous mixing after discharge, calculate the ultimate CBOD of the river water after it has received the waste. The river has a flow rate ($Q$) equal to 9 $m^3$/s and a background ultimate CBOD of 2 mg $O_2$/L upstream of the waste discharge. Also calculate the ultimate CBOD ($L_0$) and $CBOD_5$ ($y_5$) in the river 50 km downstream of the point of discharge. The river has a width ($W$) of 20 m and a depth ($H$) of 5 m.

## solution

This problem has several steps. First, calculate the ultimate CBOD of the waste before discharge:

$$L_{0,waste} = \frac{y_5}{(1 - e^{-k_L \times 5\,days})} = \frac{200\,mg\,O_2/L}{(1 - e^{-0.1/day \times 5\,days})} = 508\,mg\,O_2/L$$

Next, perform a mass balance mixing basin calculation to determine the ultimate CBOD after the waste has been discharged and mixed with the river. The general relationship for calculation of the concentration of any chemical in a mixing basin ($C_{mb}$) is

$$C_{mb} = \frac{C_{up} \times Q_{up} + C_{in} \times Q_{in}}{Q_{mb}}$$

Here the total flow, $Q_{mb}$, equals $Q_{up} + Q_{in}$, and the ultimate CBOD equals

$$L_{0,mb} = \frac{2\,mg\,O_2/L \times 9\,m^3/s + 508\,mg\,O_2/L \times 1\,m^3/s}{10\,m^3/s}$$

$$= 52.6\,mg\,O_2/L$$

This is the value of the ultimate CBOD of the river water after it has received the waste.

To answer the last two questions concerning the ultimate CBOD and 5-day CBOD 50 km downstream of discharge, first calculate the 5-day CBOD of the river water after it has received the waste:

$$y_t = L_0 \times (1 - e^{-k_L \times t})$$

$$y_{5,mb} = 52.6\,mg\,O_2/L \times (1 - e^{-0.1/day \times 5\,days})$$

$$y_{5,mb} = 20.7\,mg\,O_2/L$$

Next, calculate the ultimate CBOD 50 km downstream of the point of discharge. As the waste travels downstream, it will decay and deplete oxygen according to first-order kinetics. The river downstream of the mixing zone can be modeled as a plug flow reactor (PFR). Therefore,

$$L_t = L_0 \times e^{-k_L \times t}$$

However, the time of travel needs to be calculated. The river velocity (U) is given by

$$U = \frac{Q}{A} = \frac{Q}{W \times H} = \frac{10\,\text{m}^3/\text{s}}{20\,\text{m} \times 5\,\text{m}} = 0.1\,\text{m/s} \times \frac{86{,}400\,\text{s}}{\text{day}} \times \frac{\text{km}}{1{,}000\,\text{m}}$$

$$= 8.64\,\text{km/day}$$

Then, to determine the time of travel, divide the distance by the river velocity:

$$t = \frac{x}{U} = \frac{50\,\text{km}}{8.64\,\text{km/day}} = 5.78\,\text{days}$$

This value can then be used to determine the ultimate CBOD 5.78 days downriver:

$$L_{0,50\,\text{km}} = L_{0,\,\text{mb}} \times e^{-k_L \times t} = 52.6 \times e^{-0.1/\text{day} \times 5.78\,\text{days}} = 29.5\,\text{mg}\,O_2/\text{L}$$

and a 5-day CBOD of

$$y_t = L_0 \times (1 - e^{-k_L \times t})$$

$$y_{5,\,50\,\text{km}} = 29.5 \times (1 - e^{-0.1/\text{day} \times 5\,\text{days}}) = 11.6\,\text{mg}\,O_2/\text{L}$$

## 8.2.2 OXYGEN SATURATION

The amount of oxygen that can be dissolved in water at a given temperature (its equilibrium or **saturation concentration**) may be determined through the Henry's law constant, $K_H$:

$$\boxed{DO_{sat} = K_H \times P_{O_2}} \tag{8.1}$$

where $DO_{sat}$ is the saturation dissolved-oxygen concentration (in moles $O_2$/L), $K_H$ is the Henry's law constant ($1.36 \times 10^{-3}$ moles/L-atm at 20°C), and $P_{O_2}$ is the partial pressure of oxygen in the atmosphere (~21 percent or 0.21 atm).

The Henry's law constant varies with temperature (see Chapter 3), so the saturation concentration of dissolved oxygen varies as well. Example 8.2 illustrates the calculation of the saturation dissolved-oxygen concentration.

> ## example/8.2 Determination of Saturation Dissolved-Oxygen Concentration
>
> Determine the saturation dissolved-oxygen concentration, $DO_{sat}$, at 20°C.
>
> ## solution
>
> Determine $DO_{sat}$ from the appropriate temperature-dependent Henry's law constant and the oxygen partial pressure:
>
> $$DO_{sat} = \frac{1.36 \times 10^{-3} \, \text{mole}}{\text{L-atm}} \times 0.21 \, \text{atm} = \frac{2.85 \times 10^{-4} \, \text{mole} \, O_2}{L}$$
>
> Convert to mg $O_2$/L:
>
> $$DO_{sat} = \frac{2.85 \times 10^{-4} \, \text{mole} \, O_2}{L} \times \frac{32 \, \text{g} \, O_2}{\text{mole} \, O_2} \times \frac{1{,}000 \, \text{mg} \, O_2}{\text{g} \, O_2}$$
>
> $$= \frac{9.1 \, \text{mg} \, O_2}{L}$$
>
> Note that the phrases *dissolved-oxygen saturation concentration* and the *solubility of oxygen* are used interchangeably.

The value for $DO_{sat}$ ranges from approximately 14.6 mg $O_2$/L at 0°C to 7.6 mg $O_2$/L at 30°C. These are typical temperature extremes for natural and engineered systems. This shows why fish with high oxygen requirements are associated with colder waters and why the impacts of oxygen-demanding wastes on water quality may be greatest in the summer. In the warmer months of summer, stream flow is typically lower as well, offering less dilution of the waste. The concentration of oxygen in water also decreases as the salinity increases, which becomes important in estuarine and ocean conditions.

## 8.2.3  THE OXYGEN DEFICIT

The **oxygen deficit** ($D$, expressed in mg $O_2$/L) is defined as the departure of the ambient dissolved-oxygen concentration from saturation.

$$D = DO_{sat} - DO_{act} \qquad (8.2)$$

$DO_{act}$ is the ambient or measured dissolved-oxygen concentration (mg $O_2$/L).

Note that negative deficits may occur when ambient oxygen concentrations exceed the saturation value. This happens in lakes and rivers under quiescent, nonturbulent conditions when algae and

## example/8.3 Determining the Oxygen Deficit

Determine the dissolved-oxygen deficit, $D$, at 20°C for a river with an ambient dissolved-oxygen concentration of 5 mg $O_2$/L.

### solution

From Example 8.2, the $DO_{sat}$ at 20°C was determined to be 9.1 mg $O_2$/L. Applying Equation 8.2 yields the deficit:

$$D = 9.1 - 5 = 4.1 \, mg \, O_2/L$$

The actual DO in this case is 5 mg $O_2$/L, which is below the saturation level. Microbial oxidation of organic matter or ammonia-nitrogen in this river may be leading to oxygen depletion.

macrophytes are actively photosynthesizing and producing dissolved oxygen. This **oversaturation** is eliminated when sufficient turbulence is available—for example, due to rapids, waves, and waterfalls.

### 8.2.4 OXYGEN MASS BALANCE

Examples 8.1 to 8.3 demonstrated that BOD exertion (for example, 29.5 mg $O_2$/L over a 50 km stretch) may exceed a river's oxygen resources, even at saturation. The shortfall (oxygen present minus oxygen required) must be made up through atmospheric exchange, that is, reaeration. Where the demand by deoxygenation exceeds the supply from reaeration, oxygen levels fall, and anaerobic conditions may develop. The dynamic interplay between the oxygen source (reaeration) and sink (deoxygenation) terms can be examined through a mass balance on oxygen in the river. Deoxygenation occurs as BOD is exerted and is described by Equation 8.3. The rate of reaeration is proportional to the deficit and is described using first-order kinetics:

$$\frac{dO_2}{dt} = k_2 \times D - k_1 \times L \tag{8.3}$$

Here the *in-stream* deoxygenation rate coefficient ($k_1$, day$^{-1}$) is comparable to (and, for the purposes of this presentation, the same as) the *laboratory* or *bottle* CBOD reaction rate coefficient ($k_L$), but also includes in-stream phenomena, such as sorption and turbulence and roughness effects. The reaeration rate coefficient ($k_2$, day$^{-1}$) varies with temperature and turbulence (river velocity and depth) and ranges from ~0.1 to 1.2/day.

In practice, the mass balance is written in terms of deficit:

$$\frac{dD}{dt} = k_1 \times L - k_2 \times D \tag{8.4}$$

Note how Equation 8.4 is a simple reversal of the order of the source–sink terms presented in Equation 8.3. Equation 8.4 can be integrated, yielding an expression that describes the oxygen deficit at any location downstream of an arbitrarily established starting point, such as the point where a waste is discharged to a river:

$$D_t = \frac{k_1 \times L_0}{(k_2 - k_1)} \times (e^{-k_1 \times t} - e^{-k_2 \times t}) + D_0 \times e^{-k_2 \times t} \tag{8.5}$$

where $L_0$ is the ultimate CBOD and $D_0$ is the oxygen deficit at the starting point ($x = 0$, $t = 0$), and $D_t$ is the oxygen deficit at some downstream location ($x = x$, $t = t$). The notation $t$ refers to time of travel, defined here as the time required for a parcel of water to travel a distance $x$ downstream. Therefore, $t = x/U$, where $x$ is distance downstream and $U$ is the river velocity.

The time–distance relationship permits expression of the analytical solution for the oxygen deficit in terms of $x$, the distance downstream of the starting point:

$$D_x = \frac{k_1 \times L_0}{(k_2 - k_1)} \times \left(e^{-k_1 \times \frac{x}{U}} - e^{-k_2 \times \frac{x}{U}}\right) + D_0 \times e^{-k_2 \times \frac{x}{U}} \tag{8.6}$$

Equation 8.6 is called the **Streeter–Phelps model** and was developed in the 1920s for studies of pollution in the Ohio River.

## 8.2.5 DISSOLVED-OXYGEN SAG CURVE AND CRITICAL DISTANCE

The discharge of oxygen-demanding wastes to a river yields a characteristic response in oxygen levels that is termed the dissolved-oxygen sag curve (Figure 8.2). Figure 8.2 demonstrates that a typical dissolved-oxygen sag curve has three phases of response:

**River Dissolved Oxygen Simulator**

1. An interval where dissolved-oxygen levels fall because the rate of deoxygenation is greater than the rate of reaeration ($k_1 \times L > k_2 \times D$)

2. A minimum (termed the **critical point**) where the rates of deoxygenation and reaeration are equal ($k_1 \times L = k_2 \times D$)

3. An interval where dissolved-oxygen levels increase (eventually reaching saturation) because BOD levels are being reduced and the rate of deoxygenation is less than the rate of reaeration ($k_1 \times L < k_2 \times D$)

The location of the critical point and the oxygen concentration at that location are of principal interest, because this is where water quality conditions are at their worst. Design calculations are based on this location because if standards are met at the critical point, they will be met elsewhere. To determine the location of the critical point, first use Equation 8.7 to determine the *critical time*, and then multiply the critical time by the river velocity to determine the *critical distance*:

$$t_{crit} = \frac{1}{k_2 - k_1} \times \ln\left(\frac{k_2}{k_1} \times \left(1 - \frac{D_0 \times (k_2 - k_1)}{k_1 \times L_0}\right)\right) \qquad (8.7)$$

To find the oxygen deficit at the critical distance, substitute the critical time into Equation 8.5. Knowledge of $DO_{sat}$ then provides the actual DO concentration at the critical distance. Example 8.4 illustrates this approach and suggests opportunities for its application in river management.

---

example/8.4 **Determining Features of the DO Sag Curve**

After receiving the discharge from a wastewater treatment plant, a river has a dissolved-oxygen concentration of 8 mg $O_2$/L and an ultimate CBOD of 20 mg $O_2$/L. The saturation dissolved-oxygen concentration is 10 mg $O_2$/L, the deoxygenation rate coefficient $k_1$ is 0.2/day, and the reaeration rate coefficient $k_2$ is 0.6/day. The river travels at a velocity of 10 km/day. Calculate the location of the critical point (time and distance) and the oxygen deficit and concentration at the critical point.

solution

First determine the initial *DO* deficit at the point of discharge, using Equation 8.2:

$$D_0 = DO_{sat} - DO_{act}$$
$$= 10 - 8 = 2\,mg\,O_2/L$$

Then use Equation 8.7 to determine the critical time and knowledge of the river's velocity to determine the critical distance:

$$t_{crit} = \frac{1}{k_2 - k_1} \times \ln\left(\frac{k_2}{k_1} \times \left(1 - \frac{D_0 \times (k_2 - k_1)}{k_1 \times L_0}\right)\right)$$

$$t_{crit} = \frac{1}{0.6/day - 0.2/day} \times \ln\left(\frac{0.6/day}{0.2/day} \times \right.$$
$$\left. \left(1 - \frac{2\,mg\,O_2/L \times (0.6/day - 0.2/day)}{0.2/day \times 20\,mg\,O_2/L}\right)\right)$$

$$t_{crit} = 2.2\,days$$

$$x_{crit} = 2.2\,days \times 10\,km/day = 22\,km$$

## example/8.4 Continued

Finally, use Equation 8.5 to determine the oxygen deficit and Equation 8.2 to determine the actual dissolved-oxygen concentration for the critical time as just calculated:

$$D_t = \frac{k_1 \times L_0}{(k_2 - k_1)} \times (e^{-k_1 \times t} - e^{-k_2 \times t}) + D_0 \times e^{-k_2 \times t}$$

$$D_t = \frac{0.2/\text{day} \times 20\,\text{mg O}_2/\text{L}}{(0.6/\text{day} - 0.2/\text{day})} \times (e^{-0.2/\text{day} \times 2.2\,\text{days}} - e^{-0.6/\text{day} \times 2.2\,\text{days}}) +$$

$$2\,\text{mg O}_2/\text{L} \times e^{-0.6/\text{day} \times 2.2\,\text{days}}$$

$$D_t = 4.3\,\text{mg O}_2/\text{L}$$

$$DO = 10 - 4.3 = 5.7\,\text{mg O}_2/\text{L}$$

In this example, the deficit occurs 22 km downstream from the point of initial discharge.

## 8.3  Lake and Reservoir Water Quality

Water quality conditions in lakes and reservoirs are influenced by the magnitude and routing of the chemical and energy fluxes passing through biogeochemical cycles. The cultural perturbations of two of those cycles, phosphorus and nitrogen, result in a water quality issue of widespread interest: *eutrophication*.

**Lake Tahoe Planning Agency**
http://www/trpa.org

### 8.3.1  THERMAL STRATIFICATION OF LAKES AND RESERVOIRS

A major difference between lakes and rivers lies in the means of mass transport. Rivers are completely mixed, while in temperate latitudes, lakes undergo *thermal stratification*, dividing the system into layers and restricting mass transport. Periods of stratification alternate with periods of complete mixing with mass transport at a maximum. The restriction of mass transport during stratification influences the cycling of many chemical species (such as iron, oxygen, and phosphorus) and can have profound effects on water quality.

The process of thermal stratification is driven by the relationship between water temperature and density. The maximum density of water occurs at 3.94°C (Figure 8.3). Ice thus floats and lakes freeze from the top down, instead of from the bottom up, as they would if the maximum density were at 0°C. (Consider the implications of the opposite situation.) During summer stratification, an upper layer of warm, less dense water floats on a lower layer of cold, denser water.

The layers are assigned three names as shown in Figure 8.4: (1) the **epilimnion**, a warm, well-mixed surface layer; (2) the **metalimnion**, a region of transition where temperature changes at least 1°C with every meter of depth; and (3) the **hypolimnion**, a cold, well-mixed bottom

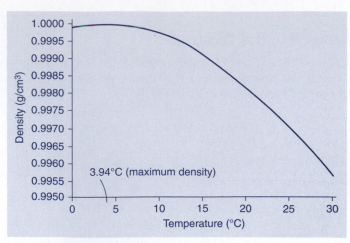

**Figure 8.3  Maximum Density of Water**  The maximum density occurs at 3.94°C. Thus, water at approximately 4°C will be found below colder waters (ice at 0°C) in winter and warmer waters (20°C) in the summer.

From Mihelcic (1999). Reprinted with permission of John Wiley & Sons, Inc.

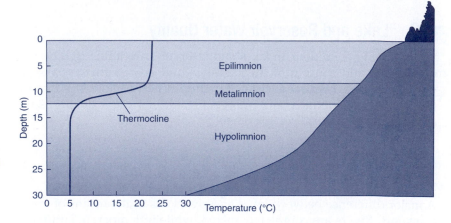

**Figure 8.4  Midsummer Temperature Profile for a Thermally Stratified Lake**  Note the epilimnion, metalimnion (with a thermocline), and hypolimnium

Adapted from Mihelcic (1999). Reprinted with permission of John Wiley & Sons, Inc.

layer. The plane in the metalimnion where the temperature–depth gradient is steepest is termed the **thermocline**.

The stratification and destratification processes (mixing) follow a predictable seasonal pattern, as shown in Figure 8.5. In winter, the lake is thermally stratified with cold (~0°C) water near the surface and warmer (2°C–4°C), denser waters near the bottom. As the surface waters warm toward 4°C in the spring, they become denser and sink, bringing colder waters to the surface to warm.

The process of mixing by convection, aided by wind energy, circulates the water column, leading to an isothermal condition termed **spring turnover**. As the lake waters continue to warm above 4°C, the lake thermally stratifies. Surface waters are significantly warmer and less dense than the lower waters during **summer stratification**. In the fall, solar input decreases, and heat is lost from the lake more rapidly than it is gained. As the surface waters cool, they become denser, sink,

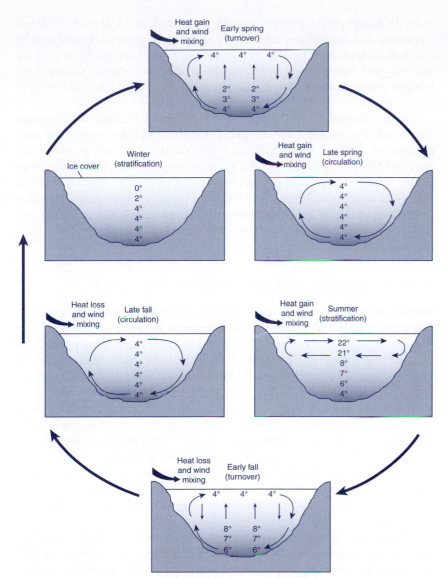

**Figure 8.5** **Annual Cycle of Stratification, Overturn, and Circulation in Temperate Lakes and Reservoirs** Variation in meteorological conditions (temperature, wind speed) may cause significant variation to the timing and extent of these events.

From Mihelcic (1999). Reprinted with permission of John Wiley & Sons, Inc.

and promote circulation through convection, aided by wind. This phenomenon, called **fall turnover**, again leads to isothermal conditions. Finally, as the lake cools further, cold, low-density waters gather at the surface, and the lake reenters **winter stratification**.

## 8.3.2 ORGANIC MATTER, THERMAL STRATIFICATION, AND OXYGEN DEPLETION

The internal production of organic matter in lakes, resulting from algal and macrophyte growth and stimulated by discharges of growth-limiting nutrients (phosphorus and nitrogen), can dwarf that supplied externally, for example, from wastewater treatment plants and surface runoff. Organic matter produced in well-lit upper waters settles to the bottom, where it decomposes, consuming oxygen. There is little resupply of oxygen under stratified conditions, and if algal and/or macrophyte growth yields a great deal of organic matter, hypolimnetic oxygen depletion may result. Oxygen concentrations in the bottom waters of productive lakes are lower than those in surface waters, and the opposite is true in unproductive waters, where cold bottom waters have a higher oxygen saturation than the warmer surface waters.

Oxygen depletion leads to acceleration in the cycling of chemicals that reside in lake sediments (especially iron and phosphorus), the generation of several undesired and potentially hazardous chemical species ($NH_3$, $H_2S$, $CH_4$), and the extirpation of fish and macroinvertebrates. Oxygen depletion is one of the most important and commonly observed water quality problems in lakes, bays, and estuaries. It is also important in drinking-water reservoirs, where intakes may encounter nuisance algal growth near the top and accumulations of noxious chemicals near the bottom.

### 8.3.3 NUTRIENT LIMITATION AND TROPHIC STATE

© jean schweitzer/iStockphoto.

**Trophy** is defined as the rate at which organic matter is supplied to lakes, both from the watershed and through internal production. The growth of algae and macrophytes in lakes is influenced by conditions of light and temperature and by the supply of growth-limiting nutrients. Because levels of light and temperature are more or less constant regionally, trophy is determined primarily by the availability of growth-limiting nutrients. As mentioned previously, phosphorus is generally found to be the nutrient limiting plant growth in freshwater environments. Because naturally occurring phosphorus minerals are sparingly soluble, anthropogenic inputs can dramatically affect the rate of algal and macrophyte growth and attendant production of organic matter. Table 8.1 shows how lakes can be classified into three groups according to their trophic state: **oligotrophic**, **mesotrophic**, and **eutrophic**.

The process of nutrient enrichment of a water body, with attendant increases in organic matter, is termed **eutrophication**. This is considered to be a natural aging process in lakes. Figure 8.6 shows the succession of newly formed water bodies to dry land. Addition of phosphorus

**Table / 8.1**

**Classification of Water Bodies Based on Their Trophic Status**

| | |
|---|---|
| Oligotrophic | Nutrient poor; low levels of algae, macrophytes, and organic matter; good transparency; abundant oxygen |
| Eutrophic | Nutrient rich; high levels of algae, macrophytes, and organic matter; poor transparency; often oxygen-depleted in the hypolimnion |
| Mesotrophic | Intermediate zone; often with abundant fish life because of elevated levels of organic-matter production and adequate supplies of oxygen |

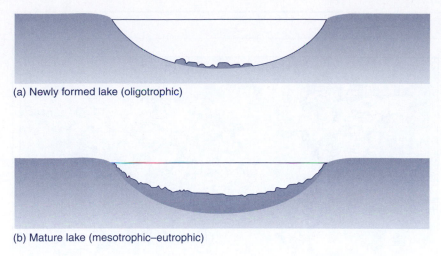

(a) Newly formed lake (oligotrophic)

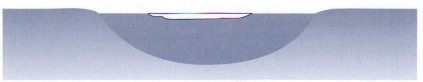

(b) Mature lake (mesotrophic–eutrophic)

(c) Meadow/marsh

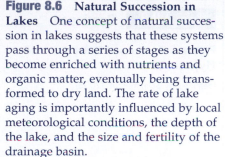

**Figure 8.6** **Natural Succession in Lakes** One concept of natural succession in lakes suggests that these systems pass through a series of stages as they become enriched with nutrients and organic matter, eventually being transformed to dry land. The rate of lake aging is importantly influenced by local meteorological conditions, the depth of the lake, and the size and fertility of the drainage basin.

From Mihelcic (1999). Reprinted with permission of John Wiley & Sons, Inc.

(d) Dry land

---

**Box / 8.1    Oxygen-Depleted Dead Zones around the World**

Over 400 coastal areas in the world are reported to experience some form of eutrophication (Figure 8.7). Of these, 169 are reported to experience hypoxia. These so-called dead zones experience very low oxygen levels (less than 2 mg/L) that can be seasonal or continual.

A small dead zone may occupy 1 km² and occur in a bay or estuary. The large dead zone located in the Gulf of Mexico has been measured at over 22,000 km² (the size of Massachusetts). This location, off the Louisiana shore, contains the most important commercial fishery in the lower 48 states and is fed by runoff from the Mississippi River.

The Mississippi drains 41 percent of the landmass of the lower 48 states and includes Corn Belt states such as Ohio, Indiana, Illinois, and Iowa. Besides using excessive amounts of fertilizers, these states have drained up to 80 percent of their wetlands, which serve as nutrient buffers. In fact, 65 percent of the nutrient input to the Gulf Coast dead zone originates in the Corn Belt, an area that provides food for a growing population and is now viewed by some as a source of energy independence through biofuels. Other nutrient inputs (that contain nitrogen and/or phosphorus) are associated with municipal wastewater and industrial discharges, urban runoff, and

**Figure 8.7**  Coastal Eutrophic and Hypoxic Zones

Identified by Selman et al. (2008).

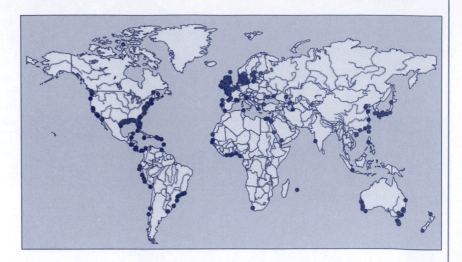

atmospheric deposition associated with fossil fuel combustion.

Atmospheric inputs can be large contributors. For example, atmospheric nitrogen inputs that originate with fossil fuel combustion contribute 25 percent of the nitrogen input to the Chesapeake Bay dead zone.

Left untouched, dead zones can cause the collapse of ecosystems and the economic and social systems that

depend on them. Fortunately, dead zones can be reversed. For example, the Black Sea once had a dead zone that occupied 20,000 km². After the 1980s collapse of many centralized economies in countries located in the watersheds that drain to the Black Sea, nitrogen inputs fell by 60 percent. This eventually resulted in the dead zone shrinking and in fact disappearing in 1996 (Larson, 2004; Selman et al., 2008).

**Class Discussion**

How would you address nutrient management using a systems approach? The average population density in coastal areas is twice the global average, while the biodiversity of coastal and freshwater aquatic ecosystems continues to decline. Increases in population and urbanization concentrate nutrients (N and P) in urban areas, where discharges from wastewater treatment plants and urban runoff can create havoc on freshwater and coastal ecosystems. Yet these nutrients are required in rural areas, where crop production is the greatest.

**Lake Trophic State Management Calculator**

through human activities and the resultant aging of the lake are termed **cultural eutrophication**. Variation in land use and population density can lead to a range of trophic states within a given region, for example, from oligotrophic Lake Superior to eutrophic Lake Erie.

### 8.3.4  ENGINEERED LAKE MANAGEMENT

The preferred option in quality management of surface water is always to prevent or eliminate discharges. The focus of lake management is typically on controlling phosphorus; however, most of the solutions presented here apply to other pollutants. (Remember that in Box 8.1, management of dead zones would also consider minimizing nitrogen inputs.) Figure 8.8 summarizes eight methods of lake management. In the case of phosphorus, great strides in treatment technologies have reduced phosphorus concentrations in municipal wastewater effluent more than two orders of magnitude from the influent. A variety of land management practices can reduce phosphorus loads from watersheds. Finally, stormwater detention basins, artificial wetlands, and low-impact development may be employed to trap phosphorus (and nitrogen and other materials, such as sediment and trace metals) washed from the land and paved surfaces.

## (a) Point source control

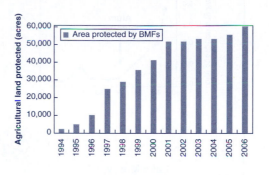

The discharge of phosphorus from a municipal wastewater treatment plant resulted in a hypereutrophic state, manifested in nuisance algal blooms, poor water clarity, and severe hypolimnetic oxygen depletion. Implementation of advanced waste treatment, in multiple stages, has reduced effluent P concentrations and led to a decline in the rate at which oxygen is consumed in the bottom waters (AHOD, areal hypolimnetic oxygen depletion).

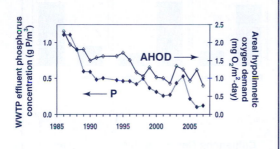

**Onondaga Lake, NY**

## (b) Nonpoint source control

This reservoir, which provides 70% of the water supply for the city of Wichita, is polluted by phosphorus and sediment originating on cropland and rangeland in its watershed. Increases in the land area protected by conservation practices in the watershed is being implemented to reduce phosphorus and sediment loading.

**Cheney Lake, KS**

## (c) Diversion

Algal blooms with attendant elimination of sensitive benthic macroinvertebrates was caused by phosphorus discharges from four municipal wastewater treatment plants. Point source loads were diverted to land application and other uses, including irrigation of citrus groves and groundwater recharge. In-lake phosphorus and chlorophyll levels fell by 50 and 30%, respectively, and Secchi disk transparency increased by 50%.

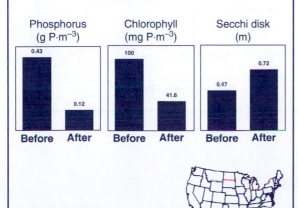

**Lake Tohopekaliga, FL**

## (d) Dredging

Prolific macrophyte growth and attendant deposition resulted in a water column of 1.4 m overlying 10 m of decomposing plant material. Dredging removed 665,000 m$^3$ of sediment from this 37 ha lake, increasing the water volume by 128% and the maximum depth to 6.6 m. Severe oxygen depletion and fish kills in winter were eliminated. Macrophytes no longer grew to nuisance levels because increased depth reduced the amount of well-lit plant habitat available. Dredging is expensive and carries with it potential side effects, largely related to sediment resuspension.

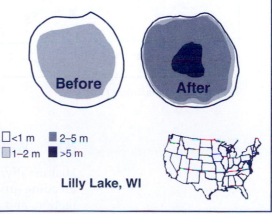

| | |
|---|---|
| ☐ <1 m | ■ 2–5 m |
| ☐ 1–2 m | ■ >5 m |

**Lilly Lake, WI**

**Figure 8.8** Examples of Engineering Lake Management

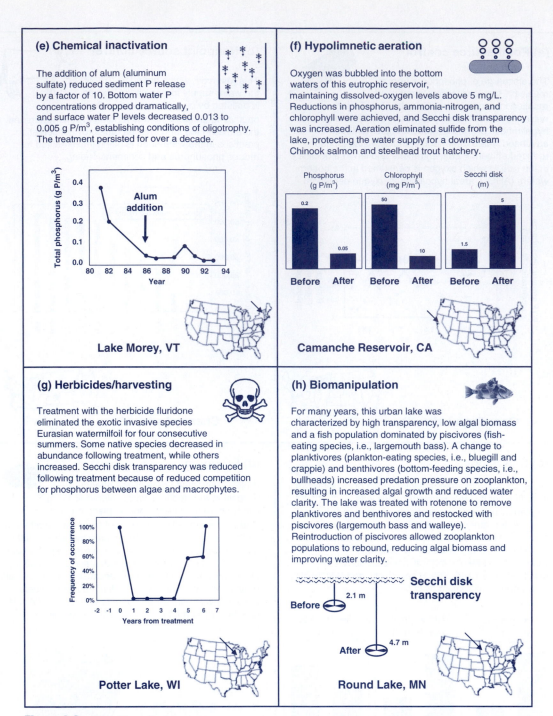

**(e) Chemical inactivation**

The addition of alum (aluminum sulfate) reduced sediment P release by a factor of 10. Bottom water P concentrations dropped dramatically, and surface water P levels decreased 0.013 to 0.005 g P/m³, establishing conditions of oligotrophy. The treatment persisted for over a decade.

Total phosphorus (g P/m³)

Alum addition

Year

**Lake Morey, VT**

**(f) Hypolimnetic aeration**

Oxygen was bubbled into the bottom waters of this eutrophic reservoir, maintaining dissolved-oxygen levels above 5 mg/L. Reductions in phosphorus, ammonia-nitrogen, and chlorophyll were achieved, and Secchi disk transparency was increased. Aeration eliminated sulfide from the lake, protecting the water supply for a downstream Chinook salmon and steelhead trout hatchery.

Phosphorus (g P/m³) — Before 0.2, After 0.05
Chlorophyll (mg P/m³) — Before 50, After 10
Secchi disk (m) — Before 1.5, After 5

**Camanche Reservoir, CA**

**(g) Herbicides/harvesting**

Treatment with the herbicide fluridone eliminated the exotic invasive species Eurasian watermilfoil for four consecutive summers. Some native species decreased in abundance following treatment, while others increased. Secchi disk transparency was reduced following treatment because of reduced competition for phosphorus between algae and macrophytes.

Frequency of occurrence

Years from treatment

**Potter Lake, WI**

**(h) Biomanipulation**

For many years, this urban lake was characterized by high transparency, low algal biomass and a fish population dominated by piscivores (fish-eating species, i.e., largemouth bass). A change to planktivores (plankton-eating species, i.e., bluegill and crappie) and benthivores (bottom-feeding species, i.e., bullheads) increased predation pressure on zooplankton, resulting in increased algal growth and reduced water clarity. The lake was treated with rotenone to remove planktivores and benthivores and restocked with piscivores (largemouth bass and walleye). Reintroduction of piscivores allowed zooplankton populations to rebound, reducing algal biomass and improving water clarity.

**Secchi disk transparency**

Before 2.1 m

After 4.7 m

**Round Lake, MN**

**Figure 8.8** (*Continued*)

## 8.4  Wetlands

Historically, **wetlands** were regarded as nuisances that provided breeding grounds for disease and stood in the way of agriculture, navigation, and urbanization. More than 50 percent of the wetlands that existed before 1700 in the lower 48 contiguous states of the United States have been destroyed; for some individual states, the percentage of wetlands lost is greater than 90 percent. Table 8.2 lists the most common contributors to wetland loss.

In 1989, the United States declared a national policy of *no net loss* of wetlands; this was followed by a goal of a net increase of 100,000 acres/year of wetlands in the Clean Water Action Plan of 1998. As a result, activities have shifted from elimination to the identification, delineation, preservation, restoration, and construction of wetlands. The majority of this activity is driven by a Clean Water Act requirement that the U.S. Army Corps of Engineers issue permits for any wetlands that are to be filled.

Under these federal initiatives, the net loss of wetlands has been slowed and potentially reversed. Rates of wetland loss dropped from 1,854 $km^2$/yr in the period 1950–1975 to 1,174 $km^2$/yr during 1975–1985 and to 237 $km^2$/yr during 1985–1997. In the latest reporting period (1998–2004), there has been a net *increase* in wetland area of 129 $km^2$/yr (Dahl, 2006).

To the extent that they are part of U.S. waters, wetlands are subject to the same protections as rivers and lakes under the Clean Water Act. The discharge of wastes to wetlands is regulated under the National Pollutant Discharge Elimination System, the same permit program applied to lakes and streams. Further, the Clean Water Act prohibits the discharge of dredge (or fill) material into wetlands without a permit from the Army Corps of Engineers. If damage to wetlands is deemed unavoidable, new or improved wetlands must be created to offset the functions lost in the damaged wetlands. The creation or restoration of wetlands to compensate for damage to other wetlands is termed **compensatory mitigation**.

## 8.4.1 TYPES OF WETLANDS

Wetlands represent an *ecotone* or transition between terrestrial and aquatic environments. Any time that soils become saturated with water, the entry of oxygen is severely restricted, and anaerobic conditions develop. Types of wetlands may be identified based on variations in the source of water (direct rainfall, groundwater discharge, or inputs from an adjacent lake or river), the position of the water table (above or below the water surface), the **hydroperiod** (seasonal pattern of water depth), and levels of productivity and enrichment with nutrients and organic matter.

Table 8.3 compares the four types of wetlands: **marshes**, **swamps**, **bogs**, and **fens**. Wetlands may also be characterized by their proximity to other habitats such as riparian wetland (adjacent to rivers) and coastal wetlands (adjacent to oceans and large lakes). Because of the wide variety in types and characteristics of wetlands, a legal definition has remained elusive. While the accepted definition can differ (even among federal agencies), a site may be characterized as a wetland if it (1) is inundated or saturated with water for at least part of the year; (2) has hydric soils; and (3) supports predominantly hydrophytic plants.

| Table / 8.2 |
|---|
| **Most Common Contributors to Wetland Loss Today** |
| Agricultural activities |
| Residential and commercial development (urbanization) |
| Construction of roads and highways |

**National Wetlands Inventory**
http://www.fws.gov/wetlands

## Table / 8.3

**Types of Wetlands**

| Wetland Type | Description |
|---|---|
| Marsh | Frequently or continually inundated with water; characterized by emergent, soft-stemmed vegetation |
| Swamp | Dominated by woody (as opposed to soft-stemmed) vegetation |
| Bog | Peat-forming wetlands that receive all water and nutrient inputs from atmospheric deposition; typically acidic and nutrient poor |
| Fen | Receives nutrients by drainage from mineral soils and groundwater input; typically less acidic and more nutrient rich than bogs |

SOURCE: Defined by EPA Office of Wetlands, Oceans and Watersheds.

**Hydric soils** exist under waterlogged, anaerobic conditions and exhibit particular chemical and physical characteristics. These include odors (for instance, from hydrogen sulfide), the presence of organic-rich layers, a gleyed or mottled appearance, and low redox potential. The roots of most plants cannot tolerate anaerobic conditions, because soil oxygen is the only source of oxygen for root tissue respiration. **Hydrophytic** plants, however, have developed adaptations that allow them to grow in oxygen-deficient soils. In practice, vegetation surveys are used to delineate the boundaries of wetlands. For an area to be considered a wetland, hydrophytic plants must be prevalent; for example, at least half of the dominant plants must be wetland species.

### 8.4.2 WETLAND FUNCTIONS

www.wiley.com/college/mihelcic **Wetland Ecosystems**

Wetlands serve several important functions: (1) water storage and flood mitigation; (2) filtration of water and removal of suspended solids, bacteria, nutrients, and toxic substances; (3) wildlife habitat; and (4) biogeochemical cycling of materials that are important on local to global scales.

Wetlands have a large capacity to store water, on the order of 1 $m^3$ per $m^2$ of surface area. They also dampen extreme rainfall events and storm surges. In fact, the loss of coastal wetlands along the Gulf of Mexico contributed to the severity of the impact of hurricane Katrina. Wetlands also remove substances that may impair water quality. Because water flow is retarded in wetlands, sediments are able to settle and collect there. Wetland plants also anchor the soil and reduce erosion that might be induced by waves. The microorganisms and vegetation present in wetlands remove nutrients from waters flowing through the system and thereby help to protect the water quality of adjacent lakes and streams. The reduced water flow rates and shallow

**Louisiana Coastal Land Loss**
http://www.nwrc.usgs.gov/special/landloss.htm

waters of wetlands promote the die-off of disease-causing bacteria, minimizing the transport of pathogens to surface waters. Wetland soils are frequently rich in organic matter or have horizons enriched in iron oxides; both substrates are capable of sorbing and removing a number of pollutants (for instance, phosphorus and benzene) from the water passing through the wetland.

It is difficult to overstate the importance of wetlands to biodiversity. Wetlands represent only 5 percent of the land area of North America but are home to 31 percent of North American plant species. It is estimated that 50 percent of North American bird species nest or feed in wetlands. Roughly 46 percent of all endangered species in the United States depend in some way on wetlands. Organic matter produced by plants in wetlands feeds not only the food web within the wetland but also may, through hydrologic export, serve as an important food source for organisms far removed from the wetland. Because wetlands are so important to wildlife, they are also critical for supporting commercial and sport fishing, as well as hunting and other leisure activities.

Wetlands are significant as global reservoirs of carbon and as sources and sinks for important trace gases, including $CH_4$, $H_2S$, and $N_2O$. Globally, 75 percent of the natural emissions of methane originate from wetlands. Rivers, estuaries, and their associated wetlands may

**Class Discussion**

How would you manage the Everglades while balancing social, economic, justice, and environment issues that will not lead to loss of opportunity for current and future generations? What are some key issues associated with water storage and quality, biodiversity, and the economy in this area? How do engineers participate in developing solutions to problems related to these issues? (For further information, see www.evergladesplan.org.)

---

## Box / 8.2 The Florida Everglades

The Everglades in the State of Florida is one of the world's truly unique ecosystems. It includes wetlands and a river once estimated to be 50 miles wide. More than 50 percent of the original wetlands have been lost to conversion to agriculture and urbanization. Efforts to protect this resource have strengthened but often conflict with increasing development demands for rapidly growing resident and tourist populations. Since 1930, the population of southern Florida has increased 25-fold, from 200,000 to more than 5 million, a growth rate approximately ten times faster than that of the United States. In 1947, Everglades National Park was formed. It now covers 1.4 million acres (nearly 5 percent of the land area of Florida) and has tripled in size since it was established.

In 2000 the U.S. Congress authorized the Comprehensive Everglades Restoration Plan, the largest environmental restoration project in history, with a 30-year time frame and a $10.5 billion budget. As part of the restoration plan, 400 km of canals and levees (installed in the 1940s to control and divert water) will be removed. Measures will be taken to control invasive and exotic species, and 6.4 billion liters a day of runoff will be treated to remove nutrients and other contami-

nants. Runoff will be stored and redirected to more closely resemble presettlement (natural) flow patterns to Florida Bay, as shown in Figure 8.9. More than 200,000 acres of land have been purchased (50 percent of the project goal) to control **land use**.

At a Web site devoted to reporting on this plan (www.evergladesplan.org), the Army Corps of Engineers spells out its promise and importance:

*Implementation of the restoration plan will result in the recovery of healthy, sustainable ecosystems in south Florida. . . . The plan will redirect how water is stored in south Florida so that excess water is not lost to the ocean, and instead can be used to support the ecosystem as well as urban and agricultural needs. . . . The ability to sustain the region's natural resources, economy, and quality of life depends, to a great extent, on the success of the efforts to enhance, protect and better manage the region's water resources. (USACE, 2008a).*

One key element of the management plan is the Florida panther, a subspecies of mountain lion. The panther once ranged in eight states over the entire southeastern United States, but now it has a range

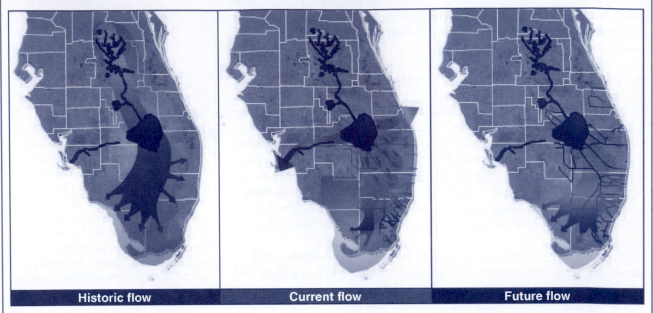

| Historic flow | Current flow | Future flow |

**Figure 8.9**   The Florida Everglades, Showing Flow of Water from Historic, Current, and Future Scenarios

From www.evergladesplan.org, an effort of the U.S. Army Corps of Engineers in partnership with the South Florida Water Management District and many other federal, state, local and tribal partners). USACE (2008 b).

reduced to approximately 10 counties in southern Florida (less than 5 percent of its native range). The current population is estimated at approximately 90 animals. The population decreased dramatically over the past century due to hunting and habitat loss. Habitat loss is primarily from urban sprawl and the conversion of woodlands to agriculture. While loss of habitat is significant, habitat degradation and fragmentation also pose major threats.

The preferred prey for the panther is white-tailed deer, the population of which has also declined.

Secondary food sources are feral hogs released for hunting, armadillos, and raccoons. In areas with low deer populations, raccoons are an increasingly important part of the panther diet. Because raccoons feed on fish and crayfish, they have increased mercury content from bioaccumulation. Increased mercury levels in panthers may be linked to decreased health and reproductive success, but conclusive scientific studies have neither proven nor disproven this hypothesis. Vehicle collisions also have been a significant source of panther mortality, with 24 deaths recorded from 1978 to 1998.

account for 20 percent of the anthropogenic emissions of another important greenhouse gas, $N_2O$ (Seitzinger and Kroeze, 1998).

### 8.4.3   BUILDING WETLANDS

As defined near the beginning of section 8.4, **compensatory mitigation** refers to the creation of wetlands to offset the permitted destruction of wetlands. **Banking** is the creation of wetlands to offset future destruction of wetlands, at which time the banked wetlands will become mitigation wetlands. The term **created wetland** refers to a wetland built for mitigation purposes, and **constructed wetland** denotes a wetland designed for pollutant removal (for example, treatment of municipal wastewater, agricultural runoff, stormwater, and mining wastes). Although created and constructed wetlands may share many common features, the guidelines for construction and the regulations for operation differ; for this reason, the two categories are often considered separately. Created wetlands are treated in the next section; constructed wetlands are described in Chapter 11.

## 8.4.4  CREATED WETLANDS: AN OPTION FOR MITIGATION

Under the U.S. environmental regulatory system, wetland destruction is to be avoided whenever less damaging practicable alternatives are available. When adverse impacts are unavoidable, those impacts are to be minimized to the extent practicable by modification of the proposed activity. Compensatory mitigation is to be permitted only after the first two options have been fully implemented. The system to be destroyed should be inventoried to characterize and quantify the types of wetland and functions that will be lost. Arguably, to fulfill the same functions of water storage, water filtration, and biogeochemical cycling, created wetlands should be in similar hydrologic settings (for instance, water table height, water flow rates, and hydroperiod) within their watersheds as those degraded or destroyed.

Compensatory mitigation is intended to offset the loss of wetland functions by creating, restoring, or enhancing wetlands that will play a similar role. Restoring historical wetlands that have been drained is the method thought best to ensure restoration of natural functions. Despite this goal, the measure used to quantify wetland loss is not typically function, but area. When a created wetland is less efficient in performing a function than the wetland it replaces, the new wetland can be made larger. To avoid degradation of the watershed, the functions lost by destruction of one wetland should be replaced in the same watershed. This goal is difficult to meet, though, when compensatory mitigation is practiced on a regional basis.

Creation of a wetland includes two phases: site selection and site development. Table 8.4 presents some guidelines for site selection: seeking to ensure that the wetland functions properly, is in harmony with its watershed, and offers the greatest environmental benefit. Site development will be specific to the location and planned functions of the wetland, with attention given to the desired hydroperiod and wetland vegetation. Site development components relating to the hydrologic regime include grading to ensure water flow (slopes of 0.0001 to 0.01), installation of an impermeable layer where not naturally present, creation of berms for confinement, and construction of a device to regulate or constrict outflow.

Artificial wetlands should have irregular geometries and nonuniform properties to promote the fulfillment of multiple functions, including the maintenance of biodiversity. It is also important to plan on adequate capacity for accumulation of introduced sediment. Often it is useful to have multiple wetland *cells*, or units with different flow paths. This variety may help the wetland fulfill several functions as well as promoting biodiversity of plants and animals.

**Wetland Facts**
http://www.epa.gov/owow/wetlands

## Table / 8.4

### Site Selection Guidelines for a Created Wetland

| Created Wetlands Should Have: | This Means the Site Should: |
|---|---|
| Function | • Have an adequate and reliable water supply<br>• Have soils that are hydric or hydrologically modified<br>• Have associated upland areas to serve as a wetland buffer |
| Harmony | • Have previously had wetlands<br>• Be connected to existing wetlands<br>• Be distant from incompatible land uses<br>• Be near an area of wildlife significance<br>• Not damage other ecologically significant areas |
| Benefit | • Be near designated priority areas such as public management lands or preserves<br>• Enhance biodiversity, outdoor recreation, and/or scientific value |

SOURCE: Adapted from ELI, 2002.

As with wetland delineation, vegetation surveys can be used to gauge the success of created wetlands. Because wetland vegetation and hydrologic functioning cannot be established immediately, post-creation monitoring should be maintained over a suitable time interval. A five- to ten-year monitoring period is recommended by the Army Corps of Engineers.

## 8.5    Low-Impact Development

**Low Impact Development**

www.wiley.com/college/mihelcic

One major problem of the **built environment** is the impact of *nonpermeable surfaces* on the natural hydrologic cycle and water quality. Figure 8.10 shows how covering natural surfaces with buildings, roofs, roads, and parking lots decreases the amount of precipitation that infiltrates to groundwater. In natural systems, approximately 50 percent of precipitation recharges to the subsurface, but in a highly urbanized environment, this declines to 15 percent, with only 5 percent reaching deep groundwater.

The impact of urbanization on runoff also affects a city's ability to store freshwater. For example, in urbanized coastal areas, precipitation that becomes runoff will travel quickly to saline seawater, where it then becomes energy intensive and expensive to treat to drinking or agricultural standards. This problem is significant, because 21 of world's 33 megacities are located in coastal areas, and the average population density in coastal areas is twice the global average. Promoting groundwater recharge by preferring permeable surfaces should thus be considered a method to store freshwater for later use by ecosystems and humans. This is especially important because groundwater makes up over 30 percent of the world's freshwater reserves.

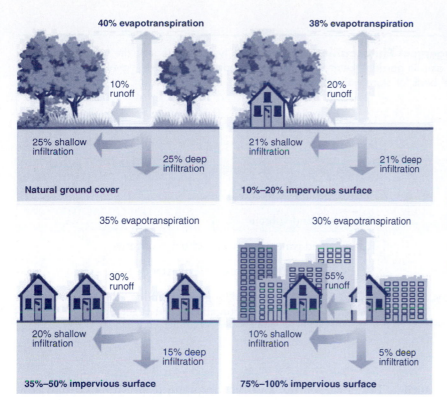

**Figure 8.10** **How Nonpermeable Surfaces Associated with the Built Environment Change Natural Hydrological Cycles** As natural ground cover is removed and replaced with nonpermeable surfaces such as buildings, roads, and parking lots, there is increased runoff and significantly less recharge of groundwater. Also, evapotranspiration is reduced in an area with large amounts of nonpermeable coverings. The process of evapotranspiration results in a cooling process (much as your skin cools when you perspire), which negates the impact of urban heat island.

EPA (2000).

Because of the water problems associated with nonpermeable surfaces, Phase II of the National Pollutant Discharge Elimination System (NPDES) requires small municipalities with separate storm sewer systems (called MS4s) to address stormwater runoff from their site with the recommended use of structural **best management practices (BMPs)**. Examples of BMPs related to this regulation are grassed swales, permeable pavements, and bioretention cells (EPA, 2000). Fortunately, BMPs can be easily integrated into a new or existing development, even at the household level.

Traditional **stormwater** management utilizes concrete catch basins and pipes that transport stormwater to a larger detention basin, where the water is discharged by an orifice or discharge pipe at a specific peak rate. This traditional technique is intended solely to reduce the peak flow rate. **Low-impact development** mimics the natural hydrology that existed before development occurred. It not only reduces the peak flow rate, but also considers the timing of the discharge off the site as well as retention of rainwater. It also integrates principles of biodiversity, green space, water storage, groundwater recharge, and water quality improvements into the overall plan. Table 8.5 compares the philosophies of traditional stormwater management with low-impact development.

## 8.5.1 GREEN ROOFS

Rooftop runoff can be a substantial contributor to municipal storm sewer systems. To provide an idea of the extent of urban roof areas, the estimated roof area in metropolitan Chicago is 680 km$^2$, and in cities

**Stormwater Basics**
http://cfpub.epa.gov/npdes/
stormwater/swbasicinfo.cfm

## Table / 8.5

**Comparison of Stormwater Management Philosophies**   Site plans will typically integrate the use of several best management practices (BMPs) such as rain gardens, permeable pavement (pavers and gravel), green roof, bioswales, and even underground gravel beds for stormwater detention.

| Traditional Management | Low-Impact Development |
|---|---|
| Catchbasins and pipes | Swales |
| Mowed detention basin | Planted basin as in bioretention cells |
| Large central detention pond | Small, distributed detention areas |
| Runoff rate | Runoff rate and volume |
| Flooding is primary concern | Flooding and water quality are primary concerns |
| Time of concentration reduced significantly | Time of concentration maintained or extended |
| Runoff has nutrients, suspended particles, and hazardous materials | Runoff constituents treated by gravity settling, filtration, sorption, and interaction with microorganisms and vegetation |

SOURCE: Ward, 2007.

such as Phoenix, Seattle, and Birmingham, it is estimated that connected residential roofs account for 30 to 35 percent of annual runoff volume.

Compared with traditional blacktop and metal roofs, **green roofs** provide many private and public benefits, listed in Table 8.6. Some of these benefits include enhanced stormwater management, reduction in building energy costs, increase of green space and habitat, and reduction in urban heat island.

There are three types of green roofs:

1. *Intensive green roofs* have a thicker soil layer (150–400 mm) and weigh more, so they require more structural support. They work well with existing concrete roofs (for example, on parking decks).

2. *Extensive green roofs* have a thinner soil layer (60–200 mm), so they require less structural support.

3. The *semi-intensive green roof* has some components of both the extensive and the intensive roof systems.

Table 8.7 summarizes design and maintenance criteria for each type of green roof.

Native vegetation is always the preferred alternative. Whether or not the plants are native plays an important role in determining the degree of maintenance and irrigation. Extensive green roofs tend to be the more manicured with lots of mulch coverage and sedum vegetation, implying the need to weed regularly throughout the lifetime of the roof. Intensive green roofs tend to incorporate more native vegetation.

**Table / 8.6**

**Private and Public Benefits of Green Roofs**

| Benefit | Description |
|---|---|
| Increased roof life | Life expectancy of a "naked" flat roof is 15–25 years because of high surface temperatures and UV radiation degradation. Green roofs increase roof life by moderating these impacts. |
| Reduced noise levels | Sound is reflected by up to 3 dB, and sound insulation is improved by up to 8 dB. |
| Thermal insulation | Provides additional insulation, which reduces heating and cooling costs. |
| Heat shield | Transpiration during growing season results in a cooler building climate. |
| Use of space | Can be incorporated into personal, commercial, and public space. |
| Habitat | Provides habitat for plant and animal species. |
| Stormwater retention | Runoff can be reduced 50%–90%, especially important during peak precipitation events. |
| Urban heat island | Transpiration results in cooler roof surfaces, which reduces building contribution to heating the local built environment. |

SOURCE: Information courtesy of the International Green Roof Association, Berlin.

**Table / 8.7**

**Design and Maintenance Criteria for Green-Roof Types**   Each has different characteristics in terms of planting depth and structural loading requirements.

| | System buildup height | Weight | Costs | Possible use |
|---|---|---|---|---|
| **Extensive Green Roof** | 60–200 mm | 60–150 kg/m$^2$ <br> 13–30 lb./sq. ft. | Low | Ecological protection layer |
| **Semi-Intensive Green Roof** | 120–250 mm | 120–200 kg/m$^2$ <br> 25–40 lb./sq. ft. | Middle | Designed green roofs |
| **Intensive Green Roof** | 150–400 mm on underground garages > 1,000 mm | 180–500 kg/m$^2$ <br> 35–100 lb./sq. ft. | High | Parklike garden |

SOURCE: Information courtesy of the International Green Roof Association, Berlin.

The green roof on Chicago's City Hall is planted with native prairie vegetation. Native vegetation can grow more densely (so it requires little to no mulch), and it more closely replicates a native ecosystem. The use of native species in the design of an intensive green roof requires heavy weeding of invasive species during the first 2 to 3 years but little maintenance after this initial period. In contrast, a manicured sedum-planted extensive green roof will require maintenance over

**Chicago Green Roofs**
www.chicagogreenroofs.org/

its lifetime. Native vegetation also stands up better to climatic variation than ornamentals and introduced species. This implies a lesser need for irrigation on an intensive green roof planted with native species.

Note that irrigation needs and the relationship to the type of vegetation will depend on the depth of the soil. For example, irrigation requirements for a succulent nonnative plant could be equal to those for native vegetation in the shallow soils of a green roof. This is because the deep-root advantage of some native plants is lessened in the shallow soil layers of green roof.

Figure 8.11 shows the components of a typical green roof. Such a roof consists of a multilayered system that sits above the roofing deck and provides water and root protection (items 3–7 in Figure 8.11). These layers sit under a drainage system (item 9). On top of the drainage layer is some growing material (such as soil), which is planted (items 11 and 12, respectively).

The volume of water that a green roof can store after a rainfall event (V) is determined as

$$V = P \times A \times C \tag{8.8}$$

where $P$ is the precipitation (mm), $A$ is the roof area, and $C$ is a measure of the water-holding capacity of the growing medium (ranges from 0 to 1). This volume of stored water can be compared with the volume of stormwater generated by a conventional roof ($P \times A$).

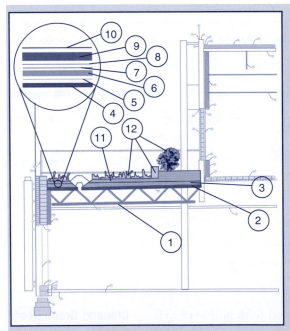

1. Steel joist

2. Metal roof dack

3. 5 in. R-30 foam insulation

4. 1/2 in. gypsum protection board

5. 75 mil ethylene propylene diene monomer (EPDM) membrane

6. 1/2 in. foam protection board

7. 40 mil hight-density polyethylene (HDPE) root barrier

8. Protection fabric

9. 1 in. drainage layer

10. Filter fabric

11. 3–9 in. lightweight growing medium

12. Stone features, sedum, native perennials, and shrubs

**Figure 8.11  Components of a Green Roof**  This particular roof uses a combination of extensive (3–4 in. soil) and intensive (4–9 in. soil) planting areas. It reduced the nonpermeable surface area 3,626 sq. ft., required an engineered support of 62 lb./sq. ft., and cost $31.80/sq. ft. in 2005.

Courtesy of Wetlands Studies and Solutions, Inc., Gainesville, Va.

## 8.5.2 PERMEABLE (OR POROUS) PAVEMENTS

**Permeable (or porous) pavement** is pavement that allows for the vertical passage of water. This pavement not only reduces runoff by enhancing groundwater recharge but also reduces the urban heat island. Permeable pavements are also believed to be more skid resistant and quieter than conventional pavements. They also eliminate the need for some drainage systems (as with other BMPs).

Permeable pavement can be constructed from grass, gravel, crushed stone, concrete pavers, concrete, and asphalt (see Table 8.8). Porous pavement can be as simple as grassy surfaces and interlocking paving stones that allow vegetation to grow between the block edges. Paver stones can also be placed along trees because they will not harm root systems like traditional pavement does. It can also refer to a multilayer system that includes a permeable course of paver stones (or gravel) at the surface, a compacted sand subbase immediately below the permeable cover, a fabric filter, and a compacted base on the bottom.

In situations where strength of the paving material is of concern, permeable pavement can be mixed with traditional pavement. In this scenario, parking areas and pedestrian walkways are specified as porous pavement, and traditional paving materials are used in limited-use areas where heavy loads are brought (e.g., truck travel to a loading dock). In this case, pavement markings or signage can guide larger delivery trucks so they remain off the porous pavement that surrounds the traditional pavement.

*Permeable concrete* consists of conventional concrete materials with the coarse aggregate being limited in its range of sizes, and the presence of fine aggregate being minimal or nonexistent (PCA, 2004). *Porous asphalt* is sometimes referred to as open-graded coarse aggregate, bonded together by asphalt cement, with sufficient interconnected voids to make it highly permeable to water (EPA, 1999).

## Table / 8.8

### Examples of Permeable (Porous) Pavement

| Permeable (Porous) Pavement | Examples |
| --- | --- |
| Grass | Excellent option for situations where commercial parking is needed in winter months when ground cover is frozen or is only needed several times a year (e.g., outside football stadium or on edge of big box stores) |
| Gravel or crushed stone | Common on many driveways and roads |
| Grid | Consists of washed gravel, plastic grid, and filter fabric; has high infiltration rate and removes sediments |
| Paver stones | High infiltration rates and filter sediments; easy to plow and maintain |
| Porous concrete or asphalt | For porous concrete, cement, and water with little to no sand or aggregate; high permeability (15%–25% voids, flow rates around 480 in. water depth per hour); filter sediment; easy to plow and maintain |

### 8.5.3 BIORETENTION CELLS

**Bioretention cells** are shallow depressions in the soil to which stormwater is directed for storage and to maximize infiltration. They are sometimes referred to as *bioinfiltration cells, vegetated biofilters,* and *rain gardens.* They are most often mulched (for aesthetic value and water treatment) and planted with native vegetation that promotes evapotranspiration. The design objective of maximizing infiltration will reduce the volume of water that needs to be stored and/or treated (hence the name bioinfiltration cell).

Figure 8.12 shows the detailed design of a bioretention cell. It includes the use of an underdrain, substantial aggregate backfill, and geotextile lining. Bioretention cells are often incorporated along streets to capture road runoff, which requires the inclusion of a curb cutoff. One novel aspect is that they can be designed to incorporate a wide variety of uses, from high infiltration to pretreatment of urban runoff and removal of nitrogen (see Figure 8.13 for examples). They can also be sized and installed by homeowners (WDNR, 2003).

As the vegetative cover grows, bioretention cells are expected to have increased capacity to accept water as the root network of the plants evolves and increases transpiration. In contrast to septic tank drainage fields, where biomats can develop due to relatively high organic and nutrient loadings, no studies have yet found loss of infiltration performance in bioretention cells. If needed, the soil may be carefully loosened (and aerated) to restore the infiltrative capacity and break up the biomat that may develop on the upper horizontal plane of the cell where water enters.

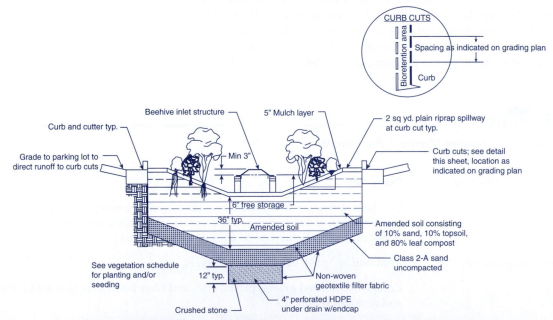

**Figure 8.12** **Typical Design of a Commercial Bioretention Cell** Note the location of amended soils, underdrains, outlet structures, vegetation, and inlet protection. These types of bioretention cells typically include the use of an underdrain, substantial aggregate backfill, or geotextile lining.

Courtesy of Spicer Group, Inc., Saginaw, Mich. Detail developed in 2006.

One common myth is that bioretention cells attract mosquitoes. Mosquitoes require 7 to 12 days to lay and then hatch their eggs. The standing water in a properly designed bioretention cell will be present for only several hours after a rain event. Also, the plants will attract dragonflies, which prey on mosquitoes.

**DESIGN AND CONCEPT OF FIRST FLUSH** Many states now recommend or require that the first flush of a storm event be captured and treated. Bioretention cells can be sized based on the concept of first flush. **First flush** is defined as the first 0.5 to 1 inch of runoff that is associated with a rain event and is calculated over the entire nonpermeable area of a site.

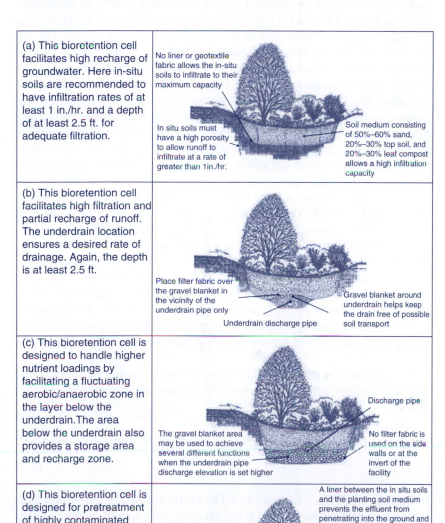

| | |
|---|---|
| (a) This bioretention cell facilitates high recharge of groundwater. Here in-situ soils are recommended to have infiltration rates of at least 1 in./hr. and a depth of at least 2.5 ft. for adequate filtration. | No liner or geotextile fabric allows the in-situ soils to infiltrate to their maximum capacity / In situ soils must have a high porosity to allow runoff to infiltrate at a rate of greater than 1in./hr. / Soil medium consisting of 50%–60% sand, 20%–30% top soil, and 20%–30% leaf compost allows a high infiltration capacity |
| (b) This bioretention cell facilitates high filtration and partial recharge of runoff. The underdrain location ensures a desired rate of drainage. Again, the depth is at least 2.5 ft. | Place filter fabric over the gravel blanket in the vicinity of the underdrain pipe only / Underdrain discharge pipe / Gravel blanket around underdrain helps keep the drain free of possible soil transport |
| (c) This bioretention cell is designed to handle higher nutrient loadings by facilitating a fluctuating aerobic/anaerobic zone in the layer below the underdrain. The area below the underdrain also provides a storage area and recharge zone. | Discharge pipe / The gravel blanket area may be used to achieve several different functions when the underdrain pipe discharge elevation is set higher / No filter fabric is used on the side walls or at the invert of the facility |
| (d) This bioretention cell is designed for pretreatment of highly contaminated water before discharge at an outlet pipe. The liner prevents groundwater contamination. | A liner between the in situ soils and the planting soil medium prevents the effluent from penetrating into the ground and reduces the likelihood of groundwater contamination / By capping the underdrain pipe, this facility type may be used to capture accidental spills and contain the level of contamination |

**Figure 8.13** **Bioretention Cells Designed for Different Purposes** From top to bottom: (a) infiltration and recharge facility for enhanced infiltration; (b) filtration and partial recharge facility; (c) infiltration, filtration, and recharge facility; and (d) filtration-only bioretention cell.

Redrawn from *The Bioretention Manual*, developed by Prince George's County Government, Md., 2006.

In a residential home where a yard slopes toward the bioretention cell, the nonpermeable area (for example, the roof) would be increased by the area of yard that is draining into the bioretention cell. The reason for including the area of the yard that drains to the cell is that there is a misconception that residential yards constitute high-infiltration green space. In fact, the effects of soil compaction during development are substantial, which is why land disturbance should be minimized during any land development. For similar reasons of soil compaction, heavy equipment should never travel across the cell during construction.

Because climates vary among states, some states have guidelines in terms of the release of the first flush. For example, in Michigan, this water volume must be released over a 1 to 2-day period or infiltrated into the ground within 3 days. For regional detention applications, Michigan suggests treatment of the 90 percent non-exceedance storm (the storm event for which 90 percent of all runoff-producing storms are smaller than or equal to the specified storm) (Ward, 2007).

Bioretention systems are sized based on several different methods, including the Prince George's County manual method, the runoff frequency method, and the rational method. Modeling programs such as EPA SWMM, WIN-TR-55, HEC-HMS, and HydroCad are applicable for modeling stormwater at the small site to regional scale. These tools have limitations in terms of simulating specific hydrologic mechanisms in bioretention at the site scale, but knowledgeable users have applied them to develop conservative designs. Two good examples of widely used models are the Prince George's County (Maryland) BMP module and RECARGA (from the University of Wisconsin–Madison).

Bioretention cells can also be designed based on the first flush and whether the water is stored below grade or above grade. Sizing the cell based on storage below grade will require an estimate of the cell's porosity. Sizing the cell based on the volume of water that can be stored above grade requires an understanding of how deep a water level the plant species requires to survive for a short period of time, along with accounting for the above-grade volume taken up by the plants.

To estimate the volume of a bioretention cell required to store the first flush below grade, determine the volume of rainwater generated at a site:

$$\text{volume of rainwater to be stored} = \text{first flush} \times \text{nonpermeable area}$$

$$(8.9)$$

In Equation 8.9, the first flush is 0.5 to 1.0 in. rainwater generated during a precipitation event, and the nonpermeable area includes the area being drained to the cell. The maximum volume of the bioretention cell below grade can be determined as follows:

$$\text{volume of bioretention cell} = \frac{\text{volume of rainwater to be stored}}{\text{soil porosity}}$$

$$(8.10)$$

The porosity is defined as

$$n = \frac{\text{volume of the void space}}{\text{total volume}} = \frac{V_v}{V_t} \qquad (8.11)$$

Equation 8.10 assumes that the soil porosity is saturated with precipitation associated with the first flush. Because soil porosity has units of volume voids divided by total volume, the area of the bioretention cell can be written as:

$$A = \frac{\text{volume of bioretention cell}}{\text{depth}} \qquad (8.12)$$

where the below-grade depth is typically 3 to 8 in.

When designing a bioretention cell to store water above grade, the designer should consider the volume of water generated by non-permeable surfaces and the depth of water that can submerge a portion of the planted vegetation for a short duration, along with the above-grade volume occupied by the plants. The volume required above grade for the bioretention cell to store the first flush precipitation event is

$$\begin{bmatrix} \text{total volume required} \\ \text{above grade for the} \\ \text{bioretention cell} \end{bmatrix} = \begin{bmatrix} \text{volume of} \\ \text{rainwater to} \\ \text{be stored} \end{bmatrix} + \begin{bmatrix} \text{volume taken} \\ \text{up by} \\ \text{vegetation} \end{bmatrix} \qquad (8.13)$$

In Equation 8.13, the volume of rainwater associated with the first flush that will need to be stored is determined from Equation 8.9. The volume above grade taken up by vegetation is

$$V = \begin{bmatrix} \text{number} \\ \text{of plants} \end{bmatrix} \times \begin{bmatrix} \text{cross-sectional} \\ \text{area of the} \\ \text{plant stem} \end{bmatrix} \times \begin{bmatrix} \text{allowable depth that} \\ \text{a plant can be} \\ \text{submerged for} \\ \text{a short duration} \end{bmatrix} \qquad (8.14)$$

With the information provided by Equation 8.14, the total volume required above grade for the bioretention cell can be determined from Equation 8.13. This volume can be divided by the vegetative-specific allowable depth that a plant can be submerged for a short duration (variable in Equation 8.14) to determine the required area. This area may be restrained by site constraints. Also, permeable pavements can be used to reduce the volume of the first flush that is generated.

You can easily design and construct a rain garden for your home. A residential bioretention cell—like commercial ones—is usually 4 to 8 inches deep. The area of a residential bioretention cell that treats runoff from a homeowner's roof typically ranges from 100 to 300 sq. ft. In all cases, bioretention cells (especially those without an overflow drain) are graded so that when they overflow, water flows away from buildings. Parking lot bioretention systems between rows of parking sometimes have a deeper slope, but the level of such an outlet would

**Residential Rain Garden Manual**
http://cee.eng.usf.edu/peacecorps

**Size Factors for Sizing Residential Bioretention Cells**

| Soil Type | Depth of Cell | | |
|---|---|---|---|
| | 3–5 in. | 6–7 in. | 8 in. |
| Sandy soil | 0.19 | 0.15 | 0.08 |
| Silty soil | 0.34 | 0.25 | 0.16 |
| Clayey soil | 0.43 | 0.32 | 0.20 |

SOURCE: WDNR, 2003.

still reflect the specific inundation requirements of the plants and/or reflect the desired volume captured (for example, a first-flush design or 90 percent non-exceedance design).

Residential bioretention cells can also be sized based on understanding the specified depth of the cell (based upon vegetation inundation tolerances) along with knowledge of the soil type. In this case, the area of the cell is determined as follows:

$$\begin{bmatrix} \text{area of residential} \\ \text{bioretention cell} \end{bmatrix} = \text{nonpermeable area} \times \text{size factor} \qquad (8.15)$$

The size factor is related to the porosity of the soil and its ability to infiltrate rainwater. Table 8.9 provides size factors for residential bioretention cells as a function of cell depth and soil type.

### 8.5.4 BIOSWALES AND OTHER LAND USE TECHNIQUES

**Bioswales** (also referred to as *grass swales* and *infiltration trenches*) are engineered conveyance channels that consist of native vegetation. They are not lined with a material like concrete. To the untrained eye, a bioswale will appear to be a grassy channel with a ridge of higher soil placed on either side of the channel, perhaps planted with trees or shrubs.

Bioswales are designed with the longest route of conveyance in mind. Thus, as the water flows through the bioswale, it either transpires through plants or infiltrates through the soil. A meandering conveyance channel encourages infiltration. When a meandering flow path is impossible, porous rock check dams may be used at intervals along bioswales for additional filtration and reduction in the rate of runoff. Bioswales are usually installed along highways or between rows of a parking lot. They have sloped walls and are also sloped along the length for conveyance. When they are placed along roads, sometimes curbs are completely removed. (To take a virtual tour of bioswales, review the Seattle SEAStreets project at www2.cityofseattle.net/util/tours/seastreet/slide1.htm.)

Another way to reduce the impact of nonpermeable coverings is to retain natural vegetation and preserve wetlands. As previously mentioned, grass cover is a permeable pavement, especially in applications where the space is needed for winter months (as in the case of holiday shopping season), when ground is frozen, or is needed in frequently (for example, surrounding sports events and fairs). **Green space** areas can also be preserved to provide seasonal flood storage. Here the green space may flood during seasonal heavy-rain events or after snowmelt. During these times, the space will also provide wildlife habitat and recreation such as bird-watching. When the space dries out later in summer and fall, it can be used as recreational green space.

### 8.5.5 SELECTION OF VEGETATION

Native plants are always preferred over nonnative plants, due to their adaptation to the regional climatic trends. For example, in Midwest bioretention cells, mesic prairie species would be preferred over other

**Class Discussion**

How would your class manage the stormwater generated from the classroom's building roof and an outside paved area? Where does the stormwater currently go? How would you maximize the benefit of low-impact development while accounting for movement of students, faculty, staff, services, water, and biodiversity? How would you design a system that uses the least energy and materials? What economic, social, and environmental benefits would be preserved for future generations?

**Figure 8.14** **Root Systems of Prairie Plants** This drawing demonstrates the proportional length of turf grass (shown on the far left with little root depth) and various native prairie plants. The longest roots shown are 15 ft. long.

Heidi Natura, Living Habitats © 1995.

plant species. (*Mesic* is a designation for an ecosystem's level of dryness; it falls midway between wet and dry.) For other U.S. regions, the Brooklyn Botanic Garden provides starter lists of plants for bioretention cells (BBG, 2007).

Plant selection is critical to any plant-based engineered system. A professional with expertise in plants should be consulted, and native plants should be used whenever possible. Not only do native plants increase biodiversity, but they are adapted to regional climatic trends and will require less irrigation and long-term maintenance.

When selecting native plants, it is important to consider the plant root depth. Figure 8.14 visually shows the proportional root depth of a variety of native prairie species relative to turf grass, which is common in urban environments and penetrates only a few inches. In comparison, the longest roots in Figure 8.14 extend downward 15 ft. Long roots and large root masses enable the plants to locate water in periods of drought and can serve as conduits to transfer oxygen into the subsurface.

## 8.6 Groundwater Quality

For many years, the prevailing view was that contaminants released into the subsurface would be cleansed by aquifer materials or diluted in the aquifer water. However, a wide range of toxic substances, including synthetic organic chemicals, trace metals, and pathogenic microorganisms, have been detected at potentially harmful levels in groundwater, and concern over subsurface contamination has grown.

### 8.6.1 SOURCES AND CHARACTERISTICS OF GROUNDWATER CONTAMINANTS

**Groundwater contaminants** originate from a wide variety of natural and anthropogenic sources. Table 8.10 lists sources of contamination, all of which exhibit different areal and temporal characteristics. For example, pathogens released from a septic tank may originate over a small area, while pesticide applications may involve more than 1,000 km². Contaminant releases may occur over a short time span, such as those

## Table / 8.10

### Selected Sources of Groundwater Contamination

| | |
|---|---|
| Septic tanks | Leaking underground storage tanks |
| Agricultural activities | Industrial activities |
| Landfills: domestic and hazardous | Mining activities |
| Chemical spills | Petroleum extraction |
| Improper hazardous-waste disposal | Home lawn care |

**Global Arsenic Crisis**

http://www.who.int/topics/arsenic/en/

**Underground Storage Tanks in Your Area**

http://www.epa.gov/OUST/wheruliv.htm

associated with a single, essentially instantaneous chemical spill, or over a longer time span (even decades), such as leakage from improperly stored chemical wastes. Knowledge of the areal and temporal extent of a contaminant release is required in the design of remediation plans and preventive measures.

Groundwater contaminants include pathogens, naturally occurring inorganic chemicals (for example, arsenic), radionuclides (for example, uranium) and anthropogenic inorganic (for example, nitrates and metals) and organic chemicals (pesticides, solvents, and petroleum products). Pathogens usually originate from improperly designed or maintained septic tanks. Radionuclide contamination, while sometimes associated with the mineral composition of aquifer materials, is most often associated with nuclear processing facilities. Nutrients (for example, nitrate) and pesticides are associated with agricultural and lawn care activities. Organic chemicals may originate from chemical spills, improper hazardous waste disposal, leaking underground storage tanks, and industrial activities.

**Nonaqueous phase liquids (NAPLs)** are a specialized class of organic contaminants that do not mix readily with water (the meaning of nonaqueous). After NAPLs are released to an aquifer, they remain more or less separate from the groundwater. Lighter-than-water NAPLs, such as gasoline, tend to collect on top of the water table. Denser-than-water NAPLs (DNAPLs), such as chlorinated solvents, will penetrate the water table and migrate vertically until they encounter an aquifer material with a low-hydraulic conductivity (for example, clay or rock). NAPLs are poorly soluble, so a small discharge can contaminate large volumes of water; even at these low solubilities, however, NAPLs often achieve concentrations exceeding toxicity limits (Mayer and Hassanizadeh, 2005).

### 8.6.2 POLLUTANT FATE AND TRANSPORT IN GROUNDWATER

Groundwater contaminants are transported by two processes: advection and dispersion. **Advection** occurs when the contaminant is transported with the bulk flow, that is, moving groundwater. Since movement in the subsurface may occur in all three dimensions simultaneously and can change in time and space, prediction of advective transport can be quite complicated. However, in a simple, one-dimensional system where groundwater flow is constant in time and space, the travel time of a contaminant over a given length can be estimated as

$$t = \frac{L}{v} \tag{8.16}$$

where $t$ is the travel time (days), $L$ is length (m), and $v$ is the pore (interstitial or seepage) velocity (m/day).

*Specific discharge* is the rate of groundwater flow per unit of cross-sectional area of the aquifer ($m^3/m^2$-day or m/day). Specific discharge is related to the aquifer's *hydraulic conductivity* ($K$, $m^3$/m-day) and the *hydraulic gradient* ($dh/dx$, dimensionless) by **Darcy's law**. Note, however, that these terms express groundwater flow normalized to the entire cross section, that is, both pore space and solid material. Because flow actually occurs only through the pore space, porosity must be accommodated in calculating the rate of groundwater transport. This is done by normalizing specific discharge and Darcy's law for aquifer's porosity ($n$, dimensionless):

$$v = \frac{q}{n} = -\frac{K}{n}\frac{dh}{dx} \tag{8.17}$$

Because $0 < n < 1$, the pore velocity will be always be greater than the specific discharge.

---

## example/8.5 Calculation of Groundwater Travel Time

A toxic contaminant is released 1 km upstream of a drinking-water supply well. The relevant aquifer properties are as follows: hydraulic conductivity $K = 10^{-5}$ m/s; porosity $n = 0.3$; and hydraulic gradient $dh/dx = -10^{-2}$. Determine how long it will take for the contaminant to reach the drinking-water supply well.

### solution

The pore velocity can be determined from Equation 8.17.

$$v = \frac{q}{n} = -\frac{K\,dh}{n\,dx} = -\left(\frac{10^{-5}\,\text{m/s}}{0.3}\right) \times -10^{-2} = 3.33 \times 10^{-7}\,\text{m/s}$$

The travel time is then determined from Equation 8.16:

$$t = \frac{L}{v} = \frac{1,000\,\text{m}}{3.33 \times 10^{-7}\,\text{m/s}} = 3.00 \times 10^{9}\,\text{s} = 95\,\text{yr}$$

The long travel time calculated here underscores the fact that groundwater transport can be slow and may suggest that we do not need to be concerned about the contamination for a long time. However, by the time the contamination might be detected at the well, a large portion of the aquifer and a significant volume of groundwater may be contaminated as the plume spreads.

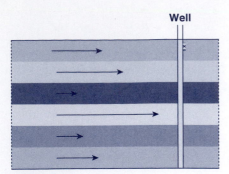

**Well**

**Figure 8.15** **Aquifer Scale Dispersion Processes That Occur in Groundwater** The shading in layers indicates the relative value of hydraulic conductivity, with lighter shades indicating higher values. The length of the arrows indicates the relative groundwater velocity in each layer. (Pore scale dispersion processes were discussed in Chapter 4.)

Contaminants tend to spread out or mix (disperse) as they are advected with the flowing groundwater, becoming more dilute and occupying an increasingly larger volume. This second transport process, **dispersion**, results from the effects of two mechanisms: diffusion and mechanical mixing. Diffusion is driven by concentration gradients, while mechanical mixing results from interactions with the aquifer's solid matrix. The contribution of mechanical mixing to dispersion dominates that of diffusion except at low pore velocities (for example, in aquifers with low hydraulic conductivity). Both these terms were discussed in Chapter 4.

The travel time calculation illustrated in Example 8.5 assumes that once the contaminant is released, the contaminant molecules experience identical velocities as they migrate through the aquifer. In fact, each molecule will be subject to velocities that can range over many orders of magnitude; from Equation 8.16, if $v$ varies by many orders of magnitude, then the travel times will vary similarly. Variations in pore velocity are caused by mixing phenomena occurring over a wide range of length scales. Velocity differences at the pore scale (described in Chapter 4) result from friction with the solid media, an effect manifested to a different degree at different positions over the cross section. Pore scale velocity differences are also influenced by the flow paths that molecules travel as they migrate through the aquifer medium. At the aquifer scale, Figure 8.15 shows how variations in velocity are caused by differences in hydraulic conductivity resulting in contaminants' arrival at a particular location at different times.

To visualize the effects of advection on contaminant transport, consider a column packed with a porous medium (Figure 8.16a). The inflow to the column originates from one of two reservoirs, one containing clear water and the other a tracer dye. Up until time $t = 0$, the pump is delivering water ($C_{in} = 0$), and the dye concentration in the effluent is zero. At time $t = 0$, the pump switches to the reservoir containing dye ($C_{in} = C_0$), and the "contaminant" enters the column. The effluent dye concentration ($C_{out}$) is presented in the inset of Figure 8.16a. Because transport is assumed to be only by advection, the effluent concentration

(a) Assuming transport by advection only

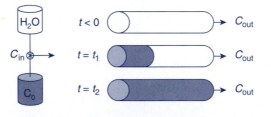

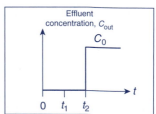

(b) Assuming transport by advection and dispersion

**Figure 8.16** **Transport of a Tracer in a Column Study** (a) The first case assumes transport by advection only. (b) The second case assumes transport by advection and dispersion.

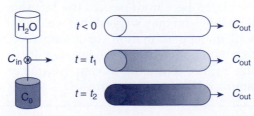

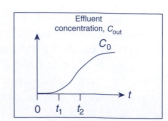

remains zero ($C_{out} = 0$) until the dye front reaches the end of the column ($t < t_2$). At time $t = t_2$, the front reaches the end of the column, and the effluent concentration immediately climbs to 100 percent of the influent chemical concentration ($C_{out} = C_0$).

Figure 8.16b shows the effects of both advection and dispersion. Here, the dye travels at a range of velocities due to dispersion and reaches the end of the column earlier and at a lower concentration than with advective transport alone. The effluent dye (shown in the inset in Figure 8.16b) climbs slowly following arrival of the front, eventually reaching 100 percent of the influent concentration ($C = C_0$). The time to reach that maximum concentration is significantly later than $t = t_2$ (the time required for $C = C_0$ in the advection-only case).

The practical significance of this is that the travel time (for arrival of the front) is shorter with dispersion than with advection alone. Referring again to the travel time calculation in Example 8.5, when the effects of dispersion are included, the contaminant reaches the drinking-water supply well at potentially toxic levels orders of magnitude more quickly than with transport only by advection.

Groundwater contaminants rarely behave conservatively, as they may be degraded by microorganisms or chemically react in the aquifer. Reaction rates vary widely in groundwater systems and depend on many factors, such as the nature of the chemical of interest and the aquifer's biogeochemistry. In the case of microbially mediated reactions, rates are influenced by the types and abundance of microorganism present, the chemical properties of the contaminant, and the concentrations of electron donors and acceptors (for redox reactions), nutrients, and energy sources. In many cases, contaminant reactions are beneficial, reducing contaminant levels. However, in some cases (especially chlorinated solvents), the reaction products can be at least as toxic as the chemical originally released. Contaminants may also interact with solid materials in the aquifer through sorption (discussed in Chapter 3).

Assuming that the chemical–solid interactions are at equilibrium, the effects of sorption can be quantified through the **retardation coefficient ($R_f$)**:

$$R_f = 1 + \frac{\rho_b}{\eta} K_p \qquad (8.18)$$

where $K_p$ is the soil–water partition coefficient (L/kg or cm$^3$/mg), $\eta$ is the porosity (unitless), and $\rho_b$ is bulk density of the aquifer material (cm$^3$/g).

As the term implies, the retardation coefficient reflects the apparent slowing of chemical migration due to sorption. If the retardation factor equals ten, the average velocity of the chemical would be ten times slower than the average velocity of the groundwater. The calculation of travel time can be modified as follows to account for chemical sorption:

nonsorbing (conservative) chemical: $\quad t = \dfrac{L}{v} \qquad (8.19)$

sorbing chemical: $\quad t = \dfrac{L}{v} \times R_f \qquad (8.20)$

**Groundwater Assessment in Africa**
http://www.unep.org/groundwater
project/

### 8.6.3  GROUNDWATER REMEDIATION STRATEGIES

Contaminated groundwater is expensive and time-consuming to clean up. This is true for several reasons. First, pollution problems are often discovered years after the initial contamination, offering the chemical time to affect millions of liters of water and extend kilometers from the source. Second, the subsurface is physically and chemically heterogeneous. This heterogeneity leads to slow rates of removal or in-place degradation of contaminants, even when remediation is aggressive. Finally, little information is usually available concerning the physical and chemical characteristics of the contaminated aquifer system. The lack of information, coupled with aquifer heterogeneity and the fact that the subsurface is essentially invisible, means that the design and implementation of economical and effective remediation systems presents a challenge.

Groundwater remediation may be accomplished by physical, biological, or chemical means. Physical remediation usually involves pumping the contaminated groundwater out of the aquifer, treating it aboveground, and disposing of the treated groundwater back to the impacted aquifer, to a surface water, or to a wastewater collection system. Soil vapor extraction and air sparging involve pumping and recovering air into and from the unsaturated (vadose) and saturated zones, respectively. Biological remediation consists of injecting electron acceptors and nutrients. Chemical remediation also involves adding chemicals, either to enhance abiotic degradation or to immobilize contaminants through chemical precipitation.

The concept of **natural attenuation** has been applied frequently in recent years. Natural attenuation implies that as the contaminant plume moves in the general direction of groundwater flow, biological and chemical degradation and dilution processes will reduce concentrations to acceptable levels at the "front" or leading edge of the plume.

Specialized remediation strategies are required when aquifers are contaminated by NAPLs. The chemical properties that cause NAPLs to remain separate from water make it difficult to recover them from aquifers. The injection of surfactants or alcohols that tend to partition in both water and NAPLs (much as detergents do in the laundry) has been somewhat successful in enhancing the efficiency of NAPL recovery. Heating the subsurface to enhance volatilization, decrease interfacial tension, and reduce viscosity also has been shown to improve NAPL recovery.

**Innovative Remediation Technologies**
http://www.epa.gov/tio/

## Key Terms

- advection
- best management practices (BMPs)
- bioretention cells
- bioswales
- bogs
- built environment

- Clean Water Act
- compensatory mitigation
- constructed wetland
- created wetland
- critical point
- cultural eutrophication
- Darcy's Law

- deoxygenation
- dispersion
- dissolved oxygen (DO)
- epilimnion
- eutrophic
- eutrophication
- fall turnover

- fens
- first flush
- green roof
- green space
- groundwater contaminants
- hydric soils
- hydroperiod
- hydrophytic
- hypolimnion
- land use
- load
- low-impact development
- macropollutants
- marsh

- mesotrophic
- metalimnion
- National Pollutant Discharge Elimination System (NPDES)
- natural attenuation
- nonaqueous phase liquids (NAPLs)
- nonpoint source
- oligotrophic
- oversaturation
- oxygen deficit
- permeable (or porous) pavement
- point sources

- reaeration
- retardation coefficient ($R_f$)
- saturation concentration
- spring turnover
- stormwater
- Streeter–Phelps model
- summer stratification
- swamp
- thermocline
- trophy
- wetland banking
- wetlands
- winter stratification

**8.1** A stream at 25°C has a dissolved-oxygen concentration of 4 mg/L. What is the dissolved-oxygen deficit (mg/L)?

**8.2** The oxygen concentration of a stream is 4 mg/L, and DO saturation is 10 mg/L. What is the oxygen deficit?

**8.3** Calculate the dissolved-oxygen deficit for a river at 30°C and a measured dissolved-oxygen concentration of 3 mg/L. The Henry's law constant at that temperature is $1.125 \times 10^{-3}$ mole/L-atm, and the partial pressure of oxygen is 0.21 atm.

**8.4** A wastewater treatment plant discharges an effluent containing 2 mg/L of dissolved oxygen to a river that has a dissolved-oxygen concentration of 8 mg/L upstream of the discharge. Calculate the dissolved-oxygen deficit at the mixing basin if the saturation dissolved oxygen for the river is 9 mg/L. Assume that the river and plant discharge have the same flow rate.

**8.5** A river traveling at a velocity of 10 km/day has a dissolved-oxygen content of 5 mg/L and an ultimate CBOD of 25 mg/L at distance $x = 0$ km, that is, immediately downstream of a waste discharge. The waste has a CBOD decay coefficient $k_1$ of 0.2/day. The stream has a reaeration rate coefficient $k_2$ of 0.4/day and a saturation dissolved-oxygen concentration of 9 mg/L. (a) What is the initial dissolved-oxygen deficit? (b) What is the location of the critical point, in time and distance? (c) What is the dissolved-oxygen deficit at the critical point? (d) What is the dissolved-oxygen concentration at the critical point?

**8.6** The wastewater treatment plant for Pine City discharges $1 \times 10^5$ m³/day of treated waste to the Pine River. Immediately upstream of the treatment plant, the Pine River has an ultimate CBOD of 2 mg/L and a flow of $9 \times 10^5$ m³/day. At a distance of 20 km downstream of the treatment plant, the Pine River has an ultimate CBOD of 10 mg/L. The state's Department of Environmental Quality (DEQ) has set an ultimate CBOD discharge limit for the treatment plant of 2,000 kg/day. The river has a velocity of 20 km/day. The CBOD decay coefficient is 0.1/day. Is the plant in violation of the DEQ discharge limit?

**8.7** An industry discharges 0.5 m³/s of a waste with a 5-day CBOD of 500 mg/L to a river with a flow of 2 m³/s and a 5-day CBOD of 2 mg/L. Calculate the 5-day CBOD of the river after mixing with the waste.

**8.8** A waste having an ultimate CBOD of 1,000 mg/L is discharged to a river at a rate of 2 m³/s. The river has an ultimate CBOD of 10 mg/L and is flowing at a rate of 8 m³/s. Assuming a reaction rate coefficient of 0.1/day, calculate the ultimate and 5-day CBOD of the waste at the point of discharge (0 km) and 20 km downstream. The river is flowing at a velocity of 10 km/day.

**8.9** A new wastewater treatment plant proposes a discharge of 5 m³/s of treated waste to a river. State regulations prohibit discharges that would raise the ultimate CBOD of the river above 10 mg/L. The river has a flow of 5 m³/s and an ultimate CBOD of 2 mg/L. Calculate the maximum 5-day CBOD that can be discharged without violating state regulations. Assume a CBOD decay coefficient of 0.1/day for both the river and the proposed treatment plant.

**8.10** A river flowing with a velocity of 20 km/day has an ultimate CBOD of 20 mg/L. If the organic matter has a decay coefficient of 0.2/day, what is the ultimate CBOD 40 km downstream?

**8.11** A river traveling at a velocity of 10 km/day has an initial oxygen deficit of 4 mg/L and an ultimate CBOD of 10 mg/L. The CBOD has a decay coefficient of 0.2/day, and the stream's reaeration coefficient is 0.4/day. What is the location of the critical point: (a) in time; (b) in distance?

**8.12** A paper mill discharges its waste ($k_L = 0.05$/day) to a river flowing with a velocity of 20 km/day. After mixing with the waste, the river has an ultimate carbonaceous BOD of 50 mg/L. Calculate the 5-day carbonaceous BOD at that location and the ultimate carbonaceous BOD remaining 10 km downstream.

**8.13** For each of the following cases, assuming all other things unchanged, describe the effect of the following parameter variations on the magnitude of the maximum oxygen deficit in a river. Use the following symbols to indicate your answers: increase (+), decrease (−), or remain the same (=).

| Parameter | Magnitude of the Deficit |
|---|---|
| Increased initial deficit | _____ |
| Increased ultimate CBOD @ $x = 0$ | _____ |
| Increased deoxygenation rate | _____ |
| Increased reaeration rate | _____ |
| Increased ThOD @ $x = 0$ | |

**8.14** Contact a wastewater treatment plant located in a nearby urban area to find out the average daily flow rate and average concentration of phosphorus in the effluent. Use census data to determine the projected population of your area in 2025 and 2050. If nothing is done in how the plant treats phosphorus, what is the current and future P loading (kg P/day)? Develop a technical and nontechnical solution to reduce future phosphorus loading to the wastewater treatment plant.

**8.15** Use the library or Internet to research one dead zone located in the United States and one located overseas, such as the Baltic Sea, northern Adriatic Sea, Yellow Sea, or Gulf of Thailand. Write a two-page report discussing environmental, social, and economic issues associated with the dead zones. What management solutions would you propose to reverse the dead zones?

**8.16** Describe some differences and similarities among the three types of green roofs: (a) extensive, (b) semi-intensive, and (c) intensive. Identify a location on your campus (or community) where an extensive roof and intensive roof could be installed. (Explain your answer.)

**8.17** Go to the Brooklyn Botanic Gardens Web site, www.bbg.org, and search for "rain garden plants" (use the quotation marks). Follow the link to the page titled "Rain Garden Plants." Use the information there to identify plant species that would be applicable for a bioretention cell installed in your community (or hometown).

**8.18** Assume that the roof area for a residential home is 12 ft. × 30 ft. If a green roof is placed on the home, what percentage of a 0.5 in. rain event will be stored on the roof if the growing medium has a water-holding capacity of 0.25? What is the volume of water (in gallons) that is stored during this rain event?

**8.19** What area (in sq. ft.) is needed for two bioretention cells used to collect rainwater coming from a household roof? The roof has dimensions of 30 ft. × 40 ft. It drains to two downspouts, each of which will be routed to a bioretention cell. Assume the soil surrounding the home is silty and the cell will be dug to a depth of 6 in.

**8.20** A 1-acre paved parking lot measures 50 ft. × 20 ft. What volume of bioretention cell is required (in cu. ft.) that can handle a first flush from the nonpermeable pavement of 0.5 in.? Assume the soil porosity is 0.30.

**8.21** Select a specific location on your campus that has a building and associated parking lot. Redesign this area, incorporating at least three low-impact development techniques. Besides thinking about management of stormwater, also consider the movement of people and vehicles, and the use of native plant species.

**8.22** Size a rain garden for your current home, apartment, or dormitory to treat stormwater that originates from the roof.

**8.23** Two groundwater wells are located 100 m apart in permeable sand and gravel. The water level in well 1 is 50 m below the surface, and in well 2 the water level is 75 m below the surface. The hydraulic conductivity is 1 m/day, and porosity is 0.60. What is: (a) the Darcy velocity; (b) the true velocity of the groundwater flowing between wells; and (c) the period it takes water to travel between the two wells, in days?

**8.24** The hydraulic gradient of groundwater in a certain location is 2 m/100 m. Here, groundwater flows through sand, with a hydraulic conductivity equal to 40 m/day and a porosity of 0.5. An oil spill has caused the pollution of the groundwater in a small region beneath an industrial site. How long would it take the polluted water from that location to reach a drinking-water well located 100 m downgradient? Assume no retardation of the pollutant's movement.

**8.25** An underground storage tank has discharged diesel fuel into groundwater. A drinking-water well is located 200 m downgradient from the fuel spill. To ensure the safety of the drinking-water supply, a monitoring well is drilled halfway between the drinking-water well and the fuel spill. The difference in hydraulic head between the drinking-water well and the monitoring well is 40 cm (with the head in the monitoring well higher). If the porosity is 39 percent and hydraulic conductivity is 45 m/day, how long after the contaminated water reaches the monitoring well would it reach the drinking-water well? Assume the pollutants move at the same speed as the groundwater.

**8.26** Spills of organic chemicals that contact the ground sometimes reach the groundwater table, where they are then carried downgradient with the groundwater flow. The rate at which they are transported with the groundwater is decreased by sorption to the solids in the groundwater aquifer. The contaminated groundwater can reach wells, which is dangerous if the water is used for drinking. (a) For a soil having $\rho_b$ of 2.3 g/cm$^3$ and a porosity of 0.3, and for which the percent organic carbon equals 2 percent, determine the retardation factors of trichloroethylene (log $K_{ow}$ = 2.42), hexachlorobenzene (log $K_{ow}$ = 5.80), and dichloromethane (log $K_{ow}$ = 1.31). (b) Which compound would be transported farthest, second farthest, and least far with the groundwater if these chemicals entered the same aquifer?

# References

Brooklyn Botanic Garden (BBG). 2007. "Rain Garden Plants." BBG Web site, www.bbg.org/gar2/topics/design/2004sp_raingardens .html, accessed June 22, 2007.

Dahl, T. E. 2006. States and Trends of Wetlands in the Conterminous United Status 1998 to 2004, U.S. Department of the Interior, Fish and Wildlife Service, Washington, D.C., 112 pp.

Environmental Law Institute (ELI). 2002. "Banks and Fees: The Status of Off-Site Wetland Mitigation in the United States," Washington, D.C. http://www.eli.org/Program_Areas/WMB/bankfees.cfm, accessed Nov 18, 2008.

Environmental Protection Agency (EPA). 1999. *StormWater Technology Fact Sheet: Porous Pavement*. EPA 832-F-99-023. Washington, D.C.: Environmental Protection Agency, Office of Water.

Environmental Protection Agency (EPA). 2000 (revised 2005). "Stormwater Phase II Final Rule: Small MS4 Stormwater Program Overview." EPA Web site, www.epa.gov/npdes/pubs/fact2-0.pdf, accessed July 7, 2006.

Larson, J. 2004. "Dead Zones Increasing in the World's Coast Waters." Earth Policy Institute (June 16). http://www.earth-policy.org/updates/update41.htm, accessed Nov 18, 2008.

Mayer, A. S., and S. M. Hassanizadeh, ed. 2005. *Contamination of Soil and Groundwater by Nonaqueous Phase Liquids (NAPLs)*. Water Resources Monograph Series. Washington, D.C.: American Geophysical Union.

Mihelcic, J. R. 1999. *Fundamentals of Environmental Engineering*. New York: John Wiley & Sons.

Portland Cement Association (PCA). 2004. "Pervious Concrete Mixtures and Properties." *Concrete Technology Today* 25 (3).

Prince George's County, Md. (PGCM), Department of Environmental Resources. 2006. *The Bioretention Manual*; Watershed Protection Branch, Largo: Maryland Department of Environmental Resources. Available at www.goprincegeorgescounty.com/government/ agencyindex/der/esd/bioretention/bioretention.asp, accessed December 3, 2007.

Seitzinger, S. P., and C. Kroeze. 1998. "Global Distribution of Nitrous Oxide Production and N Inputs in Freshwater and Coastal Marine Ecosystems." *Global Biogeochemical Cycles* 12 (1): 93–113.

Selman, M., S. Greenhalgh, R. Diaz, and Z. Sugg. 2008. "Eutrophication and Hypoxia in Coastal Areas: A Global Assessment of the State of Knowledge." WRI Policy Note. Washington, D.C.: World Resources Institute (March).

United Nations Educational, Scientific, and Cultural Organization (UNESCO) and World Water Assessment Programme (WWAP). 2003. *Water for People, Water for Life: The United Nations World Water Development Report*. New York: UNESCO / Berghahn Books.

U.S. Army Corps of Engineers (USACE). 2008a. "FAQs: What You Should Know about the Comprehensive Everglades Restoration Plan (CERP)." Comprehensive Everglades Restoration Plan Web site, www.evergladesplan.org, accessed Nov 18, 2008.

U.S. Army Corps of Engineers (USACE). 2008b. "Water Flow Maps of the Everglades: Past, Present & Future." Comprehensive Everglades Restoration Plan Web site, www.evergladesplan.org, accessed Nov 18, 2008.

Ward, A. S. 2007. "A Review of the Practice of Low Impact Development: Bioretention Design, Analysis, and Lifecycle Assessment." (M.S. report, Civil & Environmental Engineering, Michigan Technological University).

Wisconsin Department of Natural Resources (WDNR). 2003. DNR Publication PUB-WT-776. Available from the University of Wisconsin Extension, www.clean-water.uwex.edu/pubs/ raingarden, accessed June 7, 2007.

# chapter/Nine Water Supply, Distribution, and Wastewater Collection

Brian E. Whitman, James R. Mihelcic, Alex S. Mayer

*In this chapter, readers will review the availability, sources, and usage of water. The chapter provides an overview of the hydrologic cycle and groundwater hydrology. It examines water availability and its usage at a global scale, discusses water usage and reuse in the United States, and finally focuses on water demand and wastewater generation for communities. Concepts for sizing and designing treatment facilities, water distribution systems, and wastewater collection systems are also introduced.*

## Major Sections

9.1 Introduction
9.2 Water Availability
9.3 Water Usage
9.4 Municipal Water Demand
9.5 Water Distribution and Wastewater Collection Systems

## Learning Objectives

1. Identify the quantities and sources of freshwater on a global basis.
2. List the components that make up groundwater hydrology.
3. Identify the major users of water and the percent of water use associated with types of users.
4. Describe the perspective of the global population that lives in areas not equally served by global water resources, and discuss the challenges experienced by people living in these areas.
5. Articulate how you as an engineer can help achieve more equitable distribution of water resources in your community and the world.
6. Associate sources of water with issues of water quality and usage.
7. Identify a local or regional water reuse or water reclamation project, and describe its environmental and social benefits.
8. Articulate how you as an engineer can help achieve more sustainable management of water resources in your region.
9. Estimate water and wastewater flow rates for residents and communities.
10. Distinguish daily demand cycles for industry, residential, and commercial uses.
11. Estimate demand factors and household water usage rates from historical records.
12. Determine water demand associated with fire protection and losses due to items such as leakage and unmetered use.
13. Calculate wet-weather flows based on inflow and infiltration.
14. Project future water demand using extrapolation methods and the type of customer.
15. Lay out a water distribution system.
16. Size a wastewater collection pipe based on the full-pipe design flow velocity and design carrying capacity.
17. Size a wet well based on pump characteristics.

# 9.1 Introduction

The fundamental science that deals with the occurrence, movement, and distribution of water on the planet is **hydrology**. The **hydrologic cycle** is defined as the pathways for how water moves and is distributed on the planet. It is depicted in Figure 9.1. The quantity and quality of water greatly varies as it moves through the hydrologic cycle.

Groundwater systems can extend from Earth's surface to meters to kilometers below the surface. These systems are composed of porous solid materials (for example, sand, gravel, and clay) containing water and air. Solid and fractured rock also can be thought of as porous materials that compose groundwater systems.

Figure 9.2 shows a typical groundwater system consisting of two **aquifers** (**confined aquifers** and **unconfined aquifers**). Aquifers are subsurface formations that yield water at sufficient amounts to provide practical sources of water for human usage. The **vadose zone** extends from the ground surface to the water table. In this zone, air and water are found in the pore spaces. The uppermost layer of the vadose zone consists of a thin (centimeters to meters) layer of soil, which is capable of supporting intense biological activity. Water in this zone consists of thin films surrounding the surfaces of the porous materials. The pressure in the pore water in the vadose zone is less than atmospheric pressure.

Water generated by precipitation infiltrates the vadose zone and, if sufficient volume is available, migrates to the water table. The *water table* is the depth at which the pressure in the pore water equals atmospheric pressure. In more practical terms, we can imagine finding the water table by digging through the vadose zone until we find standing water. The *saturated zone* begins near the location of the water table. A thin, saturated layer extends above the water table, drawn upward by *capillary forces*. This saturated layer is referred to as the *capillary fringe*. Below the water table, the porous media are saturated, and the pore water is under positive pressure.

**Public Water Information**
http://water.usgs.gov/

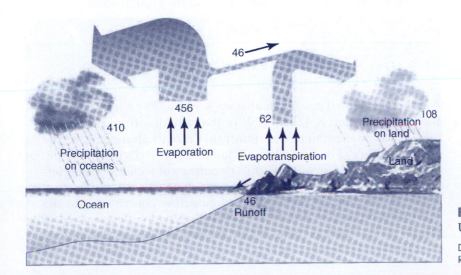

46

456    62          Precipitation    108
410                on land

Precipitation    Evaporation    Evapotranspiration
on oceans                                          Land

Ocean          46
               Runoff

**Figure 9.1    The Hydrologic Cycle**
Units of water transfer are $10^{12}$ m$^3$/yr.

Data from Budyko (1974); from Mihelcic (1999). Reprinted with permission of John Wiley & Sons, Inc.

**Figure 9.2  Typical Groundwater System**   Vertical scale is exaggerated. The aquitard/aquiclude will not yield significant water to a well. An aquiclude is relatively impermeable (e.g., clay). Similarly, an aquitard is not very permeable; however, it may allow water to move between adjoining aquifers.

Figure adapted from Hemond and Fechner, *Chemical Fate and Transport in the Environment*, Copyright Elsevier (1994).

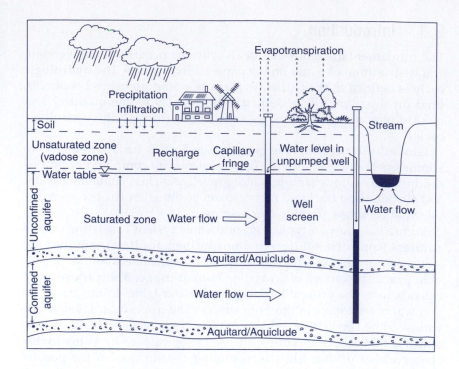

The role of groundwater systems in the hydrologic cycle is shown in Figure 9.2. A portion of the water that infiltrates the vadose zone is lost to the overlying atmosphere via evaporation from the pore spaces or transpiration through plant roots. The combination of these processes is referred to as *evapotranspiration*. Groundwater systems are recharged by the remaining, infiltrated water. The water that reaches the saturated zone moves in accord with pressure gradients and gravity and can migrate from one aquifer to another. As indicated in Figure 9.2, horizontally flowing groundwater can reach surface water, such as rivers, lakes, and oceans. Groundwater contributed to flows in streams and rivers is termed **baseflow** and usually constitutes a significant portion of surface water flow. In some circumstances, the direction of flow is reversed, and water is lost from the surface water body to the groundwater.

## example/9.1  Using a Water Balance to Determine Baseflow

The average annual rainfall in an area is 90 cm. The average annual evapotranspiration is 42 cm. Half of the rainfall infiltrates into the subsurface; the remainder is runoff that moves along or near the ground surface. The underlying aquifer is connected to a stream. Assuming there are no other inputs or outputs of water to the underlying aquifer and the aquifer is at steady state (it neither gains nor loses water), what is the amount of baseflow contributed to the stream from the groundwater?

## solution

To solve this problem, first determine the inputs and outputs to the aquifer. The input is

$$I = \text{infiltration} = 90 \text{ cm/yr} \times 0.5 = 45 \text{ cm/yr}$$

The outputs are

$$ET = \text{evapotranspiration} = 42 \text{ cm/yr}$$

$$BF = \text{baseflow}$$

A water balance equation (inputs = outputs) can be written for the system as follows:

$$I = ET + BF$$

Solve this expression for baseflow:

$$BF = I - ET = 45 \text{ cm/yr} - 42 \text{ cm/yr} = 3 \text{ cm/yr}$$

The units for the baseflow (length per time) may seem strange. If you knew the plan (bird's-eye view) area of the aquifer, you could multiply the precipitation and evapotranspiration by that area, resulting in units of volume per time. For example, if the area under consideration were 100 km$^2$, the baseflow (in units of volume/time) would be

$$BF = \frac{3 \text{ cm}}{\text{yr}} \times \frac{1 \text{ m}}{100 \text{ cm}} \times 100 \text{ km}^2 \times \left(\frac{1,000 \text{ m}}{\text{km}}\right)^2 = 3 \times 10^6 \text{ m}^3/\text{yr}$$

This example does not consider the possibility that water is being withdrawn from the groundwater system. If we assume there are 20 pumping wells within the 100 km$^2$, each of which is pumping at an average of 2 $\times$ 10$^{-3}$ m$^3$/s, the total pumping rate is then

$$Q = 20 \times \frac{2 \times 10^{-3} \text{ m}^3}{\text{s}} \times \frac{60 \text{ s}}{\text{min}} \times \frac{60 \text{ min}}{\text{hr}} \times \frac{24 \text{ hr}}{\text{day}} \times \frac{365 \text{ days}}{\text{yr}}$$

$$= 1.26 \times 10^6 \text{ m}^3/\text{yr}$$

Note here how the total pumping rate is approaching the magnitude of the baseflow. This might indicate that this level of pumping is not sustainable in the long term. It could reduce the amount of baseflow that contributes to the stream flow and supports ecological functions.

## 9.2 Water Availability

The total volume of the world's water is estimated to be $1.386 \times 10^9 \, km^3$. Oceans hold 96.5 percent of this total volume, and the atmosphere contains only $1.29 \times 10^4 \, km^3$ of water (which is only 0.001 percent of the total hydrosphere). Table 9.1 shows the world's freshwater reserves.

The total amount of **freshwater** on our planet is approximately $3.5 \times 10^7 \, km^3$. In terms of freshwater availability, only 2.5 percent of the world's total water budget is estimated to be freshwater, and of this, almost 70 percent is currently present as glaciers and ice sheets. As shown in Table 9.1, a large percentage of the world's freshwater is available as a groundwater resource, much of which has a renewal period of over 1,000 years. All of this information shows that very little of the total freshwater budget is available as surface water (lakes, rivers) or as groundwater that is recharged over a short duration.

Figure 9.3 shows the relationship of global water availability to population. The Americas are relatively rich in available water resources relative to their population. North and Central America combined have 8 percent of the world's population and 15 percent of the world available water resources. In contrast, Asia contains 36 percent of the world's available water resources but houses 60 percent of the global population.

## 9.3 Water Usage

Water is required by a wide variety of human users, including residential homes, commercial entities, industry, agriculture, and importantly, ecosystems. The concept of the ecological footprint (discussed in

**State of the World's Water**

www.unep.org/vitalwater

**Class Discussion**

Discuss some regional and global challenges you expect over the next century related to the distribution of population and water. How do demographics (for example, income, level of education, age, and gender) relate to your discussion?

### Table / 9.1

**Percent of World's Total Freshwater in Different Locations**  The total amount of freshwater on Earth is approximately $3.5 \times 10^7 \, km^3$

| Location | Percent of World's Freshwater |
|---|---|
| Glaciers and permanent snow cover | 68.7 |
| Groundwater | 30.1 |
| Lakes | 0.26 |
| Soil moisture | 0.05 |
| Atmosphere | 0.04 |
| Marshes and swamps | 0.03 |
| Biological water | 0.003 |
| Rivers | 0.006 |

SOURCE: Data from UNESCO–WWAP, 2003.

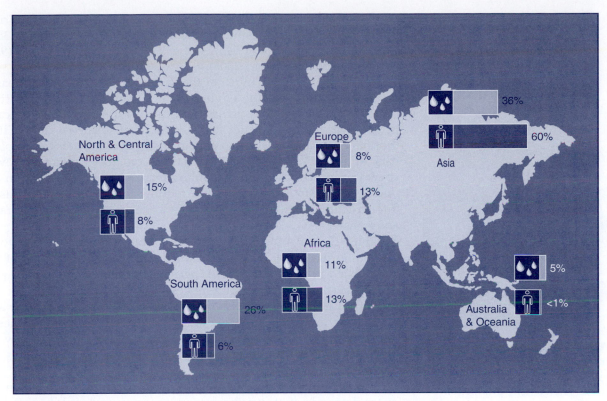

**Figure 9.3    Global Overview of Water Availability versus Population**    Continental disparities exist, particularly on the Asian continent. About 60% of the world's population resides in Asia, yet only 36% of the world's water resources are located there.

Figure adapted from UNESCO–WWAP (2003).

Chapter 5) assumed that a minimum of 12 percent of the world's ecological capacity should be preserved for biodiversity protection. This ecological capacity requires water. In addition, much of the world's poor (those living on less than $1 or $2 per day) depend on ecosystems for their economic livelihood. The water requirements of ecosystems must therefore be accounted for when managing how water is distributed among various users.

Globally, 3,800 km$^3$ of water are withdrawn every year. Of this, 2,100 km$^3$ are consumed. **Consumed water** is evapotranspirated or incorporated into products or organisms (UN-Habitat, 2003). The difference of 1,700 km$^3$ is returned to local water bodies, usually as wastewater that comes primarily from domestic and industrial users. This large volume returned to the local water system may not be available for easy reuse, though, depending on its next use and, importantly, whether it has been contaminated and/or treated prior to discharge.

Similar to the **ecological footprint** that calculated the land area required to support human activities, a **water footprint** determines the water required to support human activities. Eight countries (in order of consumption) are responsible for half of the world's water footprint: India, China, the United States, Russia, Indonesia, Nigeria, Brazil, and Pakistan. On a per capita basis, for 1997–2001, the United States had the highest footprint: 2,483 m$^3$/capita-yr. In comparison, for the same

**World Water Assessment Programme**

http://www.unesco.org/water/wwap

**Class Discussion**

Online, visit www.waterfootprint.org/, and calculate your personal water footprint and the footprint of several countries. Discuss how changes in technology, policy, and human behavior can reduce the water footprint at the household and country level. Which changes are most effective at the household and national level? Which are most equitable?

**Calculate Your Water Footprint**

http://www.waterfootprint.org

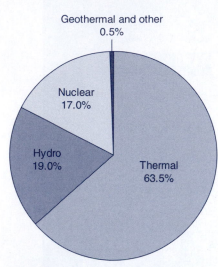

Geothermal and other
0.5%

Nuclear
17.0%

Hydro
19.0%

Thermal
63.5%

**Figure 9.4    Breakdown of Global Electricity Production**   Thermal electricity generation accounts for two-thirds of electricity production worldwide. Hydropower is the most widely used renewable source of electricity.

period, the global water footprint was 1,243 $m^3$/capita-yr. Footprints of other countries ($m^3$/capita-yr) are Australia (1,393), Brazil (1,381), China (702), Germany (1,545), India (980), and South Africa (931) (Hoekstra and Chapagain, 2006).

As stated previously, much of the domestic water in urban areas is discharged back into the environment. (Think about the multiple reuses along a long river such as the Ohio or Mississippi.) Historically urban areas met their water needs from local surface waters. Cities thus have an important role to play in ensuring that water they replace back to the environment does not harm ecological systems or downstream users. They are also becoming increasingly dependent on the *interbasin transfer* of water, which requires tremendous amounts of infrastructure investment and associated energy for collection, storage, and transfer.

Table 9.2 shows the primary uses of water throughout the world (excluding the use of water in production of **electricity**). Most of the usage listed in Table 9.2 involves **agricultural use** (globally at 70 percent). There are regional differences, however, particularly in the **industrial use** of water in higher-income areas such as Europe and North America. The percent of annual water withdrawals for **domestic use** ranges from 6 to 18 percent of the total withdrawals. Note how the level of development for a particular region of the world affects the distribution of water use.

Table 9.2 excludes the use of water for the energy sector. When the water demands of electricity generation are considered, over half of the water use is for power generation. Figure 9.4 shows that thermal electricity production accounts for two-thirds of global electricity

## Table / 9.2

**Volume of Annual Water Withdrawals and Percent of Withdrawals Associated with Agricultural, Industrial, and Domestic Sectors**

|  | Annual Water Withdrawals (km$^3$) | Agriculture (%) | Industry (%) | Domestic (%) |
|---|---|---|---|---|
| **World** | **3,317** | **70** | **20** | **10** |
| Africa | 152 | 85 | 6 | 9 |
| Asia and Pacific | 1,850 | 86 | 8 | 6 |
| Europe | 456 | 36 | 49 | 15 |
| Latin American and Caribbean | 263 | 73 | 9 | 18 |
| North America | 512 | 39 | 47 | 13 |
| West Asia | 84 | 90 | 4 | 6 |

SOURCE: Adapted from UN–Habitat, 2003, with permission of Earthscan Publications Ltd., London.

production. Hydropower provides 19 percent of the total electricity generation, and nuclear 17 percent. Other sources such as geothermal, tidal, wave, solar, and wind energy (which are not associated with large water usage) account for less than 0.5 percent of the world's electricity production.

A major benefit of **hydropower** is that each additional terawatt of hydropower produced per hour that displaces coal-generated electricity annually offsets 1 million tons of $CO_2$ equivalents. Hydropower has other benefits, such as low operation and maintenance costs, few atmospheric emissions, and no production of hazardous solid wastes. However, large-scale hydropower has problems, including large investment costs, issues related to fish entrainment and restriction of passage, loss and modification of fish habitat, and displacement of human and wildlife populations.

**World Commission on Dams**
http://www.dams.org

**Class Discussion**
What are some of the social and environmental impacts of large dams? The World Commission on Dams has further information at its Web site, www.dams.org.

---

**Box / 9.1   Social and Environmental Concerns of Large-Scale Hydropower Systems**

Many of the people displaced by large-scale hydropower systems are poor, less educated, and indigenous. Remember our discussion on **environmental justice** in Chapter 6? As just one example, the 18.2-gigawatt (GW) Three Gorges dam project in China is estimated to have already displaced more than 1 million people that reside in over 1,200 villages and many cities. The dam will submerge 632 $km^2$, which includes burial sites, historic sites of cultural significance, and environmental treasures. Three hundred species of fish live in the Yangtze River, and many will be separated from their spawning grounds. Expected to be completed in 2009, the dam is estimated to provide one-ninth of China's electricity needs.

In contrast, **micro-hydropower** systems (generating less than 100 kW) and *mini-hydropower* systems (100 kW to 1 MW) have a much lower negative impact on the environment and society than large hydropower systems. They are usually decentralized and not connected to the electric grid. In terms of environmental benefits, compared with an equivalent coal plant, a 1 MW mini-hydropower system that produces 6,000 MWh every year would supply the electricity needs of 1,500 families and avoid emissions of 4,000 tons of carbon dioxide and 275 tons of sulfur dioxide (UNESCO-WWAP, 2003).

---

### 9.3.1  U.S. WATER USAGE

Total water withdrawals in the United States exceed 400,000 million gallons per day (gpd). Table 9.3 shows the breakdown (by use) of fresh and saline water withdrawals in the U.S. The largest use of water is for the production of electricity.

Figure 9.5 shows the volume of *freshwater* withdrawals in the United States since 1950, broken down by withdrawals from groundwater and surface water. California and Texas withdraw the most surface water. California and Florida withdraw the most groundwater. The total volume of withdrawals has remained relatively constant since the mid-1980s, varying by less than 3 percent, even though the population has increased over this time period. Surface water has made up 80 percent of the total, and groundwater has made up 20 percent of the total over the past 50 years. In contrast, the percent of withdrawals associated with public water supplies has tripled since 1950.

www.wiley.com/college/mihelcic

**Water Usage**

### Total Fresh and Saline Water Withdrawals in the United States

These withdrawals total 408,000 million gpd. Freshwater accounts for 85% of this total, and surface water accounts for 79% of the total.

| User | Percent of Total Fresh and Saline Water Withdrawals |
| --- | --- |
| Thermoelectric | 48 |
| Irrigation | 34 |
| Public Supply | 11 |
| Industrial | 5 |
| Domestic | <1 |
| Livestock | <1 |
| Aquaculture | <1 |
| Mining | <1 |

SOURCE: Data from Hutson et al., 2004.

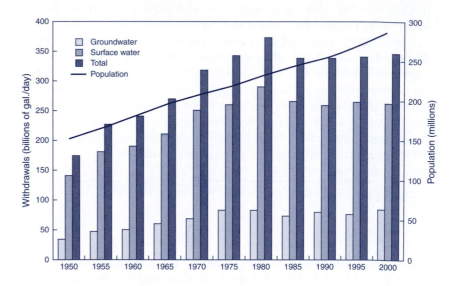

**Figure 9.5** U.S. Freshwater Withdrawals by Source and Population, 1950–2000

Courtesy of U.S. Geological Survey; Hutson et al. (2004).

In terms of industrial withdrawals, Figure 9.6 shows the geographic distribution of industrial water withdrawals in the United States. Louisiana, Texas, and Illinois account for 38 percent of total industrial water withdrawals. Over 80 percent of this is from surface waters. The states of Georgia, Louisiana, and Texas account for 23 percent of groundwater withdrawals associated with industrial use.

Table 9.4 reviews various sources of water. Most domestic and industrial users obtain their water from **surface waters** (streams, rivers, lakes, reservoirs) and **groundwater**. However, desalination plants allow **seawater** to be used. Providing safe **reused water** (reclaimed water) is technically feasible. Reclamation is becoming an increasingly

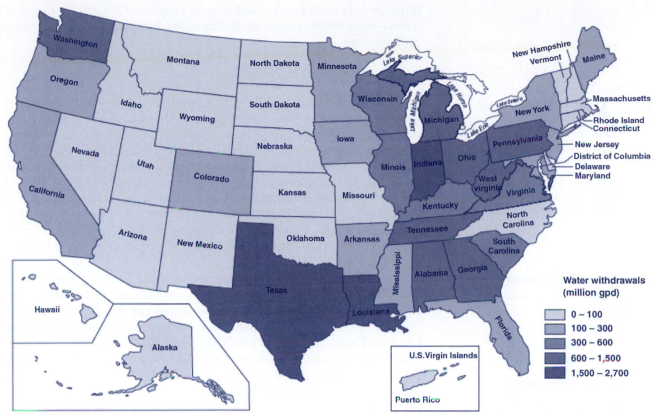

**Figure 9.6** Geographic Distribution of Industrial Water Withdrawals in the United States and Its Territories

Courtesy of U.S. Geological Survey; Hutson et al. (2004).

## Table / 9.4

### Sources of Water and Issues Associated with the Source

| Source of Water | Issues |
|---|---|
| Surface water | High flows, easy to contaminate, relatively high suspended solids (TSS), turbidity, and pathogens. In some parts of the world, rivers and streams dry up during the dry season. |
| Groundwater | Lower flows but natural filtering capacity that removes suspended solids (TSS), turbidity, and pathogens. May be high in dissolved solids (TDS), including Fe, Mn, Ca, and Mg (hardness). Difficult to clean up after contaminated. Renewal times can be very long. |
| Seawater | Energy intensive to desalinate, so costly compared with other sources, and disposal of resulting brine must be considered. Desalination can occur by distillation, reverse osmosis, electrodialysis, and ion exchange. Of these, multistage distillation and reverse osmosis are the two technologies most commonly used (they account for approximately 87 percent of worldwide desalination capacity). There are more reverse osmosis plants in the world; however, they are typically smaller in capacity than distillation plants. |
| Reclaimed and reused water | Technically feasible. Currently used for irrigating agricultural crops, landscaping, groundwater recharge, and potable water. Includes decentralized use of gray water (wastewater produced from baths and showers, clothes and dishwashers, lavatory sinks, and drinking-water fountains). |

important source of water and is now employed by a wide range of users—for domestic use, agriculture, landscaping, and recharging groundwater.

### 9.3.2 PUBLIC WATER SUPPLIES

In the United States, **public water** supplies are those that serve at least 25 people and have a minimum of 15 connections. They can be owned by the public or a private organization. This water can serve domestic, commercial, industrial, and even thermoelectric users. Figure 9.7 shows the total water withdrawals associated with public water supplies for every state and several territories of the United States. Public water usage is strongly dependent on population. For example, large states that account for 38 percent of the U.S. population (for example, California, Texas, New York, Florida, and Illinois) account for 40 percent of the total withdrawals from public water supplies.

Single-family U.S. households average 101 gallons of water use per day per capita (gpdc), indoor and outdoor. In multifamily dwellings such as apartments, water use can be as low as 45 to 70 gpdc, because these households use less water, have fewer fixtures and appliances, and use little or no water outdoors (Vickers, 2001). Outdoor use may exceeed natural rainfall in some locations. It ranges from 10 to 75 percent of total residential demand, depending on location. Table 9.5 shows a breakdown of this **household use** by different activities.

The table also shows what the same breakdown would be if households installed more efficient water fixtures and performed regular leak detection. If every U.S. household installed water-efficient features, water use would decrease by 30 percent. This would not only save money, but would also eliminate the demand to identify and

**Water Sources**

The largest desalination plant in the United States provides 10% of Tampa's (FL) water needs. Capacity is up to 25 mgd.

**Learn More about Desalination**

http://ga.water.usgs.gov/edu/drinkseawater.html

**Class Discussion**

What type of water-efficient features are present in or missing from your house, your apartment or dormitory, and your university campus? Does use of this technology require any behavioral changes by users or maintenance staff?

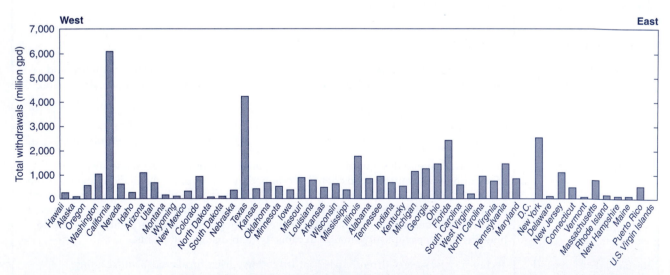

**Figure 9.7** **Public Supply Water Withdrawals in the United States, 2000** The states and territories in the figure are arranged from west to east.

Courtesy of U.S. Geological Survey; Hutson et al. (2004).

## Table / 9.5

**Water Usage in U.S. Households: Typical and Efficient Alternatives** Typical water usage in a U.S. household is far greater than when water-efficient fixtures are installed and households pay attention to leak detection. Percentages are based on total use.

| Activity | Typical Water Usage, gpdc (% total use) | Water Usage with Water-Efficient Fixtures and Leak Detection, gpdc (% total use) |
|---|---|---|
| Showers | 11.6 (16.8%) | 8.8 (19.5%) |
| Clothes washing | 15.0 (21.7%) | 10.0 (22.1%) |
| Dishwashing | 1.0 (1.4%) | 0.7 (1.5%) |
| Toilets | 18.5 (26.7%) | 8.2 (18.0%) |
| Baths | 1.2 (1.7%) | 1.2 (2.7%) |
| Leaks | 9.5 (13.7%) | 4.0 (8.8%) |
| Faucets | 10.9 (15.7%) | 10.8 (23.9%) |
| Other domestic uses | 1.6 (2.2%) | 1.6 (3.4%) |

SOURCE: Data from Vickers, 2001.

secure new water sources, saving energy and materials associated with collecting, storing, transporting, and treating water.

### 9.3.3 WATER RECLAMATION AND REUSE

Because of items such as increasing population, demographic shifts in where people live and industry locates, and current and future issues of water scarcity, **water reclamation** and **reuse** are beginning to be increasingly important. For example, states such as California, Arizona, Georgia, and Florida already incorporate water reclamation and reuse into the water management decisions. This is due to their climate, expanding populations, and extensive water-dependent agricultural production. However, with water scarcity now becoming a common occurrence, areas that were once thought to be water rich must *always* incorporate water reuse and reclamation into water management decisions.

Wastewater treatment plants have typically been sited to take advantage of gravity to transport collected wastewater and to make the disposal of treated effluent easy. These are two reasons why many treatment plants are located near surface water bodies. Treatment plants are also situated closer to urban areas, where the vast majority of wastewater is generated. One key to successful water reclamation and reuse is to pair the quality of wastewater effluent with the water quality requirements of new users.

Most reclaimed water today is used in industrial or agricultural settings that may be located far away from a large wastewater treatment

**Water Reclamation and Reuse**
http://www.epa.gov/region09/water/recycling/

## Table / 9.6

**Examples of Water Reclamation and Reuse**  All sources of reclaimed water are treated domestic wastewater. On a global scale, water reuse capacity is expected to increase from 19.4 million m³/day to 54.5 million m³/day by 2015.

| Location | Use of Reclaimed Water | Issues Solved by Engineers through Technical, Policy, and Outreach Solutions |
|---|---|---|
| Hampton Roads Sanitation District, Virginia | Service water and boiler feed water at oil refinery | Needed to treat ammonia during cold weather and produce water with more consistent turbidity levels. |
| Irvine Ranch Water District, California | Landscape irrigation for public and businesses; dual-plumbed office buildings use water for toilet and urinal flushing; commercial office cooling towers; agricultural irrigation | TDS builds up in recycled water. Seasonal demands for landscape irrigation need to be balanced with storage limitations of urban environment. |
| San Antonio Water System, Texas | Power plant cooling water; industrial cooling; river maintenance; landscape irrigation | Water quality deterioration in distribution system can occur from higher solids content of reclaimed water. Concerns of cross connection with potable water and impact of higher TDS levels on vegetation. |
| South Regional Water Reclamation Facility, Florida | Agricultural and landscape irrigation; freeze protection of citrus crops; groundwater recharge | Possible impacts on irrigating with reclaimed water. |
| Orange County Water District, California | Groundwater recharge that subsequently supplements potable water supply; landscape irrigation | Emerging contaminants such as low-molecular-weight organics, pharmaceuticals, and endocrine-disrupting chemicals in reclaimed water. |

SOURCE: Crook, 2004; GWI, 2005.

### Class Discussion

Many technical issues associated with water reclamation and reuse have been successfully addressed. Are energy and material inputs high or low for this technology? What are some social challenges to reusing water for domestic use? What challenges might you encounter as an engineer working with a community on a water reuse project? How would you overcome these challenges in a fair and equitable manner?

plant. Thus, any reclaimed water needs to be transported over large distances before it can be reused. To counter this problem, small decentralized **satellite reclamation plants** are being designed and constructed. These plants combine primary, secondary, and/or tertiary treatment processes to treat a portion of a wastewater stream close to where it can be used, thus eliminating the need to transport reclaimed water over large distances.

Table 9.6 provides several examples of successful water reclamation and reuse. In each case, the source of the reclaimed water is treated domestic wastewater. The reclaimed water has a variety of uses, including domestic, industrial, and agricultural use, as well as groundwater recharge and landscape irrigation.

## 9.3.4  WATER SCARCITY

One of the most pressing global security problems in the future is likely to be **water scarcity**, a situation where there is insufficient water to satisfy normal human requirements. A country is defined as experiencing **water stress** when annual water supplies drop below 1,700 m$^3$ per person. When annual water supplies drop below 1,000 m$^3$ per person, the country is defined as **water scarce**. By one measure, nearly 2 billion people currently suffer from severe water scarcity. This number is expected to increase substantially as population increases and as standards of living (and therefore consumption) rise around the world.

Climate change is expected to have an impact on precipitation (see Table 9.7). Some areas may benefit from 10 to 40 percent increases in rainfall, but others are likely to suffer from 10 to 30 percent decreases in rainfall. Some regions that will see an increase in rainfall will also become vulnerable to extreme rain events associated with flooding and erosion. The population that will be most vulnerable to climate change

### Table / 9.7

**Example of Possible Impacts of Climate Change on Water Resources Projected for the Mid to Late Century**

| Phenomenona and Direction of Trend | Likelihood of Future Trends Based on Projections for 21st Century | Major Impact(s) |
|---|---|---|
| Over most land areas, warmer and fewer cold days and nights, warmer and more frequent hot days and nights | Virtually certain | Effects on water resources relying on snowmelt; effects on some water supplies |
| Warm spells/heat waves; frequency increases over most land areas | Very likely | Increased water demand; water quality problems, e.g., algal blooms |
| Heavy precipitation events; frequency increases over most areas | Very likely | Adverse effects on quality of surface water and groundwater; contamination of water supply; water scarcity may be relieved |
| Increase in area affected by drought | Likely | More widespread water stress |
| Increase in intense tropical cyclone activity | Likely | Power outages, causing disruption of public water supply |
| Increased incidence of extreme high sea level (excludes tsunamis) | Likely | Decreased freshwater availability due to saltwater intrusion |

SOURCE: Used with permission of the Intergovernmental Panel on Climate Change, *Climate Change 2007: Impacts. Adaption and Vulnerability,* Summary for Policymakers, from Table SPM.1.

is poor and depends on rain for agricultural water and on local water resources for health and economic livelihood. These people also tend to live in areas that are prone to water-associated disasters of drought and flooding.

## 9.4 Municipal Water Demand

The amount of water used (or needed) is critical in the planning and design of a municipal water system. The estimated water usage rate is commonly called the **municipal water demand**. In general, the source(s), water facility location and size, and the piping to connect these facilities to the customers all depend upon demand. Although estimating water demand is critical to the planning of a system, there is no single method to measure or estimate it.

The amount of municipal water use is based on the type and number of customers in the system. The design and sizing of a water (or wastewater) treatment plant is based on an estimate of the current and potential future water usage by the customers served by the system. Other factors, such as additional water for fire protection and wet-weather inflows into a sewer system, increase the actual volume of water to be treated.

The design and sizing of the piping network to deliver water or collect wastewater is based not only on the estimated water usage, but also on the location of the specific customers relative to the treatment facilities. For example, the location of a large industrial user may not greatly affect the total water to be processed by the treatment plant, but the sizing of the piping network to connect the user to the treatment plant will be greatly affected. An industrial user very near the treatment plant would need a much shorter length of large-diameter pipe than if the user were far from the treatment plant.

### 9.4.1 CREATING MODELS TO ESTIMATE DEMAND

Estimating the water demand generally involves creating a model of the system that mimics the real system. Decisions made to create an accurate model depend upon the intended purpose of the model and the information required from it. Therefore, the purposes for the model are defined at the outset, so that the appropriate type of model is selected. In general, **modeling objectives for water demand** are twofold:

1. *Existing systems*: Develop a model to accurately simulate the operation of the existing system.

2. *Proposed systems*: Develop a model that will become a planning tool that will guide the design of a future system.

The type of model and the associated detail are defined based on the specific model objectives. The detail of a model can be classified as

© Eric Delmar/iStockphoto.

either a *macroscale model* or *microscale model*. Macroscale models are used to estimate the overall water demand, sizing of treatment facilities, and system storage required to account for daily water usage cycles. The detail that a specific-size pipe is connected to five commercial customers residing in a LEED-certified green building on Main Street would not be required in this case. However, a microscale model of the specific pipe diameter and nearby surrounding system might be used to size a pump in a pumping station used for fire protection. For this case, the water demand would be the water needed to put out the fire.

Table 9.8 summarizes the variety of data used in **estimating water demand**. Depending on the model's objectives and detail, the data actually required may be just one or two types, or additional data may needed beyond what is listed in Table 9.8. The availability and accuracy of the data will vary greatly. Maps and drawings of the existing system, historical climate data, and operational data are usually readily available and generally accurate. Data required for future planning, such as demographic changes, future land use, and projected climate, can have significant uncertainty. Once any data are located, they must be critically reviewed for suitability to meet the model's needs. Any historical documents must be evaluated and verified to assure the accuracy of the data.

## Table / 9.8

### Types of Data That May Be Needed to Create a Water Demand Model

| Type of Data | Description of Data |
|---|---|
| System data | This is the physical layout of the system. Examples of data include process drawings for a treatment plant, piping network for water distribution or sewer collection, or the layout of a new development. This includes the system dimensions (length, width, height) and elevations. |
| Operational data | This is the information about the system when it is in operation. Examples of data include water levels in the treatment plant tanks or storage tanks, pumping rates for pumps, or wet-well water levels. Much of this information is known by the system operators. |
| Consumption data | This is estimated water use by the customers. Data includes the per capita daily water demand, water demand value for specific customers such as a large building, and fire protection demand. Also there are typically estimates of changing water use patterns, such as water conservation strategies. |
| Climate data | This consists of the seasonal temperature and rainfall data. Temperature and rainfall can have a great influence on the estimated water demand. Other climate data could be used if needed, and climate forecasting also may be used. |
| Demographic and land use data | This is data about the customers and how the customers use their property. This includes population numbers and expected future growth (or decline), the types of customers (residential, commercial, industrial, etc.), and the location of these customers. Transportation planning has large influence on this. |

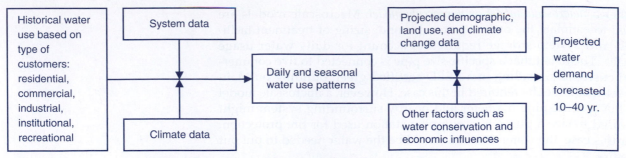

**Figure 9.8**  General Process Used to Create a Model to Estimate Future Water Demand

The general **modeling process to estimate water demand** is provided in Figure 9.8. The process begins with collecting and evaluating historical information about the type of customers served. From this, the daily and seasonal cycles can be determined. Including expected future demographic and land use data (and even climate change forecasting) makes it possible to estimate the future water demand.

Because of data availability and uncertainty, it is common to use different methods to estimate future demand. For example, when estimating the water demand for a residential area to size a treatment plant, a simple analysis of the current water use per household multiplied by the projected number of households might be as good an estimate as a complicated, detailed model of the entire system.

### 9.4.2  ESTIMATING WATER AND WASTEWATER FLOWS

**Water Demand: Demand Discharge Simulator**

Obtaining water and wastewater data is a fundamental step in designing a water distribution system or sewer collection system, or in sizing a treatment plant. Flow rates and patterns vary greatly from system to system and are highly dependent on the type and number of customers served, climate, and local economics. Figure 9.9 shows the daily metered **wastewater flows** for a population of approximately

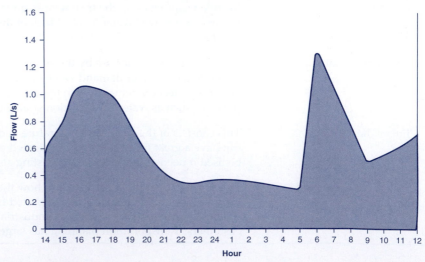

**Figure 9.9**  20-hour Wastewater Flow Distribution in San Antonio, Bolivia
Approximately 420 individuals live in San Antonio. Peak flows occur in the early morning and later in the afternoon/ early evening.

420 people living in San Antonio, Bolivia. Note how wastewater production (and corresponding water usage) are time dependent as households and other users of the system incorporate water into their daily lifestyle. Demographics specific to a region will change the shape of this figure. For example, in the United States, the morning peak is usually higher than the afternoon peak. And in bedroom communities, there is usually a very early morning peak as people wake up early for long commutes into urban work areas.

The best source of information for estimating demand is usually recorded flow data. For existing systems, **historical records** of water usage are generally available. Generally, water supply and treatment facilities have information on water level changes in reservoirs and the rate of water pumped and treated (so water balances can be used to estimate hourly flows). For customers, there are typically billing records, which may have metered flows associated with them, but these will provide only averages over one or two months.

Typically, more detailed information is available about **water usage rates** than **wastewater generation rates**. One approach to estimating wastewater generation is to estimate the water usage rate and then assume that between 60 and 90 percent of the water becomes wastewater (Tchobanoglous et al., 2003; Walski et al., 2004). However, this range will change greatly with climate, season, and type of customer.

Indoor water use is generally equivalent to wastewater generation, because water used outdoors generally does not enter the wastewater collection system. A simple model to estimate indoor residential water use is expressed as follows (Mayer et al., 1999):

$$Y = 37.2X + 69.2 \qquad (9.1)$$

where $Y$ is the indoor water use per household (gpd) and $X$ is the number of people per household.

For new customers, historical water use or wastewater generation rates can be used to estimate the new water demand. The historical records are used to estimate values for per unit water use or wastewater generation. Table 9.9 provides typical values and expected ranges. An estimate of the number of new units added to the system is determined based on the information for the projected type of new customers. Then, to determine the additional water demand in gallons per day, the typical flow values are multiplied by the number of additional units.

Water usage rates for industrial sources are highly site specific and should be based on historical records or the design flow rates for new industrial customers. For example, a Ford Taurus (including tires) requires over 147,000 L of water for production and delivery to the marketplace, a pair of blue jeans requires over 6,800 L, and a Sunday newspaper requires 568 L. The wastewater generated from industrial sources also greatly varies depending upon how much water is used and what the water is used for. For many industries, water is used in processes where much of the water can be lost as evaporation. When possible, industries should be metered to determine the actual flows.

## Table / 9.9

**Typical Water Use and Wastewater Generation Values**   Values are shown for different customers and can be used for estimating future scenarios.

| Source | Water Usage Flow (gal./unit day) | | | Wastewater Generation Flow (gal./unit day) | |
|---|---|---|---|---|---|
| | Unit | Range | Typical | Range | Typical |
| Apartment | Person | 100–200 | 100 | 35–80 | 55 |
| | Bedroom | | | 100–150 | 120 |
| Department store | Rest room | 400–600 | 550 | 350–600 | 400 |
| Hotel | Guest | 40–60 | 50 | 65–75 | 70 |
| Individual residence: | | | | | |
|   Typical home | Person | 40–130 | 95 | 45–90 | 70 |
|   Luxury home | Person | | | 75–150 | 95 |
|   Summer cottage | Person | | | 25–50 | 40 |
| Office building | Employee | 8–20 | 15 | 7–16 | 13 |
| Restaurant | Customer | 8–10 | 9 | 7–10 | 8 |
| School: | | | | | |
|   With cafeteria, gym, and showers | Student | 15–30 | 25 | 15–30 | 25 |
|   With cafeteria only | Student | 10–20 | 15 | 10–20 | 15 |

SOURCE: Data obtained from Tchobanoglous and Burton, 1991; Tchobanoglous et al., 2003.

Water usage and wastewater generation rates can also be estimated using **land use** data. Although this method is primarily used when designing water distribution or wastewater collection systems for future development (Walski et al., 2003, 2004; AWWA, 2007), it can also be used to estimate the expected flows for sizing water and wastewater treatment facilities. The land use is classified based on customer type (residential, commercial, industrial) and customer density (households per area, light commercial, dense commercial, and so on). From analyzing historical records or using assumed values, a water usage rate or wastewater generation rate per land area can be determined. Then the local zoning regulations for the proposed development provide information for determining a water usage rate and wastewater generation rate.

### 9.4.3   TIME-VARYING FLOWS AND SEASONAL CYCLES

The methods described in the previous section generally provide average flow rates for water use and wastewater generation. The average flow rate provides an idea of the amount of water that needs to be treated or transported in the pipe network, but the actual design needs to be able to handle the expected daily and seasonal variations in water flow. A properly designed treatment plant must be able to handle a range of expected flows. Storage facilities such as storage tanks for water distribution or wet wells for wastewater collection can be used to minimize the daily flow fluctuations into a treatment plant, but seasonal variations can have a great impact on the treatment facility and how it is operated. Also, the piping network should be sized to handle the expected maximum flow rate but also work effectively for very small flow rates.

The **variation in flows** for municipal systems typically follows a 24-hour cycle. However, this cycle can gradually change throughout the week (weekday flow versus weekend flow) and seasonally. The water use in the afternoon of a hot summer weekend day can be enormous due to outside water use such as watering lawns and filling pools. At the same times however, wastewater generation may follow a typical day pattern because the inside water use would be typical. Alternatively, on a cool, rainy day, water use would be typical, but wastewater flows could be high because of stormwater entering the collection system. For most days in communities, there is very little flow during the night, an increase in flow during the morning hours, and close to average flow during the day, followed by a second increase in flow during the evening hours.

The daily cycle for individual users also can be determined. Figure 9.10 shows examples of the daily water usage rate for different types of customers. Note how the individual demand pattern can be very different from the demand pattern for the entire community. Most of the time, an individual demand pattern is insignificant and has little effect on the entire community's pattern. However, a large water user (such as a large industry) may affect the demand pattern in the local water distribution or sewer collection system, especially in a small community with a large single user.

For municipal systems, a **demand factor (DF)** is determined from historical records to estimate the typical maximum and minimum daily

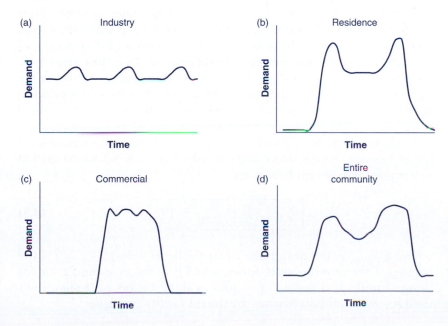

**Figure 9.10  Daily Demand Cycles** Cycles differ, depending on the type of customers: (a) industry, (b) residential, (c) commercial, and (d) the entire community.

## Table / 9.10

### Commonly Determined Demand Factor Events for Communities

| Event | Description | Demand Factor Range |
|---|---|---|
| Maximum-day demand | The average rate of all recorded annual maximum-day demand | 1.2–3.0 |
| Minimum-day demand | The average rate of all recorded annual minimum-day demand | 0.3–0.7 |
| Peak-hour demand | The average rate of all recorded annual maximum-hour demand | 3.0–6.0 |
| Maximum day of record | The highest recorded maximum-day demand | <6.0 |

SOURCE: Adapted from WEF, 1998; Walski et al., 2003.

flow rates. Determining demand factors for entire communities is relatively easy, because flow rate records exist at the treatment facilities. The demand factor for different conditions is determined from the average flow rate and extreme-condition flow rate:

$$\text{DF} = \frac{Q_{\text{event}}}{Q_{\text{average}}} \qquad (9.2)$$

where $Q_{\text{event}}$ is the event flow rate (volume/time), $Q_{\text{average}}$ is the average flow rate (volume/time), and DF is the demand factor (unitless). Table 9.10 shows how demand factors are associated with particular events. Historical records can be used to determine the annual average, maximum, and minimum recorded water usage rates or wastewater generation values. Since these values are system specific, the actual demand factors should be determined for each evaluated system.

To determine the **maximum and minimum design flow rates** for a treatment facility or piping network, a peaking factor (similar to the demand factor) is applied to the average daily flow rate. The **peaking factor (PF)** is a multiplier that is used to adjust the average flow rate to design or size components in a water or wastewater treatment plant, or components of a water distribution or wastewater collection system (pipes, pumps, storage tanks, and so forth). Equation 9.3 can be used to determine these design flow rates:

$$Q_{\text{design}} = Q_{\text{average}} \times \text{PF} \qquad (9.3)$$

where $Q_{\text{design}}$ is the design flow rate (volume/time), $Q_{\text{average}}$ is the average flow rate (volume/time), and PF is the peaking factor for design (unitless). Table 9.11 provides peaking factors or design flows used for water and wastewater treatment facility processes.

## Table / 9.11

**Design Flows and Peaking Factors Used for Sizing Drinking-Water and Wastewater Treatment Plants**

| Treatment Process of Facility Operation | Water Treatment Plant | Wastewater Treatment Plant |
|---|---|---|
| Plant hydraulic capacity | $Q_{\text{max day}} \times (1.25 \text{ to } 1.50)$ | $Q_{\text{maximum instantaneous}}$ |
| Treatment processes | $Q_{\text{max day}}$ | $Q_{\text{average}} \times (1.4 \text{ to } 3.0)$ |
| Sludge pumping | $Q_{\text{max day}}$ | $Q_{\text{average}} \times (1.4 \text{ to } 2.0)$ |

SOURCE: Adapted from Crittenden et al., 2005; Chen, 1995.

Details on using peaking factors for the design of treatment facilities and piping networks are discussed elsewhere for water treatment facilities (Crittenden et al., 2005), wastewater treatment facilities (Chen, 1995; Tchobanoglous et al., 2003), water distribution systems (Walski et al., 2003), and wastewater collection systems (Walski et al., 2004).

example/9.2 **Using Historical Records to Estimate Demand Factors and Household Water Usage Rate**

Estimate the maximum- and minimum-day demand factors, using data gathered from the annual water reports for a small water treatment plant. Then estimate the average household usage rate, using the data for all years. The metered flow values for each year are summarized in Table 9.12.

## Table / 9.12

**Metered Flow Values for Example 9.2**

| Year | Average (gpd) | Maximum Day (gpd) | Minimum Day (gpd) | Households Served |
|---|---|---|---|---|
| 2001 | 834,514 | 1,325,486 | 324,851 | 5,567 |
| 2002 | 843,842 | 1,354,826 | 314,584 | 5,603 |
| 2003 | 854,247 | 1,334,287 | 300,145 | 5,671 |
| 2004 | 837,055 | 1,341,024 | 365,454 | 5,789 |
| 2005 | 828,103 | 1,362,487 | 298,764 | 5,894 |
| 2006 | 858,076 | 1,356,214 | 325,141 | 5,969 |
| 2007 | 861,003 | 1,384,982 | 336,954 | 6,002 |
| 2008 | 868,150 | 1,368,920 | 310,247 | 6,048 |

## solution: Demand Factors

Determine the demand factor every year for the extreme events. For year 2001:

$$\text{DF}_{\text{max day}} = \frac{Q_{\text{max day}}}{Q_{\text{average}}} = \frac{1,325,486 \text{ gpd}}{834,514 \text{ gpd}} = 1.59$$

$$\text{DF}_{\text{min day}} = \frac{Q_{\text{min day}}}{Q_{\text{average}}} = \frac{324,851 \text{ gpd}}{834,514 \text{ gpd}} = 0.39$$

In the same way, the average demand factor can be determined for the other years, using all the data. From the annual averages, overall average can be determined as follows:

| Year | $\text{DF}_{\text{max day}}$ | $\text{DF}_{\text{min day}}$ |
|---|---|---|
| 2001 | 1.59 | 0.39 |
| 2002 | 1.61 | 0.37 |
| 2003 | 1.56 | 0.35 |
| 2004 | 1.60 | 0.44 |
| 2005 | 1.65 | 0.36 |
| 2006 | 1.58 | 0.38 |
| 2007 | 1.61 | 0.39 |
| 2008 | 1.58 | 0.36 |
| **Average** | **1.60** | **0.38** |

The maximum-day demand factor is 1.60, and the minimum-day demand factor is 0.38.

Our results compare very well with the established DF values provided in Table 9.10, where the maximum-day demand factor ranged from 1.2 to 3.0 and the minimum-day demand factor ranged from 0.3 to 0.7.

## solution: Usage Rates

Next, we estimate the average household usage rate, using the data for all years. The household water usage rate for each year can be determined for each year of data. For year 2001:

$$\text{usage rate} = \frac{Q_{\text{average}}}{\text{metered households}} = \frac{834,514 \text{ gpd}}{5,567 \text{ households}}$$

$$= 150 \text{ gpd/household}$$

## example/9.2 Continued

Using the same formula, the average household user rate can be determined for the remaining years, as shown in the following table:

| Year | Household Usage Rate (gpd/household) |
|---|---|
| 2001 | 150 |
| 2002 | 151 |
| 2003 | 151 |
| 2004 | 145 |
| 2005 | 140 |
| 2006 | 144 |
| 2007 | 143 |
| 2008 | 144 |
| **Average** | **146** |

The average household water usage rate is 146 gpd/household.

Remember that average residential water usage is approximately 101 gpdc. So it appears this community is averaging approximately 1.5 individuals per household. Leak detection, incorporation of water-saving technologies, public reminders to conserve water, and promotion of the use of native vegetation that requires little water are some methods that can reduce water usage and eliminate the need to develop additional sources of water that are expensive and ecologically or socially destructive.

### 9.4.4 FIRE FLOW DEMAND AND UNACCOUNTED-FOR WATER

A water system must be able to supply water quickly for societal needs to ensure adequate protection due to fire emergencies. Also, a portion of the supplied water will be lost due to system leakage, unmetered use (fire protection and maintenance), theft, or other causes.

During a fire emergency, the **fire protection water demand** can have a large effect on supply and distribution. In a community, water used for fire protection is generally pulled from nearby hydrants, which can greatly lower the available water pressure to local customers. For large industries, water is sometimes stored on site for fire protection. Generally, the amount of water required for fire protection depends on the size of the burning structure, the way the structure was constructed, the amount of combustible material in the structure, and the proximity of other buildings.

## Table / 9.13

**Needed Fire Flow (NFF) for Small Family Residences**

| Distance between Buildings (ft) | NFF (gpm) |
|---|---|
| <11 | 1,500 |
| 11–30 | 1,000 |
| 31–100 | 750 |
| >100 | 500 |

SOURCE: Values from ISO, 1998.

## Table / 9.14

**Recommended Needed Fire Flow (NFF) Duration for Fire Protection**

| NFF (gpm) | Duration (hr) | Storage (gal) |
|---|---|---|
| <2,500 | 2 | ~300,000 |
| 3,000–3,500 | 3 | 540,000–630,000 |
| >3,500 | 4 | >840,000 |

SOURCE: Values from Walski et al., 2003.

In the United States, community fire protection is rated and evaluated by the Insurance Services Office (ISO) using the Fire Protection Rating System (ISO, 1998, summarized in AWWA, 1998). For a municipal system, the ISO will evaluate the water supply source, treatment plant and pumping capacities, water distribution piping network, and placement and spacing of hydrants. The water system should have the available storage, pumping capacity, and piping to deliver the maximum daily demand plus the fire flow demand at any time during the day. In many situations, the **fire flow demand** is equal to the ISO's determined **needed fire flow** (NFF) for residential, commercial, and industrial properties. Table 9.13 provides the NFF for small family residences. The NFF is determined based on the spacing of the residential dwellings.

For commercial or industrial structures, the NFF is based on the size of the building, construction class (for example, wood frame), type of occupancy (for example, department store), exposure to adjacent buildings, and what is known as a communication factor (location and types of fire protection doors). The general ISO equation for a minimum NFF being 500 gallons per minute (gpm) is

$$NFF = 18 \times F \times A^{0.5} \times O \times (1 + \Sigma(X + P)) \qquad (9.4)$$

where NFF is the needed fire flow (gpm), $F$ is the construction factor (0.6 to 1.5), $A$ is the building effective area (sq. ft.), $O$ is the occupancy factor (0.75 to 1.25), $X$ is the exposure factor (0 to 0.25), and $P$ is the communication factor (0 to 0.25). The full procedure to determine the NFF for a structure can be found in ISO's *Fire Suppression Rating Schedule* (ISO, 1998) and the AWWA's Manual M-31 (AWWA, 1998). In addition to the NFF, Table 9.14 provides values for the recommended storage for fire protection along with the duration that water should be supplied.

Water that is produced by a treatment facility is delivered to a user in a water distribution system. However, a portion of that water does not make it to the customers or is used as unmetered flow. This **unmetered flow** includes (1) what is lost in the system due to leaks and breaks in pipes and joints; (2) unmetered uses such as fire protection and maintenance; (3) water theft; and (4) a variety of other minor water losses. Generally, in all cases, more water will be produced and enter the distribution system than is delivered to users. To determine this unaccounted-for water, subtract the sum of the metered water for each user from the metered water leaving the water treatment facilities:

$$\text{unaccounted-for water (\%)} = \frac{\text{water produced} - \text{metered use}}{\text{water produced}} \times 100$$

(9.5)

Many times, the unaccounted-for water is used to gauge the performance of a water distribution system. On an annual basis, it is expected that less than 10 percent of the water produced will be lost as unaccounted-for water. However, for older systems, the unaccounted-for water can be much higher due to the aging pipe network, which can have a significant amount of leakage. Closely monitoring the unaccounted-for water can also be used as an indicator of when something is wrong in the water distribution system. An increase in unaccounted-for water indicates a leak the piping network, which should be repaired, or significant loss due to another problem, such as water theft. With enough internal metering (meters for water mains, pumping station records), it is possible to determine the general location of needed repairs or problems.

## 9.4.5 WET-WEATHER FLOWS FOR WASTEWATER

Runoff water from either rain events or snowmelt can enter the wastewater collection system through manhole covers, designed inlets and catchments, and defects or cracks in the piping network. Domestic wastewater collection systems are designed to carry only domestic sewage (*sanitary system*) or to carry both domestic sewage and runoff (*combined system*). When a sanitary system is used to collect domestic sewage, many times a separate stormwater collection system is designed for runoff events.

**Wet-weather flows** are defined as the stormwater or snowmelt that directly enters a combined sewer system through designed inlets and catchments or that enters into the sanitary sewer though manholes and pipe defects. The amount of wet-weather flow that enters a sanitary sewer is generally in the same order of magnitude as the domestic (dry-weather) flows. However, the wet-weather flow can be much higher for aged sanitary systems or when there are many "leaky" manhole covers. For combined systems, the wet-weather flow is many times higher than the dry-weather flows and is be used to size the pipes in the system. Wet-weather flows are generally subdivided into two categories: **inflow** and **infiltration** (I/I), described in Table 9.15.

When the wet-weather flows are not distinguished from each other, the combination of flows is called inflow/infiltration (I/I). The I/I flow can vary widely, based on the age and condition of the sewer system, climate or season, and groundwater elevation. (If the groundwater table is above the sewer system, there can be infiltration into the sewer.)

| Table / 9.15 | |
| :--- | :--- |
| **Categories of Wet-Weather Flow** | |
| Inflow | This is the water that enters a collection system through direct connections such as stormwater catchments, roof leaders, sump pumps, yard and foundation drains, manhole covers, and other designed inlets. It is estimated by subtracting the typical dry-weather flow from the total metered flow after a storm event. |
| Infiltration | This is the water that enters a collection system through pipe defects, pipe joints, manhole walls, or other non-engineered designed inflows. It is estimated as the metered flow early in the morning when domestic use is relatively small and the sewer system has been drained of domestic sewage. The remaining flow is mostly infiltration. |

Recording typical I/I values can be based on the length of sewer, land area drained into the sewer, or the number of manholes.

I/I values for new sanitary sewers are generally between 200 to 500 gpd per inch of pipe diameter and mile of pipe length (Hammer and Hammer, 1996). An aged sewer may have higher values. I/I based on the number of manholes is generally divided into three categories based on the rainfall depth: (1) low, or 3,000 gal./in.-manhole; (2) medium, or 7,700 gal./in.-manhole; and (3) high, or 20,000 gal./in.-manhole (Walski et al., 2004). Common I/I values based on land area range from 20 to 3,000 gal./acre-day for nonrainy days. During storm events or snowmelt, I/I values could exceed 50,000 gal./acre-day for older sewer systems with considerable leaks and points for inflow (Tchobanoglous et al., 2003).

The type of sewer system can greatly influence the design I/I value. A sanitary sewer would have a relatively small I/I design flow rate where most is from infiltration. A combined system, which is designed to handle a large amount of runoff, would have a much higher design I/I flow rate. The wastewater collection system and treatment facilities should have the hydraulic capacity to handle the maximum-day domestic wastewater generation plus expected I/I.

Design infiltration values for new sewers varies greatly based on location, type of pipe material, construction practices, and whether the pipe is above or below the groundwater table. Regulatory agencies in most states have a maximum allowance for infiltration. For a sewer system designed to capture stormwater runoff, inflow values are estimated using a runoff model based on a design rain event, drainage area, and land use information. Since this inflow value is based on a design storm event, the potential for hydraulic failure (surcharging in the sewer system or overloading the treatment plant) is high for large-rainfall events.

## 9.4.6 DEMAND FORECASTING

**Alliance for Water Efficiency**
http://www.a4we.org/

Estimating future scenarios is a major part of designing a treatment facility, water distribution system, or sewer collection system. In almost all situations, there will be a level of uncertainty regarding how much water is required and wastewater is generated. The long-term planning of a community usually includes the estimation of the **future water demand** for 5, 10, 20, or more years. It is common to estimate the future demand for several different scenarios before deciding on the actual values for design. Also, the comparison of alternative future projections provides a way to understand what effects the input data or assumptions can have on the future water demand. This can be used as a sensitivity analysis to guide community leaders in the decision-making process for how to plan future developments, influence water use practices, or understand the impact of large water users on the system.

Determining the specifics of future scenarios requires a consensus that incorporates environmental, economic, and social needs for current and future generations and is developed by engineers, planners, utilities, and community stakeholders. Population and demographic trends and the location of any new users will greatly influence future water demand. Other important issues could include impact of climate

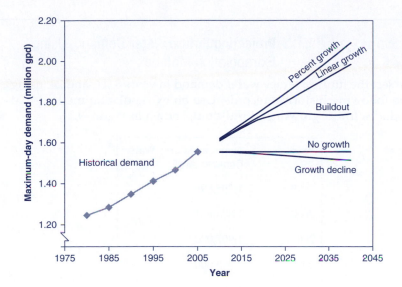

on water availability, climate change, and issues of land use within the watershed. Often more data are collected than what is actually used in the analysis.

Figure 9.11 shows some of the potential scenarios that can be evaluated to estimate the future water demand. This approach analyzes and extrapolates historical data to future scenarios. Caution should be used when simply extrapolating forward (linear growth), because past influences may not hold in the future.

Rather than basing projections using an extrapolation method, the forecaster can develop a more in-depth analysis of the potential causes for projected changes in water demand. This analysis is based on estimating the future population, demographics, water use, land use, number of large users (for example, industry), future climate scenarios, and technological and societal issues that affect water conservation. In this case, the water demand is separated into **disaggregated segments** to determine the volume of water used (or wastewater generated) per unit. Then the projected change for each segment is predicted, and the resulting new water demand is determined (AWWA, 2007; Walski et al., 2003).

These disaggregated segments are typically based on population estimates or equivalents, or on land use designations. A **population equivalent** is a method of converting the water use (or wastewater generation) of commercial or industrial users into the equivalent amount of water used by a population number. For example, an industrial unit may use the water equivalent of 150 people in a residential area. The projected equivalent population (real population plus population equivalents) is estimated to determine the future water demand.

Water use (or wastewater generation) can also be categorized based on land use such as light residential, dense residential, heavy commercial, and industrial. After determining the water use for each category and knowing any proposed new development, the forecaster determines a projected water demand. Another important issue is that users do not all require the same quality of water in relation to the source and level of treatment.

**Class Discussion**

Develop future scenario(s) specific to your location that can be added to Figure 9.11 for sustainable water use. In your scenario(s), incorporate water reuse as a source of your regions's water portfolio and account for your location's specific population growth; changes in demographics such as age, education, and wealth; changes in climate; estimation of future precipitation; and expected changes in use (shift from agriculture to domestic uses for example).

## example/9.3 Projecting Future Water Demand, Using Extrapolation Methods

Project the future average water demand in years 8, 14, and 24, using the following historical records. Use an extrapolation method that includes **linear growth** and **buildout**, shown in Figure 9.11.

| Year | Average Metered Water Demand (gpd) |
|------|------------------------------------|
| 1999 | 1,797,895 |
| 2000 | 1,843,661 |
| 2001 | 1,907,000 |
| 2002 | 1,813,000 |
| 2003 | 1,890,000 |
| 2004 | 1,901,145 |
| 2005 | 1,891,860 |
| 2006 | 2,012,201 |
| 2007 | 2,058,492 |
| 2008 | 2,051,339 |

## solution

First, graph the recorded values to visualize the historical trend (see Figure 9.12a). Based on these observations, we can make assumptions of expected growth (or decline).

In this case, we will project the future trend using the extrapolation method (Figure 9.11). Both future scenarios are shown in Figure 9.12b. The following table contains the details of the projected water usage. For linear growth, linear regression analysis was used to project the future water demand. The buildout extrapolation method requires additional assumptions. In Figure 9.12b, we assume 8 years of linear growth followed by buildout starting in year 9. Note that many other assumptions can be made to extrapolate the future water demand.

| **Projected Average Water Demand** | | |
|------|------|------|
| Year | Linear Growth Assumption | Buildout Assumption |
| 2016 | 2,190,000 gpd | 2,190,000 gpd |
| 2022 | 2,328,000 gpd | 2,239,000 gpd |
| 2032 | 2,604,000 gpd | 2,239,000 gpd |

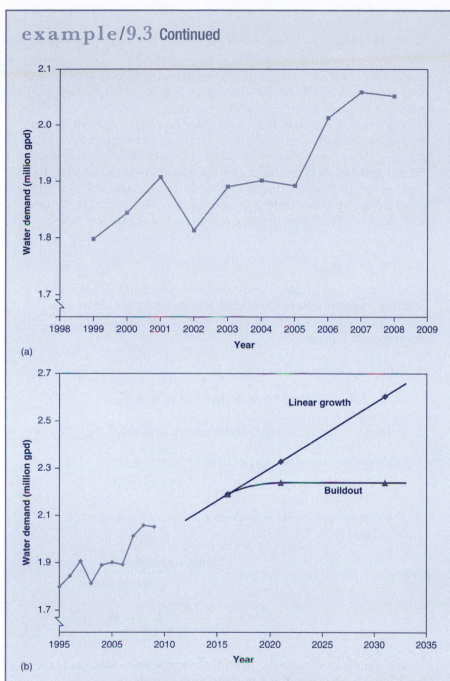

**Figure 9.12** **Historical and Projected Water Demand**
(a) Recorded values of water demand graphed to visualize the historical trend of data used in Example 9.3. (b) Projected future trend of water demand using the extrapolation method and data for Example 9.3.

---

example/9.4 Projecting Wastewater Generation Based on Types of Customers

A residential community with a population of 10,000 is planning to expand its wastewater treatment plant and sewer system. In 15 years, the population is estimated to increase to 17,000 residents, and a new

## example/9.4 Continued

500-person apartment complex is to be constructed. A new industrial park, also planned, will contribute an average flow of 550,000 gpd and maximum-day flow of 750,000 gpd. It is expected that approximately 9 mi. new sewer pipe will be needed.

The present average daily flow into the plant is 1.0 MGD with 16.5 mi. sewers. The current average inflow and infiltration (I/I) is 2,800 gpd/mi., and maximum-day I/I expected on a rainy day is 53,000 gpd/mi. Residential per capita water use is expected to be 8 percent less in 15 years, due to in-house water-saving strategies. The maximum-day wastewater generation typically occurs on a rainy day. Estimate the future average and maximum-day flow rates.

## solution

Based on the information provided, we first need to estimate the various contributions to the overall wastewater generation. The current estimated per capita generation of wastewater from different sources is as follows:

1. Estimate the current average I/I:

$$I/I = 2,800 \, gpd/mi. \times 16.5 \, mi. = 46,200 \, gpd$$

2. Determine the current average domestic wastewater generation:

$$domestic \; wastewater \; generation = metered \; flow - I/I$$
$$= 1,000,000 \, gpd - 46,200 \, gpd = 953,800 \, gpd$$

3. The current per capita wastewater generation can then be calculated as:

$$per \; capita \; wastewater \; generation = \frac{domestic \; wastewater \; generation}{population \; served}$$

$$= \frac{953,800 \, gpd}{10,000 \, people} = 95.4 \, gpdc$$

We can now estimate the future average wastewater flow rate to be 2,140,900 gpd from the following four contributions:

1. Future average domestic flow (including 8 percent wastewater reduction):

$$95.4 \, gpdc \times 0.92 \times 17,000 \, people = 1,492,000 \, gpd$$

2. Apartment complex average flow (using values from Table 9.9):

$$55 \, gpdc \times 500 \, people = 27,500 \, gpd$$

## example/9.4 Continued

3. Industrial park average flow (from problem statement): 550,000 gpd.

4. Average infiltration/inflow (I/I):

$$(16.5 + 9) \text{ mi.} \times 2{,}800 \text{ gpd/mi.} = 71{,}400 \text{ gpd}$$

In similar fashion, we can also estimate the future maximum-day wastewater flow rate (assuming the dry-day domestic wastewater generation is equal to the wet-day domestic wastewater generation) as 3,621,000 gpd from the following four contributions:

1. Future domestic maximum flow (including 8 percent wastewater reduction):

$$95.4 \text{ gpdc} \times 0.92 \times 17{,}000 \text{ people} = 1{,}492{,}000 \text{ gpd}$$

2. Apartment complex maximum flow (using values from Table 9.9):

$$55 \text{ gpdc} \times 500 \text{ people} = 27{,}500 \text{ gpd}$$

3. Industrial park maximum flow (from problem statement): 750,000 gpd.

4. Maximum infiltration/inflow (I/I):

$$(16.5 + 9) \text{ mi.} \times 53{,}000 \text{ gpd/mi.} = 1{,}351{,}500 \text{ gpd}$$

Note that the values used in this example are "typical" and not overly aggressive in terms of water conservation. Readers are encouraged to think much bigger in terms of sustainable water management. For example, can the projected 8 percent reduction in wastewater generation be raised much higher to achieve more sustainable use of existing water? What would happen if the apartment complex were designed to be "green" in terms of aggressively reducing water usage, not only through use of water-saving technology and education of the building occupants, but also with innovative use of gray water in cooling and landscaping? How about if the design of the industrial complex selected some businesses that had an integrated reuse of others' generated wastewater? And how will water conservation impact the "strength" of the wastewater?

## 9.5 Water Distribution and Wastewater Collection Systems

Before the design of a **water distribution** or **wastewater collection** system, a comprehensive investigation of the proposed service area takes place. Part of this investigation includes forecasting water demand throughout the service area, as previously discussed. Other factors such as pipe material, location of appurtenances (manholes, junctions, inlets, and other structures), hydrant placement, type and location of valves, design and locations of pumping stations, and water storage locations all need to be determined and integrated to design a working efficient system.

This process typically uses computer software. (See Walski et al. 2003, 2004 for a full description of such modeling software.) The remainder of this section will discuss system layout, design flow velocities for estimating pipe size, and system needs related to pumping and storage.

### 9.5.1   SYSTEM LAYOUT

A water distribution system and a wastewater collection system both carry water. However, they do so very differently. Generally, the piping is placed under or along roadways dedicated for public use or through lands where the utility has a right-of-way easement. It is common practice for the wastewater piping be at least 10 ft. from, and 18 in. below, the water service lines, to minimize the possible contamination of potable water (Hammer and Hammer 1996). Large piping systems are connected to each customer by small laterals or supply pipes. The piping capacity should be designed to meet needs of the customer without excessive costs. Customers should therefore not notice or have to worry about a well-designed and well-maintained system.

A water distribution system has a layout that contains many loops where pressurized flow occurs throughout the system (see Figure 9.13). For large systems, many **looped** subsystems are connected to large pipe force mains or transmission lines. These loops allow water to be delivered to the customers by many different routes. When demand at a specific location is high, as during a fire emergency, water needs to be delivered to that location as efficiently as possible. Hydrants are placed along roads at intersections and spaced so fire personnel can pull water from multiple hydrants if necessary. Also, many shutoff valves are placed throughout the system. Thus, if a break occurs or maintenance is scheduled, isolating the problem area by using shutoff valves will minimize the number of customers that have to go without water. If designed correctly, a water distribution system should provide the needed water for a variety of demand scenarios with adequate pressure and good water quality throughout the whole system.

A wastewater collection system has a layout that is **branched** (or dendritic), with very few or no loops (see Figure 9.13). Most of the time, wastewater flows by gravity in one direction. The smallest pipes are located at the "end" of the branches and progressively get larger as the wastewater flows toward the treatment plant. Since the wastewater flows by gravity, an adequate pipe slope must be maintained throughout the system. There can be sections where wastewater is pumped in a force main to lift the wastewater to a higher elevation to again flow by gravity, or to force wastewater uphill due to the topography. Generally manholes are placed at every pipe diameter change, slope change, junction (two or more pipes joining), and upstream pipe ends. Manholes are spaced no more than 300 to 500 ft. (90 to 150 m) between each other to provide enough entry points for maintenance (ASCE, 2007).

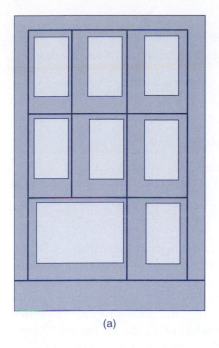
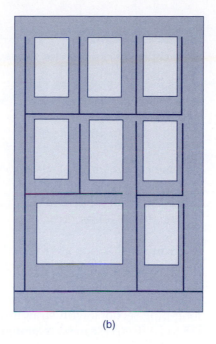

(a)                                  (b)

**Figure 9.13** Examples of a Water Distribution and Wastewater Collection System Layout for a Residential Area. (a) This "looped" system is typical for a water distribution system. (b) This "branched" system is typical for a wastewater collection system.

## 9.5.2 DESIGN FLOW VELOCITIES AND PIPE SIZING

The analysis of water flow in pipes is based on conservation of mass, conservation of energy, and conservation of momentum. Fluid flow in pipes can be classified as either *full pipe* (pressure flow) or *open channel* (gravity flow). In both cases, modeling software is used to analyze the flow in pipes for water distribution and wastewater collection systems.

Pressure flow conditions are found in water distribution systems where the delivered water must have both the capacity and the pressure to suit the needs of the customer. The **design water velocity in a pipe** is typically between 2.0 ft./sec. (0.6 m/s) to 10 ft./sec. (3.1 m/s) for peak flow conditions. Smaller design velocities tend to size pipes to be bigger than what is really economically needed. Higher velocities tend to cause excessive head loss throughout the system and to increase the potential for water transients (for example, water hammer).

Typical service pressure requirements are provided in Table 9.16. In residential communities, it is necessary to maintain a minimum pressure for each customer that allows for "normal" water use when more than one water-using device is in service on the second story. Also, the pressure at the residence should not exceed 80 psi; otherwise, the potential for in-house leaks increases, excessive flows in showers and faucets could occur, and the hot-water pressure relief valve could discharge.

**Open-channel conditions** are commonly found in wastewater collection systems where the water flows by gravity. Since the piping is sized for infrequent high-flow conditions, the pipes are only partially filled most of the time. During low-flow conditions, this causes the wastewater velocity to be too small to move solids, which then can be deposited in the sewer. It is usual practice to design sewer slopes so that the minimum full-pipe velocity is 2.0 ft./sec. (0.6 m/s) or greater

© Lacy Rane/iStockphoto.

**Typical Required Service Pressures in Residential Communities**

| Condition | Service Pressure (psi) |
|---|---|
| Maximum pressure | 65–75 |
| Minimum pressure during maximum-day demand | 30–40 |
| Minimum pressure during peak-hour demand | 25–35 |
| Minimum pressure during fires | 20 |

SOURCE: Values from Mays, 2000.

to ensure that solids are flushed out during peak flow hours of the day. Also, it is common practice for the maximum velocity to be about 10 ft./sec. (3.0 m/s) to avoid damaging the sewer (ASCE, 2007) and causing excessive "sewer spray" in manholes. However, for simple long-pipe sections with straight-through manholes, higher velocities (15 to 20 ft./sec.) can be tolerated (Walski et al., 2004). Table 9.17 provides the minimum slope for circular concrete sewer pipe based on

Table / 9.17

**Minimum and Maximum Slope for Gravity Flow in Circular Concrete Sewer Pipe for a Manning's Roughness Value Equal to 0.013**

| Pipe Diameter | | Minimum Slope (2 ft./sec., 0.6 m/s) | Maximum Slope (10 ft./sec., 3.1 m/s) |
|---|---|---|---|
| in. | mm | | |
| 4 | 100 | 0.00841 | 0.21030 |
| 6 | 150 | 0.00490 | 0.12247 |
| 8 | 200 | 0.00334 | 0.08345 |
| 10 | 250 | 0.00248 | 0.06197 |
| 12 | 300 | 0.00194 | 0.04860 |
| 15 | 375 | 0.00144 | 0.03609 |
| 18 | 450 | 0.00113 | 0.02830 |
| 21 | 525 | 0.00092 | 0.02304 |
| 24 | 600 | 0.0008* | 0.01929 |
| 27 | 675 | 0.0008* | 0.01648 |
| 30 | 750 | 0.0008* | 0.01432 |
| 36 | 900 | 0.0008* | 0.01123 |
| 42 | 1,050 | 0.0008* | 0.00655 |

*Minimum practical slope for construction.

the minimum and maximum full-pipe velocity for a Manning's roughness value equal to 0.013.

Equation 9.6 can be used to make an initial estimate of a pipe size based on the full-pipe design flow velocity and design carrying capacity (design peak flow):

$$D = k\sqrt{\frac{Q}{v}}$$

(9.6)

where $D$ is the pipe diameter (in., mm), $Q$ is the design flow rate (gpm, L/s), $v$ is the design velocity (ft./sec., m/s), and $k$ represents a constant that is 0.64 for U.S. customary units or 35.7 for SI units.

For example, if the design flow rate is 6,000 gpm (380 L/s) and the design velocity is 5 ft./sec. (1.5 m/s), then the estimated pipe diameter would be 22.2 in. (568 mm). The next largest nominal pipe size would be selected as a starting point—in this case, a 24 in. (600 mm) pipe. Then a model would be used to evaluate whether this pipe is the best choice for a variety of different scenarios.

### 9.5.3 PUMPING STATIONS AND STORAGE

A **pumping station** is located where wastewater needs to be lifted or increased pressure is required. Pumps are characterized by their capacity, pumping head, efficiency, and power requirements. The design of a pumping station requires that each pump's performance curve (capacity versus head curve) match the system head curve. The system curve is developed by adding the static head to the head loss (friction plus minor losses) of the system for varied flow rates. The pump characteristic curves should be available from the utility for existing pumps or can be obtained from the manufacturer of the pump. An old pump may not perform as it did when it was new. In fact, older pumps should be checked to verify that the real pump performance matches what is shown on the pumping head performance curve.

Like most mechanical machines, a pump operates best at its **best efficiency point (BEP)**. As a pump operates further from its BEP, the life of the pump or impeller decreases considerably, and the energy costs (and resulting $CO_2$ emissions) to run the pump greatly increase. The typical operating range of a centrifugal pump is the flow rate between 60 and 120 percent of the flow rate at the BEP. When designing and sizing a pumping station, it is common to have multiple pumps working together to maximize the pumping efficiency and minimize the pumping costs while enabling the facility to deliver the required water for a wide range of flow conditions.

In the example shown in Figure 9.14, the actual pump efficiency is near the BEP, which is at the peak of the pump efficiency curve. Therefore, this pump configuration is working near its maximum efficiency. As the system curve shifts because of changes in the system (water elevation changes or required pumping rate changes), the match point would move right or left along the performance curve. If it moves too far to the right or left, the pump efficiency can decrease to the point where this configuration may no longer be very efficient or may be outside the working range of the pumps. In this case, additional pumps could be turned on or

**Distribution, Collection, Pump Control, and Optimization**

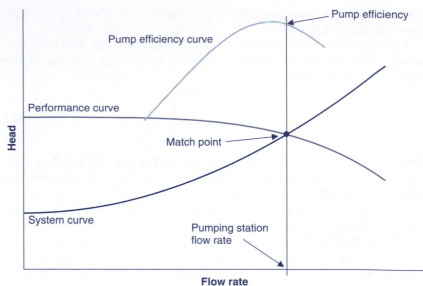

**Figure 9.14** **Pump Characteristic Curves** The match point between a pumping station's performance curve and system curve indicates the flow rate and pumping head leaving the pumping station.

off to better match the needed pumping rate while maximizing efficiency, or different pumps should be installed that have efficiency curves where the match point is closer to the pump's BEP. These changes will lessen the amount of associated energy consumption.

Often a pumping station is near and controlled by a storage tank or wet well. The water levels in the tank or wet well control when pumps are on or off and how many pumps are needed. In the case of a variable-speed pump, the water level controls the speed of the pump motor.

A **storage tank** is commonly used to set an elevated hydraulic grade line across a water distribution system. For flat topography, storage tanks are easily seen as elevated water towers. For areas with hilly terrain, many times the storage tanks are placed at ground level in the hills above the community. The water level in a storage tank should be sized so the tank will routinely fill and drain during the daily demand cycle. This is because stagnant water in a tank becomes "stale" as it ages.

The volume of a storage tank is determined as the sum of the operational (equalization) storage, fire flow storage, and other emergency storage if needed. As shown in Figure 9.15, the *operational storage* is determined by estimating the additional volume of water needed for the maximum-day demand above the volume of water produced at the tank location. The *estimated fire flow storage* was provided previously in Table 9.14. Again, modeling software is used to complete the actual design of storage tanks and the pumping stations that feed them.

A **wet well** provides storage so the actual pumping rate does not need to match the inflow rate into the pumping station. For constant-speed pumps, the wet-well volume is determined based on the **pump cycle time**, which is defined as the time between successive pump starts. It is generally recommended that pumps be started six or fewer times per hour; the pump manufacturer should be consulted. The minimum wet-well volume can then be estimated as

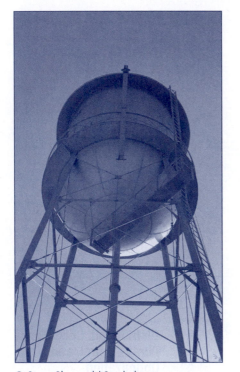

© Steve Shepard/iStockphoto.

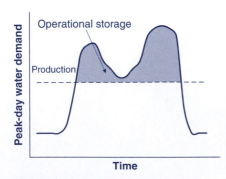

**Figure 9.15** **Operational Storage for a Storage Tank in a Water Distribution System**

$$V_{min} = \frac{Q_{design} \times t_{min}}{4}$$ (9.7)

where $V_{min}$ is the active wet-well volume, $Q_{design}$ is the pump's design flow rate, and $t_{min}$ is the minimum pump cycle time. Derivation of Equation 9.7 is provided by Jones and Sanks (2008).

For example, if the pump's design flow rate is 500 gpm and the pump cycle is six times per hour (so that each pump's cycle time would be 10 min.), the estimated minimum wet-well volume would be 1,250 gal.:

$$V_{min} = \frac{500 \text{ gpm} \times 10 \text{ min.}}{4} = 1,250 \text{ gal.} \qquad \textbf{(9.8)}$$

This value is known as the active volume in the wet well or the amount of water between the "pump on" and "pump off" water levels in the wet well. An additional amount of volume is needed in the wet well to maintain a suction head for the pumps, and more volume is added as a margin of safety.

## Key Terms

- agricultural use
- aquifer
- baseflow
- best efficiency point (BEP)
- branched
- buildout
- confined aquifer
- consumed water
- demand factor
- demand forecasting
- design water velocity in a pipe
- disaggregated segments
- domestic use
- electricity
- energy use
- environmental justice
- estimating water demand
- fire flow demand
- fire protection water demand
- freshwater
- future water demand
- groundwater
- historical records

- household use
- hydrologic cycle
- hydrology
- hydropower
- industrial use
- inflow
- inflow/infiltration (I/I)
- land use
- linear growth
- looped
- maximum and minimum design flow rates
- micro-hydropower
- modeling objectives for water demand
- modeling process to estimate water demand
- municipal water demand
- needed fire flow
- open-channel conditions
- peaking factor (PF)
- population equivalent
- public water
- pump cycle time

- pumping station
- reused water
- satellite reclamation plants
- seawater
- storage tank
- surface waters
- unconfined aquifers
- unmetered flow
- vadose zone
- variation in flows
- wastewater collection
- wastewater flows
- wastewater generation rates
- water distribution
- water footprint
- water reclamation and reuse
- water scarce
- water scarcity
- water stress
- water usage rates
- wet-weather flows
- wet well

# chapter/Nine Problems

**9.1** The average annual rainfall in an area is 60 cm. The average annual evapotranspiration is 35 cm. Thirty percent of the rainfall infiltrates and percolates into the underlying aquifer; the remainder is runoff that moves along or near the ground surface. The underlying aquifer is connected to a stream. Assuming there are no other inputs or outputs of water to the underlying aquifer and the aquifer is at steady state (neither gains nor loses water), what is the amount of baseflow contributed to the stream from the groundwater?

**9.2** Go to the United Nations Environment Programme's Global Environment Outlook Web page, http://geodata.grid.unep.ch/. Look up two countries located in different hemispheres. What are their current amounts of water withdrawals and freshwater withdrawals? Are these countries currently experiencing water scarcity or expected to experience water scarcity?

**9.3** Go to the Web site of the U.S. Geological Survey (USGS), www.usgs.gov, and navigate to "Water Use in the United States." Look up the total water withdrawals associated with the following uses in your state: thermoelectric, irrigation, public supply, industrial, domestic, livestock, aquaculture, and mining. Place the eight uses in a table in order of greatest to least water withdrawals. Determine the percent of the total water withdrawals associated with each of these uses. Compare these percentages to the national percentages listed in Table 9.3. Discuss how your state compares with the national average.

**9.4** Go to the Web site of the U.S. Geological Survey (USGS), www.usgs.gov, and navigate to "Water Use in the United States." Look up the total surface water and groundwater withdrawals associated with your state. Determine the percent of surface water and groundwater withdrawals relative to total withdrawals in your state. Compare these percentages with the national distribution of surface water and groundwater use. Discuss how your state compares with the national average.

**9.5** Contact your local water or wastewater authority. Ask for the annual water usage rates (average, maximum-day, minimum-day, and so on). For a water authority, ask how much is unaccounted-for water; for a wastewater authority, ask for how much is wet-weather flows. Use these numbers to estimate a demand factor and per capita (or metered connections) water usage rates. Discuss how your local values compare with the expected range of values described in this chapter.

**9.6** Estimate your own actual water use during a typical day. List your water use activities, and estimate the volume of water used for each activity. Compare your water usage rate with that of an average per person rate such as 101 gpdc. Explain why your rate may be more or less than the average rate. How much of your water use do you think was discharged as wastewater? Did you do any water use activity that did not create any wastewater?

**9.7** Estimate the daily water demand and wastewater generation for a small commercial area that has the following buildings. Clearly indicate all assumptions and the estimated water usage for each building: (a) a 200-room hotel with 35 employees and one kitchen; (b) three restaurants, one being an organic restaurant with regionally produced foods, another an all-you-can-eat buffet (dinner only), and the third a vegan deli open from 5:00 A.M. to 3:00 P.M.; (c) a newsstand that sells magazines, refreshments, and snacks with one lavatory used only by the employees; and (d) a three-story office building with basement employing 140 people and with two sets of men's and women's lavatories per floor.

**9.8** Estimate the maximum-day demand plus fire flow for a residential area. The residential area is 400 acres divided into .25-acre lots with yards that are 75 ft. wide. Assume the average population density is 2.8 people per home and the maximum-day demand factor is 2.1.

**9.9** A 2.5 MGD wastewater treatment plant is currently running at 80 percent capacity during the annual maximum day, servicing a city of 38,500 people with 26.7 mi. sewers. During the next 10 years, it is expected that new residential developments for 15,000 people will be built, along with an additional 6.5 mi. sewers. The sewer is assumed to leak 8,500 gpd/mi. (a) Project the maximum daily demand for the wastewater treatment plant after the new development is built. (b) Should the wastewater treatment plant's capacity be increased?

**9.10**  You are hired to upgrade the existing water treatment plant for Nittany Lion City. Using the historical records provided in Table 9.18, forecast the water demand through 2024. The population is expected to increase about 1.8 percent per year. (a) Create a graph that has the historical average, minimum-day, and maximum-day water demand in gpd for each year. Extrapolate trend lines for the projected future water demand through 2015. (b) Use your graph to predict the average, minimum-day, and maximum-day water demand for years 2014, 2019, and 2024. (c) Estimate the 2009, 2014, 2019, and 2024 per capita water use by calculating the average water use divided by the population served. (d) Determine a demand factor for the minimum-day and maximum-day demand, using the historical records.

### Table / 9.19

**Historical Records Used to Solve Problem 9.11**

| Year | Metered Flow from Treatment Plant (gpd) | Metered Flow Based on Billing Records (gpd) | | |
|------|------|------|------|------|
| | | Domestic | Commercial | Industrial |
| 2003 | 1,687,517 | 824,247 | 423,229 | 92,676 |
| 2004 | 1,789,453 | 837,055 | 465,232 | 102,707 |
| 2005 | 1,745,658 | 828,103 | 476,429 | 76,916 |
| 2006 | 1,728,750 | 858,076 | 454,928 | 79,029 |
| 2007 | 1,779,854 | 861,003 | 461,669 | 87,422 |
| 2008 | 1,826,650 | 875,548 | 475,254 | 91,214 |
| 2009 | 1,872,456 | 899,545 | 479,451 | 90,248 |

### Table / 9.18

**Historical Records Used to Solve Problem 9.10**

| Year | Water Demand (gpd) | | | Population Served |
|------|------|------|------|------|
| | Average | Minimum | Maximum | |
| 2003 | 1,707,190 | 1,018,655 | 2,624,414 | 14,251 |
| 2004 | 1,713,230 | 1,086,201 | 2,817,674 | 14,352 |
| 2005 | 1,820,602 | 1,094,415 | 3,003,411 | 14,354 |
| 2006 | 1,901,145 | 1,248,011 | 2,945,221 | 14,598 |
| 2007 | 1,891,860 | 1,068,574 | 3,038,157 | 14,587 |
| 2008 | 1,948,648 | 1,124,125 | 3,076,542 | 14,684 |
| 2009 | 1,923,458 | 1,184,214 | 3,067,821 | 14,857 |

into account the additional industry water demand in 2016. (b) Estimate the percent of produced water that is unaccounted-for water based on the historical records. (c) Use the graph and estimated percent unaccounted-for water to predict the total water demand for years 2014, 2019, and 2024.

**9.12**  The recorded metered flow for June 3 at the Wilkes City wastewater treatment plant is shown in Figure 9.16. (a) Estimate the average flow rate for June 3. (b) Estimate the sewer infiltration flow rate, assuming no rain fell during this day.

**9.11**  You are hired to upgrade the existing water treatment plant for USF City. Using the historical records in Table 9.19, forecast the water demand through 2024. A new industry is expected to require 65,000 gpd starting in 2016.

(a) Create a graph that has the historical domestic, commercial, and industrial demand for each year. Extrapolate trend lines for the projected future water demand through 2024 for each category. Take

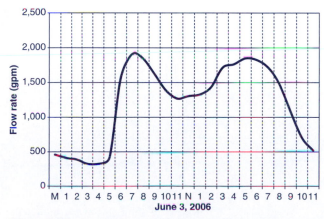

**Figure 9.16**  Wilkes City Wastewater Metered Flow Rate (Used in Problem 9.12.)

**9.13** A storage tank is designed to supply water for fire protection at a small industry. The NFF for this industry is 3,400 gpm. (a) Estimate the volume of water that would be needed for fire protection. (b) Estimate the nominal pipe size of a single pipe supplying the fire protection water from the tank if the design velocity for the pipe is 9.5 ft./sec.

**9.14** A pumping station with wet well is to be sized in a wastewater collection system for a design pumping rate of 1,200 gpm. (a) Estimate the minimum active wet-well volume with a pump cycle of 4 times per hour. (b) Size the force main (pumping station discharge pipe) with a design velocity of 7.5 ft./sec.

**9.15** Identify one regional and one global water scarcity issue. Develop a long-term sustainable solution that protects future generations of humans and the environment.

**9.16** Go to the U.S. Green Building Council Website (http://www.usgbc.org) and research the LEED credits associated with new commercial construction and major renovation (Version 2.2, from U.S. Green Building Council). A project can obtain a maximum of 69 points. (a) How many possible points directly relate to the category water efficiency? (b) What are the specific credits provided for the category water efficiency?

**9.17** The average annual rainfall in an area is 100 cm. The average annual evapotranspiration is 40 cm. 55% of the rainfall infiltrates into the subsurface; the remainder is runoff that moves along or near the ground surface. The underlying aquifer is connected to a stream. Assuming there are no other inputs or outputs of water to the underlying aquifer and the aquifer is at steady-state (neither gains nor loses water), what is the amount of baseflow contributed to the stream from the groundwater (cm/yr)? (b) If the area under consideration is 50 km$^2$, what is the baseflow in m$^3$/yr? (c) What percentage of the baseflow could be modified by human activity assuming there are 15 nearby pumping wells within the 50 km$^2$, each of which is pumping at an average rate of $2 \times 10^{-3}$ m$^3$/s.

**9.18** How does the answer to part (c) in Problem 9.17 change if climate change results in reduced precipitation by 20% and increased population and water consumption increases the pumping rate of the 15 wells in this area to $3 \times 10^{-3}$ m$^3$/s? Assume the rate of evapotranspiration does not change.

**9.19** Go the following web site (www.unesco.org/water/wwap) and look up the report, *The 1$^{st}$ UN World Water Development Report: Water for People, Water for Life.* (a) Of the 11 challenge areas, list the ones related to "life and well being" and the ones related to "management". (b) Access the link on "facts and figures on securing the food supply". Develop a table with columns of product, unit equivalent, and water in m$^3$ per unit for the following products: cattle, sheep and goats, fresh beef, fresh lamb, fresh poultry, cereals, citrus fruits, palm oil, and roots and tubers. Use this table to answer the question, on a per kg basis, does providing meat or grains/fruits use more water?

**9.20** Go to the following web site to learn how you can save water at home (http://www.epa.gov/p2/pubs/water.htm). For the following three areas (in the bathroom, in kitchen/laundry, outdoors) list a minimum of 3 items you can do at home to conserve water.

# References

American Society of Civil Engineers (ASCE). 2007. *Gravity Sanitary Sewer Design and Construction*, ASCE MOP #60 and WEF MOP #FD-5. Reston, Va.: American Society of Civil Engineers; Alexandria, Va.: Water Environment Federation.

American Water Works Association (AWWA). 1998. *Distribution System Requirements for Fire Protection*. AWWA Manual M-31. Denver, Colo.: AWWA.

American Water Works Association (AWWA). 2007. *Water Resources Planning*, 2nd ed. AWWA Manual M-50. Denver, Colo.: AWWA.

Budyko, M. I. 1974. *Climate and Life*. New York: Academic Press.

Chen W. F. 1995. *The Civil Engineering Handbook*. Boca Raton, Fla.: CRC Press, Inc.

Crittenden, J. C., R. R. Trussell, D. W. Hand, K. J. Howe, and G. Tchobanoglous. 2005. *Water Treatment: Principles and Design*, 2nd ed. Hoboken, N.J.: John Wiley & Sons, Inc.

Crook, J. 2004. *Innovative Applications in Water Reuse: Ten Case Studies*. Alexandria, Va.: WaterReuse Foundation.

Global Water Intelligence (GWI). 2005. Water Reuse Markets 2005–2015: A Global Assessment & Forecast. http://www.globalwaterintel.com, accessed November 18, 2008.

Hammer, M. J., and M. J. Hammer Jr. 1996. *Water and Wastewater Technology*, 3rd ed. Englewood Cliffs, N.J.: Prentice Hall.

Hemond, H. F., and E. J. Fechner. 1994. *Chemical Fate and Transport in the Environment*. San Diego: Academic Press.

Hoekstra, A. Y., and A. K. Chapagain. 2006. "Water Footprints of Nations: Water Use by People as a Function of Their Consumption Pattern." *Water Resource Management*, DOI 10.1007/s11269-006-9039-x.

Hutson, S. S., N. L. Barber, J. F. Kenny, K. S. Linsey, D. S. Lumia, and M. A. Maupin. 2004. *Estimated Use of Water in the United States in 2000*. U.S. Geological Survey Circular 1268. U.S. Geological Survey.

Insurance Services Office (ISO). 1998. *Fire Suppression Rating Schedule*. New York: ISO. http://www.iso.com.

Intergovernmental Panel on Climate Change (IPCC). 2007. "Summary for Policymakers." In *Climate Change 2007: Impacts, Adaptation and Vulnerability. Contribution of Working Group II to the Fourth Assessment Report of the Intergovernmental Panel on Climate Change*, ed. M. L. Parry, O. F. Canziani, J. P. Palutikof, P. J. van der Linden, and C. E. Hanson, 7–22. Cambridge: Cambridge University Press.

Jones, G. M., and R. L. Sanks. 2008. *Pumping Station Design*, 3rd ed. Woburn, MA: Butterworth Heinemann.

Mayer, P. W., W. B. DeOreo, E. M. Opitz, J. C. Kiefer, W. Y. Davis, B. Dziegielewski, and J. O. Nelson. 1999. *Residential End Uses of Water*. Denver: American Water Works Association Research Foundation.

Mays, L. W. 2000. *Water Distribution Systems Handbook*. New York: McGraw-Hill, Inc.

Mihelcic, J. R. 1999. *Fundamentals of Environmental Engineering*. New York: John Wiley & Sons, Inc.

Tchobanoglous, G., and F. L. Burton. 1991. *Wastewater Engineering: Treatment, Disposal, and Reuse*, 3rd ed. New York: McGraw-Hill, Inc.

Tchobanoglous, G., F. L. Burton, and H. D. Stensel, 2003. *Wastewater Engineering, Treatment and Reuse*, 4th ed. New York: Wiley Interscience.

United Nations Educational, Scientific, and Cultural Organization (UNESCO) and World Water Assessment Programme (WWAP). 2003. *Water for People, Water for Life: The United Nations World Water Development Report*. New York: UNESCO / Berghahn Books.

United Nations Human Settlements Programme (UN-Habitat). 2003. *Water and Sanitation in the World's Cities: Local Action for Global Goals*. London: Earthscan.

Vickers, A. 2001. *Handbook of Water Use and Conservation*. Denver: American Water Works Association.

Walski, T. M., D. V. Chase, D. A. Savic, W. Grayman, S. Beckith, and E. Koelle. 2003. *Advanced Water Distribution System Modeling*. Waterbury, Conn.: Haestad Press.

Walski, T. M., T. E. Barnard, E. Harold, L. B. Merritt, N. Walker, and B. E. Whitman. 2004. *Wastewater Collection System Modeling and Design*. Waterbury, Conn.: Haestad Press.

Water Environment Federation (WEF). 1998. *Design of Municipal Wastewater Treatment Plants*. MOP no. 8, vol. 1, 4th ed. Alexandria, Va.: Water Environment Federation.

# chapter/Ten Water Treatment

David W. Hand,
Qiong Zhang, James R.
Mihelcic

*In this chapter, readers will learn about the constituents in untreated water, their concentration, and water quality standards associated with them. Mass balance concepts, stoichiometry, and kinetics are employed to develop expressions describing physical–chemical treatment unit processes used to remove constituents. These unit processes include coagulation and flocculation, sedimentation, granular filtration, disinfection, hardness removal by lime–soda softening, removal of dissolved iron and manganese by aeration, and removal of other dissolved organic and inorganic chemicals by activated carbon and membrane techniques. Readers will also learn about energy requirements of collection and treatment.*

## Major Sections

## Learning Objectives

1. Identify physical, chemical, and biological constituents that exist in untreated water and typical concentration ranges for the major constituents.

2. Match major raw water constituents with the unit process(es) that remove a significant amount of each constituent.

3. Identify the difference between maximum contaminant level goals (MCLGs) and maximum contaminant levels (MCLs) and relate these values to treatment objectives at a water treatment plant.

4. Related specific biological pathogens to specific human health impact.

5. Develop viable solutions that address the magnitude of global water issues related to improving human health.

6. Use a systems approach to discuss the relationship between human health, watershed management issues associated with water quality, and design and performance of water treatment plants.

7. Size and understand the operation of water treatment unit processes used for coagulation and flocculation, sedimentation, granular filtration, disinfection, hardness removal by lime–soda softening, and removal of dissolved organic and inorganic chemicals by activated carbon and membrane techniques.

8. Apply several appropriate water treatment processes to design ways to improve water sources for the 1 billion people in the world who do not have access to an improved source of water.

9. Identify sources of energy use during the life stages of water supply and treatment, and thoughtfully discuss how to reduce energy use during treatment.

# 10.1    Introduction

Freshwater is a finite resource, and readily accessible supplies are becoming less abundant. With water scarcity a reality in many parts of the world, increases in population and income along with the impacts of climate change are expected to further exacerbate this issue. Achieving sustainable solutions is compounded by the energy demands of obtaining, storing, and producing a safe water supply, from pumping water to manufacturing the chemicals and materials used throughout the process. And as society develops less desirable sources of water to meet increasing demand, the amount of embodied energy in our water supply is expected to increase. Consequently, there is a need to develop innovative water management strategies (for example, sustainable watershed management, water conservation, and water reuse practices) to meet the global demand for safe drinking water.

The purpose of **water treatment** is to provide potable water that is palatable. **Potable water** refers to water that is healthy for human consumption and free of harmful microorganisms and organic and inorganic compounds that either cause adverse physiological effects or do not taste good. **Palatable** describes water that is *aesthetically* acceptable to drink or free from turbidity, color, odor, and objectionable taste. Water that is palatable may not be safe.

In developed countries, water is treated to be both potable and palatable. However, some people do not like the palatability of municipal waters, and this has given rise to the increased use of point-of-use treatment systems and bottled water. Bottled water has added an additional layer of embodied energy to water, because petroleum is used to produce the water container, and there are recycling or disposal costs for the bottles during their end-of-life life stage.

**Test Your Water Sense**
www.epa.gov/watersense.quiz

# 10.2    Characteristics of Untreated Water

Most consumers expect drinking water to be clear, colorless, odorless, and free of harmful chemicals and pathogenic microorganisms. Natural waters usually contain some degree of dissolved, particulate, and microbiological constituents, which are obtained from the surrounding environment. Table 10.1 summarizes many of the important chemical and biological constituents found in water.

Natural geologic weathering processes can impart dissolved inorganic ions into water, and these can cause problems related to color, hardness, taste, and odor. Dissolved organic matter in water, which is derived from decaying vegetation, can impart a yellowish or brownish color to the water. The surrounding terrestrial environment can cause small or colloidal clay particles to be suspended in the water, which can make the water appear turbid or cloudy. Naturally occurring microorganisms such as bacteria, viruses, and protozoa can make their way into natural waters and cause health issues. Synthetic organic chemicals can be released into the environment and cause chronic or acute health problems to humans and aquatic life. Consequently, a

**Drinking Water at the WHO**
http://www.who.int/topics/
drinking_water/en/

**Concentration of Major Constituents Found in Water**

| General Classification | Specific Constituents | Typical Concentration Range |
|---|---|---|
| Major inorganic constituents | Calcium ($Ca^{2+}$), chloride ($Cl^-$), fluoride ($F^-$), iron ($Fe^{2+}$), manganese ($Mn^{2+}$), nitrate ($NO_3^-$), sodium ($Na^+$), sulfur ($SO_4^{2-}$, $HS^-$) | 1–1,000 mg/L |
| Minor inorganic constituents | Cadmium, chromium, copper, lead, mercury, nickel, zinc, arsenic | 0.1–10 μg/L |
| Naturally occurring organic compounds | naturally occurring organic matter (NOM) that is measured as total organic carbon (TOC) | 0.1–20 mg/L |
| Anthropogenic organic constituents | synthetic organic chemicals (SOCs) and emerging chemicals of concern used in industry, households, and agriculture (e.g., benzene, methyl tert-butyl ether, tetra-chloroethylene, trichloroethylene, vinyl chloride, alachlor) | Below 1 μg/L and up to the tens of mg/L |
| Living organisms | Bacteria, algae, viruses | Millions |

water's physical, chemical, and microbiological characteristics need to be considered in the design and operation of a supply and treatment system.

## 10.2.1 PHYSICAL CHARACTERISTICS

Several aggregate physical characteristics of natural water (also referred to as *raw* or *untreated water*) are used to quantify the appearance or aesthetics of the water. These parameters, described in Table 10.2, are turbidity, number and type of particles, color, taste and odor, and temperature.

**Turbidity** measurements of natural waters vary depending upon the water source. Low turbidity measurements (less than 1 NTU) are typical for most groundwater sources, while surface water turbidity varies depending upon the source. In lakes and reservoirs, turbidity is usually stable and ranges from 1 to 20 NTU, but some waters can vary seasonally due to turnover, storms, and algal activity. Turbidity in rivers is highly dependent on precipitation events and can range from less than 10 NTU to more than 4,000 NTU. Because climate change is expected to increase severe storm events in some parts of the world (including the United States), the resulting runoff and erosion may result in increased seasonal turbidity of some raw water supplies.

Turbidity measurements are primarily used for process control, regulatory compliance, and comparison of different water sources. They are also used as an indicator of increased concentrations of microbial water constituents, such as bacteria, *Cryptosporidium* ocysts, and *Giardia* cysts.

## Table / 10.2

**Physical Characteristics of Natural Water**

| | |
|---|---|
| Turbidity | Turbidity measures the optical clarity of water. It is caused by the scattering and absorbance of light by suspended particles in the water. |
| | A turbidimeter is used to measure the interference of the light passage through the water. Turbidity is reported in terms of **nephelometric turbidity units (NTU)**. |
| | The World Health Organization reports that a turbidity of <5 NTU is usually acceptable but may vary depending upon the availability and resources for treatment. In the United States, many water utilities aim to treat the water to <0.1 NTU. |
| Particles | Particles in natural waters are solids larger than molecules but are generally not distinguished by the unaided eye. They may adsorb toxic metals or synthetic organic chemicals. |
| | Water treatment considers particles in the size range 0.001–100 µm. Particles larger than 1 µm are called **suspended solids**, while particles between about 0.001 and 1 µm can be considered **colloidal particles** (though some researchers go as low as 0.0001 µm). Constituents smaller than 0.001 µm are called **dissolved particles**. |
| | **Natural organic matter (NOM)** comprises colloidal particles and **dissolved organic carbon (DOC)**. The DOC is the portion of NOM that can be filtered through a 0.45 µm filter. It is not classified in terms of size. |
| Color | Color is imparted to water by dissolved organic matter, natural metallic ions such as iron and manganese, and turbidity. |
| | Most people can detect color at greater than 15 true color units in a glass. |
| Taste and odor | Taste and odor can originate from dissolved natural organic or inorganic constituents and biological sources present in raw waters. They can also be an outcome of the water treatment process. |
| Temperature | Surface water temperatures may vary from 0.5°C to 3°C in the winter and 23°C to 27°C in the summer. Groundwater can vary from 2.0°C to 25°C depending upon location and well depth. |

**Particles** found in natural waters can be measured in terms of their numbers and size. Particle counters can measure the number of suspended particles in size ranges generally from 1.0 to 60 µm. Typical particle size distributions for an untreated water source and treated water are shown in Figure 10.1. Note how the size distribution shifts to smaller particles after treatment. Particle removal is important because it has been suggested as an indicator of removal of *Giardia* and *Cryptosporidium* cysts from water (LeChevallier and Norton, 1995). Accordingly, many treatment facilities employ online particle counters to evaluate process performance and to aid in process-control decisions.

Suspended particles such as algae, organic debris, protozoa cysts, and silt can be removed by conventional sedimentation and depth filtration methods. Coagulation and flocculation processes can remove colloidal particles. However, many dissolved constituents will remain in solution, such as the lower-molecular-weight natural organic matter

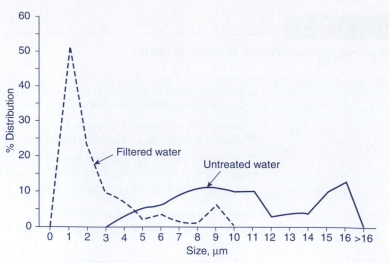

**Figure 10.1** **Particle Size Distribution of a Raw-Water Source and Treated Water** The raw water has 44,000 particles per mL, and the filtered water has 328 particles per mL

From Davis and Cornwell, *Introduction to Environmental Engineering*, 2004, with permission of the Mc-Graw Hill Companies.

(NOM) (for example, humic and fulvic acids) and synthetic organic compounds. Other treatment methods, such as activated carbon adsorption and reverse osmosis, may be used to remove these constituents.

**Color** is categorized as apparent or true color. *Apparent color* is measured on unfiltered samples, so it includes the color imparted by turbidity. *True color* is measured on a water sample passed through a 40 μm filter, so it is a measure of the color imparted by dissolved constituents. While color is not a regulated health concern, it can be an aesthetic problem for some individuals and communities, and treatment is usually provided.

**Taste and odor** threshold concentrations have been established as guidelines for determining when constituents can be detected. Table 10.3 summarizes several constituents and the concentration of each that causes taste and odor in water. The most prevalent natural odor-causing compounds in surface waters come from the decay of algae (for example, geosmin and methyl isoborneal, which impart a musty odor). Water treated with excess chlorine will have a chlorine odor. Groundwaters that have a low redox potential may contain a dissolved gas such as hydrogen sulfide, which smells like rotten eggs. Waters containing dissolved inorganic compounds such as iron, manganese, and copper may have a metallic taste. Some natural or synthetic organic compounds will impart an objectionable taste to water. Examples of these include phenol and 2-chlorophenol.

Water **temperature** is very important because it affects many physical and chemical parameters of water, such as density, viscosity, vapor pressure, surface tension, solubility, and reaction rates, which are used in the design and operation of a treatment plant and associated conveyance system.

## Table / 10.3

### Taste and Odor Thresholds of Substances Found in Water

| Substance Causing Odor and Taste Problems | Concentration (mg/L) |
|---|---|
| **Odor** | **Faint Odor** |
| Chlorine | 0.010 |
| Hydrogen sulfide | 0.0011 |
| Ozone | 0.001 |
| Chloroform | 20.0 |
| Geosmin | 0.000005 |
| Methylisoborneal | 0.000005 |
| Toluene | 0.014 |
| **Taste** | **Taste Threshold** |
| Phenol | 1.0 |
| 2-Chlorophenol | 0.004 |
| Fluoride | 10 |
| $Fe^{2+}$ | 0.04–0.01 |
| $Mn^{2+}$ | 0.4–30 |

SOURCE: Data from Crittenden et al., 2005.

## 10.2.2 MAJOR AND MINOR INORGANIC CONSTITUENTS

Table 10.4 summarizes the major dissolved inorganic constituents found in water. Calcium is one of the most abundant cations found in water and is the major constituent of water **hardness** (along with magnesium). Calcium concentrations greater than about 60 mg/L are considered a nuisance by some. Chloride ($Cl^-$) concentrations in terrestrial waters vary from 1 to 250 mg/L depending on the location, and typical surface water is usually less than 10 mg/L $Cl^-$. However, waters affected by saltwater intrusion and groundwater that contains trapped brine may have chloride concentrations similar to the ocean. Fluoride exists in natural waters primarily as the anion $F^-$ but can also be associated with ferric ion, aluminum, and beryllium. Some water utilities add fluoride in the form of sodium fluoride or hydrofluorosilic acid at concentrations of about 1.0 mg/L.

Iron is abundant in geological formations and is frequently found in water. If not removed, it can impart a brownish color to laundry and bathroom fixtures. At concentrations around 0.2 to 0.4 mg/L, manganese can impart an unpleasant taste to the water and stain laundry and fixtures.

Surface water can contain high concentrations of nitrate (and other forms of nitrogen) from urban and agricultural runoff. Groundwater can also contain high concentrations of nitrate, especially in agricultural areas, where ammonia fertilizers are biochemically converted to

**Lead in Drinking Water**
http://www.epa.gov/safewater/lead

## Table / 10.4

### Major Dissolved Constituents Found in Water

| Constituent | Source | Problem in Water Supply | Range in Natural Waters |
|---|---|---|---|
| Calcium and magnesium | Surface water and groundwater | Above 60 mg/L can be considered nuisance as hardness. | For calcium, less than 1 mg/L to more than 500 mg/L<br><br>Surface water concentrations of magnesium are less than 10 up to 20 mg/L. Groundwater concentrations are less than 30 up to 40 mg/L. |
| Chloride | Surface water and groundwater; saltwater intrusion | Above 250 mg/L can impart salty taste. Below 50 mg/L can be corrosive to some metals. | Typical surface water is usually less than 10 mg/L. |
| Fluoride | Surface water and groundwater<br><br>Some water utilities add fluoride in the form of sodium fluoride or hydrofluorosilic acid at doses of about 1.0 mg/L. | Toxic to humans at concentrations of 250–450 mg/L; fatal at concentrations above 4.0 g/L. | For surface water with total dissolved solids (TDS) concentrations less than 1,000 mg/L, fluoride is usually less than 1.0 mg/L. |
| Iron and manganese | Surface water and groundwater | Taste threshold of iron for many consumers is around 0.01 mg/L. Iron can impart a brownish color to laundry and bathroom fixtures.<br><br>Manganese ion can impart a dark brown color. At concentrations around 0.4 mg/L, manganese can impart an unpleasant taste to the water and can stain laundry and fixtures. | In oxygenated surface waters, the concentration of total iron is usually less than 0.5 mg/L. In groundwater that has low bicarbonate and dissolved oxygen, iron concentrations can range from 1.0 to 10.0 mg/L.<br><br>The concentration of manganese ion in surface water and groundwater may be less than 1.0 mg/L. |
| Nitrate | Surface water and groundwater can contain high concentrations of nitrate from runoff from fertilizers found in urban and agricultural watersheds. | Very high nitrate concentrations may produce infant methemoglobinemia. | |
| Sulfur | Surface water and groundwater | Groundwater low in dissolved oxygen can contain reduced sulfur compounds, which impart objectionable odors such as that of rotten eggs. Sulfates are also corrosive in concrete structures and pipes. | Sulfate concentrations in freshwater can approach 10 mg/L. |

nitrate in the soil. Nitrate is regulated because high concentrations may produce infant methemoglobinemia.

Sulfur can occur as sulfates ($CaSO_4$, $Na_2SO_4$, $MgSO_4$) and reduced sulfides ($H_2S$, $HS^-$). Sulfides can be found in water where there is significant organic decomposition that results in anoxic conditions. Groundwater low in dissolved oxygen can contain reduced sulfur that imparts objectionable odors. Sulfates are also corrosive in concrete structures and pipes.

Several minor inorganic constituents are sometimes a significant health concern or diminish water quality. Examples include copper, chromium, nickel, mercury, strontium, and zinc. Some of these constituents are the result of the surrounding natural environment, while others are present due to human activities. For example, some industrial sites that use arsenic as a wood preservative have contaminated water supplies, while naturally occurring arsenic is widespread

**Global Arsenic Crisis**

http://www.who.int/topics/arsenic/en/

---

**Box / 10.1    The Global Arsenic Crisis**

As Table 10.5 shows, naturally occurring arsenic is widespread in water supplies throughout the world. Unfortunately, when humans are exposed over long periods to low concentrations of arsenic from consuming contaminated water, several forms of cancer can develop. The World Health Organization (WHO) has accordingly set a drinking-water guideline for arsenic of 10 µg/L (10 ppb$_m$).

The magnitude of the problem is most serious in Bangladesh and West Bengal (India). In the 1970s and 1980s, 4 million hand-pump wells were installed in Bangladesh and India to provide people there with a pathogen-free drinking water supply. The presence of arsenicosis began to appear in the 1980s, shortly after the well installation program. By the early 1990s, it was determined that the arsenic poisoning was originating from these wells. The arsenic is naturally occurring.

Today it is estimated that, every day in Bangladesh, up to 57 million people are exposed to arsenic concentrations greater than 10 µg/L. In West Bengal, an estimated 6 million people are exposed to arsenic concentrations between 50 and 3,200 µg/L. The magnitude of the problem shows why some have called this the greatest mass poisoning of humans that has ever occurred.

---

**Table / 10.5**

**Global Locations Where Naturally Occurring Arsenic Has Been Detected in Drinking-Water Supplies**

| Region | Specific Countries |
|---|---|
| Asia | Bangladesh, Cambodia, China, India, Iran, Japan, Myanmar, Nepal, Pakistan, Thailand, Vietnam |
| Americas | Argentina, Chile, Dominica, El Salvador, Honduras, Mexico, Nicaragua, Peru, United States |
| Europe | Austria, Croatia, Finland, France, Germany, Greece, Hungary, Italy, Romania, Russia, Serbia, United Kingdom |
| Africa | Ghana, South Africa, Zimbabwe |
| Pacific | Australia, New Zealand |

SOURCE: Petrusevski et al., 2007.

Box / 10.1   Continued

The most commonly used arsenic removal systems in both the developed and developing world are based on coagulation–separation and adsorption processes. Membrane filtration (such as reverse osmosis and nanofiltration) also is effective at removing arsenic from water; however, it is not practical in much of the world because of high costs. Accordingly, appropriate technologies have been developed to treat this water. Figure 10.2 shows one such technology.

This unit is installed directly at the hand-pumped wells that were installed in the 1970s and 1980s. It requires no electricity or chemical addition. The unit is packed with granular activated alumina, which removes the arsenic from the water. The unit can be regenerated with caustic soda about every four months. The community is instructed to dispose of the arsenic-laden sludge in a pit lined with bricks. After ten years of typical operation, it is estimated that the volume of sludge generated will occupy 56 cu. ft.

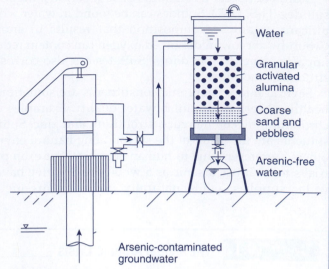

**Figure 10.2   Well Head Arsenic Removal Unit Developed by Dr. Arup Sangupta and Others at Lehigh University**

throughout much of the world. In this latter case, arsenic is found mostly as a solid in the mineral form. However, it can be found dissolved in groundwater in the form of arsenite ($H_3AsO_3$) and arsenate ($H_2AsO_4^-$, $HAsO_4^{2-}$) species. In contrast, lead contamination is usually associated with human activities that even include leaching from old distribution systems.

### 10.2.3   MAJOR ORGANIC CONSTITUENTS

Organic constituents found in water can either be naturally occurring or associated with human activities. Natural organic matter (NOM) in water is the result of the complexation of soluble organic material derived from biochemical degradation of vegetation in the surrounding environment. NOM occurs in all waters and is measured as total organic carbon (TOC). Typical TOC concentrations in natural waters range from less than 0.1 to 2.0 mg/L in groundwater, 1.0 to 20 mg/L in surface waters, and 0.5 to 5.0 in seawater. Table 10.6 summarizes the impact NOM can have on drinking-water treatment processes.

Anthropogenic organic constituents found in water are associated with industrial activity, land use by agriculture, urban runoff, and municipal effluents from wastewater treatment plants. Most of these organic contaminants are classified as **synthetic organic chemicals (SOCs)**. Representative SOCs are found in fuels, cleaning solvents,

**Pharmaceuticals and Personal Care Products**
http://epa.gov/ppcp/faq.html

## Table / 10.6

### Effect of Natural Organic Matter (NOM) on Water Treatment Processes

| Water Treatment Process | Effect |
|---|---|
| Disinfection | NOM reacts with, and consumes, disinfectants, which increases required dose to achieve effective disinfection. |
| Coagulation | NOM reacts with, and consumes, coagulants, which increases required dose to achieve effective turbidity removal. |
| Adsorption | NOM adsorbs to activated carbon, which depletes adsorption capacity of the carbon. |
| Membranes | NOM adsorbs to membranes, clogging membrane pores and fouling surfaces. This leads to decline in water passed through the membrane. |
| Distribution system | NOM may lead to corrosion and slime growth in distribution systems (especially when oxidants are used during treatment). |

SOURCE: Adapted from Crittenden, 2005.

chemical feedstocks, and herbicides and pesticides. **Emerging chemicals of concern** are now found in water and wastewater from the use of personal-care and pharmaceutical products.

## 10.2.4   MICROBIAL CONSTITUENTS

Potable water must be free from pathogenic microorganisms. As just one example of the global magnitude of the problem, the World Health Organization reports that diarrhea contributed 4.3 percent of the global burden of disease in 2002. Of that 4.3 percent, 88 percent was caused by unsafe water, sanitation, and hygiene.

**Pathogens** are microorganisms that cause sickness and disease. Pathogens include many classes of microorganisms, among them viruses, bacteria, protozoa, and helminths.

Details about some representative pathogenic organisms found in untreated water and associated health effects are provided in Table 10.7. Because there are many different water-based pathogens, monitoring and detecting all of them would require a prohibitive amount of resources. Consequently, **indicator organisms** (such as **coliforms**) have been identified and are used to monitor the microbial water quality.

At present, the EPA requires water utilities to monitor their water distribution system monthly for **total coliforms**. The total coliform rule maximum contaminant level is based on frequency of detection (no more than 5 percent for systems collecting at least 40 samples per month) or the combination of a positive *E. coli.* sample (or fecal coliforms) with a positive total coliform sample. While the total coliform test can provide a good indication of fecal contamination, it cannot prove that the source water is safe. Other methods must be used to confirm the absence of longer-surviving organisms such as viruses and spores.

**How's Your Local Drinking Water?**
http://www.epa.gov/safewater/dwinfo/index.html.

## Table / 10.7

### Representative Pathogenic Organisms in Raw-Water Supplies

| Pathogen(s) | Type | Health Effects in Healthy Persons | Normal Habitat |
|---|---|---|---|
| *Vibrio cholerae* <br> Shape: wormlike <br> Size: 0.5 by 1–2 μm | Bacteria | Classic cholerae—explosive diarrhea and vomiting without fever followed by dehydration; abnormally low blood pressure and temperature; muscle cramps; shock; coma followed by death | Human stomach and intestines |
| Salmonella (several species) <br> Shape: rod <br> Size: 0.6 μm | Bacteria | *S. typhi* specie causes enteric fever, headaches, malaise and abdominal pain | Intestines of warm-blooded animals |
| *Shigella dysenteriae* <br> Shape: round <br> Size: 0.4 μm | Bacteria | Bacillary dysentery: abdominal pain, cramps, diarrhea, fever, vomiting, blood and mucus in stools | Human stomach and intestines |
| *Escherichia coli* (EPEC) <br> Shape: rod <br> Size: 0.3–0.5 by 1–2 μm | Bacteria | Diarrhea | Intestines of warm-blooded animals |
| Poliovirus Types 1, 2, 3 <br> Shape: round <br> Size: 28–30 nm | Virus | Fever; severe headache; stiff neck and back; deep muscle pain; skin sensitivity | Human intestinal tract |
| Human adenovirus Type 2 <br> Shape: 12 vertices <br> Size: 70–90 nm | Virus | Severe infections in lungs, eyes, urinary tract, genitals; some strains affect intestines | Human intestinal tract |
| Rotavirus A <br> Shape: round <br> Size: 80 nm | Virus | Severe diarrhea and dehydration | Human intestinal tract |
| *Cryptosporidium parvum* Type 1 <br> Oocyst: <br> Shape: ellipsodial, <br> Size: 3–5 μm <br> Sporozoite and merozoites: <br> Shape: wormlike, <br> Size: 10 by 1.5 μm | Protozoa | Severe diarrhea, abdominal pain, nausea or vomiting, and fever | Human intestinal tract |
| *Giardia lamblia* <br> Shape: single-celled flagellated protozoa <br> Size: 9–15 μm long, 5–15 μm wide, 2–4 μm thick | Protozoa | Sudden diarrhea, abdominal cramps, bloating, cramps, and weight loss | Human intestinal tract |
| *Schistosoma haematobium* a wormlike organism | Helminthic | Squamous cell carcinoma of the bladder; urolithiasis; ascending urinary tract infection; urethral and ureteral stricture with subsequent hydronephrosis; renal failure | Blood vessels of the human bladder and in mammals |

## 10.3 Water Quality Standards

In 1974, the U.S. Congress passed the Safe Drinking Water Act (Public Law 93–523). Under this law, the primary responsibility of setting the water quality regulations was moved from the states to the EPA. To protect public health, EPA established primary drinking-water standards by setting health-based **maximum contaminant level goals (MCLGs)** and **maximum contaminant levels (MCLs)** for a large number of anthropogenic pollutants.

The MCL is the enforceable standard and is based not only on health and risk assessment information, but also on costs and the availability of technology. The MCLGs are based solely on health and risk assessment information.

Table 10.8 provides the MCLGs and MCLs for several important chemicals found in water supplies and their potential health effects. A

**Safe Drinking Water Act**
http://www.epa.gov/safewater/sdwa/index.html

 **Regulations**

| Table / 10.8 | | |
|---|---|---|

**Representative Chemicals Found in Water and Their Maximum Contaminant Level Goals (MCLGs) and Maximum Contaminant Levels (MCLs)**   MCLGs are based solely on health and risk assessment information. MCLs are based not only on health and risk assessment information but also on costs and the availability of technology.

| Chemical | Maximum Contaminant Level Goal (mg/L) | Maximum Contaminant Level (mg/L) |
|---|---|---|
| **Synthetic Organic Chemicals (SOCs)** | | |
| 2,4-D | 0.07 | 0.07 |
| Alachor | 0 | 0.002 |
| Atrazine | 0.003 | 0.003 |
| Benzo(*a*)pyrene | 0 | 0.0002 |
| Chlordane | 0 | 0.002 |
| Lindane | 0.0002 | 0.0002 |
| Polychlorinated biphenyl (PCB) | 0 | 0.0005 |
| **Volatile Organic Chemicals (VOCs)** | | |
| 1,1-Dichloroethylene | 0.007 | 0.007 |
| 1,1,1-Trichloroethane | 0.2 | 0.2 |
| 1,2-Dichloroethane | 0 | 0.005 |
| Benzene | 0 | 0.005 |
| Carbon tetrachloride | 0 | 0.005 |
| cis-1,2-Dichloroethylene | 0.07 | 0.07 |
| Dichloromethane | 0 | 0.005 |
| Ethyl benzene | 0.7 | 0.7 |
| Toluene | 1 | 1 |
| Tetrachloroethylene | 0 | 0.005 |
| Trichloroethylene | 0 | 0.005 |
| Vinyl chloride | 0 | 0.002 |
| Xylenes (total) | 10 | 10 |

## Table / 10.8

(Continued)

**Inorganics**

| | | |
|---|---|---|
| Arsenic | 0 | 0.01 |
| Cadmium | 0.005 | 0.005 |
| Chromium (total) | 0.1 | 0.1 |
| Cyanide | 0.2 | 0.2 |
| Mercury | 0.002 | 0.002 |
| Nitrate (as N) | 10 | 10 |
| Nitrite (as N) | 1 | 1 |
| Selenium | 0.05 | 0.05 |

SOURCE: Data from Davis and Cornwell, 2004; Crittenden et al., 2005.

**Class Discussion**

At present, EPA has developed MCLs for over 90 contaminants, even though there are tens of thousands of commonly used chemicals in commerce. It has thus been a daunting task for government regulators to keep pace with the introduction of new chemicals into commerce. How would more widespread implementation of green chemistry and green engineering affect issues of regulation and treatment?

**Water Treatment Overview**

similar set of MCLs has been established by the World Health Organization (see www.who.org).

## 10.4 Overview of Water Treatment Processes

The typical *unit processes* used for the treatment of surface water and brackish waters are shown in Figure 10.3. Table 10.9 summarizes the unit processes associated with significant removal of particular water constituents. Treatment of surface waters (Figure 10.3a) mostly requires the removal of particulate matter and pathogens. Removing particles also assists in pathogen removal, because most pathogens either are particles or are associated with particles.

If the water source contains dissolved constituents, then additional unit processes can be added to remove them as well. Membrane processes are now widely used for drinking-water treatment, and a schematic diagram for the treatment of brackish water is shown in Figure 10.3b. This treatment process will become more important in the future as increases in population and demand, along with climate change, force society to search for waters of poorer quality. These sources are sometimes high in total dissolved solids (TDS) and are present as brackish groundwater, seawater, and **reclaimed water**.

## 10.5 Coagulation and Flocculation

The most common method used to remove particles and a portion of dissolved organic matter is a combination of coagulation and flocculation followed by sedimentation and/or filtration. **Coagulation** is a charge neutralization step that involves the conditioning of the suspended, colloidal, and dissolved matter by adding coagulants. **Flocculation** involves the aggregation of destabilized particles and formation of larger particles known as floc.

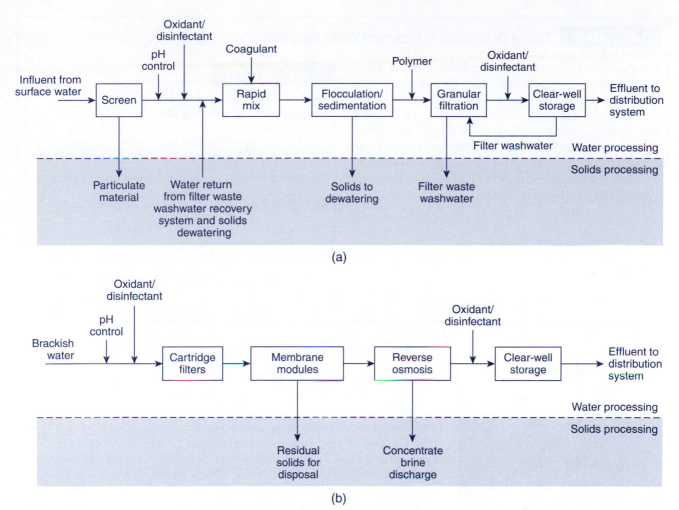

**Figure 10.3** **Typical Water Treatment Unit Processes and Their Arrangement** These processes are typically used for: (a) treatment of surface water and (b) treatment of water with high levels of dissolved constituents.

From Crittenden et al. (2005). Redrawn with permission of John Wiley & Sons, Inc.

**Table / 10.9**

**Unit Processes That Remove a Significant Amount of Raw-Water Constituents**

| Constituent | Unit Process(es) |
|---|---|
| Turbidity and particles | Coagulation/flocculation, sedimentation, granular filtration |
| Major dissolved inorganics | Softening, aeration, membranes |
| Minor dissolved inorganics | Membranes |
| Pathogens | Sedimentation, filtration, disinfection |
| Major dissolved organics | Membranes, adsorption |

For those working on engineering projects in the developing world, a water supply can be improved with many types of projects and appropriate technology, from protecting a water source to building a distribution system. Table 10.10 describes how the World Health Organization (WHO) defines *unimproved* and *improved water supplies*. Bottled water is considered unimproved because of possible problems of sufficient quantity, not quality (WHO 2000).

### Table / 10.10

**Global Definitions of Improved and Unimproved Water Supplies**

| Unimproved Water Supplies | Improved Water Supplies |
|---|---|
| Unprotected well | Household connections |
| Unprotected spring | Public standpipes |
| Vendor-provided water | Boreholes |
| Bottled water | Protected dug wells |
| Tanker-truck-provided water | Protected springs |
| | Rainwater collection |

## 10.5.1  PARTICLE STABILITY AND REMOVAL

Surface charge is the primary contribution to particle stability. Stable particles are likely to remain suspended in solution (and measured as turbidity or TSS). Suspended colloids and fine particles are relatively stable and cannot flocculate and settle in a reasonable period of time. The stability of particles in natural waters primarily depends on a balance of the repulsive and attractive forces between particles. Most particles in natural waters are negatively charged, and a *repulsive electrostatic force* exists between particles of the same charge.

Counteracting these repulsive forces are *attractive forces* between particles, known as *van der Waals forces*. The potential energy from the combined repulsive electrostatic force and attractive van der Waals forces is related to the distance between two particles. Because the net attractive force is very weak at long distances, flocculation usually will not occur. At very short distances, an energy barrier exists, and the kinetic energy arising from Brownian motion of particles is not high enough to overcome the energy barrier. After a coagulant is added, the repulsive forces are reduced, particles will come together, and rapid flocculation can occur. Table 10.11 explains the combined mechanisms of coagulation and flocculation in greater detail.

## 10.5.2  CHEMICAL COAGULANTS

A **coagulant** is the chemical that is added to destabilize particles and accomplish coagulation. Table 10.12 provides examples of commonly used coagulants. Selection of the proper coagulant depends upon (1) the characteristics of the coagulant; (2) concentration and type of particulates; (3) concentration and characteristics of NOM; (4) water temperature; (5) water quality (for example, pH); (6) cost and availability; and (7) dewatering characteristics of the solids that are produced. Natural coagulants are being promoted in many parts of the world

Photo courtesy of Brooke Tyndell Ahrens.

## Table / 10.11

**Mechanisms of Coagulation and Flocculation**

| | |
|---|---|
| **Compression of the electrical double layer (EDL)** | Most particles in water have a net negative surface charge. The *electric double layer* (EDL) consists of a layer of cations bound to the particle surface and a diffuse set of cations and anions that extend out into the solution. When the ionic strength is raised, the EDL shrinks (the repulsive forces are reduced). |
| **Charge neutralization** | Since most particles found in natural waters are negatively charged at neutral pHs, they can be *destabilized* by adsorption of positively charged cations or polymers, such as hydrolyzed metal salts and cationic organic polymers. The dose (in mg/L) of such salts or polymers is critical for subsequent flocculation process. With the proper dose, the charge will be neutralized, and particles will come together. However, if the dose is too high, the particles, instead of being neutralized, will attain a positive charge and become stable once again. |
| **Adsorption and interparticle bridging** | With the further addition of *nonionic polymers* and long-chain low-surface-charge polymers, a particle can be adsorbed on the chain, and the remainder of the polymer may adsorb on available surface sites of other particulates. This results in formation of a bridge between particles. Again, an optimum dose (in mg/L) of the nonionic polymer exits. If too much polymer is added, particles will be enmeshed in a polymer matrix and will not flocculate. |
| **Precipitation and enmeshment** | Enmeshment (also referred to as *sweep floc*) occurs when a high enough dose of aluminum (and iron salts) is added and they form various hydrous polymers that will precipitate from solution. As the amorphous precipitate forms, particulate matter is trapped within the floc and swept from the water with the settling floc. This mechanism predominates in water treatment applications where aluminum or iron salts are used at high concentrations and pH is maintained near neutral. |

## Table / 10.12

**Types of Coagulants Commonly Used in the Field**

| Coagulant Type | Examples |
|---|---|
| Inorganic metallic coagulants | Aluminum sulfate (also referred to as alum, $Al_2(SO_4)_3 \cdot 14H_2O$); sodium aluminate ($Na_2Al_2O_4$); aluminum chloride ($AlCl_3$); ferric sulfate ($Fe_2(SO_4)_3$) and ferric chloride ($FeCl_3$) |
| Prehydrolyzed metal salts | Made from alum and iron salts and hydroxide under controlled conditions; include polyaluminum chloride (PACl), polyaluminum sulfate (PAS), and poly-iron chloride |
| Organic polymers | Cationic polymers, anionic polymers, and nonionic polymers (for synthetic polymers, molecular weight in the range of $10^4$–$10^7$ g/mole) |
| Natural plant-based materials | *Opuntia* spp. and *Moringa oleifera* (used in many parts of the world, especially the developing world) |

because they are considered renewable, they can be used as food and fuel, and their production relies on local materials and labor.

Coagulant and flocculant aids are substances that enhance the coagulation and flocculation processes. Coagulant aids are typically insoluble particulate materials, such as clay, diatomite, powdered activated carbon, or fine sand, that form nucleating sites for the formation of larger flocs. They are used in conjunction with the primary coagulants. Flocculant aids such as anionic and nonionic polymers are used to strengthen flocs. They are added after the addition of coagulants and the destabilization of the particles.

The most commonly used coagulant is alum (molecular weight of 594 g/mole). Addition of $Al^{3+}$ in the form of alum (or $Fe^{3+}$ in the form of the iron salts such as $FeCl_3$) at concentrations greater than their solubility limits results in the formation of the hydroxide precipitate that is typically used in the sweep floc mode of operation. The overall stoichiometric reaction for addition of alum in the formation of a hydroxide precipitate is as follows:

$$Al_2(SO_4)_3 \cdot 14H_2O + 6(HCO_3^-) \rightarrow$$

$$2Al(OH)_{3(s)} + 3SO_4^{2-} + 14H_2O + 6CO_2 \qquad \textbf{(10.1)}$$

In Equation 10.1, alkalinity (expressed as $HCO_3^-$) is consumed with the addition of alum. This is because alum and the other iron salts are weak acids. Based on stoichiometry, 1 mg/L of alum will consume approximately 0.50 mg/L of alkalinity (as $CaCO_3$). If the natural alkalinity of the water is not sufficient, it may be necessary to add lime or soda ash ($Na_2CO_3$) to react with the alum to maintain the pH in the appropriate range. The pH range for operating region of alum is 5.5 to 7.7, and for iron salts is 5 to 8.5.

**Jar testing** is widely used for screening the type of coagulant and the proper coagulant dosage. A jar test apparatus is shown in Figure 10.4.

**Coagulation and Flocculation**

www.wiley.com/college/mihelcic

**Figure 10.4** Jar Test Apparatus Used to Screen Coagulants for Correct Coagulant and Proper Dosage

Photo courtesy of David Hand.

It consists of six square batch reactors, each equipped with a paddle mixer that can turn at variable speeds. In a jar test, batch additions of various types and different dosages of coagulants are added to the water sample. A rapid mixing stage is combined with the addition of the coagulant. This stage is followed by a slow mixing stage to enhance floc formation. The samples are then allowed to settle under undisturbed conditions, and the turbidity of the settled supernatant is measured and plotted as a function of coagulant dose in order to determine the proper coagulant dosage.

example/10.1 Use of Jar Testing to Determine the Optimal Coagulant Dosage

A jar test was conducted on untreated water with an initial turbidity of 10 NTU and a $HCO_3^-$ concentration of 50 mg/L as $CaCO_3$. Using the following data obtained from a jar test, estimate the optimum alum dosage for turbidity removal and the theoretical amount of alkalinity that will be consumed at the optimal dosage. Alum is added as dry alum (molecular weight of 594 g/mole).

| Alum dose, mg/L | 5 | 10 | 15 | 20 | 25 | 30 |
|---|---|---|---|---|---|---|
| Turbidity, NTU | 8 | 6 | 4.5 | 3.5 | 5 | 7 |

solution

The data are graphed as shown in Figure 10.5. The graph shows that the turbidity reaches the lowest value when the alum dose is 20 mg/L. This is the optimum alum dosage.

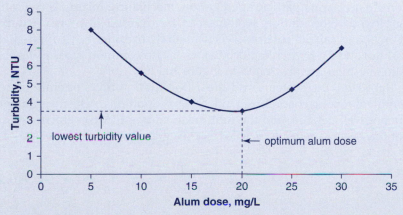

**Figure 10.5** Jar Test Results That Will Aid Identification of Proper Coagulant Dosage

Next, we must determine the amount of alkalinity consumed and check this value against the naturally occurring alkalinity to determine whether additional alkalinity needs to be added. Use the stoichiometry of Equation 10.1 to determine the alkalinity consumed:

$$\text{alkalinity consumed} = (20 \, \text{mg/L} \, Al_2(SO_4)_3 \cdot 14H_2O) \left( \frac{1 \, g}{1{,}000 \, \text{mg}} \right)$$

$$\times \left( \frac{1 \, \text{mole/L} \, Al_2(SO_4)_3 \cdot 14H_2O}{594 \, g/\text{mole} \, Al_2(SO_4)_3 \cdot 14H_2O} \right) \times \left( \frac{6 \, \text{mole} \, HCO_3^-}{1 \, \text{mole} \, Al_2(SO_4)_3 \cdot 14H_2O} \right)$$

$$\times \left( \frac{1 \, \text{eqv alkalinity}}{1 \, \text{mole} \, HCO_3^-} \right) \times \left( \frac{100 \, g \, CaCO_3}{2 \, \text{eqv alkalinity}} \right)$$

$$= 0.01 \, g/L \, \text{as} \, CaCO_3 = 10 \, \text{mg/L} \, \text{as} \, CaCO_3$$

Thus, 10 mg/L of alkalinity as $CaCO_3$ are consumed, and the water sample had an initial alkalinity of 50 mg/L as $CaCO_3$. Therefore, the alkalinity in the raw water is sufficient to buffer the acidity produced after alum addition.

### 10.5.3 OTHER CONSIDERATIONS

Coagulants are dispersed into the water stream via **rapid mixing** systems: (1) pumped mixing (for example, pumped flash mixing, which can be simple and reliable); (2) hydraulic methods (for example, in-line static mixer, which is simple, reliable, and nonmechanical); and (3) mechanical mixing (conventional stirred tanks being the most common). Several of the devices are illustrated in Figure 10.6.

The *gentle mixing* of the water is the major mechanism for flocculation. To accelerate particle aggregation, mechanical mixing is typically employed. Flocculation systems can be divided into two groups: (1) mechanical flocculators (vertical-shaft turbine, horizontal-shaft paddle) and (2) hydraulic flocculators. Three common types of flocculation systems are illustrated in Figure 10.7.

Rapid-mixing systems and most flocculation units operate under turbulent mixing conditions. Under these conditions, velocity gradients are not well defined, and the **root mean square (RMS) velocity gradient** has been widely adopted for assessing energy input:

$$\overline{G} = \sqrt{\frac{P}{\mu V}} \tag{10.2}$$

where $\overline{G}$ is the global RMS velocity gradient (energy input rate, in $s^{-1}$; $P$ is the power of mixing input to the vessel (J/s); $\mu$ is the dynamic viscosity of water ($N \cdot s/m^2$); and $V$ is the volume of the mixing vessel ($m^3$). The value of $\overline{G}$ assumes that all elemental cubes of liquid in the volume $V$ are sheared at the same rate (on average) and $\overline{G}$ is a time-averaged value.

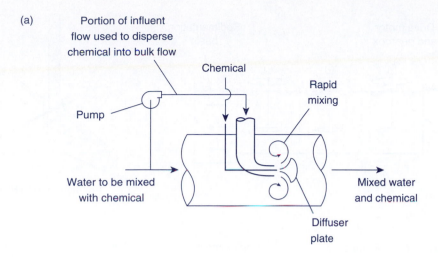

(a)

Portion of influent
flow used to disperse
chemical into bulk flow

Chemical

Rapid
mixing

Pump

Water to be mixed
with chemical

Mixed water
and chemical

Diffuser
plate

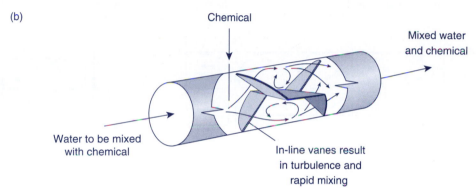

(b)

Chemical

Mixed water
and chemical

Water to be mixed
with chemical

In-line vanes result
in turbulence and
rapid mixing

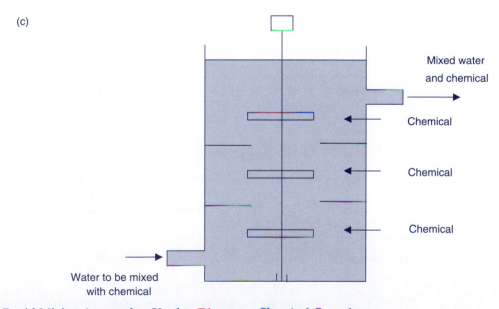

(c)

Mixed water
and chemical

Chemical

Chemical

Chemical

Water to be mixed
with chemical

**Figure 10.6** **Rapid Mixing Approaches Used to Disperse a Chemical Coagulant during Water Treatment** The approaches are (a) pumped flash mixing; (b) in-line static mixer; and (c) conventional stirred tank.

Illustrations (a) and (b) from Crittenden et al. (2005). Redrawn with permission of John Wiley & Sons, Inc. Illustration (c) from M. L. Davis and D. A. Cornwell, *Introduction to Environmental Engineering*, 1998, McGraw-Hill. Redrawn with permission of the McGraw-Hill Companies.

(a)

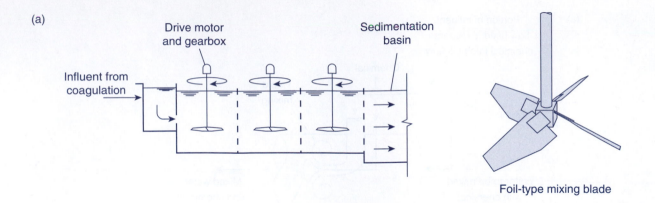

(b)

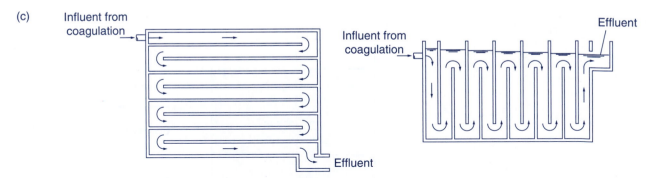

(c)

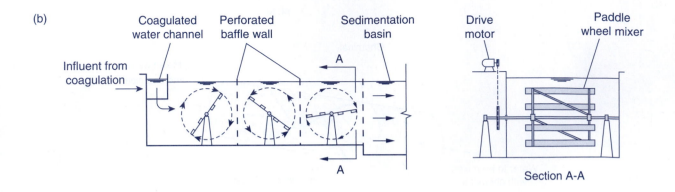

**Figure 10.7** **Common Types of Gentle Mixing Employed in Flocculation Systems**
The drawing show (a) a vertical shaft turbine flocculation system, (b) a horizontal shaft paddle wheel flocculation system, and (c) a hydraulic flocculation system. Note how the hydraulic flocculation system requires no energy input during the use life stage.

Crittenden et al. (2005). Redrawn with permission of John Wiley & Sons, Inc.

The mixing performance depends not only on the velocity gradient ($\overline{G}$) but also the hydraulic detention time ($t$) and the product of $\overline{G}$ and $t$, a measure of degree of mixing. In practice, the values of $\overline{G}$ and $\overline{G}t$ are used as design criteria. Typical values for $\overline{G}$, $t$, and $\overline{G}t$ in the design of rapid-mix systems and flocculation tanks are provided in Table 10.13.

## Table / 10.13

**Typical Values Used in Design of Rapid-Mixing and Flocculation Systems**

| System Category | RMS Velocity Gradient, $\bar{G}$ (sec$^{-1}$) | Detention Time, $t$ | $Gt$ Values |
|---|---|---|---|
| Mechanical mixing | 600–1,000 | 10–120 s | $5.0 \times 10^4$ to $5.0 \times 10^5$ |
| In-line mixing | 3,000–5,000 | 1 s | $1.0 \times 10^3$ to $1.0 \times 10^5$ |
| Horizontal-shaft paddle flocculator | 20–50 | 10–30 min | $1.0 \times 10^4$ to $1.0 \times 10^5$ |
| Vertical-shaft turbine flocculator | 10–80 | 10–30 min | $1.0 \times 10^4$ to $1.0 \times 10^5$ |

example/10.2 Design of a Mechanical Rapid-Mix Tank

A conventional stirred tank is used for rapid mixing in a water treatment plant with a flow of $100 \times 10^6$ L/day. The water temperature is 10°C. Determine the tank volume and power requirement.

solution

The tank volume equals the flow rate ($Q$) times the hydraulic detention time ($\theta$). Table 10.13 provides appropriate detention times. Conventional stirred tanks are considered a form of mechanical mixing, and we will select a detention time value of 60 s:

$$V = Q \times \theta = \frac{100 \times 10^6 \, \text{L}}{\text{day}} \times 60 \, \text{s} \times \frac{1 \, \text{min}}{60 \, \text{s}} \times \frac{1 \, \text{day}}{1{,}440 \, \text{min}}$$

$$\times \frac{\text{m}^3}{1{,}000 \, \text{L}} = 69 \, \text{m}^3$$

To determine the power requirement, use Table 10.13 to select an appropriate RMS velocity gradient. Here we will select a $\bar{G}$ value of 900/s, and the product of $\bar{G}$ and $t$ (900/s × 60 s = $5.4 \times 10^4$) is within the range ($5 \times 10^4$ to $5 \times 10^5$) provided in Table 10.13. At 10°C, $\mu = 0.001307$ N·sec/m$^2$. To obtain power consumption, rearrange Equation 10.2 to solve for $P$:

$$P = \bar{G}^2 \times \mu \times V = (900/\text{s})^2 \times 0.001307 \, \frac{\text{N} \cdot \text{s}}{\text{m}^2} \times 69 \, \text{m}^3$$

$$\times \frac{1 \, \text{kN}}{1{,}000 \, \text{N}}$$

$$= 73 \, \frac{\text{kN} \cdot \text{m}}{\text{s}} = 73 \, \text{kW}$$

Northbrook, Illinois, is the first community in Illinois and one of only a few in the United States to offset the energy used to run its water plant. The Northbrook water treatment plant provides 2.1 billion gal. water per year for 34,000 residents. It began purchasing 155 MWh/y in renewable-energy certificates from wind farms in north central Illinois several years ago to offset electricity derived from coal (Figure 10.8).

After seeing the movie *An Inconvenient Truth*, Northbrook's director of public works recommended the village increase its purchase of renewable-energy certificates to 4,500 MWh/y, enough energy to operate the water treatment plant.

**Figure 10.8    A Wind Farm in North Central Illinois**   Wind turbines provide energy that the water treatment plant for Northbrook, Illinois, purchases through renewable energy certificates. The community buys enough energy to operate the water treatment plant.

Photo courtesy of Iberdrola Renewables.

## Class Discussion

*What energy efficiency techniques or use of renewable energy can you envision for your local water or wastewater treatment plant, public works department, or community?*

## 10.6    Hardness Removal

Water hardness is caused by divalent cations, primarily calcium and magnesium ions ($Ca^{2+}$ and $Mg^{2+}$). When $Ca^{2+}$ and $Mg^{2+}$ are associated with alkalinity anions (for example, $HCO_3^-$), the hardness is defined as **carbonate hardness**. The term **noncarbonated hardness** is used if the $Ca^{2+}$ and $Mg^{2+}$ are associated with nonalkalinity anions (for example, $SO_4^{2-}$) The distribution of hard waters in the United States is shown in Figure 10.9.

Complexation agents can be added to prevent divalent cations from precipitating, or hardness can be removed. A process flow diagram for a two-stage excess **lime–soda ash softening process** is shown in Figure 10.10. Lime is sold commercially in forms of quicklime (90 percent CaO) and hydrated lime (70 percent CaO). Granular quicklime is usually crushed in a slaker and fed to slurry containing about 5 percent calcium hydroxide. Powdered hydrated lime is prepared by fluidizing in a tank containing a turbine mixer. The soda ash (approximately 98 percent $Na_2CO_3$) is a grayish-white powder that can be added either with lime or following the addition of lime. Carbon

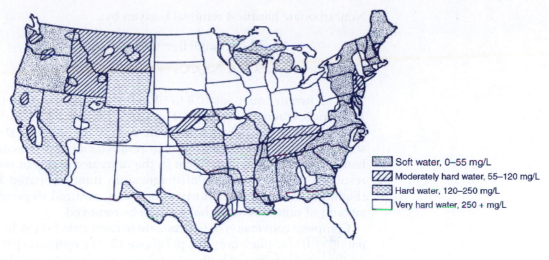

**Figure 10.9** **Distribution of Hard Water in the United States** Units are mg/L as $CaCO_3$. The areas shown define approximate hardness values for municipal water supplies.

Legend:
- Soft water, 0–55 mg/L
- Moderately hard water, 55–120 mg/L
- Hard water, 120–250 mg/L
- Very hard water, 250 + mg/L

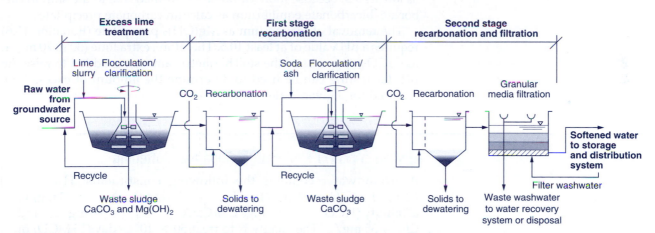

**Figure 10.10** **Process Flow Diagram for a Two-Stage Excess Lime–Soda Treatment Process Used to Treat Hard Waters** Considerations should be made for reusing all waste streams.

From Crittenden et al. (2005). Redrawn with permission of John Wiley & Sons, Inc.

dioxide is used for recarbonization to reduce the pH and precipitate excess calcium of the lime-softened water.

When lime slurry ($Ca(OH)_2$) is added to water, it first reacts with free carbon dioxide, because the $CO_2$ is a stronger acid than $HCO_3^-$. (Remember that, by definition, $HCO_3^-$ can act as an acid or a base.) The chemical reactions for the removal of carbonate and noncarbonate hardness are provided in Equations 10.3 to 10.7:

$$CO_2 + Ca(OH)_2 \rightarrow CaCO_{3(s)} + H_2O \qquad (10.3)$$

Carbonate hardness removal is given by:

$$Ca(HCO_3)_2 + Ca(OH)_2 \rightarrow 2CaCO_{3(s)} + 2H_2O \qquad (10.4)$$

$$Mg(HCO_3)_2 + 2Ca(OH)_2 \rightarrow 2CaCO_{3(s)} + Mg(OH)_{2(s)} + 2H_2O \qquad (10.5)$$

Noncarbonate hardness removal is given by:

$$MgSO_4 + Ca(OH)_2 \rightarrow Mg(OH)_{2(s)} + CaSO_4 \qquad (10.6)$$

$$CaSO_4 + Na_2CO_3 \rightarrow CaCO_{3(s)} + Na_2SO_4 \qquad (10.7)$$

As shown in Equations 10.4 to 10.7, lime will remove $CO_2$ (Equation 10.3) and carbonate hardness (Equations 10.4 and 10.5) and replace magnesium with calcium (Equation 10.6). Equation 10.7 shows that soda ash ($Na_2CO_3$) is then used to remove calcium noncarbonated hardness, which may be present in the untreated water or may be the result of the precipitation of magnesium noncarbonated hardness (Equation 10.6). The amount of soda ash required depends on the amount of noncarbonated hardness to be removed.

Complete conversion of bicarbonate to carbonate for calcium precipitation will take place only at a pH above 12. The optimum pH depends on the concentration of both calcium and bicarbonate ions. In practice, the optimum pH for maximum calcium carbonate precipitation may be as low as 9.3, because more carbonate is formed due to the shift in carbonate–bicarbonate equilibrium as calcium carbonate precipitates.

The removal of magnesium as $Mg(OH)_2$ precipitate (Equation 10.6) requires a pH value of at least 10.5. Therefore, extra lime (30 to 70 mg/L as $CaCO_3$) in excess of the stoichiometric amount is added to raise the pH. Jar testing can be used to determine the amount of excess lime required for a given water source.

example/10.3 Lime–Soda Ash Softening

A groundwater contains the following constituents: $H_2CO_3^* = 62 \text{ mg/L}$, $Ca^{2+} = 80 \text{ mg/L}$, $Mg^{2+} = 36.6 \text{ mg/L}$, $Na^+ = 23 \text{ mg/L}$, alkalinity ($HCO_3^-$) $= 250 \text{ mg/L}$ as $CaCO_3$, $SO_4^{2-} = 96 \text{ mg/L}$, and $Cl^- = 35 \text{ mg/L}$. The facility is to treat $50 \times 10^6 \text{ L/day}$ (15 MGD) of water from this source using lime–soda ash to reduce the hardness.

1. Determine the total, carbonate, and noncarbonated hardness present in the raw water.

2. Determine the lime and soda ash dosages for softening (units of kg/day). Assume the lime is 90 percent CaO by weight and the soda ash is pure sodium carbonate.

solution

This problem requires several steps. First construct a table of the chemical constituents and their concentrations in terms of mg/L as $CaCO_3$ (see Table 10.14).

The second step determines the total hardness, carbonate hardness, and noncarbonated hardness. The total hardness is the sum of the calcium and magnesium ions as $CaCO_3$:

$$\text{total hardness} = (200 + 150) = 350 \text{ mg/L as } CaCO_3$$

## Table / 10.14

**Table Constructed to Solve Example 10.3**

| Chemical Constituent | Concentration (mg/L) | Equivalents | Molecular Weight (g/mole) | Equivalent Weight (eqv/mole) | Concentration (meqv/L) | Concentration (mg/L as CaCO₃) |
|---|---|---|---|---|---|---|
| $H_2CO_3^*$ | 62 | 2 | 62.0 | 31.0 | 2.0 | 100 |
| *Cations* | | | | | | |
| $Ca^{2+}$ | 80 | 2 | 40.0 | 20.0 | 4.0 | 200 |
| $Mg^{2+}$ | 36.6 | 2 | 24.4 | 12.2 | 3.0 | 150 |
| $Na^+$ | 23.0 | 1 | 23.0 | 23.0 | 1.0 | 50 |
| | | | | **Total** | 9.0 | 400 |
| *Anions* | | | | | | |
| Alk ($HCO_3^-$) | 250.0 | 2 | 100.0 | 50.0 | 5.0 | 250 |
| $SO_4^{2-}$ | 96.0 | 2 | 96.0 | 48.0 | 2.0 | 100 |
| $Cl^-$ | 35 | 1 | 35.5 | 35.5 | 1.0 | 50 |
| | | | | **Total** | 9.0 | 400 |

The carbonate hardness is the sum of the calcium and magnesium ions associated with bicarbonate ions. Because the total hardness (350 mg/L as $CaCO_3$) is greater than bicarbonate alkalinity (250 mg/L $CaCO_3$), all the bicarbonate is associated with calcium (200 mg/L $CaCO_3$) and magnesium (50 mg/L $CaCO_3$). The carbonate hardness is thus equal to bicarbonate alkalinity as $CaCO_3$:

$$\text{carbonate hardness} = 200 + 50 = 250\,\text{mg/L as } CaCO_3$$

The noncarbonated hardness equals the magnesium ions not associated with carbonate hardness ($MgSO_4$):

$$\text{Noncarbonate hardness} = (150 - 50) = 100\,\text{mg/L as } CaCO_3$$

The second question requests we determine the daily mass of lime and soda ash required for softening. The lime that is required will react with $CO_2$ ($H_2CO_3^*$), $Ca(HCO_3)_2$, $Mg(HCO_3)_2$, and $MgSO_4$. The stoichiometric amount of $Ca(OH)_2$ can be calculated based on Equations 10.3 to 10.6.

$$Ca(OH)_2 \text{ required to react with } CO_2 = 100\,\text{mg/L as } CaCO_3$$
$$Ca(OH)_2 \text{ required to react with } Ca(HCO_3)_2 = 200\,\text{mg/L as } CaCO_3$$
$$Ca(OH)_2 \text{ required to react with } Mg(HCO_3)_2 = 2 \times Mg(HCO_3)_2$$
$$= 2 \times 50 = 100\,\text{mg/L as } CaCO_3$$

## example/10.3 Continued

$Ca(OH)_2$ required to react with $MgSO_4 = 100\ mg/L$ as $CaCO_3$

Remember that 30 to 70 mg/L of extra lime must be added to raise the pH above 10.5 to ensure that magnesium is removed as $Mg(OH)_2$. Assume 30 mg/L (as $CaCO_3$) of additional lime is required in the process.

The total lime requirement is thus determined from the summation of all five individual lime requirements:

$$
\begin{aligned}
\text{lime required} = &\left( \frac{100\ mg\ CaCO_3}{L} + \frac{200\ mg\ CaCO_3}{L} \right. \\
&\left. + \frac{100\ mg\ CaCO_3}{L} + \frac{100\ mg\ CaCO_3}{L} + \frac{30\ mg\ CaCO_3}{L} \right) \\
&\times \frac{56\ mg\ CaO/mmole}{100\ mg\ CaCO_3/mmole} \times \frac{kg}{10^6\ mg} \times \frac{50 \times 10^6\ L}{day} \\
&\times \frac{kg\ bulk\ lime}{0.9\ kg\ CaO} \\
= &\ 16,500\ kg/day
\end{aligned}
$$

The soda ash ($Na_2CO_3$) required is also determined by reaction stoichiometry:

$$
\begin{aligned}
\text{soda ash} &= \text{noncarbonate hardness} \\
&= 100\ mg\ CaCO_3/L \times \frac{106\ mg\ Na_2CO_3/mmole}{100\ mg\ CaCO_3/mmole} \\
&\times \frac{kg}{10^6\ mg} \times \frac{50 \times 10^6\ L}{day} = 5,300\ kg/day
\end{aligned}
$$

## 10.7   Sedimentation

**Sedimentation** is the process in which the majority of the particles will settle by gravity within a reasonable time and be removed. Particles with densities greater than $1,000\ kg/m^3$ will eventually settle, and particles with densities less than $1,000\ kg/m^3$ will float to the water surface. In water treatment, there are common types of settling: discrete particle settling and flocculant settling.

### 10.7.1   DISCRETE PARTICLE SETTLING

**Discrete particle settling** occurs when particles are discrete and do not interfere with one another as they settle. For this type of settling, the movement of a particle in water is determined by a balance of a downward gravitational force, an upward buoyancy force, and an upward drag force.

The settling velocity of particles in a liquid such as water can be described by either **Stokes' law** or **Newton's law**. Table 10.15 describes each of these laws in greater detail. Stokes' law was derived in Chapter 4. It is applicable to spherical particles when the Reynold's number is less than or equal to 1 (laminar flow). Newton's law is used to determine the settling velocity of particles when the Reynold's number is greater than 1 (transition and turbulent flow). The dimensionless **Reynold's number (Re)** is defined as

$$Re = \frac{\rho \, d_p \, v_s}{\mu} = \frac{d_p \, v_s}{\upsilon} \approx \frac{\text{inertial forces}}{\text{viscous forces}} \qquad (10.8)$$

where $\rho$ is the density of the liquid (kg/m$^3$), $d_p$ is the particle diameter (m), $v_s$ is the settling velocity of the particle at any point in time (m/s), $\mu$ is the dynamic viscosity of the liquid (N-s/m$^2$), and $\upsilon$ is the kinematic viscosity of the liquid (m$^2$/sec).

## Table / 10.15

### Determination of Settling Velocity of Particles Using Stokes' and Newton's Laws

| Applicable Law | Settling Velocity (m/sec) | Terms | Drag Coefficient | Applicability |
|---|---|---|---|---|
| Stokes' Law | $v_s = \dfrac{g(\rho_p - \rho)d_p^2}{18\mu}$ | $g$ is the acceleration due to gravity (m/s$^2$); $\rho_p$ is the density of the particle (kg/m$^3$); $\rho$ is the density of the liquid (kg/m$^3$); $d_p$ is the particle diameter (m); and $\mu$ is the dynamic viscosity of the liquid (N-s/m$^2$) | For laminar flow: $C_d = \dfrac{24}{Re}$ | Applicable for spherical particles when the Reynold's number $\leq 1$ (laminar flow). Has limited application in water treatment because conditions in most treatment facilities are not laminar. |
| Newton's law | $v_s = \sqrt{\dfrac{4g(\rho_p - \rho)d_p}{3C_d\rho}}$ | $g$ is the acceleration due to gravity (m/s$^2$); $\rho_p$ is the density of the particle (kg/m$^3$); $\rho$ is the density of the liquid (kg/m$^3$); $C_d$ is the drag coefficient. | For the transition regime: $C_d = \dfrac{24}{Re} + \dfrac{3}{\sqrt{Re}} + 0.34$ $C_d$ becomes constant in turbulent regime (Re > 10,000) | Applicable for particles when the Reynold's number > 1 (transition and turbulent flow). |

## example/10.4 Application of Stokes' Law

Calculate the terminal settling velocity for a sand particle that has a diameter of 100 $\mu$m and a density of 2,650 kg/m$^3$. The water temperature is 10°C.

### solution

The terminal settling velocity of the particle can be calculated using Stokes' law (Table 10.15). For water at 10°C, $\rho$ = 999.7 kg/m$^3$, $\mu$ = 1.307 × 10$^{-3}$ N·s/m$^2$, and $\upsilon$ = 1.306 × 10$^{-6}$ m$^2$/s.

$$v_s = \frac{g(\rho_p - \rho)d_p^2}{18\mu}$$

$$= \frac{9.81 \text{ m/s}^2 \times (2,650 - 999.7 \text{ kg/m}^3) \times (1.0 \times 10^{-4} \text{ m})^2}{18 \times 1.307 \times 10^{-3} \text{ N·s/m}^2}$$

$$\times \frac{3,600 \text{ s}}{\text{hr}} = 24.8 \frac{\text{m}}{\text{hr}}$$

We must verify the flow conditions to ensure Stokes' law is applicable. The Reynold's number is calculated to verify that the particle is settling under laminar conditions:

$$\text{Re} = \frac{d_p\, v_s}{\upsilon} = \frac{1.0 \times 10^{-4} \text{ m} \times 24.8 \text{ m/hr} \times (1 \text{ hr}/3,600 \text{ s})}{1.306 \times 10^{-6} \text{ m}^2/\text{s}} = 0.53$$

Because Re < 1, laminar flow exists, and Stokes' law is applicable.

## example/10.5 Application of Newton's Law

Calculate the terminal settling velocity for a sand particle that has a diameter of 200 $\mu$m and a density of 2,650 kg/m$^3$. The water temperature is 15°C.

### solution

The terminal velocity of the particle is calculated using Stokes' law (Table 10.15). For water at 15°C, $\rho$ = 999.1 kg/m$^3$, $\mu$ = 1.139 × 10$^{-3}$ N·s/m$^2$, and $\upsilon$ = 1.139 × 10$^{-6}$ m$^2$/s.

$$v_s = \frac{g(\rho_p - \rho)d_p^2}{18\mu}$$

$$= \frac{9.81 \text{ m/s}^2 \times (2,650 - 999.1 \text{ kg/m}^3) \times (2.0 \times 10^{-4} \text{ m})^2}{18 \times 1.139 \times 10^{-3} \text{ N·s/m}^2}$$

$$\times \frac{3,600 \text{ s}}{\text{hr}} = 113.8 \frac{\text{m}}{\text{hr}}$$

Check the Reynold's number (Equation 10.8) to verify the particle is settling under laminar conditions:

$$Re = \frac{d_p v_s}{v} = \frac{(2.0 \times 10^{-4}\,m) \times (113.8\,m/hr) \times (1\,hr/3{,}600\,s)}{(1.139 \times 10^{-6}\,m^2/s)} = 5.55$$

Because Re > 1, Stokes' law is not valid. The equation for the drag coefficient is provided in Table 10.15, and the settling velocity can be calculated using Newton's law (Table 10.15).

Because $v_s$ cannot be determined explicitly, a trial-and-error solution must be used. Using the value of Re just obtained (5.55), the drag coefficient can be calculated as

$$C_d = \frac{24}{5.55} + \frac{3}{\sqrt{5.55}} + 0.34 = 5.94$$

The terminal settling velocity also can be recalculated:

$$v_s = \sqrt{\frac{4(9.81\,m/s^2) \times (2{,}650 - 999.1\,kg/m^3) \times 2.0 \times 10^{-4}\,m}{3 \times 5.94 \times 999.1\,kg/m^3}} \times 3{,}600\frac{s}{hr}$$

$$= 97.1\,\frac{m}{hr}$$

The Reynold's number is calculated again, and then the drag coefficient and terminal settling velocity are recalculated. After several iterations, a convergent answer is obtained, as shown in Table 10.16. The settling velocity begins to converge at the sixth or seventh trial and has a value of 86.5 m/hr.

## Table / 10.16

**Iterative Process Used in Example 10.5 to Determine Settling Velocity Using Newton's Law** After several iterations, a convergent answer is obtained, as shown here.

| Trial | Re (dimensionless) | $C_d$ (dimensionless) | $v_s$ (m/hr) |
|-------|--------------------|-----------------------|--------------|
| 0 | 5.55 | 5.94 | 97.1 |
| 1 | 4.74 | 6.78 | 90.9 |
| 2 | 4.43 | 7.18 | 88.3 |
| 3 | 4.31 | 7.36 | 87.3 |
| 4 | 4.26 | 7.43 | 86.8 |
| 5 | 4.23 | 7.47 | 86.6 |
| 6 | 4.23 | 7.48 | 86.5 |
| 7 | 4.22 | 7.49 | 86.5 |

## 10.7.2 PARTICLE REMOVAL DURING SEDIMENTATION

Figure 10.11 shows particle trajectories in a rectangular sedimentation basin. Here it is assumed that particles move horizontally at the same velocity as the water and are removed by gravity once they reach the bottom of the basin. Particle trajectories in the basin depend upon the particle settling velocity ($v_s$) and the fluid velocity ($v_f$).

The settling velocity for discrete particles is constant, because particles will not interfere with one another, and the size, shape, and density of particles is assumed to not change as they move through the reactor. A particle (particle 2 in Figure 10.11) that enters at the top of the basin and settles just before it flows out of the basin is called a *critical particle*. Its settling velocity is defined as the **critical particle settling velocity**, determined as follows:

$$v_c = \frac{h_o}{\theta} \tag{10.9}$$

where $v_c$ is the critical particle settling velocity (m/hr), $h_o$ is the depth of the sedimentation basin (m), and $\theta$ is the hydraulic detention time of the sedimentation basin (hr).

The critical particle settling velocity is also called the **overflow rate (OR)** because it is equal to the ratio of process flow rate to surface area:

$$v_c = \frac{h_o}{\theta} = \frac{h_o Q}{V} = \frac{h_o Q}{h_o A} = \frac{Q}{A} = OR \tag{10.10}$$

where $A$ is the surface area of the top of the settling basin (m$^2$) and $Q$ is the process flow rate (m$^3$/hr). Important to our discussion is the term OR, the overflow rate. Note in Equation 10.10 that the OR is not a function of the tank depth.

**Sedimentation**

The OR (m$^3$/m$^2$-hr, also written as m/hr) is equal to the critical settling velocity, $v_c$. Any particles with a settling velocity ($v_s$) greater than or equal to $v_c$ (or the OR) will be removed. Particles with a settling

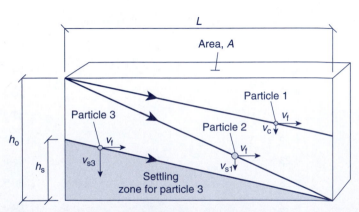

**Figure 10.11** Discrete Particle Trajectories in a Rectangular Sedimentation Basin

velocity ($v_s$) less than $v_c$ (particle 3 in Figure 10.11) can also be removed, depending on their position at the inlet, as shown in Figure 10.11. In Figure 10.11, particle 1 will not be removed, because its settling velocity is not high enough relative to its entry point into the sedimentation basin.

The percentage of particles removed is determined as follows:

$$\text{percent of particles removed} = \frac{v_s}{OR} \times 100 \qquad \textbf{(10.11)}$$

## example/10.6 Determining Particle Removal

A treatment plant has a horizontal flow sedimentation basin with a depth of 4 m, width of 6 m, length of 36 m, and process flow rate of 450 m³/hr. What removal percentage should be expected for particles that have settling velocities of 1.0 m/hr and 2.5 m/hr? What is the minimum size of particles that would be completely removed? Assume the particle density is 2,650 kg/m³. The water temperature is 10°C.

## solution

First, determine the sedimentation basin overflow rate (critical settling velocity), using Equation 10.10:

$$OR = v_c = \frac{Q}{A} = \frac{450\,\text{m}^3/\text{hr}}{36\,\text{m} \times 6\,\text{m}} = 2.1\,\text{m}^3/\text{m}^2\text{-hr}$$

The percent removal of particles for each particle size can then be calculated as follows.

For particles with a settling velocity of 1.0 m/hr, because $v_s$ equals 1.0 m/hr (which is less than the $v_c$ of 2.1 m/hr), the percentage of removal is calculated using Equation 10.11:

$$\text{fraction of particles removed} = \frac{v_s}{OR} = \frac{1.0\,\text{m/hr}}{2.1\,\text{m/hr}} = 0.48$$

For particles with a settling velocity of 2.5 m/hr, because $v_s$ equals 2.5 m/hr (which is greater than $v_c$ of 2.1 m/hr), all particles with this settling velocity will be removed:

$$\text{fraction of particles removed} = 1.0$$

The final question is to determine the minimum particle size that would result in complete removal. This particle size can be determined using Stokes' law. The settling velocity of the minimum-size particle is equal to the critical settling velocity, $v_s = v_c = 2.1$ m/hr. Plugging $v_s$ into Stokes' law will provide the correct particle size.

## example/10.6 Continued

For water at 10°C, $\rho = 999.7 \, \text{kg/m}^3$, $\mu = 1.307 \times 10^{-3} \, \text{N} \cdot \text{s/m}^2$, and $\upsilon = 1.306 \times 10^{-6} \, \text{m}^2/\text{s}$.

$$v_s = \frac{g(\rho_p - \rho)d_p^2}{18\,\mu} = \frac{9.81 \, \text{m/s}^2 \times (2{,}650 - 999.7 \, \text{kg/m}^3) \times (d_p)^2}{18 \times 1.307 \times 10^{-3} \, \text{N} \cdot \text{s/m}^2}$$

$$\times \frac{3{,}600 \, \text{s}}{\text{hr}} = 2.1 \frac{\text{m}}{\text{hr}}$$

$$d_p^2 = 8.48 \times 10^{-10} \, \text{m}^2$$

$$d_p = 2.9 \times 10^{-5} \, \text{m} = 2.9 \times 10^{-3} \, \text{cm}$$

To verify the flow conditions, check the Reynold's number (Equation 10.8) to see if the particle is settling under laminar conditions:

$$\text{Re} = \frac{d_p v_s}{\upsilon} = \frac{8.48 \times 10^{-10} \, \text{m} \times 2.1 \, \text{m/h} \times 1 \, \text{h}/3{,}600 \, \text{s}}{1.306 \times 10^{-6} \, \text{m}^2/\text{s}}$$

$$= 3.8 \times 10^{-7}$$

The Reynold's number is much less than 1. Therefore, laminar flow exists, and Stokes' law is valid.

Readers should note that horizontal settling tanks can be designed and operated without the use of mechanical rakes and other moving parts that require energy, maintenance, and funds to purchase spare parts. They are thus a more appropriate technology for many applications throughout the world.

Typical detention times for a rectangular sedimentation basin with horizontal flow are in the range of 1.5 to 4 hours. Other important design criteria are considered in the design of such basins, including the number of basins (more than 2), the horizontal mean flow velocity (0.3–1.1 m/min), overflow rate (1.25–2.5 m/hr), length-to-depth ratio (greater than 15), and length-to-width ratio (greater than 4) (Kawamura, 2000).

### 10.7.3 OTHER TYPES OF SETTLING

Table 10.17 summarizes the various types of settling observed during treatment of water and wastewater. The type of discrete particle settling discussed in Example 10.6 is also referred to as Type I settling. In **Type I settling**, particles settle discretely at a constant settling velocity.

When particles flocculate during settling due to the velocity gradient of fluids or differences in the settling velocities of particles, their size is increasing, and they settle faster as time passes. This type of settling is known as flocculant or **Type II settling**, which is found in coagulation processes and most conventional sedimentation basins. At high particle concentrations (greater than 1,000 mg/L), a blanket of particles is formed, and a clear interface is observed between the blanket

## Table / 10.17

**Types of Particle Settling Encountered during Drinking-Water and Wastewater Treatment**

| Type of Settling | Description | Where Used in Treatment Process |
|---|---|---|
| Type 1 | Particles settle discretely at a constant settling velocity. | Grit removal |
| Type 2 | Particles flocculate during settling due to velocity gradient of fluids or differences in the settling velocities of particles. Their size is increasing, and they settle faster as time passes. | Coagulation processes and most conventional sedimentation basins |
| Type 3 | Blanket of particles is formed at high particle concentrations (above 1,000 mg/L), and a clear interface is observed between the blanket and clarified water above it. | Lime-softening sedimentation and in wastewater sedimentation and sludge thickeners |

and clarified water above it. This type of settling is hindered or **Type III settling**, which occurs in lime-softening sedimentation, activated-sludge sedimentation, and sludge thickeners.

## 10.8 Filtration

**Filtration** is widely used for removing small flocs or precipitated particles. It may be used as the primary turbidity removal process such as direct filtration of raw water with low turbidity. It is also used for removal of pathogens, such as *Giardia lamblia* and *Cryptosporidium*. Two types of filtration employed in water treatment facilities include **granular media filtration** (discussed next) and membrane filtration (discussed in a later section).

### 10.8.1 TYPES OF GRANULAR FILTRATION

Granular media filtration can operate at either a high hydraulic loading rate (5–15 m/hr) or a low rate (0.05–0.2 m/hr). (The units on the hydraulic loading rates are really $m^3/m^2$-hr, which reduces to m/hr.) In both processes, the influent water is driven by gravity flow through a bed of granular material, and particles are collected within the bed.

**High-rate filtration** (also known as **rapid filtration**) is the process used by nearly all U.S. filtration plants. **Slow sand filtration** is an appropriate water treatment technology for rural communities due to its simplicity and low energy requirement. It is also commonly employed in

## Table / 10.18

**Comparison of Typical Ranges for Design and Operating Parameters for Slow Sand Filtration and Rapid Filtration** Some filters are designed and operated outside of these ranges.

| Process Characteristic | Slow Sand Filtration | Rapid Filtration |
|---|---|---|
| Filtration rate | 0.05–0.2 m/hr | 5–15 m/hr |
| | (0.02–0.08 gpm/sq. ft.) | (2–6 gpm/sq. ft.) |
| Media diameter | 0.3–0.45 mm | 0.5–1.2 mm |
| Bed depth | 0.9–1.5 m (3–5 ft.) | 0.6–1.8 m (2–6 ft.) |
| Required head | 0.9–1.5 m (3–5 ft.) | 1.8–3.0 m (6–10 ft.) |
| Run length | 1–6 mo | 1–4 days |
| Pretreatment | None required | Coagulation |
| Regeneration method | Scraping | Backwashing |
| Maximum raw-water turbidity | 10–50 NTU | Unlimited with proper pretreatment |

SOURCE: Crittenden et al., 2005. Reprinted with permission of John Wiley & Sons, Inc.

community and household point-of-use systems implemented in the developing world. Table 10.18 compares the process characteristics of slow sand filtration and rapid filtration.

Figure 10.12 shows a schematic of a dual-media granular filtration system. Filter media in rapid filtration must be fairly uniform in size to allow the filters to operate at a high loading rate. Coagulation pretreatment is necessary before filtration because particles have to be properly destabilized for effective removal.

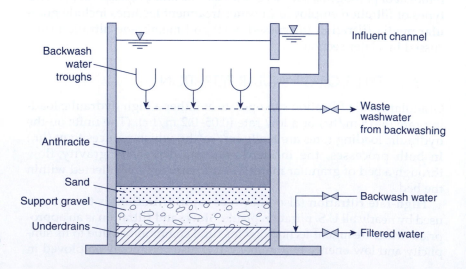

**Figure 10.12** Dual-Media Granular Media Filter

Crittenden et al. (2005). Adapted with permission of John Wiley & Sons, Inc.

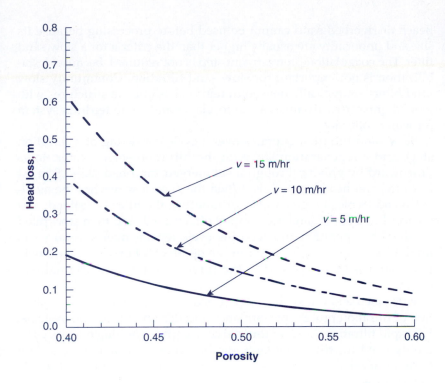

**Figure 10.13** **Effect of Porosity and Filtration Rate ($v$) on Clean-Bed Head Loss through Clean Granular Filter Bed** The clean-bed head loss increases as the porosity decreases or as the filtration rate increases.

Crittenden et al. (2005). Redrawn with permission of John Wiley & Sons, Inc.

Rapid filtration operates over a cycle consisting of a filtration stage and a backwash stage. During the *filtration stage*, water flows downward through the filter bed, and particles are captured within the bed. During the *backwash stage*, water flows upward and flushes captured particles up and away from the bed. Typically, the filtration step lasts from 1 to 4 days, and the **backwashing** takes 15 to 30 min.

The **head loss** during the filtration and backwashing stages is important for proper design of granular filtration systems. For a given depth of filter media and the effective size of media, the **clean-bed head loss** depends on the bed porosity, filtration rate, and water temperature. The effect of the porosity and filtration rate on head loss is illustrated in Figure 10.13. The clean-bed head loss increases as the porosity decreases or as the filtration rate increases. The figure also shows that the clean-bed head loss is more sensitive to filtration rate at a lower porosity. The clean-bed head loss also increases as temperature decreases, because fluid viscosity increases. For example, the clean-bed head loss at 5°C is 60 to 70 percent higher than it is at 25°C.

The **backwash rate** must be above the minimum fluidization velocity of the largest media. The largest media is typically taken as the $d_{90}$ diameter. The minimum fluidization velocity can be calculated by inputting the fixed bed porosity into design equations. Backwash rates range from approximately 20 to 56 m/hr, with a typical rate being 45 m/hr. The targeted expansion of the bed is approximately 25 percent for anthracite and 37 percent for sand.

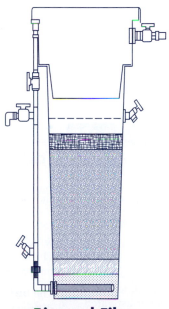

**Biosand Filter**

### 10.8.2 MEDIA CHARACTERISTICS

The sand used for slow sand filtration is smaller and less uniform than the media used in rapid filtration. Only washed sand should be used.

Beach or riverbed sand cannot be used before processing because its size and uniformity are usually higher than the criteria for a slow sand filter. The coagulation pretreatment step is not required, because destabilization is not important for slow sand filtration. Community slow sand filters are typically housed in reinforced-concrete structures with graded gravel (0.3–0.6 m) as a support layer and an underdrain system for water collection.

Slow sand filtration operates over a cycle consisting of a filtration stage and a regeneration stage. In the filtration stage, water flows downward by gravity through a submerged sand bed (0.9–1.5 m) at low rate, and head loss builds. When the head loss reaches the available head (typically after weeks or months), the filter is drained, and the top 1 to 2 cm of sand are scraped off, cleaned, and then stockpiled on site. The operation and scraping cycle repeats over several years until the sand bed reaches a minimum depth of 0.4 to 0.5 m. The stockpiled sand is then replaced in the filter to restore the original bed depth.

Filter media and bed characteristics are very important for evaluation of filtration process performance and design of filtration systems. For rapid filtration, sand, anthracite coal, garnet, and ilmenite are commonly used for filter media. Sometimes granular activated carbon (GAC) is used when filtration is combined with adsorption in a single unit process.

Sand is the granular media used in *slow sand filtration*. Important media characteristics include media size (described by the effective size), size distribution (described by a uniformity coefficient), density, shape, and hardness. Table 10.19 compares important filter and bed characteristics for rapid and slow sand filters. Figure 10.14 shows how

**Filtration**

---

## Table / 10.19

### Filter Media and Bed Characteristics for Rapid and Slow Sand Filters

| | Important Media Characteristics |
|---|---|
| **Rapid filtration**<br><br>Uses granular materials such as sand, anthracite coal, garnet, ilmenite, and granular activated carbon (GAC) | The **effective size (ES or $d_{10}$)** is determined by sieve analysis. It is the media diameter at which 10% of the media by weight is smaller. The effective size is determined by sieve analysis. The typical effective size for rapid filtration media are: sand 0.4–0.8 mm, anthracite coal 0.8–2.0 mm, garnet 0.2–0.4 mm, ilmenite 0.2–0.4 mm, GAC 0.8–2.0 mm.<br><br>The **uniformity coefficient (UC)** is the ratio of the 60th-percentile media diameter (the diameter at which 60% of the media by weight is smaller) to the effective size ($d_{10}$):<br><br>$$UC = \frac{d_{60}}{d_{10}}$$<br><br>UC is an important parameter in the design of rapid filters because it directly affects the overall effectiveness of the filter bed because of media stratification during backwash. During backwash, coarse grains (more weight) will settle to the bottom of the bed and are difficult to fluidize for effective cleaning. Fine grains (less weight) will collect at the top of the filter, which in turn will cause excessive head loss during the filtration stage. Typical ranges of uniformity coefficients for rapid filtration media are as follows: sand 1.3–1.7, anthracite coal 1.3–1.7, garnet 1.3–1.7, ilmenite 1.3–1.7, GAC 1.3–2.4. |

## Table / 10.19

(Continued)

| | Important Media Characteristics |
|---|---|
| **Slow sand filtration** Uses sand | Slow sand filter media has a smaller **effective size (ES or $d_{10}$)** to achieve a lower filtration rate. It also has a higher **uniformity coefficient (UC)** because backwash is not involved in slow sand filtration operation; therefore stratification is not a concern. The typical effective size for slow sand filtration is between 0.3 and 0.45 mm, and the uniformity coefficient is less than 2.5. |
| | **Filter bed porosity** is like soil porosity. Porosity has a strong influence on the head loss and filtration effectiveness. If the porosity is too small, the head loss will be high, and the filtration rate will decrease rapidly over operation time. If the porosity is too high, the filtration rate will exceed the criterion, and the effluent will not meet the treatment objective. Typical values for porosity are in the range of 40%–60%. |

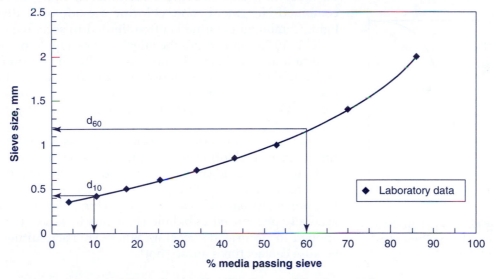

**Figure 10.14** Analysis of the suitability of a sample of native sand for slow sand filtration

a sand sample is analyzed by a sieve analysis to determine appropriate size for use in filtration.

In Figure 10.14, a 1,000 g sample of naturally occurring sand was sifted through a stack of sieves, and the weight retained on each sieve was recorded and plotted. Here $d_{10} = 0.43$ mm , and $d_{60} = 1.18$ mm. Therefore, the uniformity coefficient equals 2.7 (see Table 10.19). See how the uniformity coefficient of this naturally occurring sand is much higher than typical values used in rapid filters (1.3 to 1.7). The higher uniformity coefficient will result in severe media stratification during backwash, causing excessive head loss and reducing overall effectiveness of the filter. Therefore, the sample sifted for this example needs to be processed to a fairly uniform size.

# 10.9   Disinfection

Pathogens can either be removed by treatment processes such as granular filtration or inactivated by disinfection agents. The term *disinfection* in drinking-water practice refers to two activities:

1. *Primary disinfection:* the inactivation of microorganisms in the water

2. *Secondary disinfection:* maintaining a disinfectant residual in the treated-water distribution system (also called residual maintenance).

## 10.9.1   CURRENT DISINFECTION METHODS

Generally, **disinfectants** can be classified as *oxidizing agents* (for example, chlorine and ozone), *cations of heavy metals* (silver or copper) and *physical agents* (heat or UV radiation). The most commonly used disinfectant is free chlorine. Four other common disinfectants are combined chlorine, ozone, chlorine dioxide and ultraviolet (UV) light. Combined chlorine is often limited to secondary disinfection.

Table 10.20 summarizes the effectiveness, regulatory limits, typical application, and chemical source of the five most common disinfectants. Table 10.21 provides detailed information on the important disinfection chemistry and application considerations for the five common disinfectants (including the importance of pH, as shown in Figure 10.15).

Reactors used for disinfection are usually called *contactors*. Free chlorine, combined chlorine, and chlorine dioxide are most often used in contactors close to ideal plug flow reactors, such as baffled, serpentine contactor chambers. Both types of contactors can be designed so that they are highly efficient, closely approaching ideal plug flow. Ozone is generally introduced in bubble chambers in series. UV light is often applied in proprietary reactors where short-circuiting is a concern because contact times are so short.

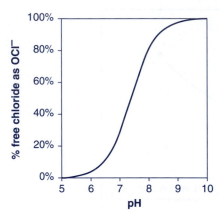

**Figure 10.15**   **Effect of pH on the Fraction of Free Chlorine Present as Hypochlorite Ion (OCl⁻).**

## 10.9.2   DISINFECTION KINETICS

The mechanisms for pathogen inactivation during disinfection are complex and not well understood. Therefore, kinetic models have been developed that are based on laboratory observations. **Chick's law** (Equation 10.27) is the most straightforward model to describe the disinfection process. It assumes the rate of the disinfection reaction is **pseudo first order** with respect to the concentration of the pathogens being inactivated:

$$\frac{dN}{dt} = -K \times N \qquad (10.27)$$

where $dN/dt$ is the rate of change in the number of organisms with time (organisms/volume/time), $N$ is the concentration of organisms

## Table / 10.20

**Effectiveness, Regulatory Limits, Typical Application, and Chemical Source of the Five Most Common Disinfectants**

| Issue | Disinfectant | | | | |
|---|---|---|---|---|---|
| | Free Chlorine | Monochloramine | Chlorine Dioxide | Ozone | Ultraviolet Light |
| Effectiveness | | | | | |
| Bacteria | Excellent | Good | Excellent | Excellent | Good |
| Viruses | Excellent | Fair | Excellent | Excellent | Fair |
| Protozoa | Fair to Poor | Poor | Good | Good | Excellent |
| Endospores | Poor to Good | Poor | Fair | Excellent | Fair |
| Frequency of use as primary disinfectant | Most common | Common | Occasional | Very common | Increasingly common |
| Regulatory limit on residuals | 4 mg/L | 4 mg/L | 0.8 mg/L | — | — |
| Formation of chemical by-products | | | | | |
| Regulated by-products | Forms 5 THMs and 4 HAAs | Traces of THMs and HAAs | Chlorite | Bromate | None |
| By-products that may be regulated in the future | Several | Cyanogen halides, NDMA | Chlorate | Biodegradable organic carbon | None known |
| Typical application Dose, mg/L | 1–6 | 2–6 | 0.2–1.5 | 1–5 | 20–100 mJ/cm$^2$ |
| Chemical source | Delivered as liquid gas in tank cars, 1 ton and 68 kg cylinders as calcium hypochlorite powder for very small applications. On-site generation from salt and water using electrolysis. | Same sources as for chlorine. Ammonia is delivered as aqua ammonia solution, liquid gas in cylinders, or as solid ammonium sulfate. Chlorine and ammonia are mixed in treatment process. | Same sources as for chlorine. Chlorite as powder or stabilized liquid solution. ClO$_2$ is manufactured with an on-site generator. | Manufactured on site using a corona discharge in very dry air or pure oxygen. Oxygen is usually delivered as a liquid. Oxygen is also manufactured on site in some very large plants. | Uses low pressure or low pressure–high intensity UV (254 nm) or medium pressure UV (several wavelengths) lamps in the contactor. |
| Typical contactor | In the past was added at beginning of plant and residual carried through. Increasingly, individual contactors are used. | In the past was added at beginning of plant and residual carried through. Increasingly, individual contactors are used. | In the past was added at beginning of plant and residual carried through. Increasingly, individual contactors are used. | Has always been added in specially engineered contactors. These contactors are using more compartments. | Lamps are placed in gravity channels or in specially manufactured UV reactors. Because the contact time is so short, reactors must be tested. |

SOURCE: Crittenden et al., 2005. Adapted with permission of John Wiley & Sons, Inc.

**Important Chemical Reactions Associated with Common Disinfectants**

| Disinfectant | Important Chemical Reactions | Considerations |
|---|---|---|
| **Free chlorine**<br><br>When gaseous chlorine is added to water, it rapidly reacts with water to form hypochlorous acid (HOCl) and hydrochloric acid (HCl).<br><br>Hypochlorous acid and hypochlorite ion together are often referred to as *free chlorine* (free chlorine = HOCl + OCl$^-$).<br><br>Both of these chemical species are active disinfecting agents; however, hypochlorous acid (HOCl) is much more effective than OCl$^-$ for disinfection. | $Cl_{2(g)} + H_2O \rightarrow HOCl + HCl$ **(10.12)**<br><br>$HOCl \rightarrow H^+ + OCl^-$ $K_a = 10^{-7.5}$ **(10.13)** | As can also be seen from the equilibrium constant for Equation 10.13 and Figure 10.15, HOCl is the predominant form of free chlorine at a pH less than 7. Consequently, a treatment plant operator will attempt to maintain the pH at 7 or slightly lower to increase the disinfection power of the added chlorine.<br><br>Although chlorine disinfection is very effective and cost-advantageous, use of chlorine has some concerns. One of them is by-product formation; chlorine will react with dissolved organic matter that occurs naturally in waters to form carcinogenic trihalomethanes (THMs). |
| **Combined chlorine**<br><br>When chlorine and ammonia (NH$_3$) are both present in water, they react to form three *chloramine compounds* (NH$_2$Cl, NHCl$_2$, NCl$_3$) according to the three reactions on the right.<br><br>These three chloramines together are referred to as **combined chlorine**. (combined chlorine = NH$_2$Cl + NHCl$_2$ + NCl$_3$). The **total chlorine residual** is the sum of the combined chlorine and any free chlorine residual. | *Monochloramine formation:*<br>$HOCl + NH_3 \rightarrow NH_2Cl + H_2O$ **(10.14)**<br><br>*Dichloramine formation:*<br>$HOCl + NH_2Cl \rightarrow NHCl_2 + H_2O$ **(10.15)**<br><br>*Trichloramine formation:*<br>$NHCl_2 + HOCl \rightarrow NCl_3 + H_2O$ **(10.16)** | The formation of these species depends upon the ratio of Cl$_2$ to NH$_3$-N. At a high ratio of Cl$_2$ to NH$_3$-N, the oxidation of ammonia to nitrogen gas and nitrate ion occurs.<br><br>$3HOCl + 2NH_3 \rightarrow N_{2(g)} + 3H_2O + 3HCl$ **(10.17)**<br><br>$4HOCl + NH_3 \rightarrow HNO_3 + H_2O + 4HCl$ **(10.18)**<br><br>The reactions are also dependent upon chlorine dosage, temperature, pH, and alkalinity. At low pH values, other reactions become more significant, such as these:<br><br>$NH_2Cl + H^+ \rightleftharpoons NH_3Cl^+$ **(10.19)**<br><br>$NH_3Cl^+ + NH_2Cl \rightleftharpoons NHCl_2 + NH_4^+$ **(10.20)** |
| **Chlorine dioxide**<br><br>This chemical does not produce significant amounts of **trihalomethanes (THMs)** as by-products from reactions with organics. | $2NaClO_2 + Cl_{2(g)} \rightarrow 2ClO_{2(g)} + 2NaCl$ **(10.21)**<br><br>$2NaClO_2 + HOCl \rightarrow 2ClO_{2(g)} + NaCl + NaOH$ **(10.22)**<br><br>$5NaClO_2 + 4HCl \rightarrow 4ClO_{2(g)} + 5NaCl + 2H_2O$ **(10.23)** | Chlorine dioxide is explosive at elevated temperatures, on exposure to light, or in the presence of organic substances. Chlorine dioxide does not produce THMs; however, it produces inorganic products such as chlorite (ClO$_2^-$) and chlorate (ClO$_3^-$), which have health concerns at certain levels of exposure. |

| | | |
|---|---|---|
| **Chlorine dioxide (*continued*)** Chlorine dioxide also has a higher oxidizing power than chlorine; however, at neutral pH values typical of most waters, it has only about 70% of the oxidizing capacity of chlorine. | | $ClO_2$ is generated on site using sodium chlorite with gaseous chlorine ($Cl_2$), aqueous chlorine (HOCl), or acid (usually hydrochloric acid, HCl). |
| **Ozone** Ozone is a stronger oxidant than the other three disinfectants. | $3O_2 \rightarrow 2O_3$          (10.24) <br><br> Ozone can decompose to the hydroxyl radical (HO·), which is formed by the reactions with high concentrations of hydroxide ion ($OH^-$) or natural organic matter (NOM). <br><br> $3O_3 + OH^- + H^+ \rightarrow 4O_2 + 2HO·$ <br>          (10.25) <br><br> $O_3 + NOM \rightarrow HO· + \text{by-products}$ <br>          (10.26) | Ozone is a highly reactive gas and decays rapidly under ambient conditions. Therefore, it has to be generated on site, commonly by electrical discharges in the presence of $O_2$. <br><br> Ozone reacts with microbes by direct oxidation or through the action of hydroxyl radicals generated as in Equations 10.25 and 10.26. Hydroxyl radicals are scavenged by carbonate species ($HCO_3^-$, $CO_3^{2-}$) and reduced metal ions (e.g., $Fe^{2+}$, $Mn^{2+}$). High pH conditions or high concentrations of organic matter favor hydroxyl radical oxidation reactions. Ozone disinfection is primarily dependent on its direct reactions, and the residual of ozone is important. Low pH, high alkalinity, low concentrations of organic matter, and low temperature will increase the stability of aqueous ozone residuals. Unfortunately, use of ozone will not result in a residual that can continue the disinfection process in the water distribution system. In addition, ozonation of waters containing bromide produces bromate ($BrO_3^-$), which is believed to be a carcinogen to humans. |
| **Ultraviolet Radiation** This is electromagnetic radiation having a wavelength between 100 and 400 nm. | Light in the UV spectrum can be divided into vacuum UV, short-wave UV (UV-C), middle-wave UV (UV-B), and long-wave UV (UV-A). <br><br> The region of the wavelength of 200–300 nm is a germicidal range where deoxyribonucleic acid (DNA) absorbs UV. | The photons in UV light react directly with the nucleic acids in the form of DNA and damage DNA. This will inhibit further transcription of the genetic code of the cell and prevent its successful reproduction. However, microorganisms may evolve to repair UV-induced damage through photoreactivation and dark repair mechanisms. Therefore, reactivation is an important consideration in UV disinfection. <br><br> The performance of UV disinfection systems is highly influenced by dissolved substances and particulate matter in waters, which have to be considered in UV reactor design. |

(organisms/volume), and $K$ is the Chick's law rate constant (time$^{-1}$). Integrating Equation 10.27 results in

$$\ln\left(\frac{N}{N_0}\right) = -K \times t \qquad (10.28)$$

where $N_0$ is the initial concentration of organisms (organisms/volume).

The rate of disinfection can be determined by plotting the log organism concentration ratio ($N/N_0$) versus time. To better estimate the rate of disinfection, several measurements should be made at each time step. Due to inaccurate measurement, the best fit often may not pass through zero. The disinfection rate from Equation 10.28 is related to disinfectant concentration, and the reaction has a different rate constant for each concentration.

## example/10.7 Application of Chick's Law

Given the data in the first three columns of Table 10.22, graph the data for the inactivation of *Poliovirus* type 1, using the disinfectant bromine. Determine the Chick's law rate constant for each of the two disinfectant concentrations.

## solution

Calculate the log of the survival value ($N/N_0$) for each data point. The results are shown in the two right columns of Table 10.22. Then plot log($N/N_0$) as a function of time ($t$), and fit a linear line through the data. Figure 10.16 shows this graph.

The value of the slope of the line in Figure 10.16 corresponds to the Chick's law rate constant. For a disinfectant concentration of 21.6 mg/L, the rate constant $K$ (base 10) is 1.8333/s; thus, $K$ (base $e$) is 4.22/s. For the disinfectant concentration of 4.7 mg/L, the

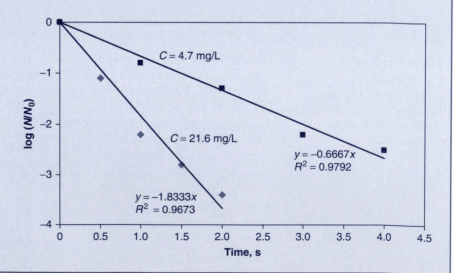

**Figure 10.16** Log ($N/N_0$) as a Function of $t$ for Example 10.7

## Table / 10.22

**Data Used to Solve Problem Applying Chick's Law in Example 10.7**

| C (mg/L) | t (min) | N (number of organism/L) | $N/N_0$ | log ($N/N_0$) |
|---|---|---|---|---|
| 21.6 | 0.0 | 500 | 1 | 0 |
| 21.6 | 0.5 | 40 | 0.080 | −1.1 |
| 21.6 | 1.0 | 3 | 0.006 | −2.2 |
| 21.6 | 1.5 | 1 | 0.002 | −2.7 |
| 21.6 | 2.0 | 0.2 | 0.0004 | −3.4 |
| 12.9 | 0.0 | 500 | 1 | 0 |
| 12.9 | 1.0 | 79 | 0.158 | −0.8 |
| 12.9 | 2.0 | 25 | 0.050 | −1.3 |
| 12.9 | 3.0 | 3 | 0.006 | −2.2 |
| 12.9 | 4.0 | 1.5 | 0.003 | −2.5 |

rate constant $K$ (base 10) is 0.6667/s; thus, $K$ (base $e$) is 1.54/s. Remember that, as we discussed previously, this is a pseudo first order rate constant, so the rate constant differs for different disinfectant concentrations.

The **Ct approach** uses the product $C \times t$, which can be viewed as the dosage of disinfectant. $C$ is the concentration of a chemical disinfectant (mg/L), and $t$ is the time required to achieve a level of inactivation. A similar concept is the product of the UV light intensity ($I$, mW/cm$^2$) and the time of exposure, $t$. $I \times t$ (units of mW/cm$^2 \times$ s or mJ/cm$^2$) is used during UV disinfection to compute the dose of UV light.

The $Ct$ approach is a useful way to compare the relative effectiveness of different disinfectants and the resistance of different organisms. Figure 10.17 illustrates this by comparing the $Ct$ and $It$ required for a 99 percent inactivation of several microorganisms using the five common disinfectants. Figure 10.17 shows that ozone requires a lower $Ct$ for most microorganisms than the other three chemical oxidants need; therefore, ozone is a stronger disinfectant. Also, the microorganism called *C. parvum* requires the highest $Ct$ among four chemical oxidants, so this microorganism is most resistant to these disinfectants. The $It$ data for UV disinfection are included in Figure 10.17. However, not much can be concluded by comparing the $It$ values with the $Ct$ values for the same organism.

www.wiley.com/college/mihelcic **Disinfection**

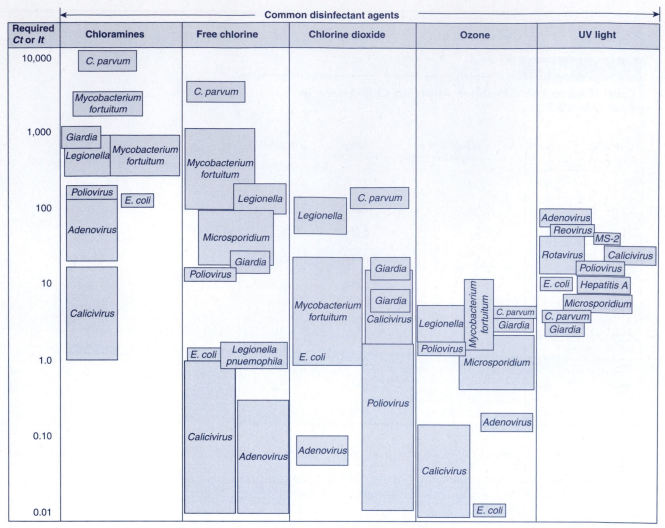

**Figure 10.17  Overview of Disinfection Requirements for 99% Inactivation**  This chart compares the *Ct* or *It* required for a 99% inactivation of several pathogens, using five common disinfectants (listed on the top). Ozone requires a lower *Ct* for most pathogens than the other three chemical oxidants, so ozone is a stronger disinfectant. The microorganism *C. parvum* requires the highest *Ct* among the four chemical oxidants, so this microorganism is most resistant to these disinfectants.

Crittenden et al. (2005). Redrawn with permission of John Wiley & Sons, Inc.

example/10.8 Application of *Ct* Value

The Surface Water Treatment Rule (SWTR) requires at least 99.99 percent (4-log) removal and/or inactivation of viruses. A water treatment plant provides disinfection using free chlorine with a contact time of 30 min. Determine the concentration of free chlorine required for disinfection at pH of 7 and water temperatures of 5°C

## example/10.8 Continued

and 20°C. The *Ct* values are provided: $Ct = 3$ min-mg/L at pH = 7 and temperature of 20°C; $Ct = 8$ min-mg/L at pH = 7 and temperature of 5°C.

## solution

The required concentration of free chlorine (mg/L) for the given contact time of 30 min is determined as follows:

$$C_{20} = \frac{3 \text{ min-mg/L}}{30 \text{ min}} = 0.1 \text{ mg/L}$$

$$C_5 = \frac{6 \text{ min-mg/L}}{30 \text{ min}} = 0.2 \text{ mg/L}$$

Note that at the lower temperature, higher *Ct* values are required. Therefore, a higher concentration of free chlorine may have to be applied during winter months.

---

## Box / 10.4   Pathogen Removal and Disinfection in the Developing World

In the developing world, inactivation of microbiological contaminants is accomplished through the addition of disinfection agents such as free chlorine, combined chlorine, ozone, chlorine dioxide, UV light, or heat (Table 10.23). Water storage needs to be carefully considered as well, because water safe at the point of collection is often recontaminated with fecal matter during collection, transport, and use in home. Vessels with narrow mouths and taps reduce the chance of recontamination, and open containers should always be covered (Mihelcic et al., 2009).

Hand-dug wells can be chlorinated by lowering a chlorination pot (Figure 10.18) into the well or directly injecting chlorine into the well daily. Some communities take the intermediate measure of periodically injecting chlorine into a well to reduce contamination. However, disinfection will be short-lived with this method if the pathogens are present in the groundwater source.

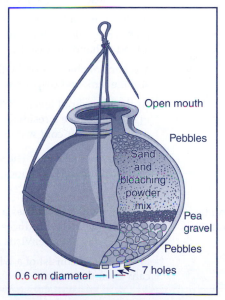

**Figure 10.18**   Example of a Chlorination Pot

From Mihelcic et al. (2009). Drawing courtesy of Linda Phillips.

## Table / 10.23

### Disinfection Agents and Processes Used at the Household Level in the Developing World

| Disinfection Agent or Process | Examples |
|---|---|
| Heat | Boiling is still the most common means of treating water in the developing world but requires valuable wood (which can be expensive), produces air pollution, and has safety concerns related to scalding of children. |
| | Water has a flat, deaerated taste. |
| | Most experts suggest attaining a "rolling boil" for 1–5 min. The World Health Organization recommends reaching a rolling boil. This gets you in excess of temperatures needed for sterilization. |
| | Heating to pasteurization temperatures (60°C) for 1–10 min will destroy many waterborne pathogens. |
| Solar | Solar systems take time and work with limited volumes of water. |
| | Uses combination of UV and pasteurization processes for disinfection. Heating to pasteurization temperatures (60°C) for 1–10 min will destroy many waterborne pathogens. |
| | *Solar disinfection (SODIS)* is a simple treatment method that takes advantage of the bacterial destruction potential of sunlight. Treatment involves placing clear bottles of water to be treated in direct sunlight for a determined amount of time. |
| | Solar cookers or reflectors (made of cardboard or aluminum foil) can attain pasteurization temperatures of 65°C. |
| Chlorination | Imparts taste to water. Requires turbidity be below 20 NTU. The World Health Organization guideline for chlorine is 5 mg/L, which is above the taste threshold of 0.6–1.0 mg/L. |
| | Liquid laundry bleach is a readily available source of chlorine in many developing communities. The powder form of chlorine is calcium hypochlorite, which includes chlorinated lime, tropical bleach, bleaching powder, and HTH. Calcium hypochlorite can be found in 30%–70% solutions. Liquid bleach is in the form of sodium hypochlorite and contains 1%–18% chlorine. |
| | At the community level, *chlorination pots* are added to hand-dug wells and use bleaching powder or chlorinated lime (Figure 10.18). Pressed calcium hypochlorite tablets can also be added to hand-dug wells. |
| | Mechanical chlorinators with no moving parts are sometimes constructed on top of a water storage tank and drip a bleach solution into the tank. |
| | At the household level, a 1% chlorine stock solution can be used. Chlorine is added at 1–5 mg/L to achieve a residual of 0.2–0.5 mg/L after contact time of 30 min. This increases to 1 hr contact time for cold waters. |
| Iodine | Imparts taste to water. The World Health Organization recommends a dose of 2 mg/L with a contact time of 30 min. Iodine is better at penetrating particulate matter than chlorine but is more expensive. |
| Sedimentation and filtration | The three-pot system uses a series of three pots where water is transferred between pots daily, and gravity settling reduces helminth ova and some protozoans by more than 90% especially with 1–2 day storage. Virus and bacteria removal are much smaller. |
| | Filtration can consist of a wide variety of media: granular media, slow sand, fabric, paper, canvas, and ceramics. |

SOURCE: Mihelcic et al., 2009.

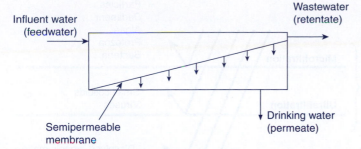

**Figure 10.19   Diagram of Membrane Separation Process**   The influent is separated into a permeate stream (for drinking water) and a retentate stream, which currently becomes waste. One challenge to the water treatment industry is to find beneficial uses for the retentate.

Crittenden et al. (2005). Redrawn with permission of John Wiley & Sons, Inc.

## 10.10   Membrane Processes

By 2007, over 20,000 membrane (including reverse osmosis) plants were operating worldwide. This number is expected to grow at a significant pace as population and water consumption increase. **Membrane processes** involve water pumped under pressure, called *feedwater*, into a housing containing a semipermeable membrane, where some of the water filters through the membrane and is called *permeate*. The rest of the water containing the filtered constituents passes by the filter and is called *retentate*, as shown in Figure 10.19.

The pressure required for constant filtration or flux through the membrane is called the *transmembrane* pressure. This pressure provides the driving force for filtration to take place. As the pore size of the membrane decreases, the transmembrane pressure increases. The **flux rate** is the rate at which the permeate flows through the membrane area and is expressed as L/m²-day (gpm/ft.²). The permeate is usually stabilized if needed, disinfected, and sent to the distribution system.

### 10.10.1   CLASSIFICATION OF MEMBRANE PROCESSES

Selection of the type of membrane systems depends upon the constituents to be removed. Four types of membrane systems are used in water treatment: **microfiltration (MF)**, **ultrafiltration (UF)**, **nanofiltration (NF)**, and **reverse osmosis (RO)**. Figure 10.20 classifies the four membrane processes used in drinking-water treatment. Table 10.24 provides more information about the four types of membrane filters. Microfiltration and ultrafiltration are sometimes classified as membrane filtration because they primarily remove particulate constituents. Nanofiltration and reverse osmosis are sometimes classified as reverse osmosis processes because they remove dissolved constituents.

The **energy** use required to pass water through a membrane can be determined as follows:

$$Hp = \frac{Q \times \Delta P}{1,714} \tag{10.29}$$

**Figure 10.20** **Classification of Four Pressure-Driven Membrane Processes**
The processes are microfiltration, ultra-filtration, nanofiltration, and reverse osmosis.

Adapted from Crittenden et al. (2005). Redrawn with permission of John Wiley & Sons, Inc.

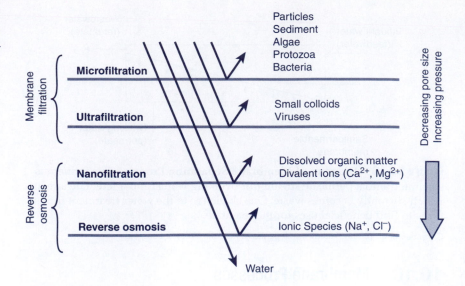

## Table / 10.24

**Types of Membrane Filter Systems**

| Type of Membrane Filter | Considerations |
|---|---|
| Microfiltration (MF) | Has membrane pore sizes $\approx 0.1\ \mu m$ in nominal diameter. |
| | Removes particles, algae, bacteria, and protozoa that have sizes larger than the nominal diameter. |
| | Transmembrane pressure operating ranges are 0.2–1.0 bar (2–15 psig). |
| Ultrafiltration (UF) | Has membrane pore sizes as small as $0.01\ \mu m$ in nominal diameter. |
| | Can remove constituents such as small colloids and viruses. |
| | Operating pressure ranges are 1–5 bar (15–75 psig). |
| Nanofiltration (NF) | Has membrane pore sizes as small as $0.001\ \mu m$ in nominal diameter. |
| | Removes dissolved organic matter and some divalent ions such as calcium and magnesium ions. |
| | Transmembrane pressure operating ranges are 5.0–6.7 bar (75–100 psig). |
| Reverse osmosis (RO) | Filters are considered nonporous, and usually only constituents the size of water molecules can pass through the filter. |
| | Transmembrane operating pressure ranges are 13.4–80.4 bar (200–1,200 psig). |

where Hp is the power required to pass a million gallons of water per day through the membrane (kWh/MGD), $Q$ is the flow rate (gpm), and $\Delta P$ is the required feed pressure (psi).

### 10.10.2 MEMBRANE MATERIALS

Membranes are manufactured in tubular form, flat sheets or as fine hollow fibers. They are composed of either natural or synthetic materials. Modified natural types consist of cellulose acetate, cellulose diacetate, cellulose triacetate, and a blend of both di- and triacetate materials. Synthetic membrane materials may be composed of polyamide, polysulfone, acrylonitrile, polyethersulfone, Teflon, nylon, and polypropylene polymers.

The membrane thickness can range from 0.1 to 0.3 μm. Some membrane materials, such as cellulose acetate and polyamide, are manufactured as thin film composites (TFC). TFC membranes usually have a very thin active layer that is bonded to a thicker porous support material for strength.

Some membrane materials are sensitive to temperature, pH, and oxidants. Cellulose acetate membranes have an operating temperature range of 15°C to 30°C and are not tolerant above 30°C. They are also sensitive to hydrolysis at high and low pH. The triacetate and di/triacetate blends have improved hydrolytic stability in water with high and low pH. Polyamide, polysulfone, nylon, Teflon, and polypropylene membranes have high physical stability, do not hydrolyze in the pH range of 3 to 11, and are immune to bacterial degradation. However, some polyamide membranes may be subject to oxidation by disinfectants such as chlorination.

## 10.10.3 MEMBRANE PROCESS TYPES AND CONFIGURATIONS

The most common types of membrane configurations used in water treatment are *spiral-wound* and *hollow-fiber modules*. A spiral-wound element consists of sheets of the membrane, permeate material, channel spacer, and outer membrane wrap wound around a porous permeate tube. Water enters one end of the element and flows across the wound filter surface area, and the water permeates through the membrane into the collection tube while the concentrate or retentate flows out the other end of the filter element. Spiral-wound elements can range from 100 to 300 mm in diameter and 1 to 6 m in length. Bundles of elements, usually 30 to 50 in number, are skid mounted (called skids or arrays) and operated in parallel mode.

For RO operation, skid-mounted units or arrays can also be operated in parallel and in series operation. For example, two arrays may operate in parallel (first stage), and the retentate from each array is fed to two other arrays (second stage) to increase water recovery. Figure 10.21 displays an RO plant with four skid-mounted units.

Perhaps the most widely used membrane is the hollow-fiber type. **Hollow fibers** have outside diameters ranging from 0.5 to 2.0 mm wall thickness ranging from 0.07 to 0.60 mm, and lengths of about 1 m. Typically, 7,000 to 10,000 fibers are housed in stainless-steel or fiberglass pressurized modules with diameters ranging from 100 to 300 mm and lengths ranging from 1 to 2 m.

In addition to pressure vessels, microfiltration and ultrafiltration membranes can operate submerged in tanks. Figure 10.22 is a schematic of what is called a *submerged-membrane system*. In the submerged mode, the membranes are submerged in a completely mixed tank, where the feed water enters the tank. Permeate is drawn through the membrane by applying a vacuum on the permeate side of the membrane. The retentate is usually drawn off in a semibatch mode or in a

Figure 10.22  Membranes Submerged
in Tanks

Crittenden et al. (2005). Redrawn with permission of
John Wiley & Sons, Inc.

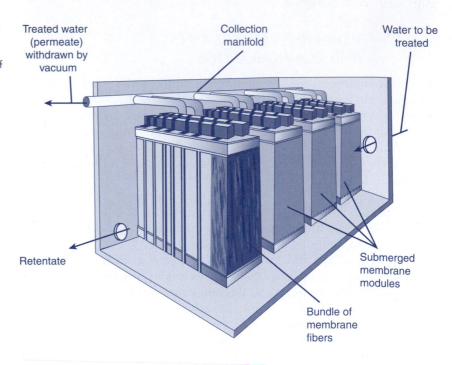

continuous mode if the reactor is operating as a completely mixed-flow reactor.

### 10.10.4   MEMBRANE SELECTION AND OPERATION

Membrane operation is highly dependent upon the influent water characteristics and the type of constituents in the water that need to be removed. If fresh surface water requires removal of particulate constituents, then the required process is membrane filtration (microfiltration or ultrafiltration). If the removal of dissolved constituents is

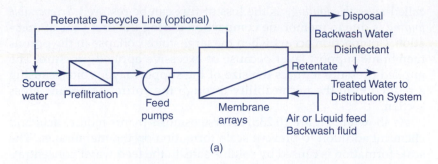

(a)

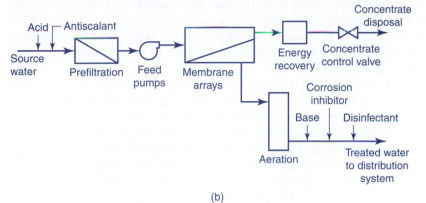

(b)

**Figure 10.23**  **Typical Components of a Membrane Drinking-Water Treatment Plant**   This schematic compares (a) a microfiltration plant and (b) reverse osmosis plant.

Crittenden et al. (2005). Redrawn with permission of John Wiley & Sons, Inc.

required, then the process should be nanofiltration or RO. Figure 10.23 depicts the layouts of microfiltration and RO water treatment plants.

As shown in Figure 10.23, microfiltration plants typically require a prefiltration step to remove coarse particles with diameters in excess of 200 to 500 μm. If dissolved constituents (such as manganese, iron, or hardness) or colloidal constituents are present and interfering with membrane operation, chemical pretreatment steps may be required. Feed pumps are utilized to provide the transmembrane pressure necessary for filtration to occur. Over time, particulate and dissolved constituents build up on the surface of the membrane, causing the permeate flux to decrease or increase the transmembrane pressure.

At some point in the operation, the membranes require backwashing to restore their performance. Backwashing a microfiltration filter is accomplished by reversing fluid flow through the use of air or permeate at a pressure higher than normal operating pressure. Permeate is then disinfected and sent to the distribution system. If there is retentate, it can be returned directly to the feed line or to a mixing basin upstream of the membranes. The backwash water containing cleaning chemicals and solids may be sewered for treatment at the wastewater treatment plant, treated on site, or permitted for discharge into source water.

For microfiltration membranes, the backwash cycle occurs approximately every 30 to 90 minutes, depending upon the influent water quality. After a period of time, chemical cleaning is required to reverse the loss of flux and restore the permeability. The loss of flux is due to fouling of the membrane caused by particulates, dissolved organic matter, and biological fouling from microorganism growth. This is called *reversible fouling*, as the loss of flux can be restored. *Irreversible fouling* is due to membrane compaction during the first year of operation. Compaction occurs when the large voids collapse in the porous membrane support layer because of excessive applied pressure. The applied pressure reduces the size of the support layer voids, causing a reduction in the permeability through the entire membrane cross section.

As shown in Figure 10.23b, reverse osmosis plants require acid and chemical addition to prevent scale formation on the membranes. The scale formation is caused by soluble salts in the feed water concentrating in the membrane to a degree that they exceed their solubility product and begin to precipitate as a solid on the membrane surface. Adding acid will change the solubility of the salts, and the antiscalant will help prevent the formation of the scale or at least slow down the rate of precipitation.

Prefiltration is used to prevent particles from clogging RO membranes. For surface waters, cartridge filters, granular filtration, and/or microfiltration may be necessary to remove particles before RO filtration. Disinfection may also be required to prevent microbial fouling of the RO membranes. Care must be taken to ensure that the RO membranes are not susceptible to oxidation by the disinfectant.

Permeate water quality is usually acidic (pH $\approx$ 5.0), low in alkalinity, and corrosive, and depending upon the water source, permeate may contain dissolved gas (for example, hydrogen sulfide if the water source is reduced groundwater). Some permeates may require aeration to remove unwanted dissolved gases. The corrosiveness of the water requires adjustments to pH and alkalinity by using a basic solution and adding a corrosion inhibitor (sodium silicate and sodium hexametaphosphate). A disinfectant is usually added to the water before it enters the clear well and distribution system.

Since the retentate stream is under high pressure, a concentrate control valve is used to capture the energy and reduce the pressure of the retentate stream. The concentrate control valve is an energy recovery system and can be used to reduce energy use at the RO plant. The retentate may be further treated, as with the use of evaporation ponds to further concentrate the salt, or may be discharged into the ocean, brackish estuary, or river, or to a municipal sewer. Obviously, the retentate stream is easier to dispose of in coastal areas, where the discharge does not present as many problems, than at inland locations where freshwater is present. However, permitting will be required to ensure protection of coastal ecosystems.

### 10.10.5   MEMBRANE PERFORMANCE

Table 10.25 summarizes typical operating characteristics and general performance parameters for membrane systems. For microfiltration systems, typical water recovery exceeds 95 percent and suspended-solids removal is around 90 to 98 percent, depending upon the membrane's nominal pore size. Microbiological removal can be as high as 7 log removal for protozoa such as *Giardia lamblia* cysts and *Cryptosporidium parvum* oocysts. Bacteria removal has been reported from 4 to 8 log removal, and virus removal from 7 to 10 log removal.

**More on Desalination**

http://www.coastal.ca.gov/desalrpt/dsynops.html

## Table / 10.25

**Operational Parameters for Membranes**

| Parameter | Microfiltration | Ultrafiltration | Nanofiltration | Reverse osmosis |
|---|---|---|---|---|
| Flux rate, $L/m^2 \cdot d$ | 400–1,600 | 400–800 | 10–35 | 12–20 |
| Operating pressure, kPa | | | | |
|   Freshwater | 0.0007–.01 | 0.007–0.7 | 350–550 | 1,200–1,800 |
|   Seawater | | | 500–1,000 | 5,500–8,500 |
| Energy consumption, $kWh/m^3$ | | | | |
|   Freshwater | 0.4 | 3.0 | 0.6–1.2 | 1.5–2.5 |
|   Seawater | | | | 5–10 |

SOURCE: Data from Crittenden et al, 2005; and Asano et al., 2006.

For RO systems, water recoveries range from 50 to 90 percent, depending upon the influent water type and the staging. TDS removals can be in the range of 90 to 99.5 percent removal. Microbiological removal is excellent, with greater than 7 log removal for protozoa and between 4 and 7 log removal for bacteria and virus removal (Asano et al., 2006). RO membranes are also very effective in removing SOCs.

### example/10.9 Calculation of Microfiltration Membrane System

A municipality is upgrading its microfiltration plant by replacing membranes of 0.2 μm pore size with membranes that have 0.1 μm pore size. The plant consists of eight arrays with 90 modules per array, and the plant capacity is 29,214 $m^3$/day. The modules are 119 mm in inside diameter and 1,194 mm in length, and they have an available filtration surface area of 23.4 $m^2$. The new hollow fibers have 1.0 mm outside diameter and length of 1,194 mm. Calculate the following:

1. Total surface area available for filtration

2. Membrane flux rate in $L/m^2 \cdot hr$ and gpm/sq. ft.

3. Total number of membrane fibers for the plant and each module

### solution

The total surface area available for filtration is determined as follows:

$$\text{total surface area} = 8\text{ arrays} \times \frac{90\text{ modules}}{\text{array}} \times \frac{23.4\text{ m}^2}{\text{module}} = 16{,}848\text{ m}^2$$

**example/10.9** Continued

The membrane flux rate is determined as follows:

$$\text{flux rate} = \frac{\text{total plant flow rate}}{\left(\begin{array}{c}\text{total surface area}\\\text{available for filtration}\end{array}\right)}$$

$$= \left(\frac{29{,}214\ \text{m}^3/\text{day}}{16{,}848\ \text{m}^2}\right) \times \left(\frac{\text{day}}{24\ \text{hr}}\right) \times \left(\frac{1{,}000\ \text{L}}{\text{m}^3}\right)$$

$$= 72.25\ \text{L/m}^2\text{-hr}$$

$$\text{flux rate} = 72.25\ \text{L/m}^2\text{-hr} \times \frac{\text{gal.}}{3.785\ \text{L}} \times \frac{\text{m}^2}{(3.28\ \text{ft.})^2}$$

$$\times \frac{\text{hr}}{60\ \text{min}} = 0.0296\ \text{gpm/sq. ft.}$$

The total number of membrane fibers required for the plant and each module is

$$\text{total number of fibers} = \frac{\text{total plant surface area available for filtration}}{\text{external surface area of a single fiber}}$$

$$= \frac{16{,}848\ \text{m}^2}{2\pi rL} = \frac{16{,}848\ \text{m}^2}{2\pi \times (0.001\ \text{m}) \times (1.194\ \text{m})}$$

$$= 2{,}246{,}000$$

$$[\text{number of fibers per module}] = \frac{\text{total number of fibers}}{\text{number of modules}}$$

$$= \frac{2{,}246{,}000}{8\ \text{arrays} \times \left(\frac{90\ \text{modules}}{\text{array}}\right)} = 3{,}120$$

## 10.11 Adsorption

### 10.11.1 TYPES OF ADSORPTION PROCESSES

Adsorption processes are widely used to remove organic and inorganic constituents from water and air. For example, **granular activated carbon (GAC)** and **powdered activated carbon (PAC)** are widely used for removing synthetic organic chemicals and odor compounds from drinking-water supplies. Under the Safe Drinking Water Act, the EPA has designated GAC as the best available technology for removing many organic and inorganic constituents from water supplies (for example, SOCs, arsenic, and radionuclides). GAC is also widely used for treating indoor air (for example, removing formaldehyde, toluene, and radon).

## Table / 10.26

**Examples of Synthetic Organic Compounds That Are Favorably and Unfavorably Adsorbed onto Activated Carbon from Water**

| Compound | Favorably Adsorbed | Unfavorably Adsorbed |
|---|---|---|
| 1-aminobutane | | ✕ |
| Acetone | | ✕ |
| Atrazine | ✕ | |
| Benzene | ✕ | |
| Carbon tetrachloride | ✕ | |
| Chloroform | ✕ | |
| Ethanol | | ✕ |
| Geosmin | ✕ | |
| Lindane | ✕ | |
| Methanol | | ✕ |
| Methyl isoborneal | ✕ | |
| Tert-butyl alcohol | | ✕ |
| Tetrachloroethylene | ✕ | |
| Toluene | ✕ | |
| Trichloroethylene | ✕ | |

**Adsorption** is a process whereby molecules are transferred from a fluid stream and concentrated on a solid surface by physical forces. A dissolved molecule in the fluid stream (gas or aqueous) that is attracted to and adsorbs to the solid surface is called *adsorbate*. The solid surface onto which the adsorbate adsorbs is called the *adsorbent*. The physical attraction is controlled by van der Waals–type forces, nonspecific binding mechanisms (attractive forces on a molecular level), or the attractive forces between the adsorbate and the adsorbent surface. SOCs that are amenable to adsorption are those that are hydrophobic (water hating). Those that are hydrophilic (water loving) prefer to stay in water, so their affinity to adsorption is low. Table 10.26 provides examples of SOCs that are favorably and unfavorably adsorbed onto activated carbon.

### 10.11.2 ADSORBENT TYPES

The most widely used adsorbent is activated carbon. Activated carbon comes in both granular (GAC) and powdered (PAC) form. Table 10.27

## Table / 10.27

**Comparison of Activated Adsorbents Used in Water Treatment**  Granular ferric hydroxide can be employed to remove arsenic and selenium.

| | Primary Use | Total Specific Surface Area (m²/g) | Particle Size (mm) | Design Considerations |
|---|---|---|---|---|
| Granular activated carbon (GAC) | Removal of organic constituents, inorganic constituents such as mercury, fluoride, per-chlorate, and arsenic | 500–2,500 | 0.3–2.4 | Carried out in a fixed-bed mode of operation either as gravity feed or under pressure. The three common modes of fixed-bed operation are: (1) single adsorber operation, (2) adsorbers operated in parallel, and (3) adsorbers operated in series. |
| Powdered activated carbon (PAC) | Primarily for removal of seasonal taste and odor; also used for periodic problems with agricultural pollutants such as pesticides and herbicides (e.g., during spring runoff) | 800–2,000 | 0.044–0.074 | Injected at head of the conventional water treatment plant and removed during sedimentation and filtration processes. Key design parameter is required dosage, obtained using jar tests. |

**Adsorption**

**Small Communities**
http://www.nesc.wvu.edu/drinkingwater.cfm

compares the use of GAC and PAC and a few physical properties of these two activated carbon adsorbents. Another adsorbent is granular ferric hydroxide (GFH), which is used primarily for removing arsenic and selenium. In comparison to GAC and PAC, GFH has a specific surface area of 250 to 300 m²/g, with particle sizes ranging from 0.32 to 2.00 mm. These adsorbents are porous and have a large internal surface area for adsorption to occur. The high total specific surface area of these adsorbents provides many sites for adsorption to take place.

Treatment with granular activated carbon (GAC) is carried out in a fixed-bed mode of operation either as gravity feed or under pressure. In some cases, GAC can replace granular media in filter operations. Beds in parallel consist of more then one adsorber where the influent concentration is fed to each bed and the effluent is blended.

PAC is typically added at the raw-water intake before the rapid-mix unit, or at the filter inlet before sand filtration, depending upon the other water treatment processes used. For example, PAC use can interfere with preoxidation and/or coagulation processes, resulting in higher PAC dosage requirements. The PAC membrane uses a completely mixed-flow reactor (CMFR) for adequate contact time, followed by PAC/water separation using an ultrafiltration membrane, and the PAC is recycled back to the head of CMFR. The membrane reactor configurations may be used when chronic problems are associated with taste and odor or micropollutants.

By definition, a public drinking-water system regularly supplies drinking water to at least 25 people or 15 service connections. Of the approximately 158,000 public water supply systems inventoried by the Environmental Protection Agency in 2005, almost 95 percent were classified as small (population 501–3,300) or very small (population less than 500) systems. These small systems serve a range of facilities, including small municipalities, mobile-home parks, apartment complexes, schools, and factories, as well as churches, motels, resorts, restaurants, and campgrounds. While the majority of the U.S. population is served by larger water systems, small systems play a vital role in providing safe drinking water.

Figure 10.24 shows the unit processes used to treat water at a resort located in northern Minnesota. Water is pumped from Rainy Lake and flows through several treatment process steps before being sent into the distribution system. Lake water is first chlorinated and then enters a set of pressure sand filters that remove larger particles. A 30 μm bag filter provides additional pretreatment. Water then flows through two ultrafiltration membrane units operating in parallel, providing more than the required 2 log removal of *Cryptosporidium* oocysts. Storage tanks provide chlorine contact time for inactivation of viruses and other pathogens and also provide distribution storage volume. Because of the high organic content in the source water (10–12 mg/L TOC), high levels of disinfection by-products have formed at this point in the process. Water therefore flows through a granular activated carbon filter (not shown) to remove the disinfection by-products. Because regulations require that a disinfectant residual be maintained at all points of the distribution system, a flow-paced rechlorination system (not shown) provides the final step of treatment.

Some challenges for this type of small system include the complexity of the treatment coupled with the lack of a full-time water operator, obtaining service in a remote area, and the ever-changing technology market and cost considerations.

**Figure 10.24**    **Small-Scale Water Supply System Serving a Seasonal Resort and Restaurant near Voyageurs National Park (Minnesota)**    Average water use for this system is approximately 3,000 gpd. The system ensures that the water consumed by the resort owners and visitors, some as part of the famous Long Island iced teas served at this resort, is safe to drink.

Photo and information courtesy of Anita Anderson, Minnesota Department of Public Health.

## 10.12   Energy Usage

Approximately 20 to 60 percent of the operating cost of a drinking-water utility is for energy use. The amount of energy used differs based on the source. For example, approximately 1.4 kWh of energy are needed to collect and treat 1,000 gal. surface water, and 1.8 kWh of energy are needed to collect and treat groundwater. The range of energy use varies widely, though, by geographical location and also

**Water Efficiency**

http://www.epa.gov/owm/water-efficiency/index.htm

© Jani Bryson/iStockphoto.

by treatment plant size. For example, energy use at drinking-water treatment plants is estimated to range from 300 to 3,800 kWh per million gallons of water treated and supplied (Burton, 1996; Elliot et al., 2003).

Focusing on the **life cycle stage** of manufacturing the treated water, the vast majority of this total energy use (up to 80 to 90 percent) is associated with pumping. Obviously, then, energy savings can be realized from proper design (and use) of more efficient motors and pumps, adjustable-speed drives, improvements in instrumentation and control, and installation of pipes and valves that reduce head loss in the collection, storage, and transmission system. As just one example—though not applicable in all situations—ball and flapper check valves have a much lower head loss than swing check valves. There is, of course, **embodied energy** associated with the reactors, materials, and chemicals used during pumping and treatment, and this is not accounted for in the manufacturing life stage.

Matching pump flows can eliminate the need to turn on additional pumps. Also, engineers often oversize motors. In these cases, it is quite cost-effective to replace oversized motors with more properly sized, energy-efficient motors. Modern high-efficiency motors may cost 15 to 25 percent more than a standard motor, but the payback period can be as short as 1 year (Elliot et al., 2003).

Two newer technologies that are seeing increasing usage in the water treatment industry are ozone disinfection and membrane filtration. Both are associated with higher energy demand but are known to provide a safe and sufficient water supply. For example, addition of ozone disinfection is expected to increase energy use by 0.12 to 0.55 kWh per 1,000 gallons of water treated, and microfiltration may increase energy use up to 0.70 kWh per 1,000 gallons of water treated (Elliot et al., 2003). Renewable (or off-peak-generated) energy can be used to lift water to where it can then be passed through membranes, using the elevation head to replace a mechanical pump.

## Key Terms

- adsorption
- backwash rate
- backwashing
- carbonate hardness
- charge neutralization
- Chick's law
- clean-bed head loss
- coagulant
- coagulation, coliform

- color
- combined chlorine
- critical particle-settling velocity
- *Ct* approach
- disinfectants
- dissolved organic carbon (DOC)
- dissolved particles
- effective size (ES or $d_{10}$)
- embodied energy

- emerging chemicals of concern
- energy
- enmeshment
- filter bed porosity
- filtration
- flocculation
- flux rate
- granular activated carbon (GAC)
- granular media filtration

- hardness
- head loss
- high-rate filtration
- hollow fibers
- indicator organisms
- interparticle bridging
- jar testing
- life cycle stage
- lime–soda ash softening process
- maximum contaminant level goal (MCLGs)
- maximum contaminant levels (MCLs)
- membrane processes
- microfiltration (MF)
- nanofiltration (NF)
- natural organic matter (NOM)
- nephelometric turbidity unit (NTU)
- Newton's law
- noncarbonated hardness
- overflow rate
- palatable
- particles
- pathogens
- potable water
- powdered activated carbon (PAC)
- precipitation
- pseudo first order
- rapid filtration
- rapid mixing
- reclaimed water
- reverse osmosis (RO)
- Reynold's number (Re)
- root mean square (RMS) velocity gradient
- sedimentation
- slow sand filtration
- Stokes' law
- suspended solids
- synthetic organic chemical (SOC)
- taste and odor
- temperature
- total chlorine residual
- total coliforms
- trihalomethanes (THMs)
- Type I, II, and III settling
- turbidity
- ultrafiltration (UF)
- uniformity coefficient (UC)
- water treatment

# chapter/Ten Problems

**10.1** The EPA provides reports (sometimes referred to as consumer confidence reports) that explain where your drinking water comes from and whether any contaminants are in the water. Go to this information at the "Local Drinking Water Information" page of EPA's Web site, (http://www.epa.gov/safewater/dwinfo/index.html). Look up the utility that serves your university and the largest city near your hometown. What is the source of water? Are there any violations? If so, are they for physical, biological, or chemical constituents?

**10.2** Jar testing was performed using alum on a raw drinking-water source that contained an initial turbidity of 20 NTU and an alkalinity of 35 mg/L as $CaCO_3$. The optimum coagulant dosage was determined as 18 mg/L with a final turbidity of 0.25 NTU. Determine the quantity of alkalinity consumed as $CaCO_3$.

**10.3** Jar tests were performed on untreated river water. An optimum dose of 12.5 mg/L of alum was determined. Determine the amount of natural alkalinity (mg/L as $CaCO_3$) consumed. If $50 \times 10^6$ gal./day of raw water are to be treated, determine the amount of alum required (kg/yr).

**10.4** A source water mineral analysis shows the following ion concentrations in the water: $Ca^{2+} = 70$ mg/L, $Mg^{2+} = 40$ mg/L, and $HCO_3^- = 250$ mg/L as $CaCO_3$. Determine the water's carbonate hardness, noncarbonated hardness, and total hardness.

**10.5** Calculate the lime dosage required for softening by selective calcium removal for the following water analysis. The dissolved solids in the water are $CO_2 = 17.6$ mg/L, $Ca^{2+} = 63$ mg/L, $Mg^{2+} = 15$mg/L, $Na^+ = 20$mg/L, Alk $(HCO_3^-) = 189$ mg/L as $CaCO_3$, $SO_4^{2-} = 80$ mg/L, and $Cl^- = 10$ mg/L. What is the finished-water hardness?

**10.6** A municipality treats $15 \times 10^6$ gal./day of groundwater containing the following: $CO_2 = 17.6$ mg/L, $Ca^{2+} = 80$ mg/L, $Mg^{2+} = 48.8$ mg/L, $Na^+ = 23$mg/L, Alk $(HCO_3^-) = 270$ mg/ L as $CaCO_3$, $SO_4^{2-} = 125$ mg/L, and $Cl^- = 35$ mg/L. The water is to be softened by excess lime treatment. Assume that the soda ash is 90 percent sodium carbonate, and the lime is 85 percent weight CaO. Determine the lime

and soda ash dosages necessary for precipitation softening (kg/day).

**10.7** Water contains 7.0 mg/L of soluble ion ($Fe^{2+}$) that is to be oxidized by aeration to a concentration of 0.25 mg/L. The pH of the water is 6.0, and the temperature is 12°C. Assume the dissolved oxygen in the water is in equilibrium with the surrounding atmosphere. Laboratory results indicate the pseudo first-order rate constant for oxygenation of $Fe^{2+}$ is 0.175/min. Assuming steady-state operations and a flow rate of 40,000 $m^3$/day, calculate the minimum detention time and reactor volume necessary for oxidation of $Fe^{2+}$ to $Fe^{3+}$. Perform the calculations for both a CMFR and a PFR. (You should be able to work this out from information provided in Chapters 3 and 4.)

**10.8** A mechanical rapid-mix tank is to be designed to treat 50 $m^3$/day of water at a temperature of 12°C. Using typical design values in the chapter, determine the tank volume and power requirement.

**10.9** Calculate the settling velocity of a particle with 100 μm diameter and a specific gravity of 2.4 in 10°C water.

**10.10** Calculate the settling velocity of a particle with 10 μm diameter and a specific gravity of 1.05 in 15°C water.

**10.11** Research the use of a filtration method that provides household (point-of-use) treatment in the developing world. Write a one-page essay that is clearly referenced. In your essay, address these issues: Is the technology affordable to the local population? Does it use local materials and local labor for its construction? What are the observed health improvements after implementation of the treatment system? What specific training do you believe is required to ensure proper operation of the technology?

**10.12** Research the global arsenic problem in the United States or Europe, and in Bangladesh or West Bengal (India). In a two-page essay, identify, compare, and contrast the extent of the problem (spatially and in terms of population affected). What are the current methods of treatment employed to remove the arsenic, and what are the associated costs? What are current treatment standards in the two locations?

**10.13** Given the following data, graph the data for the *Poliomyelitis* virus, using hypobromite as a disinfectant. Determine the Chick's law rate constant and the time required for 99.99 percent (4 log removals) inactivation of this virus.

| Time (s) | N (number of organisms/L) |
|----------|---------------------------|
| 0.0 | 1,000 |
| 2.0 | 350 |
| 4.0 | 78 |
| 6.0 | 20 |
| 8.0 | 6 |
| 10.0 | 2 |
| 12.0 | 1 |

**10.14** Graph the following data for the inactivation of a virus using hypochlorous acid (HOCl). Determine the coefficient of specific lethality and the time required to obtain 99.99 percent inactivation using 1.0 mg/L of HOCl.

| Time (min) | log (N/N$_o$) |
|------------|---------------|
| 1.0 | −0.08 |
| 3.0 | −0.64 |
| 5.0 | −1.05 |
| 9.0 | −1.87 |
| 15.0 | −3.23 |

**10.15** Define the meaning of Ct product. In addition to C and t, what factors influence the rate of chemical disinfection? What kind of microorganism is most readily inactivated by free chlorine? What kind is most difficult to inactivate?

**10.16** Visit the Web site for the World Health Organization (www.who.org). Write a referenced essay of up to two pages about a particular pathogen and associated disease that is transmitted through contaminated water. What is the global extent of the public health crisis in terms of spatial and population effects? Does the disease affect wealthy or poor communities? What are the social and engineered barriers that can be used to reduce human exposure to the specific pathogen?

**10.17** Research a professional engineering society that works on drinking-water treatment in the developing world. Examples include Water for People, Engineers without Borders, and Engineers for a Sustainable World. Determine how you could contribute to such a professional society as a student and after graduation. Provide specific detail about membership requirements, costs, and how you might be involved.

**10.18** A city is upgrading its water supply capacity by 81,378 m$^3$/day, using microfiltration. The new plant will consist of 25 arrays with 90 modules per array. The modules have an inside diameter of 119 mm, a length of 1,194 mm, and an available surface area of 23.4 m$^2$. The hollow fibers have an outside diameter of 1.0 mm, and a length of 1,194 mm. Determine: (a) the total surface area available for filtration; (b) the membrane flux rate in L/m$^2$·hr; and (c) the total number of membrane fibers for the plant and each module.

**10.19** Use the Internet to determine the number of people in the world who do not have access to an improved water supply. Then look up the Millennium Development Goals (MDGs) at the United Nations Web site, www.un.org. MDG 7 states that by 2015 the world will decrease by half the number of people without access to an improved water supply. Select a country in Africa and another country in Asia or Latin America. Compare the progress of these two countries in meeting goal 7 in terms of the number of people still not served by an improved water supply.

**10.20** What is the source of drinking water in the town where you currently live (groundwater, surface water, reclaimed water, or a mixture)? Sketch the unit processes used to treat this water in order of occurrence as currently practiced. What water constituent(s) does each unit process remove?

**10.21** Identify three significant water users in your town or city that could benefit from the use of reclaimed water. What economic, social, and environmental challenges do you see that might need to be overcome before you can implement your plan for these users to use reclaimed water? What would you do as engineer to overcome these barriers?

**10.22** Go to the World Health Organization's Web site, www.who.org. Research a particular chemical water constituent, and write a one-page report on the global magnitude of the problem and what engineered and non-engineering solutions are needed to deal with this problem.

**10.23** A 1,000-g sample of naturally occurring sand was sifted through a stack of sieves and the weight retained on each sieve is recorded as shown in the table below. Determine the effective size and uniformity coefficient for the media.

| Sieve Designation | Sieve opening, mm | Weight of retained media, g |
|---|---|---|
| 10 | 2.000 | 140 |
| 14 | 1.400 | 160 |
| 18 | 1.000 | 170 |
| 20 | 0.850 | 100 |
| 25 | 0.710 | 90 |
| 30 | 0.600 | 85 |
| 35 | 0.500 | 80 |
| 40 | 0.425 | 70 |
| 45 | 0.355 | 65 |

**10.24** A drinking water treatment plant processes 1 million gallons of water per day. The chlorinator feed rate is 4.0 mg/L and the free chlorine residual is 1.5 mg/L. (a) What is the chlorine demand (mg/L)? (b) How many kg of chlorine must be fed to the plant on a daily basis?

**10.25** Methyl tert butyl ether (MTBE) adsorption isotherm was performed on an activated carbon at 15°C using 0.250-L amber bottles with an initial MTBE concentration, $C_o$, of 150 mg/L. The isotherm data for each experimental point is summarized below. Calculate the adsorbed-phase concentration,

$q_e$, for each isotherm point, plot the log ($q_e$) versus log ($C_e$) and determine the Freundlich isotherm parameters K and 1/n (this problem will require you review information presented in Chapter 3).

| Mass of GAC, g | MTBE equilibrium liquid-phase concentration, $C_e$, mg/L |
|---|---|
| 0.155 | 79.76 |
| 0.339 | 42.06 |
| 0.589 | 24.78 |
| 0.956 | 12.98 |
| 1.71 | 6.03 |
| 2.4 | 4.64 |
| 2.9 | 3.49 |
| 4.2 | 1.69 |

**10.26** Powdered activated carbon (PAC) is to be added to a water treatment plant to remove 10 ng/L of MIB which is causing odor problems in the finished water. A standard jar test is performed to evaluate the impact of PAC dosage on the removal of MIB. The results are shown Figure 10.25. If 60 percent MIB removal is required, determine the dosage of PAC required and quantity of PAC needed for 3 months (90 days) of treatment if the plant flow rate for is 40,000 m³/d.

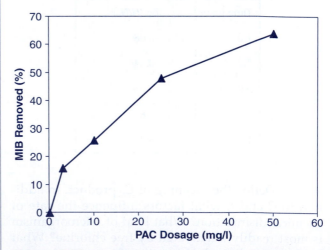

**Figure 10.25** Results from Jar Test Investigating Powdered Activated Carbon Usage at a Water Treatment Plant.

# References

Asano, T., F. L. Burton, H. L. Leverenz, R. Tsuchihashi, and G. Tchobanoglous. 2006. *Water Reuse: Issues, Technologies, and Applications*. Wakefield, Mass.: Metcalf and Eddy.

Burton, F. L. 1996. "Water and Wastewater Industries: Characteristics and Energy Management Opportunities." Report prepared for Electric Power Research Institute (EPRI) CR-106941.

Crittenden, J. C., R. R. Trussell, D. W. Hand, G. Tchobanoglous, and K. Howe. 2005. *Water Treatment Principles and Design*, 2nd ed. New York: John Wiley & Sons.

Davis, M. L., and D. A. Cornwell. 2004. *Introduction to Environmental Engineering*, 3rd ed. New York: McGraw-Hill.

Elliot, T., B. Zeier, I. Xagoraraki, and G. Harrington. 2003. "Energy Use at Wisconsin's Drinking Water Facilities." ECW Report Number 222-1. Wisconsin Focus on Energy & Energy Center of Wisconsin, Madison (July).

Kawamura, S. 2000. *Integrated Design and Operation of Water Treatment Facilities*, 2nd ed. New York: Wiley-Interscience.

LeChevallier, M. W., and W. D. Norton. 1995. "*Giardia* and *Cryptosporidium* in Raw and Finished Water." *AWWA* 87 (9): 54–68.

Mihelcic, J. R., E. A. Myre, L. M. Fry, L. D. Phillips, and B. D. Barkdoll. 2009. *Field Guide in Environmental Engineering for Development Workers: Water, Sanitation, Indoor Air*. Reston, Va.: American Society of Civil Engineers Press.

Petrusevski, B., S. Sharma, J. C. Schippers, and K. Shordt. 2007. "Arsenic in Drinking Water." Thematic Overview Paper 17. IRC International Water and Sanitation Centre, Delft, The Netherlands (March).

# chapter/Eleven Wastewater Treatment

James R. Mihelcic,
David W. Hand, Martin
T. Auer

*In this chapter, readers will learn about the composition of wastewater and the various unit processes employed to remove the various water quality constituents. Mass balance and biochemical kinetics are employed to develop expressions to size a biological reactor used to remove biochemical oxygen demand. Less mechanized natural treatment processes such as free surface wetlands and lagoons are discussed. The treatment of nitrogen and phosphorus also is discussed, along with processing of sludge produced at the treatment plant. Readers will also learn about energy requirements (and energy sources) in terms of wastewater treatment technologies and how operation of a plant influences energy usage.*

## Learning Objectives

1. Identify distinct hydrological, physical, chemical, and biological inputs that make up municipal wastewater, and list typical concentrations of the major constituents.

2. Match major wastewater constituents with the unit process(es) that remove a significant amount of each constituent.

3. Design a grit chamber, flow equalization basin, activated-sludge biological treatment system, and stabilization pond.

4. Integrate mass balances with biological growth kinetics to develop activated-sludge design equations and relate solids retention time, food-to-microorganism ratio, sludge wasting, and growth kinetics to plant design and operation.

5. Empathize with the magnitude of global sanitation coverage and its relationship to human health.

6. List advantages and disadvantages of membrane bioreactors.

7. List the components of wastewater solids management, calculate sludge volume index, and relate this value to the settling characteristics of sludge.

8. Relate the configuration and operation of a wastewater treatment plant to the removal of nitrogen and phosphorus via biochemical and chemical processes.

9. Describe specific removal processes that occur in different treatment zones in facultative lagoons and free surface wetlands.

10. Identify the magnitude of energy usage during wastewater treatment and the greenhouse gases emitted from different unit processes.

11. Identify sources of energy that can be obtained from wastewater treatment.

# 11.1 Introduction

**Municipal wastewater treatment plants**, also referred to as **publicly owned treatment works (POTWs)**, receive inputs from many domestic and industrial sources. These distinct hydrological inputs are depicted in Figure 11.1. There are four components of **domestic wastewater**: (1) wastewater from domestic, commercial, and industrial users; (2) stormwater runoff; (3) infiltration; and (4) inflow. Infiltration and inflow were discussed in Chapter 9. With population increase, climate change, nutrient cycling, and water scarcity becoming common, sustainable treatment of wastewater must address issues beyond treatment performance and cost.

Industrial wastewaters vary in quantity, composition, and strength, depending on the specific industrial source. The Environmental Protection Agency (EPA) has identified 129 *priority pollutants*. Industrial wastes include conventional pollutants found in domestic wastewater but may also contain heavy metals, radioactive materials, and refractory organics. Industries may choose to treat their waste on site, following specific guidelines for *best available treatment* and *effluent guidelines* for priority pollutants. They may also choose to sewer their waste to a municipal wastewater treatment plant, after first providing pretreatment to protect operation of the municipal treatment plant and prevent discharge of pass-through pollutants.

**Wastewater Introduction**

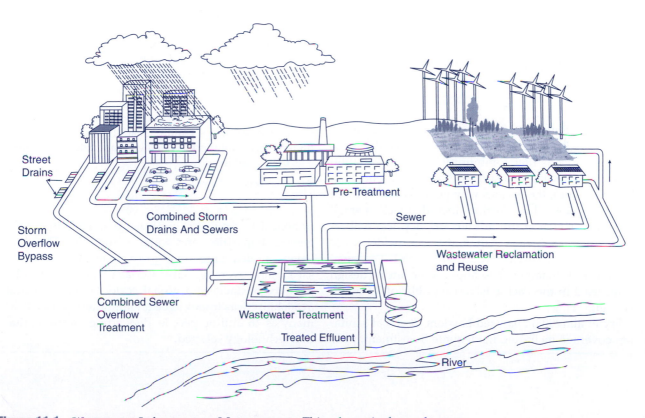

**Figure 11.1  Wastewater Infrastructure Management**  This schematic shows the many possible hydrological and chemical pollutant contributors to municipal wastewater.

Table 11.1 provides the global definition for *improved* and *unimproved* sanitation technologies. Currently, 2.5 billion people in the world are without access to improved sanitation technology (including 1.2 billion who have no facilities at all), and lack of sanitation has a large negative impact on human health and the environment (UNICEF and WHO, 2008).

## Table / 11.1

**Improved and Unimproved Sanitation Technologies: Global Definitions**

| Improved | Unimproved |
| --- | --- |
| Connection to a public sewer | Service or bucket latrines (excreta are removed manually) |
| Connection to a septic system | Public latrines |
| Pour-flush latrines | Open latrines |
| Ventilated improved pit latrines | |
| Composting latrines Simple pit latrines | |

**Figure 11.2**   **Double-Vault Compost Latrine Being Constructed in Vanuatu**   These latrines can be designed to separate urine and feces. The urine is either routed to a soak pit or collected in a pot or routed to a garden and used as a fertilizer. One side of the latrine is used for up to 12 months, while the other side is composting. Desiccants such as wood ash and sawdust are added to reduce odors and kill pathogens. The privacy shelter is being constructed of local wood and woven plant material. Compost latrines require no addition of water, unlike other sanitation technologies, and allow the nutrients to be used locally.

Photo courtesy of Eric Tawney.

The perception of sanitation varies significantly from culture to culture. Improvements in health are not the only reasons that communities accept sanitation projects. According to a survey of rural households in the Philippines, the reasons why people were satisfied with their newly built latrines included lack of smell and flies, cleaner surroundings, privacy, less embarrassment when friends visit, and fewer incidences of gastrointestinal disease (Cairncross and Feachem, 1993). For an example of an appropriate latrine design being used in the Pacific island republic of Vanuatu, see Figure 11.2.

Fry et al. (2008) analyzed barriers to global sanitation coverage including inadequate investment, poor or nonexistent policies, governance, too few resources, gender disparities, and water availability. The challenges studied were found to be significant barriers to sanitation coverage, but water availability was not a primary obstacle at a global scale. However, water availability was found to be an important barrier to as many as 46 million people, depending on the sanitation technology selected.

One purpose of municipal wastewater treatment is to prevent pollution of a receiving surface water or groundwater. Examples of pollutants associated with untreated wastewater include dissolved oxygen depletion (measured as BOD and COD), unsightly and oxygen-depleting solids (TSS), nutrients that cause eutrophication (N and P), chemicals that exert toxicity ($NH_3$, metals, organics), emerging chemicals of concern, and pathogens (bacteria and viruses). Aesthetic problems include visual pollution and odor. In terms of pathogens, humans produce on average $10^{11}$ to $10^{13}$ coliform bacteria per day. While treatment processes are very efficient at removing pathogens and other pollutants, in the near future, treatment plants will need to be concerned with removing other chemicals now found in our wastewater. These include fragrances, surfactants found in soaps and detergents, pharmaceutical chemicals, endocrine-disrupting chemicals, and other **emerging chemicals of concern**.

Through the Federal Water Pollution Control Act of 1972 (commonly known as the **Clean Water Act**), the U.S. Congress established a national strategy to reduce water pollution. The objectives of the Clean Water Act are to restore and maintain the chemical, physical, and biological integrity of the nation's waters by achieving a level of water quality that provides for the protection and propagation of fish, shellfish, and wildlife and for recreation in and on the water, and to eventually eliminate the discharge of pollutants into U.S. waters (zero discharge). This is accomplished through the **National Pollutant Discharge Elimination System (NPDES)**, which issues permits defining the types and amounts of polluting substances that may be discharged. The NPDES permitting system is administered and enforced at the state level. Violation of the technology-based and water-quality-based effluent standards may result in civil penalties (fines) and criminal penalties (imprisonment).

## 11.2 Characteristics of Domestic Wastewater

*Raw* (that is, untreated) wastewater is considered highly polluted, yet the amount of contaminants it contains may seem small. For example, 1 $m^3$ municipal wastewater weighs approximately 1 million g, yet it may contain only 500 g pollutants. This small fraction of pollution can have serious ecological and health impacts, however, if discharged untreated.

Domestic wastewater appears gray and turbid and has a temperature of 10°C to 20°C. Table 11.2 provides the composition of an **average-strength municipal wastewater** and shows the most common **municipal wastewater constituents**. We will not go into great detail about each constituent at this time because their measurement and importance were described in previous chapters, as noted in Table 11.2. As we discuss the various unit processes for treatment, you may want to refer back to this table, because specific processes remove distinct wastewater constituent(s).

© david pullicino/iStockphoto.

**Concentration of Major Constituents Found in Average-Strength Wastewater**

| Constituent | Discussed previously in | Average Concentration | Comments |
|---|---|---|---|
| Biochemical oxygen demand (BOD) | Chapters 2, 3, and 5 | 200 mg/L | Oxygen-demanding materials can deplete the oxygen content of receiving waters. |
| Suspended solids | Chapter 2 and 10 | 240 mg/L (total solids typically 800 mg/L) | Cause water to be turbid; may contain organic matter and thus contribute to BOD; may contain other pollutants or pathogens. |
| Pathogens | Chapter 5 and 10 | 3 million coliforms per 100 mL | Disease-causing microorganisms usually associated with fecal matter. |
| Nutrients such as nitrogen and phosphorus | Chapters 3 and 5 | Total nitrogen: 35 mg N/L Inorganic nitrogen: 15 mg N/L Total phosphorus: 10 mg P/L | Can accelerate growth of aquatic plants, contribute to eutrophication; ammonia is toxic to aquatic life, can contribute to NBOD. |
| Toxic chemicals | Chapters 3, 5, 6, and 10 | Variable | Heavy metals such as mercury, cadmium, and chromium; organic chemicals such as pesticides, solvents, fuel products. |
| Emerging chemicals of concern | Chapter 6 and 10 | Unknown or variable | Pharmaceuticals, caffeine, surfactants, fragrances, perfumes, other endocrine-disrupting chemicals |

## 11.3 Overview of Treatment Processes

**Wastewater Treatment Overview**

Design and operation of a wastewater treatment facility require an understanding of unit operations that employ fundamental physical, chemical, and biological processes (see Chapters 3, 4, and 5) to remove specific water quality constituents. Assembly of the correct process removal train requires the accomplishment of four tasks: (1) identifying the characteristics of the untreated wastewater; (2) identifying treatment objectives and assessing community involvement; (3) integrating unit operations into a complete process that recognizes the appropriateness and limits of each unit process and how they complement each other; and (4) integrating concepts of green engineering, life cycle thinking, and sustainability to incorporate issues beyond end-of-pipe treatment standards and capital and operating costs (for example, water reuse and energy consumption).

Figure 11.3 provides an aerial view of a typical municipal wastewater treatment plant. The schematic drawing in Figure 11.4 shows how the different unit processes can be integrated. Different processes

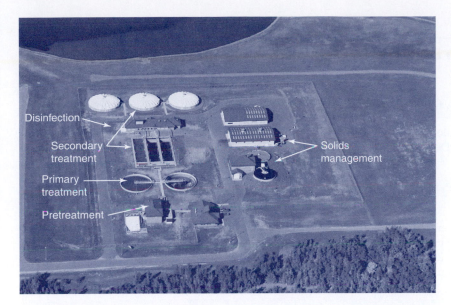

**Figure 11.3** **Overhead View of Wastewater Treatment Plant** This plant serves approximately 14,000 people.

Photo courtesy of the Portage Lake Water and Sewage Authority.

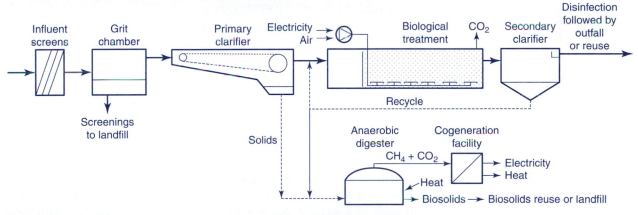

**Figure 11.4** **Typical Layout of Conventional Wastewater Treatment** Preliminary treatment with screens and grit removal is followed by a primary clarifier, biological treatment, a secondary clarifier, and anaerobic treatment of the sludge.

Adapted from figure provided by Dr. Diego Rosso, University of California–Irvine.

occur for the treatment of liquid and solid waste streams. The steps involved in conventional wastewater treatment are: (1) pretreatment, (2) primary treatment, (3) secondary treatment, (4) tertiary treatment to remove nutrients (N, P), and (5) disinfection. With wastewater now viewed as a resource by many communities, the flow chart for conventional wastewater treatment is now changing to accommodate issues of water reuse.

Pursuant to Section 304(d) of Public Law 92-500, the EPA published its definition of minimum standards for secondary treatment. Table 11.3 provides an overview of these treatment standards and the specific unit processes that remove significant amounts of specific wastewater constituents.

**Minimum Standards for Treatment and Unit Processes That Remove a Significant Amount of Major Wastewater Constituents**

| Constituent and EPA Standard for Minimum Treatment | Unit Process(es) That Remove Significant Amount of Constituent |
|---|---|
| **Biochemical oxygen demand (BOD):** Allowable 30-day $BOD_5$ is 30 mg/L, and 7-day $BOD_5$ is 45 mg/L. | BOD can be in a dissolved or particulate form. Biological reactor; primary and secondary sedimentation. |
| **Suspended solids:** Allowable 30-day TSS is 30 mg/L, and 7-day TSS is 45 mg/L. | Primary and secondary sedimentation. |
| **Pathogens:** Depends upon the NPDES permit based on the receiving water (e.g., at the plant depicted in Figure 10.3, fecal coliforms <200 counts/mL monthly average or <400 counts/mL 7-day average). | Primary and secondary sedimentation; disinfection. Predation also occurs in the biological reactor. |
| **Nutrients such as nitrogen and phosphorus:** Depends upon the NPDES permit. | Nutrients can be in a dissolved or particulate form. Sedimentation; biological reactor; addition of chemicals to precipitate phosphorus. |
| **Toxic chemicals** | Some are removed via sedimentation (if they are sorbed to or complexed by particles), some are biodegradable, and some pass through the treatment plant. |
| **pH:** Discharge must be in range of 6.5–10.0. | Not applicable. |

## 11.4 Preliminary Treatment

**Wastewater Treatment Primer**
http://www.epa.gov/owm/primer.pdf

**Preliminary treatment** prepares the wastewater for further treatment. It is used to remove oily scum, floating debris, and grit, which may inhibit biological processes and/or damage mechanical equipment. Equalization tanks are employed to balance flows or organic loading. Industrial effluents may additionally require physical–chemical pretreatment for removal of ammonia-nitrogen (air stripping), acids/bases (neutralization), heavy metals (oxidation-reduction, precipitation), or oils (dissolved air flotatation).

### 11.4.1 SCREENING

**Bar racks** (parallel bars or rods, 20–150 mm) and **bar screens** (perforated plates or mesh, 10 mm or less) retain the coarse solids (large objects, rags, paper, plastic bottles, etc.) present in wastewater, preventing damage to the piping and mechanical equipment that follows this treatment step (Figure 11.5). They are cleaned by hand in some older, smaller plants, but most are equipped with automatic cleaning rakes. Screenings are typically disposed of by landfilling or incineration.

As an alternative to screens, some plants utilize a **comminutor**, which grinds up (comminutes) coarse solids without removing them from the wastewater flow. This reduction in size makes the solids easier to treat in subsequent operations that employ settling. Comminutors eliminate the need for handling and disposal of coarse solids removed during screening.

## 11.4.2 GRIT CHAMBERS

**Grit** consists of particulate materials in wastewater that have specific gravities of approximately 2.65 and a temperature of 15.5°C. Particles with specific gravities between 1.3 and 2.7 have also been removed, based on field data. Grit can consist of inorganic sand or gravel (about 1 mm in diameter), eggshells, bone fragments, fruit and vegetable pieces and seeds, and coffee grounds.

Grit is primarily removed to prevent abrasion of piping and mechanical equipment. During grit removal some organic materials are removed along with the grit. Sometimes grit-washing equipment is added to remove organic materials and return them to the wastewater.

For horizontal-flow systems, grit is removed through sedimentation by gravity (using Stokes' law or Newton's law). In an aerated **grit chamber** air is introduced along one side of the tank, which provides a helical flow pattern of the wastewater through the chamber, enabling the grit to settle out while keeping the smaller organic material suspended in the wastewater. The *aerated grit chamber* has the added advantage that it keeps the wastewater fresh by adding oxygen to the wastewater. In a *vortex grit chamber*, the wastewater enters and exits tangentially, creating a vortex flow pattern where the grit settles to the bottom of the tank.

Both aerated and vortex systems are designed based on typical design parameters. For example, Table 11.4 provides design information used to size an aerated grit chamber. These chambers are normally designed to remove particles with diameters of at least 0.21 mm. Detention times range from 2 to 5 minutes based on peak hourly flow, and air flow rates range from 0.2 to 0.5 m$^3$ of air per minute per length of tank. Example 11.1 illustrates the design of an aerated grit chamber.

**Grit Chamber Simulator**

www.wiley.com/college/mihelcic

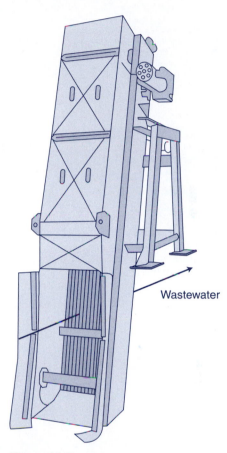

**Figure 11.5** **Bar Screens Used to Remove Coarse Solids from Wastewater** If not removed, these solids may damage piping and mechanical equipment that follows in the treatment process.

## Table / 11.4

### Design Information Used to Size Aerated Grit Chambers

| Parameter | Range |
|---|---|
| Peak flow detention time (min) | 2–5 |
| Depth (m) | 2–5 |
| Length (m) | 7.5–20 |
| Width (m) | 2.5–7 |
| Width-to-depth ratio | 1:1 to 5:1 |
| Length-to-width ratio | 3:1 to 5:1 |
| Air requirement per length of tank (m$^3$/m-min) | 0.2–0.5 |
| Quantity of grit (m$^3$/10$^3$ m$^3$) | 0.004–0.20 |

SOURCE: Values obtained from Tchobanoglous et al. (2003).

## example/11.1 Design of an Aerated Grit Chamber

Design an aerated grit chamber system to treat a 1 day sustained peak hourly flow of 1.5 $m^3$/s with an average flow of 0.6 $m^3$/s. Determine: (a) the grit chamber volume (assuming two chambers will be used); (b) the dimensions of the two grit chambers; (c) the average hydraulic retention time in each grit chamber; (d) air requirements, assuming 0.35 $m^3$/m · min of air; and (e) the quantity of grit removed at peak flow, assuming a typical value of 0.015 $m^3/10^3$ $m^3$ of grit in the untreated wastewater.

### solution

Much of this problem can be solved by using the design guidance provided in Table 11.4.

1. The volume of the grit chambers is determined assuming a detention time of 3 min:

$$\text{total grit chamber volume} = 1.5 \, m^3/s \times 3 \, min \times 60 \, s/min$$
$$= 270 \, m^3$$

$$\text{volume of each grit chamber} = \frac{1}{2} \times 270 \, m^3 = 135 \, m^3$$

2. Assuming a width-to-depth ratio of 1.5:1 and a depth of 3 m, the dimensions of the two grit chambers are

$$\text{grit chamber width} = 1.5 \times 3 \, m = 4.5 \, m$$

$$\text{grit chamber length} = \frac{\text{volume}}{\text{width} \times \text{depth}} = \frac{135 \, m^3}{4.5 \, m \times 3 \, m} = 10 \, m$$

3. The average hydraulic detention time in each grit chamber is based on the average flow rate:

$$\text{detention time} = \frac{\text{volume}}{\text{flow}} = \frac{135 \, m^3}{(0.6 \, m^3/s)/(2 \, tanks)} \times \frac{1 \, min}{60 \, s} = 7.5 \, min$$

4. The air requirements, assuming 0.35 $m^3$/m-min of air, are

$$\text{total air requirement} = (2 \, tanks) \times (10 \, m \, length)$$
$$\times (0.35 \, m^3 \, air/m{\cdot}min) = 7.0 \, m^3/min$$

5. Finally, solve for the amount of grit to be disposed of, assuming peak flow conditions:

$$\text{grit volume} = (1.5 \, m^3/s) \times (0.015 \, m^3/10^3 \, m^3)$$
$$\times (86,400 \, s/day) = 1.94 \, m^3/day$$

Similar design methods are employed to size horizontal-flow and vortex grit removal devices.

### 11.4.3 FLOTATION

**Flotation** is the opposite of sedimentation, utilizing buoyancy to separate solid particles such as fats, oils, and greases, which would not settle by sedimentation. The separation process is accomplished by introducing air at the bottom of a floatation tank. The air bubbles rise to the surface, where they are removed by skimming. A popular variation on this scheme is termed *dissolved-air flotation*. Recycled effluent is retained in a pressure vessel, where it is mixed and saturated with air. The effluent is then mixed with the raw wastewater, and as the pressure returns to atmospheric, dissolved air comes out of solution, carrying floatable solids to the surface, where they may be skimmed and collected.

Today, fats, oil, and grease (termed FOG)—specifically, material generated at local restaurants—do not have to become part of the wastewater stream. They can be easily converted to biodiesel (which could help fuel your municipality's vehicle fleet) and used to generate energy, when combined with digester gas, or used as a supplementary fuel in solid-waste-to-energy plants.

### 11.4.4 EQUALIZATION

**Flow equalization** is implemented to dampen the flow and organic loading rate to a wastewater treatment facility. Remember from Chapter 9 that large variations occur in the flow for many reasons. Implementing flow equalization in some instances can overcome operational problems associated with large flow variations and improve the performance of the downstream unit processes. For example, the biological processes used during wastewater treatment can be more easily controlled with a steady flow rate and near-constant BOD loading. In addition, implementation of flow equalization can reduce the size of the downstream treatment processes and in some cases improve performance at plants that are overloaded.

Figure 11.6 compares the diurnal flow rate and BOD loading variation with an equalized flow and BOD loading pattern. BOD

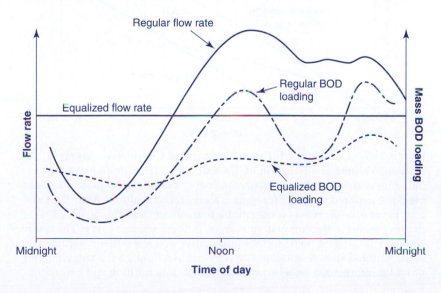

**Figure 11.6   Changes in Regular Flow Rate and Regular Mass BOD Loading over a Typical Day**   The equalized flow rate is shown as a constant, and the equalized BOD loading is dampened, so larger variations over the day are removed.

loading is equal to the flow times the concentration of BOD in the wastewater and has units of kg BOD/m$^3$ of wastewater per day. As Figure 11.6 shows, the dampening of the flow rate and BOD can be considerable.

Flow equalization can be accomplished in two ways: in-line or off-line equalization. *In-line equalization* is the process where all the flow passes through the equalization basin. In contrast, with *off-line equalization*, only a portion of the flow is diverted through the equalization basin. Off-line flow equalization requires that the diverted flow be pumped and mixed with the plant influent when the influent flow rate to the plant is reduced. This typically occurs late at night. In this case, the flow can be equalized, but the change in BOD loading is reduced less than with in-line flow equalization. Therefore, in-line equalization is typically used when stringent dampening of both the flow and organic loading is required.

Mass diagrams like the one shown in Figure 11.7 can be used to determine the required *equalization storage volume*. Figure 11.7 shows the *cumulative influent volume* and the *cumulative average influent volume* as a function of the time of day. To determine the equalization volume required, take the vertical distance between the average volume and the parallel line that is tangent to the cumulative inflow volume curve for Figure 11.7. The tangent point on the cumulative inflow curve is where the equalization tank is empty. As time increases, the slope of the cumulative inflow curve is greater than the average inflow curve. The equalization tank will fill up to about midnight, when the slopes of both curves are about the same. From midnight to a little past noon,

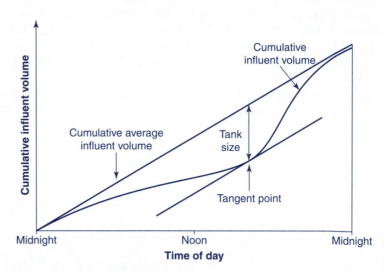

**Figure 11.7  Cumulative Influent Volume and Cumulative Average Influent Volume as a Function of Time of Day**  The cumulative average influent volume and the cumulative influent volume can be plotted to determine the required tank size for equalization storage volume. After the tangent point is determined on the cumulative influent volume curve, a line is drawn parallel to the cumulative average influent volume curve. The distance between the tangent point and the cumulative average influent volume curve is the required storage volume. The curve drawn through the tangent point provides information on when the storage volume is filling and emptying.

the slope of the cumulative inflow curve is less than the average curve, meaning the equalization tank is losing volume (is draining).

For Figure 11.7, the equalization volume would be found at the tangent of both parallel lines on the cumulative influent volume curve. During the time between the two tangent points, approximately 1 P.M. to about midnight, the equalization basin is filling, as the slope of the cumulative influent volume curve is greater than the average influent volume curve. From about midnight to 1 P.M., the slope of the cumulative influent volume curve is less than the average influent volume curve, and the equalization tank is draining.

---

example/11.2 Sizing a Flow Equalization Tank

Given the data for average hourly flows shown in Table 11.5 (in the two left columns), determine the required on-line flow equalization volume ($m^3$).

## Table / 11.5

**Data and Results for Flow Equalization Problem in Example 11.2**  The average cumulative influent flow (not shown, units of $m^3/hr$) is determined by dividing the cumulative influent volume by 24 hr.

| Time Period | Volume of Flow during Period ($m^3$) | Cumulative Influent Volume ($m^3$) |
|---|---|---|
| midnight–1 a.m. | 1,090 | 1,090 |
| 1–2 | 987 | 2,077 |
| 2–3 | 701 | 2,778 |
| 3–4 | 568 | 3,346 |
| 4–5 | 487 | 3,833 |
| 5–6 | 475 | 4,308 |
| 6–7 | 532 | 4,840 |
| 7–8 | 838 | 5,678 |
| 8–9 | 1,375 | 7,053 |
| 9–10 | 1,565 | 8,618 |
| 10–11 | 1,630 | 10,248 |
| 11–noon | 1,649 | 11,897 |
| noon–1 | 1,640 | 13,537 |
| 1–2 | 1,545 | 15,082 |

*(Continued)*

## Table / 11.5

| Time Period | Volume of Flow during Period (m³) | Cumulative Influent Volume (m³) |
|---|---|---|
| 2–3 | 1,495 | 16,577 |
| 3–4 | 1,490 | 18,067 |
| 4–5 | 1,270 | 19,337 |
| 5–6 | 1,270 | 20,607 |
| 6–7 | 1,290 | 21,897 |
| 7–8 | 1,424 | 23,321 |
| 8–9 | 1,548 | 24,869 |
| 9–10 | 1,550 | 26,419 |
| 10–11 | 1,476 | 27,895 |
| 11–midnight | 1,342 | 29,237 |

## solution

This solution requires several steps. First, determine the cumulative hourly flow during the period. The answers are shown in the right column of Table 11.5. For one time period, the cumulative hourly flow is

$$\begin{bmatrix} \text{cumulative hourly} \\ \text{flow, 1–2 A.M.} \end{bmatrix} = V_{M-1} + V_{1-2} = 1,090 + 987 = 2,077 \text{ m}^3$$

Then, to determine the cumulative average influent volume (not listed in the table), divide the cumulative flow (listed in the table) by 24 hours:

$$\text{average flow} = \frac{\text{cumulative flow}}{24 \text{ hr}} = \frac{29,237 \text{ m}^3}{24 \text{ hr}} = 1,218 \text{ m}^3/\text{hr}$$

The solution now requires a graph of the cumulative influent volume and cumulative average influent volume. (The figure is not shown, so readers should consult Figure 11.7 and complete this on their own.) From this graph, the required equalized flow volume is found to be approximately 4,100 m³.

## 11.5 Primary Treatment

The goal of **primary treatment** is to remove solids through quiescent, gravity settling. Typically, domestic wastewater is held for a period of approximately 2 hr. **Settling tanks**, also referred to as **sedimentation tanks** or **clarifiers**, can be either rectangular or circular. During sedimentation, solids settle to the bottom of the tank, where they are collected as a liquid–solid sludge. Figure 11.8 shows a cross section of a circular clarifier.

Primary treatment removes about 60 percent of the suspended solids (TSS), 30 percent of the BOD, and 20 percent of the phosphorus (P). The BOD and phosphorus removed in this stage are primarily in the particulate phase (that is, part of the TSS). Any dissolved BOD, N, or P will pass through primary treatment and enter secondary treatment. Coagulants can be added to improve the removal of particulate matter. This may reduce the overall energy costs required during second treatment to biologically convert these particles to $CO_2$, water, and new biomass.

The clarified effluent that exits primary treatment is routed to secondary treatment, and the solids (the sludge) removed during settling are segregated for further treatment. Primary sludge is malodorous, may contain pathogenic organisms, and has high water content (perhaps less than 1 percent solids). These characteristics make it difficult to dispose of. Secondary clarifiers are designed to remove much smaller particles because, as we will explain in the next section, most of the particulate matter at this point in the treatment plant consists of microorganisms.

This chapter does not go into great detail on the design of settling tanks for wastewater treatment. Chapter 10 (Section 10.7) provided a detailed discussion of sedimentation theory and design principles, including how to use established **overflow rates** to size settling tanks.

**Primary Treatment**

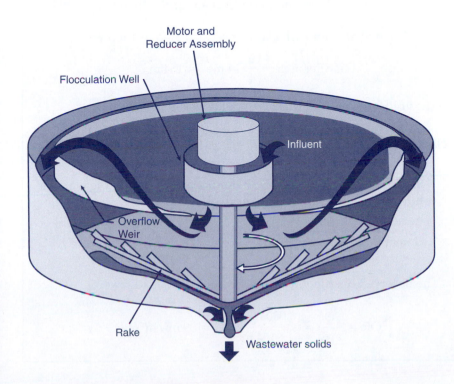

**Figure 11.8** Cross Section of a Circular Sedimentation Tank

## example/11.3 Sizing a Primary Settling Tank

A municipal wastewater treatment plant treats an average flow of 12,000 m³/day and a peak hourly flow of 30,000 m³/day. Two circular clarifiers are to be designed, using a depth of 4 meters and overflow rate of 40 m³/m²-day. Calculate the area, diameter, volume, and detention time required for each clarifier.

## solution

To calculate the surface area required for clarification, divide the average flow rate ($Q$) by the overflow rate (OR):

$$\text{total clarifier area} = \frac{Q}{\text{OR}} = \frac{12,000 \text{ m}^3/\text{day}}{40 \text{ m}^3/\text{m}^2\text{·day}} = 300 \text{ m}^2$$

Since there are two clarifiers, the area for each clarifier would be

$$\text{clarifier area} = \frac{300 \text{ m}^2}{2 \text{ clarifiers}} = 150 \text{ m}^2$$

The tank diameter can be calculated from the area as follows:

$$\text{clarifier diameter} = \sqrt{\frac{\text{clarifier area}}{\frac{\pi}{4}}} = \sqrt{\frac{150 \text{ m}^2}{\frac{\pi}{4}}} = 13.8 \text{ m}$$

The diameter will be rounded up to 14 m for the final design.

The actual area for each clarifier is calculated as follows:

$$\text{clarifier area} = \frac{\pi}{4}(14 \text{ m})^2 = 154 \text{ m}^2$$

The volume of each clarifier is calculated as follows:

$$\text{clarifier volume} = \text{Area} \times \text{Depth} = \left(\frac{\pi}{4}(14 \text{ m})^2\right) \times (4 \text{ m}) = 616 \text{ m}^3$$

To determine the hydraulic detention time, divide the clarifier volume by the flow rate ($Q$) to each clarifier:

$$\text{detention time} = \frac{\text{volume}}{Q} = \frac{616 \text{ m}^3 \times 24 \text{ hr/day}}{6,000 \text{ m}^3/\text{day}} = 2.46 \text{ hr}$$

The observed overflow rate (OR) is calculated as follows:

$$\text{OR} = \frac{Q}{\text{area}} = \frac{6,000 \text{ m}^3/\text{day}}{154 \text{ m}^2} = 39 \text{ m}^3/\text{m}^2\text{·day}$$

## example/11.3 Continued

Determine the detention time and overflow rate at peak flow:

$$\text{OR at peak flow} = \frac{(Q \text{ at peak flow})/2}{\text{area}}$$

$$= \frac{15{,}000 \text{ m}^3/\text{day}}{154 \text{ m}^2} = 97.4 \text{ m}^3/\text{m}^2 \cdot \text{day}$$

$$[\text{detention time at peak flow}] = \frac{\text{clarifier volume}}{Q \text{ at peak flow}}$$

$$= \frac{616 \text{ m}^3 \times 24 \text{ hr}/\text{day}}{30{,}000 \text{ m}^3/\text{day}/2} = 0.99 \text{ hr}$$

At the average flow, the calculated values of detention time and overflow rate are within the ranges we discussed in the previous chapter (Section 10.7). At peak flow, the calculated value of detention time is fine, but the overflow rate is slightly lower than desired. The final clarifier design may need to have an increased surface area to provide enough detention time for sufficient solids to settle.

## 11.6 Secondary Treatment

The wastewater that exits the primary clarifier has lost a significant amount of the particulate matter it contained, but it still has a high demand for oxygen due to an abundance of dissolved organic matter (measured as BOD). **Secondary treatment** (which is a form of biological treatment) utilizes microorganisms to decompose these high-energy molecules.

There are two basic approaches to biological treatment, differing in the manner in which the waste is brought into contact with the microorganisms. In *suspended-growth reactors*, the organisms and wastewater are mixed together, while in *attached-growth reactors*, the organisms are attached to a support structure, and the wastewater is passed over the organisms.

**Secondary Treatment**

### 11.6.1 SUSPENDED-GROWTH REACTORS: ACTIVATED SLUDGE

The most common biological treatment system is a **suspended-growth** system called the **activated-sludge** process. Effluent from the primary clarifier is routed to an **aeration tank** (also referred to as an **aeration basin**), usually by gravity, and mixed with a diverse mass of microorganisms comprising bacteria, fungi, rotifers, and protozoa. This mixture of liquid, waste solids, and microorganisms is called the **mixed liquor**. A measurement of TSS obtained from the aeration basin is termed the **mixed liquor suspended solids (MLSS)**, expressed in mg/L. Volatile suspended solids (VSS) can be used as a

**Figure 11.9** Food Web of the
Activated-Sludge Process

From Mihelcic (1999). Reprinted with permission of
John Wiley & Sons, Inc.

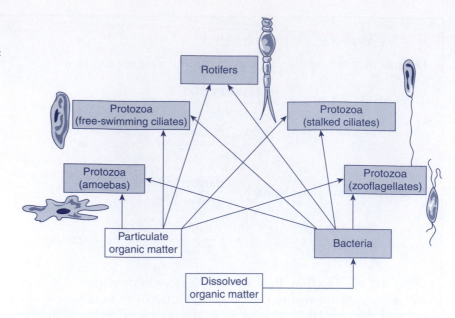

surrogate to describe the reactor's biomass. This is because most of
the solids are microorganisms that have a high carbon content in their
cell structure. Typically the volatile fraction **mixed liquor volatile
suspended solids (MLVSS)**, expressed in mg/L, is 60 to 80 percent of
the MLSS.

The food web of the activated-sludge process is shown in Figure 11.9.
The food web is somewhat truncated, both laterally (primary producers
are unimportant because the waste provides a source of organic matter)
and vertically (higher consumers are absent because the system is engineered to top out at a point where the remaining particulate matter is
easily removed by sedimentation).

Different organism groups predominate depending on the degree of
stabilization of the waste. At first, amoeboid and zooflagellate proto-
zoans dominate, utilizing the dissolved and particulate organic matter
initially present. Next zooflagellate and free-swimming ciliate proto-
zoans increase in numbers, feeding on developing populations of bacteria. Finally, stalked ciliates and rotifers become most abundant, feeding
from the surfaces of activated sludge floc. Plant operating practices—
such as the solids retention time (SRT), which will be discussed later—
dictate the degree of stabilization and thus the successional position of
the microbiology community. Molecular biology techniques are currently being used to further understand the unique microbial ecology of
wastewater treatment systems.

The majority of the BOD is degraded in the presence of oxygen, so
air is added to the reactor to supply oxygen, which must be transferred
to the aqueous phase. This requires energy inputs. In practice, dissolved-oxygen concentrations in the aeration tank are maintained at
1.5 to 4.0 mg/L, with 2 mg/L being a common value. Levels greater
than 4 mg/L do not significantly improve operation but raise operating costs because of the energy associated with forcing air into the system. Low oxygen levels can lead to *sludge bulking*, an abundance of
filamentous organisms with poor settling characteristics.

Bacteria are primarily responsible for assimilating the dissolved organic matter in wastewater, and the rotifers and protozoa are helpful in removing the dispersed bacteria, which otherwise would not settle out. This would cause the plant's effluent to not meet permit requirements for suspended solids. The energy derived from the decomposition process is primarily used for cell maintenance and to produce more microorganisms. Once most of the dissolved organics have been used up, the microorganisms are routed to the secondary (or final) clarifier for separation.

In the secondary clarifier, two streams are produced: (1) a clarified effluent, which is sent to the next stage of treatment (usually disinfection); and (2) a liquid–solid sludge largely comprising microorganisms (but perhaps 2 to 4 percent solids). Lying at the bottom of the secondary clarifier, without a food source, these organisms become nutrient-starved or *activated*. A portion of the sludge is then pumped to the head of the tank (**return activated sludge**), where the process starts all over again. The remainder of the sludge is removed from the system and is processed for disposal (**waste activated sludge**). As we will see in the next several sections, it is necessary to continuously waste sludge from the systems to balance the gains in biomass that occur through microbial growth.

**DESIGN OF THE ACTIVATED-SLUDGE SYSTEM** A set of equations allow sizing of the biological reactor and, importantly, understanding relationships in the activated-sludge system between microorganism concentration, solids removal, and influent organic matter. Figure 11.10 shows a schematic of the activated-sludge process with a control volume added for our mass balance.

Because the suspension of wastewater and suspended organisms appears mixed, the aeration basin is modeled as a *completed mixed flow reactor* (CMFR). In this reactor, the organisms convert dissolved organic matter (measured as CBOD and NBOD) into gaseous $CO_2$, water, nitrate, and particulate organic matter (more microorganisms). A settling tank (called the *secondary clarifier*) that follows the biological reactor captures the particulate matter (sludge).

The typical life of a microorganism in a wastewater plant is first to feed in the aeration basin for several hours (4–6 hrs, for example), then

**The National Small Flows Clearinghouse provides technical assistance to help small communities and homeowners with their wastewater treatment**
*http://www.nesc.wvu.edu/wastewater.cfm*

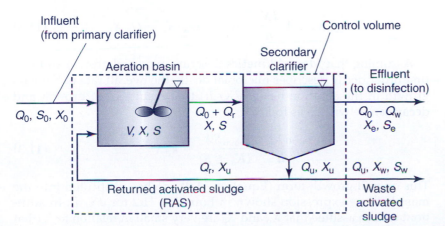

**Figure 11.10**  Schematic of the Activated-Sludge Process

From Mihelcic (1999). Reprinted with permission of John Wiley & Sons, Inc.

flow to the secondary clarifier for several additional hours, where the organism rests while it settles to the bottom of the tank. When the organisms are hungry again, they are recycled back to the biological reactor to seed the biological reactor with a metabolically active (hungry) group of organisms. For a microorganism in the process, this process is repeated several times (feed and rest, feed and rest, feed and rest, and so on).

Because the microorganism population is increasing due to the presence of substrate (CBOD and NBOD) and the plant operator needs to maintain a constant concentration of microorganisms in the aeration basin, some organisms must be removed from the secondary clarifier. These organisms are *wasted* from the process; hence the term **wasting of sludge** is used to describe the removal of solids from the activated-sludge system via the secondary clarifier.

To develop a master design equation, we will first set up and analyze two mass balances, conducted on dissolved organic matter (substrate) and solids (biomass). This analysis, when combined with our understanding of microbial growth, will allow us to determine the volume of the aeration basin. In all these expressions, $Q$ represents flow, expressed in $m^3/day$; $S$ is substrate concentration (usually measured as mg BOD or COD per L); $X$ is solids (biomass) concentration, measured as mg SS/L or mg VSS/L; and $V$ is the volume of the aeration tank, expressed in $m^3$. The subscripts refer to influent (o), effluent (e), recycle (r), underflow from clarifiers (u), and wasted solids (w).

## MASS BALANCE ON SOLIDS (BIOMASS)

A mass balance on microorganisms (solids) within the stated control volume (the dashed line in Figure 11.10) can be expressed as follows:

$$\begin{bmatrix} \text{biomass entering} \\ \text{aeration basin} \end{bmatrix} + \begin{bmatrix} \text{biomass produced} \\ \text{due to growth in} \\ \text{aeration basin} \end{bmatrix} = \begin{bmatrix} \text{biomass leaving} \\ \text{system} \end{bmatrix}$$

**(11.1)**

Using Equation 11.1, Figure 11.10, and the stated control volume, the mathematical expression that describes the *solids mass balance* is

$$Q_o X_o + V \frac{dX}{dt} = (Q_o - Q_w)X_e + Q_w X_w \qquad \textbf{(11.2)}$$

Assuming that **Monod kinetics** describe microbial growth and that first-order decay describes microbial death, the overall rate of biomass growth in the aeration basin ($dX/dt$) that results from growth and decay can be written as follows:

$$\frac{dX}{dt} = \frac{\mu_{max} S X}{(K_s + S)} - k_d X \qquad \textbf{(11.3)}$$

This overall growth term (Equation 11.3) can be substituted into the mass balance expression shown in Equation 11.2 for $dX/dt$. In addition, we can assume that $X_o$ and $X_e$ are very small in relation to $X$ (that is, $X_o \approx X_e \approx 0$). This is a good assumption because the concentration of biomass in the reactor, $X$, is maintained at approximately 2,000 to 4,000 mg TSS/L, while the influent solids into the aeration basin ($X_o$) might be 100 mg SS/L, and the effluent solids ($X_e$) less than 25 mg/L. (Note that $X$ is the MLSS or MLVSS, defined previously.)

After performing this substitution and making the assumptions of $X_o$ and $X_e$ being negligible in terms of solids concentration, the resulting expression can be rearranged to yield

$$\frac{\mu_{max}S}{K_s + S} = \frac{Q_w X_w}{VX} + k_d \qquad (11.4)$$

Equation 11.4 is important. It will be revisited in the next section, so readers should become familiar with it.

**MASS BALANCE ON SUBSTRATE (BOD)** Next, a mass balance is performed on the dissolved organic material (the BOD), which is substrate for the organisms. A mass balance on substrate (food) within the stated control volume (the dashed line in Figure 11.10) can be worded as follows:

$$\begin{bmatrix} \text{substrate entering} \\ \text{aeration basin} \end{bmatrix} - \begin{bmatrix} \text{substrate consumed} \\ \text{by microorganisms} \end{bmatrix} = \begin{bmatrix} \text{substrate leaving} \\ \text{system} \end{bmatrix}$$

$$(11.5)$$

Using Equation 11.5, Figure 11.10, and the stated control volume, we can write the mathematical expression that describes the *substrate mass balance*:

$$Q_o S_o - V\frac{dS}{dt} = (Q_o - Q_w)S + Q_w S \qquad (11.6)$$

Here the effluent substrate, $S_e$, and the substrate in the waste sludge, $S_w$, are assumed to equal $S$ ($S = S_e = S_w$). This should make sense because the secondary clarifier's purpose is to remove solids, not to biologically transform substrate to carbon dioxide and water.

The yield coefficient ($Y$) relates the change in substrate concentration to the change in biomass concentration. Thus, the change in substrate concentration with time, $dS/dt$, can be written as follows:

$$\frac{dS}{dt} = \left(\frac{1}{Y}\right)\left(\frac{\mu_{max}S}{K_s + S}\right)X \qquad (11.7)$$

This expression for $dS/dt$ can be substituted into the substrate mass balance (Equation 11.6). Rearranging the overall expression then results in

$$\frac{\mu_{max}S}{K_s + S} = \frac{Q_o Y}{VX}(S_o - S) \qquad (11.8)$$

## Table / 11.6

### Explanation of Terms Used in Expression for Activated-Sludge Design (Equation 11–9)

| Term(s) in Final Design Expression | Explanation |
|---|---|
| Yield coefficient ($Y$) and decay coefficient ($k_d$) | Biokinetic coefficients, which are either measured or estimated (see Chapter 5 for example values) |
| Substrate concentration in reactor ($S$) | Plant effluent concentration, which is set by the state through the NPDES permitting process |
| Initial substrate concentration entering reactor ($S_o$) and flow entering reactor ($Q_o$) | Independent parameters that are a function of community demographics such as population, community wealth, water conservation measures, and commercial and industrial activity in a community |
| Rate of wasting sludge ($Q_w$) and concentration of solids in the wasted sludge ($X_w$) | Items a plant operator can control by wasting sludge, especially the rate at which solids are removed (wasted) from the system |

Note that the left side of the two final rearranged expressions obtained from the solids (Equation 11.4) and substrate mass balances (Equation 11.8) are the same. Thus, these two expressions can be set equal to provide a design expression:

$$\frac{Q_w X_w}{V X} = \frac{Q_o Y}{V X}(S_o - S) - k_d \tag{11.9}$$

Equation 11.9 can be used to solve for the volume of the aeration basin ($V$). This is because all the other terms either can be measured or are fixed. Table 11.6 provides a real-world explanation of the terms in this activated-sludge design expression (Equation 11.9). However, one problem arises with using Equation 11.9 to solve for $V$; the volume term appears on both sides of the equation. Fortunately, as we will see in the next section, the terms on the left side of Equation 11.9 can be combined into one term, referred to as the solids retention time (SRT).

SOLIDS RETENTION TIME The term on the left side of Equation 11.9 is an important expression for design and operation of an activated-sludge plant. It is the inverse of a term referred to as the **solids retention time (SRT)**. SRT is sometimes also referred to as **sludge age** and **mean cell retention time (MCRT)**. SRT is defined as

$$\boxed{SRT = \frac{V X}{Q_w X_w}} \tag{11.10}$$

If you look closely at the units of this expression, you will see that it has units of time (typically days). The sludge age in most treatment plants typically ranges from 2 to 30 days.

The SRT is not the hydraulic retention time ($V/Q$). Sludge age refers to the average time a microorganism spends in the activated sludge process before it is expelled, or wasted, from the system.

Remember that the organisms feed in the aeration basin and then rest in the secondary clarifier until they are recycled back into the aeration basin to feed, and then are sent back to the secondary clarifier to rest. This process is repeated many times until the organism is finally removed from the process, or wasted. The SRT thus refers to the number of days that an average microorganism undergoes this feed-and-rest cycle.

Equation 11.10 can be substituted into Equation 11.9 to yield our final design equation:

$$\frac{1}{\text{SRT}} = \frac{Q_o Y}{VX}(S_o - S) - k_d \qquad (11.11)$$

Equation 11.11 can be used to size the aeration basin (solve for $V$) for a given (or range of) SRT.

## example/11.4 Design of the Aeration Basin Based on Solids Retention Time

Given the following information, determine the design volume of the aeration basin and the aeration period of the wastewater for an activated-sludge treatment process: population = 150,000; flow rate is $33.75 \times 10^6$ L/day (equals 225 L/person-day); and influent $BOD_5$ concentration is 444 mg/L (note this is high-strength wastewater). Assume that the regulatory agency enforces an effluent standard of $BOD_5 = 20$ mg/L and a suspended-solids standard of 20 mg/L in the treated wastewater.

A wastewater sample is collected from the biological reactor and is found to contain a suspended-solids concentration of 4,300 mg/L. The concentration of suspended solids in the plant influent is 200 mg/L, and that which leaves the primary clarifier is 100 mg/L. The microorganisms in the activated-sludge process can convert 100 g $BOD_5$ into 55 g biomass. They have a maximum growth rate of 0.1/day and a first-order death rate constant of 0.05/day, and they reach half of their maximum growth rate when the $BOD_5$ concentration is 10 mg/L. The design solids retention time is 4 days, and sludge is processed on the belt filter press every 5 days.

### solution

1. *For the aeration basin volume.* This problem provides a lot of information. To solve for the aeration basin volume ($V$), you need to know what information is important and what is not required. Look closely at Equation 11.11. Here, $S_o$ equals the substrate (or $BOD_5$) entering the biological reactor, so assume that some $BOD_5$ is particulate and is removed in the primary clarifier. Assuming

## example/11.4 Continued

that 30 percent of the plant influent $BOD_5$ is removed during primary sedimentation, this means that $S_o = 0.70 \times 444\,\text{mg/L} = 310\,\text{mg/L}$. Accordingly,

$$\frac{1}{\text{SRT}} = \frac{Q_o Y}{VX}(S_o - S) - k_d$$

$$\frac{1}{4\,\text{days}} = \frac{\left(33.75 \times 10^6 \dfrac{\text{L}}{\text{day}}\right) \times \left(0.55 \dfrac{\text{gm SS}}{\text{gm BOD}_5}\right)}{V \times \left(4,300 \dfrac{\text{mg SS}}{\text{L}}\right)}$$

$$\times \left(310 \frac{\text{mg}}{\text{L}} - 20 \frac{\text{mg}}{\text{L}}\right) - \frac{0.05}{\text{day}}$$

Solve for $V = 4.173 \times 10^6\,\text{L}$.

2. *For the aeration period.* The plant's aeration period is the number of hours that the wastewater is aerated during the activated-sludge process. This equals the hydraulic detention time of the biological reactor:

$$\theta = \frac{V}{Q} = \frac{4.173 \times 10^6\,\text{L}}{33.75 \times 10^6 \dfrac{\text{L}}{\text{day}}} = 0.12\,\text{day} = 3\,\text{hr}$$

## example/11.5 Use of Solids Retention Time to Calculate Solids Processing

Using data provided from Example 11.4, how many kg of primary and secondary dry solids need to be processed daily from the treatment plant?

## solution

Assume that the amount of solids processed from the primary sedimentation tanks equals the difference in suspended-solids concentrations (influent minus effluent) measured across the sedimentation tanks multiplied by the plant flow rate:

$$33.75 \times 10^6 \frac{\text{L}}{\text{day}} \times \left(200 \frac{\text{mg TSS}}{\text{L}} - 100 \frac{\text{mg TSS}}{\text{L}}\right) \times \left(\frac{\text{kg}}{1,000,000\,\text{mg}}\right)$$

$$= 3,375\,\text{kg primary solids per day}$$

We are not provided with the concentration difference of suspended solids across the secondary sedimentation tanks, so we cannot determine the amount of secondary solids produced daily in the same manner that we used for primary solids. However, careful examination of the expression of solids retention time (SRT = 4 days) shows the term $Q_w X_w$ equals the answer. Therefore:

$$4 \text{ days} = \frac{VX}{Q_w X_w} = \frac{4.173 \times 10^6 \text{ L} \times \left( 4,300 \frac{\text{mg SS}}{\text{L}} \right)}{Q_w \times X_w}$$

Solve for $Q_w X_w$, which equals 4,486 kg secondary dry solids per day.

## RELATING SOLIDS RETENTION TIME TO MICROBIAL GROWTH RATE

You may have already noted that the inverse of SRT has the units of $\text{day}^{-1}$. Remembering from Chapter 5 the definition of the *specific growth rate* of microorganisms,

$$\frac{dX}{dt} = \mu X \tag{11.12}$$

This expression can be arranged to solve for the specific growth rate, $\mu$:

$$\mu = \frac{dX/dt}{X} \tag{11.13}$$

Equation 11.13 shows that the specific growth rate (units of $\text{day}^{-1}$) equals the mass of biomass produced in the aeration basis (kg MLSS produced per day) divided by the mass of biomass present in the reactor (kg MLSS).

Remember that SRT refers to the average time a microorganism spends in the activated-sludge process before it is expelled, or wasted, from the system. If a treatment plant operator wants to maintain the same concentration of biomass in the biological reactor (i.e., $X$), the operator would have to waste the same volume of solids per day ($Q_w X_w / V$) that are produced by microbial growth ($dX/dt$).

Equation 11.13 and Equation 11.10 are thus related. In fact, for a completely-mix activated sludge process, the SRT (which is controlled by wasting solids) is the inverse of the average of the specific growth rate of the microorganisms:

$$\boxed{\frac{1}{\text{SRT}} = \mu} \tag{11.14}$$

The relationship shown in Equation 11.14 is important for a design engineer and plant operator because it tells us there is a *critical* value of SRT. Below this **critical SRT** value (sometimes referred to as $\text{SRT}_{min}$) the microbial cells in the activated-sludge process will be **washed out** or removed from the system faster than they can reproduce. This would not be good, because if specific types of microorganisms wash out of the system, the activated-sludge process will lose its ability to degrade particular pollutants. For example, washing out of nitrifying organisms will result in poor removal of ammonia nitrogen, and washing out of heterotrophic organisms will result in poor removal of BOD.

Fortunately, the $SRT_{min}$ can be approximated as

$$\frac{1}{SRT_{min}} \approx \mu_{max} - k_d \tag{11.15}$$

$\mu_{max}$ and $k_d$ were defined previously in Chapter 5. Never design a biological treatment process where the SRT is equal to the $SRT_{min}$! In fact, many treatment plants are designed for an SRT that is 2 to 20 times greater than the $SRT_{min}$.

FOOD-TO-MICROORGANISM RATIO The rate of food introduction (BOD loading) is largely fixed by the flow rate ($Q_o$) and BOD ($S_o$) of the influent. The size of the microbial population is equal to the product of the MLSS (or MLVSS) concentration in the biological reactor ($X$) and the reactor volume ($V$). Earlier we stated that operating experience in waste treatment plants suggests that MLSS concentrations in the reactor should be maintained at levels ranging from 2,000 to 4,000 mg/L. Too low concentrations (less than 1,000 mg/L) may lead to poor settling, and too high concentrations (greater than 4,000 mg/L) may result in solids loss in the secondary clarifier overflow and excessive oxygen requirements.

Another key process design parameter (besides SRT) is referred to as the **food-to-microorganism (F/M) ratio**. It also can be used to estimate the required tank volume. Essentially a feeding rate, the food-to-microorganism (F/M) ratio is equivalent to the BOD loading rate divided by the mass of MLSS in the reactor. The **BOD loading rate** (kg BOD/$m^3$-day) is the mass of food that enters the biological reactor per day divided by the volume of the reactor.

Using the terminology from Figure 11.10

$$F/M = \frac{S_o Q_o}{XV} \tag{11.16}$$

F/M has units of kg BOD/kg MLSS-day. Remembering from Chapter 4 the definition of **hydraulic retention time** ($\theta = V/Q$), the food-to-microorganism ratio can also be written as

$$F/M = \frac{S_o}{\theta X} \tag{11.17}$$

Referring back to Table 11.6, $S_o$ and $Q_o$ are largely fixed by local demographics such as population, wealth, and the mix of residential and commercial establishments in a community. Note that water conservation measures will not reduce the mass of food entering the

## example/11.6 Calculating the F/M Ratio

Determine the F/M ratio (in units of lb. $BOD_5$/lb. MLSS-day), using data provided from Example 11.4.

### solution

Remember by definition,

$$F/M = \frac{Q \times S_o}{X \times V} = \frac{\left(33.75 \times 10^6 \dfrac{L}{day}\right) \times \left(310 \dfrac{mg}{L}\right)}{\left(4,300 \dfrac{mg\ SS}{L}\right) \times (4.173 \times 10^6\ L)}$$

$$= 0.58 \frac{kg\ BOD_5}{kg\ MLSS\text{-}day}$$

$$= 0.58 \frac{lb.\ BOD_5}{lb.\ MLSS\text{-}day}$$

Note that converting units of F/M ratio from metric to English units requires no conversion factor because the mass unit is in both the numerator and denominator. Also, be careful in your units for F/M, becase the denominator can have units of MLSS or MLVSS.

system. Water conservation will reduce $Q_o$, but the resulting $S_o$ will increase proportionally. Remember also that the concentration of microorganisms in the biological reactor ($X$) is controlled by how much solids the operator wastes. Thus, a particular reactor volume ($V$) can be selected to achieve the desired F/M ratio.

Equation 11.16 shows the F/M ratio is really a feeding rate. The lower the F/M ratio, the lower the feeding rate, the hungrier the microorganisms, and the more efficient the removal. Likewise, if the SRT is lowered, the operator will be building up the solids inventory (increasing $X$). Examination of Equation 11.16 shows that in this case, with lower SRT and higher $X$, the F/M ratio will increase.

Our examination of F/M will not go into additional detail. However, readers need to understand that SRT and F/M are related (Table 11.7). This relationship includes the efficiency of the BOD removal and some of microbial parameters that were discussed in Chapter 5 ($Y$ and $k_d$). Higher SRTs equate to lower F/M, and lower SRTs equate to higher F/M. This should make sense. At a higher SRT, more cells are being wasted from the system, so the concentration of microorganisms in the biological reactor ($X$) will decrease. Because the incoming food ($S_o$ times $Q_o$) is not controlled by the plant operator, examination of Equation 11.16 shows that the F/M ratio would decrease.

At low F/M ratios, the microorganisms are maintained in the **death growth phase** or **endogenous growth phase**, meaning they are starved

### Table / 11.7

**Relationship of Solids Retention Time (SRT) and Food-to-Microorganism (F/M) Ratio**

| SRT (days) | F/M (gm BOD/gm VSS-day) |
|---|---|
| 5–7 | 0.3–0.5 |
| 20–30 | 0.10–0.05 |

SOURCE: Values obtained from Tchobanoglous et al. (2003).

and thus very efficient at BOD removal. Because $S_o$ is relatively constant for domestic wastes and because there are limits on the levels of $X$ that a reactor can support, maintenance of a low F/M ratio requires either a very small flow or a very large tank volume. In either case, this leads to a long hydraulic residence (aeration) time.

Operating an activated-sludge plant at low F/M ratios is termed **extended aeration**. The cost of operation and maintenance is high for large tank volumes, so extended aeration is largely limited to systems with small organic loads (for example, at mobile-home parks and recreational facilities).

At high F/M ratios, the microorganisms are maintained in the exponential-growth phase. These organisms are more food saturated, meaning there is an excess of substrate, so BOD removal is less efficient. This approach is termed *high-rate activated sludge*. In this approach, higher MLSS concentrations are employed, so a shorter hydraulic residence time is achieved, and smaller aeration tank volumes are required.

In addition to influencing BOD removal efficiency, the selection of an F/M ratio affects the settleability of the sludge and thus the efficiency of TSS removal. In general, as the F/M ratio decreases, the settleability of the sludge increases. Starving microorganisms flocculate and thus settle well, while those maintained at high F/M ratios form buoyant filamentous growths, which settle poorly, a condition termed *sludge bulking*.

## SETTLING CHARACTERISTICS OF ACTIVATED SLUDGE

Design of settling clarifiers is similar to that of primary clarifiers, except that the hydraulic detention rates and overflow rates reflect

### example/11.7 Relating F/M to Aeration Tank Volume

The suspended-solids concentration is 220 mg/L in the plant influent; 4,000 mg/L in the primary sludge; 15,000 mg/L in the secondary sludge; and 3,000 mg/L exiting the aeration basin. The concentration of total dissolved solids in the plant influent is 300 mg/L, and the concentration of total dissolved solids exiting the aeration basis is 3,300 mg/L. The $BOD_5$ is 150 mg/L measured after primary treatment and 15 mg/L exiting the plant. Total nitrogen levels in the plant are approximately 30 mg N/L.

If the F/M ratio is 0.33 lb. $BOD_5$/lb. MLSS-day, estimate the hydraulic retention time of the aeration basins if the total plant flow is 5 million gal./day.

### solution

Note that this problem statement provides a lot of extra material, so readers must understand the order of various unit processes in a wastewater treatment plant, as well as the definition of F/M ratio. The mass of food that the microorganisms see equals the plant flow rate multiplied by the concentration of $BOD_5$ exiting the primary sedimentation tank (which thus enters the aeration basin). This value is $S_o$.

$$F/M = 0.33 \frac{\text{lb. BOD}_5}{\text{lb. MLSS-day}} = \frac{Q \times S_o}{V \times X} = \frac{Q \times \left(150 \frac{\text{mg}}{\text{L}}\right)}{V \times \left(3{,}000 \frac{\text{mg MLSS}}{\text{L}}\right)}$$

$Q/V = 6.6/\text{day}$, and because the hydraulic detention time equals $V/Q$, the detention time ($\theta$) equals 0.15 day or 3.6 hr. The size of the aeration basins can then be found from knowledge of the design flow (or $Q$) of the plant: $V = Q \times \theta$.

the fact that the particles are smaller than in primary settling. The efficiency of settling in the secondary clarifier is influenced by the degree to which the MLSS flocculates (that is, the ability of the MLSS to stick together to form a larger mass of particles). This **flocculation** capacity is lowest in exponential-growth phase populations (high F/M ratio), increases in declining-growth phase populations (inter-mediate F/M ratio), and is highest in endogenous-growth phase pop-ulations (low F/M ratio).

There are two reasons for this behavior. First, flocculation is aided by the presence of microbial-produced slime (polysaccharide gums), which helps the particles stick together. Slime is produced through the attrition of slime layers on the zoogleal masses. Slime layers are most abundant in populations grown under the endogenous phase and are least abundant in exponential-phase populations. Second, as we described in the previous chapter, flocculation is greatest under condi-tions where particles can be easily brought together. Inactive, endoge-nous-phase cells behave as simple colloids and flocculate well. Active, highly motile particles characteristic of exponential-phase populations (high F/M ratio) tend to flocculate poorly.

If you placed some activated sludge in a 1,000 mL graduated cylin-der and watched the sludge settle over time, you would observe four zones of settling. Near the top of the cylinder, you would observe **dis-crete settling**, which is typical of low particle concentrations. In this zone, the particles settle alone according to **Stokes' law**. Beneath this, you would observe **flocculent settling**. Here the particles coalesce dur-ing sedimentation, and the sample is still relatively dilute. The third zone is called **hindered settling**. Here, as the solids concentrations increases, interparticle forces hinder settling of neighboring particles. You can actually observe the particles settling as a unit, and you observe a solid–liquid interface. The final zone, called **compression settling**, occurs near the bottom of the cylinder and is visible as time passes. Here the concentration of solids is now large, so the downward movement of solids is opposed by the upward movement of water. In

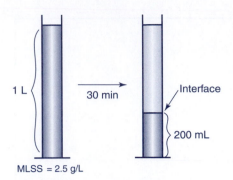

1 L

30 min

Interface

200 mL

MLSS = 2.5 g/L

**Figure 11.11 Determination of Sludge Volume Index** The sludge volume index (SVI) is measured to determine the settleability of the sludge. The SVI is the volume (in mL) occupied by 1 g MLSS (dry weight) after settling for 30 min in a 1,000 mL graduated cylinder. In the situation depicted here, 2.5 g MLSS occupy 200 mL of volume after 30 min settling, so the SVI is 80.

| Table / 11.8 | |
| --- | --- |

| **Interpretation of Sludge Volume Index (SVI)** | |
| --- | --- |
| Value of Sludge Volume Index (SVI) | Settleability of the Sludge |
| 0–100 | Good |
| 100–200 | Acceptable |
| >200 | Poor |

this case, no settling can occur until water is compressed from the sludge.

A test termed the **sludge volume index (SVI)** is performed at the treatment plant to determine the settleability of the sludge. The SVI is the volume (in mL) occupied by 1 g MLSS (dry weight) after settling for 30 min in a 1,000 mL graduated cylinder (see Figure 11.11). The units of the SVI are mL/g.

Another way to think of the SVI is that it is the percent volume occupied by sludge in an MLSS sample. It thus tells you how well your sludge is currently settling. Table 11.8 provides a method to interpret the value of SVI in terms of how well sludge will settle.

Good settling characteristics (low SVI) are one indication of a properly operating wastewater treatment plant. A person at such a plant who collects a sample from the biological reactor would first observe a moderate amount of brown **foam** in the sample. After the sample has been allowed to sit for a few minutes, the resulting supernatant should appear clear and have a low BOD. Further examination of the contents of the biological reactor under a microscope would show a well-flocculating sludge with large numbers of free-swimming ciliates and bacteria.

During extended aeration, **endogenous respiration** takes place, because there is not enough food (BOD) to support the population of microorganisms. The microorganisms will thus begin to utilize food they have stored in their cellular structure, and some organisms will begin to die. A dead organism's cells begin to lyse and subsequently become food for other, higher organisms such as stalked ciliates and rotifers. During endogenous respiration, the biological solids appear very dense (so the SVI is very low).

A common occurrence in an activated-sludge plant is the site of a poorly settling brown foam that is high in TSS. The microorganisms found in this foam have hydrophobic cell walls. Most of these organisms belong to a group called **nocardioform** organisms. Because of the hydrophobic cell wall, air bubbles from the aeration process can attach to the microorganism and the associated biological floc. As a result, the biological floc rises to the surface. At the surface, the air bubble will eventually collapse but leave behind the floated solids, which appear as foam.

Fortunately several design and operating techniques can eliminate or reduce the formation of foam. For example, treatment plants need to watch their inputs of toxic chemicals that can change the ecology of the reactor and may favor the presence of these dispersed bacteria. Also, activated sludge plants can be run at low SRT in an attempt to wash out the nocardioform organisms. Recycling foam that contains these organisms back to the front end of the plant should be discouraged. In addition, it has been found that activated-sludge systems containing subsurface aeration basin draw-off and secondary scum baffles have higher nocardioform levels than systems with aeration basin draw-off and no clarifier surface baffles. Apparently, reactor features that encourage the selection of dispersed nocardioforms over a clumped and settleable biological floc result in greater chance for foaming (Jenkins, 2007).

**Examples of Process Modifications Made to the Conventional Activated-Sludge System**

| Process | Description |
| --- | --- |
| Conventional activated sludge | Primary effluent and return activated sludge (RAS) are introduced at head of aeration basin. The aeration is provided in a nonuniform manner over the length of the tank as more aeration is required at the beginning of the tank, since the organic loading is higher there because the BOD is removed along the length of the aeration basin. |
| Step feed aeration | Modification where primary clarifier effluent is introduced at several points along the beginning of the aeration basin. The peak oxygen demand is thus more evenly distributed throughout the aeration tank. Aeration is uniform along the length of the aeration basin. |
| Contact stabilization | The aeration basin is separated into a stabilization zone followed by a small contact zone. Primary clarifier effluent is routed to the contact zone first. Return activated sludge is recycled back into the stabilization zone. |
| Extended aeration | Similar to conventional activated sludge except primary clarifier is usually eliminated, SRT is very long (20–30 days), and hydraulic detention times are close to 1 day. Used primarily by smaller communities, schools, resorts. |
| Oxidation ditch | Oval reactor where wastewater moves at relatively high velocities. Return activated sludge is recycled back to beginning of the reactor. |
| Sequencing batch reactor | Fill and draw reactors where a minimum of two reactors are used. While one reactor is being filled, the other reactor is overseeing the biological reactions, settling of solids, and removal of settled wastewater. |

## 11.7 Modifications to the Activated-Sludge Process

Operation of the activated-sludge process at midrange F/M ratios with microorganisms in the declining-growth phase is termed **conventional activated sludge**. This option offers a balance between removal efficiency and cost of operation. Various reactor configurations are available (see Table 11.9), each with its own set of advantages and disadvantages. The two basic types are plug flow and completely mixed flow reactors. Plug flow reactors offer higher treatment efficiency than completely mixed flow reactors but are less able to handle spikes in the BOD load. Other modifications of the process are based on the manner in which waste and oxygen are introduced to the system.

### 11.7.1 MEMBRANE BIOREACTORS (MBRs)

One of the fastest-growing segments of biological wastewater treatment processes is the use of **membrane bioreactors (MBRs)**. MBRs combine the suspended-growth activated-sludge process previously described with the microfiltration membrane process (discussed in

**Figure 11.12** **Two Bioreactor Membrane Processes** (a) Membrane bioreactor where the membranes are immersed into the aeration basin. (b) Membrane bioreactor where the membranes are external to the aeration basin.

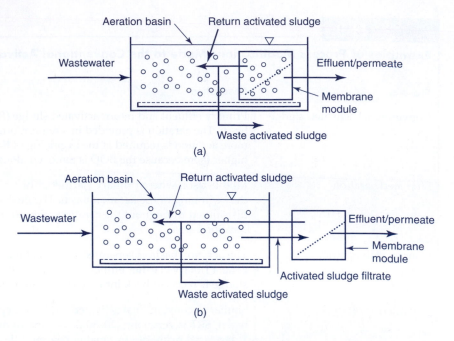

## Wastewater Infrastructure on the U.S. Mexican Borderlands

http://www.epa.gov/owm/mab/mexican/mxsumrpt.htm

Chapter 10). Figure 11.12 displays two types of process configurations: (1) *the submerged-membrane process* and (2) *the external-membrane process*.

For both process flow configurations, the mixed liquor in the aeration basin is filtered through the membrane, separating the biosolids from the effluent water. In the submerged-membrane process, a vacuum of less than 50 kPa is applied to the membrane that filters the water through the membrane while leaving the biosolids in the aeration basin. For the external membrane system, a pump is used to pressurize the mixed liquor at less than 100 kPa, and the water is filtered through the membrane while the biosolids are sent back to the aeration basin. In both systems, wasting of solids is done directly from the aeration basin.

MBRs have several advantages over conventional activated-sludge systems (Tchobanoglous et al., 2003). The use of MBRs eliminates the need for secondary clarifiers or filters. The MBRs can operate at much higher MLSS loadings, which decreases the size of the aeration basin; can operate at longer SRTs, leading to less sludge production; and can operate at lower DO concentrations with the potential of nitrification/denitrification at long SRTs. The MBR requires about 40 to 60 percent less land use footprint than a conventional activated-sludge plant, which is especially important in urban areas, where land is a premium and populations are expanding. Also, the effluent quality is much better in terms of BOD, low turbidity, TSS, and bacteria.

In addition, MBRs do not have the problematic issues such as sludge bulking, filamentous-organism growth, and pinpoint floc, sometimes experienced in many conventional activated-sludge facilities. However, they do contain more dispersed microorganisms, and the resulting biological flocs tend to be smaller than obtained from gravity settling. For an example of a facility that has realized some of these benefits, see Figures 11.13 and 11.14 and Table 11.10.

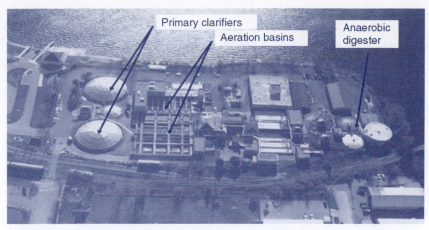

**Figure 11.13** **Aerial View of the Traverse City, Michigan, Wastewater Treatment Facility, Highlighting Some Major Unit Processes** The plant employs one-membrane bioreactors. The plant consists of 8 trains containing 13 cassettes, and each cassette consists of 32 membrane modules. The plant was initially designed to treat 19,000 m$^3$/day (maximum monthly flow) and was upgraded with the bioreactor membrane system to treat 32,000 m$^3$/day (maximum monthly flow) to 68,000 m$^3$/day (peak-day flow) of wastewater. With addition of the membrane bioreactors, the physical footprint of the plant was decreased by approximately 40 percent, as the two secondary clarifiers were no longer required. Photo courtesy of David Hand.

MBRs do have some disadvantages, including (1) higher capital costs; (2) the potential for short membrane life due to membrane fouling and higher energy costs resulting from module aeration; and (3) the operational issue that membranes need to be cleaned on a cyclic basis. In addition, a more skilled operation staff is needed in case of problems with plant disturbances, which can upset the plant operation very quickly.

**Figure 11.14** **Membrane Cassette Being Lowered into the Aeration Basin at the Traverse City, Michigan, Wastewater Plant**

Photo courtesy of David Hand.

**Comparison of Effluent Characteristics of Membrane Bioreactor (MBR) and Conventional Wastewater Treatment at the Traverse City (Michigan) Wastewater Treatment Plant**

| Parameter | Performance of Conventional Activated-Sludge Plant Before Installation of Membrane Bioreactor | After Installation of Membrane Bioreactor |
|---|---|---|
| $BOD_5$ | 2–5 mg/L | <2 mg/L |
| TSS | 8–20 mg/L | <1 mg/L |
| $NH_3$-N | <0.03–20 mg/L | <0.03 mg/L |
| $PO_4$-P | 0.6–4.0 mg/L | <0.5 mg/L |
| Fecal coliforms | 50–200 cfu (colony forming units)/100 ml | <1 cfu/100 ml |

## 11.8 Attached-Growth Reactors

In contrast to suspended-growth treatment systems, discussed in the previous section, microorganisms can also be attached (or fixed) to a surface during biological treatment. Table 11.11 provides an overview of specific types of **attached-growth** processes, identifying some advantages and disadvantages associated with each of them.

In the **trickling filter**, primary effluent is "trickled" over and percolates through a 1 m to 3 m deep tank filled with stone media, slag, or plastic media (termed the filter bed). An active biological growth forms on the solid surfaces (the active biofilm thickness ranges from 0.07 mm

**Attached-Growth Processes Used to Treat Wastewater**   The three processes are configured in several specific types of reactors, which have different advantages and disadvantages.

| Attached-Growth Processes | Specific Type | Advantages and Disadvantages |
|---|---|---|
| Nonsubmerged attached-growth systems | Trickling filters; biotowers; rotating biological contactors (RBCs) | Less energy required; simpler operation and fewer equipment maintenance needs than suspended-growth systems; better recovery from shock toxic loads than suspended-growth systems; difficult to accomplish biological N and P removal compared with suspended designs |
| Suspended-growth processes with fixed-film packing | Submerged RBCs; aeration basins with submerged packing materials | Increased treatment capacity; greater process stability; reduced solids loadings on the secondary clarifier; no increase in operation and maintenance costs |
| Submerged attached-growth aerobic processes | Upflow and downflow packed-bed reactors and fluidized-bed reactors that do not use secondary clarification | Small physical footprint with an area requirement one-fifth to one-third of that needed for activated-sludge treatment. |

to 4.0 mm) and dissolved organic matter (BOD) diffuses from the water phase into the biofilm as the wastewater trickles down. Design of trickling filters is based on a maximum allowable hydraulic loading (5–10 $m^3/m^2$-d) and a maximum organic loading (250–500 g BOD $m^3$/d). The organic loading must be limited so that the substrate uptake capabilities of the system are not saturated (which would result in poor removal efficiency). The hydraulic load must be limited so that the filter is not flooded. In this case, ponding could occur, which would limit oxygen transfer into the system.

---

### Box / 11.2    Biofiltration for Control of Odorous Air

Emission and control of odorous gases is a growing concern in municipal wastewater collection and treatment and some industrial processes. The odor is primarily caused by hydrogen sulfide ($H_2S$) and other reduced sulfur compounds such as methyl mercaptan and dimethyl sulfide. The presence of these chemicals can result in complaints from community members who live near a treatment plant (or pump station). These chemicals may also damage human health and corrode infrastructure and equipment.

This problem can be addressed with **biofiltration** units (see Figure 11.15), which can be classified into two categories: *biotrickling filters* and *biofilters*. Biofiltration utilizes microorganisms that are attached to a packing material (synthetic or natural) to break down pollutants in a contaminated airstream that is passed through the packing material. If water is added consistently, the system is called a biotrickling filter. If the amount of water applied is minimal, the system is called a biofilter. Packing media typically consist of a combination of wood chips and compost, lava rock, or synthetic packing material. Biofiltration units can also be constructed above or below the ground surface.

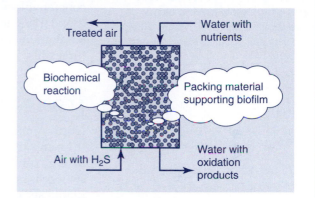

**Figure 11.15   Biofiltration of Odorous Air**
Biofiltration units consist of packing media in which odorous air is passed through the column. In this case, the odorous air contains hydrogen sulfide ($H_2S$). Water is either added intermittently (as with a biofilter) or consistently (as in a biotrickling filter). The microorganisms oxidize reduced sulfur to sulfate ($SO_4^{2-}$) and in the process produce acidity ($H^+$). See Martin et al. (2004) for further information on factors influencing design and performance.

---

## 11.9   Removal of Nutrients: Nitrogen and Phosphorus

Secondary treatment is sometimes inadequate to protect the receiving water. Additional removal of pollutants, especially nitrogen (N) and phosphorus (P), is accomplished through a variety of physical, chemical, and biological processes collectively termed advanced or **tertiary wastewater treatment.** As we will see, advanced removal of nutrients such as N and P is now being incorporated into the existing biological process.

Today suspended- or attached-growth biological systems can treat inorganic nitrogen down to 1 to 1.5 mg/L and phosphorus to as low as 0.1 mg/L (after filtration). Dissolved organic nitrogen concentrations will still remain at the range of 0.5 to 1.5 mg/L. An excellent review of

### Class Discussion

Think about the advantages and disadvantages of separating urine from feces at the household and community scale. What environmental, social, and economic issues would need to be addressed if the established methods to collect and treat wastewater were changed?

the history of biological removal of nitrogen and phosphorus is available elsewhere (Barnard, 2006).

A significant fraction of phosphorus was removed in some parts of the world via source reduction by eliminating it from soaps and detergents. Interestingly, approximately 90 percent of the total nitrogen and 75 percent of the total phosphorus discharged from a household into a sanitary sewer is found in urine. Much of this is diluted by the excessive water usage found in North American homes and commercial districts. Some countries are now proposing the development of toilets that separate urine from feces and dual-sewer systems that handle each waste stream.

Composting toilets can also be designed to separate urine from feces. They are used throughout the world and can be constructed from concrete block or modular plastic units that resemble ceramic toilets (Box 11.1). The lead author of this book has a working composting toilet. Both the urine and composted feces can be used to amend agricultural soils.

### 11.9.1 NITROGEN

North American treatment plants typically receive influent nitrogen in the range of 25 to 40 mg N/L. This value can reach the low hundreds of mg/L in areas where wastewater is primarily sewage and the sewage is not diluted by excessive water use. At a minimum, treatment plants seek ammonia removal, but it is becoming more common to totally remove nitrogen from a wastewater effluent.

Approximately 10 percent of the nonwater portion of a microbial cell is nitrogen. Therefore, the growth of biological solids removes some nitrogen from the dissolved to the particulate phase. However, this removal of nitrogen is nowhere sufficient to protect receiving water bodies from pollution.

**Nitrification**, the conversion of ammonia ($NH_4^+$) to nitrite ($NO_2^-$) and then nitrate ($NO_3^-$), is accomplished during secondary treatment by specialized genera of lithotrophic bacteria. *Nitrosomonas* (and *Nitrosococcus*) bacteria convert ammonia ($NH_4^+$) to nitrite ($NO_2^-$), and *Nitrobacter* (and *Nitrospira*) bacteria convert nitrite to nitrate ($NO_3^-$). The overall reaction for these two processes can be written as follows:

$$NH_4^+ + 2O_2 \rightarrow NO_3^- + 2H^+ + H_2O \qquad \textbf{(11.18)}$$

If you work out the stoichiometry in Equation 11.18, you will find that 4.57 gm of oxygen are required to oxidize every 1 gm of nitrogen (as N).

Equation 11.19 not only includes nitrification of ammonia but also the incorporation of dissolved inorganic carbon and some of the ammonia into biomass (written as $C_5H_7NO_2$):

$$NH_4^+ + 1.863O_2 + 0.098CO_2 \rightarrow 0.0196C_5H_7NO_2 + 0.98NO_3^-$$
$$+ 0.0941H_2O + 1.98H^+ \qquad \textbf{(11.19)}$$

In Equation 11.19, only 4.52 g oxygen are required to oxidize every gram of nitrogen (as N). This stoichiometric value is lower than Equation 11.18 because Equation 11.19 accounts for some ammonia being used for synthesis of new cells. Equation 11.19 also shows that lithotrophic microorganisms obtain carbon for their cellular mass not from dissolved organic carbon, but by converting inorganic carbon (dissolved $CO_2$). Both Equations 11.18 and 11.19 show that some alkalinity is consumed for every mole of ammonia oxidized.

The nitrification reactions proceed slowly, require adequate oxygen (more than 0.5 mg/L) and alkalinity, and are sensitive to temperature, pH (preferring pH near 7), and the presence of toxic chemicals. For complete nitrogen removal, a variety of bacteria, including those of the genus *Pseudomonas*, can convert nitrate to nitrogen gas ($N_2$). Some of the nitrogen gas produced in this process (not shown in Equation 11.20) is actually $N_2O$, a **greenhouse gas.** (See how increased populations will produce increased amounts of greenhouse gases because of constituents found in domestic waste streams.) In this reaction, the nitrate serves as the electron acceptor (as oxygen does in carbonaceous oxidation), and the organic material in the wastewater is the electron donor.

Assuming the biodegradable organic material (measured as CBOD) in wastewater can be written as $C_{10}H_{19}O_3N$, the removal of nitrate to nitrogen gas can be written as follows:

$$C_{10}H_{19}O_3N + 10NO_3^- \rightarrow 5N_2\,(gas) + 10CO_2$$

$$+ 3H_2O + NH_3 + 10OH^- \qquad \textbf{(11.20)}$$

In Equation 11.20, alkalinity is produced (written as $OH^-$), and no dissolved oxygen is written in the expression. In fact, the presence of dissolved oxygen will inhibit the nitrate-reducing enzymes required for the denitrification reaction. This occurs at dissolved-oxygen concentrations as low as 0.1 or 0.2 mg/L.

During the design and operation of the biological reactor to denitrify nitrate, the amount of CBOD in the wastewater is a critical design parameter. If organic carbon is limiting in the waste stream, a waste organic stream (for example, dairy waste) or chemicals (for example, methanol) can be added in carefully controlled amounts to support **denitrification**. However, the amount added must be carefully controlled to ensure that no untreated residual BOD remains.

More than nine patented methods are available to configure the biological reactor such that nitrogen can be removed via nitrification and denitrification reactions. Figure 11.16 shows the most commonly used process, termed the **modified Ludzak–Ettinger (MLE) process**. Here an anoxic zone is located at the beginning of the biological reactor. All the processes use a combination of aerobic and anoxic zones where the biological reactor is configured to remove organic carbon (in the aerobic and anoxic zones), convert inorganic ammonia nitrogen (in the aerobic zone), and remove inorganic nitrate nitrogen (in the anoxic zone).

In the anoxic zone, the internal carbon of the wastewater (measured as CBOD) is oxidized to carbon dioxide and new biomass, while the

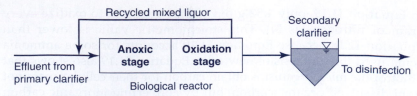

**Figure 11.16** **Modified Ludzak–Ettinger (MLE) Process for Configuring a Biological Reactor to Remove Nitrogen** In the second oxygenated compartment (Oxidation), nitrification of ammonia to nitrate occurs and the nitrate contained in the mixed liquor is recycled back to the first anoxic stage (Anoxic) for denitrification.

nitrate, serving as the electron acceptor, is reduced to nitrogen gas. The nitrate is produced in the aerobic zone (by nitrification reactions that convert ammonia to nitrate) and is recycled to the anoxic zone of the biological reactor.

One key to biological nitrogen removal is to prevent the washout of the slower-growing autotrophic bacteria (for example, *Nitrosomonas*) that convert ammonia nitrogen to nitrate. This requires that the solids retention time (SRT) be greater than the inverse of the growth rate of the nitrifying bacteria. The nitrifiers are also more sensitive to inhibition by toxic chemicals that can find their way into a treatment plant.

Compared with traditional wastewater treatment or a plant that treats BOD and ammonia nitrogen, the nitrifying/denitrifying processes have been found to have lower total operating costs if one considers the cost of aeration, sludge disposal costs, and credits for releasing less methane. Table 11.12 summarizes this information in detail.

## Table / 11.12

**Operational Costs for Activated-Sludge Practices** Assumptions: plant flow of 20,000 $m^3$/day; influent of 350 mg BOD/L; and effluent of 20 mg BOD/L.

| | Conventional Activated Sludge | Conventional Activated Sludge with Nitrification | Conventional Activated Sludge with Nitrification and Denitrification |
|---|---|---|---|
| Range of solids retention time (SRT) (days) | 1.2–8.5 | 12–21 | 4.7–22 |
| Sludge disposal cost ($/day) | 140 | 78 | 69 |
| Oxygen requirement (kg $O_2$/day) | 3,800 | 5,034 | 3,469 |
| Aeration cost ($/day) | 39 | 52 | 36 |
| Methane production credit ($/day) | 96 | 36 | 32 |
| Total cost (sludge disposal cost + aeration cost – methane production credit) ($/day) | 83 | 94 | 73 |

SOURCE: Results from Rosso and Stenstrom (2005).

Sludge disposal costs (Table 11.12) are greater in conventional activated sludge because the SRT is lower. Remember that when the SRT is high, there is greater endogenous respiration, so more of the resulting waste sludge is oxidized during the treatment process, and less sludge is produced. When fewer solids are produced from the biological reactor, there will be less sludge produced that will require sludge digestion.

As we will see later in this chapter, anaerobic sludge digestion produces **methane**, and if less sludge is produced, less methane also will be produced. Of course, this methane can be utilized as an energy source to supply heat or electricity. This is why the conventional process listed in Table 11.12 shows a higher methane production credit than the other processes.

Readers should investigate whether their local wastewater treatment plant recovers methane that is produced to use for heating or electricity generation. With future energy needs and climate change upon us, treatment plants should have as a first rule the successful recovery and use of methane generated at the treatment plant. However, technology to convert methane to electricity still produces $CO_2$ because it is an end product of methane combustion with oxygen.

In all three processes compared in Table 11.12, oxygen is required to oxidize organic carbon (CBOD) and ammonia nitrogen (NBOD). Table 11.12 demonstrates how conventional activated sludge with nitrification increases the oxygen requirement (and thus the aeration cost) compared with conventional activated sludge. However, with the addition of nitrification and denitrification, the oxygen requirements are lowered, because the nitrate that is produced during the nitrification process can be used as an electron acceptor in the denitrification process (where it displaces oxygen as the electron acceptor).

**Class Discussion**

Given all the choices in Table 11.12, how would you operate a municipal wastewater treatment plant to balance environmental, social, and economic issues in a sustainable manner?

## 11.9.2 PHOSPHORUS

Traditionally, chemicals such as alum ($Al_2(SO_4)_3$, ferric sulfate ($Fe_2(SO_4)_3$), and ferric chloride ($FeCl_3$) have been added to remove **phosphorus** by precipitation. All three chemicals precipitate dissolved polyphosphates (as illustrated here for alum):

$$Al^{3+} + PO_4^{3-} \rightarrow AlPO_{4\,(s)} \qquad \textbf{(11.21)}$$

Chemicals are typically added either during primary or secondary treatment. They thus generate a chemical sludge in addition to the sludge associated with the respective treatment processes. Of course, pollution prevention activities such as preventing phosphorus use in detergents can reduce the need for a portion of this treatment requirement. Biological processes also can be used to remove significant amounts of phosphorus and are gaining widespread usage. The phosphorus content of a typical dry cell is approximately 1 percent, so some phosphorus is removed simply by the growth and subsequent wastage of sludge. Typically, some hybrid of chemical addition and enhanced biological uptake is used to reach low P concentration in the effluent and minimize chemical usage.

**Figure 11.17  Configuration of a Biological Reactor to Remove Phosphorus**  In the anaerobic compartment, phosphate is stored internally by phosphate-accumulating microorganisms. Phosphorus is thus removed by converting it from dissolved phosphate to particulate phosphorus stored in biological cells, which are removed in the secondary clarifier.

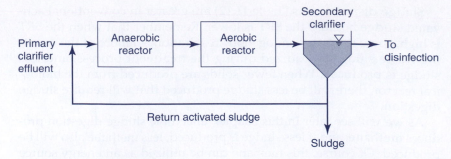

In the 1970s, it was determined that enhanced biological phosphorus removal was possible by use of organisms called **phosphate-accumulating organisms (PAOs)**. These organisms were found to uptake phosphorus well above the 1 percent typical of most microorganisms. Fortunately PAOs are present in wastewater treatment systems. They have the ability to take up volatile fatty acids such as acetic acid when oxygen and nitrates are not present. It was found that if PAOs were first exposed to an anaerobic zone (no oxygen present) followed by an aerobic zone, they would take up extra phosphate as they grew on the organic carbon in the aerated zone of the biological reactor (Figure 11.17). This extra phosphate is stored in energy-rich polyphosphate chains, which are subsequently used to take up the volatile fatty acids. This ability to transfer phosphorus in the dissolved aqueous phase to the particular microbial phase does not require chemical addition, and the particulate phosphorus can be removed as sludge.

## 11.10   Disinfection and Aeration

The final step before flow measurement and discharge to the receiving water is **disinfection**. The purpose of disinfection is to ensure removal of pathogenic organisms. This is most commonly accomplished by the addition of liquid sodium hypochlorite, chlorine dioxide, or chlorine gas; on-site hypochlorite generation; ozonation; or exposure to ultraviolet light. Disinfection was covered in Chapter 10, so it is not discussed further in this chapter.

During aeration, oxygen is transferred from a gaseous phase to the liquid phase. Remember in Chapter 3 that we explored the solubility of oxygen in water. Though oxygen makes up approximately 21 percent of Earth's atmosphere, only 8 to 11 or so parts per million (mg/L) of oxygen can dissolve into water at equilibrium if air is used as the source.

Table 11.13 compares the three methods commonly used to aerate wastewater: (1) *surface aeration*, (2) *fine-pore diffusion*, and (3) *coarse-bubble diffusion*. Table 11.13 also provides information on the energy usage associated with each of these aeration technologies. Fine-pore diffusers reduce energy costs by 50 percent over coarse-bubble diffusers, but they are more easily fouled by constituents found in wastewater. Figure 11.18 shows examples of fine-pore diffusers placed in an aeration basin.

**Disinfection**

**Chlorine Disinfection of Combined Sewer Overflows**
http://www.epa.gov/OWM/mtb/chlor.pdf

## Table / 11.13

### Aeration Devices Used During Wastewater Treatment

| Aeration Device | Description |
|---|---|
| Surface aerator | Shear wastewater surface with a mixer or turbine to produce a spray of fine droplets that land on the wastewater surface over a radius of several meters. Can be connected to a solar-powered pump. |
| Diffusers (fine pore and coarse bubble)<br><br>Fine bubbles have a diameter less than 5 mm; coarse bubbles have diameters as large as 50 mm. | Nozzles or porous surfaces are placed in the tank bottom, where they release bubbles that travel upward toward the tank's surface.<br><br>Fine-pore diffusers are used more in the United States and Europe than coarse-bubble diffusers. Fine-pore diffusers reduce energy costs by 50 percent over coarse-bubble diffusers.<br><br>Fine-pore diffusers can be fouled or experience scale buildup, so they need more cleaning. Cleaning typically consists of either emptying a tank, using a hose to clean the diffusers, or scrubbing with a 10–15 percent HCl solution. Note that emptying a tank works best in the situation where a plant has excess capacity (typically at larger plants). Periodic cleaning will maintain the diffuser's efficiency, which will reduce energy requirements.<br><br>The transfer efficiency of gas to liquid in the presence of wastewater contaminants (e.g., surfactants, dissolved organic matter) is quantified by the $\alpha$ factor. This alpha factor is lower for fine-pore diffusers, which suggests the presence of contaminants inhibits the transfer of oxygen to a greater extent than for coarse bubbles. |

SOURCE: Rosso and Stenstrom (2006).

(a)

(b)

**Figure 11.18** **Fine-Pore Diffusers** (a) Fine-pore diffusers at bottom of rectangular aeration basin. (b) Fine-pore diffusers in operation in same aeration basin.

Photo courtesy of James Mihelcic.

Conservation, improvement in efficiency of use, and **water reclamation and reuse** are all methods that can be used to bridge the gap between demand for and supply of water. Treatment technology is so advanced today that wastewater can be converted to a water source for use in a variety of residential, commercial, industrial, and agricultural applications. Globally, water reuse capacity is increasing rapidly. One challenge, though, is matching issues of quality, demand, and supply. For example, the extensive water needs of agriculture are typically located far away from treatment plants that collect wastewater, which are located in urban areas. In this case, it makes little sense to use energy to pump treated wastewater back to rural areas. One solution to this problem is the use of small-scale **satellite reclamation processes**. These types of small plants treat and reclaim wastewater where it is needed.

An example of using technology discussed in our drinking-water and wastewater chapters to reclaim wastewater is the Gippsland Water Factory, located in Traralgon (Victoria, Australia). Every day this facility treats 16,000 m³ of municipal wastewater and 19,000 m³ of industrial wastewater. Figure 11.19 provides an example of how the Gippsland Water Factory (Australia) has combined treatment processes used to produce the reclaimed water. In this case, the reclaimed water is being used to supplement a current source of groundwater (Daigger et al., 2007).

**The WateReuse Association**
http://www.watereuse.org/

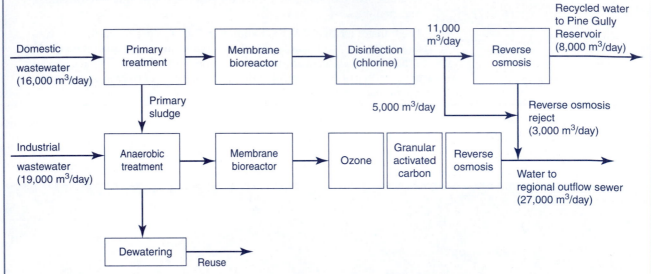

**Figure 11.19   Combination of Drinking-Water and Wastewater Unit Processes Employed to Produce Usable Water from a Domestic and Municipal Wastewater Source**   This example is from the Gippsland Water Factory located in Victoria, Australia.

## 11.11   Sludge Treatment and Disposal

The sludge generated through primary and secondary treatment has three characteristics that make its direct disposal difficult: (1) it is aesthetically unpleasing in terms of odor; (2) it is potentially harmful because of the presence of pathogens; and (3) it contains too much water, which makes it difficult to process and dispose of. The first two problems are often solved by *sludge stabilization*, and the third by *dewatering*.

## 11.11.1   SLUDGE STABILIZATION

The objective of **sludge stabilization** is to reduce problems associated with sludge odor and putrescence and the presence of pathogenic organisms. The first stabilization alternative, **aerobic digestion**, is simply an extension of the activated-sludge process. Waste activated sludge is pumped to dedicated aeration tanks for a much longer time period than with the activated-sludge process. The concentrated solids are allowed to progress well into the endogenous respiration phase, in which food is obtained through the destruction of viable organisms. The result is a net reduction in organic matter.

Another method of sludge treatment is **anaerobic digestion**. It is more commonly employed, because it does not require energy-intensive aeration. It is primarily a three-step biochemical process mediated by specialized groups of microorganisms (Table 11.14).

A pH near neutral is preferred for anaerobic digestion, and at a pH below 6.8, the methane formers begin to be inhibited. If the digester is not operated properly, the methane-forming bacteria will not be able to use the hydrogen that is produced at a fast enough rate. In this case, the pH of the reactor may drop due to accumulation of volatile fatty acids from the fermentation step. The methane formers may further become inhibited by the low pH; however, the acid formers keep on mediating the second step. This further lowers the pH, which may *sour* the digester and stop the process. Lime is often added to correct this problem.

The resulting gas of an anaerobic digester is approximately 35 percent $CO_2$ and 65 percent $CH_4$. In terms of greenhouse gas emissions,

## Table / 11.14

**Three-Step Biochemical Process during Anaerobic Digestion of Wastewater Solids**

| Step | Description |
|---|---|
| Step 1: hydrolysis | Microorganisms produce extracellular enzymes, which solubilize particulate organics in the presence of water. |
| Step 2: fermentation | Sometimes referred to as acidogenesis. A specialized group of bacteria termed the acid formers convert the soluble organics (things like sugars, amino acids, fatty acids) to volatile fatty acids (e.g., weak organic acids such as acetate and propionate). In this process, hydrogen and carbon dioxide are also formed. The microorganisms that mediate the hydrolysis and fermentation steps are facultative and obligate anaerobic bacteria. |
| Step 3: methanogenesis | A group of specialized bacteria, the methane formers, convert the organic acids that the acid formers produced to the end products **methane** and **carbon dioxide**. The methane-forming bacteria are termed strict obligate anaerobes. |

remember that in Chapter 2 we learned that the radiative forcing of gases differs. Thus, 1 ton of methane emissions equates to 25 tons of carbon dioxide emissions in terms of their equivalent potential as greenhouse gases. The methane that results from anaerobic digestion should be viewed as a valuable gas that should not be emitted directly into the atmosphere.

Methane generated at a wastewater treatment plant can be converted to electricity or used for heating. Combusting methane still results in production of greenhouse gases, according the following equation:

$$CH_4 + 2O_2 \rightarrow CO_2 + 2H_2O \qquad \textbf{(11.22)}$$

Equation 11.22 shows that, on an equivalent basis, 2.75 kg $CO_2$ are produced for every 1 kg of $CH_4$ that is combusted. But although some greenhouse gas is produced by combusting methane, there is not only a benefit from converting methane to carbon dioxide, but also some carbon offset associated with generating electricity from the methane. Recall the methane production credits compared in Table 11.12. These are related to the quantity of sludge produced during the different biological configurations of the activated-sludge process used to remove nitrogen.

---

**Box / 11.4**  **Energy Production from Wastewater and Solid Waste**

The potential for electricity production from wastewater treatment is not trivial. Besides the real possibility to use solar, wind, or microhydro power, wastewater also should be viewed as a source of energy. As just one example, the Los Angeles Sanitation District treats approximately 520 million gal. wastewater every day and manages the final disposal of half the 40,000 tons/day of nonhazardous solid waste generated in Los Angeles County. From this, the Sanitation District currently obtains 23 MW of electricity from digester gas,

63 MW from landfill gas, and 40 MW from combusting solid waste. In comparison to needs of the Sanitation District, this 126 MW of electricity production dwarfs the 41 MW of electricity required by the district (McDannel and Wheless, 2007). Methane from digester gas is even used in a fuel cell after it has been transformed to hydrogen upstream. Fuel cells generate electricity by using electrochemical reactions between hydrogen and oxygen.

---

**Sludge Treatment and Disposal**

## 11.11.2  DIGESTERS

In the past, a two-stage **digester** was employed. In this process, the primary tank was covered, heated to 35°C, and kept well mixed to enhance the reaction rate. The secondary tank had a floating roof cover. The secondary tank would not be mixed or heated and was used for gas storage and for concentrating the solids by settling. Settled solids (termed **digested sludge)** were routed to a dewatering process, and the liquid supernatant was recycled to the beginning of the treatment plant. The reason two-stage digesters have fallen out of favor is the cost associated with building a second tank to be used primarily for storage.

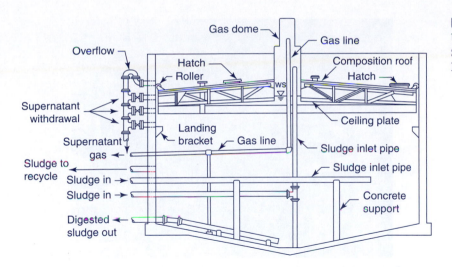

In the single-stage digester (see Figure 11.20), sludge is pumped to the reactor every 30 to 120 min to maintain constant conditions within the digester. The digester is typically sized based on a design solids retention time. In this case, the solids retention time equals the mass of solids in the reactor divided by the mass of solids removed every day. Typical design SRTs for anaerobic digesters range from 15 to 30 days.

## 11.11.3    DEWATERING

After stabilization, solids are typically dewatered before disposal. *Dewatering* is generally the final method of volume reduction before ultimate disposal. Sludge pumped from the primary and secondary clarifiers has a solids content of only 0.5 to a few percent. Dewatering can improve this to 15 to 50 percent.

The simplest and most cost-effective dewatering method if land is available and labor costs are low is to use *drying beds*. The beds consist of tile drains in gravel covered by about 10 in. of sand. Liquid is lost by seepage into the sand and by evaporation. A typical drying time is 3 mo. If dewatering by sand beds is considered impractical, mechanical techniques may be employed. One mechanical dewatering method is a belt filter press (Figure 11.21), where the sludge is introduced to a moving belt and squeezed to remove water, producing a sludge cake. A second mechanical method is the centrifuge, a solid bowl where solids are moved to the wall by centrifugal force and scraped out by a screw conveyor. The performance of mechanical dewatering devices may be improved by some type of chemical pretreatment. Here, polymers are added to improve dewatering.

## 11.11.4    DISPOSAL

Dewatered sludge is typically incinerated, applied to agricultural land, composted, provided to the public or a municipality as a soil conditioner, or disposed of in a landfill. Recently, burial in a landfill was not considered a preferred disposal method because of the lack

**Figure 11.21** **Belt Filter Press** This is one example of a commonly used mechanical dewatering method.

Photo courtesy of Woodridge Greene Valley Treatment Plant, Illinois.

**Biosolids**

http://www.epa.gov/owm/mtb/biosolids

of landfill space. However, burial in a landfill is now being considered as a disposal option in order to sequester carbon. Use of dewatered sludge (also referred to as sludge or biosolids) in agricultural areas can be seen as a way to return nutrients back into the environment. The resulting material can then be used to support agricultural production of food or as a soil conditioner that can be used in community gardens.

On a global basis, readily available phosphorus (P) that is used to promote agricultural activity on phosphorus-deficient soils is expected to run out in the next 50 to 100 years. The phosphorus that is accumulated in biosolids is one source of P. A simple way to determine the large mass of P that is potentially available to society from wastewater treatment is to do the mass balance based on plant flow rate and influent P concentrations. Of course, the sludge should not be contaminated with industrial and household hazardous wastes. Therefore, an integrated program to use solids from a municipal wastewater treatment plant should be coordinated with well-publicized household hazardous-waste collection program and aggressive monitoring and enforcement of industrial pretreatment standards where industries discharge to municipal sewers.

Two concerns about biosolids are runoff and groundwater contamination associated with chemical constituents in the sludge and the presence of pathogens. Table 11.15 shows the survival times of pathogens in soil. The survival times for some pathogens range from days to several years. This is one reason for either treating biosolids before application or minimizing the risk by preventing human exposure by placing the solids in an area with little human contact.

Sludge that is applied to land is broken down into *Class A* and *Class B biosolids*. Class A solids can be applied in areas open to the public. These biosolids may even be provided (or sold) to the public in a small bag. Therefore, Class A solids must be further treated by heat or chemical treatment to reduce the presence of pathogens to undetectable

## Table / 11.15

**Survival Times in Soil for Commonly Occurring Pathogens Found in Domestic Wastewater**

| Pathogen | Absolute Maximum | Common Observed Maximum |
|----------|------------------|--------------------------|
| Bacteria | 1 yr | 2 mo |
| Viruses | 6 mo | 3 mo |
| Protozoa | 10 days | 2 days |
| Helminths | 7 yr | 2 yr |

SOURCE: Data from Kowal (1995).

levels. Thermal treatment can consist of heat drying or composting. Chemical treatment typically involves a combination of increased pH and temperature.

Class B solids are processed to a point where pathogens may still be present, but land restrictions are in place to limit the public's

---

### Box / 11.5    Dealing with Household Food Wastes

Garbage disposals have become a standard feature in many homes because of consumer perceptions of status and convenience. Garbage disposals result in a larger fraction of food wastes entering the wastewater collection system. This can increase the BOD and TSS of wastewater by 10 to 20 percent. As explained in this chapter, removal of these pollutants requires additional plant capacity (more or larger reactors) and energy to pump and aerate the wastewater. Also, some of the organic carbon found in the food wastes will be converted to solids, which will require treatment and handling. Another part of the organic carbon will be converted to climate-change-causing greenhouse gases such as $CO_2$ and $CH_4$.

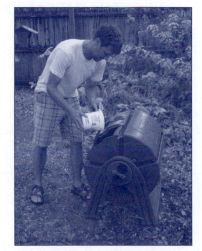

Photo courtesy of James R. Mihelcic

Other options to deal with food scraps include depositing them in the solid-waste stream, where they might be landfilled; reusing them as animal feed for local farms; collecting them at the community level for composting, and treating them with backyard composting. Landfills produce the greenhouse gases $CH_4$ and $CO_2$. Composting can occur at the home or be arranged at the community level. But even with composting, biological processes will break down the organic matter and produce $CO_2$. Some people might have a negative bias against household composting, but backyard composting does not require transportation costs or mechanical energy in the form of mechanically aerating the compost.

#### Class Discussion

Each disposal option for food scraps has different impacts on the environment, aesthetic concerns, and levels of societal acceptance. Thinking over the whole life cycle of the food scraps (from production by a farmer and in the house to final treatment or disposal), if you were trying to minimize the environmental impact of dealing with food wastes, what would you suggest to: (1) your family and (2) a municipal client? If you think carefully, you can see that the answer requires a broad systems approach to come up with a more sustainable solution. One solution being used by some communities is to collect food scraps and add them to anaerobic biodigesters that can produce energy.

exposure. The most common example of restricted land use where human exposure is limited would be to land apply Class B solids to an agricultural field. Restrictions are further placed on the timing of when root crops or aboveground crops may be harvested after the final land application of municipal wastewater sludge. There are also restrictions on application of solids to agricultural land to ensure that runoff from the field does not cause problems with water quality in local steams and lakes.

## 11.12 Natural Treatment Systems

Natural waste treatment systems are discussed in this section, which emphasizes the treatment technologies of lagoons and wetlands. These technologies not only use more natural methods to treat wastewater, but also have lower capital costs because they do not employ aboveground reactors constructed from steel-reinforced concrete, metal, or plastic. They also typically have lower operation costs because they may rely on natural aeration methods (versus mechanical aeration) and may utilize nonoxygenated biological processes. Natural wastewater treatment systems are also employed in decentralized treatment systems. Figure 11.22 shows one such system, the Living Machine®, which can be scaled to homes, dormitories, offices, and schools.

### 11.12.1 STABILIZATION PONDS

**Stabilization ponds** are referred to as **lagoons** or **oxidation ponds**. The lagoon is essentially an engineered hole-in-the-ground design to confine wastewater for treatment before discharge to a natural watercourse. Lagoons are typically found in smaller communities. Table 11.16 describes the various types of stabilization ponds. Each type of stabilization pond supports different biological processes. The type of biology is influenced by the depth of the lagoon and whether the

**25% of U.S. Homes Have Septic Systems**
http://cfpub.epa.gov/owm/septic/index.cfm

**Figure 11.22** **The Living Machine®**
This example of engineered natural treatment processes uses methods that incorporate bacteria, protozoa, plants, and snails to treat wastewater. EPA reports the largest of these systems can treat 80,000 gpd. They have produced effluent with $BOD_5$, TSS, and total nitrogen less than 10 mg/L. Phosphorus removal is reported to be 50 percent with effluents in the range of 5–11 mg/L.

Adapted from EPA (2001).

### Types of Waste Stabilization Ponds and Associated Design Information

| Type of Stabilization Pond | Comments | Water depth (m) | Detention Time (days) |
|---|---|---|---|
| Facultative lagoon | Uses combination of aerobic, anoxic, and anaerobic processes. Not typically mixed or aerated. Does not function well in colder climates. | 1.2–2.4 | 20–180 |
| Aerated lagoon | Typically placed in series, in front of a facultative pond. Aeration consists of either mechanical surface aerators or submerged diffused aeration systems. Requires less area than a facultative lagoon and can operate effectively in the winter. | 1.8–6 | 10–30 |
| Anaerobic pond | Usually used to pretreat high-strength wastewaters. Deep, non-aerated and no mixing. Performance decreases at temperatures below 15°C. | >8 | ≤50 |
| Tertiary pond | Typically treats treated effluent from an activated-sludge process or a trickling filter. Also referred to as a maturation or polishing pond. | <1 | 10–5 |

lagoon is mixed and aerated. The capability of lagoon treatment systems to meet water quality guidelines and protect public health and environmental integrity, coupled with their cost-effectiveness and ease of operation and management, make them a highly desirable technology, particularly for smaller communities and developing countries. They are designed in such a way as to remove primary wastewater constituents, including total suspended solids (TSS), biological oxygen demand (BOD), nutrients, and pathogens. Major pathogen removal mechanisms in ponds include sedimentation, adsorption to particles, lack of food and nutrients, solar ultraviolet radiation, temperature, pH, predators, straining/filtration in sediments, toxins and antibiotics excreted by some organisms, and natural die-off.

Figure 11.23 shows the various zones found in a **facultative lagoon** and the respective biological processes that occur. An *aerobic zone* is located near the surface. It is aerated due to oxygen transfer from the overlying air to the water and also by algal photosynthesis. The amount of **photosynthetic oxygen** production can be significant and does not require any energy input, except from the sun. In the presence of oxygen, CBOD is converted to $CO_2$, and NBOD is converted to nitrate (along with the production of biomass solids). An *anaerobic zone* forms on the bottom of the lagoon, where the solids settle. This part of the lagoon supports anaerobic biological fermentation processes discussed in the section on sludge digestion and converts CBOD into $CH_4$ and $CO_2$. Between these two layers is an *anoxic layer*, also termed the facultative zone. In this zone, denitrification reactions take place where nitrate can be reduced to nitrogen gas and in the process oxidize CBOD.

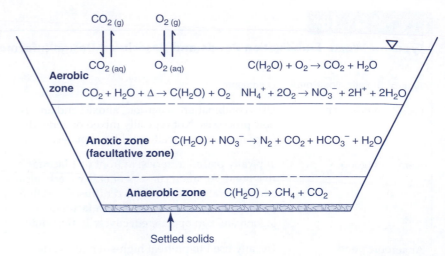

**Figure 11.23** **Zones of a Facultative Lagoon** There are three different zones for wastewater treatment. Oxygen is transferred to the upper aerobic zone via gaseous diffusion or algal photosynthesis. Here CBOD, written as $C(H_2O)$, is oxidized, and NBOD, written as $NH_4^+$, is oxidized to $NO_3^-$. In the middle anoxic zone, CBOD is oxidized, and $NO_3^-$ is reduced via denitrification reactions. In the lower anaerobic zone, solids accumulate, and fermentation reactions break down CBOD to $CH_4$ and $CO_2$.

Typical organic loading values of facultative ponds range from 15 to 80 kg/ha/day. One possible problem with these ponds is that they cool off quickly and can experience a large reduction in biological activity during the winter months of northern climates. In addition, algae may accumulate in the pond effluent, which will cause problems with an effluent TSS going well above 20 to 100 mg/L in a poorly designed lagoon.

## 11.12.2 WETLANDS

Natural ecosystems such as wetlands are the prototype for raising water quality by using natural energy (sunlight) and ambient temperature and pressures, without adding materials or requiring large amounts of human labor. They can also provide the public with open

**Figure 11.24** **A Solar-Powered Aerator** Solar-powered aerators can provide oxygen to lakes, reservoirs, stormwater ponds, and wastewater lagoons. The aerator in this photo can move up to 10,000 gal. water per min from depths up to 100 ft. These types of aerators can displace up to 30 hp (25,000 W) of electrical-grid-connected equipment every day, which equates to 148 tons of carbon dioxide per year.

Photo courtesy of SolarBee, Inc.

green space. A wastewater treatment system designed with that proto-type in mind can be sustainable in terms of energy and material input/output as well as social and environmental benefits. Wastewater treatment technologies that combine the soil–water–air–vegetation environment include constructed wetlands and evapotranspiration beds. Both require pretreatment of the influent solids load with either a septic tank, oxidation pond, or other primary treatment structure for settling solids. The two types of constructed wetlands are the free water surface (FWS) wetland and the subsurface flow (SSF) wetland (Figure 11.25). Only free water surface wetlands will be described in detail in this chapter.

In some literature, the term **created wetland** refers to a wetland built for mitigation purposes, and **constructed wetland** denotes a wetland designed for treatment of wastes. Wastes that are commonly treated with constructed wetlands include domestic sewage, agricultural runoff, stormwater runoff, and mining wastes. Although created and constructed wetlands may share many features in common, the

**Decentralized Wastewater Treatment**
http://www.epa.gov/seahome/decent.html

**Natural Treatment**

**Figure 11.25**   **Subsurface Flow Wetland (before Planting of Vegetation) That Serves Over 200 Students at the Pisgah All Age School (Jamaica)**
Subsurface flow (SSF) wetlands typically use gravel as the aquatic plant rooting media, and the water level is intentionally maintained below the surface of the gravel. Several hydraulic regimes may be employed: horizontal flow, vertical upflow, and vertical downflow. The septic tanks and wetland treat only black water from the toilets and no gray water. The wetland has two parallel plastic-lined rock media beds (separated by the black plastic in center of bed). Each bed has inside dimensions of 19.1 m length by 4.7 m width by 0.5 m depth of media. Media are irregularly shaped washed stones ordered as "1/2 inch river shingle," and the bulk of the stone ranges in size from 1/4 to 1 in. (6–25 mm) diameter. The media porosity was measured to be 37.7 percent without plant roots.

Photo courtesy of Edward Stewart.

**Figure 11.26** Zones of a Free Water Surface Wetland

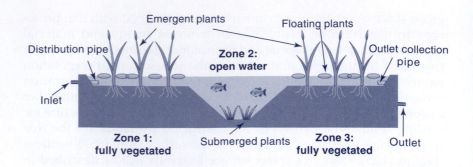

Distribution pipe

Inlet

Zone 1: fully vegetated

Emergent plants

Zone 2: open water

Submerged plants

Floating plants

Outlet collection pipe

Zone 3: fully vegetated

Outlet

guidelines for construction as well as the regulations that are applied are different; for this reason, the two categories are often considered separately. Created wetlands were discussed in Chapter 8; constructed wetlands are considered in the following section.

## FREE WATER SURFACE (FWS) WETLANDS Free water surface (FWS) wetlands, also called surface flow wetlands, are similar to natural open-water wetlands in appearance and treatment mechanisms (see Figure 11.26). The majority of the wetland surface area has aquatic plants rooted in soil or sand below the water surface. The wastewater travels in deep sheet flow over the soil and through the plant stems (zones 1 and 3). The area (zone 2) with no surface vegetation is exposed to sunlight and open to the air to increase the potential for oxygen transfer from the gaseous to aqueous phase. Zone 2 may also have submerged aquatic plants to enhance the dissolved-oxygen content.

The first vegetated zone (at water depth of approximately 1 ft.) acts as an anaerobic settling chamber so that only a 1 or 2 day hydraulic retention time is necessary to achieve the required reactions. The hydraulic residence time of the open zone (zone 2, at water depth of approximately 3 ft.) should be less than the amount of time required for algae to form and will depend on climate and temperature as well as nutrient limitations. In the United States and Canada, this time is typically between 2 and 3 days. The hydraulic residence time for the second vegetated zone (zone 3, at water depth of approximately 1 ft.) is 1 day to achieve denitrification.

There may be multiple vegetated and open zones to achieve the desired treatment objectives. The calculation of head loss along the length of an FWS wetland is usually not necessary, since a typical FWS wetland with a recommended 5:1 to 10:1 aspect ratio (L:W) may have a hydraulic slope gradient of only 1 cm in 100 m (EPA, 2000). The use of hydraulic residence time and the maximum areal loading guidelines (see Table 11.17) makes sizing for TSS, BOD, and nitrogen an iterative design process. The values provided in Table 11.17 represent maximum monthly mass loading rates that should reliably keep a FWS wetland's effluent below the noted concentration.

Table 11.18 compares free water surface (FWS) wetlands with subsurface flow wetlands (SSF). One advantage of FWS wetlands is that they can be designed to provide long-term nitrogen removal because of the aerobic open-water zones that allow biological nitrification. However, this is only the case for wetlands that are shallow or well

## Table / 11.17

**Maximum Wetland Areal Mass Loading Values and Typical Resultant Effluent Concentrations** Data were obtained for a variety of applications, from sewage to stormwater, and cover a range of temperate climate locations, from Florida to Canada.

| Constituent | Free Water Surface Wetland Loading | Subsurface Flow Wetland Loading | Effluent Concentration |
|---|---|---|---|
| BOD | 60 kg/ha-d | 60 kg/ha-d | 30 mg/L |
| TSS | 50 kg/ha-d | 200 kg/ha-d | 30 mg/L |
| TKN | 5 kg/ha-d | Not applicable | 10 mg/L |

SOURCE: EPA (2000).

## Table / 11.18

**Comparison of Free Water Surface and Subsurface Flow Wetlands**

| Characteristics | Free Water Surface | Subsurface Flow |
|---|---|---|
| Wastewater exposure | Open-water aerobic zones enhance biological nitrification and provide wildlife habitat. | Wastewater remains 2–4 in. below the media surface, so there is no surface water to attract aquatic birds, little risk of human exposure, and no mosquito breeding. |
| Hydraulics | Not likely to overflow from an accumulation of solids at the inlet. | Surface flooding will occur at the inlet if there is excessive accumulation of solids. |
| Bedding | The sand or sandy loam plant rooting media have a lower materials cost than the gravel media used in SSF wetlands. | The plant rooting rock media should have diameters of 0.25–1.5 in. and be relatively free of fines. This is more expensive than FWS wetland media. |
| Dimensions | Recommended aspect ratio (L:W) is 5:1 to 10:1. Depth of water may range from a few inches in vegetated zones to 4 feet in open-water zones. | Recommended aspect ratio is in the range of 1:1 to 0.25:1. Depth of gravel may be 1 to 2 ft. |

mixed. In deeper water zones, good top-to-bottom mixing may not occur, so deeper water may not achieve nitrification levels. SSF wetlands will not typically provide long-term nitrogen removal without plant harvesting or some oxygenation of the water by means of cascading or mechanical aeration.

# 11.13 Energy Usage during Wastewater Treatment

**Energy** use during wastewater treatment is not trivial. In fact, drinking-water and wastewater treatment combined account for 3 percent of electricity use in the United States, and energy costs can account

**Energy Demands on Water Resources: Report to Congress**

http://www.sandia.gov/energy-water

for up to 30 percent of a treatment plant's total operation and maintenance costs. Mechanical systems have been discussed in much of this chapter and have been preferred in highly populated areas. Land treatment systems utilize soil and plants without significant need for reactors and operational labor, energy, and chemicals. Lagoon treatment systems are also less mechanized, as was discussed previously. While large treatment plants (in excess of 100 MGD) serve many cities and a large percentage of the U.S. population, most treatment plants in the United States serve small communities. In fact, the EPA reports there are over 16,000 wastewater treatment plants, and over 80 percent of existing plants have a capacity less than 5 MGD.

Operating and maintenance costs associated with wastewater treatment include labor, purchase of chemicals and replacement equipment, and energy to aerate, lift water, and pump solids. Mechanized plants obviously cost more to run than less-mechanized forms of treatment. Table 11.19 shows a breakdown of energy use at a 7.5 MGD treatment plant. As expected, the activated-sludge process accounts for more energy use than gravity settling and sludge dewatering. For smaller plants, the operational life stage of the life cycle has been found to have the highest energy consumption (95 percent), compared with the construction and refurbishment/demolition stages. This is significant, because energy production and its use are associated with many environmental problems, including release of airborne pollutants and global warming.

Figure 11.27 shows an example of energy requirements to treat a million gallons per day (MGD) of wastewater with mechanical, lagoon, and terrestrial treatment systems. Note the much higher energy costs associated with mechanical treatment. This is primarily due to the mechanical aeration of water, which accounts for 45 to 75 percent of a treatment plant's energy costs. However, mechanical treatment

## Table / 11.19

**Breakdown of Energy Use at a 7.5 MGD Wastewater Treatment Plant**

| Unit Process/Activity | Percent of Total Energy Use |
|---|---|
| Activated sludge | 55 |
| Primary clarifier | 10 |
| Heating | 7 |
| Dewatering solids | 7 |
| Pumping raw wastewater | 5 |
| Secondary clarifier (returned activated sludge) | 4 |
| Other | 12 |

SOURCE: California Energy Commission.

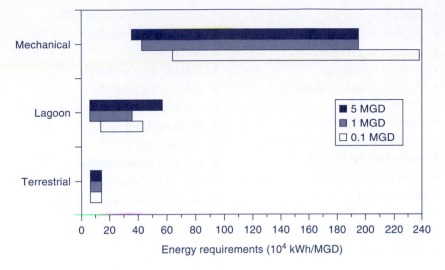

**Figure 11.27** **Total Energy Requirements for Various Sizes and Types of Wastewater Treatment Plants Located in Intermountain Areas of the United States** Total electricity requirements are measured in kWh/MGD at flow rates of 0.1, 1, and 5 MGD.

This figure was published in *Journal of Environmental Management* 88, H. E. Muga and J. R. Mihelcic, "Sustainability of Wastewater Treatment Technologies," 437–447. Copyright Elsevier (2008).

systems are well documented to be very effective at treating wastewater constituents to specified levels, especially given the smaller land area required per unit of wastewater treated. Obviously, the future of wastewater treatment needs to look beyond treatment objectives and integrate issues of energy and materials use throughout the process life cycle. For just one small example, pumps selected for wastewater treatment and drinking water are typically purchased based on initial costs, not pumping efficiencies.

## Key Terms

- activated sludge
- aeration basin
- aeration tank
- aerobic digestion
- anaerobic digestion
- attached growth
- average-strength municipal wastewater
- bar rack
- bar screen
- biofiltration
- BOD loading rate
- carbon dioxide
- clarifiers
- Clean Water Act
- comminutor
- compression settling
- constructed wetland
- conventional activated sludge

- created wetland
- critical SRT
- death growth phase
- denitrification
- digested sludge
- digester
- discrete settling
- disinfection
  domestic wastewater
- emerging chemicals of concern
- endogenous growth phase
- endogenous respiration
- energy
- extended aeration
- facultative lagoon
- flocculation
- flocculent settling
- flotation
- flow equalization

- foam
- food-to-microorganism (F/M) ratio
- free water surface (FWS) wetland
- greenhouse gas
- grit
- grit chamber
- hindered settling
- hydraulic retention time
- lagoon
- mean cell retention time
- membrane bioreactors (MBRs)
- methane
- mixed liquor
- mixed liquor suspended solids (MLSS)
- mixed liquor volatile suspended solids (MLVSS)

- modified Ludzak–Ettinger (MLE) process
- Monod kinetics
- municipal wastewater constituents
- municipal wastewater treatment plants
- National Pollutant Discharge Elimination System (NPDES)
- nitrification
- nocardioform
- overflow rate
- oxidation ponds phosphate-accumulating organisms (PAOs)
- phosphorus
- photosynthetic oxygen
- preliminary treatment
- primary treatment
- publicly owned treatment works (POTWs)
- return activated sludge
- satellite reclamation processes
- secondary treatment
- sedimentation tanks
- settling tanks
- sludge age
- sludge stabilization
- sludge volume index (SVI)
- solids retention time (SRT)
- stabilization ponds
- Stokes' law
- surface flow wetlands
- suspended growth
- tertiary wastewater treatment
- trickling filter
- washed out
- waste activated sludge
- wasting of sludge
- water reclamation and reuse

**11.1**  Research an emerging chemical of concern that might be discharged to a wastewater treatment plant or household septic systems. Examples would include pharmaceuticals, caffeine, surfactants found in detergents, fragrances, and perfumes. Write an essay of up to three pages on the concentration of this chemical found in wastewater influent. Determine whether the chemical you are researching is treated in the plant, passes through untreated, or accumulates on the sludge. Identify any adverse ecosystem or human health impacts that have been found for this chemical.

**11.2**  Research whether there are state or regional pollution prevention programs to keep mercury out of your local municipal wastewater treatment plant. This mercury might come from your university laboratories or local dental offices and hospitals. What are some of the specifics of these programs? How much mercury has been kept out of the environment since the program's inception?

**11.3**  A wastewater treatment plant receives a flow of 35,000 $m^3$/day. Calculate the required volume ($m^3$) for a 3 m deep horizontal flow grit chamber that will remove particles with a specific gravity of more than 1.9 and a size greater than 0.2 mm diameter.

**11.4**  A wastewater treatment plant will receive a flow of 35,000 $m^3$/day. Calculate the surface area ($m^2$), diameter (m), volume ($m^3$), and hydraulic retention time of a 3 m deep circular, primary clarifier that would remove 50 percent of suspended solids. Assume the surface overflow rate used for the design is 60 $m^3/m^2$-day.

**11.5**  Assume a plant flow of 12,000 $m^3$/day. Determine the actual detention time observed in the field of two circular settling tanks with depth of 3.5 m that were designed to have an overflow rate not to exceed 60 $m^3/m^2$-day and a detention time of at least 2 hr.

**11.6**  A wastewater treatment plant has a flow of 35,000 $m^3$/day. Calculate the mass of sludge wasted each day ($Q_w X_w$, expressed in kg/day) for an activated-sludge system operated at a solids retention time (SRT) of 5 days. Assume an aeration tank volume of 1,640 $m^3$ and an MLSS concentration of 2,000 mg/L.

**11.7**  You are provided with the following information about a municipal wastewater treatment plant.

This plant uses the traditional activated-sludge process. Assume the microorganisms are 55 percent efficient at converting food to biomass, the organisms have a first-order death rate constant of 0.05/day, and the microbes reach half of their maximum growth rate when the $BOD_5$ concentration is 10 mg/L. There are 150,000 people in the community (their wastewater production is 225 L/day-capita, 0.1 kg $BOD_5$/capita-day). The effluent standard is $BOD_5 = 20$ mg/L and TSS = 20 mg/L. Suspended solids were measured as 4,300 mg/L in a wastewater sample obtained from the biological reactor, 15,000 mg/L in the secondary sludge, 200 mg/L in the plant influent, and 100 mg/L in the primary clarifier effluent. SRT is equal to 4 days. (a) What is the design volume of the aeration basin ($m^3$)? (b) What is the plant's aeration period (days)? (c) How many kg of secondary dry solids need to be processed daily from the treatment plants? (d) If the sludge wastage rate ($Q_w$) is increased in the plant, will the solids retention time go up, go down, or remain the same? (e) Determine the F/M ratio in units of kg $BOD_5$/kg MLVSS-day. (f) What is the mean cell residence time?

**11.8**  Using information provided in Example 11.7, determine the critical SRT value (sometimes referred to as $SRT_{min}$). This term refers to the SRT where the cells in the activated-sludge process would be washed out or removed from the system faster than they can reproduce.

**11.9**  In the following sentences, circle the correct term in boldface. If the solids retention time (SRT) is low (for example, 4 days), which conditions exist? (a) The F/M ratio is **low/high**. (b) The power requirements for aeration will be **less/greater**. (c) The microorganisms will be **starved/saturated** with food. (d) The mean cell retention time is **low/high**. (e) The sludge age is **low/high**. (f) The sludge wastage rate may have been recently **increased/decreased**. (g) The MLSS may have been **increased/decreased**.

**11.10**  Determine the sludge volume index (SVI) for a test where 3 g MLSS occupy a 450 mL volume after 30 min settling.

**11.11**  A 2 g sample of MLSS obtained from an aeration basin is placed in a 1,000 mL graduated cylinder. After 30 min settling, the MLSS occupies 600 mL.

Does the following sludge have good, acceptable, or poor settling characteristics?

**11.12** Figure 11.16 shows the modified Ludzak–Ettinger (MLE) process, which is used to configure a biological reactor to remove nitrogen. Explain the role of the two compartments in terms of: (a) whether they are oxygenated; (b) whether CBOD is removed in the compartment; (c) whether ammonia is converted in the compartment; (d) whether nitrogen is removed from the aqueous phase in the compartment; and (e) the primary electron donor(s) and electron acceptor(s) in each compartment.

**11.13** Investigate the specific mechanisms by which ammonia nitrogen, total nitrogen, and phosphorus are removed at your local municipal wastewater treatment plant. Are the processes chemical or biochemical (or a combination)? Discuss your answer.

**11.14** Investigate the specific mechanisms that your local municipal wastewater treatment plant uses for aeration. Is it surface aeration, fine- or coarse-bubble aeration, or natural aeration (via a facultative lagoon or attached growth system)?

**11.15** A wastewater treatment plant will receive a flow of 35,000 $m^3$/d (~10 MGD) with a raw wastewater $CBOD_5$ of 250 mg/L. Primary treatment removes ~25% of the BOD. Calculate the volume ($m^3$) and approximate hydraulic retention time (hr) of the aeration basin required to run the plant as a "high rate" facility (F/M = 2 kg BOD/kg MLSS-day). The aeration basin MLSS concentration will be maintained at 2,000 mg MLSS/L.

**11.16** Table 11.20 provides suspended solids concentrations in several different wastestreams at a municipal wastewater treatment plant. The $BOD_5$ is measured in the sewer located just before the treatment plant as 250 mg/L, after primary treatment is 150 mg/L, and after secondary treatment is 15 mg/L. Total nitrogen levels in the plant are approximately 30 mg N/L.

(a) If the design hydraulic retention time of each of four aeration basins operated in parallel equals 6 hours and the total plant flow is 5 million gallons per day, what is the F/M ratio in units of lbs $BOD_5$/lbs MLVSS-day. (b) Suppose the plant engineer wishes to increase the concentration of microorganisms in the biological reactor because she expects the substrate

| Table / 11.20 | |
|---|---|
| **Suspended solids concentration for different process streams in Problem 11.16** | |
| Process stream | Suspended Solids Concentration (mg SS/L) |
| Plant influent | 200 |
| Primary sludge | 5,000 |
| Secondary sludge | 15,000 |
| Aeration basin effluent | 3,000 |

level to increase. What would she command the operator to do to accomplish this goal?

**11.17** The community of San Antonio is located in the Caranavi Province, Bolivia. According to the 2005 year survey, there are 420 habitants of this community. The population is estimated to increase to 940 by the year 2035. The average peak flow is currently 1.2 L/sec and is expected to increase to 2.14 L/sec by 2035 The organic load is estimated to be 45 gram $BOD_5$/capita-day. The community is considering a free surface wetland or facultative lagoon to treat their wastewater. (a) What is the $BOD_5$ loading generated in the year 2035 (kg/day)? (b) Use the BOD loading to estimate the maximum surface area (ha) required for a free surface wetland that would serve the community in 2035 and remove BOD and TSS to 30 mg/L. (c) What is the required surface area ($m^2$) for a facultative lagoon to handle the peak flow in 2035, assuming a design water depth of 4 m and a detention time is 20 days (see Table 11.16)?

**11.18** Table 11.1 showed that pit latrines are considered an improved technology for treating wastewater. Determine the depth required for a pit latrine 1 m × 1 m in area that serves a household of 7 people and has a design life of 10 yr. Assume the pit is dug above the water table and the occupants use bulky or nonbiodegradable materials for anal cleansing (e.g., corncobs, stones, newspaper); therefore, the solids accumulation rate is assumed to be 0.09 $m^3$/person/year. Allow a 0.5 m space between the ground surface and the top of the solids at the end of the design life which is the point when the pit is filled.

# References

Barnard, J. L. 2006. "Biological Nutrient Removal: Where We Have Been, Where We Are Going" *Proceedings of the 79th Annual Water Environment Federation Technical Exhibition and Conference*, Dallas, Texas.

Cairncross, S., and R. G. Feachem. 1993. *Environmental Health Engineering in the Tropics: An Introductory Text*. New York: John Wiley.

Daigger, G. T., A. Hodgkinson, and D. Evans. 2007. "A Sustainable Near-Potable Quality Water Reclamation Plant for Municipal and Industrial Wastewater." *Proceedings of the 80th Annual Water Environment Federation Technical Exhibition and Conference*. San Diego.

Environmental Protection Agency (EPA). 2000. *Manual: Constructed Wetlands Treatment of Municipal Wastewaters*. EPA 625/R-99/010. Office of Research and Development, Cincinnati, Ohio.

Environmental Protection Agency (EPA). 2001. *The "Living Machine" Wastewater Treatment Technology: An Evaluation of Performance and System Cost*. EPA 832-R-01-004.

Fry, L. M., J. R. Mihelcic, D. W. Watkins. 2008. "Water and Non-Water-Related Challenges of Archieving Global Sanitation Coverage," *Environmental Science & Technology*, 42 (4): 4298–4304.

Jenkins, D. 2007. "From TSS to MBTs and Beyond: A Personal View of Biological Wastewater Treatment Process Population Dynamics." *Proceedings of the 80th Annual Water Environment Federation Technical Exhibition and Conference*, San Diego.

Kowal, N. E. 1985. "Health Effects of Land Application of Municipal Sludge." EPA/600/1–85/015. Health Effects Research Laboratory, EPA, Cincinnati, Ohio.

Martin Jr., R. W., J. R. Mihelcic, J. C. Crittenden. 2004. "Design and Performance Characterization Strategy Using Modeling for Biofiltration Control of Odorous Hydrogen Sulfide." *Journal of Air & Waste Management Association*, 54:834–844.

McDannel, M., and E. Wheless. 2007. "The Power of Digester Gas: A Technology Review from Micro to Megawatts." *Proceedings of the 80th Annual Water Environment Federation Technical Exhibition and Conference*, San Diego.

Mihelcic, J. R. 1999. *Fundamentals of Environmental Engineering*. New York: John Wiley & Sons.

Mihelcic, J. R., E. A. Myre, L. M. Fry, L. D. Phillips, B. D. Barkdoll. 2009. *Field Guide in Environmental Engineering for Development Workers: Water, Sanitation, Indoor Air*. Reston, VA: American Society of Civil Engineers (ASCE) Press.

Muga, H. E., and J. R. Mihelcic. 2008. "Sustainability of Wastewater Treatment Technologies." *Journal of Environmental Management* 88 (3): 437–447.

Rosso, D., and M. K. Stenstrom. 2005. "Comparative Economic Analysis of the Impacts of Mean Cell Retention Time and Denitrification on Aeration Systems." *Water Research* 39:3773–3780.

Rosso, D., and M. K. Stenstrom. 2006. "Surfactant Effects on $\alpha$-Factors in Aeration Systems." *Water Research* 40:1397–1404.

Tchobanoglous, G., F. L. Burton, and H. D. Stensel. 2003. *Wastewater Engineering*. Boston: Metcalf & Eddy / McGraw-Hill.

World Health Organization and United Nations Children Fund Joint Monitoring Programme for Water Supply and Sanitation (JMP). 2008. Progress on Drinking Water and Sanitation: Special Focus on Sanitation. UNICEF, New York and WHO, Geneva.

# chapter/Twelve Air Resources Engineering

Kurtis G. Paterson
and James R. Mihelcic

*In this chapter, readers will learn about the constituents in outdoor and indoor air and the magnitude of the concentrations of those constituents. Global, regional, and local transport of air pollutants is discussed and integrated into how transport relates to air quality. Mechanical and natural ventilation in the indoor environment are discussed as they relate to air quality. The chapter explains estimation of emissions by the use of direct measurement, indirect measurements using mass balances, process modeling, and emission factor modeling. This is followed by a presentation of four methods to control emissions: emission-control technologies and regulatory, market-based, and voluntary solutions. Information is provided on ways to design several emission-control technologies for gases: thermal oxidization, packed-bed absorption scrubber, biofilter, and for particles: cyclone, venturi scrubber, baghouse, and electrostatic precipitator.*

## Learning Objectives

1. Describe the ambient air and indoor air environment in a North American and global context.

2. Discuss the impact of air pollution on the environment, human health, and other social costs.

3. Explain the role that atmospheric transport has on air quality from a local, regional, and global perspective.

4. List important air pollutants for the ambient and indoor environments, and describe the concentrations of these constituents.

5. Identify major health and environmental impacts associated with key air pollutants.

6. Identify the six criteria air pollutants, discuss their importance in terms of adverse impact, and understand the current status of controlling them.

7. Estimate air pollutant emissions using direct measurement, indirect measurement, process modeling, and emission factor modeling techniques.

8. Create effective strategies to improve air quality that include regulatory solutions, market-based solutions, voluntary solutions, and emission-control technologies.

9. Summarize the extent of environmental risk associated with women and children exposed to indoor air pollution in the developing world.

10. Access information resources, and design tools for air resource engineering.

11. Size and list major advantages and disadvantages of the following unit processes for air pollution treatment: thermal oxidization, packed-bed absorption scrubber, biofilter, cyclone, venturi scrubber, baghouse, and electrostatic precipitator.

## 12.1 Introduction

Of all Earth's natural resources, air is the most shared. Air moves throughout the atmosphere, connecting sources of pollution to points of impact, often far apart. Three other features of air make it an unusually challenging resource to deal with:

1. Air has traditionally not been viewed as a resource like water or land resources, so there are no private or public claims on it. This has resulted in the atmosphere being viewed as a commons, with resulting abuse by pollutant emitters. (Remember our discussion in Chapter 1 related to the **Tragedy of the Commons**.)

2. Air does not obey geopolitical boundaries, so transboundary air pollution problems are common and especially difficult to resolve.

3. Polluted ambient air cannot be remediated with technology. While there are some natural cleansing processes for the atmosphere, they are often insufficient for returning the atmosphere to desirable conditions within reasonable time frames of interest to society and the environment. This places unique constraints on control strategies.

The quality of the air resource at any given location is a function of a complex interplay among numerous processes (Figure 12.1). The examination of any particular problem usually begins with an understanding of the system and its components. Additional analyses may then focus on components of this system to resolve a particular issue.

## 12.2 Human Health Impacts and Defenses

Primary public health impacts from air pollution include increased mortality rates, increased health care costs, decreased productivity, and decreased quality of life. In fact, decreased productivity due to air-pollution-related illness is estimated to cost U.S. society $4.4 billion to $5.4 billion per year (EPA, 1999). In addition, many diseases are associated with air pollution (Table 12.1).

**Access Air Pollution Data in the U.S.**

http://www.epa.gov/air/data/geosel.html

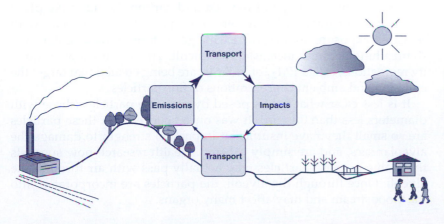

**Figure 12.1** **Main Components of an Air Quality Systems Analysis Applied to the Ambient Environment** Emissions are transported (and possibly transformed) and result in air quality impacts. When significant impacts are experienced, solutions are created. This same systems analysis approach is effective for investigating air resource issues in ambient and indoor environments.

## Table / 12.1

### General Categories of Human Disease Caused by Air Pollutants

| Category of Disease | Description |
|---|---|
| Irritation | Low-level burning of the respiratory surface tissues results in several adverse outcomes, from breathing difficulty to respiratory failure. Ozone is one pollutant that causes considerable irritation. |
| Cell damage | Arises from the introduction of pollutants into the cell. Ultrafine particles contribute to these diseases. |
| Allergies | Pollutants can trigger a release of histamines to fight the invader, resulting in breathing difficulty and irritation of sensitive tissues (e.g., skin, eyes). Pollens cause these effects. |
| Fibrosis | Pollutants permanently scar the tissues of the respiratory system, resulting in increased breathing difficulty and often death. Asbestos is a pollutant that causes this disease. |
| Oncogenesis | Pollutants can trigger abnormal, malignant tissue growth that leads to cancer. Many hazardous air pollutants are cancer-causing. |

The **respiratory system** has several defenses against air pollution, particularly large particles. When someone is breathing particle-contaminated air, the human respiratory system works a bit like a sieve, preferentially removing certain size fractions at various stages, as shown in Figure 12.2. The nose is pointed downward so particles cannot settle into the nostrils. Nose hairs trap large particles (similar to fibers in a baghouse); mucus in the nose, mouth, and throat increases the removal efficiency of larger particles (in a manner similar to a scrubber). The skin also offers protection to the internal organs from many pollutants. Particulate matter (PM) is a nonuniform combination of different compounds. Collectively, all particles 10 $\mu$m and less in diameter are denoted as $PM_{10}$. All particles 2.5 $\mu$m and less in diameter are denoted as $PM_{2.5}$.

In general, smaller particles can travel deeper into the respiratory system (Figure 12.2). Larger particles are trapped in the upper respiratory system, producing a clogged nose and scratchy throat. Fine particles (diameters less than 10 $\mu$m) make it to the alveoli, the air sacs where the exchange of oxygen and carbon dioxide take place within the cardiovascular system. Fine particles accumulate with repeated exposure, fouling the exchange surface in the alveolar sacs, so breathing becomes increasingly difficult. This is why new air quality regulations, called $PM_{2.5}$ standards, are being created that target the emission and ambient concentrations of fine particles.

It is less clear what risk is posed by ultrafine particles (those with diameters less than 0.1 $\mu$m). It was once believed that these particles are so small they have insufficient impaction strength to damage the alveolar sacs, and are simply exhaled. Health research now suggests that some of these particles may actually pass with air through the alveoli. Once through the alveoli, the particles are incorporated into the bloodstream and may affect many organs.

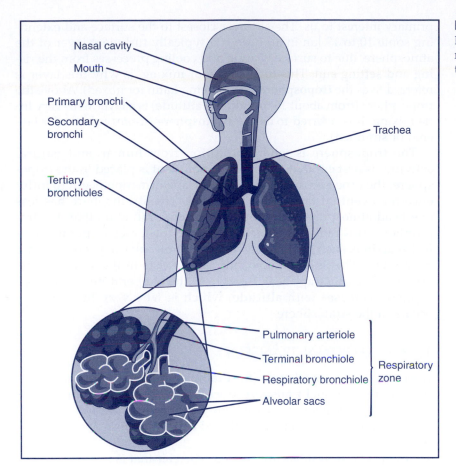

**Figure 12.2** **Major Parts of the Human Respiratory System** The main subsections are nasopharyngeal, tracheobronchial, and pulmonary.

One of the cardiovascular system's defenses is to mobilize white blood cells to metabolize foreign objects. These particles may explain the association between heart attacks and concentrations of particulate matter (PM). Accounting for all other risk variables, it has been observed that increased PM in air results in more heart attacks. The connection may be the ultrafine particles. Preliminary evidence suggests that such particles alter the blood's properties (for example, its viscosity and composition) either directly or through the metabolism process, precipitating sometimes fatal cardiovascular distress.

## 12.3 Transport of Air

In the context of engineering analyses of air quality, two distinct systems are of interest: the outdoor environment (**ambient air**) and **indoor** environment. Remember from Figure 12.1 that air emissions are transported, which can worsen air quality.

### 12.3.1 GLOBAL TRANSPORT

The physics of Earth's atmosphere are such that several distinct layers (or spheres) are present. The two layers closest to Earth's surface are of

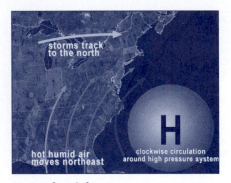

**Bermuda High**

primary interest to us. The air layer closest to the surface and extending some 10 to 15 km in altitude is a typically turbulent layer of the atmosphere, due to surface heating and cooling processes from the rising and setting sun. This turbulence, or mixing, is why this layer is referred to as the **troposphere** (*tropos* being Latin for mixed). Above the troposphere (from about 15 to 50 km in altitude) is a layer with very little mixing. It is referred to as the **stratosphere** (*stratos is* Latin for layered or stratified).

The troposphere is where pollution from human and natural activities is first emitted. Once an air pollutant is placed in the troposphere, the processes of transport and transformation begin to influence the eventual fate of the pollutant. The stratosphere has less overhead atmosphere, so it has more intense radiation than the troposphere. This radiation, much of it ultraviolet, creates **photochemical reactions** such as the conversion of molecular oxygen ($O_2$) into **ozone** ($O_3$). This is why the **ozone layer** exists in the stratosphere. Ozone effectively traps heat, though, so the temperature of the stratosphere increases with altitude, which is why very little mixing occurs in the stratosphere.

### 12.3.2 REGIONAL TRANSPORT

Operating at a regional scale (for example, hundreds of kilometers in size) are the **regional transport** processes familiar from their appearance on daily weather forecasts: high- and low-pressure systems. Both are created from the interaction of cold and warm air masses near the surface.

**Low-pressure systems** are created when air that is warmed by surface heating begins to rise. **High-pressure systems** are typically created when air is descending to the ground. The pressure systems thus have unique vertical transport tendencies—upward for low pressure, downward for high pressure. There are more forces, however, than the rising or falling of air. With pressure changes at different points in the atmosphere, air also begins to move laterally from high pressure to low pressure. Table 12.2 compares the resulting atmospheric consequences (including air quality) that result from low- and high-pressure systems in the Northern Hemisphere. Note how low-pressure systems can result in improved air quality from enhanced dispersion and cleansing due to precipitation.

### 12.3.3 LOCAL TRANSPORT

Much of the human population resides along the coast. In fact, 21 of the world's 33 megacities are located in coastal areas, and the average population density in coastal areas is twice the global average. Coastal locations are prone to additional transport patterns created by the **land–sea interface** and the resulting temperature differences that develop due to the differential heating that occurs. This results in sea–land breezes (Figure 12.3). In the morning, land heats up faster than the water. The air over the land subsequently warms faster and rises through convection (much as a hot-air balloon rises). As the air over land rises, air moves in from over the water to replace it. This

## Table / 12.2

### Atmospheric Consequences of Low- and High-Pressure Systems in the Northern Hemisphere

| Consequence | Low Pressure | High Pressure |
| --- | --- | --- |
| Lateral air movement | Inward | Outward |
| Rotational air movement | Counterclockwise | Clockwise |
| Vertical air movement | Upward | Downward |
| Typical weather conditions | Windy, cloudy, precipitation, storms | Calm, clear skies |
| Air quality | Improving due to enhanced dispersion and cleansing | Worsening due to stagnation and sunshine |

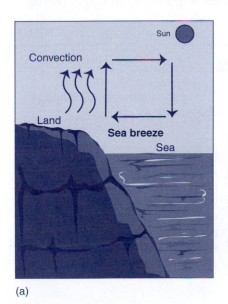

(a)

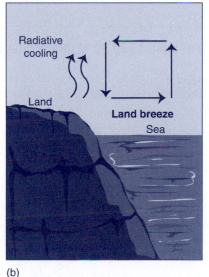

(b)

**Figure 12.3   Impact of the Land–Sea Interface on Temperature**   (a) In the morning, the sun heats up the land surface more quickly, resulting in a sea breeze, the movement of air from sea to land. (b) In the evening, the land cools more quickly and results in surface air moving from land to sea, the land breeze. Air pollutants will be carried along with these winds.

movement of air is referred to as a sea breeze, a persistent wind from the ocean or extremely large lakes, to land.

As shown in Figure 12.3, these movements create a circulation cell during the morning. In the late afternoon and early evening, the sun is less intense, and due to less radiative heating, the surface begins to cool. The land cools faster, cooling the air over it more quickly. Cooler air becomes denser and descends toward the surface. Ultimately, a circulation cell opposite to that of the morning results, with a persistent breeze, known as a land breeze, blowing from the land to the ocean.

These circulation patterns can strongly influence air quality over the course of a day, particularly in coastal urban environments. In the morning, relatively clean air is carried from the sea to the land, but it is quickly polluted by human activities. Pollutant concentrations can build as the air slowly moves inland. In the evening, the emissions over a city can be blown out to sea, decreasing the urban air pollution but with potential impacts to the aquatic ecosystem. Within a city, this diurnal pattern of sea and land breeze ensures that opposite sides of the city take turns being downwind throughout the day. However, if emissions are concentrated in a particular part of the day, then the people who are downwind during that time may be exposed to an inequitable amount of pollution.

### 12.3.4 AIR STABILITY

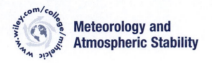

**Meteorology and Atmospheric Stability**

The vertical movement of air is strongly influenced by the stability of the atmosphere and can be assessed by measuring the vertical temperature profile of the atmosphere. The rate of change of temperature with altitude is referred to as the **lapse rate**, $\Gamma$. Typically, the temperature decreases with altitude. (A tropospheric average is $-6.5°C/km$.)

The atmospheric lapse rate is always changing. Time of day (solar heating), topographic features, and larger-scale weather influences will produce a lapse rate different from the tropospheric average. Therefore,

---

**Box / 12.1**  **Relating the Shape of a Plume to Atmosphere Stability**

A classic indicator of the current stability of the atmosphere is the behavior of pollutant plumes emanating from smokestacks. Characteristic plume shapes are suggestive of the state of the atmosphere. A few profiles are provided in Figure 12.4.

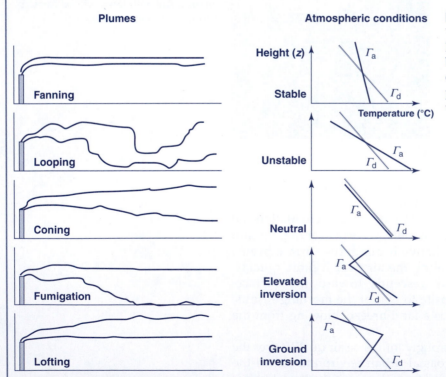

**Figure 12.4** **Plume Depictions for Various Atmospheric Conditions** Plume names and atmospheric stability conditions are in the center for each situation. The measured atmospheric lapse rate ($\Gamma_a$) is shown, along with the standard lapse rate ($\Gamma_d$).

## Table / 12.3

### General Stability Conditions Encountered in the Atmosphere

| Stability Condition | Description |
| --- | --- |
| Stable | This situation exists when the atmospheric temperature changes more dramatically with altitude than the temperature of the air parcel that contains the pollutant(s). The result of this suppression of vertical movement is less vertical mixing of the pollutant and therefore less dilution. Inversions are an example of stable conditions. |
| Unstable | This situation exists when the atmospheric temperature changes less dramatically with altitude than the temperature of the air parcel that contains the pollutant(s). Comparing the average atmospheric lapse rate ($-6.5°C/km$) to the standard lapse rate ($-9.8°C/km$), it is apparent that the atmosphere is typically in an unstable condition. Unstable conditions enhance mixing, resulting in enhanced dispersion of pollutants and lowered pollutant concentrations. |
| Neutral | The atmosphere exerts no force on pollutant emissions that move vertically. This situation occurs when the temperature changes for the atmosphere and the air parcel containing the pollutant(s) are nearly identical. |

measurements, typically through routine weather balloon launches, are used to reveal the current state of the vertical temperature profile for a given location.

The measured *atmospheric lapse rate* is denoted $\Gamma_a$, and the *standard lapse rate* is denoted as $\Gamma_d$. The atmosphere changes temperature according to the atmospheric lapse rate, and air parcels that contain pollutant emissions are assumed to change temperature according to the standard lapse rate. For the standard lapse rate, the air is assumed to be dry, behave adiabatically with the surrounding atmosphere, and have a lapse rate of $-9.8°C/km$.

By comparing the standard lapse rate with the current atmospheric lapse rate, we can assess the stability of the atmosphere, as well as the vertical movement of the emissions through the atmosphere. Table 12.3 compares the three general **stability conditions** for the atmosphere; **stable, unstable,** and **neutral**.

**Inversions** are a special category of strongly stable air. During an inversion, the temperature profile of the atmosphere (which typically decreases with altitude) is inverted; thus, the temperature profile increases with altitude. Pollutant emissions released into such an atmosphere have little ability to move upward, due to the strong stabilizing force of the atmosphere, so little dilution of pollutants occurs. Consequently, some of the worst air quality occurs when inversions develop.

Several factors (mostly geographical) increase the likelihood of inversions. Common inversion-prone locations include coastal settings with nearby hills or mountains (for example, Los Angeles), plains downwind of mountains (for example, Denver), and valleys. As stated previously, coastal locations tend to support large human populations (and more emissions) as well. Topographic features such as these can produce inversions, as shown in Figure 12.5.

**Ground Level Plume Simulator**

www.wiley.com/college/mihelcic

**Figure 12.5  Inversions Produced by Topographic Features**   (a) Surface inversion in a valley and (b) Subsidence inversion in a coastal setting with nearby mountains. The vertical line shows how temperature (*T*) varies with height above the surface (*z*). The inversion is recognized by the portion of the plot with a positive slope. Within and below the inversion, pollutant emissions are not mixed in the vertical direction, so pollutants build up until the inversion is broken. The upper-altitude winds can then transport the pollution elsewhere.

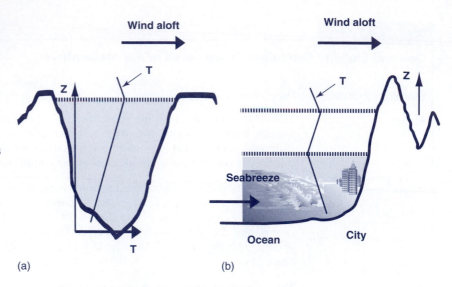

(a)

(b)

In valleys, as evening arrives, air cools and becomes denser, pooling at the bottom of the valley. Only with strong daytime heating from the sun will the surface air warm, begin to rise, and then break apart the inversion. In these valley surface inversions (Figure 12.5), it is not unusual for the inversion to extend to the top of the valley. As the inversion breaks down, pollutants that are trapped within the inversion can be carried away by wind.

Subsidence inversions arise from descending air masses. High-pressure systems can create these conditions via downward motion. Air that is also moving from high to low elevations (say, having passed over mountains and now descending into plains) also can create this type of inversion. As the air descends, it encounters high atmospheric pressure, is compressed, and warms, resulting in warm air atop cooler surface air. Because of these mechanisms, the subsidence inversion can be found well above the ground level, essentially creating a cap on the air over an area. Pollutant emissions below the inversion can mix up to, but not through, the inversion.

A critical difference between surface inversions and subsidence inversions is their persistence. Surface inversions are often due to daily heating and cooling. They can be common occurrences for certain valley locations, but they tend to break apart with the rising sun and surface heating that follows. Subsidence inversions, while typically less frequent for many locations, can be long-lived, allowing air quality to worsen over several days. The combination of influencing factors results in site-specific chances for the occurrence of inversions.

### 12.3.5  INDOOR AIR

From an exposure perspective, the indoor environment is a potential significant contributor to environmental risk. This is a critical driver in the **green building** design movement. As an engineered environment, the indoor setting influences pollutant movement and fate very differently than the atmospheric processes described previously. The movement of air in an indoor environment is achieved through the design (features or flaws) of ventilation systems. Due to the controlled nature of ventilation systems, indoor air, unlike ambient air, can be conditioned to specific temperature, moisture, and pollutant levels, if desired.

Natural systems capitalize on wind and natural indoor/outdoor temperature differences to create ventilation. **Natural ventilation** systems are best used in mild or moderate climates with minimal heating or cooling demands. It is also best used in locations with minimal pollution, as it is difficult to integrate pollution-control systems with passive ventilation. Natural ventilation systems capitalize on wind and thermally generated pressures. The design of such systems focuses on channeling these forces primarily through the clever design of openings, building geometry, and materials. Obviously, these types of systems are more sustainable because they do not require energy to move air, and also they do not require the many natural resources that go into mechanical ventilation systems (such as the metals in ductwork).

Wind pressure is created when air strikes a building, imparting a positive pressure on the windward face and negative pressure on the other faces. This causes air to enter openings and move through the building from the high-pressure to the low-pressure region, as depicted in Figure 12.6.

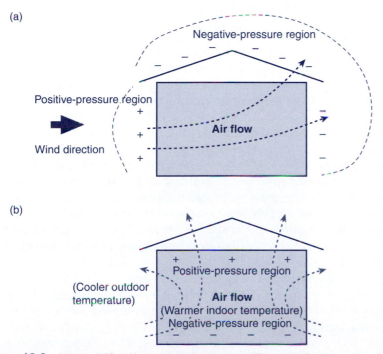

**Figure 12.6** **Natural Ventilation Mechanisms** These mechanisms may be either (a) wind-driven air flow or (b) thermal-driven air flow. The flow pattern for thermal-driven air occurs when outside temperatures are cooler than indoor temperatures. In real conditions, both wind and thermal effects are likely to occur. In general, wind-driven ventilation does not become important until wind speeds exceed 5 m/s. Below this wind speed, thermal ventilation dominates. In Chapter 14, Figure 14.13 provides examples of passive ventilation that can be used to eliminate or minimize the need for mechanical heating and cooling.

## Healthy Homes

http://www.epa.gov/region1/
healthyhomes/

Thermal pressure results from the difference between indoor and outdoor air temperatures. The temperature difference creates a density difference, with lower-density (warmer) air rising and creating a stack effect through the building. When the warmer air is in the building, it drafts cooler air through openings at the lower part of the building as it exits through higher openings. The air movement is reversed when the warmer air is outside the structure.

In most large buildings (commercial, industrial, and multiunit residential) and many residences, air flow is controlled through mechanical ventilation systems. Such systems not only provide controlled ventilation to most spaces within the structure, but often are coupled with heating, cooling, and air pollutant control subsystems. Collectively these systems are referred to as HVAC (heating, ventilation, air conditioning). Many modern systems include heat exchangers that transfer energy in exhaust air to inlet air, thereby minimizing cooling and heating costs. Natural ventilation systems are always preferred in terms of minimizing the amount of energy and materials that go into a building over its life cycle.

example/12.1 Clearing Kitchen Smoke with Natural Ventilation

More than half the world's population relies on solid fuels for use around the home. These fuels and the methods used result in extremely smoky conditions. Women and young children accompanying their mothers are at most risk for this exposure. This type of indoor air pollution is annually responsible for 1.6 million deaths, one every 20 seconds (World Bank, 2007b).

You are asked to investigate the problem of elevated indoor air pollution in an indigenous community. The goal of the project is to create a culturally appropriate solution that minimizes exposure to the smoke during cooking. In this location, it is common practice to cook with wood and charcoal in essentially an open fire, in a semi-enclosed area of the simple house structures (Figure 12.7). Your best idea is to improve the ventilation of the homes in order to reduce the fine-particulate levels ($PM_{2.5}$).

Assume the average house dimensions are: $3\,m \times 5\,m \times 2.5\,m$; therefore, $V = 37.5\,m^3$. After several days of measurements in homes around the community, you have gathered the following data: background $PM_{2.5}$ concentration, $C_a = 5\,\mu g/m^3$; average indoor $PM_{2.5}$ concentration during cooking, $C = 700\,\mu g/m^3$; peak indoor $PM_{2.5}$ concentration, $C_{max} = 2{,}000\,\mu g/m^3$; and $PM_{2.5}$ emission estimate from the fire, $E = 440\,\mu g/min = 7.33\,\mu g/s$.

You also determine, after the stove burns out, the time it takes for the indoor air to go from peak concentration back to ambient background concentrations, $\Delta t = 100\,min$. To design better ventilation, you want to calculate the current average ventilation rate for the structure, using the data you collected.

**Figure 12.7** Three-Stone Fire Typically Used for Cooking in Many Parts of the Developing World

## solution

The mass balance on $PM_{2.5}$ is written (in words and symbols) as follows:

$$\begin{bmatrix} \text{rate of change in} \\ PM_{2.5} \text{ mass} \end{bmatrix} = \begin{bmatrix} \text{mass of } PM_{2.5} \\ \text{entering the home} \end{bmatrix} - \begin{bmatrix} \text{mass of } PM_{2.5} \\ \text{leaving the home} \end{bmatrix} - \begin{bmatrix} PM_{2.5} \\ \text{decay rate} \end{bmatrix}$$

$$V\frac{dC}{dt} = QC_a + E - QC - kCV$$

Here, $Q$ is the natural ventilation rate into and out of the house, and $k$ is the $PM_{2.5}$ decay rate coefficient, given in $s^{-1}$ (from literature review, $k = 1.33 \times 10^{-4}\,s^{-1}$).

Without better data, we can estimate the rate of change ($dC/dt$) using the peak to background data:

$$V\frac{dC}{dt} = V\frac{\Delta C}{\Delta t} = 37.5\,m^3\,\frac{2{,}000\,\mu g/m^3 - 5\,\mu g/m^3}{0 - 100\,\text{min}}$$

$$= -748\,\mu g/\text{min} = -12.5\,\mu g/s$$

Solve for $Q$, the ventilation rate:

$$Q = \frac{\left(kCV - E + V\dfrac{dC}{dt}\right)}{C_a - C}$$

Substitute the other measurements:

$$Q = \frac{(1.33 \times 10^{-4}/s \times 700\,\mu g/m^3 \times 37.5\,m^3) - 7.33\,\mu g/s - 12.5\,\mu g/s}{5\,\mu g/m^3 - 700\,\mu g/m^3}$$

$$= 0.02\,m^3/s = 84.5\,m^3/hr$$

To measure indoor air quality, a common calculation is air changes per hour (ACH). This is simply the inverse of the retention time:

$$\text{ACH} = \frac{Q}{V} = \frac{84.5\,m^3/hr}{37.5\,m^3} = 2.25 \text{ per hour}$$

Essentially the ventilation purges the house volume 2.25 times over the course of an hour. This is a strong ventilation rate; probably no design alterations could be made to the home to enhance natural ventilation appreciably. Other strategies must be used to lower the $PM_{2.5}$ exposure.

## Class Discussion

What would you recommend as an appropriate solution for the community members in Example 12.1? Would you suggest a more efficient stove design, change in type of fuel use (which would cost money), installation of a chimney to divert smoke out of the home, or some behavioral change? Do you think behavioral change is easy to enact? If not, what is required to effect positive behavioral change for this particular situation that is culturally and economically appropriate?

## Clean Air Act

http://www.epa.gov/lawsregs/laws/caa
.html

## 12.4 Air Pollutants

### 12.4.1 CRITERIA POLLUTANTS

The U.S. federal government, through the Environmental Protection Agency (EPA), has identified six pollutants of special concern: ozone, particulate matter, sulfur dioxide, nitrogen dioxide, carbon monoxide, and lead. They are called **criteria air pollutants** on the basis of criteria regarding specific health and/or environmental effects. The criteria pollutants make this list for two primary reasons: (1) they are very common, so most people and the environment are exposed to them; and (2) they can do great harm.

To minimize health or environmental impacts, the EPA and the World Health Organization (WHO) have set *maximum recommended atmospheric concentrations* of the criteria pollutants. In the United States, these concentration limits are approved by Congress and detailed in the Clean Air Act (and subsequent amendments). The **Clean Air Act (CAA)** was passed in 1963 and stressed that there was scientific evidence linking air pollution to health and environmental impacts. In 1970, shortly after the formation of the EPA, the **Clean Air Act Amendments (CAAA)** were passed into law. The CAAA identified the criteria pollutants and set concentration limits that should not be

## Table / 12.4

### Air Quality Standards for Criteria Pollutants

| Criteria Pollutant | World Health Organization (WHO)* | Environmental Protection Agency (EPA) | California Air Resources Board (CARB) |
|---|---|---|---|
| Ozone (O$_3$) | 0.05 ppm$_v$ (8 hr average) | 0.075 ppm$_v$ (8 hr average) | 0.07 ppm$_v$ (8 hr average) |
| Carbon monoxide (CO) | 9 ppm$_v$ (8 hr average) 26 ppm$_v$ (1 hr average) | 9 ppm$_v$ (8 hr average) 35 ppm$_v$ (1 hr average) | 9 ppm$_v$ (8 hr average) 20 ppm$_v$ (1 hr average) |
| Nitrogen dioxide (NO$_2$) | 0.023 ppm$_v$ (annual average) | 0.053 ppm$_v$ (annual average) | 0.030 ppm$_v$ (annual average) |
| Sulfur dioxide (SO$_2$) | 0.007 ppm$_v$ (24 hr average) | 0.14 ppm$_v$ (24 hr average) | 0.04 ppm$_v$ (24 hr average) |
| Lead (Pb) | 0.5 $\mu$g/m$^3$ (annual average) | 1.5 $\mu$g/m$^3$ (quarterly average) | 1.5 $\mu$g/m$^3$ (30 day average) |
| Respirable particulate matter (PM$_{10}$) | 50 $\mu$g/m$^3$ (24 hr average) | 150 $\mu$g/m$^3$ (24 hr average) | 50 $\mu$g/m$^3$ (24 hr average) |
| Fine particulate matter (PM$_{2.5}$) | 25 $\mu$g/m$^3$ (24 hr average) 10 $\mu$g/m$^3$ (annual average) | 35 $\mu$g/m$^3$ (24 hr average) 15 $\mu$g/m$^3$ (annual average) | 35 $\mu$g/m$^3$ (24 hr average) 12 $\mu$g/m$^3$ (annual average) |

*Recommended Limits.

exceeded in any region, particularly urban environments. These concentration limits are referred to as *air quality standards*. The standards targeting the criteria air pollutants are so important they are designated **National Ambient Air Quality Standards (NAAQS)** and vary with each pollutant, as shown in Table 12.4.

The NAAQS are legally enforceable. In the United States, each state is required to adopt the federal standards, but the states also have the freedom to set more stringent standards. For example, California, through its California Air Resources Board (CARB), has established stricter standards. This is in recognition of California's large population, sensitive environmental resources, numerous pollution sources, and unique topographical features that enhance inversion formation. The CARB air quality standards also are legally enforceable. WHO guides nations with or without regulatory agencies, so its guidelines are recommended limits. Table 12.5 provides an overview of the source(s), health concerns, and current status of the six criteria air pollutants.

There is no similar set of enforceable standards for indoor air. However, suggested limits have been promoted by several organizations, including the Occupational Safety and Health Administration (OSHA), the National Institute for Occupational Safety and Health (NIOSH), and the American Society of Heating, Refrigeration, and Air-conditioning Engineers (ASHRAE).

**Spatial Distribution of Air Quality**

NITROGEN DIOXIDE **Nitrogen dioxide** is produced when air (which contains more than 78 percent N$_2$) is used to combust fuels. The high temperature (and sometimes pressure) that can exist during combustion converts the oxygen and nitrogen to NO, which is then quickly transformed to NO$_2$. Their combined presence is denoted as **NO$_x$**.

## Table / 12.5

### Source(s), Health Concern, and Current Status of the Criteria Air Pollutants

| Criteria Pollutant | Source | Concern | Status |
|---|---|---|---|
| Nitrogen oxide ($NO_x$) | Produced when $N_2$ in air reacts with $O_2$ during fuel combustion. | Imparts brownish haze to atmosphere.<br><br>Can form small particles that are respiratory irritants, thereby producing breathing problems in sensitive individuals.<br><br>Critical ingredient in the formation of ozone.<br><br>Some $NO_2$ may react with water vapor to form nitric acid, thus forming acidic rain, snow, or fog. | The average U.S. nitrogen concentration has been declining fairly consistently since 1980.<br><br>The current national average concentration is about two thirds the 1980 level, and all urban areas are now in compliance with the nitrogen dioxide NAAQS. However, ozone and haze problems persist in most urban areas. |
| Sulfur dioxide ($SO_2$) | Produced when sulfur-containing fuels are burned or metals are extracted from sulfur-containing ore. The sulfur becomes oxidized, producing $SO_2$.<br><br>In the United States about two-thirds of emissions come from burning coal to generate electricity. | Dissolves easily in water vapor, becoming sulfuric acid, thus forming acidic precipitation.<br><br>Can form small sulfate particles that elevate rates of respiratory disease and death, particularly in children and the elderly.<br><br>In addition to haze created by the sulfate particles, $SO_2$ can accelerate the decay of building materials, including historical sculptures and architecture. | In the United States the current national average concentration is about one-third the 1980 level. All urban areas meet the NAAQS for $SO_2$.<br><br>Overseas, $SO_2$ damage is extensive downwind of major coal-burning facilities. In China, acid rain now affects 25% of all land area and 33% of agricultural land. On top of this, China is expected to double its consumption of coal by 2020. |
| Carbon monoxide (CO) | Produced from incomplete combustion of fuels.<br><br>In the United States more than half of emissions come from on-road vehicles. | Is absorbed easily into blood. At low levels, the health threat is most serious for those with cardiovascular problems. A single exposure may impair that person's ability to function and could lead to other cardiovascular problems.<br><br>While high levels are rarely encountered in the ambient atmosphere, asphyxiation can occur in indoor environments, often through a combination of poorly functioning heating systems and inadequate ventilation. | The national average concentration in the United States is nearly 75% lower than in 1980.<br><br>While most urban areas are well below the NAAQS, there still remain several counties out of compliance. |
| Lead (Pb) | Lead is added to some products. Historically, a major source of lead was motor vehicles burning gasoline with lead added to improve the engine performance. High levels of atmospheric lead are | Accumulates in the body and produces most significant impacts to developing bodies of children, even at low levels.<br><br>Excessive exposure affects the nervous system, resulting in seizures, mental retardation, memory problems, and behavioral issues. | Leaded fuel has been banned in the United States since 1986. Average U.S. atmospheric lead concentrations are 96% less than in 1980, a direct result of gasoline reformulation. (Some leaded fuel is still available for limited off-road vehicle use.) |

(Continued)

| Criteria Pollutant | Source | Concern | Status |
|---|---|---|---|
| | found near lead smelters, waste incinerators, and lead-acid battery manufacturers. | Low levels have been shown to impair brain development in the fetus and young children, resulting in lowered IQs and learning difficulties. Can also increase heart disease and anemia.<br><br>Unlike many other air pollutants, most health impacts from lead are permanent. | There are still two lead nonattainment areas left in the United States (located in Missouri and Montana), and people exposed to high levels of lead still live with the health consequences. |
| Ozone ($O_3$) | Ozone (smog, urban ozone) has no direct sources. It is created by a complex sequence of chemical reactions driven by sunlight. | Known to be a strong respiratory irritant. Individuals with compromised respiratory systems are particularly at risk. This includes people with lung disease, children, older adults, and active people.<br><br>Specific impacts are lung inflammation, breathing difficulties, aggravated asthma, increased susceptibility to respiratory diseases such as pneumonia, and permanent lung damage from repeated exposure. | Ozone is proving to be a difficult issue to solve. Figure 12.8 shows the several hundred counties in the United States that exceed the 8-hr NAAQS level.<br><br>A large part of the problem is the inability to reduce the emissions of ozone precursors ($NO_x$ and RH) using conventional approaches. |
| Fine particulate matter | Particles are not a uniform combination of compounds; some are acids, organic chemicals, metals, or soils.<br><br>Some particles are created through atmospheric reactions of gases, as in the case of sulfur dioxide or nitrogen dioxide.<br><br>Collectively, all particles 10 $\mu$m and less in diameter are denoted as $PM_{10}$. All particles 2.5 $\mu$m and less in diameter are denoted as $PM_{2.5}$. | Health research suggests that particles <10 $\mu$m in diameter are problematic for the respiratory system. Primary particles can be of this size, but most are larger. Most secondary particles are <10 $\mu$m, many are much smaller than that. | The NAAQS for $PM_{10}$ was created in the 1980s. In the late 1990s, EPA established a NAAQS for $PM_{2.5}$. Progress has been made on both fronts. $PM_{10}$ is down 30% since 1987. $PM_{2.5}$ is down 10% since 2002.<br><br>While there are some $PM_{10}$ nonattainment areas, areas that exceed $PM_{2.5}$ NAAQS are widespread (Figure 12.9).<br><br>Most $PM_{10}$ exceedances are found in rural, dry areas, whereas most $PM_{2.5}$ violations are in large metropolitan areas. |

**CARBON MONOXIDE** **Carbon monoxide** is produced from the incomplete combustion of fuels. This usually arises from an insufficient amount of air for the amount of fuel. The cause of this inadequate air-to-fuel ratio can be poorly operated or maintained equipment, airflow limitations, or low temperatures. Major sources of carbon monoxide include vehicles, power plants, and wood-burning stoves. Natural sources, (such as wildfires) can also produce significant emissions.

**OZONE** Unlike the other criteria pollutants, ozone has no direct sources. Instead ozone is created by a complex sequence of chemical reactions driven by sunlight. The overall process of this photochemical system of reactions is summarized as follows:

$$RH + NO_x \xrightarrow{\text{sunlight}} O_3 + \text{other} \tag{12.1}$$

where RH is an abbreviation for reactive hydrocarbons. The reactive hydrocarbons are a class of compounds including many commercial, industrial, and personal products (benzene, propane, components of gasoline, and others). Some of these materials escape to the atmosphere accidentally, and some may be released in small quantities through permitted discharges. As we previously discussed, the nitrogen oxides (notably NO and $NO_2$) are produced during the combustion of fuel with air. When these compounds are present in the air and the sun rises, a sequence of chemical reactions begin, and the formation of ozone commences.

Typically, the concentration of ozone builds through the day, peaking in early afternoon in sync with the peak in sunlight strength and air temperature, both of which strengthen the reactions producing ozone. After the sun sets, the photochemical sequence summarized in Equation 12.1 stops. This allows reactions that destroy ozone as well as physical processes that remove ozone to dominate and reduce the ozone concentrations through the evening.

Ozone transport is a problem for downwind receptors. At typical wind speeds, it is easy for ozone (and its precursors) to be transported hundreds of kilometers in a day. In remote locations (such as national parks and over the ocean), most of the ozone is transported from upwind urban sites. The ozone concentration in most urban settings is the combination of locally generated ozone plus upwind sources. In especially dense urban corridors, such as the U.S. East Coast, there are substantial upwind contributions (Figure 12.8).

Ozone also has detrimental effects on ecosystems. Sensitive plants can be damaged by ozone, making them more susceptible to diseases, insects, weather stresses, and competition. Such plant damage has certain ripples through the food web of ecosystems. As not all plants are similarly sensitive to ozone, such pollution can decrease biodiversity in an ecosystem. In agricultural systems, ozone reduces crop yields and forest productivity. Aesthetic impacts also are evident, as ozone damages the appearance of vegetation in urban green space, national parks, and recreational areas.

In the European Union, ozone exposure is estimated to cause 20,000 premature deaths every year (EEA, 2007). The increased need for respiratory medicines (for example, inhalers) results in 30 million additional medication-use days per year. While it is impossible to completely represent this cost to society, annual health impacts are estimated to cost $16.7 billion and annual crop damage to cost $3.9 billion (EEA, 2007). Some species are especially sensitive to ozone (for example, wheat), so these economic impacts are not distributed equally across the agricultural sector.

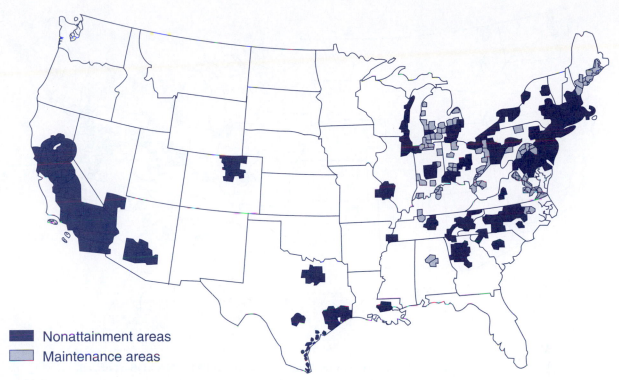

**Figure 12.8 Nonattainment and Maintenance Areas for Ozone** As of 2007, several hundred U.S. counties exceed the ozone 8-hr NAAQS (designated nonattainment areas). Over 100 countries have recently improved to a point where they are no longer in nonattainment status (termed maintenance areas).

From EPA (2007).

**PARTICULATE MATTER** Particulate matter (PM) is due to a complex mixture of small airborne solids and liquids. The primary source of fine particulate matter is combustion. Today, vehicles, power plants, industry, and agricultural practices provide abundant emissions around the world. Fuel combustion (for power generation), transportation, and industrial processes each provide about one-third of all fine-particulate emissions

Some particles are generated directly from a source and are known as *primary particles*. Sources of primary particles include construction sites, agricultural fields, unpaved roads, fires, and smokestacks. These particles often arise from forces that grind or abrade solid materials into particles small enough to become entrained in air. Other particles form from reactions in the atmosphere and are called *secondary particles*. Major sources of these precursors include power plants, industry, and vehicles. Figure 12.9 shows the locations of counties in the United States that exceed the $PM_{2.5}$ standard.

Table 12.6 provides concentrations of $PM_{10}$ in several major cities of the world. Remember that WHO guidelines for $PM_{10}$ advise limiting it to 50 $\mu g/m^3$ (24 hr average). This pollution has negative health consequences. For example, a recent World Bank study estimates the air pollution consequences for China are 400,000 to 450,000 premature deaths

**Figure 12.9** U.S. Counties That Exceeded the PM$_{2.5}$ NAAQS as of 2007
These counties are referred to as nonattainment areas.

From EPA (2007).

## Table / 12.6

**Annual Average Concentrations of Particulate Matter in Selected Major Cities of the World** PM$_{10}$ data are presented because only a few countries routinely measure PM$_{2.5}$. The WHO guideline for PM$_{10}$ is 50 μg/m$^3$ (24 hr average).

| Country | City | Population (thousands) | PM$_{10}$ (μg/m$^3$) |
|---|---|---|---|
| Brazil | Sao Paolo | 18,333 | 40 |
| Canada | Toronto | 5,312 | 22 |
| China | Beijing | 10,717 | 89 |
| | Chongqing | 6,363 | 123 |
| | Tianjin | 7,040 | 125 |
| Ghana | Accra | 1,981 | 33 |
| India | Kolkata | 14,277 | 128 |
| | Madras | 6,916 | 37 |
| | Delhi | 15,048 | 150 |
| Indonesia | Jakarta | 13,215 | 104 |
| Ireland | Dublin | 1,037 | 19 |

## Table / 12.6

(Continued)

| | | | |
|---|---|---|---|
| Japan | Tokyo | 35,197 | 40 |
| Kenya | Nairobi | 2,773 | 43 |
| Mexico | Mexico City | 19,411 | 51 |
| Norway | Oslo | 802 | 14 |
| Philippines | Manila | 10,686 | 39 |
| Thailand | Bangkok | 6,593 | 79 |
| Turkey | Istanbul | 9,712 | 55 |
| Ukraine | Kiev | 2,672 | 35 |
| United Kingdom | London | 8,505 | 21 |
| United States | Chicago | 8,814 | 25 |
| | Los Angeles | 12,298 | 34 |

SOURCE: World Bank, 2007b. Data for 2004.

each year (World Bank, 2007a). By comparison, the number of deaths associated with waterborne disease in China is estimated to be 66,000 per year.

**SULFUR DIOXIDE** Sulfur is present in many raw materials, including oil, coal, aluminum, copper, and iron. When such fuels are burned or metals are extracted from ore, the sulfur can be oxidized, producing several sulfur oxide gases. The most common of these is **sulfur dioxide, $SO_2$**.

**LEAD** Like sulfur, **lead** also is found naturally in the environment, but it is also added to some products. Historically, a major source of air pollution from lead was motor vehicles burning gasoline with lead added to improve the engine performance. More recently, because the United States and many other countries have eliminated leaded gasoline, the processing of metals is becoming the major source of lead emissions. Leaded gasoline is still used in many developing nations (see Figure 12.10), which is a major reason why exposure to lead contributes 11 percent of the global environmental risk. (For more on this subject, refer back to Chapter 1.)

**Lead Air Pollution**
http://epa.gov/air/lead/

### 12.4.2 HAZARDOUS AIR POLLUTANTS

**Hazardous air pollutants (HAPs)** are not covered by ambient air quality standards. They tend not to be uniformly present in the global, regional, or local atmosphere. Hence, unlike the criteria pollutants, these pollutants are not routinely subject to ambient air monitoring.

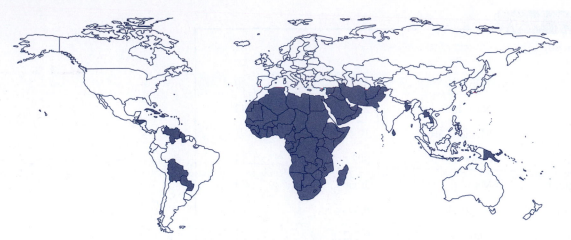

**Figure 12.10** **Availability of Leaded Gasoline Worldwide** Use of leaded gasoline is associated with high levels of lead in ambient air. Countries using only leaded gasoline are shaded.

Data from UNEP (2007).

Instead, emissions are monitored at the source. EPA compiles a list of HAPs. Because of their potential impacts, hazardous air pollutants are also called *air toxics*. EPA currently lists 188 HAPs targeted for emission reduction efforts. These pollutants tend to fall into four general categories: (1) pesticides, (2) metals, (3) organics, and (4) radionuclides.

Sources of HAPs include a few natural sources, for example, volcanoes and forest fires. Also, many organic HAPs are present in vehicle exhaust. Factories, refineries, and power plants can release metals and organics. Pesticides may drift from agricultural fields. Radionuclides may be released accidentally from nuclear power plants or intentionally from the detonation of radioactive weapons. In addition to their emission to the ambient atmosphere, considerable potential exists for the release of HAPs in homes, the workplace, and other indoor settings, often through poor choices of building materials or poor maintenance practices.

Exposure to air toxics occurs in several ways. Breathing contaminated air introduces the HAP deep inside the body, where it may stick to the lining of the lung or be carried along with the air into the bloodstream. Food may be contaminated by air toxics settling on produce or contaminated feed being consumed by livestock. Air toxics may attach to particulate matter that eventually is removed from the atmosphere via precipitation or gravitational settling. Some of this material falls onto water bodies, introducing the toxics into the water ecosystem. Some air toxics will settle onto land, thereby contaminating soil. People ingest soil accidentally (dust in the mouth, for example) or intentionally (on children's dirty hands), creating another path of entry.

Most reduction of HAPs has been achieved through substitution of green chemicals or materials for hazardous ones in the manufacturing or use of products. These practices are also being adopted in the home. Such efforts are necessary if overall exposures to air toxics are to be lowered appreciably.

**Where are the HAPs High?**
http://www.scorecard.org/
env-releases/hap/

**Common Indoor Air Pollutants, Sources, and Permissible Exposure Limits**

| Pollutant | Source | Permissible Exposure Limit (PEL) | Short-Term Exposure Limit (STEL) |
|---|---|---|---|
| Carbon monoxide | Stoves, furnaces | 50 ppm$_v$ | 400 ppm$_v$ |
| Formaldehyde | Carpets, particleboard, plywood, finishes, subflooring, paneling, furniture | 0.75 ppm$_v$ | 2 ppm$_v$ |
| Particulate matter | Materials processing, cooking, carpets | 5 mg/m$^3$ | |
| Volatile organic compounds (VOCs) | Solvents, cleaning supplies, paints, home furnishings, personal-care products | Compound specific | |
| Radon | Diffusion from radium in soil | 100 pCi/L* | |
| Ozone | Photocopiers, printers, air-cleaning devices | 0.1 ppm$_v$ | 0.3 ppm$_v$ |
| Biological agents | Mold, fungi, pets | N/A | |
| Environmental tobacco smoke (ETS) | Cigarettes, cigars, pipes | N/A | |
| Asbestos | Heating system and acoustic insulation, floor and ceiling tiles, structural fireproofing | 0.1 fiber/cm$^3$ | |

*Maximum permissible concentration, not PEL.
SOURCE: Exposure limits from OSHA.

## 12.4.3 INDOOR AIR POLLUTANTS

Indoor air pollution differs from ambient pollution in several regards. First, the sources are different. Also, pollutants may linger longer in the air, due to limited ventilation rates. A third difference is that, unlike ambient pollution, indoor air quality is rarely monitored. Finally, there is limited enforcement of air quality in most indoor settings.

The indoor environment may possess any of the criteria or hazardous air pollutants. The Occupational Safety and Health Administration (OSHA) has established **permissible exposure limits (PELs)** for some pollutants in the indoor workplace (Table 12.7). Workers' exposures may not exceed these levels, as OSHA does have the power to warn, cite, and fine violating companies. The OSHA Act required OSHA to set these standards to protect workers and help businesses provide safer working environments. The PELs are to be agreed upon through negotiations with industry, so progress has been slow since 1973. Until more PELs can be established, OSHA provides guidance on other pollutants through *threshold limit values (TLVs)*.

Both PELs and TLVs are set at a concentration not to be exceeded by a time-weighted average (TWA) exposure over the course of an 8 hr workday. The following equation expresses this TWA:

$$\text{TWA} = \frac{\sum_i C_i t_i}{\sum_i t_i} \tag{12.2}$$

where $t_i$ is the period of time during which an air sample is taken for analysis, and $C_i$ is the average concentration of the pollutant during the sample period.

As workers may change environments or sources change emissions, the TWA is a simple calculation that estimates the average exposure level over the course of a day. To protect workers from especially hazardous compounds, some pollutants also have short-term exposure limits (STELs), which are TWAs for a 15 min exposure. Regardless of whether a STEL exists, it is generally recommended that short-term exposures never exceed three times the TWA.

If a PEL or TLV is exceeded, OSHA can force the employer to take several steps, including employee health monitoring, air sampling, and control measures. Each of the PELs includes recommendations for air-sampling procedures, record keeping, control methods, and labeling of work areas.

example/12.2 Time-Weighted Average (TWA) Exposure

A worker is exposed to a chemical over the course of her 8 hr shift as shown in Figure 12.11. The PEL for this chemical is 6 ppm$_v$. Estimate the TWA, and check for safety violations.

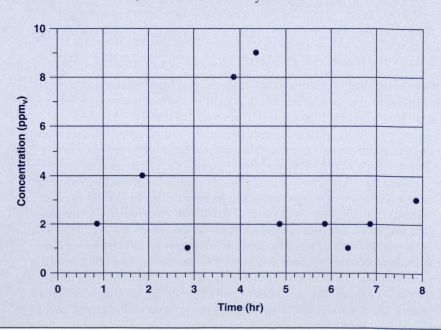

Figure 12.11   Exposure to Chemical over 8 hr Shift Used in Example 12.2

## solution

Using the data provided and Equation 12.2, the TWA can be determined as follows:

$$
TWA = \frac{\begin{pmatrix}(2\text{ ppm} \times 60\text{ min}) + (4\text{ ppm} \times 60\text{ min}) + (1\text{ ppm} \times 60\text{ min}) + (8\text{ ppm} \times 60\text{ min}) \\ + (9\text{ ppm} \times 30\text{ min}) + (2\text{ ppm} \times 30\text{ min}) + (2\text{ ppm} \times 60\text{ min}) + (1\text{ ppm} \times 30\text{ min}) \\ + (2\text{ ppm} \times 30\text{ min}) + (3\text{ ppm} \times 60\text{ min})\end{pmatrix}}{(60 + 60 + 60 + 60 + 30 + 30 + 60 + 30 + 30 + 60\text{ min})}
$$

$$= 3.4\text{ ppm}_V$$

The TWA of 3.4 $ppm_V$ is below the given PEL of 6 $ppm_V$.

Best practices suggest the short-term exposure should be less than 3 × TWA. The highest concentration observed in our study was 9 $ppm_V$, which is less than 3 × 3.4 = 10.2 $ppm_V$. Thus, the work environment meets the guidelines. However, use of a green manufacturing process would eliminate the need (and cost) of expensive workplace monitoring and potential liability from workplace exposure.

## 12.5    Emissions

**Emissions** are the amount of pollutant a source puts into the air, usually over a fixed amount of time; hence, emission rates are expressed as mass per time. Typical emission rate units depend on the source, pollutant, and problem of interest, but common units are g/s, kg/day, and tonnes/yr. In a few places, tons/day and lb./hr might still be used.

Emission rates can be highly variable. Emissions from the same source can change with time, and emissions from similar sources can be very different. Also, the composition of pollutants in the emission air can change dramatically with operating conditions.

**Air Quality in Your Neighborhood**
http://www.scorecard.org

### 12.5.1    EMISSION CLASSIFICATION

Table 12.8 reviews the many methods available to classify air emissions: (1) by specific pollutant, (2) by type of source, and (3) type of release (a) by natural processes, (b) as permitted emissions, (c) as fugitive emissions, and (d) as accidental emissions.

Classified by type of source, emissions may come from *stationary sources*, *point sources*, and *mobile sources*. For accounting purposes, these emissions are measured at the smallest region possible and then totaled as needed by the states. Figure 12.12 shows an example of a county-level presentation of VOC emission densities (annual emissions divided by the area of the county) for several eastern states. Such data sets help communicate emission patterns and may offer better input into computer models of air quality.

## Table / 12.8

### Classification of Air Emissions

| Ways to Classify Emissions | Explanation |
|---|---|
| Specific pollutant | Impacts are usually specific to a particular pollutant, as are emission reduction strategies. The criteria pollutants receive the most attention. A common list would include sulfur dioxide ($SO_2$), nitrogen oxides ($NO_x$), carbon monoxide (CO), particulate matter ($PM_{2.5}$, $PM_{10}$), ammonia ($NH_3$), and volatile organic compounds (VOCs). <br><br> Ammonia and VOCs are not criteria pollutants but are included because they are important precursors to the formation of criteria pollutants, fine particulate for ammonia, ozone for VOCs. |
| Type of source | One critical classification is whether the source is stationary or mobile. *Stationary sources* are fixed to a specific location. Stationary sources are usually further identified as point or area sources. <br><br> *Point sources* are large, often with the emissions entering the atmosphere from a stack (e.g., coal-fired power plants). *Area sources* are often many similar small sources that are better addressed collectively, rather than individually (e.g., gas stations that emit VOCs during storage and fueling). Individually, each gas station has a small amount of emissions, and it would be challenging to measure each around a city. Instead, average emission values can be used for each station and multiplied by the number of stations in the region to arrive at a total contribution. An area source can also be a process that emits over a larger region (e.g., windblown particle emissions from a construction site or from an agricultural field). <br><br> *Mobile sources* are subdivided into on-road mobile sources and nonroad mobile sources. The major contributors to on-road mobile source emissions are automobiles and trucks. Nonroad mobile sources include ships, airplanes, construction equipment, trains, and recreational vehicles. |
| Natural processes | In some areas, natural emissions may dominate the total for certain pollutants, and in these cases, there will be little impact from engineered emission-control strategies. Identifying such regions is done only through a careful auditing of emissions. <br><br> A major natural source is the emission of VOCs by certain trees. The Smoky Mountains in the southeastern United States are so named because of the natural haze generated from the forest VOC emissions. While this region is greatly influenced by anthropogenic emissions transported from urban areas, the natural VOCs are persistent and contribute substantially. Solving ozone problems in the area thus becomes even more challenging due to this natural source of VOC emissions and rapid population growth that is concentrated in vehicle-oriented communities. |
| Permitted emissions | Another important emission classification scheme, particularly to industry, is whether an emission is permitted, fugitive, or accidental. Permitted emissions are those granted by an agreement with the state air resource regulatory agency. Based on the process or activity proposed by the company, the state will allow a certain amount of emissions to be released to the atmosphere, the amount being determined so as to not result in degradation of the environment. The emissions you see coming from a stack are permitted releases, assuming the company is adhering to the emission rate prescribed by the state permit. |
| Fugitive emissions | Pollutant releases from an unconfined source. These sources are often area sources and outdoors. Heavy construction equipment kicking up dust on an unpaved road, earth-moving operations, or agricultural pesticide spraying are all |

<br/>

## Table / 12.8

(Continued)

| | |
|---|---|
| | examples of activities that create fugitive emissions. These are especially difficult to control. Generally, practices must be redesigned to solve such problems.<br><br>The vast majority of fugitive emissions are particulate-matter releases from large open sources; earth-moving, transport, and storage operations are common culprits. Fugitive-emission controls from these activities require changes in design and operation, often creating ways to minimize forces that liberate the material into the atmosphere, notably wind-driven and mechanical agitation. Wind barriers (e.g., vegetative rows and berms) and moisture control (e.g., routine surface-wetting programs) are two general engineered strategies to control these emissions. |
| Accidental emissions | Releases due to accidents or poor operation and maintenance of equipment. Most of these emissions are best controlled by diligent maintenance of equipment, including ducts, valves, and storage vessels. Engineering efforts in this area not only reduce air quality impacts, but for many industries, offer financial gains due to the elimination of inefficiencies. For some air quality issues, fugitive and accidental releases can rival permitted releases. Due to the nature of their release, it is much more difficult to quantify these emissions, though. |

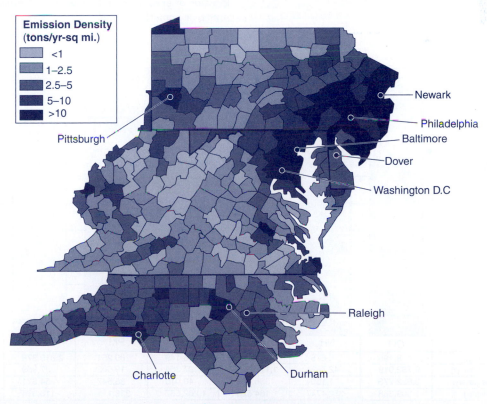

**Figure 12.12  Annual Average VOC Emission Densities (Excluding Biogenic Emissions) at the County Level in Several Eastern States, 2002**  For accounting purposes, emissions are estimated at the smallest region possible, then totaled as needed by the states.

From MARAMA (2005). Figure prepared under the direction of Dr. Serpil Kayin (MARAMA), using data obtained from MARAMA member state and local agencies. Dr. Jeffrey W. Stehr of the University of Maryland at College Park was the principal author of the report.

Stationary and mobile sources require very different control strategies. Additionally, some air quality issues are of such large scale that only a regional strategy will effectively improve the situation. Along the eastern coast of the United States, states in the heavily urbanized corridor often work together to address air quality issues. It is important to understand how much of a pollutant originates from each source type, as done in this 2002 total emission graph shown in Figure 12.13 for important pollutants in the Mid-Atlantic states.

Upon further examination of $SO_2$ emissions, Figure 12.13 suggests that most $SO_2$ emissions originated from stationary point sources (in the United States, two-thirds of $SO_2$ emissions originate from combustion of coal to generate electricity). Regional variability could also be assessed to determine whether emission-control solutions should be applied uniformly or selectively (see Figure 12.14).

**Class Discussion**

Note how each of the criteria pollutants (or the ammonia precursor) has a different mix of source contributions. Understanding each mix is the start in designing effective mitigation strategies. How might solutions for $SO_2$ differ from those for CO, for example?

**Class Discussion**

Carefully examine Figure 12.14. What causes the large differences in $SO_2$ emissions among the seven states and District of Columbia? Is it population, consumption, demographics (age, income, education), presence (and use) of nearby resources, location of specific industrial sectors, location of urban areas, or something else?

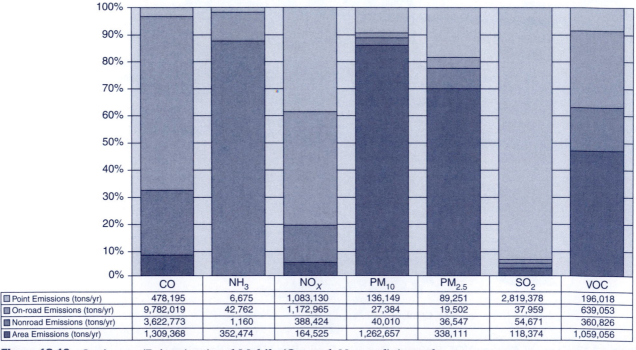

| | CO | NH₃ | NOₓ | PM₁₀ | PM₂.₅ | SO₂ | VOC |
|---|---|---|---|---|---|---|---|
| ☐ Point Emissions (tons/yr) | 478,195 | 6,675 | 1,083,130 | 136,149 | 89,251 | 2,819,378 | 196,018 |
| ☐ On-road Emissions (tons/yr) | 9,782,019 | 42,762 | 1,172,965 | 27,384 | 19,502 | 37,959 | 639,053 |
| ☐ Nonroad Emissions (tons/yr) | 3,622,773 | 1,160 | 388,424 | 40,010 | 36,547 | 54,671 | 360,826 |
| ☐ Area Emissions (tons/yr) | 1,309,368 | 352,474 | 164,525 | 1,262,657 | 338,111 | 118,374 | 1,059,056 |

**Figure 12.13**  Stationary (Point, Area) and Mobile (On-road, Nonroad) Annual Emissions for Five Criteria Pollutants and Ammonia (NH₃) in the Mid-Atlantic United States, 2002

Data from MARAMA (2005).

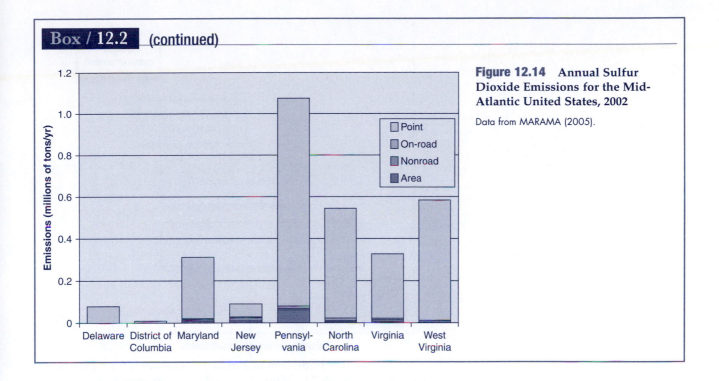

**Figure 12.14** **Annual Sulfur Dioxide Emissions for the Mid-Atlantic United States, 2002**

Data from MARAMA (2005).

## 12.5.2 ESTIMATING EMISSIONS

Four methods are used to quantify the magnitude of air pollutant emissions: (1) direct measurement, (2) the mass balance approach, (3) process modeling, and (4) emission factor modeling. Figure 12.15 shows the trade-offs for estimating emissions with these four methods. For example, while emission factor modeling has its place as an engineering tool, the reliability of the information produced with this technique is of the lowest quality compared with the other estimation techniques.

**DIRECT MEASUREMENT** The **direct measurement** of emissions poses several unique challenges. The sampling occurs in the stack discharge airstream. Besides being a difficult place to obtain such samples, the environment in these airstreams can be extreme with regard to temperature, moisture, and discharge velocity. This requires specialized and expensive equipment. The sensors on such equipment are specific to the source facility and the pollutants emitted.

Typically the emission-monitoring equipment is integrated into the operations of the facility, with the sensors permanently mounted into the stack. Such an arrangement enables continuous emissions monitoring (CEM). CEM systems are required for some sources, including waste incinerators. Other facilities may install CEM systems for the ease in acquiring data—for example, from extremely tall stacks at coal-fired power plants. Where safe conditions exist, mobile stack samplers can be used. These devices, operated by hand, can be carried to the top of a stack. An air-sampling line is extended into the

**Air Pollution and Children**
http://www.who.int/ceh/risks/cehair/en/

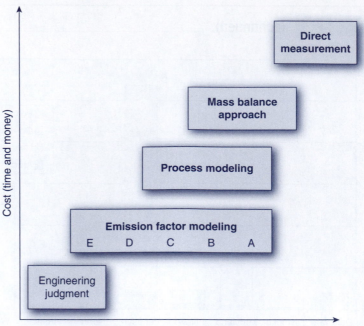

**Figure 12.15** **Relative Trade-Offs of Estimating Emissions from Emission Factor Modeling, Process Modeling, Mass Balance Measurements, and Continuous Emission Monitoring** In general, greater reliability in emission estimates requires more time and money. The easiest, and least reliable, method would be an engineer's best judgment.

emission airstream, and emissions are measured with a portable detector.

MASS BALANCE APPROACH The mass balance approach can be used to indirectly determine the emission rate from some sources. In its most basic form, this indirect measurement is a simple but useful accounting tool for pollutant tracking. It may also be the easiest way to identify the existence of fugitive or accidental emissions.

example/12.3 Mass Balance Approach Applied to Indirect Measurement of Air Emissions

A benzene storage tank has a capacity of 500,000 kg. The mass of benzene supplied to it from a chemical plant is 10,000 kg/day, and the mass pumped out to transport vehicles is 9,980 kg/day.

Accidental losses are suspected from poor seals on the tank but cannot be measured easily. Use a mass balance to assess the daily average accidental emissions (the mass out via accidental air emissions) in units of kg/day.

## solution

The mass balance on benzene is written (in words and symbols) as follows:

$$\begin{bmatrix} \text{rate of change in} \\ \text{benzene mass} \end{bmatrix} = \begin{bmatrix} \text{mass of benzene entering} \\ \text{tank per unit time} \end{bmatrix} - \begin{bmatrix} \text{mass of benzene leaving} \\ \text{tank per unit time} \end{bmatrix} - \begin{bmatrix} \text{benzene} \\ \text{decay rate} \end{bmatrix}$$

$$\frac{dm}{dt} = \dot{m}_{in} - \dot{m}_{out} \pm \dot{m}_{rxn}$$

To solve this problem, make two assumptions: First, the tank capacity is not changing over time, so it is at steady state (therefore, $dm/dt = 0$). Second, the benzene does not react inside the tank (therefore, $\dot{m}_{rxn} = 0$).

The total mass rate is from the chemical plant (and was given). Therefore,

$$\dot{m}_{in} = 10{,}000 \text{ kg benzene/day}$$

The total mass rate out of the tank is from loading benzene in the transport vehicles (which was given) plus the accidental emissions we are determining:

$$\dot{m}_{out} = \dot{m}_{transport} + \dot{m}_{emissions} = 9{,}980 \text{ kg/day} + \dot{m}_{emissions}$$

Combining everything we know along with our assumptions results in the following expression, which is our original mass balance:

$$0 = 10{,}000 - (9{,}980 + \dot{m}_{emissions}) \pm 0$$

Solve for the accidental emissions:

$$\dot{m}_{emissions} = 20 \text{ kg benzene/day}$$

Thus, 20 kg/day (of the 10,000 kg/day supplied to the tank every day) is lost. This value of 20 kg/day was obtained from an indirect method of determining the fugitive air emissions of benzene from the tank. This represents 0.2 percent of the benzene supply being lost daily to accidental emissions, a cost that will add up.

**PROCESS MODELING** **Process models** attempt to describe emissions as mathematical functions of relevant process information. In general, process models try to incorporate the physics or chemistry of a process that creates the specific pollutant(s). Typical output is pollutant-specific emission rates. While some process models are publicly available, most are created in-house by process owners or researchers and thus are proprietary.

An example of a commonly used process model is EPA's **MOBILE6** (EPA, 2003). MOBILE6 calculates emission factors (see next section) for 28 vehicle types in different operating conditions. The emission factors depend on such operating conditions as ambient temperature, vehicle travel speed, vehicle operating mode, fuel volatility, and mileage rates, among many others. These variables are specified by the model user. The model can estimate emissions for hydrocarbons, carbon monoxide, nitrogen oxides, exhaust particulate matter, tire wear particulate matter, brake wear particulate matter, sulfur dioxide, ammonia, six hazardous air pollutants, and carbon dioxide for both gasoline and diesel-fueled on-road vehicles including cars, trucks, buses, and motorcycles. There are also options for specialized vehicles such as natural-gas-fueled or electric vehicles.

MOBILE6 also has the ability to calculate other emissions, for example, accidental emissions from the fuel tank generated by the rise in temperature of a car sitting in the sun, from refueling, or from emissions occurring after the end of a trip, due to heating of the fuel lines. The output from MOBILE6 is emission factors in grams or milligrams of pollutant per vehicle per vehicle mile traveled (for example, g/vehicle mi). With knowledge of the fleet of vehicles in a region (say a city), driving patterns, and the emission factors generated from MOBILE6, an estimate of the total vehicular emissions can be created, perhaps for input to air quality models.

**EMISSION FACTOR MODELING** Emission factor modeling relates the air pollution released by a source activity to the magnitude of that activity. The proportionality between emissions and activity is called an **emission factor**. Emission factors are the best estimates of the most likely values based on literature reviews of process emission measurements of acceptable quality. These factors are expressed as the mass of pollutant emitted per unit weight, volume, distance, or duration of the source activity. These measurements are then used to create the emission factors, which can be implemented in emission factor models to predict the likely emissions from a new source activity.

In most cases, the emission factors are generally assumed to be representative of the long-term averages for emissions from facilities in the source category. As suggested by this methodology, the emission factors are specific to each source activity. Equation 12.3 shows the general emission factor model:

$$E = A \times EF \times (1 - \eta) \tag{12.3}$$

where $E$ is the pollutant emission (mass per time), $A$ is the activity rate (process events per time), EF is the emission factor (mass per process event), and $\eta$ is the emission-control efficiency (1 is complete control of emissions, and 0 is no control of emissions).

Emission factors are compiled and available for use in an EPA document referred to as AP-42 (EPA, 1995). AP-42 provides emission factors for over 250 source activities in 15 source groupings, covering a wide range of engineering processes, human activities, and even a few natural processes. Due to their literature-derived nature, emission factors are rated to provide some indication of the quality and robustness of the factor for estimating average emissions from a source. The factor ratings range from A (excellent) to E (poor). In general, factors determined from more observations and approved testing methodologies receive the highest ratings.

While there are many emission factors for common industrial, community, and natural sources, no factors exist for indoor air pollutant sources or for activities based on emerging technologies.

example/12.4 Using an Emission Factor Model to Estimate Carbon Monoxide Emissions from Oil Combustion

An industrial boiler burns 120,000 L distillate oil per day. Estimate the carbon monoxide emissions from this boiler (kg/day), assuming no emission-control devices are used.

solution

Chapter 1, Section 3, of AP-42 covers fuel oil combustion. The emission factor, EF, for CO emissions from industrial boilers burning distillate oil is reported to be 0.6 kg CO per 1,000 L oil burned. This emission factor has an A rating (excellent). The CO emission is then determined using Equation 12.3:

$$E = A \times \text{EF} \times (1 - \eta)$$

where $A$ (the oil burned) = 120,000 L/day, EF = 0.6 kg CO per 1,000 L oil, and $\eta = 0$.

Solve for $E$:

$$E = (120,000 \text{ L oil/day}) \times (0.6 \text{ kg CO}/1,000 \text{ L oil}) \times (1 - 0)$$

$$= 72 \text{ kg CO/day}$$

Due to the high rating for the emission factor (A for excellent), this emission estimate should be reasonable for a long-term average.

example/12.5 Using an Emission Factor Model to Estimate Particulate Matter Emissions from a Coal Pile

Freshly processed coal is delivered daily to a coal-fired power plant, moved with a conveyor, and stored in piles. The coal has an average moisture content of 4.5 percent. Average wind speed at the storage yard is 3 m/s. What is the emission of $PM_{10}$ associated with a 100,000 kg coal delivery under these conditions? Express your answer in g/day.

## solution

Chapter 13, Section 2.4, of AP-42 deals with aggregate handling and storage piles. The emission factor includes process variables and for storage piles is given as

$$EF = 0.0016 \, k \, \frac{\left(\dfrac{U}{2.2}\right)^{1.3}}{\left(\dfrac{M}{2}\right)^{1.4}} \text{ kg particles per Mg material transferred}$$

where EF is the emission factor, $k$ is a particle size multiplier, $U$ is the mean wind speed (m/s), and $M$ is the material moisture content (%).

The emission factor has an A rating as long as the material moisture is within the acceptable range provided for coal in this section of AP-42 (that range is 2.7%–7.4%). As also stated in this section of AP-42, $k = 0.35$ for $PM_{10}$. Remember that $U = 3$ m/s and $M = 4.5\%$. Solve for EF:

$$EF = (0.0016)(0.35) \frac{\left(\dfrac{3.0}{2.2}\right)^{1.3}}{\left(\dfrac{4.5}{2}\right)^{1.4}} = 2.7 \times 10^{-4} \text{ kg } PM_{10}/\text{Mg coal}$$

Use Equation 12.3 to estimate the emission of $PM_{10}$:

$$E = A \times EF \times (1 - \eta)$$

The activity, $A$, is the coal delivery: 100 Mg/day (1,000 kg = 1 Mg). In this situation, there is no control, so $\eta$ is 0. Substitute in these values to solve for the $PM_{10}$ emission for the coal pile under these conditions:

$$E = (100 \text{ Mg coal/day}) \times (2.7 \times 10^{-4} \text{ kg } PM_{10}/\text{Mg coal}) \times (1 - 0)$$

$$= 2.7 \times 10^{-2} \text{ kg } PM_{10}/\text{day} = 27 \text{ g } PM_{10}/\text{day}$$

The A rating (excellent) indicates this should be a reasonable average emission estimate.

**Congestion Pricing Reduces Traffic in Urban Corridors**
http://www.edf.org/page.cfm?tagID6241/

## 12.6 Control of Air Emissions

Historically, air pollution was discharged up a tall stack to take advantage of dilution and regional transport processes. Today, stacks are still used, but only as one part of a comprehensive solution.

Table 12.9 reviews five options to control air pollution: prevention, regulatory solutions, market-based solutions, voluntary solutions, and

## Table / 12.9

### Options to Control Air Pollution

| Solution | Description |
|---|---|
| Prevention | Pollution prevention, green chemistry, and green engineering can be used to reduce or eliminate air emissions and associated adverse impacts. |
| Regulatory | The foundation of U.S. air quality regulations is the Clean Air Act (passed in 1963 and amended numerous times). State governments are responsible for implementing federal regulations within their states (as with regulations related to drinking water and wastewater). |
| | *Permits* cover which pollutants may be released, their allowable emissions, control approaches to be used, monitoring requirements, and permit fees. Permit applications and the resulting permits are available for public review. |
| | Indoor environments do not have the same kind of air quality regulations as provided by the Clean Air Act. Instead, OSHA is charged with protecting workers by promoting industry-specific safety guidelines and requiring the industry to meet PELs. |
| Market-based emission reductions | The Clean Air Act Amendments (1990) introduced the concept of market-based solutions. A nationwide cap-and-trade market was subsequently established for sulfur dioxide, and a $NO_x$ market was established for the eastern states in 2003. While there is currently a European Union carbon dioxide market, to date, there are only congressional proposals for creating a similar U.S. market for carbon dioxide and other greenhouse gases. |
| | Market-based solutions work by setting a total allowable emission, or *cap*. A reasonable cap is estimated by assessing health or environmental impacts at various ambient concentrations, coupled with air quality models that predict ambient concentrations from various emission schemes. The cap is then progressively ratcheted lower over time (usually decades). Businesses determine the best way each year to meet these lowered emission totals by technological improvements to the process generating pollution, emission-control technologies, and purchases of emission credits on the market. |
| | Facilities that generate lower emissions than permitted may sell these unused emission allowances on the market, although they may also bank the unused emissions for later. Facilities that produce greater emissions than permitted can buy the emission credits from the sellers to account for their overage. |
| Voluntary emission reductions | Voluntary emission reductions are often grassroots efforts championed by individuals in communities and corporations. Such actions are also good at getting the public involved in improving air quality. In fact, many urban air quality problems have significant emissions from public activities, such as driving a motorized vehicle. |
| | In addition to making behavioral changes that reduce air pollutant emissions, the Clean Air Act Amendments created other ways for the public to participate in protecting their air resource. The public is encouraged to attend public hearings on permits. The public can sue the government or emission source's owner to get action as stated in applicable regulations. The public can also request action from the state regulatory agency or EPA against sources suspected of violating permits. |
| | Two common approaches that business uses to generate voluntary emission reductions include: (1) *Pollution prevention or minimization programs* and (2) *employee incentive programs* that create a system of recognition or financial bonuses to stimulate innovative solutions. |
| Emission control | Emission-control technologies are typically grouped by their ability to control the emissions of gaseous or particulate air pollutants. Technologies are typically pollutant specific. Therefore, understanding the composition of the airstream is important for narrowing the list of candidate technologies (see Table 12.10 for review). |

The Acid Rain Program created by the 1990 Clean Air Act Amendments instituted the first U.S. emissions trading program, for sulfur dioxide (SO$_2$). This market will also assist compliance with the Clean Air Interstate Rule for reducing fine particulate matter. SO$_2$ allowances (1 allowance = 1 ton of emissions) can be sold or purchased on the market.

The SO$_2$ market has been effective at creating a flexible approach to lowering total emissions. The market also has great volatility in allowance prices, due to a variety of factors, including changes in the price of natural gas and coal, electricity demand, and regulatory changes (see Figure 12.16). For example, prices surged more than 80 percent between October ($900) and December 2005 ($1,600), then returned to $900 just two months later. These impacts are more strongly experienced on markets that have low volume, like that for SO$_2$ emissions (EPA, 2006).

In 1995, the first year of the SO$_2$ market, emissions decreased by 24 percent compared with 1990 levels. Between 1995 and 2005, SO$_2$ emissions decreased another 14 percent, despite a 24 percent increase in power generation. As a snapshot of trading activity, analysis of average activity in the period 2000–2002 reveals 335 facilities reduced emissions (by 6.3 million tons), whereas 282 increased emissions (by 1.2 million tons) relative to 1990 values. This is a net reduction of 5.1 million tons (about 32 percent less than 1990 total emissions) (EPA, 2006).

**Market Based Regulatory Programs**
http://www.epa.gov/airmarkets/

The emission markets are executed alongside many other commodities markets, so any individual can purchase allowances. Allowances can be purchased directly from a company (or individual) that holds them, through EPA's annual auction, or through a commodities broker. Some nonprofit groups even buy emissions with the purpose of taking them off the market, thereby reducing the emissions available to industry and improving air quality in the process. Even your student professional society could purchase emissions.

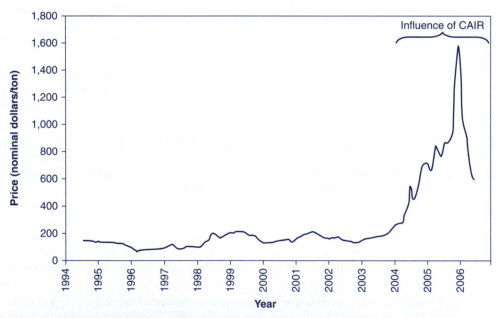

**Figure 12.16  Sulfur Dioxide Allowance Price History**   Note the influence of the Clean Air Interstate Rule (CAIR), which aims to permanently cap the emissions of sulfur dioxide (SO$_2$) and nitrogen oxides (NO$_x$) across 28 eastern states and the District of Columbia.

## Table / 12.10

**Air Pollutant Emission-Control Technologies** Detailed cost comparisons can be made using resources like EPA's *Air Pollution Control Cost Manual* (EPA, 2002).

| Technology | Pollutant Type | Operating Principle | Common Target Pollutants |
|---|---|---|---|
| Thermal oxidizer | Gaseous | Oxidization of pollutant through high-temperature combustion | VOCs, CO |
| Packed absorption bed scrubber | Gaseous | Transfer of pollutant into liquid absorbent | $NH_3$, $Cl_2$, $SO_2$, HCl |
| Adsorption tower | Gaseous | Transfer of pollutant onto solid adsorbent | VOCs, $SO_2$ |
| Biofilter | Gaseous | Metabolism of pollutants by microorganisms | Odors, VOCs |
| Cyclone | Particulate | Removal of pollutant by centrifugal force | $PM > 10 \ \mu m$ |
| Venturi scrubber | Particulate | Removal of pollutant by impaction into water droplets | $10 \ \mu m > PM > 5 \ \mu m$ |
| Baghouse | Particulate | Filtration of pollutant by fabric bags | $PM > 1 \ \mu m$ |
| Electrostatic precipitator | Particulate | Attraction of charged particles to collection plates | $PM < 1 \ \mu m$ |

emission-control technologies. Box 12.3 provides an example of a market-based solution. Table 12.10 provides information on eight gaseous and particulate emission-control technologies.

Voluntary emission reductions are cited as a method to get the public involved in improving air quality. Perhaps the best way to have an involved public is to make air quality information easily available. State and federal databases available on the Internet give the public access to a vast array of air quality data. Table 12.11 describes three online resources enabling anyone to learn about the emissions and air quality in nearly any U.S. region. These resources are developed in response to regulatory measurement requirements. In fact, nearly 4,000 sites in the U.S. monitoring network continuously measure ambient air quality. Additionally, criteria and hazardous air pollutant emissions from major source facilities are required in the permitting process. These data are collected by the state regulatory agencies and reported annually to EPA, which makes it publicly available through these online resources, reports, and other means.

To keep the public informed of air quality, the EPA also requires daily publication of an overall measure of air quality. This **air quality index (AQI)** is often reported by newspaper, TV, radio, and Internet outlets in metropolitan areas with more than 350,000

**What's the Air Quality Index Where You Live?**

*http://www.airnow.gov/*

## Table / 12.11

**Publicly Available Resources That Provide Air Quality Information**

| Resource | Developer | Web Site | Data |
|---|---|---|---|
| AirNow | EPA | www.airnow.gov | Current and forecast air quality conditions for most cities and some rural areas, expressed as air quality index |
| AirData | EPA | www.epa.gov/air/data/ | Historic emissions of criteria and hazardous air pollutants at county to national levels |
| Scorecard | Environmental Defense | www.scorecard.org | Historic emissions of criteria and hazardous air pollutants for individual facilities throughout the United States |

people. The AQI was developed to more easily communicate air quality, using a score from 0 to 500 based on the daily measurements of criteria air pollutants except lead. Table 12.12 shows how to calculate the AQI for specific pollutants and also provides information on how to interpret the AQI.

When the AQI is greater than 100, one or more of the criteria pollutants has exceeded its ambient air quality standard, and authorities

## Table / 12.12

**Air Quality Index Calculation Table**   Daily concentrations for each pollutant at a monitoring site are used to interpolate an AQI for that pollutant from this table. The reported AQI is the highest of the individual-pollutant AQIs.

| AQI Value | Corresponding Level of Health Concern (Range of AQI) | $O_3$ (8 hr) (ppb$_v$) | $O_3$ (1 hr) (ppb$_v$) | PM$_{2.5}$ ($\mu g/m^3$) | PM$_{10}$ ($\mu g/m^3$) | $NO_2$ (ppm$_v$) | $SO_2$ (ppb$_v$) | CO (ppm$_v$) |
|---|---|---|---|---|---|---|---|---|
| 0 | Good (0–50) | 0 | — | 0 | 0 | — | 0 | 0 |
| 50 | Moderate (51–100) | 64 | — | 15.4 | 54 | — | 34 | 4.4 |
| 100 | Moderate (51–100) | 84 | 125 | 40.4 | 154 | — | 144 | 9.4 |
| 150 | Unhealthy for sensitive groups (101–150) | 104 | 165 | 95.4 | 204 | 0.32 | 224 | 12.4 |
| 200 | Unhealthy (151–200) | 124 | 204 | 150.4 | 254 | 0.65 | 304 | 15.4 |
| 300 | Very unhealthy (201–300) | 374 | 404 | 250.4 | 424 | 1.24 | 604 | 30.4 |
| 500 | Hazardous (301–500) | — | 604 | 500.4 | 604 | 2.04 | 1,004 | 50.4 |

will declare an Action Day. Action Days are meant to trigger two types of action: (1) protective measures that affected individuals should take (for example, staying indoors and reducing outdoor exertion); and (2) emission reduction measures that the public can (or must) take (for example, using mass transit, not using highly polluting small-engine equipment such as lawn mowers, and reducing electricity use).

Keeping the public informed can be an effective way to reduce potential health problems resulting from poor air quality. It also engages people in improving air quality through decisions that reduce emissions. At the very least, an aware public can strengthen the political will to enact protective regulations.

**Class Discussion**

The Supreme Court ruled in April 2007 that the Clean Air Act gives EPA the authority to regulate the emissions of $CO_2$ and other greenhouse gases from cars. What political barriers at the state and federal level might cause increased or slow implementation of regulatory action on this topic? How would you overcome the barriers if you worked in the area of public policy?

## example/12.6 Calculating the Air Quality Index (AQI)

A city's monitoring site reports the following air pollutant concentrations. Use the data provided in the first two columns to determine the air quality index.

| Pollutant | Measured Concentration | Resulting AQI |
|---|---|---|
| $O_3$, 8-hr average | 80 $ppb_v$ | 89 |
| $O_3$, 1-hr average | 142 $ppb_v$ | 122 |
| $PM_{2.5}$ | 30 $\mu g/m^3$ | 79 |
| $PM_{10}$ | 93 $\mu g/m^3$ | 69 |
| $NO_2$ | 0.5 $ppm_v$ | — |
| $SO_2$ | 109 $ppb_v$ | 84 |
| CO | 5.6 $ppm_v$ | 62 |

### solution

The AQI for each pollutant measurement is determined using Table 12.12 and interpolation. (An online calculator is available at the federal government's AirNow Web site. Search for "AQI calculator," or go directly to http://airnow.gov/index.cfm?action=aqi.conc_aqi_calc) The resulting AQI values are provided in the third column of the preceding table.

The highest AQI of all pollutants is from the 1-hr ozone measurement. Therefore, the city's AQI would be reported to be 122. Because this value is greater than 100, one or more of the criteria pollutants has exceeded its ambient air quality standard, and authorities would declare an Action Day. This AQI suggests the air quality is at a level that is unhealthy for sensitive groups. The news media would therefore issue notices that children, the elderly, and individuals with respiratory diseases should limit their outdoor activities.

**Advantages and Disadvantages of Gaseous Emission-Control Technologies**

| Gaseous Emission-Control Technology | Advantages and Disadvantages |
| --- | --- |
| Thermal oxidizer | VOC removal often exceeds 99.99 percent. |
| | Because many VOCs have some fuel content, they can be burned with less auxiliary fuel, thereby potentially lowering energy use and associated operating costs. |
| | Generates $NO_x$, and possibly new hazardous pollutants, via the combustion process. |
| Packed absorption tower | VOC removal exceeds 95 percent. Although many pollutants can be effectively controlled with absorption towers, ammonia, hydrochloric acid, hydrofluoric acid, and sulfur dioxide are some of the common pollutants with adequate transfer speeds. |
| | Generated wastewater requires treatment. |
| Biofilter | Used to treat VOCs and odorous air. |
| | As with many other treatment devices, the residence time is a critical design parameter. Increased residence time results in increased pollutant removal, yet this is countered by increasing unit size (area requirements) and pressure drop across a taller bed. Increased capital costs for the larger unit and operating costs for the additional blower power are the trade-offs for seeking greater treatment efficiency. |

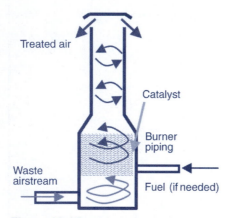

**Figure 12.17** **Thermal Oxidizer** A thermal oxidizer combines air and auxiliary fuel to create a region of high temperature in which air pollutants are oxidized. Auto-ignition temperatures can be lowered through the addition of a catalyst.

## 12.7 Gaseous Emission-Control Technologies

Several technologies are available to treat gaseous criteria and hazardous air pollutant emissions (refer back to Table 12.10). Three devices are discussed here: thermal oxidizers, packed absorption towers, and biofilters. Table 12.13 compares the advantages and disadvantages of these technologies.

### 12.7.1 THERMAL OXIDIZERS

**Thermal oxidizers** include devices that use air and auxiliary fuel to produce a region of high temperature in which pollutants are oxidized (Figure 12.17). Modern thermal oxidizer units are descendents of incinerators, devices based on similar principles but abandoned due to a poor performance history and poor public perception. Thermal oxidizers are one of the most regulated and expensive technologies. Advanced controls, continuous emissions monitoring, and extensive fuel requirements make operating costs high. The permitting process is especially difficult, often due to public concerns.

Design and operation of thermal oxidizers are based on the three $T$s (temperature, time, and turbulence), described in Box 12.4.

The complete conversion of a VOC to carbon dioxide and water vapor in a thermal oxidizer is achieved as follows:

$$C_xH_y + \left(\frac{x}{2} + \frac{y}{4}\right)O_2 \rightarrow (x)CO + \left(\frac{y}{2}\right)H_2O \qquad (12.4)$$

**New and Emerging Technologies**
http://www.epa.gov/ttn/catc/

Temperature, time, and turbulence form the basis for design and operation of a thermal oxidizer.

## Temperature

High temperature (1,200°F to 2,000°F) accelerates transformation of pollutants to more benign products. The temperature in the combustion zone is typically set several hundred degrees above the target VOC's auto-ignition temperature.

Temperature can be controlled through auxiliary fuel addition. Auto-ignition temperatures can effectively be lowered through the addition of a catalyst.

As shown in Figure 12.18, in general the higher the combustion temperature, the greater the conversion of pollutants. In the case of multiple compounds in the airstream, the highest auto-ignition temperature is used as the reference (in this case, $CH_4$).

## Time

Sufficient time (0.2 to 2.0 s) is required to achieve adequate conversion during combustion. Each VOC

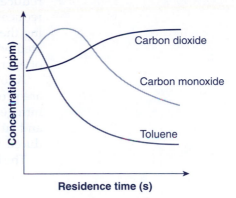

**Figure 12.19    Relationship of Residence Time to Conversion of Toluene**

requires a different burn time. The residence time of the air can be controlled either by increasing the volume (usually height) of the unit or by splitting the flow into multiple units.

Shown in Figure 12.19 is the conversion of toluene to end products (CO and $CO_2$) as a function of residence time. Short burn times do not allow complete oxidation of the toluene, resulting in high concentrations of CO. Longer burn times allow more toluene to be transformed to $CO_2$.

## Turbulence

Pass-through velocity should range from 20 to 40 ft./sec. Adequate oxygen must be provided to the combustion zone to ensure VOC oxidation.

More complete combustion can be achieved by ensuring rapid airstream movement through the thermal oxidizer, adding ambient air through blowers, and positioning baffles within the unit. The pass-through velocity must be set while considering the restrictions on residence time.

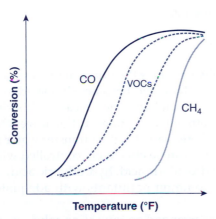

**Figure 12.18    Relationship of Combustion Temperature to Conversion of Pollutants**

$$(x)CO + \left(\frac{x}{2}\right)O_2 \rightarrow (x)CO_2 \qquad (12.5)$$

In Equations 12.4 and 12.5, $x$ and $y$ are stoichiometric coefficients used to balance the reactions (as in the toluene example depicted in Box 12.4), the VOC is oxidized to a greenhouse gas ($CO_2$). These examples suggest how green engineering activities that prevent use

**Gaseous Emission Control Technologies**

or production of VOCs can have a much broader societal and environmental return on the investment. Not only do these activities reduce the use and emissions of hazardous chemicals, they also reduce the demand for water, materials, and energy that go into treating these chemicals, as well as the greenhouse gas emissions associated with the treatment process (from oxidation of the VOC and use of energy to achieve high temperatures in the combustion zone).

Another concern is whether pollutants other than carbon dioxide are produced from less than 100 percent efficient combustion. Also, nitrogen oxides ($NO_x$) are always produced in a combustion process, and sulfur dioxide ($SO_2$) and hydrochloric acid (HCl) are often produced if the VOCs contain sulfur or chlorine, respectively.

The height of the thermal oxidizer tower (Z) is determined as follows:

$$Z = u \times \tau_r \qquad (12.6)$$

where $u$ is the face velocity through the tower (the average speed at which the air moves in the axial direction of the tower) and $\tau_r$ is the reaction time needed for adequate conversion.

The diameter of the tower (D) can also be determined:

$$D = \sqrt{\frac{4 \times Q}{\pi \times u}} \qquad (12.7)$$

where $Q$ is the exhaust flow rate.

## 12.7.2 ABSORPTION

In air pollution control, **absorption** is the physical process in which the air pollutant leaves the gas phase and becomes dissolved in the liquid phase. In many ways, this technology is identical to air strippers that are employed in water treatment, except absorption towers transfer an air pollutant into water, whereas air strippers transfer water pollutants into the air. Many pollutants can be effectively controlled with absorption towers; ammonia, hydrochloric acid, hydrofluoric acid, and sulfur dioxide are some of the common pollutants with adequate transfer speeds.

Most absorption units, arranged as towers or columns, are run in *countercurrent flow*, as shown in Figure 12.20. Clean water is used as the absorbent liquid in most implementations and is introduced at the top of the tower through a system of sprayers. The airstream to be treated enters at the bottom. As the water falls through the tower, the air pollutant is absorbed into the water droplets.

Due to cost, the scrubber liquid of choice is water. This can greatly constrain design and operation flexibility. One such application where this becomes especially problematic is the treatment of sulfur dioxide emissions from coal-fired power plants. Water alone is insufficient as a scrubber liquid for adequate emission control. To make it easier to transfer the $SO_2$ into the scrubber liquid, chemicals are added to the water, usually in a storage tank before introduction into the scrubber. The two chemical additive options available are limestone ($CaCO_3$)

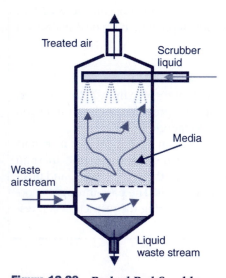

**Figure 12.20** **Packed Bed Scrubber**
This type of absorption column is usually run with countercurrent flow. The packing media is usually irregularly shaped pieces of plastic that enhance the absorption of pollutants into the scrubber liquid. This is achieved by increasing the travel distance of the air through the tower. The packing creates tortuous paths for the airflow.

and lime ($Ca(OH)_2$). Limestone is more abundant and less expensive, but lime results in better treatment efficiencies.

Note that the scrubber water in these systems will contain dissolved sulfur dioxide and the chemical additive (limestone or lime). Upon leaving the scrubber, this wastewater is typically sent to a holding tank to undergo one of the following chemical reactions:

$$\text{Limestone additive:} \quad SO_2 + CaCO_3 + \frac{1}{2}O_2 \rightarrow CaSO_{4(s)} + CO_2 \quad \textbf{(12.8)}$$

$$\text{Lime additive:} \quad SO_2 + Ca(OH)_2 + \frac{1}{2}O_2 \rightarrow CaSO_{4(s)} + H_2O \quad \textbf{(12.9)}$$

The calcium sulfate ($CaSO_4$) precipitates out of the water in the holding tank.

The formation of $CaSO_4$ can produce substantial scaling and require expensive cleaning of a fouled holding tank. However, a benefit can be seen if the resulting solids are not viewed as a waste. Calcium sulfate is better known by its common name, gypsum. Power plants have entered many reciprocal agreements to ship this material to manufacturers of gypsum-based products, such as drywall. Gypsum can also be incorporated into agricultural soils as a beneficial soil amendment. However, it is paramount to ensure that no hazardous chemicals are contained within the solids.

### 12.7.3 BIOFILTER

A **biofilter**, shown in Figure 12.21, is essentially a reactor filled with a biological support medium (lava rock, plastic media, mixture of wood chips and compost). The polluted airstream is introduced into the base of the unit. Figure 11.15 in the previous chapter also provided a schematic of the process. The air then passes through the pore space

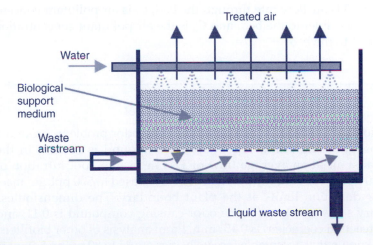

**Figure 12.21  Biofilters Used to Treat Odors and VOCs**  Contaminated air is passed through a biological support medium to bring the pollutant into contact with the microorganisms that reside there. Water is added intermittently to maintain needed moisture for the microorganisms.

between the media, bringing the pollutant into contact with the attached microorganisms that reside there.

To maintain a positive environment for the microorganisms, water is added to the bed to keep it moist (usually intermittently), and the bed is either insulated if aboveground or simply installed below grade. The moistened support medium also allows the pollutant to transfer to the water phase, where it is more readily available to the microbes.

One method to estimate the pollutant removal efficiency in a biofilter is with the following equation:

$$\eta = 1 - \exp\left(-\frac{Zk}{K_H u}\right) \tag{12.10}$$

where $\eta$ is the biofilter treatment efficiency (0 equals no treatment, and 1 equals 100 percent treatment of the air pollutant), $Z$ is the biofilter bed height, $K_H$ is the dimensionless Henry's law constant of the target pollutant, and $u$ is the air face velocity through the bed. $k$ is a reaction term, obtained through laboratory studies or empirically.

Equation 12.10 provides some insight into how to maximize pollutant removal; the goal should be to make the exponential term approach zero. Two apparent ways would be to decrease the velocity through the bed or increase the bed height. Decreasing the velocity allows the pollutant to transfer from the gaseous to liquid phase and then be taken up by the microorganisms. However, there is a maximum *elimination capacity (EC)* within the biofilter, suggesting the velocity can be lowered only to a certain point before untreated air is passing through the bed. This maximum elimination capacity is given by the following equation:

$$EC = \frac{Q \times (C_i - C_e)}{u} \tag{12.11}$$

where $Q$ is air flow rate through the bed, $C_i$ is air pollutant concentration at the inlet to the bed, and $C_e$ is the air pollutant concentration at the exit to the bed.

## example/12.7 Designing a Biofilter

A biofilter has been chosen for eliminating odor problems around a meat-rendering plant. The odorous compound is present in the exhaust air at an average flow of 12.5 m³/s and concentration of 3,810 ppb$_v$. The concentration must be decreased to 500 ppb$_v$ to meet odor detection limits at the plant boundary. The dimensionless Henry's law constant for the odor-causing compound is 0.11, and the reaction coefficient is 0.15/min. From analysis of other biofilters, it is known that acceptable face velocities are 35 to 80 m/hr. Estimate the range of bed heights needed to ensure the smell cannot be detected and the associated area of the bed.

solution

The problem is requesting the bed height ($Z$) for the range of acceptable face velocities ($u$). Rearrange Equation 12.10 to solve for $Z$:

$$Z = -\frac{K_H u}{k} \ln(1 - \eta)$$

The problem provided some of the values: $K_H = 0.1$, $k = 1.2/\text{min}$, and $u = 55$ to $120$ m/hr (which equals $0.92$ to $2.0$ m/min). The biofilter treatment efficiency can be determined as follows:

$$\eta = \frac{C_i - C_e}{C_i} = \frac{3{,}810 \text{ ppb}_v - 500 \text{ ppb}_v}{3{,}810 \text{ ppb}_v} = 0.87$$

For $u = 0.92$ m/min,

$$Z = -\frac{0.11 \times (0.92 \text{ m/min})}{0.15/\text{min}} \times \ln(1 - 0.87) = 1.38 \text{ m}$$

For $u = 2.0$ m/min,

$$Z = -\frac{0.11 \times (2.0 \text{ m/min})}{0.15/\text{min}} \times \ln(1 - 0.87) = 3.0 \text{ m}$$

The area can now be determined, using the airflow rate (given as $12.5 \text{ m}^3/\text{s}$) and face velocities:

$$A = \frac{Q}{u} = 12.5 \text{ m}^3/\text{s} \times \frac{60 \text{ s/min}}{u}$$

For the deep bed:

$$A = 12.5 \text{ m}^3/\text{s} \times \frac{60 \text{ s/min}}{2.0 \text{ m/min}} = 375 \text{ m}^2$$

For the shallow bed:

$$A = 12.5 \text{ m}^3/\text{s} \times \frac{60 \text{ s/min}}{0.92 \text{ m/min}} = 815 \text{ m}^2$$

Note that the deeper bed requires less area but will have greater pressure loss and higher fan power costs. The shallower bed will require more area but carry fewer operating costs. Deeper beds may not result in greater removal. We have also found that most removal of odorous compounds occurs in the beginning section of the bed (Martin et al., 2002). The final choice depends on space availability at the site compared with the additional anticipated power costs.

## 12.8 Particulate Emission-Control Technologies

**Partnership for Clean Indoor Air and Improved Cookstove Technology**

*www.PCIAonline.org*

Several technologies are available to treat particulate air emissions (refer back to Table 12.10). Four devices are discussed here: cyclones, scrubbers, baghouses, and electrostatic precipitators. Table 12.14 compares the advantages and disadvantages of these technologies, including the appropriate particle size on which they are effective.

### Table / 12.14

**Advantages and Disadvantages of Particulate Emission-Control Technologies**

| Particulate Emission-Control Technology | Appropriate Particle Size (Diameter) | Advantages and Disadvantages |
|---|---|---|
| Cyclone | >5 μm | Low capital costs but relatively high operating costs. <br><br> Relatively low collection (removal) efficiencies. The centrifugal force is the limiting factor; it is difficult to get fine particles to impact the cyclone walls with great effectiveness. |
| Scrubber | 5–10 μm | Can collect particulate and gaseous pollutants in the same unit. <br><br> Due to their moist environment, they can also handle flammable, explosive, or corrosive particles, mists (liquid particles), and hot airstreams. <br><br> Corrosion moisture can be produced in the scrubber liquid, which may damage equipment. The units must also be protected against freezing if placed in cold climates. The effluent liquid creates substantial wastewater and, ultimately, issues of sludge disposal. |
| Baghouse | 1–100 μm | High removal efficiencies even for small particles and over variable flow rates. Operating costs are primarily for fan power, replacement bags, and labor. <br><br> Major disadvantage is due to the bags. Hot and corrosive airstreams can degrade the bag fabric. More heat-tolerant fabrics are available, but they are more costly. Wet airstreams impair the effectiveness of the bags, particularly on particle release during the cleaning step. <br><br> Require a large area for installation, so they are difficult to site at space-limited locations. <br><br> If the particles or gases in the airstream are potentially flammable or explosive, baghouses are not a good treatment choice, because these compounds may accumulate within the unit, creating serious safety issues. |
| Electrostatic precipitator (ESP) | <1 μm | High removal efficiencies, even for extremely small particles. <br><br> Can handle large gas flow rates and create little pressure drop. Dry or wet airstreams can be handled, as can a wide range of air temperatures. <br><br> Operating costs are manageable with proper design but escalate dramatically if extremely high collection efficiencies are warranted. Design issues with ESPs include their high capital costs, lack of ability to remove gaseous pollutants, inflexibility to changes in airstream conditions once installed, need for large installation area, and inability to collect particles with high resistivities. |

## 12.8.1 CYCLONE

The **cyclone** is a simple, low-cost technology for removing larger particles from an airstream. As depicted in Figure 12.22, the technology is named after the centrifugal force generated in the device, used to separate the particles from the airstream. The contaminated air is introduced into the cyclone tangentially to its circular cross-sectional area. This creates a swirling path within the cyclone. As the particles have much greater density than the air, the particles impact the inside walls of the cyclone, accumulating to a point where they simply slide down the unit into the conical hopper, where they can be collected and transported away.

The effectiveness at which this centrifugal force results in particle impaction on the cyclone wall is directly related to the size of the particles. Larger particles have a larger mass and thus greater centrifugal force and greater impaction rates.

The collection efficiency of the cyclone is determined as follows:

$$\eta_j = \frac{1}{1 + (d_{p,c}/d_{p,j})^2} \qquad (12.12)$$

where $\eta_j$ is the removal efficiency of particles with diameter $d_{p,j}$, and $d_{p,c}$ is referred to as the cut diameter. The *cut diameter* is the size of particles that are removed with 50 percent efficiency. The cut diameter is a function of air and particle properties, as well as the geometry of the cyclone:

$$d_{p,c} = \sqrt{\frac{9\mu W}{2\pi N u_i(\rho_p - \rho_a)}} \qquad (12.13)$$

where $\mu$ is the viscosity of the airstream, $\rho_a$ is the density of the airstream, $\rho_p$ is the average density of the particles, $W$ is the width of the cyclone inlet, $u_i$ is the velocity of the airstream through the cyclone inlet, and $N$ is the number of revolutions the air makes within the cyclone. $N$ is either estimated by simple calculations of the cyclone's geometry (not shown here) or obtained from the manufacturer.

It is typical to encounter a range of particle sizes within the airstream. Each size will be removed at a different efficiency, according to Equation 12.12. To assess the overall removal rate, $\eta_o$, calculate the cut diameter for the cyclone of interest (Equation 12.13), and then estimate the removal efficiency for each particle size, using Equation 12.12. These removal efficiencies, along with knowledge of the mass fraction, $m_{p,j}$, that each particle size constitutes of the total particle mass, are used to determine the overall performance, as shown in the following equation:

$$\eta_o = \sum \eta_j m_{p,j} \qquad (12.14)$$

Some design configurations attempt to improve the collection efficiency of finer particles. One system, called the *multicyclone*, positions numerous smaller cyclones in parallel. The airflow is split evenly among these cyclones, and due to their smaller size, each generates much higher centrifugal force. To appreciate this, reflect on the motion of your body as you drive fast around a long curve compared with a tight curve. One downside for this arrangement is the cost associated

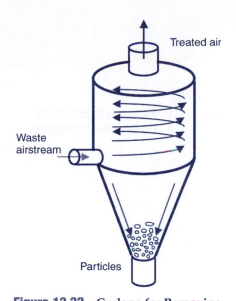

**Figure 12.22  Cyclone for Removing Particles from an Airstream**
Contaminated air is introduced into the cyclone tangentially to its circular cross-sectional area. Particles are removed by impacting the inside walls and falling to the bottom.

with the increase in number of units. Another issue is the greater pressure drop across a small cyclone, which requires higher blower costs to move the air through the system.

**Particulate Emission Control Technologies**

## 12.8.2 SCRUBBERS

Similar to the absorption technology for gaseous emission control, scrubbers can also be used to collect particles from airstreams. These systems work by impacting particles into water. The three key scrubber technologies used for particulate emission control are the spray chamber, packed-bed scrubber, and venturi scrubber.

A spray chamber consists of a tower with water droplets sprayed through nozzles, mixing with the airstream. These chambers can be designed for countercurrent, concurrent, or cross-current flows. This technology is the simplest but also least effective.

In a **packed-bed scrubber** the particles impact a liquid film found on the packing media and then drain away. One operating challenge is avoiding clogging of the packed bed.

A **venturi scrubber** employs a venturi (a duct with a decreasing, then increasing cross-sectional area) to increase the velocity of the airflow. Near the point of minimum cross section, called the throat, water droplets are injected, resulting in the most effective impaction of the particles into the water droplets. These combined (larger) water-solid particles are then typically collected in a cyclone as shown in Figure 12.23.

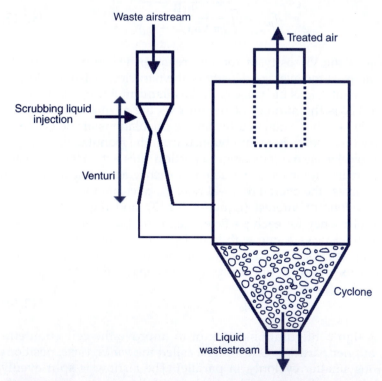

**Figure 12.23  Venturi Scrubber**  A venturi (left side of figure) is used to increase the velocity of the airflow. Water droplets are injected into the venturi, where they combine with particles in the contaminated airstream. The resulting combined water–solid particles are then typically separated in a cyclone.

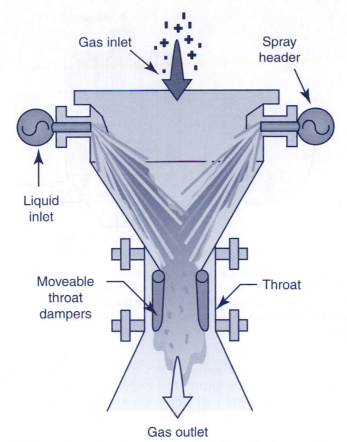

Figure 12.24 **Adjustable Throat Venturi** Adjustable dampers in a venturi are used to adjust the open cross-sectional area to affect the speed of the particles entrained in the inlet gas stream (from EPA).

## 12.8.3 BAGHOUSES

The **baghouse** particle emission-control technology uses filtration to remove particles from the airstream (Figure 12.25). These devices get their name because of the presence of many fabric bags within the housing. The bags work like vacuum cleaner bags. Air is drawn into the baghouse structure and then split evenly into the bags. Particles are trapped on the outside of the fabric filter, while the air easily passes through the filter.

The collection efficiency improves as more particles are trapped within the bags. The formation of this dust cake (see Figure 12.25) provides an additional filtering layer. When designed and operated well, baghouses offer close to 100 percent particle removal efficiencies. With proper bag fabric selection, baghouses can control a wide range of particle sizes, 1 to 100 μm in diameter.

Several important design steps are involved in designing and operating a baghouse: selecting the bag fabric, estimating the bag surface area required, determining the number of compartments within the baghouse, calculating the run and clean times, choosing the best cleaning method, and optimizing the filter velocity. Selection of the bag material depends upon the particle size and composition;

**Air Pollution from Burning Municipal Solid Waste (Santa Cruz, Bolivia)** (photo courtesy of Heather Wright Wendel).

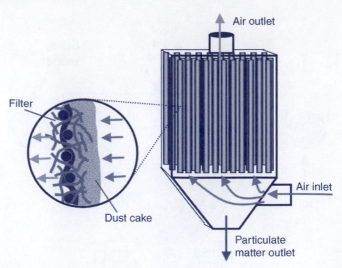

**Figure 12.25 Use of a Baghouse to Remove Particles from an Air Stream**
Air is drawn into a baghouse structure and then split evenly into the bags. Particles are trapped on the fabric filter, while the air easily passes through. The addition of *dust cake* on the filter fabric provides an additional barrier for particle removal.

the airstream temperature, flow rate, and composition; and fabric properties for particle capture and release. Many bag fabrics are available, including cotton, nylon, wool, and fiberglass. Less expensive bags typically have a lower temperature threshold and less chemical resistivity.

At some point, the bags will become full of particles. This is usually sensed by monitoring the pressure drop across the baghouse (Figure 12.26). As the bags fill, more fan power is needed to draw the air through the baghouse. The baghouse must be cleaned at this point. So that the whole unit need not be shut down, most baghouses are subdivided into a few compartments. Each compartment contains many bags and is cycled through an operation and cleaning cycle out of phase with the other units. In this way, one compartment can be taken offline for cleaning while all the others are in operation. Table 12.15 describes specific cleaning methods used for different types of baghouses.

**Figure 12.26 Variation of Pressure Drop ($\Delta P$) with Operating Time in a Baghouse** The time to reach the maximum permissible pressure drop ($\Delta P_m$) governs the operation cycle (run time $t_r$ plus cleaning time $t_c$) for each compartment in the baghouse. Typical values for run time are 1 hr, and cleaning time is 3 min. In general, most baghouse compartments can be cleaned and returned to service fairly quickly.

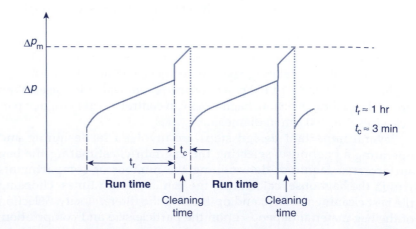

## Table / 12.15

### Baghouses Differentiated by Cleaning Method

| Type of Baghouse | Cleaning Method |
| --- | --- |
| Reverse-flow | Air flowing in the opposite direction is used during the cleaning phase. This flow essentially blows the particles out of the bags into a collection hopper. |
| Shaker | A shaker rack is used to dislodge particles from the bags during the cleaning phase. Each bag is mounted to the shaker mechanism, which is mechanically agitated on a predetermined schedule. |
| Pulse jet | High-pressure puffs of air dislodge the particles. This can be done more seamlessly with an operating baghouse, effectively reducing the downtime of any compartment. |

The number of compartments is determined by examining the total airflow rate to be treated, the permissible pressure drop (bags can rupture if the pressure drop becomes too large), the required cleaning time for the type of baghouse, and the filtration time, $t_f$ (the time between two cleanings of the same compartment):

$$t_f = N(t_r + t_c) - t_c \qquad (12.15)$$

Here $N$ is the number of compartments, $t_r$ is the run time when particles are being collected in the baghouse compartment, and $t_c$ is the cleaning time required for a compartment.

The filter velocity, $u_f$, is sometimes called the *air-to-cloth ratio* and can be estimated by considering the type of particles to be collected and the bag fabric:

$$u_f = \frac{Q_a}{A_T} \qquad (12.16)$$

$Q_a$ is volumetric flow rate of the airstream into the baghouse (sometimes referred to as the air), and $A_T$ is the total surface area of all the bags in the baghouse.

The filter velocity is typically within 0.01 to 0.02 m/s and is usually available from bag vendors. Because the airstream flow is primarily set by the upstream source processes that generate the particles, Equation 12.16 can then be used to estimate the total bag area required. Knowing this and the bag sizes available from vendors, the number of bags for the baghouse ($N_b$) can be determined as follows:

$$N_b = \frac{A_T}{A_b} \qquad (12.17)$$

where $A_b$ is the surface area of one bag. This is generally the area of the sidewalls only, so $A_b = \pi \times$ bag diameter $\times$ bag height.

Baghouses have high removal efficiencies even for small particles and over variable flow rates. Operating costs are primarily for fan power, replacement bags, and labor.

The major disadvantages of a baghouse are due primarily to the bags. Hot and corrosive airstreams can degrade the bag fabric. More heat-tolerant fabrics are available, but they are more costly. Wet airstreams impair the effectiveness of the bags, particularly on particle release during the cleaning step. Baghouses also require a large area for installation, so they are difficult to site at space-limited locations. Last, if the particles or gases in the airstream are potentially flammable or explosive, baghouses are not a good treatment choice, because these compounds may accumulate within the unit, creating serious safety issues.

### 12.8.4 ELECTROSTATIC PRECIPITATOR

With the particle-control technologies presented so far, ultrafine particles (those measuring less than 1 μm in diameter) would pass mostly unaffected. But because very small particles exacerbate health problems, control methods are especially crucial for this size range. The one technology effective at controlling ultrafine particles is the **electrostatic precipitator (ESP)** (Figure 12.27).

The ESP works by creating a *corona field*, which imparts a charge to particles opposite that of collector plates. As shown in Figure 12.28, the system is designed with the following important components: (1) an inlet flow straightener to minimize turbulence through the unit; (2) a main compartment with vertically hanging collector plates creating airflow channels; (3) discharge wires hanging between the plates; (4) a cleaning mechanism; and (5) a particle collection hopper. As the particle-laden airstream flows through the plate channels, the discharge wires create the corona field, and the particles become charged as they pass through. The plates carry an opposite charge, so the particles drift to the plate surfaces. As the particles continue to cover the plates, these *particle mats* slide off the plates, sometimes with the help of mechanical vibration. The collected particles are discharged from the hopper and transported away for disposal or secondary use.

**Green Power Options in Your State**

http://www.epa.gov/greenpower/pubs/gplocator.htm

**Figure 12.27**  **Electrostatic Precipitator**

Photo from California Air Resources Board.

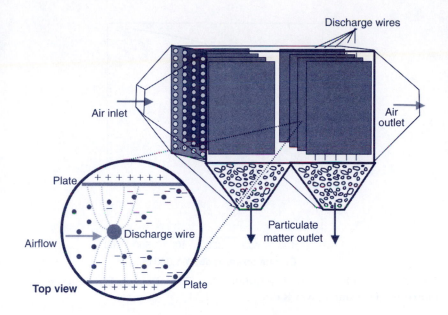

**Figure 12.28** **Electrostatic Precipitator (ESP)** The main compartment has vertically hanging collector plates that create airflow channels. As the particle-laden air flows through the plate channels, they become charged. The plates carry an opposite charge, so the particles drift to the plate surfaces.

The design of the ESP unit focuses on the particle removal (collection) efficiency ($\eta$):

$$\eta = 1 - \exp\left(-\frac{w_e A_p}{Q_a}\right) \tag{12.18}$$

where $A_p$ is the total surface area of the plates (plates have two sides), $Q_a$ is airstream flow rate, and $w_e$ is the effective drift velocity.

Typical values for the effective drift velocity are in the ranges of 2 to 20 cm/s for coal fly ash, 6 to 7 cm/s for cement dust, and 6 to 14 cm/s for blast furnace dust. The effective drift velocity can be determined from continuously measured operating properties given by

$$w_e = \left(\frac{k I_c V_{avg}}{A_a}\right) \tag{12.19}$$

where $V_{avg}$ is average voltage through the discharge wires, $I_c$ is the discharge corona current, and $k$ is an experimentally determined coefficient. This coefficient is obtained from pilot plant studies to accommodate inefficiencies inherent in the system, including irregular particles, non-uniform field strength and airflow, and reentrainment of particles off the collector plates from turbulence (specific to particle composition and size).

The impact of corona power ($I_c \times V_{avg}$) on collection efficiency is dramatic, as suggested by Equations 12.18 and 12.19 and shown in Figure 12.29. While the removal efficiency equation (12.18) suggests improvements could also be obtained by decreasing the airflow rate, this is not possible in practice. Altering the current and voltage can happen continuously and offers the surest way to maintain a collection efficiency necessary for meeting permitted emission limits. Electricity costs temper the strategy to increase the power too dramatically,

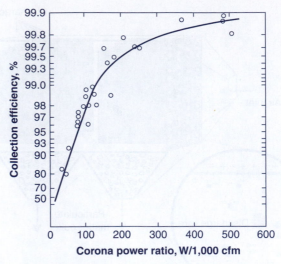

**Figure 12.29** Electrostatic Precipitator Particle Collection Efficiency as a Function of Corona Power Ratio

though. As the design equations and Figure 12.29 show, gains in collection efficiency are incremental with increases in corona power.

Several components of the ESP must be sized, including the plates and the housing. Table 12.16 provides the design equations to determine the following component requirements: (1) number of plate channels within the ESP; (2) number of collector sections along the length of the ESP; (3) overall length of the precipitator; and (4) total collection plate area. One major issue in selecting an electrostatic precipitator is the significant electrical demands for maintaining the corona.

## Table / 12.16

### Equations Used to Size Components of an Electrostatic Precipitator (ESP)

| Component of ESP That Must Be Sized | Equation | Equation Number |
|---|---|---|
| Number of plate channels within the ESP ($N_d$) | $$N_d = \frac{Q_a}{uWZ}$$ $Q_a$ is the airstream flow rate, $u$ is the air velocity through the channels, $W$ is the channel width, and $Z$ is the plate height. | (12.20) |
| Number of collector sections along the length of the ESP ($N_s$) | $$N_s = \frac{RZ}{L_p}$$ $R$ is the plate aspect ratio (total plate length to plate height), and $L_p$ is the length of one plate. | (12.21) |
| Overall length of the precipitator ($L_o$) | $$L_o = N_s L_p + (N_s - 1)L_s + L_{en} + L_{ex}$$ $L_s$ is the spacing between plate sections, $L_{en}$ is the entrance section length, and $L_{ex}$ is the exit section length. | (12.22) |
| Total collection plate area ($A_p$) | $$A_p = 2 \times Z \times L_p \times N_s \times N_d$$ | (12.23) |

# example/12.8 Designing an Electrostatic Precipitator (ESP)

A 98 percent efficient ESP is being designed to treat a gas stream flowing at 5,000 $m^3$/min. The effective drift velocity is 6.0 cm/s. Calculate the plate area in $m^2$ and the number of plates. Each plate is 6 m ($L$) by 10 m ($Z$), and there are three sections in the direction of flow.

## solution

To determine the total plate area, rearrange Equation 12.18 to solve for $A_p$:

$$A_p = -\frac{Q_a}{w_e} \ln(1 - \eta)$$

Substitute the given values for $\eta = 0.98$, $w_e = 6.0$ cm/s (or $6.0 \times 10^{-2}$ m/s), and $Q_a = 5,000$ $m^3$/min (83.33 $m^3$/s), and solve for total area:

$$A_p = \frac{83.33 \text{ m}^3/\text{s}}{6.0 \times 10^{-2} \text{ m/s}} \ln(1 - 0.98)$$

$$= 5,433 \text{ m}^2, \text{ which can be rounded up to 5,500 m}^2$$

To determine the number of plates ($N_p$) required, use the given values of $L_p = 6$ m, $Z = 10$ m, and $A_{\text{one plate}} = 2 \times Z \times L$ (multiply by 2 because the plate has two sides). Divide the total plate area by the area of one plate:

$$N_p = \frac{A_p}{A_{\text{one plate}}}$$

Substitute in the known values and solve:

$$N_p = \frac{5,500 \text{ m}^2}{2 \times 10 \text{ m} \times 6 \text{ m}} = 45.83$$

$N_p = 45.83$ plates, which is rounded up 48 because the number needs to be divisible by 3, the number of sections specified in the problem.

Because this ESP has three sections, it should then be designed to have 16 (that is, 48/3) plates in each section.

## Key Terms

- absorption
- air quality index (AQI)
- ambient air
- baghouse
- biofilter
- carbon dioxide
- carbon monoxide
- Clean Air Act (CAA)
- Clean Air Act Amendments (CAAA)
- criteria air pollutants
- cyclone
- direct measurement
- electrostatic precipitator (ESP)
- emission factor
- emissions
- fugitive emissions
- green building

- hazardous air pollutants (HAPs)
- high-pressure systems
- indoor environment
- inversions
- land–sea interface
- lapse rate
- lead
- low-pressure systems
- market-based emission reductions
- MOBILE6
- National Ambient Air Quality Standards (NAAQS)
- natural ventilation
- neutral
- nitrogen dioxide ($NO_2$)
- $NO_x$
- ozone
- ozone layer

- packed bed scrubber
- particulate matter (PM)
- permissible exposure limits (PELs)
- photochemical reactions
- $PM_{2.5}$
- $PM_{10}$
- process models
- regional transport
- respiratory system
- stability conditions
- stable
- stratosphere
- sulfur dioxide ($SO_2$)
- thermal oxidizer
- Tragedy of the Commons
- troposphere
- unstable
- venturi scrubber
- voluntary emission reductions

# chapter/Twelve Problems

**12.1** Identify specific locations near where you live that experience frequent inversions. Are there major pollutant sources in that area that might be trapped by the inversion layer?

**12.2** Identify a household cleaning product you have around your dormitory, apartment, or home. Referring to the list of chemical ingredients on the package, go to the OSHA Web site (www.osha.gov) and find the PEL for one of the chemicals listed. See if you can also find information on possible health impacts of the chemical. Next determine a nonhazardous substitute for the product you have identified. The following Web site provides information on natural substitutes for items such as household cleaning, pest-control, and personal care hygiene products: www.care2.com/greenliving/.

**12.3** Define NAAQS, and identify NAAQS pollutants.

**12.4** Go to the World Health Organization's Web site (www.who.org). Select a particular air pollutant, and write a report of up to two pages on the global magnitude of the problem and what engineered or policy solutions are needed to deal with this problem.

**12.5** Investigate sources of hazardous air pollutants emitted near your community. Go to Scorecard (www.scorecard.org) or a similar Web site to gather data about air pollutant emissions. The Scorecard site makes the Toxics Release Inventory easily searchable; by entering your zip code, you can find a list of major air polluters in your area. For your area, identify the top three polluters, their emissions (pollutant, annual emission rate, emission trend over the past several years). What percent of the emissions are to air, relative to other media?

**12.6** Emission trends are not the same everywhere. Reductions in $SO_2$ have been widespread but not universal from 1980 to 2000. Hypothesize why some states have decreased emissions while others have not.

**12.7** Examine today's national air quality index (AQI) pattern at the urban area closest to your school (see www.airnow.gov). What conditions are causing any patterns you observe?

**12.8** Assume you are employed at a large utility company that provides electricity via a network of coal-fired power plants. Your demographic forecasts suggest that the region to which your company provides power should experience a 10 percent increase in population over the next decade. Unfortunately, the region frequently has ambient concentrations of sulfur dioxide near the NAAQS, and you know that historically population is proportional to emissions, and those emissions are proportional to ambient concentrations. The bottom line is that regional emissions cannot increase. Identify and briefly explain three different sustainable strategies to deal with this problem.

**12.9** Imagine you are the environmental manager of a utility company in West Virginia that, because of population increases, needs to serve more customers. What options could you exercise so your $SO_2$ emissions do not increase? Identify one effective option within each of the four types of solution categories (regulatory, market-based, voluntary, and control technology).

**12.10** You purchase a house with a known radon problem. The house has a dirt-floor crawl space (1 m high). The emission rate of radon into the crawl space is reported to be 0.6 pCi/m$^2$ − s. An exhaust system was added to provide 1 air exchange per hour (ACH) in the crawl space. The outdoor concentration of radon is 1 pCi/L. Estimate the radon concentration in the crawl space (in pCi/L).

**12.11** An electric utility claims it is combating the $CO_2$ buildup in the atmosphere by purchasing rain forest in Belize and managing it for $CO_2$ removal, rather than seeing it clear-cut for farming. The utility indicates that it will manage 120,000 acres and in doing so could reduce greenhouse gas emissions by 5.2 million tons over 40 years. Estimate the emission reduction in tons of $CO_2$ per acre per year.

**12.12** Between 1980 and 2000, the average CO emission factor of the vehicle fleet in Hillsborough County, Florida, dropped by almost half, from about 65 to 34 g/vehicle-mil. However, the total miles driven in the county by all vehicles increased by 60 percent during this same time period. Did countywide emissions of CO go up or down, and by how much? Vehicle exhaust is getting cleaner through a combination of engine improvements, emission control technologies, auto re-design, and fuel reformation, but what other solutions can you think of?

**12.13**   A biofilter at a pressboard manufacturer in Vermont needs to be 70 percent efficient in its control of a 30 m$^3$/s dryer exhaust that contains 100 ppb$_v$ xylene. Preliminary design work suggests the biofilter is to have a below-grade 2 m deep bed and be located in an adjacent field that has an area of 8 m by 10 m available for excavation. The dimensionless Henry's constant ($K_H$) is 0.3, and the average reaction coefficient ($k$) is 0.05/sec. A pilot study suggests the bacteria in the unit will need at least 5 sec contact time to be effective at this efficiency. (a) Is the design sufficient for the minimum microbial time limit? (b) Estimate the surface area required for the unit (m$^2$).

**12.14**   A thermal oxidizer removes 99.99 percent of incoming VOC and has 1 μg/m$^3$ of VOCs in the exhaust. Estimate the VOC concentration in the inlet air (μg/m$^3$).

**12.15**   Two cyclones are available to treat an airstream with particles. The units can be placed in series or in parallel. When in series, though, the collection efficiency decreases due to the much higher flow rates. (The parallel arrangement gets a 50/50 split.) In series, each unit has a 90 percent collection efficiency; in parallel, each has a 98 percent efficiency. Which arrangement should be used, if overall efficiency is the primary design consideration?

**12.16**   A shaker-type baghouse is to be designed to filter particles from an exhaust airstream. The airflow is 150,000 cu. ft./min at 220°F and 1.1 atm. Specify the following requirements: (a) superficial filtering velocity; and (b) number of bags needed for the baghouse if the bags are $D = 0.5$ ft. by $L = 10$ ft.

**12.17**   A 20 m$^3$/s airstream from a cement kiln carries 120 mg/m$^3$ of particulate matter to an electrostatic precipitator for control. The ESP has four plates, each measuring 8 m by 5 m. The particles have a drift velocity of 0.15 m/s. Estimate the particle concentration in the ESP exhaust (mg/m$^3$).

**12.18**   A flow of 50 m$^3$/s of air exits a manufacturing facility, containing particles whose drift velocity is 0.121 m/s. Assume that 99.2 percent removal efficiency is required. (a) Calculate the total plate surface area for an electrostatic precipitator. (b) Estimate how many plates are needed. Assume the plates available for this unit measure 2 m by 5 m.

**12.19**   An electrostatic precipitator (ESP) is being designed to collect flue gas particles at a power plant. The following information has been collected: electric field intensity = 100,000 V/m, and average particle charge = 0.30 fC (1 fC = $10^{-15}$ C). Characterization of the airstream yields the following information: air viscosity = 18.5 × $10^{-6}$ Pa − sec; average particle diameter = 1 μm; airflow rate = 1.5 m$^3$/sec; air temperature = 120°C; and average inlet particle concentration = 100 mg/m$^3$. The state regulators are granting this facility a permit to have an average stack (effluent) concentration of 0.5 mg/m$^3$. (a) Estimate the required removal efficiency of particles in the ESP. (b) Estimate the required plate area to achieve this removal efficiency.

**12.20**   Using Example 12.5, estimate the PM$_{10}$ emissions when the conditions at the coal pile change to 2 percent moisture and wind speed of 10 m/s.

# References

Environmental Protection Agency (EPA). 1995. *AP-42: Compilation of Air Pollutant Emission Factors,* Vol. 1, *Stationary Point and Area Sources,* 5th ed, Washington, D.C.

EPA. 1999. *The Benefits and Costs of the Clean Air Act, 1990 to 2010: EPA Report to Congress.* EPA-410-R-99-001. Washington, D.C.

EPA. 2002. *EPA Air Pollution Control Cost Manual,* 6th ed. EPA/452/B-02-001. Washington, D.C.

EPA. 2003. *User's Guide to MOBILE6.1 and MOBILE6.2.* Mobile Source Emission Factor Model. EPA420-R-03-010. Washington, D.C.

EPA. 2006. *Acid Rain Program 2005 Progress Report.* EPA-430-R-06-015. Washington, D.C.

EPA. 2007. *Green Book: Nonattainment Areas for Criteria Pollutants.* Available at www.epa.gov/oar/oaqps/greenbk/index.html, accessed November 18, 2007. Washington, D.C.

European Environment Agency (EEA). 2007. "Ozone Pollution across Europe." EEA Web site, www.eea.europa.eu/maps/ozone/welcome, accessed November 18, 2007.

Martin, R. W., H. Li., J. R. Mihelcic, J. C. Crittenden, D. R. Lueking, C. R. Hatch, and P. Ball. 2002. "Optimization of Biofiltration for Odor Control: Model Verification and Applications." *Water Environment Research* 74 (1): 17–27.

Mid-Atlantic Regional Air Management Association (MARAMA). 2005. *A Guide to Mid-Atlantic Regional Air Quality.* Towson, Md.

United Nations Environment Programme (UNEP). 2007. *Toolkit for Clean Fleet Strategy Development.* Available at http://www.unep.org/tnt-unep/toolkit/Actions/Tool9/index.html, accessed November 18, 2007.

World Bank. 2007a. *Cost of Pollution in China: Economic Estimates of Physical Damage.* Washington, D.C.: World Bank.

World Bank. 2007b. *2007 World Development Indicators.* Washington, D.C.: World Bank.

# chapter/Thirteen Solid-Waste Management

Mark W. Milke and
James R. Mihelcic

*In this chapter, readers will learn about management of municipal solid wastes. The types of solid waste, their quantities, composition, and physical properties are first described. Then the chapter addresses the storage, collection, transport, treatment, and disposal of solid wastes, including recycling and materials recovery, composting, incineration, and landfilling. The chapter concludes with an introduction to public consultation, public policy, and cost estimation. Emphasis is placed on the application of basic mass balance concepts to solid-waste problems, with a mixture of quantitative problems and discussion of broader management topics.*

## Major Sections

## Learning Objectives

1. Describe the key components of a solid-waste management system.
2. Identify the objectives of solid-waste management.
3. Describe key relevant U.S. legislation related to solid wastes.
4. Distinguish between municipal solid waste and other solid wastes.
5. Calculate dry and wet weight generation rates for specific solid-waste components from available data.
6. Discuss the differences between solid-waste management in developing and developed countries.
7. Explain the issues associated with design and operation of successful solid-waste subsystems (collection, transfer stations, materials recovery facilities, composting facilities, incinerators, and landfills).
8. Solve mixing problems to determine an appropriate C/N ratio for composting.
9. Calculate oxygen requirements for aerobic biological or thermal treatment processes, as well as methane generation rates from landfills, using stoichiometry and mass composition data.
10. Explain the concerns with landfill gas and leachate and how they are addressed.
11. Calculate the size of an area landfill based on daily cell construction.
12. Empathize with community stakeholders related to siting a landfill or incinerator, and carefully decide how to reach consensus.
13. Identify key public consultation methods and their strengths and weaknesses.
14. Identify key public policy options for solid-waste management, and summarize their strengths and weaknesses.
15. Estimate costs for different-sized landfills, using economy-of-scale factors.

PAPER    PLASTIC    TRASH

# 13.1 Introduction

**Solid wastes** include paper and plastic generated at home, ash produced by industry, cafeteria food wastes, leaves and cut grass from parks, hospital medical wastes, and demolition debris from a construction site. These materials are considered a **waste** when owners and society believe they no longer have value.

**Solid-waste management** varies greatly between cultures and countries and has evolved over time. The components of solid-waste management are depicted in Figure 13.1. Solid-waste management requires an understanding of **waste generation**, **storage**, **collection**, **transport**, **processing**, and **disposal**. The end points in Figure 13.1 are recycled materials, compost, and energy recovery; these end points will become more common as society adopts more sustainable waste management practices. Referring back to Chapter 7, remember that *waste* is a human-derived word; thus, we need to identify ways to minimize the amount of waste that is generated, transported, processed, and disposed.

Solid wastes differ from liquid or gaseous wastes because they cannot be pumped or flow like fluids. However, solid wastes can be placed into solid forms (including soils) and thus can be contained more easily. These differences have led to different approaches for managing solid wastes than the approaches described in previous chapters for liquid and gaseous waste streams.

Proper management of solid wastes has four main objectives:

1. Protect public health.

2. Protect the environment (including biodiversity).

3. Address social concerns (equity, environmental justice, aesthetics, risk, public preferences, recycling, renewable energy).

4. Minimize cost.

It is common practice for individual communities to place varying weights on these objectives. Therefore, many different systems exist to manage solid waste.

**Zero Waste as a Design Principle for the 21st Century**

http://www.sierraclub.org/committees/zerowaste/
http://www.zerowaste.co.nz/index.sm

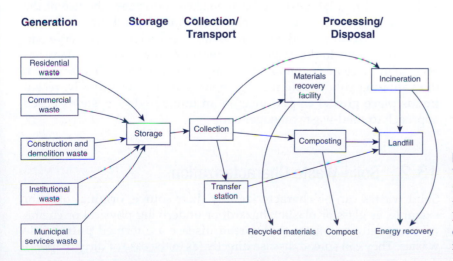

**Figure 13.1 Overview of the Solid-Waste Management System** The system consists of storage, collection and transport, processing, and disposal. Materials recovery, composting, recycling, and energy recovery are important in the processing and disposal stage.

Preindustrial communities arranged to have solid waste collected and disposed of in central locations termed *middens*. Some middens can be found today. For example, some coastal areas along Florida's Gulf Coast are home to middens derived from eating shellfish that formed small islands that can now be explored. As population, urbanization, and consumption increased, solid waste became more of a problem. Some wastes were left in streets and alleys, feeding dogs, pigs, and rats. Other wastes were hauled outside of the city and dumped in large mounds. The increasing population of cities and the large death rates experienced in many parts of the world due to plague, cholera, and other infectious diseases led to a demand to clean cities of solid waste. Organized collection, treatment, and disposal of solid waste began in the late 1800s and was closely tied to the objective of improving public health and sanitation. Early approaches to solid-waste management included feeding food wastes to farm animals, burning waste to heat city water, and creating solid-waste **dumps** (often in wetlands) to reclaim land.

During the 1900s, increased industrialization resulted in the production of different wastes—more *hazardous solid wastes*. At the same time, increasing urban populations, along with increasing affluence and consumption, increased the amount of solid waste produced. Nonrecovered solid wastes were placed into engineered facilities, which were called engineered or sanitary **landfills**. The improvement from dumps to engineered sanitary landfills has been a gradual process over the past 100 years that has led to technically advanced and sophisticated systems for protection of human health and the environment.

**RCRA**

http://www.epa.gov/awsregs/laws/rcra .html

The major piece of U.S. legislation affecting solid-waste management is the **Resource Conservation and Recovery Act (RCRA)** of 1976. RCRA mandated tracking and rigorous management of hazardous wastes. It also led to regulations to improve the design of landfills and reduce their risk. The *Pollution Prevention Act* then followed.

In the past twenty years, there has been an increased emphasis on **reduction**, **reuse** and **recycling** (the three *R*s) as part of the **pollution prevention** hierarchy discussed in Chapter 7. Numerous local, state, and national regulations have been enacted to increase the use of the three *R*s. These initiatives have been motivated by a desire to further reduce adverse social and environmental impacts and to conserve natural resources (including water and energy). At the same time, there has been an increased recognition of the importance of life cycle assessment in evaluating solid-waste management options. Accordingly, recent trends place greater importance on an integrated or systems-based approach to solid-waste management.

## 13.2 Solid-Waste Characterization

Solid wastes can be characterized by their source, original use (for example, as glass or plastic), hazard, or underlying physical or chemical composition. Wastes that spread disease are termed **putrescible** wastes. They can spread disease directly (as in the case of dirty diapers), or indirectly by providing a food source for disease vectors such as insects (flies) or animals (rats, dogs, birds).

## 13.2.1 SOURCES OF SOLID WASTE

The sources of solid waste and typical constituents are identified in Table 13.1. Some solid wastes (for example, mining wastes and most agricultural and industrial wastes) are managed by the waste generator. Smaller sources are usually managed jointly under one integrated system. The solid wastes jointly managed by a municipality are called **municipal solid waste (MSW)**. The focus of this chapter is on the management of MSW, though many of the principles and processes discussed are also relevant to the management of industrial, agricultural, and mining wastes.

## Table / 13.1

### Sources of Solid Waste and Typical Percentage That Makes Up Municipal Solid Waste

| Source | Examples | Comments | Typical Percentage of MSW |
|---|---|---|---|
| Residential | Detached homes, apartments | Food wastes, yard/garden wastes, paper, plastic, glass, metal, household hazardous wastes. | 30%–50% |
| Commercial | Stores, restaurants, office buildings, motels, auto repair shops, small businesses | Same as above, but more variable from source to source. Small quantities of specific hazardous wastes. | 30%–50% |
| Institutional | Schools, hospitals, prisons, military bases, nursing homes | Same as above; variable composition between sources. | 2%–5% |
| Construction and demolition | Building construction or demolition sites, road construction sites | Concrete, metal, wood, asphalt, wallboard, and dirt predominate. Some hazardous wastes possible. | 5%–20% |
| Municipal services | Cleaning of streets, parks, and beaches; water and wastewater treatment grit and biosolids; leaf collection; disposal of abandoned cars and dead animals | Waste sources vary among municipalities. | 1%–10% |
| Industrial | Light and heavy manufacturing, large food-processing plants, power plants, chemical plants | Can produce large quantities of relatively homogeneous wastes. Can include ashes, sands, paper mill sludge, fruit pits, tank sludge. | Not MSW |
| Agricultural | Cropping farms, dairies, feedlots, orchards | Spoiled food wastes, manures, unused plant matter (e.g., straw), hazardous chemicals. | Not MSW |
| Mining | Coal mining, uranium mining, metal mining, oil/gas exploration | Can produce vast amounts of solid waste needing specialized management. | Not MSW |

SOURCE: Tchobanoglous et al., 1993.

## 13.2.2   QUANTITIES OF MUNICIPAL SOLID WASTE

Table 13.2 provides the quantities of municipal solid waste generated and managed in the United States from 1960 to 2005. Note how generation rates have increased dramatically in just over 40 years. Though recycling rates increased in the 1990s, over the same period, the rate of waste production steadily increased as well. As a whole, U.S. citizens spent more in the 1990s, recycled more, and disposed of more waste.

The 2005 generation rate of 0.75 Mg per person per year excludes construction and demolition debris and biosolids from wastewater treatment plants (which, as listed in Table 13.1, is often included in the solid waste managed by a municipality). As a rough rule of thumb, an overall generation rate for MSW in most industrialized countries is now approximately 1 Mg per person per year. Because residential solid waste makes up 30 to 50 percent of the MSW, this converts to a residential waste generation rate of approximately 1 kg per person per day.

The amount of generated MSW can vary within a year, between urban and rural areas, geographically, with income, and among countries. The fraction managed by recycling, composting, incineration, or landfilling varies even more according to local conditions. It is common to read reports that a particular city or country produces more waste than others. However, these reports need to be read cautiously, because different authorities count different waste streams, and some define *produced* waste as waste that remains after recycling and composting. One key is to understand the differences between *generation rates* and *discard rates* (the difference being the part of the waste stream that is reused or recycled).

**Class Discussion**

EPA reports the amount of household solid waste in the United States typically increases by 25 percent between Thanksgiving and New Year's Day, from 4 million tons to 5 million tons. What specific technological, policy, and human behavioral changes can a household and community implement to reduce the volume of waste generated during the holiday season?

## Table / 13.2

### Quantities of Municipal Solid Waste in the United States over Time

|  | Mg per Person per Year[1] | | | | | |
|---|---|---|---|---|---|---|
|  | 1960 | 1970 | 1980 | 1990 | 2000 | 2005 |
| Generation | 0.44 | 0.54 | 0.61 | 0.75 | 0.77 | 0.75 |
| Recycling | 0.03 | 0.04 | 0.06 | 0.11 | 0.17 | 0.18 |
| Composting | Negligible | Negligible | Negligible | 0.01 | 0.05 | 0.06 |
| Incineration | 0.00 | 0.00 | 0.01 | 0.11 | 0.11 | 0.10 |
| Landfill[2] | 0.42 | 0.50 | 0.54 | 0.52 | 0.43 | 0.41 |

[1] These quantities exclude construction and demolition debris and wastewater treatment plant biosolids.

[2] This includes small quantities of waste incinerated without energy recovery and does not include wastes produced during recycling, composting, and incineration (e.g., ashes).

SOURCE: EPA, 2006a.

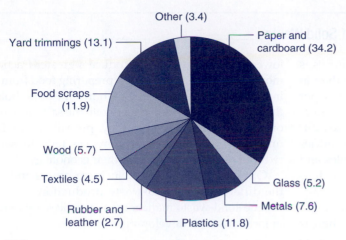

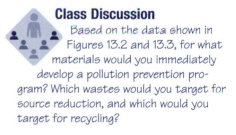

**Figure 13.2** Percentage of Various Materials (on a Mass Basis) that Compose U.S. Municipal Solid Waste, 2005

Data from EPA (2006a).

### 13.2.3 MATERIALS IN MUNICIPAL SOLID WASTE

Figure 13.2 shows the estimated percentage breakdown of the various materials found in MSW at the point of generation. This information can be used to assess which waste streams can be targeted for composting or materials recovery programs. For example, Figure 13.2 shows that a high percentage of the overall waste is yard waste (13.1 percent) and paper/cardboard (34.2 percent). The types of materials in MSW have also changed over time. Figure 13.3 shows this change over the past 45 years, especially for materials associated with packaging (plastic and paper/cardboard).

**Class Discussion**

Based on the data shown in Figures 13.2 and 13.3, for what materials would you immediately develop a pollution prevention program? Which wastes would you target for source reduction, and which would you target for recycling?

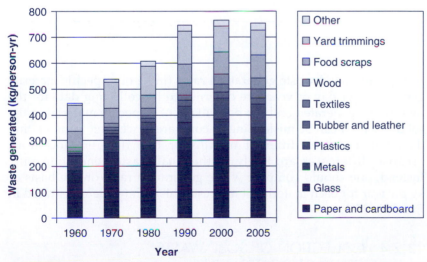

**Figure 13.3** Rate of Generation of Various Materials in U.S. Municipal Solid Waste, 1960–2005

Data from EPA, (2006a).

The solid waste generated in developing countries is much less (0.15 to 0.3 Mg per person per year) than in developed countries (0.7 to 1.5 Mg per person per year). The composition of the waste also differs throughout the world, as shown in Table 13.3. Key differences in waste composition in developing countries include the higher fraction of organic putrescibles and the lower fraction of manufactured products, such as paper, metals, and glass.

Higher-income households tend to generate more inorganic material from packaging waste, whereas lower-income households produce a greater fraction of more organic material through preparing food from base ingredients. However, some high-income households in the developing world may generate the same amount of organic material because they prepare more fresh, unpackaged food. These differences tend to become reduced as countries develop their economies.

Combined with fewer financial resources and skills, the differences in solid-waste production mean that unique solid-waste management practices are required in locations with developing economies.

**Table / 13.3**

**Solid-Waste Composition for Five Cities of the World**

| Location | Food Waste | Paper | Metals | Glass | Plastic, Rubber, Leather | Textiles | Ceramics, Dust, Ash, Stones | Generation (Mg/person-yr) |
|---|---|---|---|---|---|---|---|---|
| Bangalore, India | 75.2 | 1.5 | 0.1 | 0.2 | 0.9 | 3.1 | 19 | 0.146 |
| Manila, Philippines | 45.5 | 14.5 | 4.9 | 2.7 | 8.6 | 1.3 | 27.5 | 0.146 |
| Asuncion, Paraguay | 60.8 | 12.2 | 2.3 | 4.6 | 4.4 | 2.5 | 13.2 | 0.168 |
| Mexico City, Mexico | 59.8[*] | 11.9 | 1.1 | 3.3 | 3.5 | 0.4 | 20 | 0.248 |
| Bogota, Colombia | 55.4[*] | 18.3 | 1.6 | 4.6 | 16 | 3.8 | 0.3 | 0.270 |

* Includes small amounts of wood, hay, and straw.
SOURCE: Diaz et al., 2003.

**Waste Characterization Around the World**

Typically, solid-waste **generation rates** (in kg of a specific material generated per day or year) are determined by collecting data for the total waste generated and the percentage of a specific material in the solid waste. It can be misleading to compare percentage composition data between two different months or two different locations, because differences are likely in the total waste generation rate. Instead, one should compare waste generation rates on the basis of kg *per year* for each material of interest. This point is demonstrated in Example 13.1.

### 13.2.4   COLLECTION OF SOLID-WASTE CHARACTERIZATION DATA

The characterization of MSW is a complex task. Because solid waste varies greatly in composition and quantity within a region and over time, there will always be large uncertainty in estimates of solid-waste

## example/13.1 Calculating Solid-Waste Generation Rates

The solid-waste composition and the quantity of solid waste generated from two cities located in the developing world are as follows:

| Component | City 1 | City 2 |
|---|---|---|
| Food waste (%) | 47.0 | 65.5 |
| Paper and cardboard (%) | 6.3 | 6.5 |
| Ash (%) | 36.0 | 10.2 |
| Other (%) | 10.7 | 17.8 |
| Waste generation rate (kg/person-day) | 0.38 | 0.28 |

Which city generates more ash on a per capita basis? Which city generates more non-ash waste on a per capita basis?

## solution

To determine which city produces more ash on a per capita basis, multiply the overall waste generation rate by the percentage of the component of interest (in this case ash) in the overall wastestream:

$$\text{City 1:} \quad 0.38 \, \text{kg/person-day} \times \frac{36.0}{100}$$

$$= 0.14 \, \text{kg ash waste/person-day}$$

$$\text{City 2:} \quad 0.28 \, \text{kg/person-day} \times \frac{10.2}{100}$$

$$= 0.03 \, \text{kg ash waste/person-day}$$

City 1 produces significantly more ash on a per person basis.

To find the non-ash waste generation rate, first determine the non-ash waste percentage by subtraction, and then multiply this value by the total waste generation rate:

$$\text{City 1:} \quad 0.38 \, \text{kg/person-day} \times \frac{(100 - 36.0)}{100}$$

$$= 0.24 \, \text{kg non-ash waste/person-day}$$

$$\text{City 2:} \quad 0.28 \, \text{kg/person-day} \times \frac{(100 - 10.2)}{100}$$

$$= 0.25 \, \text{kg non-ash waste/person-day}$$

The non-ash waste generation rates are roughly equal.

The greatest difference is that one city generates much more ash than the other. One possible explanation is that it is more common for people in City 1 to burn solid fuels such as wood or coal for cooking and heat. Another possible explanation is that City 2 collects ash waste along with other solid wastes, while City 1 collects only some of the ash waste produced. Other climatic and social-economic reasons are plausible.

In any case, the example shows how waste characterization is an important aspect of a solid-waste management program because waste generation and composition can differ within a country and around the world.

composition. The generation of good data assists with making appropriate management decisions, but reduction of uncertainty in estimated composition can be very expensive.

Three methods are commonly used to characterize solid waste:

1. *Literature review*. This method relies on using past data to characterize the makeup of solid waste. It has some limitations: Definitions used to collect the data can be unclear. Solid-waste generation patterns vary with time and space. Most historical data lack an analysis of the data's uncertainty. And most historical data lack a joint estimate of total waste and percent composition.

2. *Input–output analysis*. This method relies on using data on the consumption of materials to estimate the generation of waste. The EPA data used for Figures 13.2 and 13.3 are examples of this method of analysis. This approach also has weaknesses: It requires clear boundaries so that imports and exports outside the boundary (for example, outside the United States) can be accounted for. It requires assumptions about storage and consumptive use (for example, eating) of purchased goods. It requires assumptions about wastes generated without an economic record (for example, yard wastes).

3. *Sampling surveys*. This method relies on collection of actual data and statistical methods to estimate averages and uncertainty. Many locations require periodic surveys, and a number of methods are available to aid in sampling surveys (ASTM, 1992; New Zealand Ministry for the Environment, 2002). This approach has three weaknesses: Large variability means that a large number of samples are needed, which in turn drives up costs. High variability from one season to another can mean that periodic surveys must be conducted for a year or two before useful data are obtained. Finally, because uncertainty increases as the percent decreases, the method is not effective for relatively uncommon components.

## 13.2.5 PHYSICAL/CHEMICAL CHARACTERIZATION OF WASTE

The selection of a particular management option for solid wastes depends on specific physical and chemical characteristics of the waste. Data on waste generation rates are provided in terms of mass units (mass/capita-time) and not volume units, because the density of waste can vary greatly among wastes or over time. Estimates of **waste density** are important so that space requirements can be estimated for the waste at various stages (for example, collection, transport, disposal) within the waste management system.

Table 13.4 provides typical densities for different types of solid waste during different stages of solid-waste management. The large increases in density that come with compaction of waste in a truck or landfill, and from baling recovered materials, are important in evaluating the economics of related solid-waste management options.

**Collecting combustible waste for incineration in Japan** (photo courtesy of James Mihelcic).

### Table / 13.4

**Densities of Various Municipal Solid Wastes and Recovered Materials**

| | Density Range (kg/m³) | | Density Range(kg/m³) |
|---|---|---|---|
| *Mixed MSW* | | Plastic containers | 32–48 |
| Loose | 90–180 | Miscellaneous paper | 48–64 |
| Loose (developing countries) | 250–600 | Newspaper | 80–110 |
| In compactor truck | 300–420 | Garden waste | 64–80 |
| After dumping from compactor truck | 210–240 | Rubber | 210–260 |
| In landfill (initial) | 480–770 | Glass bottles | 190–300 |
| In landfill (with overburden) | 700–1,100 | Food waste | 350–400 |
| Shredded | 120–240 | Tin cans | 64–80 |
| Baled | 480–710 | *Recovered Materials (Densified)* | |
| *Recovered Materials (Loose)* | | Baled aluminium cans | 190–290 |
| Powdered refuse-derived fuel | 420–440 | Cubed ferrous cans | 1,040–1,500 |
| Densified refuse-derived fuel | 480–640 | Baled cardboard | 350–510 |
| Aluminum scrap | 220–260 | Baled newspaper | 370–530 |
| Ferrous scrap | 370–420 | Baled high-grade paper | 320–460 |
| Cardboard | 16–32 | Baled PET plastic | 210–300 |
| Aluminum cans | 32–48 | Baled HDPE plastic | 270–380 |

SOURCE: Diaz et al., 2003.

The **moisture content** of solid waste is determined as follows:

$$\text{moisture content} = \frac{\text{mass of moisture}}{\text{total mass of waste}} \tag{13.1}$$

This definition is different from the dry weight basis, the definition commonly used in geotechnical engineering applications. However, it is the same as the wet weight basis, the definition commonly used in the soil sciences. The dry weight can be found as follows:

$$\text{dry mass} = \text{total mass of waste} \times \frac{100 - \text{moisture content (in \%)}}{100}$$

$$\tag{13.2}$$

Different amounts of moisture are associated with different solid wastes, and data are typically collected on moist solid waste *as received*. This value is then converted to dry waste mass before further calculations are made. Table 13.5 provides typical values for the moisture content of different solid-waste components, along with information on the energy content and elemental chemical composition.

## Table / 13.5

### Common Physical/Chemical Characteristics of Solid-Waste Components

| | Moisture (% by wet mass) | Energy Value as Received (MJ/kg) | Energy Value after Drying (MJ/kg) | Carbon (% by dry mass) | Hydrogen (% by dry mass) | Oxygen (% by dry mass) | Nitrogen (% by dry mass) | Sulfur (% by dry mass) | Ash (% by dry mass) |
|---|---|---|---|---|---|---|---|---|---|
| Food wastes | 70 | 4.2 | 13.9 | 48 | 6.4 | 37.6 | 2.6 | 0.4 | 5 |
| Magazines | 4.1 | 12.2 | 12.7 | 32.9 | 5 | 38.6 | 0.1 | 0.1 | 23.3 |
| Paper (mixed) | 10 | 15.8 | 17.6 | 43.4 | 5.8 | 44.3 | 0.3 | 0.2 | 6 |
| Plastics (mixed) | 0.2 | 32.7 | 33.4 | 60 | 7.2 | 22.8 | <0.1 | <0.1 | 10 |
| Textiles | 10 | 18.5 | 20.5 | 48 | 6.4 | 40 | 2.2 | 0.2 | 3.2 |
| Rubber | 1.2 | 25.3 | 25.6 | 69.7 | 8.7 | <0.1 | <0.1 | 1.6 | 20 |
| Leather | 10 | 17.4 | 18.7 | 60 | 8 | 11.6 | 10 | 0.4 | 10 |
| Yard wastes | 60 | 6.0 | 15.1 | 46 | 6 | 38 | 3.4 | 0.3 | 6.3 |
| Wood (mixed) | 20 | 15.4 | 19.3 | 49.6 | 6 | 42.7 | 0.2 | <0.1 | 1.5 |
| Glass | 2 | 0.2 | 0.2 | 0.5 | 0.1 | 0.4 | <0.1 | <0.1 | 99 |
| Metals | 4 | 0.6 | 0.7 | 4.5 | 0.6 | 4.3 | <0.1 | <0.1 | 90.6 |

SOURCE: Some data from Tchobanoglous et al., 1993.

The design of systems for **energy recovery** requires data on the energy content of the waste. Similarly, evaluation of many solid-waste treatment systems will require information on the elemental composition of wastes. The high-energy components of MSW are plastic and paper. Food waste and yard waste have high moisture content, which limits the energy they release when burned.

The moisture contents provided in Table 13.5 are typical values. These values can vary greatly depending on the specific composition of the waste component or local factors such as the weather. For example, yard waste that is generated during the summer could be predominately grass, and after collection in wet weather, the yard waste could have a moisture content of 80 percent. In late autumn, the yard waste could be predominately leaves, and after collection in dry weather, it could have a moisture content as low as 20 percent. The varying moisture content of a waste material can affect evaluation of the overall composition of a waste stream (as shown in Example 13.2) and estimates of the total waste. It is therefore preferable to work with the dry mass of waste during intermediate calculations.

example/13.2 Adjusting for Varying Moisture Content of Waste

Food wastes can make up a significant fraction of a municipal solid-waste stream. In order to design a food waste collection system, a municipality is interested in determining the amount of food waste generated. The solid waste was analyzed on one particular day when it was found that the total annual waste generation was 700 kg/person-year. The study also showed that the food waste was 20 percent of the total (wet) mass of waste generated (see the following table). Thus, the rate of food waste generation is 140 kg/person-year.

The study did not, however, measure the moisture content of the waste. Assume that the waste percentage data were collected on a dry summer day with a low moisture content. Assuming a typical moisture content, estimate the food waste generation rate and the total waste generation rate.

|  | Percentage of Total Mass | Low Moisture Content (%) | Typical Moisture Content (%) |
|---|---|---|---|
| Food waste | 20 | 50 | 70 |
| Paper waste | 30 | 3 | 10 |
| Yard waste | 30 | 20 | 60 |
| Other waste | 20 | 2 | 5 |

solution

First determine the dry generation rate of the various waste streams on a day when the waste has a low moisture content. This value can

## example/13.2 Continued

be converted to the typical dry generation rate (which is probably more applicable to the whole year). The total generation rate of each component of the solid-waste stream can be calculated and tabulated as shown in Table 13.6. To estimate conditions on a day with a typical moisture content, determine the dry generation rate by using an assumed low moisture content (provided in the preceding table). Then add the typical moisture back to the dry generation rate.

Equations 13.1 and 13.2 can be combined to solve for the total mass as a function of dry mass and moisture content:

$$\text{total mass} = \frac{\text{dry mass}}{\dfrac{100 - \text{moisture content (in \%)}}{100}}$$

The results of our analysis are as shown in Table 13.6.

The expected typical total generation rate of solid waste with typical moisture added is 1,024.1 kg/person-year, and the wet mass generation rate of food waste is 233.3 kg/person-year. These values are significantly higher than the sampled values (provided in the first column of Table 13.6) because the sampled values were determined on a relatively dry summer day.

### Table / 13.6

**Results for Example 13.2**

|  | Sampled Total Generation Rate (kg/person-yr) | Assumed Moisture Content (%) | Dry Generation Rate (kg/person-yr) | Typical Moisture Content (%) | Typical Total Generation Rate (kg/person-yr) |
|---|---|---|---|---|---|
| Food waste | 140 | 50 | 70 | 70 | 233.3 |
| Paper waste | 210 | 3 | 203.7 | 10 | 226.3 |
| Yard waste | 210 | 20 | 168 | 60 | 420.0 |
| Other waste | 140 | 2 | 137.2 | 5 | 144.4 |
| **Total** | **700** |  | **578.9** |  | **1,024.1** |

### 13.2.6   HAZARDOUS-WASTE CHARACTERIZATION

Wastes are considered **hazardous waste** when they pose a direct threat to human health or the environment. Wastes can be classified as hazardous by one of six characteristics: ignitable, corrosive, reactive, toxic, radioactive, or infectious. Table 13.7 provides more detail on each characteristic.

Most countries have laws and regulations related to hazardous wastes and provide a definition that classifies particular wastes as

## Table / 13.7

**Methods to Classify a Solid Waste as Hazardous**

| Characteristic of Waste | Question Related to Characteristic |
| --- | --- |
| Ignitable | Can the waste create a fire (e.g., waste solvents)? |
| Corrosive | Is the waste very acidic or basic and so able to corrode storage containers (e.g., battery acids)? |
| Reactive | Can the waste participate in rapid chemical reactions leading to explosions, toxic fumes, or excessive heat (e.g., lithium that can react with water explosively, explosives, cyanide sludge, strong oxidizing agents)? |
| Toxic | Can the waste cause internal damage to a person or organism (e.g., poisons causing death or blindness, carcinogens)? |
| Radioactive | Can the waste release subatomic particles that can cause toxic effects (e.g., some medical and laboratory wastes, wastes associated with nuclear energy production)? |
| Infectious | Can the waste lead to the transmission of disease (e.g., used syringes, hospital medical wastes)? |

SOURCE: Environmental Protection Agency, www.epa.gov/osw/hazwaste.htm.

legally hazardous or not. In the United States, the relevant regulations are promulgated under RCRA and focus primarily on large generators of relatively homogeneous wastes. The management of hazardous wastes is a specialized topic that is beyond the scope of this book. We have described in Chapter 7 how green chemistry and green engineering can be deployed to reduce or eliminate the production and use of hazardous chemicals (and their associated risk and wastes).

Small quantities of wastes that exhibit the characteristics of a hazardous waste are often not legally considered a hazardous waste and are managed (and thus disposed of) along with municipal waste. Table 13.8 lists common household hazardous products. The storage and use of these products in the home is a concern (especially because of the poor indoor air environments). Many municipalities provide household hazardous-waste programs that educate consumers on using green alternatives (see www.care2.com/greenliving/) while also collecting household hazardous waste.

Readers should consider the impact on the environment of their purchases. To varying degrees, household hazardous-waste products will increase environmental impacts in ecosystems and at wastewater treatment plants, materials recovery facilities, compost facilities, landfills, and waste-to-energy facilities. In addition, the use of these items in the home or workplace presents an environmental risk to humans who occupy buildings. As we will explore in Chapter 14, specifying and using degradable and nontoxic cleaning products and equipment reduces emissions and operation and maintenance costs, and improves indoor air quality and occupant productivity.

### Is It Non–Hazardous or Hazardous Waste?

http://www.epa.gov/osw/

### Class Discussion

What green alternatives could you use at your house or apartment to eliminate use of household hazardous wastes listed in Table 13.8? How would you effectively communicate this information to the public as an engineer employed by the local municipality?

### Pollution Prevention Act

http://www.epa.gov/oppt/p2home/pubs/laws.htm

## Table / 13.8

### Common Hazardous Products Found in Households

| Product | Concern |
|---|---|
| *Household Cleaning Products* | |
| Oven cleaners | Corrosive |
| Drain cleaners | Corrosive |
| Pool acids, chlorine | Corrosive |
| Chlorine bleach | Corrosive |
| *Automotive Products* | |
| Motor oil | Ignitable |
| Antifreeze | Toxic |
| Car batteries | Corrosive |
| Transmission and brake fluid | Ignitable |
| *Lawn and Garden Products* | |
| Herbicides, insecticides | Toxic |
| Wood preservatives | Toxic |
| *Indoor Pesticides* | |
| Flea repellents and shampoos | Toxic |
| Moth repellents | Toxic |
| Mouse and rat poisons | Toxic |
| *Home Maintenance/Hobby Supplies* | |
| Oil or enamel-based paints | Flammable |
| Paint solvents and thinners | Flammable |

SOURCE: Adapted from Environmental Protection Agency (www.epa.gov/msw/hhw-list.htm) and Tchobanoglous et al., *Integrated Solid Waste Management*, 1993, copyright The McGraw-Hill Companies.

---

### Box / 13.2   Solid-Waste Management in the Developing World: Case Study from Mali, West Africa

Proper solid-waste management is important in all parts of the world.* The components of solid-waste management can look quite different, though. Here we explore solid-waste management in a neighborhood of the city of Sikasso (Mali) that has a population of approximately 150,000. Figure 13.4 provides an overview of the components.

#### On-Site Storage

Women sweep their small shops and homes every morning and evening and place the waste in a trash can or corner. Residents may purchase a trash can constructed of halved metal drums, which have holes along the walls. Residential solid waste consists mainly of paper, organics (dust, leaves), and some plastic. There is also significant solid waste generated at markets, especially organics and cardboard.

#### Collection

Typically, solid-waste collection begins with privately owned organizations that collect solid waste.

#### Transport and Transfer

If your household has an account with the collection organization, a man with a donkey cart will come to your concession/house each day, empty your trash can, and take the waste to a collection area on the periphery of the city. They will not empty your trash without a trash can. Those without trash cans and agreements will ask children to carry trash to the collection areas. In some cases, the mayor or private collection organization owns a large truck and tractor and can transfer large piles of trash from central areas to a collection area. In Sikasso, there are ten collection areas on the periphery.

### Processing

Scavengers, especially children, living near the collection areas may pick through the solid waste before the collection team arrives to take the trash outside the city. Some residents may burn the trash occasionally if the piles become too large.

### Disposal

Later on, solid waste is burned or carried from the periphery to rural areas. It may be spread out over fallow fields, like compost. Currently, there are no official landfills for disposal, but plans for construction are pending.

### Source Reduction, Recycling, Composting

Much of the refuse is organic. People reuse cans, bags, and other plastics and metals as long as they can. These may be used for containers in the reselling of products (for example, juice, spices) or recycled into toys and art (for example, metal milk cans hammered into toy trucks).

### Disease Vectors

Rats, mice, maggots, cockroaches, and mosquitoes may be found in collection areas. Those living in proximity to these zones may experience increased health risks. Many people throughout the world still do not understand the connection between these vectors and diarrheal diseases or malaria.

### Cultural Considerations

Men and women consider transporting and burning solid waste children's work, unless they have an account with the collection organization.

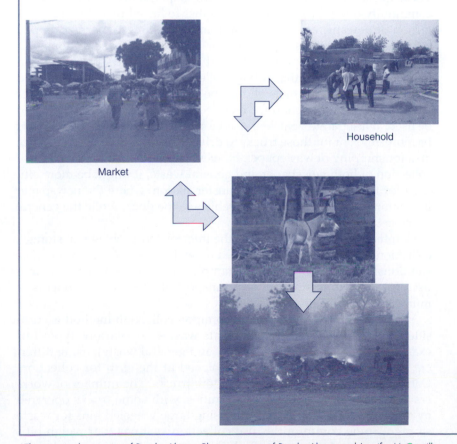

Market

Household

**Figure 13.4** Solid-Waste Management in the City of Sikasso, Mali (Population Approximately 150,000)

*This case study courtesy of Brooke Ahrens. Photos courtesy of Brooke Ahrens and Jennifer McConville.

## 13.3  Components of Solid-Waste Systems

Recall from Figure 13.1 that a solid-waste system comprises waste generation, storage, collection and transport, and processing and disposal. In this section, we consider each responsibility in greater detail, beginning with storage, collection, and transport.

### 13.3.1  STORAGE, COLLECTION, AND TRANSPORT

Storage, collection, and transport of municipal solid waste typically accounts for 40 to 80 percent of the total cost of solid-waste management. Three questions need to be considered when designing a storage, collection, and transport system:

1. Which wastes should be collected from the generator, and which should the generator transport to a processing facility?

2. To what extent should the generators be asked to separate collected waste into different fractions?

3. Should waste be transported directly to a treatment/disposal facility, or should collection vehicles transfer wastes into a more efficient vehicle first?

These questions cannot be considered independently. For example, a community might be considering whether to have house-to-house collection of newspapers versus a system where newspaper is dropped off at recycling centers. If newspaper is collected house to house, an appropriate system will be required for households to collect and deliver newspapers to the curb for collection. An assessment is needed whether a separate vehicle should be used to collect newspapers or whether existing collection for recycled goods or general waste could be used. If a separate vehicle is used for newspaper collection, it might be more efficient for those trucks to drive directly to a dockside storage area for shipping of wastepaper. However, if the same truck is used for collection of both newspaper and general waste, it might be more efficient for the truck to travel to a transfer station, where the newspaper is separated and sent in a special vehicle to the dock, while the general waste is taken to a landfill.

Common options available for the interrelated problems of storage, collection, and transport can be used to devise creative and sustainable solutions. Figure 13.5 provides examples of vehicles and a container used for collection, storage, and transport of specific components of municipal solid waste.

For residential MSW, the most common collection method is curbside. Residents are asked to segregate wastes into various types (for example, recyclables, organic waste, and general waste), using different containers or bags, which are placed at the curb for collection. Collection is performed by one or more trucks. The number of workers per truck varies among communities, with some trucks operated by only one person and residents using large wheeled bins for waste storage. Some communities use collection systems that weigh bins (or charge per bag) and thus bill the generator accordingly. For

(a) This front-end loading vehicle is commonly used for commercial collection.

(Photo courtesy of Heil Environmental.)

(b) The side-loading vehicle is commonly used for residential collection.

(Photo courtesy of Heil Environmental.)

(c) The rear-loading vehicle is suited for residential collection.

(Photo courtesy of Heil Environmental.)

(d) This is a street and footpath sweeper.

(Photo courtesy of Hako-Werke GmbH.)

(e) This bin offers a means to collect recyclables in a shopping district.

(Photo courtesy of Glasdon UK Ltd.)

**Figure 13.5** **Devices to Aid Solid-Waste Collection, Storage, and Transport**

**Solid Waste Storage in Fiji**
(Photo courtesy of James Mihelcic.)

**Recycle Your Cellphone**
© Bakaleev Aleksey/iStockphoto.

residential waste collection, the costs increase most rapidly with the number of stops, the number of workers, and the total number of trucks in service. These factors have influenced collection systems so they now typically use large waste containers for residents, more multiuse trucks, higher density of waste per truck, and fewer workers per truck.

Commercial and institutional waste generators typically use larger storage containers and have a separate system for waste collection. This system uses vehicles specifically designed for collection of large quantities of waste per stop. High-rise commercial or residential buildings are a special case. Many have specialized systems to transport waste to the bottom of the building for storage and compaction prior to collection.

Drop-off stations for recyclables are another valuable part of a MSW collection system. These can be designed for use by people who are walking or riding a bicycle, as well as driving via personal vehicles or using shared public transit. Drop-off systems can also be used for general waste in rural areas without door-to-door collection. Where vehicular access would be inappropriate or difficult (say, in a very old city or a tourist zone), pneumatic systems can be used to convey wastes by vacuum out of the urban core.

**Transfer stations** are used in larger cities and towns to reduce the costs associated with transport. They have also become common because of the movement toward regional disposal sites (versus the many local disposal sites that existed several decades ago). Collection trucks are specialized vehicles and are most efficiently used to collect, rather than transport, waste. While a typical collection truck might hold 4 to 7 Mg of MSW, a larger truck using more efficient compaction can carry 10 to 20 Mg of MSW. Stations can also be used to transfer waste to containers that are shipped by rail or sea. With increases in the distance to waste treatment and disposal sites and the amount of waste generated, urban transfer stations generally become more economical.

### 13.3.2   RECYCLING AND MATERIALS RECOVERY

Recycling requires separation of materials and removal of low-quality wastes. Successful recycling systems use a mixture of separation at the source by the waste generator, by machinery at a central location, and by trained people at a central location. Successful recycling systems require careful consideration of costs involved, and of the markets for recycled goods. Several types of **materials recovery facilities (MRFs)** can be used, with some specializing in processing of separately collected wastes (see Figure 13.6).

Expansion in recycling requires the development of new markets. Otherwise, the excess of supply over demand will lead to decreases in the value of the recovered materials, to the point that more resources are used to recover the materials than are saved by the recovery. In some cases, additional recycling can occur by distributing information between waste holders and potential users. Waste exchange systems can be operated to help one small business (or homeowner) solve a waste problem while another finds a valuable input.

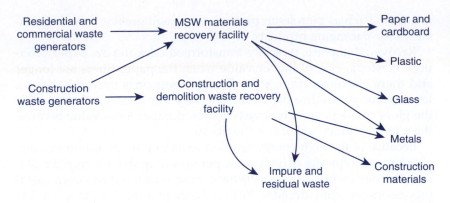

**Figure 13.6** Recycling Systems: Waste Generators, Materials Recovery Facilities, and Markets for Recovered Materials

Note that in recycling programs, the control of impurities can be critical. For example, a small amount of ceramics in glass can make the glass impractical to recycle. Limiting the number of impurities requires extensive and ongoing communication with waste generators, whether they be children, adults, business owners, or community leaders.

TYPES OF MATERIALS RECOVERED OR RECYCLED Plastics recycling is a challenge partly because the plastics industry has developed and marketed many unique types of plastics that are not necessarily compatible when recycled. To assist with plastics recycling, an international resin code is marked on most plastic consumer products (see Table 13.9). The most commonly recovered plastics are polyethylene terephthalate (PET) (type 1) and high-density polyethylene (HDPE) (type 2). As society moves away from use of nonrenewable

## Table / 13.9

**Types of Plastics Found in Commercial Products with Resin Codes Used to Aid in Recovery**

| Resin Code | Material | Sample Applications |
|---|---|---|
| 1 | Polyethylene terephthalate (PET) | Plastic bottles for soft drinks; food jars |
| 2 | High-density polyethylene (HDPE) | Bottles for milk; bags for groceries |
| 3 | Polyvinyl chloride | Blister packs; bags for bedding, pipe |
| 4 | Low-density polyethylene | Bags for dry cleaning and frozen foods |
| 5 | Polypropylene | Containers for yogurt, takeout meals |
| 6 | Polystyrene | Cups and plates; furniture and electronics packaging |
| 7 | Other plastics | Custom packaging |

resources such as petroleum plastics, there will greater use of biomaterials for packaging products.

Recovered paper is typically transformed back to new paper products. Wastepaper is of higher value when the paper fibers are longer and there are fewer impurities. Glossy magazines currently have a lower value than office paper because they use minerals that provide the gloss to the paper. Previously recycled paper loses value because the recycling process shortens the fibers.

Because of the high energy requirements to process aluminum ore, aluminum is typically of high value per unit weight of recovered material. Ferrous metals (iron, steel) have been recovered by scrap metal processors for many decades. With a developed market for waste ferrous metal, the recovery of ferrous metals from appliances, vehicles, equipment, cans, and demolition debris is now common.

The system for turning waste glass, called cullet, into new glass is well developed. However, the large cost of transportation to a glass smelter can make it impractical to turn waste glass into new glass. As a result, new markets for this material are under development.

**Construction and demolition debris** includes metals, wood, stone, and concrete. Some building materials (for example, tiles and fittings) can be reused, while others are processed for new uses. Broken stone and concrete can become aggregate for new concrete or for other building-fill purposes (see Chapter 14).

**E–Cycle Your Electronic Wastes**
http://www.epa.gov/epawaste/conserve/
materials/ecycling/index.htm

SEPARATION OF MATERIALS A wide variety of mechanical equipment is available for separation of waste materials. Magnets can separate ferrous metals, but only after any bags have been opened and the waste has been placed on conveyors. To separate papers and plastics, machinery can exploit the lower density and larger size of these materials. Methods can involve screens, sieve-like inclined shaking tables, bursts of air, and rotating sieves called trommels. In some cases, paper and plastic wastes are separated from other low-energy materials and then left in a mixed state to be used as a source of fuel. The paper/plastic mix is termed **refuse-derived fuel (RDF)**, and can be shredded and then compressed to lessen transport costs. Mechanized techniques are also available to separate aluminum from other materials, and to distinguish various colors of glass.

In many situations, people are employed to aid in the separation of waste materials. In some cases, people ensure that high-quality recovered goods are produced. In other cases, staff pick specific wastes from

---

**Box / 13.3    Role of Scavenging in the Developing World**

In developing countries, it is common for *scavengers*, the *informal sector*, to participate in solid-waste management activities. This is due primarily to inadequate municipal services, which create a large need for informal waste collection and an opportunity for income among the poor. Medina (2000) writes:

*When scavenging is supported—ending exploitation and discrimination—it represents a perfect illustration of sustainable development that can be achieved in the Third World: jobs are created, poverty is reduced, raw material costs for industry are lowered (while improving competitiveness), resources are conserved, pollution is reduced, and the environment is protected.*

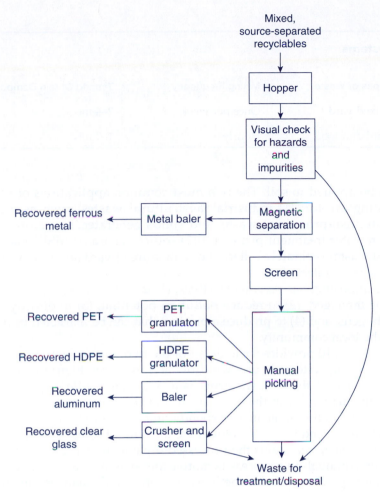

**Figure 13.7** **Process Flow Diagram for a Materials Recovery Facility** This facility is designed to process source-separated PET and HDPE plastics, metal and aluminum cans, and clear glass.

a conveyor and place them into separate containers. The health and safety of workers is obviously a primary concern in MRFs.

Design of an MRF is difficult, and creative approaches are valued. Separated materials need to be compressed for transport and safely stored. A typical process flow diagram for an MRF is shown in Figure 13.7. The MRF design needs to manage waste materials that arrive in variable quantities, as well as adapt to markets that vary in the price paid for processed materials.

### 13.3.3 COMPOSTING

**Composting** is a microbial process that treats biodegradable wastes. The reactions are similar to those employed in aerobic wastewater treatment (discussed in Chapter 11). Wastes are processed down to a suitable size, water is added, air is allowed to enter to transfer oxygen into the waste pile, and the waste is mixed to ensure even degradation. The microorganisms feed on the organic matter in the waste, producing carbon dioxide and leaving behind a solid (called *compost*) that

**Most Common Types of Composting Systems**

| System Type | Particle Size | Types of Waste | Mixing Frequency | Time to Obtain Compost |
|---|---|---|---|---|
| Windrows | 5–20 mm | Mixed yard | Once per week | 2–4 mo |
| In-vessel | 5–20 mm | Yard and food | Hourly | 1–2 mo |

can be applied to soil. The two most common applications of composting are for: (1) industrial/agricultural wastes, such as wood waste, fish-processing waste, and solids generated at a municipal wastewater treatment plant; and (2) source-separated MSW, such as yard wastes separately collected or a mixture of yard and food wastes separately collected.

Composting has several objectives: (1) to reduce the mass of waste to be managed; (2) to reduce pollution potential; (3) to destroy any pathogens; and (4) to produce a product that can be marketed or used by the local community.

Table 13.10 provides some detail of the two most common types of composting systems in use: windrow and in-vessel. Figure 13.8 illustrates each of them. A **windrow** is a trapezoidal pile of processed organic matter left in the open air. It must be occasionally turned to ensure that all organic matter spends time inside the pile, where the temperature and moisture content are ideal for decomposition. An **in-vessel** system maintains the processed organic matter in a large container. Although this process is more expensive to construct, it allows for more precise control of the process, more readily manages air emissions, and produces compost faster.

Production of compost requires maintenance of temperatures above 40°C for several days or longer. Providing the right conditions

(a) Windrows.

(Photo provided by Dr. Ian Mason, University of Canterbury.)

(b) In-vessel system.

(Photo provided by Robert Rynk).

**Figure 13.8** Two Composting Systems

## Table / 13.11

### Nutrient Content of Various Materials Used in Composting

Composting systems perform best when the carbon-to-nitrogen ratio is optimized in the range of 20–40 (C:N of 20:1 to 40:1).

| Material | Nitrogen (% dry mass) | C:N Ratio (dry mass basis) |
|---|---|---|
| Urine | 15–18 | 0.8 |
| Human feces | 5.5–6.5 | 6–10 |
| Cow manure | 1.7–2 | 18 |
| Poultry manure | 5–6.3 | 15 |
| Horse manure | 1.2–2.3 | 25 |
| Activated sludge | 5 | 6 |
| Nonlegume vegetable wastes | 2.5–4.0 | 11–12 |
| Potato tops | 1.5 | 25 |
| Wheat straw | 0.3–0.5 | 130–150 |
| Oat straw | 1.1 | 48 |
| Grass clippings | 2.4–6.0 | 12–15 |
| Fresh leaves | 0.5–1.0 | 41 |
| Sawdust | 0.1 | 200–500 |
| Food wastes | 3.2 | 16 |
| Mixed paper | 0.19 | 230 |
| Yard wastes | 2.0 | 23 |

SOURCE: Haug, 1993.

for aerobic microbial growth leads to rapid release of energy in the form of heat, and composting systems can commonly reach temperatures of 70°C for short time periods. The high temperature ensures destruction of pathogens and unwanted seeds, and leads to a faster production of the final product. The process is controlled through nutrient (especially nitrogen) content, pH, moisture content, and the air content.

The nutrient content is commonly expressed as a **carbon-to-nitrogen ratio** on a dry-weight basis (the C:N ratio). This ratio should be in the range of 20 to 40 (20:1 to 40:1) for materials entering a composting process. Table 13.11 lists the nitrogen content and C:N ratio for a variety of materials commonly composted.

When one waste is not compostable on its own, it can be mixed with other materials to ensure the proper nutrient content, pH, moisture content, and air porosity. Example 13.3 demonstrates a typical calculation used to determine the correct composition of the waste added to a composting system.

 **Composting Mix Design**

## example/13.3 Determining Proper Ingredients for Successful Composting

A poultry manure has a moisture content of 70 percent and is 6.3 percent N (on a dry mass basis). The manure is to be composted with oat straw with a moisture content of 20 percent. The desired C:N for the mixture is 30. Using Table 13.11 values for the nitrogen composition and C:N ratio for these two materials, determine the kg of oat straw required per kg of manure to attain the desired C:N ratio.

## solution

Assume 1 kg of moist poultry manure dry mass. Let $X$ = kg of moist oat straw on a dry mass basis. The mass of carbon and nitrogen obtained from each material in the mixture is

$$\text{Dry mass nitrogen from poultry manure} = 1 \times (1 - 0.7) \times 0.063$$
$$= 0.0189 \text{ kg}$$

$$\text{Dry mass carbon from poultry manure} = 1 \times (1 - 0.7) \times 0.063 \times 15$$
$$= 0.2835 \text{ kg}$$

$$\text{Dry mass nitrogen from oat straw} = X \times (1 - 0.2) \times 0.011 \text{ kg}$$
$$= 0.0088 \times X \text{ kg}$$

$$\text{Dry mass carbon from oat straw} = X \times (1 - 0.2) \times 0.011 \times 48 \text{ kg}$$
$$= 0.4224 \times X \text{ kg}$$

The overall C:N ratio is

$$30 = \frac{(\text{mass carbon from poultry manure} + \text{mass carbon from oat straw})}{(\text{mass nitrogen from poultry manure} + \text{mass nitrogen from oat straw})}$$

$$30 = \frac{(0.2835 + 0.4224 \times X)}{(0.0189 + 0.0088 \times X)}$$

Solving for $X$, we find $X = 1.8$ kg. Thus, for every 1 kg of poultry manure, 1.8 kg of oat straw must be added to obtain an optimal C:N ratio of 30. The reason for this is that the poultry manure is a better source of nitrogen and the oat straw provides a better source of carbon.

Composting can be performed by individuals, by individual businesses, or by municipalities with large amounts of organic waste. In all cases, the principles are the same. Backyard composting can reduce the costs (and associated environmental impacts) of collection, transport, and processing and disposal of organic wastes. Many municipalities now provide subsidized or free home composting units (or worm farms) to encourage the practice. To be effective, home composting must have the proper mix of nitrogen- and carbon-rich materials and have adequate airflow and moisture. Poor household composting practices will not reach the temperatures required for effective treatment, can create odor problems, and can attract animals. Municipalities need to find the proper balance of education, subsidies, and enforcement to achieve effective backyard composting systems.

At the larger scale, composting systems must match the product to suitable markets. Municipalities can return (or sell) compost back to the local residents who generated the waste. Larger markets for compost include municipal parks, golf courses, nurseries, landscapers, landfills (as daily and final cover material), and turf growers. Many users will insist on a lack of impurities in compost, and strict control of the wastes inputted to a composting system will ensure a quality, in-demand product.

**Construct Your Own Compost Bin**
http://www.edf.org/article.cfm?ContentID=2030

### 13.3.4 INCINERATION

**Incineration** (also called **waste-to-energy**) is a combustion process where oxygen is used at high temperatures to liberate the energy in waste. In the United States in 2004, waste-to-energy facilities burned 29 million tons of MSW. In addition, 380 U.S. landfills now recover methane. Incineration can reduce the amount of waste needing disposal, generate energy for a community, and also reduce MSW transportation costs. Incineration becomes more favorable for wastes that have high energy content, low moisture content, and low ash content. These wastes include paper, plastics, textiles, rubber, leather, and wood (energy values previously listed in Table 13.5).

Table 13.12 describes six incineration systems. The most commonly used method is *mass-burn incineration*. In this process, unsegregated

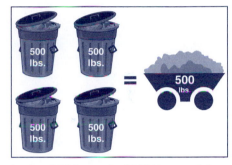

It takes 2,000 pounds of MSW to equal the heat energy in 500 pounds of coal.

## Table / 13.12

### Common Incineration Systems

| Type of Incinerator System | Explanation |
| --- | --- |
| Mass-burn | Unsegregated municipal solid waste is combusted. |
| Modular | Small incinerators focus on treatment of specific waste streams (e.g., medical waste). |
| Refuse-derived fuel (RDF) | Energy-rich waste streams can be separated from other wastes and burned, typically as a substitute for fossil fuels such as coal, in power plants. Wastewater treatment biosolids are one such waste stream. |
| Co-incineration | Specific postproduction commercial/industrial wastes, such as wood wastes from construction, can be combusted with production wastes, such as paper mill sludge or dried wastewater treatment plant biosolids to produce energy. |
| Hazardous waste | Hazardous organic wastes (e.g., solvents, pesticides) can be combusted to destroy the wastes, though this requires very close attention to air emissions. |
| Cement kilns | Cement factories may provide suitable conditions for combustion of many wastes, including waste tires and waste oils, during the production of cement. |

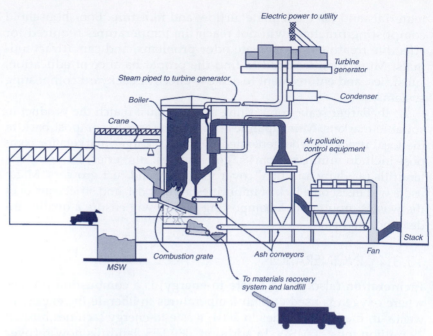

**Figure 13.9** **Mass-Burn Incineration System for Treatment of Municipal Solid Waste** Programs in reducing and recycling wastes and eliminating household hazardous wastes and heavy metals (e.g., batteries) from a waste stream can be coordinated with incineration systems to mimimize potential risk associated with pollutants found in the offgas and ash.

MSW is combusted. Figure 13.9 provides a schematic of a typical mass-burn incinerator. Incineration creates two solid by-products. *Bottom ash* is the unburnt fraction of the waste. *Fly ash* is the particulate matter suspended in the combustion air and removed by air pollution control equipment. Both of these ashes have a number of hazardous components and need to be carefully managed; therefore, a mixture of recovery and landfill processes are used for ashes.

Air pollution control is required for incinerators to limit emissions of particulates, volatile metals (such as mercury), nitrous oxides, and incomplete combustion products (such as dioxins). In the late 1980s, incinerators were the major contributor of dioxin emissions in the United States, but since 2000, the contribution of dioxin emissions from incinerators has been less than 6 percent of the total. Informal backyard burning of solid waste (for example, burn barrels) is now the major source of U.S. dioxin emissions (EPA, 2006b).

Sufficient oxygen must be provided in incineration processes to ensure complete combustion. Carbon, hydrogen, and sulfur in the waste stream are oxidized to $CO_2$, $H_2O$, and $SO_2$ during combustion. Although some oxygen is present in wastes initially, most of this oxygen comes from air. Example 13.4 shows how to estimate the air required in combustion.

Incineration systems often have high construction and operating costs. However, the costs can be offset by savings in transporting wastes to a disposal site, land requirements for disposal, and recoverable energy. Successful incineration systems are those where the waste

**Class Discussion**

How would you lead a public meeting where you were going to discuss behavioral changes that would decrease use of burn barrels in a rural community? What stakeholders need to be present at such a meeting? What specific actions might you employ to achieve consensus?

# example/13.4 Air Requirements for Combustion of Municipal Solid Waste

A refuse-derived fuel (RDF) comprises 60 percent mixed paper, 30 percent mixed plastic, and 10 percent textiles. Assume it is dried before combustion. Determine the volume of air (in L) at 20°C, 1 atmosphere pressure that is required to combust 1 kg of the RDF.

## solution

Use values of percent composition of solid waste from Table 13.5 to determine the moles of C, H, S, and O in the waste. (The molecular weights are 12, 1, 32, and 16 g/mole, respectively.) Complete a table for each of the RDF components:

| | Dry Mass (g) | Moles C | Moles H | Moles S | Moles O |
|---|---|---|---|---|---|
| Paper (mixed) | 600 | 21.7 | 34.8 | 0.038 | 16.6 |
| Plastic (mixed) | 300 | 15.0 | 21.6 | <0.01 | 4.3 |
| Textiles | 100 | 4.0 | 6.4 | 0.006 | 2.5 |
| **Total** | **1,000** | **40.7** | **62.8** | **0.05** | **23.4** |

The moles of $O_2$ required for combustion are determined for each molecular species. The balanced oxidation reactions are as follows:

$$C + O_2 \rightarrow CO_2$$
$$H + 1/4O_2 \rightarrow 1/2H_2O$$
$$S + O_2 \rightarrow SO_2$$

From the balanced reactions, 1, ¼, and 1 mole of $O_2$ are required per mole of carbon, hydrogen, and sulfur, respectively. In addition, each mole of oxygen (O) in the waste can offset the requirement for 0.5 mole $O_2$ gas. The overall moles of $O_2$ required to incinerate the waste are then

$$\left[ \left( 40.7 \text{ moles C} \times \frac{1 \text{ mole } O_2}{\text{mole C}} \right) + \left( 62.8 \text{ moles H} \times \frac{\frac{1}{4} \text{ mole } O_2}{\text{mole H}} \right) \right.$$
$$\left. + \left( 0.05 \text{ mole S} \times \frac{1 \text{ mole } O_2}{\text{mole S}} \right) \right] - \left( 23.4 \text{ moles O} \times \frac{0.5 \text{ mole } O_2}{\text{mole O}} \right)$$
$$= 44.8 \text{ moles } O_2$$

This value of $O_2$ can be converted to liters of oxygen, using the ideal gas law, after which we can determine the volume of air. Rearrange the ideal gas law ($PV = nRT$) to solve for $V$. Use a value of $R = 0.082$ L-atm/mole-K, along with the values of 20°C and 1 atmosphere pressure:

$$V = 44.8 \text{ moles } O_2 \times 0.082 \text{ L-atm/mole-°K} \times \frac{(273 + 20)\text{°K}}{1 \text{ atm}}$$

$$= 1{,}080 \text{ L } O_2/\text{kg}$$

Because air is approximately 20.9 percent $O_2$, the liters of $O_2$ can be converted to liters of air:

$$1{,}080 \text{ L } O_2/\text{kg} \times \frac{1 \text{ L air}}{0.209 \text{ L } O_2} = 5{,}200 \text{ L air/kg waste}$$

is appropriate for incineration and the cost offsets can be found. In addition, incineration systems are complex and require advanced skills in construction and operation. These two factors cause incineration systems to be preferred in more economically advanced regions of the world and where land values, energy costs, and transportation costs associated with other alternatives are high.

### 13.3.5 LANDFILL

**Landfills** are engineered facilities designed and operated for the long-term containment of solid wastes. Design of the landfill will vary greatly based on the waste and the location of the facility. Based on waste type, the four main types of landfills are listed in Table 13.13, along with the relevant federal regulations under RCRA. In landfills, wastes are placed and compacted into solid forms, then covered to limit exposure to water and air. **Leachate** is water that contacts the wastes and becomes a contaminated wastewater. As biological materials decompose in landfills, oxygen is consumed, and carbon dioxide is

**Table / 13.13**

**Major Types of Landfills**   Regulatory oversight depends on the type of waste.

| Type of Waste | Applicable Regulations in United States |
|---|---|
| Construction and demolition debris | Regulation of construction and demolition debris landfills is managed at the state level in the United States. |
| Municipal solid waste | 40 C.F.R. Part 258 |
| Industrial waste (e.g., for ash and mining waste) | 40 C.F.R. Part 257 |
| Hazardous waste | 40 C.F.R. Parts 264 and 265 |

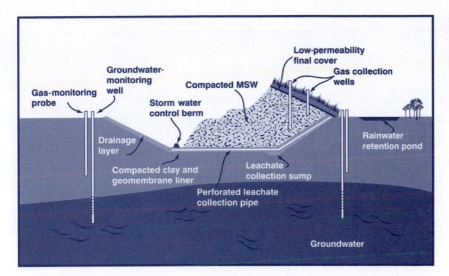

**Figure 13.10  Cross Section of a Modern Landfill, Showing Barriers Incorporated into Engineering Design**  The many barriers help protect the environment.

(Redrawn with permission of the National Solid Wastes Management Association (NSWMA).

produced. Over time, an anaerobic environment evolves that leads to the production of methane gas.

As shown in Figure 13.10, landfills are technically advanced facilities with sophisticated environmental protection systems. Environmental protection in landfills occurs through a combination of four barriers: (1) appropriate siting; (2) engineered design that is carefully implemented during construction and operation; (3) exclusion of inappropriate wastes; and (4) short- and long-term monitoring.

LANDFILL SITING  Landfills need to be located where the risks to the environment and society are low, so that even in the case of poor design, construction, or operation, the resulting risk is minimized. Negating issues associated with social and environmental justice also is critical. Table 13.14 summarizes locations to avoid when siting a landfill and other issues that need to be considered. Landfill siting is a highly contentious social issue, and the engineer's role is to provide input and analysis in an equitable manner.

**Table / 13.14**

**Items to Consider when Siting a Landfill**

**Location-Specific Items to Avoid**

Floodplains

Active geological faults

Land prone to slips or erosion

Wetlands and intertidal zones

Areas with significant ecoystems and important biodiversity

Areas of cultural or archaeological significance

Drinking-water catchments

(Continued)

**Other Items to Consider**

Minimize costs for transport of wastes by placing the landfill near centers of waste generation.

Minimize costs required for construction of transportation infrastructure required to access the site.

Identify sites less prone to extremes of rain or wind.

Identify sites with soils that can be used during construction.

Match the potential site to a final use of the landfill that will benefit the local community.

Eliminate issues of environmental justice and other social objections to the development of a landfill.

Seek collocation of users that can make beneficial use of waste materials or derived energy.

### Class Discussion

Discuss issues of environmental siting in terms of distributing risk equitably. Do you understand why many people do not want a landfill sited in the backyard? Who are the important stakeholders, and how would you address their concerns to arrive at some consensus?

A common tool for evaluating potential landfill sites is a *geographic information system (GIS)*. GIS is a valuable means of processing large amounts of data and ensuring an evaluation of all options. Figure 13.11 shows the result of a GIS evaluation of potential landfill sites. In this case, the location of existing transportation infrastructure can be integrated with the location of high-volume urban waste generators and other location issues listed in Table 13.14. GIS also allows the engineer to add demographic information that is related to other concerns such as equitable distribution of risk.

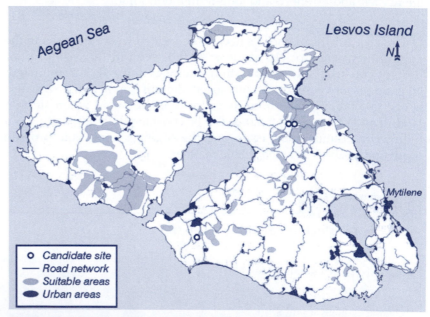

**Figure 13.11**   **Graphical Information System Evaluation of Potential Landfill Sites**   Suitable areas are highlighted in the gray shaded areas, which can be compared to existing transportation networks and waste-producing urban areas highlighted in dark shaded areas. The location is Lesvos Island, in the Aegean Sea east of Greece.

(Reproduced with permission from T. D. Kontos et al., *Waste Management and Research*, 2003, by permission of Sage Publications Ltd.)

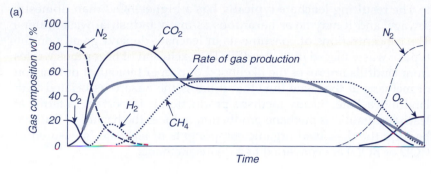

(a)

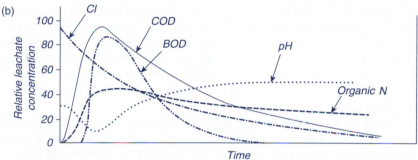

(b)

**Figure 13.12** **Typical Landfill Decomposition Pathways** The upper figure (a) shows gas composition (and production) over time. The lower figure (b) shows the relative leachate concentration of various constitutents. As the rate of gas production increases, the leachate becomes less strong in terms of the concentration of leachate constitutents.

Adapted from Farquhar and Rovers (1973)

**LANDFILL DECOMPOSITION** Wastes that are disposed into a landfill undergo a series of interrelated chemical and biological reactions. These reactions determine the quantity and composition of the gas and leachate produced by the landfill, and thus determine the management required. In terms of the biological reactions that take place, a landfill is best viewed as a batch biochemical decomposition process.

Figure 13.12 depicts gas and leachate composition over time as the waste of biological origin (for example, food waste, yard waste, paper) decomposes. In the early stages of decomposition (shown in Figure 13.12), oxygen is consumed, and carbon dioxide and organic acids are produced. Both of these products decrease the pH of the leachate. An increase in the leachate's oxygen demand (measured as COD and BOD) also is observed in the early stages of decomposition as the organics are converted from a particulate to dissolved phase. In later stages, after all the oxygen is consumed and an anaerobic environment is established, microorganisms convert the high-BOD organic acids to methane gas. Here the leachate becomes less concentrated as the dissolved constituents are converted to the gaseous phase and the readily leachable components of the waste material become less prevalent. At later stages of decomposition, leachate is still high in strength and will still require collection and treatment.

The time required to reach a steady state for methane production can vary from 1 or 2 years to as long as 2 to 20 years. Methane production may never occur if biological conditions are unfavorable. Reasons for this could be the presence of methane-inhibiting chemicals in the waste stream or infiltration of oxygen from a poorly designed cover. During steady-state methane production, the gas is roughly 50 percent methane and 50 percent carbon dioxide. Remember that methane has 25 times the global-warming potential of carbon dioxide.

The resulting leachate typically has a higher BOD than domestic sewage, and it may be as hazardous as many industrial wastewaters. The concentrations of constituents in leachate depend greatly on the type of waste placed in the landfill. The exclusion of hazardous wastes from landfills improves the likelihood of proper biological production of methane at the same time that it limits the hazard associated with landfill leachate. Steady methane production is important in terms of leachate quality, as methane production reduces the hazard of leachate because the dissolved organic components of leachate (measured as COD or BOD) are converted to gaseous methane.

LANDFILL GAS The production of landfill gas is best viewed as both a problem and an opportunity. First, consider the reasons why it is a problem: (1) It can be explosive when mixed with oxygen. (2) It can be a human health concern for site workers. (3) It can create odors. (4) It can displace oxygen in soils, which may suffocate nearby plants. (5) It can emit methane to the atmosphere, which contributes to **greenhouse gas emissions**.

The main reason that landfill gas is an opportunity is that it has the potential to produce an economical and non-fossil-fuel-derived form of electricity. In Chapter 2, Table 2.4 showed that in the United States alone, $CH_4$ emissions from landfills were 140.9 Tg of $CO_2$ equivalents in 2004. In Chapter 11 (Box 11.4), we discussed how the Los Angeles County Sanitation District currently obtains 63 MW of electricity from landfill gas.

The total amount of **methane** that can be produced will vary depending on the amount of biodegradable material and the suitability of landfill conditions to biological methane production. Typical values are 100 L $CH_4$ produced per kg MSW landfilled. The maximum amount of methane produced can be estimated from the stoichiometry of the waste decomposition:

$$C_aH_bO_cN_d + \frac{(4a - b - 2c + 3d)}{4} H_2O \rightarrow \frac{(4a + b - 2c - 3d)}{8} CH_4$$

$$+ \frac{(4a - b + 2c + 3d)}{8} CO_2 + dNH_3 \qquad \textbf{(13.3)}$$

where $a$, $b$, $c$, and $d$ are stoichiometric coefficients for a specific organic chemical. The actual amount of methane produced may be only 10 to 50 percent of the maximum amount estimated from Equation 13.3. This is because some organic waste is not degradable under anaerobic conditions and because, in some sections of a landfill, effective biodegradation may be hampered by low moisture, presence of toxins, adverse pH, or a lack of nutrients.

Figure 13.12 showed that the rate at which methane is produced (in units of L methane/kg waste-yr) is commonly described with a lag phase of zero methane production, followed by exponential decay, as follows (for $t > t_{lag}$):

$$\text{rate of } CH_4 \text{ production} = V_{gas} \times k \times e^{[-k \times (t - t_{lag})]} \qquad \textbf{(13.4)}$$

where $V_{gas}$ is the total volume of gas (L) that can be produced per kg of waste, $k$ is a first-order decay rate (time$^{-1}$), $t$ is the time measured from the point the waste is disposed, and $t_{lag}$ is the lag time required before the waste begins to produce methane.

To determine the methane produced between time $t_1$ and $t_2$, we can integrate this function to provide the following equation (with $t_1, t_2 > t_{lag}$):

$$\text{cumulative CH}_4 = V_{gas} \times \left[1 - e^{[-k(t_2 - t_{lag})]}\right] - V_{gas} \times \left[1 - e^{[-k(t_1 - t_{lag})]}\right]$$

$$(13.5)$$

For the situation when $t_1 \ll t_{lag}$, Equation 13.5 reduces to (with $t_2 > t_{lag}$)

$$\text{cumulative CH}_4 = V_{gas} \times \left[1 - e^{[-k(t_2 - t_{lag})]}\right] \qquad (13.6)$$

The first-order decay rate is related to the half-life, $t_{1/2}$ (as discussed in Chapter 3):

$$k = \frac{0.693}{t_{1/2}} \qquad (13.7)$$

Half-lives for methane production in a landfill depend on the waste's degradability and can vary from 1 to 35 yr (McBean et al., 1995; Pierce et al., 2005). They also vary with the moisture content of the waste. Dry waste will have a longer half-life than moister waste. Accordingly, some landfills recycle leachate into the waste pile or add freshwater to the waste to increase the rate of decomposition.

## example/13.5 Estimating Methane Production

Calculate the methane production (in m$^3$/person at 0°C) from land-filling one year's waste for each person. Use the landfilling rate for 2005 provided in Table 13.2. Assume the three waste components that produce methane are food wastes (15 percent of total), mixed paper (30 percent of total), and yard wastes (25 percent of total). Assume that 60 percent of the food and paper waste and 40 percent of the yard trimmings decompose and 90 percent of the methane is captured for energy recovery.

## solution

This problem requires we determine the methane produced by 1 kg landfilled waste and multiply this value by the mass of waste land-filled per person per year.

For each kg of waste, determine the moles of carbon, hydrogen, and oxygen that will degrade to methane, and then use the molar stoichiometry from Equation 13.3 to determine the moles of methane produced. As Table 13.15 shows, the solution uses not only information from Table 13.5, but also values provided in the problem statement and molecular weights, to find the required molar ratios.

## Table / 13.15

**Results for Example 13.5**

| | Wet Weight (g) | Moisture Content (%) | Dry Weight (g) | Total Carbon (g) | Total Hydrogen (g) | Total Oxygen (g) | Total Nitrogen (g) |
|---|---|---|---|---|---|---|---|
| Food wastes | 150 | 70 | 45 | 21.6 | 2.9 | 16.9 | 1.2 |
| Paper (mixed) | 300 | 10 | 270 | 117.2 | 15.7 | 119.6 | 0.8 |
| Yard wastes | 250 | 60 | 100 | 46.0 | 6.0 | 38.0 | 3.4 |
| Other | 300 | — | — | — | — | — | — |
| Total | 1,000 | | | | | | |

| | Degraded Carbon (g) | Degraded Hydrogen (g) | Degraded Oxygen (g) | Degraded Nitrogen (g) | Degraded Carbon (moles) | Degraded Hydrogen (moles) | Degraded Oxygen (moles) | Degraded Nitrogen (moles) |
|---|---|---|---|---|---|---|---|---|
| Food wastes | 13.0 | 1.7 | 10.2 | 0.7 | 1.08 | 1.71 | 0.63 | 0.05 |
| Paper (mixed) | 70.3 | 9.4 | 71.8 | 0.5 | 5.85 | 9.32 | 4.49 | 0.03 |
| Yard wastes | 18.4 | 2.4 | 15.2 | 1.4 | 1.53 | 2.38 | 0.95 | 0.10 |
| Other | — | — | — | — | — | — | — | — |
| Total | | | | | 8.46 | 13.41 | 6.07 | 0.18 |

From Equation 13.3, the moles of methane produced are determined as $(4a + b - 2c - 3d)/8$. From Table 13.15, $a = 8.46$, $b = 13.41$, $c = 6.07$, and $d = 0.18$. Using this expression, the methane produced per kg of waste is 4.32 moles.

Using the ideal gas law, at 0°C (273°K), there are 22.4 L gas per mole or 0.0224 m³ of gas per mole of gas. The volume of methane produced per kg waste is then

$$4.32 \text{ moles } CH_4 \times 0.0224 \text{ m}^3/\text{mole} = 0.0967 \text{ m}^3 \text{ per kg}$$

The problem states that 90 percent of the methane was recovered; therefore, 0.087 m³ methane is produced per kg waste.

From Table 13.2, the U.S. landfilling rate in 2005 was 0.41 Mg (or 410 kg) per person per year. Therefore, the gas production rate is

$$0.087 \text{ m}^3/\text{kg} \times 410 \text{ kg/person-yr}$$

$$= 35.7 \text{ m}^3 \text{ methane/person-yr at } 0°C$$

The production of landfill gas leads to pressure in landfills. This results in gas movement. Gas flow in landfill waste (and soil) has many similarities to groundwater flow. To control the movement of gas, gas-impermeable barriers are incorporated into the design, and outside the landfill high-permeability soils are placed, which direct gas flow into trenches. Even in past circumstances when landfill gas wasn't collected for energy, it was common to collect and combust the gas to minimize negative impacts associated with landfill gas.

Full capture of landfill gas requires installing inside the landfill waste, gas wells, and gas-permeable layers, along with pumping and piping systems (refer back to Figure 13.10). New landfills can typically capture more than 90 percent of the methane produced. Landfill gas can then provide heat, steam, or electricity. The most common method of converting gas to electricity is with large, 1 MW modular engines. Because landfill gas is a renewable energy source, many efforts are taking place around the world to expand its use. As one example, the Landfill Methane Outreach Program is a voluntary assistance and partnership program run by the EPA. Readers are encouraged to visit this site or similar ones in their country.

**LANDFILL LEACHATE** No matter what controls are implemented to minimize the movement of water into a landfill, some water will enter and produce **leachate**. Control of leachate needs to consider the quantity and quality of the leachate, as well as its potential adverse effects. Leachate concentrations will vary dramatically by location and the life of the landfill (refer back to Figure 13.12). Table 13.16 provides typical concentrations for young and old leachates. Concentrations of leachate constituents are much higher than similar constituents found in untreated municipal wastewater. The information provided in Figure 13.12 and Table 13.16 also shows how the concentration of the constituents decreases as the landfill ages and the readily leachable components are removed.

**Should EPA Update Emission Standards for Landfill Gas?**
http://www.edf.org/pressrelease.cfm?contentID=8714

**Landfill Methane Outreach Program**
http://www.epa.gov/lmop

## Table / 13.16

### Composition of Young and Old Landfill Leachates

| Constituent | Units | Young Leachate | Old Leachate |
|---|---|---|---|
| $BOD_5$ | mg/L | 10,000 | 100 |
| COD | mg/L | 18,000 | 300 |
| Organic nitrogen | mg/L as N | 200 | 100 |
| Alkalinity | mg/L as $CaCO_3$ | 3,000 | 500 |
| pH | — | 6 | 7 |
| Hardness | mg/L as $CaCO_3$ | 3,500 | 300 |
| Chloride | mg/L | 500 | 200 |

SOURCE: Data adapted from Tchobanoglous et al., 1993.

Three different strategies are used to control the volume and strength of leachate. A landfill will typically use a combination of these three strategies to limit leachate impacts.

1. *Isolation.* Waste is isolated by limiting the entrance of water and thus the production of leachate. The waste may be bound to some physical or chemical matrix to reduce its leaching potential. This latter option is most suitable for highly hazardous wastes (for example, waste that is radioactive) but is always part of any overall strategy to limit leachate impact.

2. *Natural attenuation.* The natural physical, chemical, and microbiological properties of soil treat the leachate. The method also relies on the dilution of leachate. This option is very suitable for poorly resourced communities with small quantities of nonhazardous wastes.

3. *Controlled biological degradation.* Landfill conditions are modified to optimize degradation. This option typically involves adding moisture and ensuring suitable mixing, ensuring that toxic chemicals are kept out of the waste, and maintaining proper pH, nutrients, and monitoring. It has been termed a landfill bioreactor strategy and is most suitable for landfills where advanced technical skills are available and nonhazardous wastes are disposed. This strategy accelerates stabilization of the waste material and so reduces the long-term risk from leachate.

**Leachate management** requires a series of subsystems: (1) hydraulic barriers to limit the ability of leachate to leave a landfill and for rainfall to enter the landfill; (2) collection systems to convey the leachate from the base of the landfill to an external location; and (3) a leachate treatment system. Figure 13.10 depicts these subsystems.

Hydraulic barriers are constructed from compacted clay, manufactured geomembranes, or geosynthetic clay products. A combination of these can be used to provide a system with multiple benefits. The top barrier is referred to as a **cover** or **final cover**, and the bottom barrier is called a **liner**. Hydraulic barriers can decrease the amount of leachate leaving a site by a factor of 1,000 or more over existing soils. Quality control of the installation of the barriers is critical to ensuring good performance.

Collection systems rely on gravity to convey leachate to a low point, a sump, within the landfill. Highly permeable, rounded gravel and perforated pipes are placed above the liner to ensure rapid movement of leachate and thus reduce the likelihood that pore pressures will increase to the point where leachate can seep out of the landfill or cause problems of geotechnical stability. From a low point, leachate is typically pumped out to a storage location.

Because landfill leachate is similar in many ways to a high-strength industrial wastewater, similar options are considered for treatment. The leachate could be conveyed (by pipe or truck) to a municipal wastewater treatment plant, where it would be carefully metered into the plant flow. Other options are to treat it on site and then discharge it to land or water, or else to partly treat it on site before conveyance to a

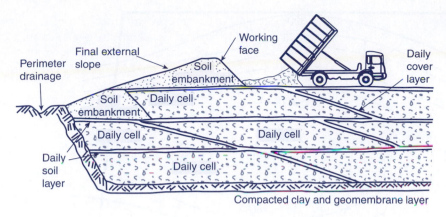

**Figure 13.13** **Landfill Construction from Daily Cells and Lifts** A daily cell of solid waste is covered with daily soil; additional cells are added vertically to the landfill, which eventually makes a lift.

central facility. The choice depends on the nature of the leachate, the environmental effects of discharge of treated leachate at the landfill, and the costs of conveying the leachate to a larger treatment facility.

**LANDFILL DESIGN** As shown in Figure 13.13, a landfill is constructed as a series of daily **cells**, where a day's waste is compacted and covered. Waste is delivered at the *working face* and then compacted against the edge of the landfill or previous daily cell. A horizontal layer of daily cells is called a **lift**. The number of lifts will depend on the topography of the site and the desired working life of the landfill.

The two principal design types are area and valley. An **area landfill** requires a relatively flat area and attempts to fit as much waste into that space as possible (moving vertically upward). This needs to be accomplished without causing geotechnical stability problems or exceeding any height limitation ($h$ in Figure 13.14) set by zoning standards. For a rectangular land area, the relevant solid is a truncated, rectangular-based pyramid as depicted in Figure 13.14.

Using the dimensions provided in Figure 13.14 and letting the slope, $G$, be the length divided by the height ($0.5 \times [L_1 - L_2]/h$), the volume of this pyramid is given as follows:

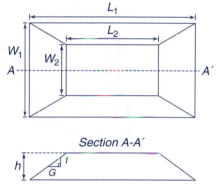

**Figure 13.14** **Geometric Design for the Area Landfill** The top figure (a) is a plan view, and the bottom figure (b) is a cross section through the upper figure.

$$V = \frac{h}{3} \times \left\{ L_1 \times W_1 + [(L_1 - 2Gh)(W_1 - 2Gh)] \right.$$
$$\left. + \sqrt{L_1 \times W_1 \times (L_1 - 2Gh)(W_1 - 2Gh)} \right\} \qquad \textbf{(13.8)}$$

Note that the volume determined in Equation 13.8 is the maximum amount of compacted solid waste that could be contained in the landfill, assuming that no volume is taken up by daily or final cover soil.

The soil located under this type of landfill can be used for covering compacted refuse or for other purposes. It is thus common to excavate soil under an area landfill before construction. Equation 13.8 can also be adapted to estimate the volume of compacted waste placed below grade.

To determine the volume available from a **valley fill landfill** design, compare the topographic contours of a site before filling and estimated contours after filling (Figure 13.15). To maximize the waste placed per unit surface area, the fill needs to increase at maximum slope from the

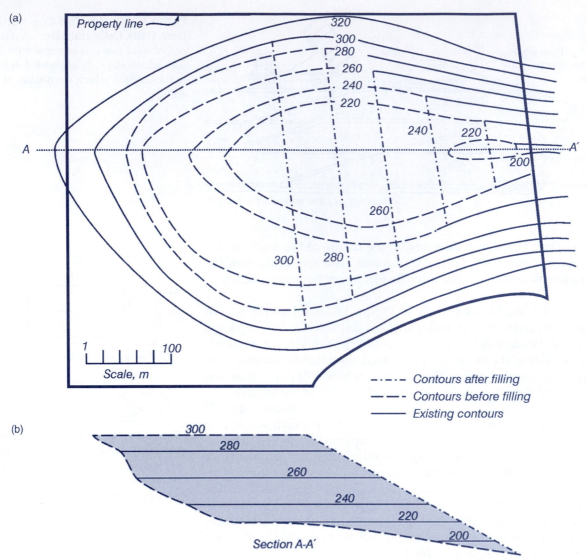

(a)

Property line

320
300
280
260
240
220

240  220

200

A ---------------------------------- A'

240  220

260

300  280

1    100
Scale, m

----- Contours after filling
--- Contours before filling
——— Existing contours

(b)

300
280
260
240
220
200

Section A-A'

**Figure 13.15** **Topographic Contours Used to Determine Available Volume for Waste Material in a Valley Fill Landfill Design** The top figure (a) shows the topographic contours. The bottom figure (b) shows a section drawn through the upper figure.

Redrawn with permission of Tchobanoglous et al., *Integrated Solid Waste Management*, 1993, copyright The McGraw-Hill Companies.

low point of the valley until it reaches its final height. Modern computer-aided design software can quickly assess volumes of valley fills.

Typical densities for compacted solid waste vary from 700 to 1,000 $kg/m^3$. Previously, Table 13.4 showed how the density of solid waste increases as the waste is collected, transported, and then landfilled. The density is 90–178 $kg/m^3$ for loose refuse and increases to 475–772 $kg/m^3$ when first placed in the landfill. This density increases further with pressure from lifts of waste above. The density assumed in design will vary with the composition of the waste (for example, construction and demolition waste is denser) and the depth of waste (greater depths lead to greater pressure and thus greater density).

In addition to compacted waste, the **landfill volume** will often include substantial amounts of **daily soil cover**. This soil is usually obtained from the landfill site. A typical thickness, $T$, of daily soil cover is 200 mm. The daily cell is designed based on the daily compacted volume of refuse ($V_r$), the slope of the cell ($G$), the daily refuse height ($H$), the refuse length ($L$), and the working face width ($W$) (different than $L_1$ and $W_1$ in Equation 13.8). The volume of daily soil cover ($V_s$) required in an idealized daily cell can be related to the volume of compacted refuse ($V_r$) as follows (Milke, 1997):

$$\frac{V_s}{V_r} = \left[\left(1 + \frac{T}{H}\right) \times \left(1 + \frac{G \times T}{L}\right) \times \left(1 + \frac{G \times T}{W}\right)\right] - 1 \quad \textbf{(13.9)}$$

The height of a daily cell ($H$) usually depends on site management factors and can vary from 2 to 5 m. $G$ is the grade ratio of length to height and is unitless. The daily volume of compacted refuse, $V_r$, is given by $H \times L \times W$.

The working face width will depend on the number of vehicles the site can accommodate at any one time under safe conditions. The working face width can be estimated by dimensional analysis. For example, the working face width can be determined, assuming a daily discard rate of 1,000 Mg/day (equals 1,000 tonnes/day) (and noting the assumptions related to truck size and operation of the landfill incorporated in Equation 13.10) as follows:

$$\frac{1{,}000 \text{ tonnes}}{\text{day}} \times \frac{1 \text{ truck}}{8 \text{ tonnes}} \times \frac{1 \text{ day}}{6 \text{ hr operation}} \times \frac{6 \text{ m}}{\text{truck}}$$

$$\times \frac{0.167 \text{ hr unloading}}{\text{truck}} = 21 \text{ m} \quad \textbf{(13.10)}$$

This value of 21 m implies that 3.5 trucks can unload at once, which is not practical. The working face length should be rounded up to the nearest multiple of truck width. In this case, we assumed a 6 m distance is required per truck to ensure safe operation, and the working face should fit four trucks, which would result in $W = 24$ m.

---

## example/13.6 Sizing a Landfill

A landfill and associated structures are to be built on flat ground. The dimensions for the landfill portion are 1,000 m by 1,000 m. The maximum height allowed (h) is 9 m aboveground (excluding final cover). The landfill will be open six days a week and accept 1,000 Mg (equals 1,000 tonnes/day) of waste per day of operation. How many years can the facility operate?

Other assumptions are as follows: thickness of daily soil cover ($T$) = 0.2 m; refuse height in daily cell ($H$) = 3 m; compacted density = 1,000 kg/m$^3$; working face width ($W$) = 24 m; and slope ($G$) = 3.

solution

The total landfill volume is obtained from Equation 13.8:

$$V = \frac{9\text{ m}}{3} \times \left\{ [1{,}000\text{ m} \times 1{,}000\text{ m} \right.$$

$$+ (1{,}000\text{ m} - 2(3)(9\text{ m}))(1{,}000\text{ m} - 2(3)(9\text{ m})]$$

$$+ \sqrt{1{,}000\text{ m} \times 1{,}000\text{ m} \times (1{,}000\text{ m} - 2(3)(9\text{ m}))(1{,}000\text{ m} - 2(3)(9\text{ m}))} \Big\}$$

$$V = 3\text{m} \times \left\{ [10^6\text{ m}^2 + 894{,}916\text{ m}^2] + 946{,}000\text{ m}^2 \right\}$$

$$= 8.5227 \times 10^6\text{ m}^3$$

The total volume of a daily cell consists of the volume occupied by compacted refuse ($V_r$) and daily cover soil ($V_s$). Using this knowledge and Equation 13.9, the volume of the daily cell can be determined as follows:

$$V_{\text{daily cell}} = V_s + V_r$$

Solve Equation 13.9 for $V_s$, and substitute it into the previous equation:

$$V_{\text{daily cell}} = V_r \left\{ \left(1 + \frac{T}{H}\right) \times \left(1 + \frac{G \times T}{L}\right) \times \left(1 + \frac{G \times T}{W}\right) \right\}$$

In the above equation, L is determined based on our previous definition of $V_r$,

$$L = \frac{V_r}{H \times W} = \frac{\dfrac{1{,}000\text{ Mg}}{1{,}000\text{ kg/m}^3} \times \dfrac{1{,}000\text{ kg}}{\text{Mg}}}{3\text{ m} \times 24\text{ m}} = 13.9\text{ m}$$

Solving for $V_{\text{daily cell}}$

$$V_{\text{daily cell}} = \frac{1{,}000\text{ Mg/day}}{1{,}000\text{ kg/m}^3} \times \frac{1{,}000\text{ kg}}{\text{Mg}} \times \left\{ \left(1 + \frac{0.2\text{ m}}{3\text{ m}}\right) \right.$$

$$\times \left(1 + \frac{3 \times 0.2\text{ m}}{13.9\text{ m}}\right) \times \left(1 + \frac{3 \times 0.2\text{ m}}{24\text{ m}}\right) \Big\}$$

$$= 1{,}140\text{ m}^3/\text{day}$$

Find the volume of all the daily cells per year, assuming 6 days' operation per 7-day week:

$$V_{\text{daily cells}} = 1{,}140 \, \text{m}^3/\text{day} \times \frac{365 \, \text{days}}{\text{yr}} \times \frac{6 \, \text{days}}{\text{7-day week}}$$

$$= 356{,}700 \, \text{m}^3/\text{hr}$$

Finally, determine the number of years of capacity in the total volume of the landfill:

$$\text{years} = \frac{8.5227 \times 10^6 \, \text{m}^3}{356{,}700 \, \text{m}^3/\text{yr}} = 24 \, \text{yr}$$

Readers may want to resolve this problem assuming that the disposal rate decreases over time, or that the population increases and the waste discard rate remains the same.

**LANDFILL MANAGEMENT** A landfill requires careful management over its lifetime. The decomposition of waste and the increase in pressure from waste added in higher lifts will cause the waste to settle. This necessitates the repair of roads, gas collection, and water drainage systems. Surface water can be easily contaminated if it contacts the waste; thus, careful separation of rainwater and waste is needed. Landfills require heavy machinery associated with earthworks and waste compaction after the waste is placed on the working face. Roads on a landfill require careful management, not only because of the use of heavy vehicles, but also because of settlement of waste. A detailed and rigorous system for protection of safety and health of workers is a key aspect of good landfill management. Landfills must also be good neighbors to the surrounding community. Therefore, vigilant measures are required to reduce impacts from noise, odor, birds, dust, and litter.

Good management is closely tied to good monitoring. Environmental protection and occupational safety require monitoring of gas, leachate, groundwater, and surface water. In addition, landfill managers will monitor the waste that arrives to ensure that inappropriate materials are not deposited and to provide data on the rate of arrival of waste.

Waste density also is often monitored to allow for planning the landfill's life. Landfills will typically charge for waste on the basis of the mass received. The charge must be high enough to pay for large construction costs and also provide for ongoing monitoring and maintenance after the landfill has completed its useful life.

**Landfill Operation**

### 13.3.6 SOLID-WASTE ENERGY TECHNOLOGIES

Because of an increasing interest in obtaining energy from waste, a number of alternative **energy technologies** are undergoing development.

**Anaerobic digestion** is a biological process of converting separated biodegradable solid wastes into methane gas and a solid residue that is suitable for turning into compost. The process works very similarly to those that produce landfill gas. The difference is that in an anaerobic digester, the wastes are mixed in a large vessel, and optimal degradation conditions are ensured, giving greater production of energy from the wastes. The technology is currently expensive, and for it to be successful, the wastes must be readily degradable (such as food wastes), and the other alternatives available, such as landfill or incineration, must be expensive.

**Gasification** is a process, similar to incineration, where less than stoichiometric amounts of oxygen are applied to the waste in the reaction vessel. The high temperatures that result from partial combustion lead to the production of high-energy gases, which can in turn be converted into power, usually in less polluting ways. Pyrolysis is a similar process, where even less oxygen or no oxygen is applied to the reaction, leading to the production of high-energy gases and a solid residue (char) that can be separated to give a high-energy residue for later combustion. Both processes have been most successfully applied to relatively homogeneous wastes, such as tires or plastics (Malkow, 2004).

## 13.4 Management Concepts

Successful solid-waste management requires a systems approach. Rather than attempt to analyze whether individual system components are better or not, society needs to evaluate, in a holistic, integrated manner, the combination of components that will maximize benefits at a given cost for current and future generations. Figure 13.16 provides an example of how municipal solid-waste management can involve a complex mix of technologies appropriate to the types of waste.

Arriving at a system like the one shown in Figure 13.16 requires focus on the overall objectives; creativity in developing new sustainable possibilities, which may include redefining the original problem; and recognition of the impact that decisions on one part of the system

The College and University Recycling Council (CURC) organizes and supports environmental program leaders at institutions of higher education in managing resource, recycling, and waste issues

http://www.nrc-recycle.org/

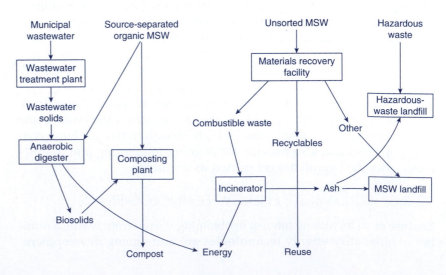

**Figure 13.16**  **Example of a Systems Approach to Solid-Waste Management** Note the many technologies and waste streams that are integrated in this approach.

can have on the functioning of the overall system. For example, operating a successful incinerator requires a steady supply of high-energy waste, such as paper. If a community chooses to invest in an incinerator, it then becomes difficult to consider other options for managing wastepaper, such as recycling through materials recovery. If a community chooses to separate food wastes and can do so at a low level of contamination, then the possibility of treating the waste along with wastewater-derived solids becomes an option that did not exist before.

Separation of waste types by waste generators, known as **source separation**, is a key part of a good system. If a community is able and willing to separate solid-waste components, the ability to create value from the wastes increases. Evaluation of the economics of source separation should consider the avoided disposal cost of separated wastes.

---

**Box / 13.4**  **Household Management and Source Reduction**

Waste management techniques such as composting, recycling, and source reduction result in the reduction of the quantity of waste bound for disposal options such as landfilling. Three waste reduction strategies were studied in the Parish of St. Ann in the small-island nation of Jamaica (Post and Mihelcic, 2009). Factors designated as incentives to waste reduction existed primarily at the household level—specifically, waste segregation, household education, environmental concern, and knowledge—whereas the barriers existed primarily at the national or regional levels, namely government policies and finances. (More informationon the incentives

Courtesy of James R. Mihelcic.

and barriers to recycling programs can be found at Troschinetz and Mihelcic, 2009.)

The greatest potential for initiating waste reduction strategies and diverting waste from a landfill in this setting was thus within the household, specifically by community-based waste reduction initiatives that build upon already existing practices and improve local solid-waste management.

Example of source reduction at the point of collection. Photo taken in Lavena, Island of Taveuni (Fiji) home of the rare orange dove.

---

### 13.4.1  CONSULTATION

Public concern about solid waste is high, as it is for many other engineering activities. As a result, engineers today must discuss proposed projects and programs with the various **stakeholders** that make up the public—an effort that includes listening to the public's concerns and ideas. A first step in a process of consultation is the identification of stakeholders who have a direct or indirect interest in the project. Stakeholders can be neighbors, the local community, the wider community, news media, elected officials, environmental and social interest groups, and cultural groups.

**Class Discussion**

What type of meaningful source reduction could be implemented in your own household? How would you implement the plan with all members of your house or apartment?

**Consultation Methods and Their Potential to Achieve Specific Outcomes** Consulation methods that appear to take more time (e.g, workshops, advisory commitees, mediation) typically result in a higher outcome.

| Consultation Method | Outcomes | | | | | | | |
|---|---|---|---|---|---|---|---|---|
| | Inform Stakeholders | Identify Values | Generate Options | Change Opinions | Resolve Conflict | Change Proposal | Costly Consultation | Lengthy Consultation |
| Information releases | High | Low | Low | Moderate | Low | Low | Moderate | Moderate |
| Field trips/site visits | Moderate | Moderate | Moderate | Moderate | Moderate | Moderate | Moderate | Low |
| Information stands/club visits | Moderate | Moderate | Moderate | Moderate | Low | Low | Low | Moderate |
| Contact person | Low | High | Low | Low | Low | Low | Low | Moderate |
| Public meetings | Moderate | High | Moderate | Low | Low | Low | Low | Low |
| Workshops | Moderate | High | High | Moderate | Moderate | Moderate | Moderate | Moderate |
| Advisory committees | Low | Moderate | High | High | High | High | High | High |
| Mediation | Low | Moderate | Moderate | Low | High | High | Moderate | High |

**Recycling Around the World**

**EPA's WAste Reduction Model (WARM) tracks greenhouse gas emissions reductions from different waste management practices**
http://www.epa.gov/climatechange/wycd/waste/calculators/Warm_home.html

Table 13.17 summarizes reasons to consult with stakeholders and identifies methods that have greater potential to match the desired outcome. In general, consultation methods that involve greater cost and longer time periods tend to result in better outcomes.

To be effective, consultation must begin early, be provided with adequate resources, be open and sincere, and involve good listening. In general, giving up a degree of power to stakeholders and allowing modifications to the project will result in the greatest chance of public acceptance. Consultation is therefore a critical part of managing any engineering project.

### 13.4.2   POLICY OPTIONS

Table 13.18 provides an overview of the policy options to meet objectives for solid-waste management. Good policy development requires assessments of costs and benefits, a focus on objectives, and a consideration of risks and unintended effects (Australian Productivity Commission, 2006).

### 13.4.3   COST ESTIMATION

Socially acceptable waste management facilities can be expensive, and equitable compromises between costs and social benefits are needed. An underlying cause of difficulty is the economy of scale of most waste

## Table / 13.18

**Policy Options for Meeting Objectives for Solid-Waste Management**

| Policy Option | Description | Example | When to Consider | When to Avoid |
|---|---|---|---|---|
| Public information campaigns | Inform the public of preferred behaviors. | Exchanges of commercial waste | Information is in individual's best interest as well as helping waste management | Changing behavior will create difficulties for individuals. |
| Ecolabeling | Inform the public which consumer goods create fewer waste problems. | Reusable shopping bags; ecolabeled detergents | Consumers lack information on waste impacts of products. | Differences between options are small or difficult to assess. |
| Waste targets | Government or industry groups set future goals. | 50% increase in paper recycling by 2020 | Society agrees on direction but lacks a focus. | Target does not consider costs, side effects, or risks. |
| Government subsidies | Government supports recycling or waste minimization efforts. | Grants to community recycling efforts | Environmental effects of waste are not reflected by costs. | Grants awarded for activities that would happen in any case. |
| User pays | Producers of waste rather than the government pay the full cost of management. | Weight-based charges for residential solid waste | Cost of charging system is small. | Users avoid charges by illegal practices. |
| Enforcement | Violators of rules pay a fine. | Tickets for littering | Behavior is clearly negative, and few violators exist. | Many violators and each causes very small impact |
| Deposit-refund | Consumers receive refund as incentive for proper management of waste. | Deposit-refund for car batteries | High negative consequences of improper waste management. | Large costs to operate system, and few environmental benefits. |
| Waste taxes | Government imposes taxes on waste. | Landfill tax | Taxes linked to environmental consequences of activities. | Large costs to operate system or undesirable side effects. |
| Producer responsibility | Producers responsible to take back goods at end-of-use life stage. | Computer return systems | Goods can be reused easily in production of new products. | Large costs in collection, storage, and transport. |
| Ban of goods and practices | Government prohibits goods or practices. | Ban of specific pesticides; ban of backyard burning | Goods/practices have high potential for harm, and other policy options are too costly. | Impacts are small or can be managed with other policies. |

management facilities, which means that a facility twice as large does not cost twice as much. For example, the costs of a landfill per tonne (Mg) over a year in the European Union (in 2003 euros) had the following form (Tsilemou and Panagiotakopoulos, 2006):

$$\text{total landfill cost} = 5{,}040 \times X^{-0.3} \tag{13.11}$$

for situations where $X$ was between 60,000 and 1,500,000 Mg/year. From Equation 13.11, the total landfill cost per Mg of discarded solid waste for 60,000 Mg per year can be estimated at 186 euros/Mg. For ten times the amount of solid waste discarded (600,000 Mg per year), the cost decreases to 93 euros/Mg.

Economy of scale means that larger landfills, incinerators, and compost plants are economically favored. However, they are more likely to face public opposition. Of course, sustainable development has shown that local solutions are often a preferred alternative.

## Key Terms

- anaerobic digestion
- area landfill
- carbon-to-nitrogen (C:N) ratio
- cell
- collection
- composting
- construction and demolition debris
- cover
- daily soil cover
- disposal
- dumps
- energy recovery
- energy technology
- final cover
- gasification
- generation rate
- greenhouse gas emissions

- hazardous waste
- incineration
- in-vessel
- landfill
- landfill volume
- leachate
- leachate management
- lift
- liner
- materials recovery facility (MRF)
- methane
- moisture content
- municipal solid waste (MSW)
- pollution prevention
- processing
- putrescible
- recycling
- reduction

- refuse-derived fuel (RDF)
- Resource Conservation and Recovery Act (RCRA)
- reuse
- solid waste
- solid-waste management
- source separation
- stakeholders
- storage
- transfer stations
- transport
- valley fill landfill
- waste
- waste density
- waste generation
- waste-to-energy
- windrow

**13.1** Design and safely perform a waste characterization on the solid waste at your residence and at an office at your university or college. (a) How does your waste characterization compare with the data in Figure 13.2? (b) Which of the three *R*s (reduce, reuse, recycle) would you implement to reduce the discard rate?

**13.2** Identify one source of solid waste on your campus that could readily be reduced, one source that could be reused, and one that could be recycled. What social, economic, and environmental benefits would come from implementing a plan to deal with the three items you identified?

**13.3** Research the energy and water savings associated with recycling 1,000 kg office paper. Which value do you consider the most reliable of the ones you found? Justify your choice, and provide a reference for your preferred source of information.

**13.4** Using the values provided in Example 13.2, estimate the low moisture content and typical moisture content for the waste as a whole.

**13.5** Waste composition has been measured for two cities. The results are summarized in Table 13.19.

**Table / 13.19**

**Data for Problem 13.5**

|  | City 1 | City 2 |
|---|---|---|
| *Wet Weight Generation Rate* |  |  |
| (kg/person-day) | 2.0 | 1.8 |
| *Wet Weight Composition (%)* |  |  |
| Food | 15 | 10 |
| Paper | 30 | 40 |
| Yard | 20 | 15 |
| Other | 35 | 35 |
| *Moisture Content of Fractions* *(% on wet weight basis)* |  |  |
| Food | 80 | 50 |
| Paper | 10 | 4 |
| Yard | 80 | 30 |
| Other | 5 | 4 |

(a) Which city generates more paper on a dry weight basis? (b) Find the percent moisture (wet weight basis) for City 1. (c) A nearby disposal site receives all of its MSW from cities 1 and 2. The average moisture content for MSW disposed of at the site is 20 percent. What fraction of the dry weight refuse comes from City 1?

**13.6** What is the dry weight percent composition for the following combined waste?

| Component | % Composition | % Moisture (Wet Weight) |
|---|---|---|
| Paper | 40 | 6 |
| Yard/food | 30 | 60 |
| Other | 30 | 3 |

**13.7** The mass composition of dry paper is 43 percent carbon, 6 percent hydrogen, 44 percent oxygen, and 7 percent other. Estimate the liters of air required to burn 1 kg dry paper. Assume carbon dioxide and water are the only products of combustion of carbon, hydrogen, and oxygen. Assume a temperature of 20°C and pressure of 1 atm.

**13.8** Estimate the oxygen demand for composting mixed garden waste (units of kg of $O_2$ required per kg of dry raw waste). Assume 1,000 dry kg mixed garden waste has a composition of 513 g C, 60 g H, 405 g O, and 22 g N. Assume 25 percent of the nitrogen is lost to $NH_{3(g)}$ during composting. The final C:N ratio is 9.43. The final molecular composition is $C_{11}H_{14}O_4N$.

**13.9** Waste of the composition shown in the following table is disposed of at a rate of 100,000 Mg/yr for 2 yr in one section of a landfill. Assume that half of the waste is disposed of at time = 0.5 yr, and half at a time of 1.5 yr. Assume that gas production follows the first-order relationship used in Equation 13.4, and use the additional information provided in the table. How long until 90 percent of the gas will be produced in this section?

|  | Initial Mass (Mg) | Half-Life (yr) |
|---|---|---|
| Slowly biodegrading | 10,000 | 10 |
| Rapidly biodegrading | 40,000 | 3 |
| Nonbiodegrading | 50,000 | Infinite |

**13.10** Assume all the waste in one section of a landfill was added at the same time. After 5 yr, the gas production rate reached its peak. After 25 yr (20 yr after the peak), the production rate had decreased to 10 percent of the peak rate. Assume first-order decay in the gas production rate after reaching its peak. Assume no gas is produced prior to the peak of 5 years. (a) What percentage of the total gas production do you predict has occurred after 25 yr? (b) How long do you predict until 99 percent of the gas has been produced?

**13.11** Determine whether your local (or regional) landfill produces energy from methane gas. If so, what is the mass of solid-waste disposed at the landfill on an annual basis, and what is the amount of $CH_4$ generated? Relate these numbers to a calculation you can perform with appropriate assumptions.

**13.12** Assume only yard waste contributes to $NH_4^+$ in leachate. Assume the waste composition provided in Figure 13.2 and Table 13.5. Assume all N in yard waste is eventually released as $NH_4^+$. What percentage reduction in yard waste would be required to give a reduction of 1 kg $NH_4^+$ released per Mg of MSW?

**13.13** Daily cells for a landfill are operated so that the following conditions are maintained: thickness of daily cover = 0.2 m; slope (horizontal:vertical) = 3:1; working face for refuse = 30 m; height of refuse = 3 m; and volume of daily refuse = 1,800 $m^3$/day. The landfill is interested in reducing requirements for daily cover soil over its 20 yr life and is considering three options. Which option would be the best? Why?

| Option 1 | Increase height of refuse to 4 m. |
|---|---|
| Option 2 | Increase daily refuse volume to 2,000 $m^3$/day. |
| Option 3 | Decrease working face to 20 m. |

**13.14** Estimate the landfill area required, in hectares, given the following specifications: daily cover thickness = 0.2 m; total final cover = 1.0 m (in addition to daily cover); height above ground before biodecay and settlement = 10 m; lift height = 3 to 5 m; depth below ground level that waste can begin to be placed = 5 m; landfill site area is square; MSW generation rate = 100,000 Mg/yr; side slopes at

3 horizontal:1 vertical for daily cells and external slopes; working face width = 8 m; open for disposal 360 days per year; 30 yr life; and in-place density of fresh MSW = 700 kg/$m^3$.

**13.15** You need to budget for a new transfer station in your district. A similar transfer station cost $1 million, but that station was 50 percent larger than yours. How much money should you budget so that your local government will have enough money to pay for the new transfer station? Assume that the economy of scale factor for transfer stations is 0.9.

**13.16** Go the World Health Organization Web site (www.who.org). Learn about a disease that is transmitted through improper disposal of solid waste. What is the extent of the disease on a global level?

**13.17** Identify an engineering professional society that you could join as a student or after graduation that deals with issues of solid-waste management. What are the dues for joining this group? What benefits would you receive as a member while working in practice?

**13.18** Determine the number of students at your university or college. Then, using the information provided in Figure 13.2 and Tables 13.2 and 13.5, determine the energy content associated with a day's worth of solid waste that would be generated by this population.

**13.19** Research the expected population of the United States in 2050. Estimate the annual mass of municipal solid waste that will be generated in the United States in 2050 and the annual amount that will require landfilling. Use information provided in Table 13.2. Justify your assumptions on changes in solid-waste generation per person and landfill disposal per person from now until 2050.

**13.20** Non legume vegetable wastes have a moisture content of 80% and are 4% N (on a dry mass basis). The vegetable wastes are to be composted with readily available sawdust. The sawdust has a moisture content of 50% and is 0.1% N (on a dry mass basis) The desired C:N for the mixture is 20. The C:N ratio for vegetable wastes is 11, and the C:N ratio for sawdust is 500. Determine the kg of sawdust required per kilogram of vegetable waste that results in an initial C:N ratio of 20.

# References

ASTM International. 1992. ASTM Standard D5231-92, "Standard Test Method for Determination of the Composition of Unprocessed Municipal Solid Waste." West Conshohocken, Pa.: ASTM International. Available at www.astm.org.

Australian Productivity Commission (APC). 2006. *Waste Management*. Report No. 38. Canberra: APC.

Diaz, L. F., G. M. Savage, L. L. Eggerth, and C. G. Golueke. 2003. *Solid Waste Management for Economically Developing Countries*, 2nd ed. Concord, Calif.: Cal Recovery.

Environmental Protection Agency (EPA). 2006a. *Municipal Solid Waste in the U.S.: 2005 Facts and Figures*. EPA530-R-06-011. Washington D.C.: EPA.

Environmental Protection Agency (EPA). 2006b. *An Inventory of Sources and Environmental Releases of Dioxin-Like Compounds in the United States for the Years 1987, 1995, and 2000*. EPA/600/P-03/002f. Washington, D.C.: EPA.

Farquhar, G. J., and F. A. Rovers. 1973. "Gas Production during Refuse Decomposition." *Water, Air, and Soil Pollution* 2:483–95.

Haug, R. T. 1993. *The Practical Handbook of Compost Engineering*. Boca Raton, Fla.: Lewis Publishers.

Kontos, T. D., D. R. Komilis, and C.P. Halvadakis. 2003. "Siting MSW Landfills on Lesvos Island with a GIS-Based Methodology." *Waste Management and Research* 21:262–77.

Malkow, T. 2004. "Novel and Innovative Pyrolysis and Gasification Technologies for Energy Efficient and Environmentally Sound MSW Disposal." *Waste Management* 24:53–79.

McBean, E. A., F. A. Rovers, and G. J. Farquhar. 1995. *Solid Waste Landfill Engineering and Design*. New York: Prentice Hall.

Medina, M. 2000. "Scavenger Cooperatives in Asia and Latin America." *Resources, Conservation, and Recycling* 31:51–69.

Milke, M. W. 1997. "Design of Landfill Daily Cells to Reduce Cover Soil Use." *Waste Management and Research* 15:585–92.

National Solid Wastes Management Association (NSWMA). 2003. *Modern Landfills: A Far Cry from the Past*. Washington, D.C. NSWMA. Available at www.nswma.org.

New Zealand Ministry for the Environment (NZME). 2002. *Solid Waste Analysis Protocol*. March, Ref. ME 430. Wellington: NZME.

Pierce, J., L. LaFountain, and R. Huitric. 2005. *Landfill Gas Generation and Modelling Manual of Practice*. Silver Spring, Md.: Solid Waste Association of North America.

Post, J. L., and J. R. Mihelcic. 2009. "Waste Reduction Strategies for Improved Management of Household Solid Waste in Jamaica." *International Journal of Environment and Waste Management*, in press.

Rynk, R., ed. 1992. *On-Farm Composting Handbook*. NRAES-54. Ithaca, N.Y.: Natural Resource, Agriculture, and Engineering Service.

Tchobanoglous, G., H. Theisen, and S. A. Vigil. 1993. *Integrated Solid Waste Management*. New York: McGraw-Hill.

Troschinetz, A. M., and J. R. Mihelcic. 2009. "Sustainable Recycling of Municipal Solid Waste in Developing Countries." *Waste Management*, 29(2):915–923.

Tsilemou, K., and D. Panagiotakopoulos. 2006. "Approximate Cost Functions for Solid Waste Treatment Facilities." *Waste Management and Research* 24:310–22.

# chapter/Fourteen Built Environment

James R. Mihelcic, Julie Beth Zimmerman, Qiong Zhang

*In this chapter, readers will learn what makes up the built environment and how planning, design, construction, operation and maintenance, and end-of-life issues related to the built environment affect society and the environment. Planning of sustainable communities is discussed, along with the environmental impacts associated with traditional use of engineering materials. Heat balance concepts are related to understanding proper design of energy-efficient structures and minimizing the impact of the urban heat island. Solutions are presented to plan, design, and construct a more sustainable built environment.*

## Learning Objectives

1. Define the built environment, and describe its impact on social, economic, and environmental systems.
2. Apply context-sensitive design to engineering design.
3. Relate LEED credits to engineering practice for new commercial construction and major renovation.
4. Integrate sustainability into a building's design, construction management, and operation and maintenance.
5. Describe magnitude and specific types of materials flows associated with the built environment and the implications of these flows for design, planning, and management.
6. Calculate heat loss from buildings through the building skin and from infiltration.
7. Relate heat loss in buildings to degree-days and the R factor of building materials.
8. Determine heat input from passive solar and storage of heat using thermal walls.
9. Relate features of the built environment, such as street and building geometry, location and number of trees and water, and nonpervious surfaces, to the urban heat island sketch.
10. Use an integrated systems approach in design and evaluation of the built environment.
11. Incorporate accessibility and mobility into design of a sustainable built environment.
12. Relate specific components of the built environment to social, economic, and environmental benefits obtained by humans.
13. Relate coatings and materials used in the urban environment to terms in the heat island energy balance.
14. Relate principles of smart growth to the design and construction of the built environment.
15. Design components of the built environment that incorporate issues such as public health, water and air quality, waste management, biodiversity, social interaction, environmental justice, and preservation of cultural and historical features.

# 14.1 Introduction

The **built environment** is the result of human intervention in the natural physical world. It is where places with different characteristics and identities are created, and the means to keep them functional and interdependent are established. It includes everything that is *built*—all types of buildings such as houses, shops, offices, factories, schools, and churches, together with engineering works such as treatment plants, storm water management systems, roads and railways, power generation facilities, bike paths, bridges, and harbors (adapted from Scottish Architecture, 2005). Clearly, the built environment includes *structures*. Also included is the re-creation of natural habitats through landscapes. These *natural habitats* are typically exterior areas, which include courtyards, parks, nature centers, gardens, farms, wetlands, lakes, and other green spaces.

The variety and scope of the built environment is organized into six interrelated components: products, interiors, structures, landscapes, cities, and regions. The sum of the six components defines the scope of the total built environment. For an example, see Figure 14.1.

The design of products for human use can have a significant impact on natural systems. These products constitute a component of the built environment. This chapter focuses on larger components of the built environment that relate to buildings and infrastructure. This includes interior spaces where components such as paint, carpet, furniture, and lighting fixtures are designed to create space and ambience for human activity.

As structures and landscapes of varying sizes and complexities are grouped, the scale of the built environment increases toward *communities* with subdivisions, neighborhoods, districts, villages, towns, and cities. The next level of the built environment is *regions* where cities and landscapes are grouped and are generally defined by common political, social, economic, or environmental characteristics (such as a

**The United Nations Human Settlements Programme**
http://www.unhabitat.org/

**Figure 14.1   Example of the Built Environment: Vancouver, British Columbia**   The built environment contains six interrelated components: products, interiors, structures, landscapes, cities, and regions. Vancouver is Canada's third largest city with a population of 1 million automobiles and more than 2 million people. The metro area occupies over 2,878 km$^2$.

iStockphoto.

**Figure 14.2** Growth of the Built Environment in the Baltimore–Washington, D.C., Metropolitan Area during the Past Century

From EPA (2001).

**Low Impact Development**

### Class Discussion

Identify some nodes and corridors in the built and natural environments of your local community. What are some of the social, economic, and environmental functions associated with each of them?

**Ecosystems and Biodiversity: The Role of Cities**

http://www.unep.org/urban_environment/

watershed). These are defined as counties, states, multistate collections, countries, or even continents.

The legacy of the U.S. built environment now includes 4.6 million commercial buildings and 101.5 million homes; 425,000 brownfield sites; and 61,000 sq. mi roads and parking lots (*Environmental Building News*, 2001; Brown, 2001). The built environment continues to grow rapidly (see, for example, Figure 14.2). The annual increase in the United States is 2.3 percent, which is 2.65 times the rate of population growth (*Environmental Building News*, 2001). And while only 2 to 3 percent of North America's land area is physically touched by the built environment, approximately 60 percent of North American land area is now affected by it (UNEP, 2002).

The built environment presents many challenges for natural ecological processes and systems. One way to consider these is to think about the patterns of the built environment as *nodes* and *corridors*. Nodes can be thought of as cities, towns, villages, and freestanding commercial and industrial areas. For economic and social functions to be carried out in these nodes, they must be linked together by transportation and utility corridors.

Similarly, patterns in the natural environment can also be thought of as nodes and corridors. In this case, nodes are large, contiguous woodlands, meadows, major groundwater recharge and source water protection areas, wetlands, streams, and lakes. For these major natural features to support natural processes and function properly ecologically, they must also be connected by corridors for the flow and transport of water, nutrients, and various plant and animal species.

As the built environment encroaches on the natural environment, habitat becomes fragmented, leading to loss of **biodiversity** and reducing water recharge capabilities. The built environment relies on the natural environment to store and assimilate wastes, purify water, and supply natural resources for construction, operation, and maintenance. For example, Figure 14.3 shows that preserving wetlands in an urban environment can reduce nitrogen loadings to an urban watershed. This new understanding of the impacts and reliance of the built environment on natural systems as well as on human health and well-being is leading to a new paradigm in development that considers these factors in addition to designing for human needs and activities.

## 14.2 Context-Sensitive Design

The term **context-sensitive design** was initially used in transportation project design. The movement from traditional design to context-sensitive design began in the 1990s (National Research Council, 2005). The Maryland State Highway Administration (1998) offers this definition:

Context sensitive design asks questions first about the need and purpose of the transportation project, and then equally addresses safety, mobility and preservation of scenic, aesthetic, historic, environmental, and other community values.

Similarly, the Minnesota Department of Transportation (2007) defines context-sensitive design as "the simultaneous balancing and advancement of the objectives of safety and mobility with preservation and enhancement of aesthetic, scenic, historic, cultural, environmental, and community values in transportation projects."

Although different definitions exist, context-sensitive design ensures that projects are not only functional and cost effective but, importantly, are also in harmony with natural, social, and cultural environments. It thus represents a new way to design projects and solve problems. Table 14.1 compares criteria used for traditional and context-sensitive design.

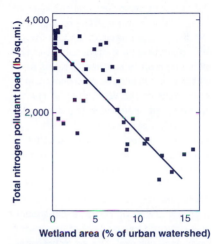

**Figure 14.3** **Impact of Amount of Wetland Area on Nitrogen Loading to Watershed** Wetlands are an important attribute of the built environment that needs to be protected for current and future generations. Traditional projects in the built environment have filled in or adversely altered the hydrology and biogeochemistry of wetlands. This figure shows that the amount of wetland area in an urban watershed has a direct impact on the mass of nitrogen loaded to the watershed. Total nitrogen pollutant loading to a watershed that decreases water quality is directly related to loss of wetlands within the urban watershed.

From EPA (2001).

### Table / 14.1

**Criteria Used for Traditional and Context-Sensitive Design**

| Traditional Design | Context-Sensitive Design |
|---|---|
| Minimize cost | Minimize cost |
| Maximize capacity | Maximize capacity |
| Maximize safety | Maximize safety |
| | Protect historic elements |
| | Protect cultural elements |
| | Protect uniqueness of community |
| | Maximize visual aesthetics |
| | Minimize environmental impacts |
| | Maximize economic benefit |
| | Incorporate local community values |
| | Protect biodiversity |
| | Provide wildlife habitat |

SOURCE: Adapted from McGowen and Johnson, 2007.

### Class Discussion

Critically think about a built-environment project that has been constructed recently in your neighborhood, hometown, or region. Did the contractor level all the vegetation, including trees that provide shading, wildlife habitat, and character, before putting up a structure? Did the local community protest the loss of historical, cultural, or environmental attributes of the development? Has the development resulted in the community losing some sense of place? That is, does the final project look like other projects you have seen in other regions of the state or country? Was there conflict between the community and the development? These are indicators of traditional design. Perhaps you thought of a project that protected important historical, cultural, and environmental attributes. Did that project seem unique and provide a positive sense of place? What was the level of conflict in the community? In either case, what are the attributes worth protecting for current and future generations?

Criteria that protect historic and cultural elements while minimizing environmental impacts have not been primary objectives for traditional design (though they may have been considered). In traditional design, engineers tended to think of historic, cultural, social, and environmental attributes as a nuisance and as something that had to be overcome during the permitting process. In context-sensitive design, such criteria are treated as primary goals and are optimized along with traditional design criteria. Context-sensitive design thus acknowledges that protection of cultural and historical elements, as well as open or wild spaces, has direct and indirect economic and social benefits for humans and the communities in which they live.

Context-sensitive design obviously requires additional thinking of the trade-offs between the criteria listed in Table 14.1. However, employing principles of context-sensitive design in project development and implementation does provide many benefits, such as increased flexibility in design, expedited project approvals, improved community acceptance of a final project, long-term economic benefits, and stronger working relationships among public agencies, communities, and citizens. Figure 14.4 identifies some of the elements and important considerations of context-sensitive design.

Engaged **stakeholders** will bring up important issues that must be incorporated into design decisions. For example, the impact of transportation infrastructure on wildlife habitat fragmentation is an issue consistently brought up by stakeholders because it affects numerous ecological and hydrologic processes across multiple spatial and temporal scales. Wildlife also provides many recreational and economic benefits to the public. As a result, **ecosystem-based mitigation** is proposed as a strategy to incorporate ecological considerations into transportation projects.

Wildlife-crossing structures and appropriately designed culverts are common components of context-sensitive design that protect habitat connectivity. Wildlife corridors are strips of natural land connecting fragmented wild habitat (Figure 14.5). They help to reduce the negative effects of habitat fragmentation and can promote species diversity (Damschen et al., 2006), affecting species as varied as butterflies, birds, fish, wolves, and grizzly bears.

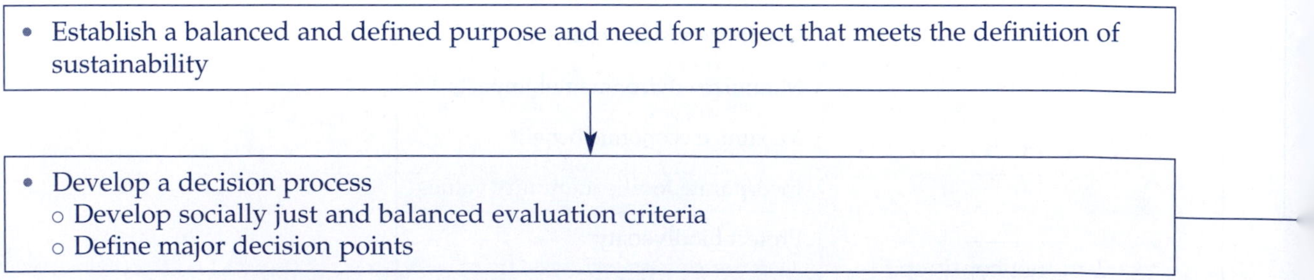

• Establish a balanced and defined purpose and need for project that meets the definition of sustainability
• Develop a decision process
o Develop socially just and balanced evaluation criteria
o Define major decision points


**Figure 14.4** **Elements and Important Considerations of Context-Sensitive Design**

Adapted from McGowen and Johnson (2007).

- Identify issues and constraints that need to be addressed in design of projects
  - Designer needs to communicate requirements (e.g., cost, capacity, safety), and impacts (e.g., environmental, cultural, community, biodiversity) along with purpose and need of project

- Involve stakeholders
  - Identify all stakeholders
  - Form project guiding committees that include key stakeholders
  - Maintain timely and coordinated stakeholder input
  - Listen to stakeholders
  - Commit to an open, creative approach to problem solving
  - Train and educate stakeholders

- Involve public throughout each stage
  - Develop a public involvement plan
  - Conduct effective public meetings
  - Have effective public notification
  - Carefully listen to the public

- Develop appropriate alternatives
  - Bring technical expertise to the decision process
  - Incorporate public input with environmental documents
  - Document project decisions
  - Ensure that community issues are addressed in design and upheld through construction
  - Present alternatives in understandable format (e.g., incorporate visual representation)
  - Engage stakeholders in identifying alternatives
  - Identify opportunities to enhance resources
  - Identify mitigation options

- Select alternatives
  - Compare alternatives based on all criteria
  - Adjust levels of detail of analysis depending on issues of importance
  - Document decisions and reasoning

- Continue context-sensitive design during construction
  - Educate contractor on importance of context-sensitive design
  - Incorporate context-sensitive design into construction specifications
  - Maintain communication with stakeholders

- Conduct continued research to determine what is acceptable (e.g., design standards and exceptions) and how that is measured

**Figure 14.4** (continued)

**Figure 14.5 Use of Wildlife Corridors to Link Ecological Areas and Wildlife Habitat** Compare (a) the existing crossing alternative and (b) proposed crossing alternative for Interstate 90 Snoqualmie Pass East Project, which links Puget Sound to eastern Washington. This area has been recognized as a significant north–south wildlife corridor for animals in the Cascades. Note how the proposed crossing provides a much wider stretch of land for terrestrial and aquatic wildlife crossing. Here wildlife habitats are connected in the proposed structure, and both large and small animals are allowed to cross under the road safely.

Adapted from U.S. Department of Transportation Federal Highway Administration, "Exemplary Ecosystem Initiatives," http://www.fhwa.dot.gov/environment/ecosystems/wa05.htm.

(a)

(b)

**Indoor Air Quality Tools for Schools**
http://www.epa.gov/iaq/schools/

## 14.3 Buildings

The average American now spends over 85 percent of his or her time indoors. This fact, along with the large material flows required to construct, operate, and maintain a building, has important consequences for engineers. In the United States, buildings use approximately one-third of the total **energy**, two-thirds of the electricity, and one-eighth of the **water** and transform land that provides valuable ecological services. Buildings

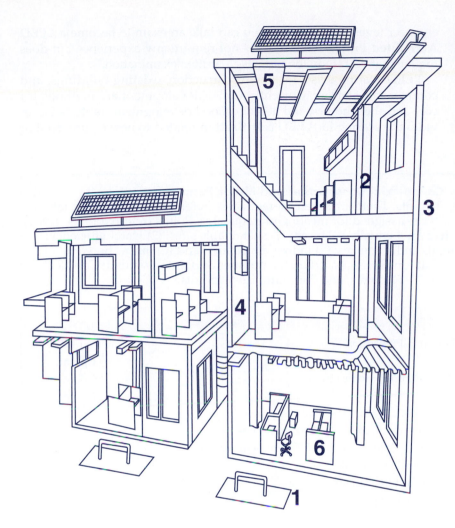

**Figure 14.6  Components of a Building**  The six components are the (1) foundation, (2) superstructure, (3) exterior envelope, (4) interior partitions, (5) mechanical systems, and (6) furnishings. For buildings to be more sustainable, engineers need to focus on integration of all six components. They also need to consider the social and environmental impacts throughout the building's entire life cycle.

From Rush (1986) with permission of John Wiley & Sons, Inc.

also account for 40 percent of global raw-materials use (3 billion tons annually).

Figure 14.6 shows the six **components of a building**: (1) foundation, (2) superstructure, (3) exterior envelope, (4) interior partitions, (5) mechanical systems, and (6) furnishings. Each of these components, during each stage of a building's life cycle, has a potential adverse impact on human health, as well as issues of energy use, water use, biodiversity, and use and release of hazardous chemicals. To develop more sustainable buildings (whether offices, manufacturing sites, schools, or homes), engineers must carefully evaluate the materials that go into the manufacture of each of these components and the way all these components are properly integrated.

## 14.3.1  DESIGN

The **U.S. Green Building Council** has developed a system to certify professionals and the design, construction, and operation of green buildings. This rating system is referred to as **Leadership in Energy and Environmental Design (LEED)**. Building owners, professionals, and operators see advantages to having their building LEED certified. As an

**U.S. Green Building Council**

http://www.usgbc.org

**LEED Construction**

engineer (even as a student), you can take an exam to become a **LEED Accredited Professional**. It does not require any experience but does require that you study for and pass a written examination.

LEED scoring exists for new construction, existing buildings, and commercial interiors. Currently under development are methods for core and shell, homes, and neighborhood development. Table 14.2 provides the scoring for **LEED certification** related to new commercial or

## Table / 14.2

### LEED Credits Associated with New Commercial Construction and Major Renovation

A project can obtain a maximum of 69 points. There are seven prerequisites (referred to as *Prereq.* in table) that all buildings must meet. No points are assigned for meeting prerequisites. Points (referred to as *credits*) are obtained in five categories, which are not equally balanced: (1) sustainable sites; (2) water efficiency; (3) energy and atmosphere; (4) materials and resources; (5) indoor air quality; and (6) innovation and design process. Certification is granted at four levels, based on the number of points awarded: Certified, 26–32 points; Silver, 33–38 points; Gold, 39–51 points; and Platinum, 52–69 points.

### Sustainable Sites (14 Possible Points)

| | | |
|---|---|---|
| Prereq. 1 | Construction Activity Pollution Prevention | (required) |
| Credit 1 | Site Selection | (1 point) |
| Credit 2 | Development Density & Community Connectivity | (1 point) |
| Credit 3 | Brownfield Redevelopment | (1 point) |
| Credit 4.1 | Alternative Transportation, Public Transportation Access | (1 point) |
| Credit 4.2 | Alternative Transportation, Bicycle Storage & Changing Rooms | (1 point) |
| Credit 4.3 | Alternative Transportation, Low Emitting & Fuel Efficient Vehicles | (1 point) |
| Credit 4.4 | Alternative Transportation, Parking Capacity | (1 point) |
| Credit 5.1 | Site Development, Protect or Restore Habitat | (1 point) |
| Credit 5.2 | Site Development, Maximize Open Space | (1 point) |
| Credit 6.1 | Stormwater Design, Quantity Control | (1 point) |
| Credit 6.2 | Stormwater Design, Quality Control | (1 point) |
| Credit 7.1 | Heat Island Effect, Non-Roof | (1 point) |
| Credit 7.2 | Heat Island Effect, Roof | (1 point) |
| Credit 8 | Light Pollution Reduction | (1 point) |

### Water Efficiency (5 Possible Points)

| | | |
|---|---|---|
| Credit 1.1 | Water Efficient Landscaping, Reduce by 50% | (1 point) |
| Credit 1.2 | Water Efficient Landscaping, No Potable Use or No Irrigation | (1 point) |
| Credit 2 | Innovative Wastewater Technologies | (1 point) |
| Credit 3.1 | Water Use Reduction, 20% Reduction | (1 point) |
| Credit 3.2 | Water Use Reduction, 30% Reduction | (1 point) |

### Energy and Atmosphere (17 Possible Points)

| | | |
|---|---|---|
| Prereq. 1 | Fundamental Commissioning of the Building Energy Systems | (required) |
| Prereq. 2 | Minimum Energy Performance | (required) |
| Prereq. 3 | Fundamental Refrigerant Management | (required) |
| Credit 1 | Optimize Energy Performance | (1–10 points) |
| Credit 2 | On-Site Renewable Energy | (1–3 points) |
| Credit 3 | Enhanced Commissioning | (1 point) |

## Table / 14.2

(Continued)

| | | |
|---|---|---|
| Credit 4 | Enhanced Refrigerant Management | (1 point) |
| Credit 5 | Measurement & Verification | (1 point) |
| Credit 6 | Green Power | (1 point) |

### Materials and Resources (13 Possible Points)

| | | |
|---|---|---|
| Prereq. 1 | Storage & Collection of Recyclables | (required) |
| Credit 1.1 | Building Reuse, Maintain 75% of Existing Walls, Floors & Roof | (1 point) |
| Credit 1.2 | Building Reuse, Maintain 95% of Existing Walls, Floors & Roof | (1 point) |
| Credit 1.3 | Building Reuse, Maintain 50% of Interior Non-Structural Elements | (1 point) |
| Credit 2.1 | Construction Waste Management, Divert 50% from Disposal | (1 point) |
| Credit 2.2 | Construction Waste Management, Divert 75% from Disposal | (1 point) |
| Credit 3.1 | Materials Reuse, 5% | (1 point) |
| Credit 3.2 | Materials Reuse, 10% | (1 point) |
| Credit 4.1 | Recycled Content, 10% (post-consumer + 1/2 pre-consumer) | (1 point) |
| Credit 4.2 | Recycled Content, 20% (post-consumer + 1/2 pre-consumer) | (1 point) |
| Credit 5.1 | Regional Materials, 10% Extracted, Processed & Manufactured Regionally | (1 point) |
| Credit 5.2 | Regional Materials, 20% Extracted, Processed & Manufactured Regionally | (1 point) |
| Credit 6 | Rapidly Renewable Materials | (1 point) |
| Credit 7 | Certified Wood | (1 point) |

### Indoor Environmental Quality (15 Possible Points)

| | | |
|---|---|---|
| Prereq. 1 | Minimum IAQ Performance | (required) |
| Prereq. 2 | Environmental Tobacco Smoke (ETS) Control | (required) |
| Credit 1 | Outdoor Air Delivery Monitoring | (1 point) |
| Credit 2 | Increased Ventilation | (1 point) |
| Credit 3.1 | Construction IAQ Management Plan, During Construction | (1 point) |
| Credit 3.2 | Construction IAQ Management Plan, Before Occupancy | (1 point) |
| Credit 4.1 | Low-Emitting Materials, Adhesives & Sealants | (1 point) |
| Credit 4.2 | Low-Emitting Materials, Paints & Coatings | (1 point) |
| Credit 4.3 | Low-Emitting Materials, Carpet Systems | (1 point) |
| Credit 4.4 | Low-Emitting Materials, Composite Wood & Agrifiber Products | (1 point) |
| Credit 5 | Indoor Chemical & Pollutant Source Control | (1 point) |
| Credit 6.1 | Controllability of Systems, Lighting | (1 point) |
| Credit 6.2 | Controllability of Systems, Thermal Comfort | (1 point) |
| Credit 7.1 | Thermal Comfort, Design | (1 point) |
| Credit 7.2 | Thermal Comfort, Verification | (1 point) |
| Credit 8.1 | Daylight & Views, Daylight 75% of Spaces | (1 point) |
| Credit 8.2 | Daylight & Views, Views for 90% of Spaces | (1 point) |

### Innovation and Design Process (5 Possible Points)

| | | |
|---|---|---|
| Credit 1.1 | Innovation in Design | (1 point) |
| Credit 1.2 | Innovation in Design | (1 point) |
| Credit 1.3 | Innovation in Design | (1 point) |
| Credit 1.4 | Innovation in Design | (1 point) |
| Credit 2 | LEED Accredited Professional | (1 point) |

SOURCE: Version 2.2, from U.S. Green Building Council.

major renovation projects. Note that the categories involve many things that engineers deal with. These include issues such as site management, storm water management and water/wastewater use, specification of building materials, indoor air quality, and energy conservation. Each of these categories can also be related to one of the six building components depicted in Figure 14.6.

## 14.3.2 CONSTRUCTION

Traditional **construction** of infrastructure tends to take a need or idea and transform it through a set of linear steps into what becomes a component of the built environment. In this linear thinking, construction is just one of the many required phases of a project. For the built environment to become sustainable, however, the construction phase must be viewed as a part of an integrated life cycle of a particular project. Figure 14.7 shows how a more sustainable nonlinear approach can be integrated into all the phases of a built-environment project.

There are three keys to implementing a **sustainable construction project** (Vanegas, 2003):

1. The delivery and management systems need to understand and support the concept of sustainability, so none of them inhibit the project's implementation.

2. All the project stakeholders shown in Figure 14.7 (for example, owners, operators, designers, construction team, vendor and suppliers, and any external parties) must have a common ground for understanding sustainability principles and concepts. These project stakeholders must operate as an integrated unit, and they must use sustainability as a fundamental criterion when making decisions, choosing among various options, or taking actions for the project at each stage of its life cycle.

3. Any process, practice, or operating procedure shown in Figure 14.7 that falls within planning, design, construction, procurement, commissioning, and the operations/maintenance phases must provide an entry point for formal and explicit input of sustainability.

As shown in Figure 14.7, a sustainable procurement phase parallels the design and construction phases. This phase provides an interface with the supply chain that ultimately provides all the systems, products, materials, and equipment that a design specifies. Specific elements that need to be considered during procurement include eliminating hazardous building materials, reducing or eliminating packaging, increasing recycled content of materials, promoting waste minimization, and minimizing environmental stressors during the manufacturing life stage.

A sustainable construction phase includes construction planning, construction operations, and start-up. Specific elements that should be considered in the construction phase include minimizing site disturbance, achieving a high-quality indoor environment, recycling construction

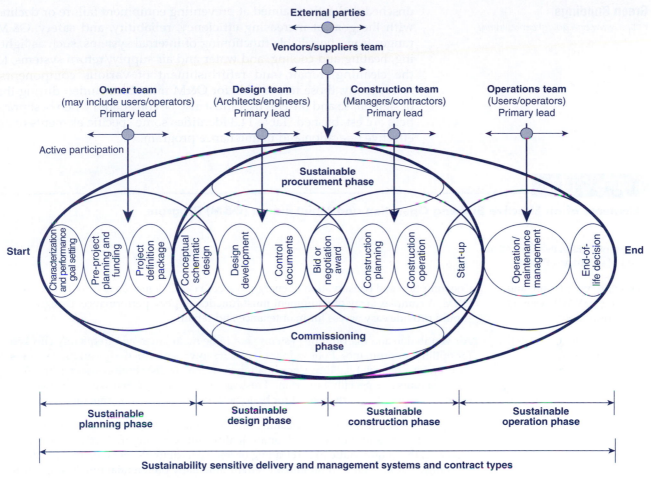

**External parties**

**Vendors/suppliers team**

**Owner team**
(may include users/operators)
Primary lead

**Design team**
(Architects/engineers)
Primary lead

**Construction team**
(Managers/contractors)
Primary lead

**Operations team**
(Users/operators)
Primary lead

Active participation

**Sustainable procurement phase**

Start

Characterization and performance goal setting

Pre-project planning and funding

Project definition package

Conceptual schematic design

Design development

Control documents

Bid or negotiation award

Construction planning

Construction operation

Start-up

Operation/ maintenance management

End-of- life decision

End

**Commissioning phase**

**Sustainable planning phase**

**Sustainable design phase**

**Sustainable construction phase**

**Sustainable operation phase**

**Sustainability sensitive delivery and management systems and contract types**

**Figure 14.7   Integration of Project Phases and Stakeholders**   Stakeholders and project phases must be integrated and fully aware of sustainable principles and practices to ensure the successful implementation of a sustainable construction project.

Reprinted with permission from Vanegas (2003). Copyright (2003) American Chemical Society.

materials along with the use of natural resources, and maintaining the health and safety of construction workers. Improper commissioning can lead to greater operation and maintenance costs because of inefficient energy and water use and can negatively affect the building occupants (Vanegas, 2003).

## 14.3.3   OPERATION AND MAINTENANCE

It is imperative that properly designed and constructed buildings be operated and maintained to perform at the highest efficiencies and effectiveness throughout the life cycle. The life cycle performance and management of the built environment must be optimized to maximize the return on investment while considering health, safety, functionality, durability, economic, environmental, and social impacts.

*Operations and maintenance (O&M)* are the activities related to the performance of routine, preventive, predictive, scheduled, and unscheduled actions aimed at preventing equipment failure or decline with the goal of increasing efficiency, reliability, and safety. O&M ranges from the reliable functioning of internal systems, such as lighting, heating and cooling, and water and air supply/return systems, to the cleaning, repair, and refurbishment of various components. Accordingly, those responsible for O&M must be included during the design phase and the end user must also be considered when best practices are established. Table 14.3 identifies some specific elements of an effective operation and maintenance program.

## Table / 14.3

### Elements of an Effective Building Operation and Maintenance (O&M) Program

| Element of Effective Operation and Maintenance Program | |
|---|---|
| Heating, ventilation, and cooling (HVAC) systems and equipment | Consumption and conservation of energy to meet HVAC demands is heavily tied to O&M. A complex array of equipment must function at peak performance. Design should provide easy access for maintenance and repair. |
| Indoor air quality systems and equipment | Air ventilation and distribution systems should be maintained and frequently checked for optimal performance. Poor indoor air quality lowers productivity, can cause illness, and can be a financial liability. Coordination between air distribution systems and furniture layouts is especially important. Fresh air should be provided where people learn and work. Regularly inspect for biological and chemical contaminants. |
| Cleaning equipment and products | Using degradable and nontoxic cleaning products and equipment reduces emissions and O&M costs, and improves indoor air quality and occupant productivity. Select materials that require little to no cleaning or less hazardous cleaners. Requirements for cleaning contracts can be built into an "environmentally preferential purchasing" program where it is a policy to choose products and services that reduce adverse effects. |
| Materials | Choices should reflect green chemistry and engineering. Materials selected for interiors (floor covering, furniture, paint, and fabrics) should be assessed not only for their impacts during manufacture and end of life but also for their O&M requirements. The same holds true for materials (sealants, gasketing, and insulation) that go into a building's superstructure and exterior envelope. |
| Water fixtures and systems | Routine inspections and maintenance program should verify that fixtures and systems are functioning effectively while also ensuring that leaks or components are quickly repaired. |
| Gray water | Can be captured and reused to flush toilets and irrigate. Reducing or eliminating consumption of potable water for nonpotable applications reduces discharge to wastewater treatment facilities and decreases energy and chemical use for treatment. O&M procedures should ensure that leaks are addressed and systems are separated from black water systems. Rainwater-harvesting systems have components for capture, storage, and delivery. Capture systems require screens and filters; maintenance of delivery systems depends on whether the system is pump- or gravity-fed. |
| Waste systems | Recycling requires consideration of space to sort materials, schedules to collect and empty containers, and methods to enforce compliance. Composting systems should ensure that estimated volumes can be accommodated and maintained in terms of turning (aerating) the material, ensuring the correct C:N ratio, and emptying and using final compost products in a way that minimizes odors. |

(Continued)

| Landscape maintenance | Use of native plantings can reduce landscape requirements and costs. After natural vegetation becomes established, there is usually no need for water, fertilizers, and pesticides. Integrated pest management can reduce the need for hazardous chemicals and pesticides. |
|---|---|
| Communication | Install a building performance feedback system that continuously reports on consumption and generation of energy and water. This information will allow rapid response to issues such as leaks or short circuits. Use information to inform occupants how their behavior (such as leaving on lights or computers or opening windows) affects the building's performance in terms of its environmental impact. Studies show that when information is delivered in real time through kiosks, display screens, or Web-enabled tools, occupants will alter their behavior to maximize the building's environmental performance. |

SOURCE: Haasl, 1999.

## 14.4   Materials

Urban populations require large material inputs for sustaining health, economy, and infrastructure. One of the most prominent features of the built environment is the large flux of materials and energy that flow through it. This material flux is dwarfed by the amount of energy, water, and air that flow through urban systems every year.

Cities also produce large outputs that have often been discarded back to the natural environment at levels beyond the assimilatory capacity of the surrounding environment. If these inputs and outputs are not sustainably balanced, large material stocks can accumulate, creating potential waste problems for future generations. Understanding these flows and stocks and how they can be related back to engineering practice and policy is vital if society hopes to improve materials and energy efficiency.

### 14.4.1   AGGREGATES, TOXIC RELEASES, CEMENT, STEEL, AND GLAZING

**AGGREGATES**   In Chapter 1 we presented a figure that showed the **flow of raw materials** in the United States. Careful examination of this figure shows that the largest materials flow (based on weight) was associated with **aggregates** (for example, crushed stone, sand, and gravel) and is now approaching 3,500 metric tons per year. Use of this material is increasing exponentially. Aggregates are mined at thousands of local and regional sites located throughout the world.

**TOXIC CHEMICAL RELEASES**   Another environmental consequence of materials use is the release into the environment of toxic materials that are associated with particular materials flows. Table 14.4 shows the large mass of **toxic chemical** releases that occur with

**Class Discussion**

What are the economic, social, and environmental consequences (for example, $CO_2$ emissions, loss of biodiversity, water quality, air quality) associated with mining aggregates that end up being incorporated into the foundations and structures that make up the built environment? In terms of flow of materials, do you see any advantages to redeveloping brownfield sites (discussed in Chapter 6) versus developing greenfield sites?

**Toxics Release Inventory Database**
http://www.epa.gov/tri/

**Total Toxic Chemical Emissions Asociated with Various Industrial Sectors That Support the Built Environment** Land disposal options are not included. (For more details, see www.epa.gov/tri.)

| Industrial Sector | Toxic Emissions measured by the 2005 Toxic Release Inventory | | | |
|---|---|---|---|---|
| | Fugitive Air Emissions (lb.) | Point Source Air Emissions (lb.) | Surface Water Discharges (lb.) | Total On-Site and Off-Site Disposal or Other Releases (lb.) |
| Metal mining | 1,496,557 | 1,981,519 | 507,435 | 1,168,724,053 |
| Textiles | 277,091 | 2,821,528 | 324,339 | 4,300,153 |
| Lumber | 2,719,680 | 24,815,519 | 13,595 | 28,759,409 |
| Chemicals | 56,644,546 | 144,953,426 | 42,424,07 | 531,452,586 |
| Plastics | 11,565,687 | 46,967,194 | 186,761 | 69,330,423 |
| Stone/clay/glass/cement | 1,497,409 | 38,631,169 | 2,361,601 | 53,597,996 |
| Primary metals | 11,983,420 | 36,104,985 | 44,039,957 | 479,223,010 |
| Transportation equipment | 12,149,617 | 44,711,942 | 376,157 | 68,563,912 |
| Electric utilities | 536,449 | 714,140,450 | 2,866,907 | 1,091,049,701 |
| **Total** | **195,896,686** | **1,315,793,746** | **240,246,101** | **4,339,463,751** |

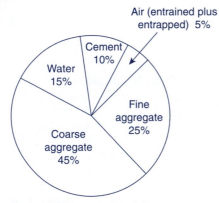

**Figure 14.8 Typical Composition of Concrete (by Mass)** The largest materials flow (by weight) in the economy is associated with aggregates (for example, crushed stone, sand, and gravel). Cement manufacturing and water usage are other large material flows in the U.S. and global economies.

industrial sectors associated with construction and operation of the built environment.

CEMENT **Concrete** is a mixture of cement, fine aggregate (sand), coarse aggregate (for example, gravel), water, and air. Figure 14.8 shows a typical composition of concrete. Note that aggregates take up 70 percent of the mass of concrete, and cement takes up 10 percent of the total mass. Aggregate primarily serves as inexpensive and inert filler.

**Portland cement** is the binding agent in concrete. It is produced from clinkering (heating to high temperatures) sources of calcium (limestone, seashells) and silica (clay, shale) in a kiln. The resulting clinker is then ground with a source of soluble sulfate (gypsum) to form the familiar gray powder.

The production of Portland cement is a major contributor to worldwide $CO_2$ production. The 1.45 billion metric tons of global cement production account for 2 percent of global primary energy use and 5 percent of anthropogenic (human-derived) $CO_2$ emissions. On average, manufacturing 1 kg cement produces 1 kg $CO_2$.

In the calcination zone of a cement kiln, the following reaction takes place between 700°C and 900°C:

$$CaCO_3\text{(limestone)} + \text{heat} \rightarrow CaO \text{ (lime)} + CO_2\text{(gas)} \quad \textbf{(14.1)}$$

As the temperature increases, the lime (CaO), silica ($SiO_2$), alumina ($Al_2O_3$), and ferric oxide ($Fe_2O_3$) enter a sintering or burning zone (between 1,200°C and 1,450°C), where several more reactions take place (and are not discussed here).

The calcination reaction shown in Equation 14.1 accounts for approximately 0.51 ton of $CO_2$ production per ton of clinker manufactured. The $CO_2$ emissions associated with the energy used for heating the kiln account for approximately 0.47 ton of $CO_2$ per ton of clinker manufacturing (van Oss and Padovani, 2003).

Some industrial by-products such as fly ash and blast furnace slag have pozzolanic properties and can be substituted for up to 25 percent of cement. Pozzolans are cementations materials that supplement Portland cement in the concrete mixture. During the hydration of Portland cement, pozzolans chemically react with calcium hydroxide ($Ca(OH)_2$) to form calcium silicate hydrates (CSHs). CSHs are the strong binders that harden concrete. Naturally occurring pozzolans such as rice husk ash, diatomaceous earth, and volcanic ash also are an option, but are not utilized as frequently.

REINFORCED STEEL Concrete is reinforced when it is used in structural applications or where cracking cannot be tolerated. Steel reinforcing bar is referred to as rebar. Steel mesh may be used as reinforcement against temperature and shrinkage cracks in concrete slabs. As indicated in Table 14.4, in 2005 the total mass of toxic emissions to air, land, and water for primary metal production exceeded 479 million pounds.

GLAZING Glazing is used on solar collectors to trap the heat associated with incoming solar radiation. It is a very important component of windows, solar hot-water heaters, greenhouses, and other technologies that incorporate passive solar heating. Materials used for glazing include glass, acrylics, polycarbonates, and polyethylene. As shown in Figure 14.9, glazing materials function similarly to the greenhouse gases that trap solar radiation and lead to climate change.

## 14.4.2 MATERIALS FLOW ANALYSIS

A materials flow analysis (MFA) measures the material flows into a system, the stocks and flows within it, and the outputs from the system. In this case, measurements are based on mass (or volume) loadings instead of concentrations. *Urban materials flow analysis* (sometimes referred to as an urban metabolism study) is a method to quantify the flow of materials that enter an urban area (for example, water, food, and fuel) and the flow of materials that exit an urban area (for example, manufactured goods, water and air pollutants including greenhouse gases, and solid wastes).

Urban metabolism studies are important because planners and engineers can use them for recognizing problems and wasteful growth, setting priorities, and formulating policy. For example, a materials flow analysis performed over ten years on the quantity of freshwater that enters and exits the Greater Toronto Area found that water inputs had grown 20 percent more than the outputs. Possible explanations for this

**Figure 14.9** **Function of Glazing Materials** Glazing allows shortwave radiation from the sun to pass through the glazing. Here, the shortwave radiation is then absorbed by surfaces. Water and masonry materials make excellent collectors of solar energy. Some longwave radiation is emitted from these surfaces. The longwave radiation cannot easily pass through the glazing material, so the collector heats up.

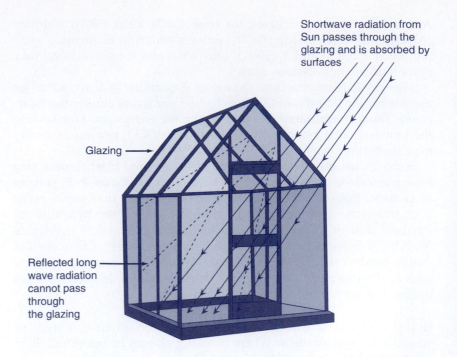

Shortwave radiation from Sun passes through the glazing and is absorbed by surfaces

Glazing

Reflected long wave radiation cannot pass through the glazing

could be leaking water distribution systems, combined sewer overflow events, and increased use of water for lawn care, all of which would allow inputted water to bypass output monitoring. The analysis also pointed to a need to further develop water conservation because of a fixed availability (or storage capacity) of freshwater.

---

**Box / 14.1   Urban Metabolism Study on Hong Kong**

Figure 14.10 shows the results of a materials flow analysis performed on the city of Hong Kong in 1997. Here, 69 percent of the building materials were used for residential purposes, 12 percent for commercial, 18 percent for industrial, and 2 percent for transport infrastructure. Also, a 3.5 percent measured increase in materials use over the 20-year period indicated that Hong Kong was still developing into a larger urban system.

During the study period, the city's economy shifted from manufacturing to a service-based center. This resulted in a 10 percent energy shift from the industrial sector to the commercial sector, yet energy consumption rose. The large increase in energy use was attributed to increases in development and residential/occupational comfort and convenience. The rate of using consumable materials also rose during the study period, plastics actually increasing 400 percent.

Overall air emissions in Hong Kong decreased; however, air pollutants associated with motor vehicle use and fossil fuel power production (such as $NO_x$ and CO) increased. Land disposal of solid waste rose by 245 percent, creating a dilemma for the space-limited city. Although a large portion of this waste is construction, demolition, and reclamation waste, municipal solid waste also rose 80 percent, with plastics, food scraps, and paper contributing the most to municipal waste.

Though the overall rate of growth for water use declined over the study (10 percent to 2 percent) from decreases in agriculture and industrial use, the per capita freshwater consumption rose from 272 to 379 L/day. Water is one of the major waste sinks for the city, due to its large volume of untreated sewage. BOD loadings increased by 56 percent. Nitrogen discharges also increased substantially. Sewage contamination in Hong Kong waters in now considered a major crisis for the city, having large harmful environmental, economic, and health effects.

One conclusion is that, at its current urban metabolic rate, Hong Kong is exceeding its own natural production and $CO_2$ fixation rates. Materials and energy consumption in the city greatly outweigh the natural assimilation capacity of the local ecosystem. High urban metabolism rates do show that, relative to other cities, Hong Kong is more efficient (on a per capita basis) in land, energy, and materials use due to lower material stocks in buildings and transportation infrastructure, has less energy and materials use (domestic consumption), and higher proportions of space dedicated to parks and open space.

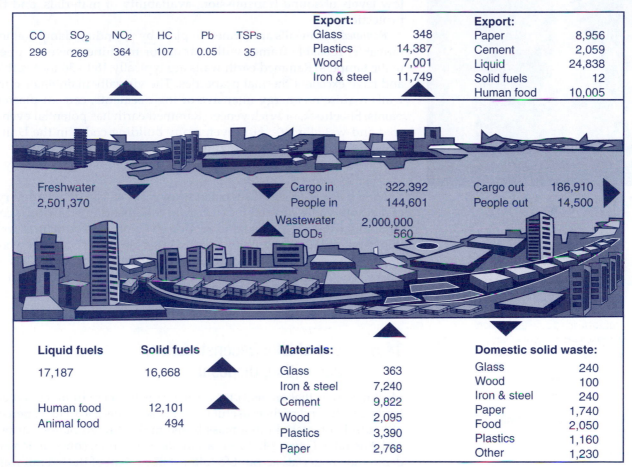

| CO | SO$_2$ | NO$_2$ | HC | Pb | TSPs |
|---|---|---|---|---|---|
| 296 | 269 | 364 | 107 | 0.05 | 35 |

**Export:**

| Glass | 348 |
|---|---|
| Plastics | 14,387 |
| Wood | 7,001 |
| Iron & steel | 11,749 |

**Export:**

| Paper | 8,956 |
|---|---|
| Cement | 2,059 |
| Liquid | 24,838 |
| Solid fuels | 12 |
| Human food | 10,005 |

| Freshwater | 2,501,370 |
|---|---|

| Cargo in | 322,392 |
|---|---|
| People in | 144,601 |
| Wastewater | 2,000,000 |
| BOD$_5$ | 560 |

| Cargo out | 186,910 |
|---|---|
| People out | 14,500 |

**Liquid fuels**

17,187

Human food

**Solid fuels**

16,668

12,101

Animal food    494

**Materials:**

| Glass | 363 |
|---|---|
| Iron & steel | 7,240 |
| Cement | 9,822 |
| Wood | 2,095 |
| Plastics | 3,390 |
| Paper | 2,768 |

**Domestic solid waste:**

| Glass | 240 |
|---|---|
| Wood | 100 |
| Iron & steel | 240 |
| Paper | 1,740 |
| Food | 2,050 |
| Plastics | 1,160 |
| Other | 1,230 |

**Figure 14.10** Important Materials Flows into and through the City of Hong Kong
All units are in tonnes per day. Arrows are intended to give some indication of the direction of flow of materials.

Adapted from *Ecological Economics*, Vol. 39, K. Warren-Rhodes and A. Koenig, "Ecosystem Appropriation by Hong Kong and Its Implications for Sustainable Development." 347–359 (2001), with permission of Elsevier.

## 14.4.3   TRADITIONAL BUILDING MATERIALS

A shift to **traditional building materials** could have revolutionary implications for the construction industry, significantly reducing the environmental impacts associated with manufacturing and transporting materials, the embodied energy associated with conventional materials, energy resource consumption during building use, and the disposal of debris during construction and later demolition.

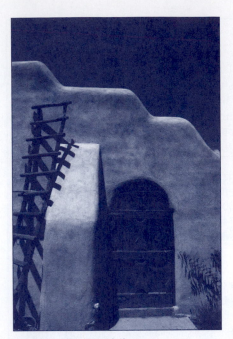

© Steven Allan/iStockphoto.

In Section 14.7, we will discuss the heat transfer properties of water that make it an advantageous material for construction of some thermal walls. Another natural material not commonly incorporated into buildings is earth. Earth construction takes many forms, including adobe and rammed earth. *Adobe* has been used all over the world in diverse regions and climates because it is strong, affordable, easy to handle, and regulates temperature well. Other benefits of using adobe are thermal storage, leading to reduced energy costs, low levels of sound transmission, availability of materials, and fire protection.

*Rammed earth* walls are formed in place by pounding damp soil into movable, reusable frames with manual or machine-powered pneumatic tampers. Rammed earth walls are typically 18 to 36 inches thick and have excellent thermal properties. The virtually maintenance-free walls require no energy-intensive or toxic sealants, cement plasters, paints, Sheetrock, or brick veneer. Rammed earth has potential even in cold and wet climates. However, many building codes in the United States require adding stabilizing agents (5 to 10 percent Portland cement) to rammed earth walls, increasing materials and energy as well as labor costs.

*Straw bale construction* uses baled straw from wheat, oats, barley, rye, and rice in walls covered by stucco. In load-bearing straw bale construction, bales are stacked and reinforced to provide structural walls that carry the roof load. With in-fill straw bale construction, a wood, metal, or masonry structural frame supports the roof, and bales are stacked to provide nonstructural insulating walls. With either alternative, the bale walls are plastered or stuccoed on both the interior and exterior.

## 14.5  End of Life: Deconstruction, Demolition, Disposal

Most materials that are used to construct the built environment enter a landfill at the end of their useful life. This accounts for 10 to 30 percent of landfilled material (on a mass basis) in the United States, Europe, and Australia. Figure 14.11a shows that in California, construction and demolition materials account for almost 22 percent of the overall waste stream. Figure 14.11b shows the composition of this construction and demolition waste stream. Note the large amount of lumber, concrete, asphalt roofing, and gypsum wallboard that make up the construction and demolition waste stream.

Many construction and demolition materials can be reused and recycled, thus prolonging the supply of natural resources and maximizing economic savings (Table 14.5). Unfortunately, only 20 to 30 percent of building-related construction and demolition waste is currently recycled in the United States (Horvath, 2004).

The concept of *demanufacturing* has been applied successfully to many consumer products, yet it is not commonly thought of as a concept that applies to the built environment. Figure 14.12 shows how demanufacturing can be applied to the built environment. Here

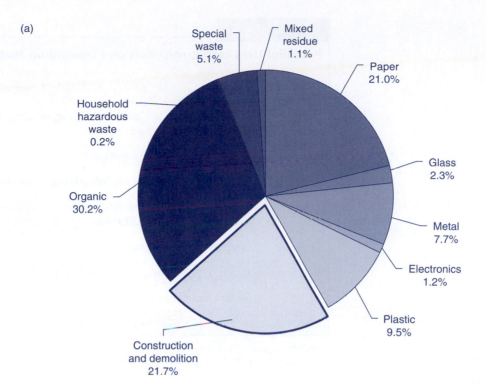

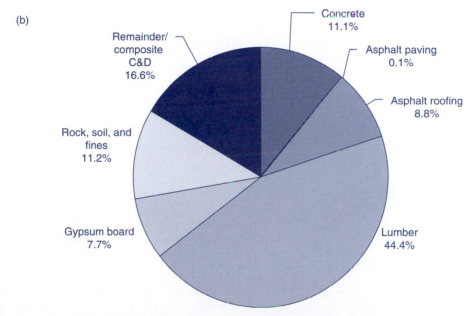

**Figure 14.11** **Composition of California's Waste Stream: Percent by Mass** (a) Types of materials found in California's *overall* waste stream (on a mass basis) that were disposed of in 2003. Note that construction and demolition materials make up approximately 22% of the waste stream. (b) Percentage of materials in California's *construction and demolition* waste (on a mass basis). Remainder/composite metal includes such items as tiles, toilets, and fiberglass insulation.

Data from Cascadia Consulting Group, Inc. (2004).

## Table / 14.5

### Potential Uses for Construction and Demolition Materials

| Material | Potential Uses |
| --- | --- |
| Wood | Reuse. Shred for fuel, animal bedding, mulch, manufactured building products, compost. |
| Bricks | Reuse. Crush to make aggregate. |
| Asphalt | Reincorporate into new asphalt paving or roadbed. |
| Concrete | Crush to make base material for roads, foundations, fill, and other aggregate applications in asphalt or concrete. |
| Drywall | Use gypsum as soil amendment. Reincorporate into drywall. |
| Roofing | Recycle asphalt shingles into asphalt paving. Reuse clay tiles. |
| Metal | Use scrap metal as feedstock. |
| Plastic | Recycle into plastic lumber, highway barriers, traffic cones. |

SOURCE: Horvath, 2004. Adapted with permission from the *Annual Review of Environment and Resources*, Volume 29, Copyright (2004) by Annual Reviews.

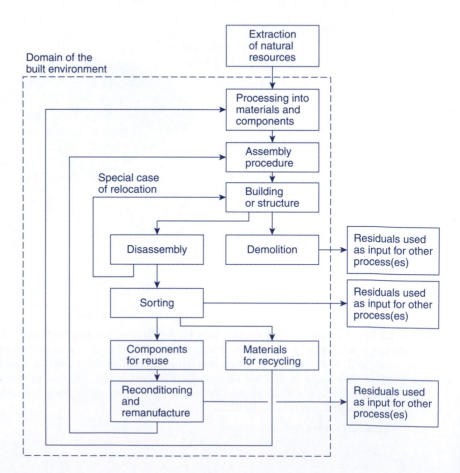

**Figure 14.12   Design for Life of Product, Showing a Demanufacturing Stage**   In demanufacturing, the built-environment structure is disassembled, reconditioned, and remanufactured into a useful component of the built environment.

Adapted from Crowther (1999) with the author's permission.

demanufacturing encompasses disassembly, recycling, reuse, reconditioning, and remanufacturing. To be successful, demanufacturing needs to be considered early in the design phase.

## 14.6 Rightsizing Buildings

### 14.6.1 MATERIALS USE IN A HOME

As Table 14.6 shows, while the average household size in the United States decreased from 3.67 members in 1940 to 2.62 in 2002, the average home size increased from 1,100 to 2,340 sq. ft. The increased size of residential dwellings has large implications for regional and global materials flows, along with materials usage and pollution production during the home's life. In terms of residential construction, Table 14.7 lists the materials used to construct a 2,082 sq. ft. U.S. home. Even appliances that are touted as being more energy efficient are consuming more and more energy, because of their larger size (think of television size).

### 14.6.2 ENERGY EFFICIENCY AND RIGHTSIZING

Previously, Figure 14.6 showed that one component of a building is the exterior envelope. **Energy efficiency** of the building envelope is a function of the building's size, how well insulated the structure is, how airtight the structure is, and how the building's glazed area (for example, its windows) is oriented to take advantage of solar heating gain.

**Rightsizing** residential, commercial, and institutional buildings is a major design tool to save materials and produce less pollution during *all* stages of the building's life cycle. As an example, in a recent study, different energy insulation scenarios were applied to 1,500 sq. ft. and 3,000 sq. ft. homes located in two North American cities with different climates (Boston and St. Louis). Table 14.8 compares the heating and cooling energy requirements associated with each building. Also compared is the past home of this book's lead author, located in the Upper Peninsula of Michigan.

**Ways to Save Energy**
http://www.energysavers.gov/

| Table / 14.6 | | |
|---|---|---|
| **Then and Now: Increasing Size of the American Home** | | |
| | **Then** | **Now** |
| Average number of occupants | 3.67 in 1940 | 2.62 in 2002 |
| Average size of home | 100 m$^2$ (1,100 sq. ft.) in the 1940s and 1950s | 217 m$^2$ (2,340 sq. ft.) in 2002 |
| Garage | 48% of single-family homes had a garage for 2 or more cars in 1967 | 82% of homes had a garage for 2 or more cars in 2002 |
| Air conditioning | 46% of new homes had central air conditioning in 1975 | 87% of new homes had air conditioning in 2002 |

**Materials Used to Construct a 2,082 sq. ft. (193 m²) House in the United States**  Larger homes are thought to consume more materials on a square-foot basis because they tend to have larger ceilings and more features.

| Component | Quantity | Component | Quantity |
|---|---|---|---|
| Framing lumber | 32.7 m² | Garage doors | 2 |
| Sheathing | 1,073 m² | Fireplace | 1 |
| Concrete | 15.35 tonnes | Toilets | 3 |
| Exterior siding | 280 m² | Bathtubs | 2 |
| Roofing | 264 m² | Shower stall | 1 |
| Insulation | 284 m² | Bathroom sinks | 3 |
| Interior wall materials | 516 m² | Kitchen sinks | 1 |
| Flooring (tile, wood, carpeting, resilient flooring) | 193 m² | Range | 1 |
| Ductwork | 69 m | Refrigerator | 1 |
| Windows | 18 | Dishwasher | 1 |
| Cabinets | 18 | Disposal unit | 1 |
| Interior doors | 12 | Range hood | 1 |
| Closet doors | 6 | Clothes washer | 1 |
| Exterior doors | 3 | Clothes drier | 1 |
| Patio door | 1 | | |

SOURCE: From Wilson and Boehland, *Journal of Industrial Ecology*, MIT Press Journals, copyright (2005).

The data in Table 14.8 show that when floor area is halved, heating costs are reduced slightly more than half, and cooling costs are reduced by about one-third. The smaller but less energy efficient, older house still uses less energy than the new and better-insulated larger house. Besides the energy required to heat and cool larger spaces, larger homes also require longer runs for ducting and hot-water pipes, which causes energy losses associated with the conveyance of warm air, chilled air, and hot water (Wilson and Boehland, 2005).

The highly insulated home located in the northern Midwest has zero cooling costs. It has no mechanical air-conditioning system, which negates the need for the materials associated with a cooling and delivery system along with energy associated to chill air. Besides being in a relatively cool geographic location, the building is designed so that insulation stores cool air obtained by opening windows during the night. Also, the strategic placement of windows that capture prevailing breezes and the use of tree shading and a shaded porch also replace the need for mechanical cooling.

**Comparative Annual Energy Use for Small versus Large Houses**  R factor is a measure of resistance to heat flow. R-19 is comparable to RSI-3.3 in the metric system.

| House | Location | Relative Energy Standard[a] | Heating (million Btu) | Cooling (million Btu) | Heating cost ($)[b] | Cooling cost ($)[c] |
|---|---|---|---|---|---|---|
| 3,000 sq. ft. | Boston, Mass. | Good | 73 | 19 | 445 | 190 |
| 3,000 sq. ft. | St. Louis, Mo. | Good | 61 | 29 | 378 | 294 |
| 1,500 sq. ft. | Boston, Mass. | Good | 35 | 13 | 217 | 131 |
| 1,500 sq. ft. | St. Louis, Mo. | Good | 29 | 20 | 181 | 198 |
| 1,500 sq. ft. | Boston, Mass. | Poor | 48 | 12 | 297 | 124 |
| 1,500 sq. ft. | St. Louis, Mo. | Poor | 40 | 21 | 247 | 206 |
| 1,500 sq. ft. | Upper Peninsula, Mich. | High | 27[d] | 0[e] | 240 | 0 |

[a] "Good" means a moderately insulated home with R-19 walls, R-30 ceilings, double-low-e vinyl windows, R-4.4 doors, R-6 insulation in air ducts, and infiltration of 0.50 air change per hour for heating and 0.25 air change per hour for cooling.
"Poor" means a poorly insulated home with R-13 walls, R-19 attic, insulated glass vinyl windows, R-2.1 doors, and infiltration of 0.50 air change per hour for heating and 0.25 air change per hour for cooling. Air ducts are not insulated.
"High" means the home is carefully designed and constructed to be airtight. It has R-25 walls, R-50 in the attic, double-low-e vinyl windows, R-14 doors, and infiltration of 0.20 air change per hour for heating.
[b] Heating costs assume natural gas costs $0.50 cents per 100,000 BTU.
[c] Cooling costs assumed to be $0.10 per kWh.
[d] Heating consumes 2 cords of hardwood and assumes 17 million usable Btu per cord.
[e] No air conditioning is installed. Building insulation stores cool air obtained during the night, and strategic placement of windows, tree shading, and use of porch contribute to no need for mechanical cooling.

SOURCE: Adapted from Wilson and Boehland, *Journal of Industrial Ecology*, MIT Press Journals, copyright (2005).

The highly insulated home also is designed to take advantage of passive solar heating in the winter, which requires no other heating source on sunny winter days. The house also incorporates use of extensive water-efficient appliances and a solar hot-water heating system. Hanging clothes outside to dry (even in the winter) is preferred over mechanical drying. Energy gains are made not only from not having to pump and treat water, but also in the energy savings associated with heating water.

Some home designers now espouse this different approach to house design—one focused not on size, but on quality and functionality, where space is designed to be used with what is termed *space efficiency*. This type of house design can use much less materials, water, and energy throughout the various life stages of building construction, occupancy, and end of life.

## 14.7 Energy Efficiency: Insulation, Infiltration, and Thermal Walls

In Chapter 4 we developed an energy mass balance expression and then applied it to heating water and thermal pollution. Equation 4.22 is rewritten here:

$$\frac{dE}{dt} = \dot{E}_{in} - \dot{E}_{out} \qquad \textbf{(14.2)}$$

Similarly, an **energy balance** can be used to describe a **heat balance** in a building to demonstrate methods to design and construct buildings that are more energy efficient. In a building, the heat balance can be written as follows:

$$\begin{bmatrix} \text{change in interal} \\ \text{plus external energy} \\ \text{per unit time} \end{bmatrix} = \begin{bmatrix} \text{heat into} \\ \text{building} \end{bmatrix} - \begin{bmatrix} \text{heat loss} \\ \text{from building} \end{bmatrix} \qquad \textbf{(14.3)}$$

In many scenarios with buildings, it is assumed that the building temperature is maintained at a constant value. Thus, the change in internal plus external energy per unit time in Equation 14.3 equals zero. In this case, after the heat loss is determined, a heating system (passive solar and/or mechanical) can be sized to counter the heat loss.

The *heat loss* from the building is related to losses through the building skin (walls, ceilings, windows, doors) and through airflow that occurs through any cracks or holes in the building (infiltration). The heat added into a conventional building is typically from conversion of nonrenewable fuels such as natural gas, oil, or electricity. Sustainable heating requires the building be oriented toward the sun, be insulated, and have a heating system designed to take advantage of the input of the sun's energy through passive solar design or use of renewable energy.

### 14.7.1 HEAT LOSS IN A BUILDING

For demonstration purposes, let us develop a heat balance on the heat loss associated with one of the 3,000 sq. ft. homes described in Table 14.8. There are several ways to perform this analysis. For our analysis, we will introduce and use a term called *degree-day*. We will also use Btu as the measure of energy (1 joule = $9.4787 \times 10^{-4}$ Btu). A Btu is defined as the amount of heat that must be added to 1 lb. water to raise its temperature by 1°F.

To simplify the calculation, we assume that the 3,000 sq. ft. home is a simple cube; thus, the four exposed walls are approximately 14.4 ft. (width) by 14.4 ft. (length) by 14.4 ft. (height). The area of each wall is then approximately 207 sq. ft. The roof area of the cube would also be 207 sq. ft. From Table 14.8, this building is assumed to have insulation specifications of R-19 walls and an R-30 ceiling, and the air infiltration rate is stated to be 0.50 air change per hour for heating.

**HEAT LOSS THROUGH BUILDING SKIN** The heat loss through the skin of the building (Btu/°F-day) is determined as follows:

$$\text{Heat loss} = \frac{1}{R} \times A \times t \qquad \textbf{(14.4)}$$

The *R value* is a measure of resistance to heat flow. The inverse of *R* (1/*R*) is defined as the flow of Btu through a 1 sq. ft. section of building skin for 1 hr., during which the temperature difference between the inside and outside of the building skin is 1°F. In Equation 14.4, *A* is the area of a particular section of the skin (wall, window, door, and ceiling), and *t* is time (usually 24 hr).

The daily total heat loss through the four walls and the ceiling can be determined as follows:

$$\text{heat loss} = \left[\left(\frac{1}{19}\frac{\text{Btu}}{\text{sq. ft. - °F-hr}}\right) \times 4 \text{ walls} \times 207 \text{ sq. ft.} \times \frac{24 \text{ hr}}{\text{day}}\right.$$
$$\left.+ \left[\left(\frac{1}{30}\frac{\text{Btu}}{\text{sq.ft. - °F-hr}}\right) \times 1 \text{ ceiling} \times 207 \text{ sq. ft.} \times \frac{24 \text{ hr}}{\text{day}}\right]\right.$$

$$(14.5)$$

Solving Equation 14.5 results in

$$\text{heat loss} = 1{,}046 \frac{\text{Btu}}{\text{°F-day}} + 6.9 \frac{\text{Btu}}{\text{°F-day}} = 1{,}053 \frac{\text{Btu}}{\text{°F-day}} \quad (14.6)$$

The unit of "°F-day" in Equation 14.6 is defined as a **degree-day**. Defined for heating, a degree-day is the number of degrees Fahrenheit below 65°F for 24 hours. Box 14.2 discusses degree-days in more detail. In our example, the value determined in Equation 14.6 can be written as 1,053 Btu/degree-day.

Once the total heat loss (in units of Btu/°F-day) is determined, that value can be multiplied by the total number of degree-days for heating in a particular location for the period of time of interest (day, month, or year). The resulting value will be the total energy requirements for heating the structure over that time period.

In our calculation, the heat loss through the actual building skin would be different if we broke the building skin down in greater detail to the area associated with the specific components of the building skin (siding, doors, and windows) and the specific R values associated with

---

**Box / 14.2    Degree-Days**

You may have seen the term *degree-day* used on your gas or electric bill. A degree-day is an index that reflects demand for energy that is used to heat or cool a building. The NOAA Climate Prediction Center provides degree-day data for almost 200 major weather stations in the United States (www.cpc.ncep.noaa.gov/). The baseline used for computations is 65°F.

A degree-day defined for heating is the number of degrees Fahrenheit below 65°F for a particular time period. Thus, if the mean daily temperature on a particular winter day was 32°F, this would equate to 33 degree-days for heating over that 24 hr period.

A degree-day defined for cooling is the number of degrees Fahrenheit above 65°F for a particular time period. Likewise, if a mean daily temperature for a summer temperature was reported as 85°F, this would equate to 20 degree-days for cooling over that 24 hr period.

Degree-days can be summed up for a week, month, or year to determine energy demand associated with heating and cooling.

example/14.1 Determining the Importance of Insulation in Minimizing Heat Loss through a Building Skin

Determine the heat loss through an insulated and uninsulated wall.[*] Each wall contains the following materials, which have the R factors given in the table:

| Component of Wall | R Factor |
|---|---|
| 1 in. stucco on outside of wall | 0.20 |
| 1/2 in. sheathing under stucco | 1.32 |
| 1/2 in. drywall on inside of wall | 0.45 |
| Inside air film along inside of wall | 0.68 |
| Outside air film along outside of wall | 0.17 |

The 3.5 in. air space in the uninsulated wall has an R factor of 1.01. If 3.5 in. fiberglass insulation is placed in this space, it will have an R factor of 11.0.

solution

Remember that Equation 14.4 allowed us to determine the heat loss through the skin of the building (Btu/°F-day) as follows:

$$\text{heat loss} = \frac{1}{R} \times A \times t$$

For the uninsulated wall, the combined R value equals

$$0.17 + 0.20 + 1.32 + 0.45 + 0.68 + 1.01 = 3.73$$

For the insulated wall, the combined R value equals

$$0.17 + 0.20 + 1.32 + 0.45 + 0.68 + 11.0 = 13.72$$

The heat loss through the uninsulated wall thus equals

$$\frac{1}{3.73} \times 100 \text{ sq. ft.} \times \frac{24 \text{ hr}}{\text{day}} = 643 \frac{\text{Btu}}{\text{°F-day}} = 643 \frac{\text{Btu}}{\text{degree-day}}$$

And the heat loss through the insulated wall equals

$$\frac{1}{13.72} \times 100 \text{ sq. ft.} \times \frac{24 \text{ hr}}{\text{day}} = 175 \frac{\text{Btu}}{\text{°F-day}} = 175 \frac{\text{Btu}}{\text{degree-day}}$$

Note that the wall with only 3.5 in. fiberglass insulation added has much less heat loss through the building skin. By knowing the number of degree-days for a particular date of the year that requires heating, we can also determine the days heating requirement.

*This example is based on Wilson (1979).

these components. In this case, we would determine the heat loss through each component of the building skin and then add up those amounts to find the total heat loss.

## HEAT LOSS FROM INFILTRATION

HEAT LOSS FROM INFILTRATION To determine the **heat loss due to infiltration**, we must know the room size. For our simplified calculation, we will assume that the 3,000 sq. ft. home is one giant room. Table 14.8 stated that the building has an air infiltration rate of 0.50 air change per hour for heating. The heat loss associated with infiltration is the amount of energy required to heat the air lost from the room every day through cracks and holes in the building envelope. For a particular volume of room or building, this can be determined as follows:

$$\begin{bmatrix} \text{heat loss from} \\ \text{infiltration} \end{bmatrix} = \text{volume} \times \begin{bmatrix} \text{air} \\ \text{infiltration} \\ \text{rate} \end{bmatrix} \times \begin{bmatrix} \text{heat to raise} \\ \text{temperature of} \\ \text{the air } 1°F \end{bmatrix}$$

$$(14.7)$$

Heat capacity is the term used to describe the heat required to raise the temperature of air. At sea level, 0.018 Btu energy is needed to increase the temperature of 1 cu. ft. air by 1°F. (At 2,000 ft. elevation, this value is 0.017; at 5,000 ft. elevation, this value is 0.015.)

Note in Equation 14.7 the importance of rightsizing a building because heat lost due to infiltration is directly related to the volume of the particular space being analyzed. (The same is true for energy requirements related to cooling.) One particular popular design feature in U.S. homes today is not only to oversize a residential home, but also to design an entry space with a large, high cathedral-type ceiling. After reading the remainder of this section, you will be able to estimate the energy required to heat such unsustainable design features.

Assuming the building in our example is located at sea level, the heat loss by infiltration is written as follows:

$$3,000 \text{ cu. ft.} \times \left( \frac{0.5 \text{ air change}}{\text{hr}} \right) \times 0.018 \frac{\text{Btu}}{\text{cu. ft.-hr}} \times \frac{24 \text{ hr}}{\text{day}}$$

$$= 648 \frac{\text{Btu}}{°F\text{-hr}} \qquad (14.8)$$

Again, using our method of degree-days, the value of 648 Btu/°F-day can be written as 648 Btu/degree-day.

Note the magnitude of this value compared with the value we determined for heat loss through the building's skin (Equation 14.6). The magnitude of this value is why it is important to make a building airtight by specification and proper installation of weather stripping, caulk, gasketing, and so on.

**Total Heat Loss** To determine the building's total heating load in our example, we can sum the heat loss through the building skin and heat loss from infiltration:

**Class Discussion**
Investigate the minimum insulation requirements for new construction in your area, and compare those requirements with the data in Table 14.8 and this example. Why haven't more consumers taken advantage of cost and energy saving strategies such as installing insulation, energy efficient windows and doors, or tankless hot water heaters?

$$1,053 \frac{\text{Btu}}{{}^{\circ}\text{F-hr}} + 648 \frac{\text{Btu}}{{}^{\circ}\text{F-hr}} = 1,701 \frac{\text{Btu}}{{}^{\circ}\text{F-hr}} = 1,701 \frac{\text{Btu}}{\text{degree-day}} \quad \textbf{(14.9)}$$

The total energy demand to meet the heat lost is found from the following expression:

$$\text{Total energy demand} = \text{total heat lost} \times \begin{bmatrix} \text{degree-days for} \\ \text{heating for} \\ \text{time period} \end{bmatrix} \quad \textbf{(14.10)}$$

Assume again that the average temperature on a particular winter day is 33°F. Remember from the earlier definition of a degree-day that the 33°F temperature would result in 32 degree days (65°F − 33°F = 32 degree-days) for that particular day. Thus, for our example, where the average temperature was 33°F, this would mean that the building would require the following amount of energy input for daily heating:

$$1,071 \frac{\text{Btu}}{\text{degree-day}} \times 32 \text{ degree-days} = 5.44 \times 10^4 \text{ Btu} \quad \textbf{(14.11)}$$

**PASSIVE SOLAR GAIN AND THERMAL WALLS** In the example used in this section to determine the energy required to make up for heat loss, we determined that for a particular winter day where the average temperature is 33°F, $5.44 \times 10^4$ Btu's of energy are required to heat the house. Equation 14.3 includes a term called *heat into building*. This added heat into the building can be derived from nonrenewable or renewable energy. Fortunately, all or part of this heat input can be derived by taking advantage of the energy provided by the sun. This heat input is called **passive solar gain**.

**Thermal walls** take advantage of passive solar energy and thermal conduction to transfer heat from warmer to cooler areas. They typically employ a large concrete or masonry wall to collect and store solar energy and then distribute this energy as heat into a building space. A masonry floor or fireplace also can accomplish this to a lesser extent. Figure 14.13 shows examples of how thermal walls can be incorporated into more sustainable building design that takes advantage of natural ventilation and overhangs. Interestingly, the Anasazi cliff dwellings of the American Southwest incorporate many of these design features.

Thermal walls can be sized to account for a particular fraction of the total heating load. The calculation first requires the determination of the heating losses, as was just performed in this section. Then, for a particular location and some assumptions related to the thermal conductivity and volumetric heat capacity of the wall material as well as the type of glazing that is placed between the wall and the sun, the percentage of the heating load that can be accounted for based on a particular area of thermal wall can be calculated. Due to space constraints, we will not go into these calculations. Readers are directed elsewhere (for example, Wilson, 1979).

© Chris Williams/iStockphoto.

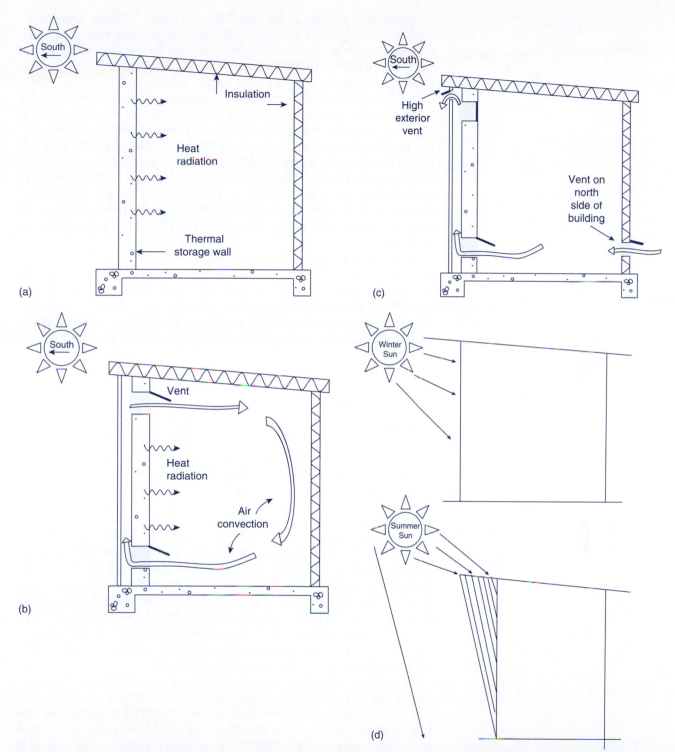

(a)

South

Insulation

Heat
radiation

Thermal
storage wall

(b)

South

Vent

Heat
radiation

Air
convection

(c)

South

High
exterior
vent

Vent on
north
side of
building

(d)

Winter
Sun

Summer
Sun

**Figure 14.13   Examples of Passive Solar Design and Ventilation Applicable to Northern Hemisphere**   These methods can be used to eliminate or minimize the need for mechanical heating and cooling. (a) Thermal walls use heat transfer to collect and dissipate heat. (b) Ventilation systems can use convection to provide natural heating. (c) Ventilation systems can provide natural cooling. (d) Overhangs take advantage of thermal properties of the sun during winter months while minimizing the sun's impact during warmer summer months.

Adapted from Wilson (1979).

As we have written in the past (Mihelcic et al., 2007), an ideal material for constructing a thermal wall would be readily available, inexpensive, nontoxic, and have optimal thermal properties (for example, heat capacity and conductivity). Water has a higher volumetric heat capacity (62 Btu/cu. ft.-°F) than wood, adobe, and concrete. These materials have heat capacity values that range in the 20s. Water is also an ideal material to release stored thermal energy as heat into a building space, because fluids can use convection to distribute heat. Also, the thermal conductivity of water (0.35 Btu-ft./sq. ft.-hr-°F) is much higher than for wood and dry adobe. Thus, a thermal wall constructed of water will provide a larger fraction of the required heating load than a similarly sized wall built of concrete or stone. Simple water-filled thermal walls can be constructed of 55 gal. drums painted black and placed on the southerly side of a building behind some type of glazing material.

Because of its high effective conductivity, water is an especially attractive material in instances where heat is required early in the day. Examples of places where heat is required early in the day are schools and offices. In residential situations where a family may be gone during much of the day and heat is needed in the evening, a conventional mass wall may be a better choice because it releases its stored energy more slowly.

## 14.8   Mobility

The largest emissions of $CO_2$ in the United States are associated with fossil fuel combustion. In fact, 85 percent of total greenhouse gas emissions in the United States are from fossil fuel combustion. The transportation sector accounts for 33 percent of these fossil-fuel-derived $CO_2$ emissions. In comparison, industry accounts for 28 percent, residential use for 21 percent, and commercial use for 17 percent.

To continue this thought process, more than 60 percent of the $CO_2$ emissions associated with the transportation sector are from personal-vehicle use. The remaining emissions come from other activities such as diesel fuel that is used in heavy-duty vehicles and jet fuel used in aircraft. Vehicle combustion is also the second largest emitter of the greenhouse gas $N_2O$. For example, the amount of $N_2O$ emitted in 2004 from vehicle combustion was 42.8 Tg $CO_2$ equivalents. This value dwarfs the $N_2O$ emitted from sources such as manure management (17.7 Tg $CO_2$ equivalents), municipal wastewater treatment (16.0), stationary combustion (13.7), and municipal solid-waste incineration (0.5).

Energy use in the transportation sector is projected to increase rapidly in the future—in fact, much faster than the industrial, residential, and commercial sectors. As noted in this chapter's introduction, the built environment contains 61,000 sq. mi. roads and parking lots. Just in the ten years of the 1990s, the number of registered U.S. cars increased six times faster than the population did from 1969 to 1995 (Alvord, 2000). If all these numbers seem enormously large, they are. If these numbers suggest that the method by which society is moving people, goods, and services is not sustainable, you are also correct.

*Transport* means to carry from one place to another. In our vehicle-dominated society, use of the word *transport* has overemphasized the use of a single-owner motorized vehicle. This form of transport has led to adverse environmental and social costs, some of which have been previously mentioned.

Differing from a vehicle-oriented transport system, **access** implies a means of approaching, entering, exiting, or making use of a passage, whereas **mobility** is a state of being capable of moving from place to place. A sustainable built environment allows people, goods and services, and wildlife that inhabit an area to move from place to place to conduct their business, be educated, take care of their health, and interact as a community, without harming the environment and other members of society.

Importantly, the built environment needs to be designed and constructed at the human and ecological scale, not at the vehicular scale. This system accounts for senior citizens, wheelchair-bound residents, and children and families who want to access schools and parks. This system also connects a mix of commercial shops, housing, services, education and recreation facilities, and other residential needs. Spread around are formal and informal gathering spots for community members. Plant and animal species also would be provided accessibility for their specific habitat needs.

Table 14.9 provides methods that community members can use to access goods and services, social interactions, and the environment. An important point about Table 14.9 is that many methods are provided besides personal-vehicle use; the community is designed to offer many options. This gives people more flexibility in their choices of mobility. It also provides options on how they can spend their household income and opportunities to improve their health.

This type of thinking also results in economic opportunities. For example, studies show that home values are higher in communities where there is less vehicular traffic. And the presence of more people on the street lowers the crash risk to pedestrians and bicyclists, because vehicle drivers become more aware of their presence. Walkable communities also provide mechanisms for community members to socially engage while improving their health.

To solve mobility problems associated with increasing population and congestion, conventional design simply adds new roads or additional

© Steve Lovegrove/iStockphoto.

**Traffic Congestion Factoids**
http://www.fhwa.dot.gov/congestion/factoids.htm

| Table / 14.9 |
| --- |
| **Methods to Access Work, Goods, Services, Recreation, and Education** |
| • Walk |
| • Bicycle |
| • Telecommute |
| • Use shared public transit (buses, shuttles, light rail) |
| • Use shared private vehicles |
| • Drive private vehicles |

## Table / 14.10

**Design Criteria for British, Australian, and U.S. Roads** U.S. roads are designed differently from roads in the rest of the world. Other countries design roads with better pedestrian accessibility and for slower vehicular speeds.

| | British Design Guide 32 | Australian Model Code | American AASHTO |
|---|---|---|---|
| **Design speed** | 20–30 mph (access roads)<br><20 mph (shared surface streets) | 18.6–24.8 mph (access roads)<br>9.3 mph (access places) | 20–30 mph |
| **Pavement width** | 12–18 ft.<br>(9 ft. 8 in. with passing bays) | 16.4–21.3 ft. (access roads)<br>11.5–16.4 ft. (access places) | 26 ft. standard |
| **Minimum curve radius** | 32.8–98.4 ft. | Maximum radius specified for traffic calming at each design speed | 100 ft. (as large as possible) |
| **Sidewalks** | Normally on both sides | At least one side of access streets | At least one side |

SOURCE: From Ewing, *Transportation Research Record* 1455, TRB, National Research Council, Washington, D.C., 1994, Table 1, p. 45. Reproduced with permission of TRB.

vehicular lanes. In fact, as shown in Table 14.10, compared with guidelines for road design in other parts of the world, U.S. guidelines are aimed at increased vehicular speeds, not to improving pedestrian accessibility or reducing runoff.

Adding lanes to roads does not solve environmental, economic, or social problems. In fact, it simply creates new sets of problems, some which are listed in Table 14.11. Two other myths commonly used in the conventional engineering design process are that placing trees alongside a road and providing access for pedestrians or bicyclists increases the liability risk. However, as will be shown later in this section, trees and the presence of pedestrians actually improve safety by slowing

---

### Box / 14.3 Traffic Congestion

The Texas Transportation Institute reported that in 2003 **traffic congestion** in the United States continued to increase, causing 3.7 billion hr travel delay and 2.3 billion gal wasted fuel. The total cost of this congestion is estimated to exceed $63 billion. This number of hours of delay would increase by 1.1 billion if public transportation were discontinued and the riders traveled in private vehicles.

In the United States, traffic congestion was the norm by 1912.[*] As early as 1907, the journal *Municipal Journal and Engineer* reported that early road-widening projects that had been expected to relieve congestion appeared to be doing the opposite. Woodrow Wilson commented in 1916 that motorists were using up roads almost as

fast as they were made. By the 1920s, congestion had decreased vehicle speed to 4 mph on New York's Fifth Avenue.

History shows that building more roads and widening existing roads will never solve the problem of traffic congestion. Providing several options for travel is a key component of a sustainable accessibility plan. This not only relieves congestion but also provides substantial savings to the public. As one example, without rail transit to and from Manhattan, New York would require 120 new highway lanes and 20 new Brooklyn Bridges. Also U.S. taxpayers recover their $15 billion in investment in public transit with congestion cost savings alone.

[*]This and the following paragraph are adapted from Alvord (2000).

**Table / 14.11**

**What Actually Happens when Additional Vehicle Lanes Are Added to Reduce Congestion**

| |
|---|
| Vehicular speeds increase |
| Pedestrian crossings increase in length and time, which makes walking less desirable |
| New signals may be needed for cross traffic |
| Congestion relief is usually temporary; lanes eventually fill up |
| New roads and lanes cause additional loss of open space because development takes place along the road corridor |

down vehicles. Myths related to liability can be overcome if engineers would take the time to address issues of liability by working more closely with risk management professionals.

**Walking** and **cycling** are healthy activities that provide society with substantial economic and social benefits in terms of reduced health costs and increased worker productivity. As shown in Table 14.12, the level of physical activity is much greater in Western Europe than in the United States. And compared with the rest of the world, Americans are more likely to use their vehicle for a trip than to use healthier activities such as walking, cycling, or using mass transportation. One reason is that engineers and planners have not provided options such as the ones listed in Table 14.9.

**Commuting to Work Trends in Florida**

http://www3.cutr.usf.edu/tdm/ commuting/FL.htm

**Table / 14.12**

**Percentage of Total Trips in Urban Areas around the World Used by Automobile, Public Transport, Bicycling, Walking, and Combined Walking and Bicycling**

| Country | Car | Public Transport | Bicycling | Walking | Walking plus Bicycling |
|---|---|---|---|---|---|
| United States | 84 | 3 | 1 | 9 | 10 |
| Canada | 74 | 14 | 1 | 10 | 11 |
| Denmark | 42 | 14 | 20 | 21 | 41 |
| France | 54 | 12 | 4 | 30 | 34 |
| Germany | 52 | 11 | 10 | 27 | 37 |
| Netherlands | 44 | 8 | 27 | 19 | 46 |
| Sweden | 36 | 11 | 10 | 29 | 49 |
| United Kingdom | 62 | 14 | 8 | 12 | 20 |

SOURCE: Data from Frank and Engelke, 2000.

Telecommuting was one mechanism mentioned in Table 14.9 for accessing work, goods, services, and education. By the end of the 1990s, 2 percent of the U.S. workforce was telecommuting to work. Options are still important, though, because relying on just one technological solution can result in problems as well. As an analogy, computers were promoted as a way to create a paperless work environment, but in practice, computer users flooded the work area with more paper (Sellen and Harper, 2002). Likewise, telecommuting that is not placed in a broader view of accessibility could actually increase personal use of a vehicle. This could happen if telecommuters become separated from their traditional car/van pool group or simply add new trips for shopping and socializing, using the time they formerly had devoted to commuting.

In terms of increasing the options of accessibility to community members, roadways can easily be designed to incorporate the needs of **pedestrians** and cyclists. In Figure 14.14, the road has been altered by adding islands that provide a safe area for pedestrians who make it only halfway across the road. Narrow roads also slow down vehicles while providing better pedestrian access.

Strategic and dense placement of **trees** along the road provides a defined edge, which guides a motorist's movement and allows drivers to assess their speed more accurately. This leads to overall speed reductions. Trees also create and frame visual walls and provide distinct edges to sidewalks, so motorists better distinguish the space of their vehicular environment and the space shared with pedestrians and cyclists. Motorists also perceive the time it takes to travel through tree-lined versus nontreed environments differently. From their perspective, a trip through a treeless environment is perceived as longer than a trip through an environment lined with trees, so they drive faster through the unlined stretch of road.

### Bicycle and Pedestrian Programs
http://www.fhwa.dot.gov/environments/bikeped/index.htm

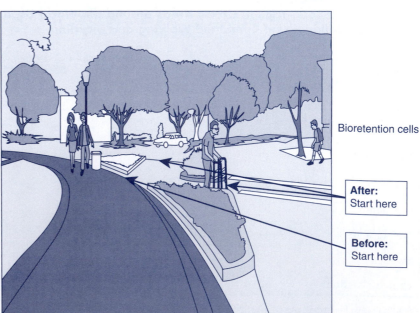

Bioretention cells

**After:** Start here

**Before:** Start here

**Figure 14.14   Roadway Design for Enhanced Walking and Bicycling**
This figure illustrates how an intersection can be retrofitted to make the crossing shorter for pedestrians. The curb has been extended into the street, bringing waiting pedestrians into better view of drivers and reducing the length of the crosswalk. Treatments like this also help slow vehicular traffic, especially turning vehicles.

Adapted from Livable Communities, Inc.

Another important consideration is to understand that walking is also a social activity for many people, who may be walking with a friend or dog. Even commuter walkways are shared by other pedestrians and possibly bicycles. For this reason, the American Association of State Highway Transportation Officials recommends that pedestrian walkways be designed with a clear travel path that is 4 to 10 ft. wide, depending on the adjacent land use. Similarly, sidewalks in commercial areas need to be wide enough to accommodate wheelchairs that might need to pass, shoppers, signage, outdoor café furniture, and public-transit shelters.

**Walkability** is the cornerstone and key to an urban area's efficient ground transportation. Every trip begins and ends with walking. Walking remains the cheapest form of transport for all people, and the construction of a walkable community provides the most affordable transportation system any community can plan, design, construct and maintain. Walkable communities put urban environments back on a scale for sustainability of resources (both natural and economic) and lead to more social interaction, physical fitness and diminished crime and other social problems. Walkable communities are more liveable communities and lead to whole, happy, healthy lives for the people who live in them. (Walkable Communities, www.walkable.org/)

Walkways need to be strategically placed, not only to connect important destination points, but also to connect pedestrian- and bicycle-friendly routes. One example is to place a walkway at the end of a dead-end street that prevents vehicle traffic but allows the passage of pedestrians and cyclists. For road crossings controlled by signals and stop signs, engineers need to decrease the distance to the location of a road crossing to provide pedestrians safe access across a road without having to walk far out of their way.

Many of the design points addressed previously for pedestrians also relate to improving accessibility for cycling. Sometimes space for a bike lane can be obtained just by making the road corridor slightly narrower.

Just as for pedestrians, buffers should be placed to separate bicycles from vehicles. The buffer strip should be a minimum of 5 ft. wide and needs to be designed so that it becomes larger as the vehicle traffic speed increases. Buffer strips are especially important for less experienced cyclists and children, as use of the bike path located in a shared corridor will increase as the buffer strip is improved.

Bicyclists are also known to be users of shared public transit. For this reason, it is critical that bike racks be placed on buses and within light-rail transit systems. Bike storage facilities should be placed near target destinations (light-rail stations, schools, commercial districts). These storage facilities should be located where there are people and associated activity to ensure safety of the bikes.

Shared **public transit** does not have to include only large public buses or light-rail transit systems. It can also include van or shuttle services to airports, schools, and shopping areas. Making public transit available to cyclists by providing racks and storage facilities is known to increase ridership.

One key issue to consider is how to integrate public transit with pedestrian and bicycle movement. As stated previously, the plan should always be to increase the options, not to limit them. The walking

© Slobo Mitic/iStockphoto.

**Class Discussion**

How many options listed in Table 14.9 are available for your commute to class? How could the built environment be better designed to provide more options and improve the performance of existing options?

## Green Highways Partnership

http://www.greenhighways.org/

distance to bus stops and light-rail stations needs to be unobstructed and account for winter changes, when snow and ice may eliminate shortcuts. Think critically about the movement of pedestrians and bicyclists, and design stations and stops to account for their movements.

General walking guidelines are that a person will walk 400 m or 5 min (assuming a walking speed of 80 m/min) to reach a bus stop. Studies in the San Francisco Bay Area and Edmonton have found that few people walk more than 1,750 m to reach a light-rail station. In Canada walking distances for light rail range from 400 to 900 m. U.S. guidelines specify a walking distance of 400 to 800 m.

People will walk farther if the bus is reliable or comes more frequently. One study in Florida found that adult usage of mass transit dropped by up to 70 percent when the walking distance increased from 200 to 400 m. Elderly passengers complained that 400 m was too far, and they disliked having to cross major arterial roads (O'Sullivan and Morrall, 1996).

In terms of improving our highways, **Green Highways Partnership** is led by the Environmental Protection Agency and the Federal Highway Administration. It uses integrated planning to incorporate environmental stewardship into all aspects of a highway life cycle (see www.greenhighways.org). Table 14.13 lists 14 characteristics of a green highway.

## Table / 14.13

### Characteristics of a Green Highway

1. Provides net increase in environmental functions and values of the watershed

2. Goes beyond minimum standards set forth by environmental laws and regulations

3. Identifies and protects important historical and cultural landmarks

4. Maps all resources in the area in order to identify, avoid, and protect critical resource areas

5. Uses innovative, natural methods to reduce imperviousness, and cleanse all runoff within the project area

6. Maximizes use of existing transportation infrastructure, provides multimodal transportation opportunities, and promotes ride sharing/public transportation

7. Uses recycled materials to eliminate waste and reduce the energy required to build the highway

8. Links regional transportation plans with local land use through partnerships

9. Controls populations of invasive species, and promotes the growth of native species

10. Incorporates post-project monitoring to ensure environmental results

11. Protects the hydrology of wetlands and stream channels through restoration of natural drainage paths

12. Results in a suite of targeted environmental outcomes based upon local environmental needs

13. Reduces disruptions to ecological processes by promoting wildlife corridors and passages in areas identified through wildlife conservation plans

14. Encourages smart growth by integrating and guiding future growth and capacity building with ecological constraints

SOURCE: Courtesy of Green Highways Partnership, www.greenhighways.org.

## 14.9    Urban Heat Island

The term **heat island** refers to urban air and surface temperatures that are higher than in nearby rural areas. Many cities and suburbs have air temperatures up to 10°F (5.6°C) warmer than the surrounding natural land cover. The **heat island sketch** (Figure 14.15) shows a typical city's heat island profile. Urban temperatures are typically lower at the urban–rural border than in dense downtown areas. The sketch also shows how parks and open land create cooler areas. This is one reason that greening the built environment provides social and environmental benefits.

Heat islands form as cities replace natural land cover with pavement, buildings, and other infrastructure (the built environment). Displacing trees and vegetation minimizes the natural cooling effects of shading and evaporation of water from soil and leaves (evapotranspiration). **Nonpervious materials** have significantly different thermal bulk properties (including heat capacity and thermal conductivity) and surface radiative properties (albedo and emissivity) than the surrounding rural areas. This initiates a change in the energy balance of the urban area, often causing it to reach higher temperatures—measured both on the surface and in the air—than its surroundings (Oke, 1982). Tall buildings and narrow streets can heat air that is trapped between them, thus reducing airflow. This is referred to as the *canyon effect*. Waste heat from vehicles, factories, and air conditioners may add warmth to their surroundings, further exacerbating the heat island effect.

**Urban heat islands** can impair a city's public health, air quality, energy demand, and infrastructure costs in several ways (Rosenfeld et al., 1997). Heat islands prolong and intensify heat waves in cities, making residents and workers uncomfortable and putting them at increased risk for heat exhaustion and heatstroke. In addition, high concentrations of ground-level ozone aggravate respiratory problems such as asthma, putting children and the elderly at particular risk. Hotter temperatures and reduced airflow increase demand for air conditioning, increasing energy use when demand is already high. This in turn contributes to power shortages and raises energy expenditures at a time when energy costs are at their highest. Urban heat islands

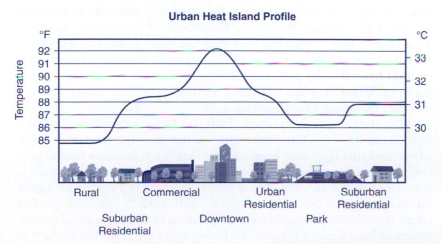

**Figure 14.15    Urban Heat Island Profile**    This profile shows that increased temperatures of up to 10°F (5.6°C) can be found in dense downtown areas, as compared with surrounding rural, suburban, and open areas. The geometry of streets and buildings, along with the built environment's dependence on masonry, concrete, and asphalt structures that have high thermal bulk properties that store the sun's energy, have helped create this problem.

Adapted from Environmental Protection Agency, Heat Island Effect Web site, www.epa.gov/heatisland.

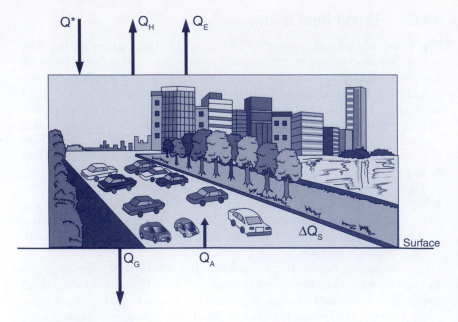

**Figure 14.16 Energy Balance Written for a Shallow Layer at the Urban Land Surface** This layer contains air and surface elements that make up the built environment. $Q^*$ is the net radiation, $Q_H$ is the sensible heat flux, $Q_E$ is the latent heat flux, $Q_G$ is the ground heat flux, $Q_A$ is the anthropogenic heat discharge, and $\Delta Q_S$ is the energy stored or withdrawn from the layer.

contribute to global warming by increasing the demand for electricity to cool our buildings.

The study of urban heat islands is complicated, though. For example, in cooler climates during the winter, the urban heat island effect can cause nighttime temperatures to be less severe, which would require less heating. Also, fewer snowfall and frost events may occur, and changes in melting patterns of snowfall may change the urban hydrology of snowmelt.

To further investigate the causes of urban heat islands, an energy balance can be written on a shallow layer at the urban land surface containing air and surface elements, as shown in Figure 14.16:

$$Q^* + Q_A - Q_H - Q_E - Q_G = \Delta Q_S \qquad (14.12)$$

Here, $Q^*$ is the net radiation, the sum of the incoming and outgoing shortwave and longwave radiation. Incoming solar shortwave radiation is a function of solar zenith angle, and a fraction of it is then reflected as outgoing shortwave radiation, which depends on the solar albedo of the surface. The higher the **albedo** of the surface, the more solar energy that is reflected back into the atmosphere and leaves the shallow layer shown in Figure 14.16. Incoming longwave radiation is emitted from the sky and surrounding environment. Outgoing longwave radiation includes both that emitted from the surface and the reflected incoming longwave radiation.

$Q_A$ is the total anthropogenic heat discharge in the box. The first two terms $(Q^* + Q_A)$ are balanced by the sensible heat flux $(Q_H)$, latent heat flux $(Q_E)$, and ground heat flux $(Q_G)$. Sensible heat is heat energy transferred between the surface and air. When the surface is warmer than the air above, heat will be transferred upward into the air and leave the box via conduction followed by convection. The latent heat flux is produced by transpiration of vegetation and evaporation of land surface water, which removes heat from the surface in the form of water vapor. The ground heat flux is the flux of heat transferred from the surface downward to subsurface via conduction. Finally, $\Delta Q_S$ is a term to account for energy that is stored or withdrawn from the layer. The ambient temperature within the layer will be influenced by $\Delta Q_S$. Later, in Table 14.14, we will investigate how these energy balance terms are related to the layout of, and materials incorporated into, the built environment.

The magnitude of the urban heat island can be described as the difference in temperature between urban (u) and rural (r) monitoring stations ($\Delta T_{u-r}$). $\Delta T_{u-r}$ will be greatest on clear, cool nights, but it also has been found to depend on street geometry. In the most dense section of the urban environment, the magnitude of this loss term (part of $Q^*$) is controlled by how well the sky is viewed at ground level. This sky view factor has been found to be approximated by the ratio of building height to street width ($H/W$). The maximum $\Delta T_{u-r}$ (in °C) can be related to the street geometry by the following expression (Oke, 1981):

$$\text{maximum } \Delta T_{u-r} = 7.45 + 3.97 \ln \left( \frac{H}{W} \right) \qquad (14.13)$$

While climate considerations related to street geometry can be designed into new urban development, in existing cities little can be done to modify the effect of the street canyon on climate. In such

**Urban Heat Island Mitigation**
http://www.epa.gov/heatisland

**Class Discussion**
Urban landscapes laid out in a more spread-out, horizontal direction (versus the densely populated vertical direction common in places like Manhattan) will have a less extreme urban heat island. Is this a more, or less, sustainable approach to populating an urban area? Obviously the answer is not easy and will require more thinking and analysis. The question does demonstrate why sustainable solutions require engineers to think beyond their individual project and take a systems approach to solving problems, which incorporates a regional and global outlook.

## example/14.2 Urban Heat Island and Street Geometry

Assume a downtown area has two 12 ft. travel lanes for vehicles, two 12 ft. bus lanes, two 12 ft. metered parking lanes, and a 12 ft. sidewalk on each side. This is all surrounded by ten-story buildings that are 125 ft. tall. What is the maximum urban heat island impact that can be expected?

### solution

The maximum urban heat island in the downtown core can be estimated using Equation 14.13. The street width includes the roadway and the sidewalk areas.

$$\text{maximum } \Delta T_{u-r} = 7.45 + 3.97 \ln (125 \text{ ft.}/96 \text{ ft.}) = 8.5°C$$

Note how this example shows the importance of **street and building geometry** (referred to as the *street canyon*) on the urban heat island. Try doing this example again for the same street size but shorter buildings. A neighborhood with the same street profile but 40 ft. tall buildings will have a maximum heat island impact of 4.0°C (Cambridge Systematics, 2005). Try doing the example again for an old historic city with narrow streets but shorter building heights. What do you discover about urban heat island intensity in the urban core as it relates to street and building geometry and population density?

cases, climate can be modified by selection of surfaces, coatings, and vegetation while also reducing the amount of mechanical waste heat that cities produce.

Table 14.14 relates many of the terms in the energy balance (Equation 14.12) with engineered features of the urban environment. Some features are related to the physical geometry of the street layout. Others

## Table / 14.14

**Features of Urban Environment Related to Terms in the Heat Island Energy Balance** Designing and modifying an urban environment to modify climate processes requires an understanding of this balance.

| Energy Balance Term | Feature of an Urban Environment That Alters the Energy Balance Term | Engineering Modifications That Reduce Intensity of Urban Heat Island |
|---|---|---|
| Net shortwave and longwave radiation, $Q^*$ | Canyon geometry of the street and building | Canyon geometry influences the way shortwave radiation enters and is absorbed by the built environment and the way longwave radiation is reflected out of the urban canopy. |
| Heat added by humans ($Q_{human}$) | Emission of waste heat from buildings, factories, and vehicles | Though this is a small term in the overall energy balance, buildings can be designed to reduce the need for mechanical cooling. Cities can be planned so they are dependent on mechanical engines to move people and goods. |
| Sensible heat flux, $Q_H$ | Types of engineering materials | Increasing the surface albedo of paints and roofing materials will limit the surface–air sensible heat flux. Albedo is a measure of the amount of solar energy reflected by the surface. Narrow canyon geometry can result in reduced air flow, which decreases the effect of $Q_H$. |
| Latent heat flux, $Q_E$ | Types of engineering materials and storm water management | The latent heat flux out of the system is the result of water evaporation. The energy is carried out in the form of water vapor (in the form of the higher energy in the water molecules in the vapor form). The heat is taken from the vegetation or water. This is the same process as sweat, where one's body is cooled with the heat going away in the form of latent heat. Impervious and nonvegetated surfaces hinder evaporative cooling (unless water is sprinkled on them). Low-impact development recognizes that leaving some standing water on the surface is not bad and vegetation such as green roofs and trees are an important feature of the urban built environment. |
| Increased storage of heat | Different abilities to store heat in different types of construction materials | The thermal conductivity of asphalt and concrete are similar (1.94 versus 2.11 J/m$^3$-K, respectively). The thermal admittance of asphalt and concrete results in increased storage of heat. Urban surfaces heat up faster than natural and impervious surfaces that retain water. Built-environment materials have a high ability to store and release heat. Paved surfaces are thick and in contact with an underlying ground surface. Buildings, though, have a thinner skin that separates indoor and outdoor air. Surfaces with higher albedo will reduce the stored heat. |

SOURCE: Adapted from Mills, 2004.

include modification of surfaces, materials choices, use of impervious pavements, preservation of wetlands, and incorporation of green roofs and low-impact development technologies for storm water control.

## 14.10   Urban Planning, Smart Growth, and Planned Communities

We began this chapter by discussing how, as the buildings and landscapes that make up the built environment are grouped, the scale of the built environment increases toward communities with subdivisions, neighborhoods, districts, villages, towns, and cities. The next level of the built environment discussed was regions where cities and landscapes are grouped and are generally defined by common political, social, economic, or environmental characteristics. When working on any component of the built environment, it is important to think how that particular piece of the built environment fits into broader community, regional, and global issues. Individual projects also need to be integrated with other pieces of the built environment, such as water, wastewater, solid waste, accessibility, housing, and commercial and education sectors. Figure 14.17 demonstrates how road layout in a residential area affects the amount of materials usage and water resources due to paved surfaces.

In terms of **urban planning**, a central core is a must. This is the facet of a city or town that brings the community members together for social gatherings, worship, business, education, and recreation. The central core also provides the occupants with a sense of place. In addition to a central core, smaller neighborhoods are needed where neighborhood residents can go for a smaller range of commercial and service-oriented businesses.

**Smart growth** is a term used to describe development of a community that protects natural resources, open space, and historical or cultural attributes of a community while also reusing land that is already

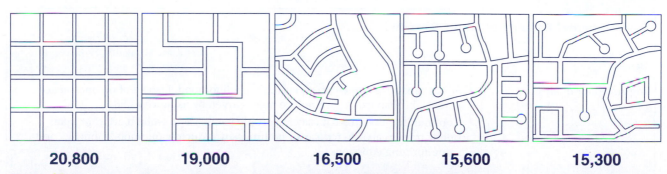

| 20,800 | 19,000 | 16,500 | 15,600 | 15,300 |

**Figure 14.17**   **Simple Features Related to Road Layout That Will Affect the Lineal Feet of Impervious Pavement**   The number given under each layout indicates the lineal feet of impervious pavement. Just adding loops and lollipops can lessen the lineal feet of pavement by 26%. Curved roads will slow traffic, and narrow pavement will reduce the paved area as well. The cul de sac shown here needs designed to connect pedestrians and cyclists.

Adapted from Prince George's County (1999).

## Table / 14.15

### Principles of Smart Growth

1. Mix land uses.

2. Take advantage of compact building design.

3. Create a range of housing opportunities and choices.

4. Create walkable neighborhoods.

5. Foster distinctive, attractive communities with a strong sense of place.

6. Preserve open space, farmland, natural beauty, and critical environmental areas.

7. Strengthen and direct development toward existing communities.

8. Provide a variety of transportation choices.

9. Make development decisions predictable, fair, and cost-effective.

10. Encourage community and stakeholder collaboration in development decisions.

SOURCE: From Smart Growth Online, available at www.smartgrowth.org.

developed. Table 14.15 lists the ten principles of smart growth. Note how the principles are not based on a vehicle-dominated community and how they promote diversity in terms of housing, land types, and commuter accessibility. Table 14.16 identifies environmental benefits of smart growth in terms of water quality, air quality, open-space preservation, and brownfields redevelopment.

## Table / 14.16

### Environmental Benefits of Smart Growth

| Benefit | Comments |
|---|---|
| Improved air quality | • Siting new development in an existing neighborhood, instead of on open space at the suburban fringe, can reduce miles driven by as much as 58%. <br> • Communities that make it easy for people to choose to walk, bicycle, or take public transit can also reduce air pollution by reducing automobile mileage and smog-forming emissions. |
| Improved water quality | • Compact development and open-space preservation can help protect water quality by reducing the amount of paved surfaces and by allowing natural lands to filter rainwater and runoff before it reaches drinking-water supplies. <br> • Runoff from developed areas often contains toxic chemicals, phosphorus, and nitrogen; nationwide, it is the second most common source of water pollution for estuaries, the third most common for lakes, and the fourth most common for rivers. |

## Table / 14.16

(Continued)

| Benefit | Comments |
|---------|----------|
| Open-space preservation | • Preserving natural lands and encouraging growth in existing communities protects farmland, wildlife habitat, biodiversity, and outdoor recreation, and promotes natural water filtration.<br>• A recent study in New Jersey found that, compared with less compact growth patterns, planned growth could reduce the conversion of farmland by 28%, open space by 43%, and environmentally fragile lands by 80%. |
| Brownfield redevelopment | • Cleaning up and redeveloping a brownfield can remove blight and environmental contamination, catalyze neighborhood revitalization, lessen development pressure at the urban edge, and use existing infrastructure. |

SOURCE: From Environmental Protection Agency, www.epa.gov/smartgrowth.

Excessive incorporation of parking into the built environment leads to problems associated with water quality, urban heat island effect, and loss of the natural environment. Many zoning codes and parking regulations require that a specific number of parking spaces be set aside for a set square footage of commercial establishment (or a particular number of housing units). However, these regulations usually overestimate the number of needed spaces because they assume all trips are by private automobile and also assume usage on some peak period such as the holiday shopping season. An engineer can work with a zoning commission and the client to eliminate this overuse of parking spaces. More accessibility choices made available to the community also will reduce the need for parking spaces (Figure 14.18). In addition, grassy open spaces that freeze over in the winter or use of

### Calculate Your Carbon Footprint from Home and Travel
http://www.fightglobalwarming.com/carboncalculator.cfm

**Figure 14.18  A Few of the World's More than 1 Billion Bicycles**  Bicycles provide human-powered transportation that can be integrated with principles of smart growth and improvements in public health.

iStockphoto.

## Table / 14.17

### Benefits of Incorporating Trees into Roads and Communities

| Benefit | Description |
|---|---|
| Reduced traffic speeds | Trees along an urban street create a vertical wall and frame the street. This provides a defined edge, which guides motorists' movement and allows them to better assess their speed. This leads to overall speed reductions. |
| Safer walking environments | Trees create and frame visual walls and provide distinct edges to sidewalks, so motorists better distinguish the space of their vehicular environment and the space shared with people. |
| Improved business | Businesses on streets landscaped with trees show 12% higher income streams, which is often the essential competitive edge needed for success of main street stores. |
| Less need for drainage infrastructure | Trees transpire the first 30% of most precipitation through their leaf system; thus, this moisture never hits the ground. Trees also facilitate natural recharge of groundwater. Both these issues reduce storm water runoff and flooding potential. |
| Lower urban air temperatures | Asphalt and concrete streets and parking lots are known to increase urban temperatures 3°F–7°F. A properly shaded neighborhood can reduce energy bills for an urban household 15%–35%. Reduced temperatures (5°F–15°F) are felt by pedestrians walking under tree-canopied streets. |
| Time-in-travel perception | Motorists perceive the time it takes to travel through tree-lined versus nontreed environments differently. A trip through a treeless environment is perceived to be longer than one on a street lined with trees. |
| Longer pavement life | Studies conducted in a variety of California environments show that the shade of urban street trees can increase the life of asphalt by reducing the daily heating and cooling extremes experienced by the pavement. |
| Improvements in air quality | Trees located close to a street take up 9 times more air pollutants than more-distant trees |

SOURCE: Adapted with permission from D. Burden, Walkable Communities, Inc; www.walkable.org.

porous pavements can accommodate larger crowds expected for a few weeks of the holiday shopping season.

Another aspect of sustainable urban planning is the incorporation of trees into the built environment. Table 14.17 describes benefits associated with planting trees along roads and within communities. Walkable Communities estimates that planting one tree costs approximately $250 to $600 over the first three yr, with the tree returning over $90,000 of direct benefits over its lifetime.

In urban areas, trees are typically spaced every 15 to 20 ft. A high density of tree placement is more desirable and will bring more benefits. In business communities where sidewalks are narrow, curbed tree wells can be spaced every 40 to 60 ft., allowing two or three parking spaces in between. This works well with parallel or angled parking (Burden, 2006).

# Key Terms

- access
- aggregates
- albedo
- biodiversity
- built environment
- components of a building
- concrete
- construction
- context-sensitive design
- cycling
- deconstruction
- degree-day
- ecosystem-based mitigation
- energy
- energy balance
- energy efficiency
- flow of raw materials
- glazing
- Green Highways Partnership
- heat balance
- heat island
- heat island sketch
- heat loss due to infiltration
- Leadership in Energy and Environmental Design (LEED)
- LEED Accredited Professional
- LEED certification
- materials flow analysis (MFA)
- mobility
- nonpervious materials
- passive solar gain
- pedestrians
- Portland cement
- public transit
- reinforced steel
- rightsizing
- smart growth
- stakeholders
- street and building geometry
- sustainable construction project
- telecommuting
- thermal wall
- toxic chemical
- traditional building materials
- traffic congestion
- trees
- urban heat islands
- urban metabolism
- urban planning
- U.S. Green Building Council
- walkability
- walking
- water

# chapter/Fourteen Problems

**14.1** List a minimum of two historical, cultural, community, and environmental attributes of your college or university that are worth preserving for future generations. Do the same for your hometown.

**14.2** List two distinct wildlife habitats located near your local college or university that are worth preserving for future generations. What components of the existing built environment are affecting this biodiversity in a positive or negative way?

**14.3** Your instructor should identify a local woods, wetlands, meadow, or historical community and provide you with a hypothetical built-environment project. Map the area before development and then after placement of the built environment project. Label all the historic, cultural, social, and environmental attributes you wish to preserve after development has occurred.

**14.4** Identify a building on your campus or in your community that has been LEED certified. Obtain the scores, and determine what percent of the total points were obtained in each of the six evaluation criteria: (1) sustainable sites, (2) water efficiency, (3) energy and atmosphere, (4) materials and resources, (5) indoor air quality, and (6) innovation and design process.

**14.5** Identify a building on your campus or in your community that has been newly constructed but not LEED certified. Walk through the building to see if it would obtain LEED credits or meet a prerequisite for the following items. For sustainable sites: (a) Credit 4.1, Alternative Transportation, Public Transportation Access; (b) Credit 4.2, Alternative Transportation, Bicycle Storage & Changing Rooms; (c) Credit 7.1, Heat Island Effect, Non-Roof, or Credit 7.2, Heat Island Effect, Roof; and (d) Credit 8, Light Pollution Reduction. For water efficiency: (e) Credit 1.2, Water Efficient Landscaping, No Potable Use or No Irrigation; (f) Credit 2, Innovative Wastewater Technologies; and (g) Credit 3.1, Water Use Reduction, 20% Reduction, or Credit 3.2, Water Use Reduction, 30% Reduction. For energy and atmosphere: (h) Credit 2, On-Site Renewable Energy. For materials and resources: (i) Prereq. 1, Storage & Collection of Recyclables. For indoor environmental quality: (j) Credit 6.1, Controllability of Systems, Lighting; and (k) Credit 8.1, Daylight & Views, Daylight 75% of Spaces, or Credit 8.2, Daylight & Views, Views for 90% of Spaces.

**14.6** Identify the specific building component (depicted in Figure 14.6) with a subset of the LEED credits your instructor provides (Table 14.2).

**14.7** (a) Determine the heat loss (in Btu/°F-day and Btu/degree-day) through a 120 sq. ft. insulated wall described in the following table. (b) Determine the heat loss through the same wall when a 3 ft. by 7 ft. door (R factor = 4.4) is inserted into the wall.

| Component of Wall | R Factor |
|---|---|
| 2 in. Styrofoam board insulation on outside of wall under siding | 10 |
| Old cedar log wall | 20 |
| Fiberglass insulation on inside of wall | 11 |
| $1/2$ in. drywall on inside of wall | 0.45 |
| Inside air film along inside of wall | 0.68 |
| Outside air film along outside of wall | 0.17 |

**14.8** Look up (a) the total degree-days for heating and (b) the total degree-days for cooling for your university town or city (or hometown).

**14.9** In Section 14.7, we worked out a problem where the combined heat loss from a hypothetical 3,000 sq. ft. building was 1,053 Btu/degree-days. Determine the total energy requirements (in Btu) to heat that hypothetical building in May 2007 for the locations in the following table.

| Location | Heating Degree-Days in May 2007 |
|---|---|
| Anchorage, AK. | 541 |
| Winslow, AZ. | 70 |
| Yuma, AZ. | 0 |
| Rochester, NY. | 237 |
| Pittsburgh, PA. | 106 |
| Rapid City, SD. | 193 |

**14.10** Which methods listed in Table 14.9 are available for you to access traveling to class? How could an engineer better design your community to make all the methods listed in Table 14.9 more available?

**14.11** You are preparing a life cycle assessment (LCA) of different transportation options for getting from your house to work (10 mi. each way). They include bicycling, one person in a car, carpooling with three or more people, or taking the bus. Write a possible goal, scope, function, and functional unit for this LCA (refer to Chapter 7).

**14.12** Select a highway located near your college or university. Visit the Web site of the Green Highways Partnership (www.greenhighways.org) to learn more about green highways. Which of the 14 characteristics of a Green Highway are incorporated into the current highway system of the highway you selected? How would you reengineer the highway to incorporate every aspect of a green highway?

**14.13** Design a bicycle–pedestrian pathway that would allow you to travel from your home to class. Provide details on the location of the pathway, details on the size and materials used to construct the pathway, signage and safety, and size and location of bicycle storage. Think about what other community attributes you could readily connect to the pathway. Are there any existing built environment systems you could incorporate into your pathway that would eliminate the need for new paving materials?

**14.14** Go to the Weather Channel Web site (www.weather.com), and look up the monthly average temperature for a major metropolitan area and nearby rural area anywhere in the world over a 12 mo. period. Use the data you looked up to estimate the magnitude of the urban heat island effect for that city. Graph your data in two figures, and determine the temperature differences in each month.

**14.15** Identify an urban core of a major metropolitan area that you are familiar with or that is close to your college or university. Calculate the magnitude of the maximum urban heat island impact in the urban core. Provide some detailed alternatives for reducing the urban heat island in this core area, and relate them to specific items in the energy balance performed on the urban canopy.

**14.16** Assume a small downtown area has two 12 ft. travel lanes with 6 ft. sidewalks on each side. This is all surrounded by buildings that are 25 ft. tall. What is the maximum urban heat island impact that can be expected?

**14.17** Using the systems analysis approach (Chapter 7), draw a systems analysis diagram for urban heat islands, including feedback mechanisms for increased energy demands for cooling and refrigeration, increased air pollution from these increased energy demands, and other effects such as global warming and public health.

**14.18** Identify a new development project in your college or university town. Does it follow principles of smart growth? If so, provide details of how it does. If it does not follow principles of smart growth, provide detailed suggestions of how you would change the design.

**14.19** Identify an important watershed on your campus or within your local community. Obtain a topographical map, and place the built environment over the watershed. (a) List at least six components of the built environment that are having an adverse impact on the local watershed. Then provide six solutions to correct these problems. (b) Increase the level of spatial scale to the problem. Identify a larger watershed that is affected by decisions made in the smaller watershed. What social and environmental attributes of the local and larger watershed are worth protecting for future generations? Are these attributes similar or different when you change spatial scales?

**14.20** Visit a commercial area near your college or university, and identify how trees could be integrated into the built environment. Describe your recommendations in detail, including size and the type of vegetation. For each recommendation, indicate social and/or environmental benefits of your plan.

**14.21** Obtain U.S. census information on the Internet to describe demographics of your community in terms of age and income. For the age and income groups in your community, suggest several options to provide accessibility, relating your suggestions to pertinent census data.

# References

Alvord, K. 2000. *Divorce Your Car!* Gabriola Island, B.C.: New Society.

Brown, L. R. 2001. *Paving the Planet: Cars and Crops Competing for Land*. Washington, D.C.: Earth Policy Institute.

Burden, D. 2006. "Urban Street Trees: 22 Benefits." State of Michigan Department of Natural Resources Web site, www.michigan.gov/documents/dnr/22_benefits_208084_7.pdf. August.

Cambridge Systematics. 2005. "Cool Pavement Report." Draft report prepared for Heat Island Reduction Initiative, U.S. Environmental Protection Agency. June.

Cascadia Consulting Group, Inc. 2004. "Statewide Waste Characterization Study." Contractor's report to the California Integrated Waste Management Board. Publication #340-04-005.

Crowther, P. 1999. "Designing for Disassembly to Extend Service Life and Increase Sustainability." Eighth International Conference on Durability of Building Materials and Components, Vancouver, Canada, May/June.

Damschen, E. I., N. M. Haddad, J. L. Orrock, J. J. Tewksbury, and D. J. Levey. 2006. "Corridors Increase Plant Species Richness at Large Scales." *Science* 313:1284–1286.

*Environmental Building News*. 2001. "Buildings and the Environment: The Numbers." 10 (5).

Ewing, R. 1994. "Residential Street Design: Do the British and Australians Know Something Americans Do Not?" *Transportation Research Record* 1455:42–49.

Frank, L. D., and P. Engelke. 2000. "How Land Use and Transportation Systems Impact Public Health: A Literature Review of the Relationship between Physical Activity and Built Form." Working Paper 1. Centers for Disease Control and Prevention, Physical Activity and Health Branch, Atlanta, Ga.

Haasl, T. 1999. "Fifteen O&M Best Practices for Energy Efficient Buildings." Portland Energy Conservation, Inc., Portland, Ore.

Horvath, A. 2004. "Construction Materials and the Environment." *Annual Review of Environment and Resources* 29:181–204.

Maryland State Highway Administration. 1998. "Thinking Beyond the Pavement." Conference Summary. May.

McGowen, P., and J. Johnson. 2007. "Habitat Connectivity and Rural Context Sensitive Design: A Synthesis of Practice." FHWA/MT-06-012/8117-31. Final report prepared for the Montana Department of Transportation.

Mihelcic, J. R., J. B. Zimmerman, and A. Ramaswami. 2007. "Integrating Developed and Developing World Knowledge into Global Discussions and Strategies for Sustainability, Part 1: Science and Technology." *Environmental Science & Technology* 41 (10): 3415–3421.

Mills, G. 2004. "The Urban Canopy Layer Heat Island." IAUC Teaching Resources, compiled for the International Association for Urban Climate Teaching Resource Committee, www.urban-climate.org/UHI_Canopy.pdf, accessed October 30, 2007.

Minnesota Department of Transportation. 2006. "Design Policy—Design Excellence through Context Sensitive Design." Technical Memorandum 06-19-TS-07. St. Paul, Minn.

National Research Council. 2005. *Does the Built Environment Influence Physical Activity? Examining the Evidence*. Washington, D.C.: Transportation Research Board, 2005.

Oke, T. R. 1981. "Canyon Geometry and the Nocturnal Urban Heat Island: Comparison of Scale Model and Field Observations." *International Journal of Climatology* 1:237–254.

Oke, T. R. 1982. "The Energetic Basis of the Urban Heat Island." *Quarterly Journal of the Royal Meteorological Society* 108 (455).

O'Sullivan, S., and J. Morrall. 1996. "Walking Distances to and from Light-Rail Transit Stations." *Transportation Research Record* 1538:19–26.

Prince George's County, Md. 1999. "Low Impact Development Design Strategies: An Integrated Design Approach." Prince George's County, Md., Department of Environmental Resources, Programs and Planning Division, June.

Rosenfeld, A., J. Romm, H. Akbari, and A. Lloyd. 1997. "Painting the Town White—and Green." *Technology Review* (February/March).

Rush, R. 1986. *The Building Systems Integration Handbook*. New York: John Wiley & Sons.

Scottish Architecture. 2005. *The Built Environment* www.scottish-architecture.com/be-home-more.html, accessed December 11, 2005.

Sellen, A. J. and R. H. R. Harper. 2002. *The Myth of the Paperless Office*. Cambridge; MA:MIT Press.

United Nations Environment Programme (UNEP). 2002. *Global Environmental Outlook 3*. Nairobi, Kenya.

United States Environmental Protection Agency (USEPA). 2001. "Our Built and Natural Environments: A Technical Review of the Interactions between Land Use, Transportation, and Environmental Quality." EPA 231-R-002, January 2001.

United States Environmental Protection Agency. 2007. Urban Heat Island Basic information. www.epa.gov/heatisland, last accessed August 31, 2007.

van Oss, H. G., and A. C. Padovani. 2003. "Cement Manufacture and the Environment." *Journal of Industrial Ecology*; 6 (1): 89–105 and 7 (1): 93–122.

Vanegas, J. A. 2003. "Road Map and Principles for Built Environment Sustainability." *Environmental Science & Technology*, 37 (23), 5363–5372.

Walkable Communities, Glatting Jackson Kercher Anglin, Inc., www.walkable.org, Orlando, Fla. accessed January 3, 2009.

Warren-Rhodes, K., and Koenig, A. 2001. "Ecosystem Appropriation by Hong Kong and Its Implications for Sustainable Development." *Ecological Economics*, 39 (3): 347–359.

Wilson, A. 1979. Thermal Storage Wall Design Manual. Santa Fe, NM:New Mexico Solar Energy Association, June.

Wilson, A., and J. Boehland. 2005. "Small Is Beautiful: U.S. House Size, Resource Use, and the Environment." *Journal of Industrial Ecology* 9 (1–2): 277–287.

# Answers to Selected Problems

## Chapter 2

**2.1** (a) 0.41 mg/L; (b) 0.21 mg/L

**2.3** 2.8 ppm

**2.5** 10.9 mg $NH_3$/L, 1.6 mg $NO_2$/L

**2.7** (a) i) 0.002 ppb, ii) 2 ppt, iii) $3.7 \times 10^{-6}$ μM; (b) i) 0.002 ppm, ii) 2 ppb

**2.9** 0.74 μg/L

**2.11** (a) 0.5 ppm, 8 ppm;
(b) $1.6 \times 10^{-5}$ moles $O_2$/L, $2.5 \times 10^{-4}$ moles $O_2$/L

**2.13** (a) $8.9 \times 10^{-2}$ ppm; (b) 0.0000089%

**2.15** concentration $= 858.9$ μg/m$^3$, so mass $= 0.7$ g

**2.17** (a) 13.6%; (b) 11.3 L

**2.19** 26 μg/m$^3$, 5,243 μg/m$^3$

**2.21** (a) 100 ppb; (b) 100 ppb > 60 ppb, therefore, sample may pose a threat

**2.23** 5 ppm

**2.25** 31.9% of total $CH_4$ emissions, 2.5% of total greenhouse emissions

**2.27** (a) 170 mg/L; (b) 70% of the solids are organic

## Chapter 3

**3.1** 5.8 g

**3.3** chloroform

**3.5** (a) H (TCE) = 0.44, H (PCE) = 1.1, H (dimethylbenzene) = 0.21,
H (parathion) = $1.6 \times 10^{-5}$; (b) PCE, TCE; 1,2-dimethylbenzene, parathion

**3.7** (a) 0.001 moles/L; (b) 33 mg/L; (c) 33 ppm

**3.9** 2.6

**3.11** (a) $3.0 \times 10^{-6}$ M; (b) $1.54 \times 10^{-6}$ M; (c) increase

**3.13** (a) $-78.4$ kJ/mole; (b) Cd $= 2.0 \times 10^3$ mg/L, so no, pH must be raised.

**3.15** 94%

**3.17** 0.7 days

**3.19** (a) first; (b) second; (c) $1.4 \times 10^{-3}$ s

**3.21** 0.055/day; 0.37/day

# Chapter 4

**4.1** 11 mg/L

**4.3** (a) 0.90 h; (b) 1.25 pCi/L

**4.5** (a) 38 mg/s; (b) 13 mg/m$^3$

**4.7** (a) 7.9 coliforms/100 mL, so no, water standard is not being met; (b) not possible, town 1 would need a concentration of $-12$ coliforms/100 mL.

**4.9** Superior 180 yr, Erie 2.6 yr

**4.11** (a) 0.15/min; (b) 30,000 gal; (c) 660,000 gal; (d) discussion; (e) $1.9 \times 10^3$ g/day

**4.13** (a) 0.090/h; (b) 0.042/h; (c) $2 \times 10^6$ gal

**4.15** (a) at x = 0.5, 3.5, 4.5 cm, the initial pollutant flux density, J = 0. At x = 1.5 and 2.5 cm, J = $10^{-8}$ mg/cm$^2$-s; (b) initial mass flux at x = 0.5, 3.5, 4.5 cm equals 0, at x = 1.5 and 2.5 cm, m = $7.1 \times 10^{-8}$ mg/s; (c) graphs; (d) The concentration profile in part (c) is changing due to the random motion of the molecules. The chemical is attempting to reach equilibrium through high concentration areas moving to areas with less concentration.

**4.17** 0.025 m/day

# Chapter 5

**5.1** 1.0/day

**5.3** (a) 91,680 mg/L; (b) 8%

**5.5** 0.2 mg biomass/mg substrate

**5.7** 0.42/day

**5.9** (a) 16 mg/L; (b) 0.3 mg/L; (c) 3 mg/L

**5.11** 457 mg/L, 240 mg/L, 697 mg/L

**5.13** 393 mg/L

# Chapter 6

**6.1** According to RCRA, toxicity is a subset of hazardous. A hazardous waste denotes a regulated waste, which is based on: 1) physical characteristics; 2) toxicity; 3) quantity generated; and 4) history of chemical (i.e., environmental damage, environmental fate).

**6.4** (a) Developed world – exposure to chemicals emitted from building materials, paints, floor coverings, and furniture; developing world – exposure to smoke associated with burning solid fuels; (b) Developed world – individuals who occupy the building for a longer period of time. At home, this could be children and a parent or adult who watches the child. At work, this would be employees; developing world – women and children.

**6.8** Hazard assessment, dose-response assessment, exposure assessment, and risk characterization

**6.9** Arsenic – human carcinogen, $3 \times 10^{-4}$ mg/kg-day, $1.5$ (mg/kg-day)$^{-1}$; methylmercury – possible human carcinogen, $1 \times 10^{-4}$ mg/kg-day; ethylbenzene – not classifiable as to human carcinogenicity, 0.1 mg/kg-day; methyl ethyl ketone – data inadequate for an assessment of human carcinogenic potential, 0.6 mg/kg-day; naphthalene – possible human carcinogen, 0.02 mg/kg-day; diesel engine exhaust – likely to be carcinogenic to human, exposure is through inhalation RfC = 5 $\mu$g/m$^3$

**6.11** (a) 0.064 mg/kg-day; (b) dose from water = 6%, dose from fish = 94%

**6.13** (a) 0.083 $\mu$g/L; (b) probable human carcinogen; (c) 1.3 $\mu$g/L

**6.15** The hazard quotient is less than 1; therefore, the concentration of the chemical in the fish will not result in adverse noncarcinogenic effects.

**6.18** (a) 40; (b) 49; (c) 18; (d) 42; (e) 23

# Chapter 7

**7.5** Example solution – Goal: determine which of the four transportation modes for commuting have the least environmental impact; scope: this LCA only considers the use phase of these transportation modes; function: provide a means of transporting a person on a 20-mile round trip commute; functional unit: 140 miles of transportation (one week of commuting).

**7.10** (a) annual cost = $2,501; (b) annual cost = $2,399; therefore, you should buy the new car (option b)

**7.12** Without solvent, E factor = 1.4; with solvent, E-factor = 23.2. These chemicals should be included if they are not recovered and recycled because they also contribute to the total waste of the process.

**7.17** $5,324

**7.18** (a) plastic; (b) no; (c) plastic

**7.19** (a) generally, paper; (b) paper becomes slightly more efficient than plastic; (c) paper; (d) paper, but gap with plastic is becoming smaller; (e) not very

# Chapter 8

**8.1** 4.7 mg/L

**8.3** 4.6 mg/L

**8.5** (a) 4 mg/L; (b) 2.6 days, 26 km; (c) 7.4 mg/L; (d) 1.6 mg/L

**8.7** 102 mg/L

**8.9** 7.1 mg/L

**8.11** (a) 0.91 days; (b) 9.1 km

**8.13** +, +, =, =, +, or = depending on biodegradability

**8.19** 150 ft$^2$ with a depth of 6 in. for each bioretention cell

**8.23** (a) 0.25 m/day; (b) 0.42 m/day; (c) 238 days

**8.25** 435 days

# Chapter 9

**9.1** 18 cm

**9.8** 672,000 gpd

**9.9** (a) 2,743,200 gpd; (b) yes

**9.12** (a) 1,193 gpm; (b) 326 gpm

**9.13** (a) 612,000 gal; (b) 12 in

**9.14** (a) 4,500 gal; (b) 8 in

**9.17** (a) 15 cm/yr; (b) 7.5 × 10$^6$ m$^3$/yr; (c) 12.6%

**9.18** 71%

# Chapter 10

**10.2** 9 mg/L as $CaCO_3$

**10.3** 862,300 kg/yr

**10.5** 61.5 mg/L as $CaCO_3$

**10.7** CMFR, t = 2.6 h, V = 4,286 $m^3$; PFR, t = 19 min, V = 528 $m^3$

**10.9** 21 m/h

**10.13** Chick's law rate: $\ln(0.0001) = -0.6092 \times t$, t = 15 s

**10.15** *Ct* product is a combined effect of disinfectant concentration and time of contact to achieve certain level of inactivation for a given microorganism; *Adenovirus* and *Calicivirus* are most readily inactivated by free chlorine; *C parvum* is the most difficult to inactivate.

**10.18** (a) 52,650 $m^2$; (b) 64.4 $L/m^2$-h; (b) number of membrane fibers for plant = 7,018,013, number for each module = 3,119

**10.23** effective size = 0.43 mm, uniformity coefficient = 2.74

**10.25** K = 5.77 $(mg/g)(L/mg)^{1/n}$, 1/n = 0.6906


# Chapter 11

**11.3** 60.8 $m^3$

**11.4** 583 $m^2$, 27.2 m, 1,749 $m^3$, 1.2 h

**11.5** Both criteria met with detention time of 2 hours.

**11.7** (a) V = 4.2 × $10^6$ L; (b) 2.9 h; (c) 4,515 kg/day; (d) SRT will decrease; (e) 0.97 lbs BOD/lbs MLVSS-day; (f) 4 days

**11.9** high, less, saturated, low, low, increased, increased

**11.11** poor because SVI = 300 mL/g

**11.15** V = 1,640 $m^3$; retention time = 1.1 hr

**11.18** 6.8 m


# Chapter 12

**12.3** National Ambient Air Quality Standards are concentration-based limits that air cannot exceed without triggering health impacts and regulatory penalties upon the region. As of 2008, NAAQS pollutants include Pb, CO, $NO_2$, $SO_2$, $O_3$, PM.

**12.7** Student specific solution; however, factors influencing the AQI could include pollutant emission activity, stability of atmosphere, weather, and transport from upwind sources.

**12.9** Regulatory: apply for a new emission permit, market-based: purchase sulfur dioxide emission credits, voluntary: set up a plan to distribute compact fluorescent light bulbs to customers, control technology: use lime instead of limestone in flue gas desulfurization unit to enhance capture efficiency

**12.11** 1.1 tons $CO_2$/acre-yr

**12.13** (a) q = 7.1 s, so yes, the design is sufficient for the minimum microbial time limit; (b) 75 $m^2$

**12.15** Series

**12.17** 66 mg/$m^3$

**12.19** 400 $m^2$

## Chapter 13

**13.5** (a) City 2, 0.69 kg/person-day compared to 0.54 kg/person-day for City 1; (b) 32.8%; (c) 31.1%

**13.7** 4,295 L

**13.8** 0.96 kg $O_2$/kg dry initial waste

**13.9** Slow biodegrading, $t_{90\%}$ = 33 yrs; rapid biodegrading, $t_{90\%}$ = 10 yrs

**13.11** Specific to student's location.

**13.13** Option 1 because Vs/Vf = 0.114

**13.14** 695,323 $m^2$

**13.15** $740,740

**13.20** 0.30 kg

## Chapter 14

**14.7** (a) 68.1 BTU/degree day; (b) 170.7 BTU/degree day

**14.9** Anchorage: 569,673 BTU, Winslow: 73,710 BTU, Yuma: 0 BTU

**14.16** 6°C

# Chapter Opener Photo Credits

**Chapter 1:** iStockphoto

**Chapter 2:** Tony Freeman/PhotoEdit

**Chapter 3:** PhotoDisc, Inc./Getty Images

**Chapter 4:** Terrance Emerson/iStockphoto

**Chapter 5:** Derek Dammann/iStockphoto

**Chapter 6:** Andrea Gingerich/iStockphoto

**Chapter 7:** Andreas Weber/iStockphoto

**Chapter 8:** Charles Taylor/iStockphoto

**Chapter 9:** Anantha Vardhan/iStockphoto

**Chapter 10:** GIPhotostock/Photo Researchers, Inc.

**Chapter 11:** Marcus Clackson/iStockphoto

**Chapter 12:** Marcus Lindstrom/iStockphoto

**Chapter 13:** Stephanie DeLay/iStockphoto

**Chapter 14:** Jeremy Edwards/iStockphoto

# Index

## A

Abiotic, 159–161, 186, 205, 348

Absorption, 61, 76–89, 136, 137, 161, 205, 207, 233, 234, 237, 244, 450–452, 553, 558, 559

Access (transportation), 634, 656–661, 667–669

Acid, 38, 39, 53, 60, 61, 68–74, 179, 192, 200–202, 204, 217, 412, 414, 419, 501, 589, 607

  Acid Rain, 68, 72–74, 436, 437, 498, 552

  Acid Rain Program, 552

Activated carbon

  Granular (GAC), 432, 450–453, 500

  Powdered (PAC), 411, 450–452

Activated Sludge, 161, 178, 475–492, 496–500, 512, 599

Activation Energy, 59, 97

Activity, 53–60

Activity Coefficient, 52, 54–56

Acute Toxicity, 225–226

Adsorption, 61, 76–89, 205, 396, 400, 404, 405, 409, 411, 432, 450–453, 507

Advection, 139, 140, 344–347

Advective Flux, 140, 141

Aeration Basin. *See* Aeration Tank

Aeration Tank, 475–480, 486, 492, 498–500

Aerobic Digestion, 501

Aerobic Respiration, 95, 184–186

Africa, 7, 11, 12, 89, 150, 181, 208, 228, 347, 359, 360, 403, 590–591

Aggregates, 19, 272, 337, 338, 596, 639, 640, 646

Air Quality Index, 553–555

Air-Water Equilibrium, 52, 60, 61, 65–68, 314–315

  Interface, 144–146

Albedo, 663–664, 666

Alien Invasive Species, 4, 209, 210

Alkalinity, 42, 71–73, 94, 412–414, 418–422, 436, 437, 448, 495, 611

Ambient Air, 279, 518, 519, 521, 527, 537, 538, 553–555, 557

American Society of Heating, Refrigeration, and Air-conditioning Engineers (ASHRAE), 531

Anaerobic Digestion, 73, 267, 465, 501–503

Anaerobic Respiration, 95, 184, 185

Anoxic, 185, 203, 403, 496, 507, 508

Apparent Concentration, 53, 58

Aquifer, 76, 81, 87, 150–152, 243, 247, 343–348, 355–357

Arrhenius equation, 97, 98

Artificial Photosynthesis, 182–183

## B

Atmosphere, 519

  Carbon dioxide, 17, 34, 36, 39, 42–45, 68, 70–72, 106, 109, 131, 137, 138, 163, 185, 193, 194, 201, 262, 270, 290, 291, 295, 296, 361, 419, 479, 495, 501, 502, 508, 520, 548, 551, 558, 597, 604, 607, 640, 641, 656

  Composition, 34

  Smog, 22, 98, 100, 275, 533, 668

  Stratosphere, 99, 100, 522

  Transport, 521–524

  Troposphere, 522

Attached Growth, 475, 492, 493

Autotrophs, 186, 191, 204, 498

## B

Baghouse, 520, 553, 562, 565–568

Bar Racks, 466–467

Bar Screens, 466–467

Base, 61, 68–73, 419

Baseflow, 356, 357

Batch Reactor, 120, 122, 165, 175–177, 413, 489

Best Management Practices (BMPs), 333–342

Bioaccumulation, 22, 205–207, 217, 221, 226, 249, 330

  Factor (BAF), 206–207

Biochemical Oxygen Demand (BOD), 46, 97, 98, 190–192, 194–201, 469–470, 473–478

Bioconcentration, 27, 205, 206, 249

  Factor (BCF), 205–206

Biodegradation, 194, 300, 608

Biodiesel, 163, 469

Biodiversity, 4, 22, 179, 180, 204, 205, 208–210, 217, 227, 311, 324, 329, 330–333, 342, 343, 359, 534, 577, 605, 628, 629, 631, 633, 639, 669

Biofilter, 121, 338, 493, 553, 556, 559–561

Biofiltration, 493, 559, 560

Biogeochemical Cycles, 201–204, 319, 328, 331

Biokinetic Coefficients, 175, 176, 480

Biomass, 5, 164–178, 186, 205, 206, 293, 326, 473, 476–483, 494, 495, 507

Biomimicry, 296

Bioretention Cells, 125, 161, 333, 334, 338–343

Biosphere, 159

Biosolids (*See* sludge), 465, 501–505

  Class A Solids, 504

  Class B Solids, 504, 505

Bioswales, 334, 342